D1269197

FORAGES

Volume I: **An Introduction to Grassland Agriculture**

The land produced vegetation [grass]: plants bearing seed according to their kinds and trees bearing fruit with seed in it according to their kinds; and God saw that it was good.

—GENESIS 1:1:12
New International Version Bible

Take not too much of a land, weare not out all the fatnesse, but leave in it some heart.

—PLINY THE ELDER (A.D. 23-79)
from his *Historiae Naturalis,* in 37 volumes

Whoever could make . . . two blades of grass to grow upon a spot of ground where only one grew before, would deserve better of mankind, and do more essential service to his country, than the whole race of politicians put together.

—JONATHAN SWIFT (1667-1765)
from *Gulliver's Travels,* 1726

It all depends upon the results. To make two blades of grass grow where one grew before is surely no achievement unless the grass is a good grass.

—HENRY SEIDEL CANBY
from *Atlantic Monthly,* June 1914

A land ethic . . . reflects the existence of an ecological conscience, and this in turn reflects a conviction of individual responsibility for health of the land. Health is the capacity of the land for self-renewal. Conservation is our effort to understand and preserve this capacity.

—ALDO LEOPOLD (1887-1948)
from *A Sand County Almanac,* 1949

FIFTH EDITION

FORAGES

Volume I: An Introduction to Grassland Agriculture

UNDER THE EDITORIAL AUTHORSHIP OF

Robert F Barnes, ASA, CSSA, SSSA

Darrell A. Miller, University of Illinois

C. Jerry Nelson, University of Missouri

WITH 54 CONTRIBUTING AUTHORS

IOWA STATE UNIVERSITY PRESS, Ames, Iowa, USA

© 1951, 1962, 1973, 1985, 1995 Iowa State University Press, Ames, Iowa 50014

All rights reserved

Copyright is not claimed for Chapters 8, 9, 10, 13, 16, 17, 18, 19, 30, 32, 33, and 36, which are in the public domain. Chapter 24 is from the Department of Agriculture, Government of Canada, and the copyright is owned by the Crown in right of Canada.

Authorization to photocopy items for internal or personal use, or the internal or personal use of specific clients, is granted by Iowa State University Press, provided that the base fee of $.10 per copy is paid directly to the Copyright Clearance Center, 27 Congress Street, Salem, MA 01970. For those organizations that have been granted a photocopy license by CCC, a separate system of payments has been arranged. The fee code for users of the Transactional Reporting Service is 0-8138-0681-X/95 $.10.

∞ Printed on acid-free paper in the United States of America

First and second editions © 1951 and 1962 by H.D. Hughes, Maurice E. Heath, Darrel S. Metcalfe, and Iowa State University Press; third edition © 1973 by Maurice E. Heath, Darrel S. Metcalfe, Robert F Barnes, and Iowa State University Press; fourth edition © 1985 by Maurice E. Heath, Robert F Barnes, and Darrel S. Metcalfe, and Iowa State University Press

Photograph used throughout frontmatter courtesy of Missouri Extension Service

Fifth edition, volume I, 1995
Second printing, 1995

Library of Congress Cataloging-in-Publication Data

Forages / under the editorial authorship of Robert F Barnes, Darrell A. Miller, C. Jerry Nelson; with 54 contributing authors.—5th ed.
p. cm.
"In memoriam: Professor Maurice E. Heath, 1910-1989"—Pref.
Includes bibliographical references and index.
Contents: v. I. An introduction to grassland agriculture.
ISBN 0-8138-0681-X (v. I: acid-free paper)
1. Forage plants. 2. Forage plants—United States. I. Barnes, Robert F. II. Miller, Darrell A. III. Nelsen, C. J. IV. Heath, Maurice E.
SB193.F64 1995
633.2—dc20 94-37719

The authors and editors dedicate this volume

To the memory of those teachers of forages, those great professionals who inspired us to learn, to discover, and to transfer knowledge and understanding about plants, soils, and livestock, and their role in agriculture.

In recognition of teachers currently in the process of keeping up with a dynamic discipline, who are helping the next generation place forages and livestock in an electronic age when environmental quality, social norms, and cultural differences have international relevance, but where genes know no bounds.

For the inspiration of today's student of forages who learns from a new perspective, within a broader context, and with a different value system; who will learn to appreciate this unique renewable natural resource and its complexity and diversity, and who will be motivated to teach the next generation.

CONTENTS

Part 2: Forage Legumes and Grasses

FARMERS

They work from dawn to darkness
With beauty on every hand.
Surrounded by Nature's bounty,
They lovingly tend the land.

They labor to well past sundown,
Taking bad days with the good.
Just to be doubly certain
We're supplied with daily food.

They're born with a love within them
For the endless hectares they tend.
A gift from their heavenly father,
An understanding, matchless friend.

The farmers all are special
With talents rare indeed.
They're endowed with heavenly caring,
For every soul they feed.

MARGARET R. LODGE
27 March 1994

PREFACE

ROLE OF FORAGES

Forages are the backbone of sustainable agriculture and contribute extensively to the US economy. Nearly a half billion hectares—more than half the land area of the US—are used for grasslands. One major role of forages is to provide feed for livestock, which convert fiber into human food such as milk and meat. There are about 6 to 10 major food crops, but hundreds of grasses, legumes, forbs, and shrubs contribute as forages. These biological resources also provide feed for wildlife and forage for winter feed, conserve our soil by reducing soil erosion, and protect our water quality by serving as water filters. These conservation measures contribute to the quality of our environment, the health of our society, and our recreational enjoyment. Economists are looking at new ways to value this multi-use resource.

FORAGE DIVERSITY

Forages and grasslands are more complex in their establishment, management, and harvesting than are feed grain crops. Forages are often grown on soils that support marginal productivity and on landscapes having both a lower productive index and a rolling topography. Many forages are perennial and, thereby, must prevail through a range of environments, including saturated soils, winter cold, and ice sheets; and summer conditions of heat, drought, and insects; and they are grazed or cut several times during the year. Many forage species are sown together or managed in natural ecosystems to retain species diversity. Plant maturity also strongly affects the quality of the forage. Therefore, management and harvesting to ensure adequate plant growth, production of nutritious feed, and persistence of desirable plants are more difficult than merely planting and harvesting a monoculture of an annual crop. In addition, ruminant livestock usually accompany forage production systems, adding biological complexity through the millions of rumen microbes and the ruminant animals themselves.

A PERISHABLE PRODUCT

Because of frequent harvests and the requirements for preservation, bulk storage and quality of product become important. Weather, labor, and quality of product are the major problems in forage management. Agricultural engineers have produced newer and more efficient machines to speed harvest and preservation in order to retain the quality at harvest. New approaches to grazing management have been developed. Animal scientists have refined nutrient needs and wildlife biologists have described ideal habitats. Thus, efficient forage production and utilization demand a high degree of managerial ability, technical information, and operational skills to meet the multiple requirements of this unique resource.

INTEGRATED TECHNOLOGIES

The objective of this edition is to integrate technologies from different scientific fields into a comprehensive production and management text that will provide knowledge to students to understand and fulfill the words of Jonathan Swift (1667–1765): *"Whoever could make . . . two blades of grass to grow upon a spot of ground where only one grew before, would deserve better of mankind, and do more essential service to his country, than the whole race of politicians put together."* However, there is also the focus on improvement in quality and value, as well as yield, as emphasized by Henry Seidel Canby in an article in the June 1914 issue of *Atlantic Monthly*, where he wrote: *"It all depends upon the results. To make two blades of grass grow where one grew before is surely no achievement unless the grass is a good grass."* This is very fitting going into the 21st century, with its greater emphasis on sustainability of productivity, conservation of resources, and expectations of economic return on investments.

VOLUME I: BASIC PRINCIPLES AND FORAGE SPECIES

The fifth edition of *Forages*, which appears as two volumes, is a major revision of the overall format presented in the fourth edition. The focus and purpose of this edition have been expanded to produce a text that facilitates classroom teaching yet retains its tradition as a strong reference. The first 15 chapters (Part 1) of Volume I are organized into the format of a college textbook, emphasizing basic principles of forages and their production, quality, and management. In this part, material from the textbook, *Forage Crops* (McGraw Hill, 1984) by D.A. Miller, has been incorporated with material from several of the principle-oriented chapters from the fourth edition of *Forages*. The next 22 chapters (Part 2, Chapters 16–37), covering important aspects of specific forage species, link technical and practical information together in a reference format. Our expectation is that the chapters in Part 2 will be used extensively to supplement the more general material in Part 1, to add emphasis to species, and to provide enrichment and examples of local and regional importance. These chapters, written by authorities on the species, will also serve as a valuable reference.

VOLUME II: NEW TECHNOLOGY AND INTEGRATED SYSTEMS

An additional 23 chapters form Volume II; they follow a format similar to the fourth edition of *Forages*. The emphasis in this volume is on new technology and the integration of forages into systems. It is envisioned that several of the chapters on specific species and forage problems in Volume II will be used to complement the general chapters in Volume I and will serve as a reference. Therefore, we believe this combination of volumes with their comprehensive coverage will provide a more versatile treatise on forages and that this fresh and relevant approach to a new edition of *Forages* will meet more diverse needs.

FORAGES AS A RESOURCE

Throughout we have encouraged emphasis on the principles of forage species and forage management. But we have also emphasized forages as valuable resources that fit into both natural and agricultural ecosystems. We have also expanded the coverage to include examples and principles of extensively managed systems as well as intensively managed systems. Chapters on conservation and wildlife management help reflect these emphases. Consistent with national trends, we have tried to incorporate contemporary issues relative to the environment, social expectations, and economic realities.

METRIC SYSTEM

We have used the metric system for expressing weights and measurements throughout both volumes. This has been the trend in our scientific societies and the federal government for several years. To facilitate interchange between units, a set of conversion tables follows the preface.

EXPANDED GLOSSARY

We have also expanded and updated the glossary. Terms continue to change, especially as the forage community becomes less regional and more international. For example, an international committee recently published a series of definitions for terms relative to grazing lands and grazing animals (J. Prod. Agric. 5:191-201, 1992). These terms plus those in the fourth edition of *Forages,* Miller's *Forage Crops,* and several from the Crop Science Society of America (CSSA) publication, *Glossary of Crop Science Terms,* have been reviewed and merged into the most comprehensive set of terminology on forages in this field today.

SCIENTIFIC NOMENCLATURE

An updated listing of scientific names is also included. New species regularly enter the forage arena, and scientific names for old species continue to change as new taxonomic findings emerge. The Crop Science Society of America periodically updates a list of approved scientific names and the USDA Genetic Resources Information Network (GRIN) system publishes regularly updated lists of adopted scientific names. Our listing is published for convenience to the reader and is current at time of publication.

AUTHORSHIP

The authors and co-authors of each chapter are recognized as authorities in that particular subject matter area. We appreciate the effort and leadership provided by each contributor. Our efforts have been to integrate the chapters to form a comprehensive treatise on this important subject. The material covered is the latest information on the subject that is synthesized and interwoven into the complex framework in which forages exist. The emphasis is on principles supported by experimental evidence that is clearly referenced.

FORAGES IN A CHANGING WORLD

There needs to be a documentation of progress in a field, especially one that is multi-disciplinary and dynamic. Forages as a science is changing. Professionals around the world are being challenged to broaden their thinking and involve all aspects of rangelands, grasslands, permanent pastures, and cultivated forages—each with its separate meaning and boundaries, many based on emotion, some on reality. Forages are a great natural resource that, when managed wisely, can help society meet public goals nationally and globally. The basic principles of plant growth, climatology, and animal management are similar over a range of climatic and social geographies. Our hope is that this edition will contribute to strengthening the relationships.

IN MEMORIAM:

PROFESSOR MAURICE E. HEATH, 1910–1989

Maurice E. Heath, Professor Emeritus, Purdue University, an international authority on the production and use of forage crops, was joint editor of the first and second editions of *Forages,* and senior editor of the third and fourth editions. He will be long remembered for his commitment to God and family, his dedication to forages, and his contributions to the forage-livestock industry and society.

Robert F Barnes

Darrell A. Miller

C. Jerry Nelson

THE METRIC SYSTEM

The International System of Units[1] (SI) is based on seven base units as follows:

Measure	Base Unit	Symbol
length	meter	m
mass	kilogram	kg
time	second	s
electric current	ampere	A
thermodynamic temperature	kelvin	K
amount of substance	mole	mol
luminous intensity	candela	cd

Following are prefixes and their meanings, suggested for use with units of the metric system:

Multiples and Submultiples	Prefix	Meaning	Symbol
1,000,000,000,000 = 10^{12}	tera	one trillion times	T
1,000,000,000 = 10^{9}	giga	one billion times	G
1,000,000 = 10^{6}	mega	one million times	M
1,000 = 10^{3}	kilo	one thousand times	k
100 = 10^{2}	hecto	one hundred times	h
10 = 10^{1}	deca	ten times	da
	The Unit = 1		
0.1 = 10^{-1}	deci	one tenth of	d
0.01 = 10^{-2}	centi	one hundredth of	c
0.001 = 10^{-3}	milli	one thousandth of	m
0.000,001 = 10^{-6}	micro	one millionth of	μ
0.000,000,001 = 10^{-9}	nano	one billionth of	n
0.000,000,000,001 = 10^{-12}	pico	one trillionth of	p

CONVERSION FROM A METRIC UNIT TO THE ENGLISH EQUIVALENT

Metric Unit		English Unit Equivalent
	Length	
kilometer (km)		0.621 mile (mi)
meter (m)		1.094 yard (yd)
meter (m)		3.281 foot (ft)
centimeter (cm)		0.394 inch (in.)
millimeter (mm)		0.039 inch (in.)
	Area	
kilometer^2 (km^2)		0.386 mile^2 (mi^2)
kilometer^2 (km^2)		247.1 acre (A)
hectare (ha)		2.471 acre (A)

[1]Additional information maybe found in the 1979 Standard for Metric Practice. *In* Annual Book of ASTM Standards, pt. 41, Designation 380-79, pp. 504-45. Philadelphia: Amer. Soc. for Testing and Materials (ASTM).

Metric Unit	English Unit Equivalent
Volume	
meter³ (m³)	1.308 yard³ (yd³)
meter³ (m³)	35.316 foot³ (ft³)
hectoliter (hL)	3.532 foot³ (ft³)
hectoliter (hL)	2.838 bushel (US) (bu)
liter (L)	1.057 quart (US liq.) (qt)
Mass	
ton (mt) = 1000 kg	1.102 ton (t)
quintal (q)	220.5 pound (lb)
kilogram (kg)	2.205 pound (lb)
gram (g)	0.00221 pound (lb)
Yield or Rate	
ton/hectare (mt/ha)	0.446 ton/acre
hectoliter/hectare (hL/ha)	7.013 bushel/acre
kilogram/hectare (kg/ha)	0.892 pound/acre
quintal/hectare (q/ha)	89.24 pound/acre
quintal/hectare (q/ha)	0.892 hundredweight/acre
Pressure	
pascal (Pa)	1.45×10^{-4} pound/inch² (psi)
megapascal (MPa)	9.87 atmosphere (atm)
megapascal (MPa)	10 bar (bar)
bar (10^6 dynes/cm²)	14.5 pound/inch² (psi)

Note: Bar is a metric term, but megapascal is the SI unit.

Temperature	
Celsius (C)	1.80C + 32 = Fahrenheit (F)
Kelvin (K) = C + 273.15	
Radiation (Light)	
lux (lx) = 1 lumen/m²	0.0929 foot-candle (ft-c)
watt (W) = J/s	none
photon (photon)	none

Note: Lux is a metric term, but it is a measure of illumination as seen by the human eye and not energy. Light energy, more correctly called *radiant energy*, is specific to the radiation source because the discrete particles (photons) at each wavelength in the output spectrum have a specific energy level. For example, about 5% of the solar energy at the earth's surface is ultraviolet (<400 nm) and about 45% is far-red and infra-red radiation (>700 nm) that is not detectable by the human eye. *Total radiation* or *radiant flux density* usually indicates the complete solar spectrum and is expressed as W/m². The portion of the spectrum between 400 and 700 nm wavelengths is active in photosynthesis and is called *photosynthetically active radiation* (PAR). The quantity of PAR available to a plant or leaf is the *photosynthetic photon flux density* (PPFD). It is expressed in photons because photosynthesis is a photon-driven process. At high noon on a clear summer day PPFD is about 2000 μmol photons/m²s.

Time	
second (s)	second (s)

Work and Energy

calorie (cal)	3.97×10^{-3} British thermal unit (Btu)

kilocalorie (kcal) = 1000 cal
megacalorie (Mcal) = 1000 kcal
erg (erg) = 2.39×10^{-8} cal
joule (J) = 1×10^{7} ergs

Power

watt (W) = J/s

PART 1

Forages in a Productive Agriculture

1 Forages in a Changing World

ROBERT F BARNES
American Society of Agronomy, Crop Science Society of America, and Soil Science Society of America

JOHN E. BAYLOR
Pennsylvania State University

EARLY RECOGNITION of the high value of grass is noted in the Book of Psalms of the Bible: "He makes *grass* grow for the cattle, and plants for man to cultivate—bringing forth food from the earth" (Psalm 104:14). This is an initial reference to the soil, plant, animal continuum to produce food. Moses promised the Children of Israel, as part of their inheritance if they obeyed the commandments of God, that He "will provide *grass* in the fields for your cattle, and you will eat and be satisfied" (Deuteronomy 11:15). Psalm 103:15-16 compares grass to man and the transitoriness of human life: "As for man, his days are like *grass,* he flourishes like a flower of the field; the wind blows over it and it is gone, and its place remembers it no more." The want of grass was recognized as the symbol of desolation: "The waters . . . are dried up and the *grass* is withered; the vegetation is gone and nothing green is left" (Isaiah 15:6). The theme of grass and grazing runs throughout the Bible.

ROBERT F BARNES is Executive Vice President of the American Society of Agronomy, Crop Science Society of America, and Soil Science Society of America in Madison, Wisconsin. He received his MS degree from Rutgers University and PhD from Purdue University. He served in the USDA, Agricultural Research Service, researching the development and application of forage evaluation methods.

JOHN E. BAYLOR is Professor Emeritus of Agronomy, Pennsylvania State University. He received his MS degree from Rutgers University and PhD from Pennsylvania State University. His 35 years in extension dealt mainly with forage production, management, and utilization. Following retirement from Penn State in 1983, he has been active in the seed industry.

THE IMPORTANT ROLE OF GRASSLANDS

From prehistory to the present, human history has been largely influenced by grassland. Civilizations began on grasslands, and civilizations have vanished with their destruction (Costello 1957). Grazing lands were vital to prehistoric peoples long before cattle (*Bos taurus*) were domesticated. The first human attempts to control their fate—to provide for future needs instead of remaining the victims of droughts or other unfortunate circumstances—must have been on grasslands, where the young calves, lambs, and kids they caught and tamed could find forage. It was also on grasslands, after they became food producers rather than food gatherers, that early peoples changed more rapidly (Smithsonian Institution 1931).

For a treatise on an intriguing and contrasting view on the stability of populations of hunters and gatherers, who have apparently "survived only in areas where agriculture has been unable to penetrate," the reader is referred to Harlan (1975). Reference is made to Marshall Sahlins' (1968) description of the hunting-gathering system as "the original affluent society." Sahlins states, "One can be affluent either by having a great deal or by wanting very little," the latter being more descriptive of hunting-gathering people. Harlan also quotes Berndt and Berndt (1970), who relate the view of the Australian aborigine:

> You people [Europeans] go to all that trouble, working and planting seeds, but we don't have to do that. All these things are there for us, the Ancestral Beings left them for us. In the end, you depend on the sun and rain just the same as we do, but the differ-

ence is that we just have to go and collect the food when it is ripe. We don't have all this other trouble.

Many of the early grassland management practices in Asia and Africa consisted of communal grazing of livestock on native forages. Such common pastures (referred to as *commons,* a term generally used today for a public place for people) were owned by everyone, and one member could not exclude animals owned by another member. Thus, overgrazing often resulted, leading to low animal, and eventually low human, nutrition.

The culture of grass as we know it in the US is mainly a product of European and American culture. In Great Britain, hay making and the scythe date from 750 B.C. Livestock survival through the winter depended upon the success of the hay harvest. The growing of hay crops and the significance of proper curing were described in detail by Columella (Roman) in about A.D. 50.

Hay making undoubtedly was an ancient agricultural practice. However, the conversion of fresh green forage into dried hay that was acceptable for storage and use over a considerable period was associated with a stable, rather than a nomadic, agriculture.

The Anglo-Saxons produced the first enclosed meadows in the Midlands of Britain about A.D. 800. The value of a change of pasture to the health of cattle and sheep (*Ovis aries*) was well known to the monks of Kelso as early as 1165. About 1400, the monks of Couper were alternating 2 yr of wheat (*Triticum* spp.) and 5 yr of grass, a practice later called *ley farming* (crop rotation).

Red clover (*Trifolium pratense* L.), one of our important legumes, was cultivated in Italy as early as 1550, in western Europe somewhat later, in England by 1645, and in Massachusetts in 1747. Its influence on civilization and European agriculture was said to have been greater than that of the potato (*Solanum* spp.) and much greater than that of any other forage plant (Heath and Kaiser 1985).

NATIVE GRASSLANDS OF NORTH AMERICA

Native grasslands in the North American Great Plains were referred to as *range* and *rangelands* soon after the turn of the 20th century. English settlers on the Atlantic Coast brought their name *meadow* for native grassland suitable for mowing. The French in Canada used *prairie* for similar grassland, and the Spanish in Florida used the word *savanna*. These various names for native grassland in North America have become a part of American vocabulary (Bidwell 1941; Gray 1958), with each term having its unique meaning (Barnes and Beard 1992).

Native Americans and Forages. The first English settlers in the American Colonies found their method of farming and producing food to be minimally successful under the harsh circumstances of those first few years of the settlement. Edwards (1940) states that it was the union of American Indian and European farming systems that produced the beginnings of American agriculture.

References to the burning of grasslands by native Americans have been found in the journals of many early explorers and settlers of the western US. These burnings of grasslands no doubt contributed to the abundance of productive grasses and control of forest and chaparral. Some appreciation of the natural grazing wealth originally provided by grassland can be gained from the numbers of herbage-consuming wildlife that roamed the great prairies in early days: 60 million buffalo (*Bison bison* L.), 40 million whitetailed deer (*Odocoileus virginianus* Boddaert), 40 million pronghorn antelope (*Antilocapra americana* Ord), and 10 million elk (*Cervus elaphus* L.). To this array must be added hundreds of millions of prairie dogs (*Cynomys* spp.), jackrabbits (*Lepus* spp.), and cottontails (*Sylvilagus floridanus*), all forage consumers (Seton 1929).

The soil-grass-fire-buffalo-Indian relationship had developed over thousands of years on the plains of North America. Grasslands were essential for the buffalo, on which the native American's existence depended. When the American buffalo herds were destroyed, the prairie native Americans were easily subdued.

Forages in American Colonial Times. Forages were equally important to the survival of early colonists. It was said there were few domestic animals in early colonies because few survived the long ocean voyage (Carrier 1923). At that time virtually all of the landmass that was settled was covered with forests or grasslands. Early settlers allowed the few livestock they had to graze the native grasses, where they did well during the summer months. However, in the North, the hard winters soon dictated shelters and supplemental harvested forages. As livestock numbers increased, the limited native pasturelands and the production of poor quality hay made it difficult to carry animals through the winter. It was then that grasses and clovers used previously in England began to make their appearance in the US.

As early as the beginning of the 18th century, the hectarage of grasses and woods and enclosed meadows did not keep pace with the increase in livestock. Lands worn out by overtillage and then abandoned to weed fallow made poor pasture. Perhaps half the average farm was a vast "pasture," mostly overrun with sourgrass, briars, and bushes. Farmers continued to cut hay chiefly from natural meadows and marshy areas (Edwards 1940).

Between 1780 and 1820 there were many trials with grassland plants in England, the most famous of them at Woburn Abbey. Various grasses and legumes were grown in small plots, and yields were determined and analyzed for nutritive value. Results were published as a book sold at home and abroad (Sinclair 1820). Although the work would be considered very crude by the modern researcher, the results were referred to in the US for about 50 yr.

Late in the 1700s, agricultural societies similar to those in England and Germany began to be formed in the US; by 1860 there were over 900 (Johnstone 1940). Among the earlier ones was the Philadelphia Society for the Promotion of Agriculture, which is still active today. During the first 100 yr following the American Revolution, agricultural technology improved more than during the previous 2000 yr. Heath and Kaiser (1985) report on a number of the historical accounts of this period. In another account, Baylor (1987) states that in 1850 haying tools consisted only of the scythe and pitchfork, along with the crude hay rake. Mechanization, as we know it, started with the cutterbar mowers. The harpoon-type fork for unloading hay in the barn came in 1864, followed by the hay loader in 1874 and the side-delivery rake in 1893. But hay making was still a time-consuming job.

The automatic pick-up baler, developed in the early 1940s, was a significant milestone in replacing the laborious, tiresome, and sweaty job of forking hay onto the wagon and into storage. The new baler replaced the big, cumbersome three- and four-person hand-tie balers that baled hay for transport and market. As late as the mid-1940s, the USDA figures show labor for harvesting and storing hay at 8.9 work h/mt. Now that figure has been cut to less that 2.8 work h/mt nationally. Studies have shown that one person can bale and load hay in as little as 15 min/mt with a modern throw-equipped baler, with unloading being done in 12 min/mt using elevators and mow conveyors to random stack bales in storage.

The Great American Prairies. Although a heavy forest growth originally covered much of the eastern US, about 40% of the total land area in the US was grassland. As the pioneer farmers pushed westward across the Alleghenies during and following the Revolution, they were confronted by the necessity of clearing heavy forest growths before crops could be grown, just as had been necessary for all farmers during the two centuries of the Colonial period. The idea had become pretty well fixed that land that grew only grass was inferior. As settlers from the eastern US entered western areas where there was a choice between forest and prairie, forest-covered soils were favored.

Forests seemingly loomed large in the thinking of pioneer farmers. The forest growth had sheltered the game that once constituted the chief source of meat, and it supplied logs for cabins, stock shelters, fuel, and fences. Fencing materials were important. The problem of fences on the prairies remained essentially unsolved until the invention of barbed wire in 1867 and its introduction into the US shortly thereafter.

The pioneers hesitated on the edge of the large prairies of seemingly endless tall grass. There was a sense of vastness about the prairie that was overwhelming and the impression of a greatness that could not be subdued. Indeed some contended at that time that the prairies would not be brought under cultivation for centuries (Edwards 1948).

George Stewart's chapter in Senate Document No. 199 provides the classic historical documentation of range resources in the US (Stewart 1936) and describes the western range thusly:

> The western range is largely open and unfenced, with control of stock by herding; when fenced, relatively large units are enclosed. It supports with few exceptions only native grasses and other forage plants, is never fertilized or cultivated, and can in the main be restored and maintained only through control of grazing. It consists almost exclusively of land which, because of relatively meager precipitation and other adverse climatic conditions, or rough topography or lack of water for irrigation, cannot successfully be used for any other form of agriculture. In contrast, the improved pastures of the East

and Middlewest receive an abundant precipitation, are ordinarily fenced, utilize introduced forage species, . . . cultivation for other crops, and are often fertilized to increase productivity, and are renewed following deterioration.

Originally there were about 283 million ha of grass-covered native prairie stretching from Ohio westward. In general, the most fertile, deep, rich black soils were formed under the vegetative growth of the prairies. The main grasslands of the central and Great Lake states were dominated by grasses so tall and stands so dense that the early settlers' cattle could be found only by the tinkling of cowbells and the waving of the grasses. So tough and thick were these prairie grass sods that some farmers preferred not to break sod until after the stand had been weakened by heavy grazing and repeated mowing. From three to seven yoke of oxen were required to break new fields (Hoover 1961).

Farther west, the short-grass prairie extended from Texas north into Canada and east from the Rocky Mountains to mid-Kansas, Nebraska, and the Dakotas. Here the native short grasses such as grama grasses (*Bouteloua* spp.) and buffalograss (*Buchloe dactyloides* [Nutt.] Engelm.) were in greatest abundance, though some of the tall grasses predominated toward the eastern margin of the area. Only in the flint hills area of eastern Kansas are there extensive native tall-grass grazing areas remaining (Johnstone 1940).

The impression of early explorers was that the growth of grasses on these vast prairie areas was such that they would endlessly support countless herds and flocks. Two factors upsetting the resiliency of the grassland resource in the western US were (1) the Spanish heritage of rearing cattle in large herds and (2) the increased demand for meat as a result of the discovery of gold on the American River in California on January 24, 1848. Livestock had previously been raised almost entirely for hides, tallow, and wool. Large herds of cattle increased in numbers most rapidly after 1870 from central Texas northward and westward (Chapline and Cooperrider 1941; Edwards 1948). The influence of the colonial Spanish New World on the use of grasslands of the southwestern US can be best summarized by Stewart (1936):

The tremendous growth in range cattle, however, carried with it a weakness that in the end proved fatal. It was based on a husbandry transplanted from Mexico, which brought to English-speaking people for the first time in history the practice of rearing cattle in great droves without fences, corrals, or feed. . . . Cattle instead of grass came to be regarded as the raw resource and the neglected forage began to give way before the heavy and unmanaged use to which it was subjected.

A publication resulting from a symposium held in 1992 promises to be another landmark publication, similar to Senate Document No. 199 with respect to an evaluation of the credence of livestock grazing on western rangelands; it is the book by Vavra et al. (1993) entitled *Ecological Implications of Livestock Herbivory in the West*. Vavra et al. state that "the westward expansion of the US was characterized by exploitation of natural resources." This exploitation was supported by the American public through federal legislation. The initial Homestead Act was passed by Congress in 1862, followed by others in 1873 and 1877, which accelerated the westward expansion through the removal from public ownership of much of the land suitable for growing crops. This influx of population into the rangeland areas resulted in an increased demand for production from the land. However, most of the rangeland was not privately owned, and the forage was grazed by animals owned by ranchers who were positioned to harvest it. There was little incentive for an individual rancher to conserve the available forage for later use (Gardner 1991). This lack of property rights led to what Hardin (1968) calls the "tragedy of the commons." The boon in unlimited livestock grazing lasted only a couple of decades, and by 1880 continued overstocking had reduced the carrying capacity of most of the western range. Dry summers coupled with severe winters resulted in the apparent loss of 30% to 80% of the cattle in the Northern Plains during the winters of 1886 and 1887 (Schlebecker 1963).

By the end of the century the plains cattle industry had been changed from an open range, exclusive use enterprise to a ranch-based industry that coexisted with cropland (Lauenroth et al. 1993). The Taylor Grazing Act of 1934 was the initial attempt to control grazing on public lands, and its implementation resulted in modifying the land management practices on such lands and ending the series of homestead acts.

Although interactions between livestock grazing and other elements of the Great Plains ecosystem are a complex issue, Lauenroth et al. (1993) contend that the management of grazing animals using current technology appears to be a sustainable resource practice. Sustainability of resource production has to be the primary goal of any forage management plan, particularly for the Great Plains, where 63% of the area is in grazing lands. Pieper (1993) similarly projects that grazing of domestic livestock at conservative levels appears to be sustainable, even on sensitive western rangelands. Many factors, in addition to livestock grazing, have had undetermined impacts on native vegetation. These include climatic shifts, increases in woody plant species, reduction of fire frequency and intensity, introduction of alien plant species, and other human activities (Pieper 1993).

GRASSLAND AGRICULTURE

Grassland agriculture is a farming system that emphasizes the importance of grasses and legumes in livestock and land management. Farmers who plan row crops and livestock production around their grassland hectares are grassland farmers. The main feature of grassland agriculture is its dependence on herbaceous plants such as grasses, legumes, and forbs and, in many range situations, on the leaves, buds, and stem tips of shrubs and woody vegetation.

One key to success in grassland farming is to recognize the importance of a healthy soil-plant-animal biological system. Land, or more specifically soil, is basic to plant production and, thereby, to all of life. Simply stated, plants convert radiant solar energy, carbon dioxide, and water to the usable form of carbohydrates. Plants also blend in nitrogen (N) with the carbohydrates to produce the amino acids found in protein. Most mineral elements required by animals and humans move from the soil to the plant. However, there is no single food plant that contains all the nutrients required for human growth and development. The primary products produced by forage plants move to the food chain through the consumption and utilization by animals, primarily ruminants. Vallentine (1990) has graphically portrayed the dual production and harvesting avenues of forage conversion by herbivores, that is, animals that subsist principally or entirely on plants or plant materials (Fig. 1.1).

Forage has been defined as "edible parts of plants, other than separated grain, that can provide feed for animals, or that can be harvested for feeding." It includes browse (i.e., buds, leaves, and twigs of woody species), herbage (leaves, stems, roots, and seeds of nonwoody species), and mast (nuts and seeds

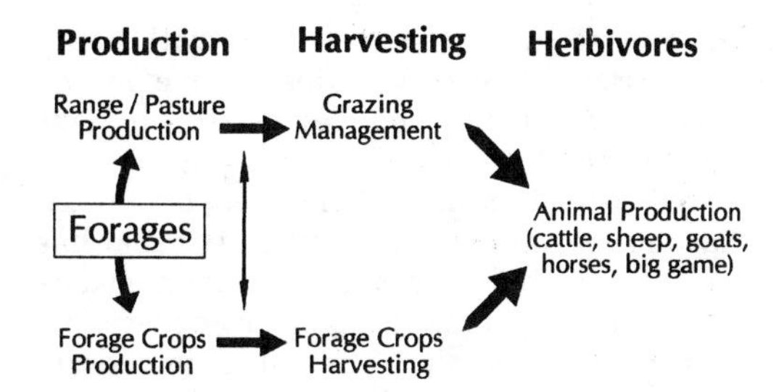

Fig. 1.1. The dual production-harvesting avenues of forage conversion by herbivores. (Adapted from Vallentine 1990.)

of woody species). *Forage* as used in the remainder of this chapter is an all-inclusive noun denoting plants and plant parts that are consumed by domestic livestock or game animals as pasturage, browse, mast, hay, haylage, silage, green chop, fodder, and certain grain crop residues.

The term *pasture* comes to us from the Latin *pastus* and is defined by *Webster's Third New International Dictionary* as "grass or other plants grown for the feeding of grazing animals." In this context, the term *pasturage* may be used as a synonym, meaning "vegetation on which animals graze, including grasses or grasslike plants, legumes, forbs, and shrubs"; however, it is not a recommended term. The preferred usage of the term *pasture* is in reference to a kind of grazing management unit, rather than to what is consumed. *Pastureland* refers to land devoted to the production of indigenous or introduced forage for harvest primarily by grazing. *Permanent pasture* involves pastureland composed of perennial or self-seeding annual plants kept for grazing indefinitely, or generally at least 10 yr or more. *Cropland* is land devoted to the production of cultivated crops, including forage crops, and *cropland pasture* in the US refers to cropland on which grazing occurs but is generally of limited duration. Cropland pastures often serve a useful role in row crop rotation schemes or in using residues after harvest of grain crops (see Chap. 14, Vol. 2).

Rangeland, a term of American origin, means "land on which the indigenous vegetation (climax or natural potential) is predominantly grasses, grasslike plants, forbs, or shrubs and is managed as a natural ecosystem." If plants are introduced to rangeland, they are also managed as indigenous species. A key element is that the forage is produced by native plants or by introduced species managed as if they were native species.

Range is difficult to define precisely and

has evolved into a collective term with broad implications. *Webster's* defines range in the broad context as "the region throughout which a plant or animal naturally lives." The preferred definition is "land supporting indigenous vegetation that is grazed or that has the potential to be grazed, and is managed as a natural ecosystem." Range encompasses all rangelands, including grazeable forestland (often referred to as *forest range*), that is, forestland that produces, at least periodically, a sufficient understory of herbaceous or shrubby vegetation that can be grazed.

The term *grassland* includes both pastureland and rangeland as grazing lands and thus, in general, denotes all plant communities on which animals are fed, annually sown crops excepted. More commonly, *grasslands* are any plant community, including harvested forages, in which grasses and/or legumes make up the dominant vegetation.

It should be noted that the terms *cropland, forestland, pastureland,* and *rangeland* provide the basis for land-use mapping units. Terminology for grazing lands and grazing animals has been addressed by the international, multisociety Forage and Grazing Terminology Committee, which has prepared a glossary of grazing terms (Forage and Grazing Terminology Committee 1992). Many of those terms are included in the glossary at the end of this book.

Land Resource. When the US was first colonized, virtually all of the landmass was covered with forests or grasslands. These lands were gradually converted into cropland and grazing land, to provide food for people and livestock, and into towns, cities, highways, railways, and industrial sites. Today, approximately 73% of the landmass of the US is covered by permanent vegetation consisting of forestland, rangeland, and pastureland (see Chap. 17, Vol. 2). An additional 2.7% is used for rotational hay and cropland pasture. The land area being maintained in vegetation has declined in recent years. The land used for crops in the US declined by 5% between 1930 and 1987, and pastureland (grassland pasture) decreased 8% during the same period. The change was reflected in major increases in what is classified as "other land," namely urban and built-up areas, roads, and areas for mining, water storage, and other miscellaneous uses (Table 1.1).

The number of farms decreased from the high of 6.8 million farms in 1935 to 2.1 million farms in 1987 (Table 1.2). Farm populations have been reported only since 1977, when 7.8 million people (about 3.6% of the total population of 218 million) lived on farms. The farm population decreased to 1.8% in 1991.

The use of federal and nonfederal land in the US is presented in Table 1.3, in which 43% of the rangeland is identified as federal land, i.e., land owned by the US government. In the late 1970s, a movement called "the sagebrush rebellion" challenged the continu-

TABLE 1.1. Land utilization, US, 1930-87 (million ha)

Major land use	1930	1978	1987
Cropland used for crops	155	149	134
Idle cropland	13	11	28
Cropland used only for pasture	27	31	26
Grassland pasture[a]	264	238	239
Forestland[b]	243	285	262
Other	69	204	227
Total land area	771	918	916

Source: USDA (1992, 359).
[a]Grassland and other nonforest pasture and range.
[b]Includes forested grazing lands.

TABLE 1.2. Statistics on farm numbers, land area, and population

Factor	1900	1930	1978	1987	1991
			millions		
Farms	5.7	6.3	2.3	2.1	2.1
Land in farms (ha)	340.0	400.0	410.0	390.0	398.0
CRP[a], total (ha)				6.3	14.3
Population					
Civilian			220.0	241.0	251.0
Farm			8.0	5.0	4.6

Source: USDA (1992, pp. 355 [farms, land in farms], 362 [population], 434 [CRP, total ha]).
[a]CRP = Conservation Reserve Program.

TABLE 1.3. Use of federal and nonfederal land in the US (million ha)

	Nonfederal land	Federal land	Total	Percentage
Cropland and pastureland[a]	214		214	23.5
Rangeland[b]	179	133	312	34.2
Transition land[c]	14		14	1.6
Forestland	166	112	278	30.4
Other land	65	30	94	10.3
Total	638	275	912	100.0

Source: USDA (1987, 15-2).

[a]Pastureland: Land used primarily for the production of adapted, introduced, or native species in a pure stand, grass mixture, or a grass-legume mixture. Cultural treatment in the form of fertilization, weed control, reseeding, or renovation is usually a part of pasture management in addition to grazing management.

[b]Rangeland: Land on which the climax vegetation (potential natural plant community) is predominantly grasses, grasslike plants, forbs, or shrubs suitable for grazing and browsing. It includes natural grasslands, savannas, many wetlands, some deserts, tundra, and certain forb and shrub communities. It also includes areas seeded to native or adapted introduced species that are managed like native vegetation.

[c]Transition Land: Land that meets the definition of forestland based on cover characteristics but where the predominant vegetation is grasses or forage plants that are used for grazing. Soil Conservation Service classification is "rangeland"; Forest Service classification is "forestland."

ing federal ownership of land and energy resources (Gardner 1991). Many western states proposed the transfer of federal land to the states. The issue centered partly on the question of who "captures the economic profits from natural resources" and partly on who "has the competence to manage resources most efficiently—the federal government or the private rancher?" The appropriate use and protection of public lands have been, and will continue to be, debated.

Changes are occuring in our national resource base. Our land and renewable natural resources are being depleted and polluted at an alarming rate. The number of farms and of people farming is declining, while the challenge remains for providing sufficient supplies of safe and wholesome food for an ever-expanding population.

Forages and Livestock in the National Economy. Grassland agriculture is highly dependent upon a reliable source of forage as the primary feed base for ruminant livestock. Resources used in cattle and sheep raising are widely scattered, both in location and in ownership because livestock raising is a land-based, forage-utilizing enterprise.

US cattle numbers peaked in 1975 with 135 million head. Ruminant livestock numbers (Table 1.4) have continued to decline, decreasing 15% from 1978 to 1991. The number of milk cows has decreased less rapidly, declining only 6%; however, a significant decrease of an additional 3.2% occurred in 1992. Sheep numbers peaked in the depression year of 1935 and declined steadily until the 1970s. A further decline of 14% has occurred for the 9-yr period from 1978 to 1987, but numbers increased an additional 4% by 1991.

Information on the use of feedstuffs for livestock and poultry in the US is presented in Table 1.5. The information was tabulated and published by the Council for Agricultural Science and Technology as part of a report on forages in 1980. The report provides a baseline of data for the proportion of the diet provided by forage for various classes of livestock. Such data are used later in the chapter in an estimate of the value of forage.

COMPLEMENTARY BENEFITS FROM FORAGE

In a grassland agriculture many complementary benefits are derived from forages. In addition to serving as the major source of feed nutrients for wild and domestic animals, forage contributes to human well-being through the following: (1) soil erosion control; (2) improvement of soil structure and fertility; (3) water conservation and resource development and protection; (4) balanced programs of wildlife and game conservation and protection; (5) potential conversion of forage bio-

TABLE 1.4. Ruminant livestock numbers

Species	1978	1987	1991	1992
		(millions)		
Cattle	116.4	102.1	98.9	
Beef cows	38.7	33.8	33.3	
Milk cows	10.9	10.5	10.2	9.9
Sheep	12.4	10.7	11.2	

Source: USDA (1992, pp. 246 [cattle], 248 [beef cows, milk cows], 269 [sheep]).

TABLE 1.5. Use of feedstuffs by livestock and poultry in the US

Type of animal	Proportion of ration[a]		Proportion of total concentrate usage (%)	Proportion of total feed usage (%)
	Concentrates	Forages		
All livestock and poultry	35.7	62.5	100.0	100.0
All dairy cattle	38.8	61.2	16.6	16.0
Beef cattle on feed	72.4	27.6	20.7	10.7
Other beef cattle	4.2	95.6	5.1	46.2
All beef cattle	17.0	83.0	25.8	56.9
Sheep and goats	8.9	91.1	0.4	1.9
Hens and pullets	100.0	0.0	12.4	4.6
Turkeys	100.0	0.0	3.3	1.3
Broilers	100.0	0.0	9.3	3.5
Hogs	85.3	14.7	30.0	13.2
Horses and mules	27.8	72.2	2.2	2.9

Source: Adapted from Council for Agricultural Science and Technology (1980).
Note: All feedstuffs are measured on a corn-equivalent basis.
[a]Per 100 kg weight gain, assuming gains in weight or other output were produced by the various feedstuffs fed as part of a properly balanced ration.

mass to energy as a renewable resource; (6) improvement and protection of the environment from pollution such as sediment, windblown soil, municipal and farm wastes, and some toxic substances; (7) outdoor recreation and pleasure; and (8) potential sources of high-quality extracted leaf proteins and other plant products.

Animals that are primary forage consumers also contribute to human well-being by providing (1) meat and milk products as rich sources of essential amino acids, fats, vitamins, and minerals; (2) hides, wool, and horns for clothing, implements, and adornments; (3) power for transportation or draft; (4) manure for fuel and fertilizer; (5) pleasure from keeping animals as pets, from training animals to perform useful or incongruous tasks, from observing animals in their natural habitats or in zoos, from the use of animals in competitive or sporting events, and from hunting; and (6) a means of harvesting both desirable and unwanted vegetation.

Estimated Value of Forage. The productivity of ruminant animals is largely dependent upon their utilization of forage, and attempts have been made to present an estimated value of forages consumed by ruminant livestock. A method outlined by Hodgson (1974) in which forage value is calculated on the basis of cash receipts for livestock and livestock products involves making an estimate of the percentage of cash receipts that represent feed costs of each class of livestock. This percentage multiplied by the percentage of feed units furnished by forage identifies the portion of cash receipts that represent the forage contributed to each class of livestock. By using estimated feed costs in livestock production, rather than feed value, the method underestimates the true value of forages in livestock production. Using food costs based on cash receipts for 1990 (USDA 1992), the calculated value of forages equals $24.027 billion and far exceeds the cash value of other crops (Table 1.6). The value of hay alone in 1990 was $10.457 billion, which was exceeded only by corn and soybeans, valued at $18.191 billion and $11.042 billion, respectively (USDA 1992).

A GRASSLAND PHILOSOPHY

A sound national grassland philosophy is needed before grassland agriculture will be practiced generally on individual farms and ranches. Soil, water, and climate, together with factors governing production and utilization of grasses and legumes, determine the intensity of grassland agriculture in different parts of the country. Forage plants common in arid and semiarid rangeland are quite different from those of humid areas, but the management principles are similar. Grassland agriculture is a longtime program directed toward increased production from improved grasslands and more efficient use of high-quality forage rich in protein, minerals, and vitamins.

As any nation tends toward self-sufficiency, the demand grows for enhanced aesthetic and recreational opportunities. Such a thrust emphasizes the need for development of fish and wildlife resources as a complement to our grasslands. In the past, conflicts have arisen in the US when traditional agronomic and grassland management practices have favored livestock use over wildlife needs. Com-

TABLE 1.6. Estimated value of forages consumed by ruminant livestock

Type animal (1)	Receipts as feed costs[a] (%) (2)	Feed units as forage[b] (%) (3)	Cash receipts as forage value[c] (4)	1990 Cash receipts[d] ($ millions) (5)	Forage value[e] ($ millions) (6)
Beef cattle	70	83	.581	30,229	17,563
Sheep and wool	70	91	.637	432	275
Dairy cattle (milk)	50	91	.302	20,495	6,189
Total forage value					24,027

[a]Receipts as feed costs from Hodgson (1974).
[b]Feed units as forage from CAST (1980) (see Table 1.5).
[c]Cash receipts as forage value obtained by multiplying column (2) by (3).
[d]Cash receipts reported in USDA (1992, 373).
[e]Forage values obtained by multiplying column (4) by (5).

munication and trust among rangeland and pastureland managers, wildlife biologists, and livestock owners must be strengthened. Only in such a way can we develop realistic management practices that emphasize the multiple use of resources, including benefits to fish and wildlife, as well as enhancing livestock productivity.

ROLE OF GRASSLAND ORGANIZATIONS

The recognition by grassland scientists of the need to make better use of grasslands resulted in the formation of grassland organizations worldwide, as well as in many countries.

International Grassland Congress. Formed in 1920 by several European scientists, the International Grassland Congress operated initially under an organizational structure involving only northern and central European countries, where, it was argued, "comparable environmental conditions prevailed" (Cardon 1948). But at the third Congress, the decision was made to extend its scope to a truly international basis. Two meetings of the International Grassland Congress have been held in the US. The sixth, in 1952, had more than 1200 delegates and members from 49 countries present. In 1981, the 14th was attended by approximately 1500 grasslanders representing 59 countries. In 1997, the 18th Congress, and the third on the North American continent, will be held in Canada.

In the foreword to the sixth Congress proceedings, Wagner et al. (1952) emphasize the world's food problems and explain the role of the Congress in helping to solve them.

> The specter of hunger stalks the world. . . . Increasing population is exerting steadily mounting pressure on food production capacities. Exploitive systems of farming are taking heavy toll of soil productivity. . . . We must find new resources for production. We must reverse the downward trend in soil productivity. And for the attainment of both of these necessary objectives, grassland farming is the most effective weapon in agriculture's arsenal. . . . More than half the total land surface of the earth is in grazing lands. Most of these enormous areas are unimproved. Improvement practices, based on research findings, can result in vast increases in production and in the utilization of livestock feed from these grasslands. And the grasslands provide the major raw material for the production of meat, milk, and other animal products. . . . Soil is the basic resource of all agriculture. . . . Sound soil management practices must be built around grasses and legumes. . . . The doctrines and teaching of science transcend national borders. . . . The . . . Congress provides the opportunity for scientists and technicians from various parts of the world to exchange information concerning the production, improvement, management, and use of grasslands.

International Rangeland Congress. The initial International Rangeland Congress was held in Denver, Colorado, in 1978. The impetus for the congress was to provide a specific focus and identity for rangelands and an understanding of worldwide rangeland ecosystems. The Society for Range Management, located in the US, undertook formation of the first congress with the expectation that future meetings, and other exchanges of knowledge, would be sponsored by a worldwide group. The pattern of development for the rangeland congresses have been complementary to the grassland congresses in that a specific attempt has been made (1) to provide coordination between the respective continuing committees and (2) to conduct one congress every 2 yr on an alternating time schedule. Following successive congresses in Australia, India, and France, the fifth International Rangeland Congress returns to the US in 1995.

Other Grassland Organizations. Through the influence of the International Grassland Congress, more grassland organizations and activities have developed within and among many countries. For example, in 1944 steps were taken in England to organize the British Grassland Society, which held its first meeting in 1945. Its main purpose is to focus greater scientific attention on all aspects of grasslands. In 1955, the first British county grassland society was organized in Surrey.

The European Grassland Federation was formed in 1963 by grassland organizations from 13 countries. Its objective is a closer working relationship and a greater interchange of scientific and applied information among the 13 groups (European Grassland Federation 1963). Grassland organizations are very active in New Zealand and Australia and many other countries.

In 1944, the Joint Committee on Grassland Farming was formed, now named the American Forage and Grassland Council (AFGC) (Baylor 1994). It brings together public service and industry (including farm) leadership to strengthen the educational and research aspects of forages and grasslands in American agriculture. By 1994, at the time of the AFGC's 50th anniversary, there were 34 affiliated state or provincial councils in the AFGC, many with local councils made up primarily of producers.

The Society of Range Management was established in 1948 to foster the science and art of range management. It was initially called the American Society of Range Management, and the name was changed to Society of Range Management in 1970. The membership generally includes forestry and range professionals in government, university faculties, and the private sector, and it has expanded to include professionals from many nations. The society publishes the results of range research through the *Journal of Range Management and Rangelands*.

Many other organizations that are more discipline oriented and focused in their scope and purpose have contributed significantly to the science of grassland agriculture. The American Society of Agronomy (ASA), organized in 1907, has played a significant role in the integrative aspects of the production, protection, and use of forage, pastureland, grassland, and rangeland resources. ASA publishes six journals, one of which, *Journal of Production Agriculture,* is cooperatively sponsored by the American Forage and Grassland Council and the Society for Range Management, as well as other discipline-oriented scientific professional societies. This journal emphasizes the publication of research findings oriented toward integrated systems and involves technology from a number of disciplinary areas.

A promising development in the US is the potential formation of a grassland and range coalition involving the major professional and scientific societies, government agencies, institutions, and professionals associated with forage, pastureland, grassland, rangeland, and wildlife resources. The initial concept of such a coalition resulted from a 1993 national workshop on "Innovative Systems for Utilization of Forage/Grassland/Rangeland Resources" (Wedin 1994) sponsored by the Cooperative States Research Service, USDA, and held at the Airlie Foundation, Airlie, Virginia. The proposed mission for the grassland and range consortium is to enhance communication among diverse groups sharing a common interest in promoting wise management, use, and protection of grassland and range resources for sustainable food production, recreation, environmental stewardship, and economic development. Grasslands play a significant role in our nation's science and economic policy concerning renewable natural resources, and such an approach for improving communication holds promise for providing a national and international understanding of forages and the science of grassland agriculture.

QUESTIONS

1. Discuss some of the early forage practices in Europe and Great Britain that were transferred to America by the colonists.
2. Discuss the role of native grasslands in the development of the western US and its implications for modern grassland agriculture.
3. What were two primary factors contributing to the deterioration of natural grasslands in the western states in the late 1800s?
4. What constitutes "sustainability of resource production" on grasslands and rangelands?
5. Explain the importance of the soil-plant-animal biological system and its significance to human health and well-being?
6. Explain what is meant by the phrase, "tragedy of the commons."
7. What is the role of forages as an overall source of feed for livestock?
8. List and discuss at least five trends related to forage production that have occurred, are occurring, or will occur.
9. Why are forages poorly appreciated in the overall food system?

REFERENCES

Barnes, RF, and JB Beard. 1992. Glossary of Crop

Science Terms. Madison, Wis.: Crop Science Society of America.
Baylor, JE. 1987. Haymakers Handbook. New Holland, Pa.: New Holland.
______. 1994. AFGC—the first 50 years—1944 to 1994. In Proc. 1994 Am. Forage Grassl. Counc., Georgetown, Tex., 1-128.
Berndt, RM, and CH Berndt. 1970. Man, Land and Myth in North Australia. East Lansing: Michigan State Univ. Press.
Bidwell, PW. 1941. History of Agriculture in the Northern United States, 1820-1860. New York: Peter Smith.
Cardon, PV. 1948. A permanent agriculture, our aim: An introduction. In Grass, USDA Yearbook of Agriculture. Washington, D.C.: US Gov. Print. Off., 1-5.
Carrier, L. 1923. The Beginnings of Agriculture in America. New York: McGraw-Hill.
Chapline, WR, and CK Cooperrider. 1941. In Climate and Man, USDA Yearbook of Agriculture. Washington, D.C.: US Gov. Print. Off., 459-76.
Costello, DF. 1957. Grasslands: America's Natural Resources. New York: Roland Press.
Council for Agricultural Science and Technology. 1980. Forages: Resources for the Future. Rep. 108. Ames, Iowa.
Edwards, EE. 1940. American agriculture—The first 300 years. In Farmers in a Changing World, USDA Yearbook of Agriculture. Washington, D.C.: US Gov. Print. Off., 171-276.
European Grassland Federation. 1963. Proc. 1st Symp. Grassl. Res. Inst. Hurley, U.K.
Forage and Grazing Terminology Committee. 1992. Terminology for grazing lands and grazing animals. J. Prod. Agric. 5:191-201.
Gardner, BD. 1991. Rangeland resources: Changing uses and productivity. In KD Frederick and RA Sedjo (eds.), America's Renewable Resources: Historical Trends and Current Challenges. Washington, D.C.: Resources for the Future, 123-66.
Gray, LC. 1958. History of Agriculture in the Southern United States to 1860. New York: Peter Smith.
Hardin, G. 1968. The tragedy of the commons. Sci. 162(13 December):1243-48.
Harlan, JR. 1975. Crops and Man. Madison, Wis.: American Society of Agronomy.
Heath, ME, and CJ Kaiser. 1985. Forages in a changing world. In ME Heath, RF Barnes, and DS Metcalfe (eds.), Forages: The Science of Grassland Agriculture, 4th ed. Ames: Iowa State Univ. Press, 3-11.
Hodgson, HJ. 1974. Importance of forages to livestock production. In HB Sprague (ed.), Grasslands of the United States. Ames: Iowa State Univ. Press, 43-56.
Hoover, H. 1961. An American Epic. Vol. 3. Chicago: Regnery.
Johnstone, P. 1940. A brief chronology of American agricultural history. In Farmers in a Changing World, USDA Yearbook of Agriculture. Washington, D.C.: US Gov. Print. Off., 1184-96.
Lauenroth, WK, DG Milchunas, JL Dodd, RH Hart, RK Heitschmidt, and LR Rittenhouse. 1993. Effects of grazing on ecosystems of the Great Plains. In M Vavra, WA Laycock, and RD Pieper (eds.), Ecological Implications of Livestock Herbivory in the West. Denver, Colo.: Society of Range Management, 69-97.
Pieper, RD. 1993. Ecological implications of livestock grazing. In M Vavra, WA Laycock, and RD Pieper (eds.), Ecological Implications of Livestock Herbivory in the West. Denver, Colo.: Society of Range Management, 177-211.
Sahlins, M. 1968. Notes on the original affluent society. In RB Lee and I DeVore (eds.), Man the Hunter. Chicago: Aldine, 85-89.
Schlebecker, JT. 1963. Cattle Raising on the Plains, 1900-1961. Lincoln: Univ. Nebraska Press.
Seton, ET. 1929. Lives of Game Animals. Vol 3. Garden City, N.Y.: Doubleday.
Sinclair G. 1820. Hortus Gamineus Woburnensis. 4th ed. London: James Ridgway.
Smithsonian Institution. 1931. Old and New Plant Lore. Sci. Ser. 11.
Stewart, G. 1936. History of Range Use. In Senate Document 199, 74th Congress.
USDA. 1987. The Second Resources Conservation Act Appraisal (Public Review Draft). Forest Service and Soil Conservation Service.
______. 1992. Agricultural Statistics. Washington, D.C.: US Gov. Print. Off.
Vallentine, JF. 1990. Grazing Management. San Diego, Calif.: Academic Press.
Vavra, M, WA Laycock, and RD Pieper. 1993. Ecological Implications of Livestock Herbivory in the West. Denver, Colo.: Society of Range Management.
Wagner, RE, WM Myers, and SH Gaines. 1952. Foreword. In Proc. 6th Int. Grassl. Congr., v-vi.
Wedin, W F. 1994. Innovative systems for utilization of forage/grassland/rangeland resources. In Proc. Natl. Workshop, 22-24 Sept., Univ. of Minnesota, St. Paul.

2 Morphology and Systematics

C. JERRY NELSON
University of Missouri

LOWELL E. MOSER
University of Nebraska

MOST cultivated forages fit into two botanical families, Poaceae (Gramineae), the grasses, and Fabaceae (Leguminosae), the legumes. Many forbs, which are herbaceous (nonwoody), nonlegume, dicotyledonous plants, exist in mixtures with grasses and legumes in extensively managed pastures and prairies. In addition, several trees and shrubs, especially woody legumes, contribute to nutritional requirements of browsing ruminants (see Chap. 23), and some brassicas (Chap. 36) are used for grazing. The objective of this chapter is to introduce plant nomenclature, describe growth properties of grasses and herbaceous legumes, and evaluate how management practices can be used to improve plant adaptation and productivity.

COMMON AND BOTANICAL NAMES

A plant is known scientifically by its species name, which consists of two Latin words. The initial letter of the first word, the genus, is always capitalized; the second, the species epithet, is written with a lowercase initial letter. The genus corresponds roughly to a last name and the species epithet to a first name, as *Medicago sativa* would to Brown, John. The scientific name includes the abbreviated name of the person or persons (authority) who correctly named the species. Thus, *Medicago sativa* L., the scientific name for alfalfa, indicates the species was named by Linnaeus, the Swedish botanist who developed the nomenclature system. If a plant is reclassified, the original authority is placed in parentheses, and the new one follows. For example, indiangrass, *Sorghastrum nutans* (L.) Nash, was first classified by Linnaeus and later reclassified by Nash.

Common names of forage plants often differ among regions of the US and the world (Bailey 1949; Fernald 1950; Hitchcock 1951; Leithead et al. 1971). For example, the name *prairie beardgrass* is occasionally used, but it is not the approved common name for little bluestem (*Schizachyrium scoparium* [Michx.] Nash). Similarly, in Great Britain, *Medicago sativa* L. is called *lucerne* instead of *alfalfa,* and *Dactylis glomerata* L. is called *cocksfoot* instead of *orchardgrass.* Common and scientific names for many grasses, legumes, and other forage plants discussed in this book are listed in the appendix.

THE GRASSES

Grasses are grouped into about 650-785 genera containing about 10,000 species (Watson and Dallwitz 1992). About 170 genera and 1400 species are found growing in the US according to Hitchcock (1951). Grasses have a wider range of adaptation to temperature and

C. JERRY NELSON is Curators' Professor in the Department of Agronomy, University of Missouri. He received the MS degree from the University of Minnesota and the PhD from the University of Wisconsin. He conducts research on genetic and environmental effects on growth of grasses and persistence of legumes.

LOWELL E. MOSER is Sunkist Fiesta Bowl Professor of Agronomy, University of Nebraska-Lincoln. He received the MS degree from Kansas State University and the PhD from Ohio State University. He teaches forages, range management, and forage physiology classes and conducts research on growth, development, and management of grasses.

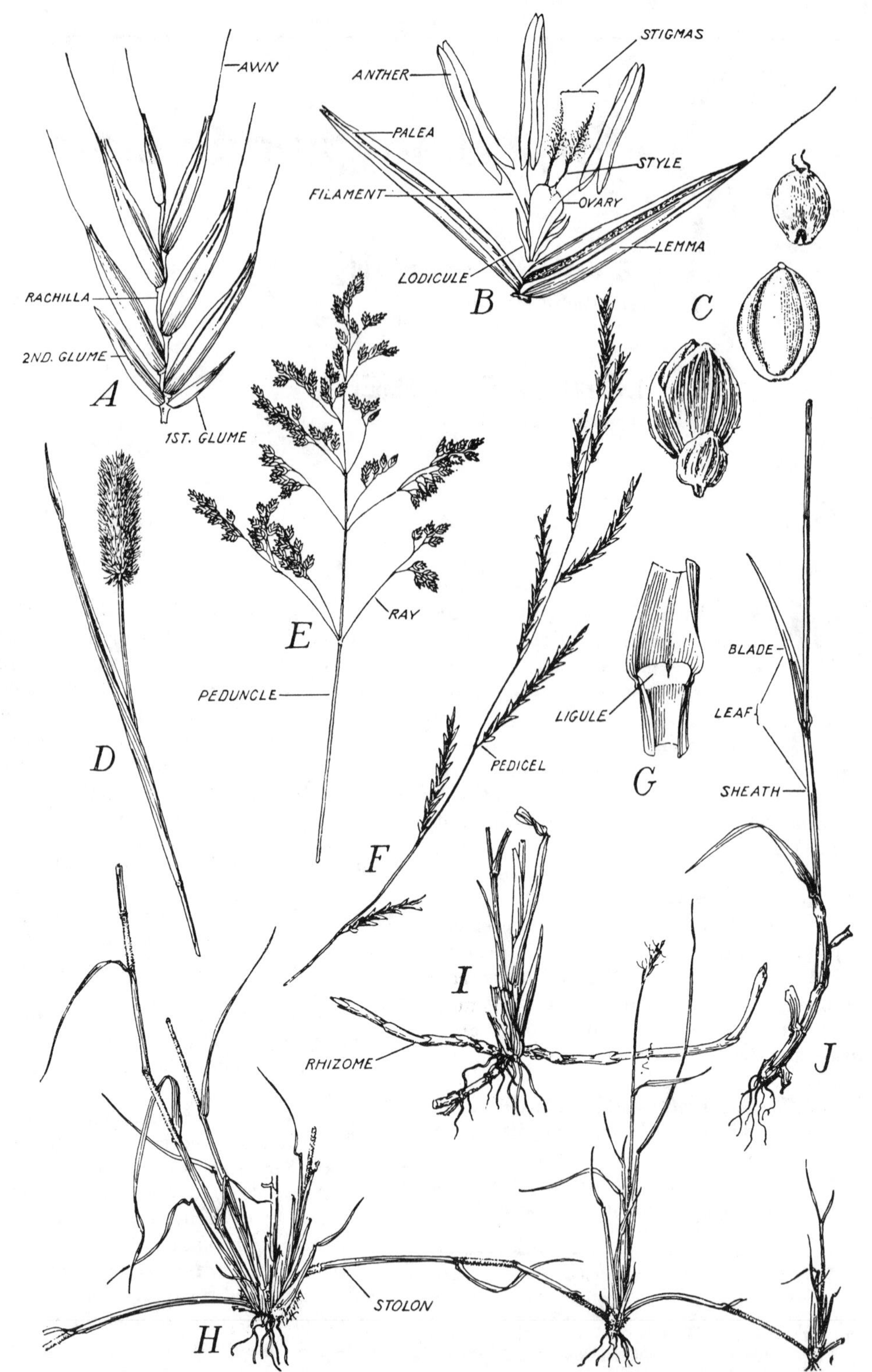

Fig. 2.1. Characteristic morphology of grass plant parts: (*A*) florets on a spikelet arranged on a central axis originally enclosed by two glumes; (*B*) floret is opened to show different parts of a grass flower that were enclosed by the lemma and the palea; (*C*) fruit (caryopsis) develops within the lemma and palea; caryopsis is shown successively enclosed by the glumes, with the lemma and palea closely adhering, and with actual caryopsis being free; (*D,E,F*) spikelets arranged in a spike, panicle, and raceme, respectively; (*G*) ligule at junction of leaf blade and leaf sheath; (*H,I,J*) means of propagating or spreading—stolon, rhizome, and bulb, respectively. (Dayton 1948)

rainfall than any other family of flowering plants, with members being found in the humid tropics, arid areas, and alpine peaks. All the cereal crops and about 75% of the species cultivated as forage crops are grasses.

Description. Grasses are either annuals or perennials, and almost all are herbaceous. The embryo within the seed has one cotyledon, sometimes called the *scutellum,* as distinguished from legumes, which have two cotyledons. At maturity the grasses range in height from a few centimeters to more than 20 m. Bamboo (*Bambusa bambos* Druce) attains the greatest size, but sugarcane (*Saccharum officinarum* L.), corn (*Zea mays* L.), and forage sorghum (*Sorghum bicolor* [L.] Moench) are also representative of larger grasses. Many warm-season perennial grasses such as big bluestem (*Andropogon gerardii* Vitman) and indiangrass often exceed 2 m in height, even in temperate zones in the northern US and Canada. Tropical forage grasses such as elephantgrass (*Pennisetum purpureum* Shumach.) in Florida may exceed 3 m in height. The basic organs of grasses are stems, roots, and leaves. The inflorescences and fruits are specialized parts consisting of modified stems and leaves (Hitchcock 1951; Gould 1968; Leithead et al. 1971).

Morphology. Despite having certain morphological characteristics in common, many modifications from the typical structure allow grass species to be adapted to specific environmental conditions and management practices. For example, some grasses tiller more actively and keep more leaf area near the soil level, allowing them to tolerate closer and more frequent grazing than species that produce few tillers and have leaf area displayed farther above the soil surface. Modifications in morphology also provide a convenient means for identification (Hitchcock 1951).

LEAVES. Leaves are borne on the stem, one at each node, but are projected alternately in two rows on opposite sides of the stem (Fig. 2.1*J*). The leaf consists of a sheath, blade, and ligule (Fig. 2.1*G* and *J*). The sheath surrounds the stem above the node where it is attached. The margins of the sheath usually are overlapping (open), though in some species they are united (closed) into a cylinder for a part or all of the distance between the node and blade (Fig. 2.2).

Leaf blades are parallel veined and typically flat and narrow. Some species have auri-

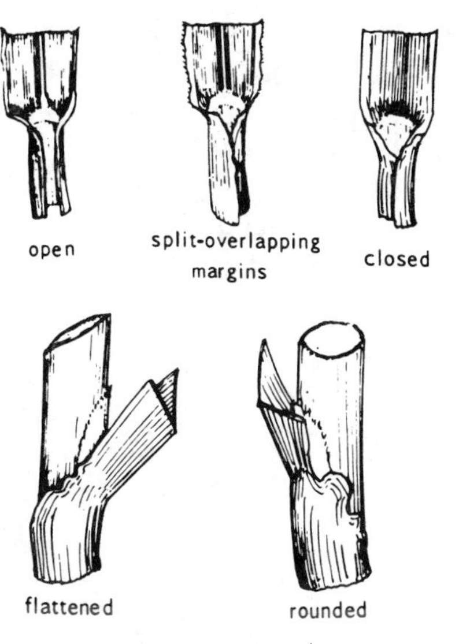

Fig. 2.2. Sheaths of grasses for specific species are characteristically open; split, with overlapping margins; or closed, with the margins fused. New leaves emerging from the whorl of preceding leaves are characteristically flattened or folded (orchardgrass) or rounded or rolled (smooth bromegrass and tall fescue).

cles, small earlike appendages, projecting from the blade near the collar area at the junction of the sheath and blade. The ligule, an appendage located where the sheath and blade join (Fig. 2.1*G*), helps in identification since it may be absent, a papery membrane of a range of sizes and shapes, or a fringe of hairs. The collar is the hardened region at the junction of the sheath and blade. The blades are oriented vertically as they emerge through the whorl of sheaths to the top of the plant. Depending on the species, they may be rolled tightly (rounded) or folded (flattened) during emergence (Fig. 2.2). Once emerged, the blade opens (flattens) and, depending on the species, decreases in angle.

STEMS. Forage grass stems have two distinct forms, depending on the stage in the life cycle. Stems of seedlings and nonreproductive or vegetative tillers tend to be short, consisting of nodes and unelongated internodes (Fig. 2.3). This adaptive mechanism keeps the terminal meristem, which is located at the top of the short stem, enclosed within the whorl of older leaf sheaths and near ground level, where it is protected from removal by grazing or cutting. When flowering begins, the terminal meristem differentiates into the reproductive structure (inflorescence), and the internodes below it elongate by action of

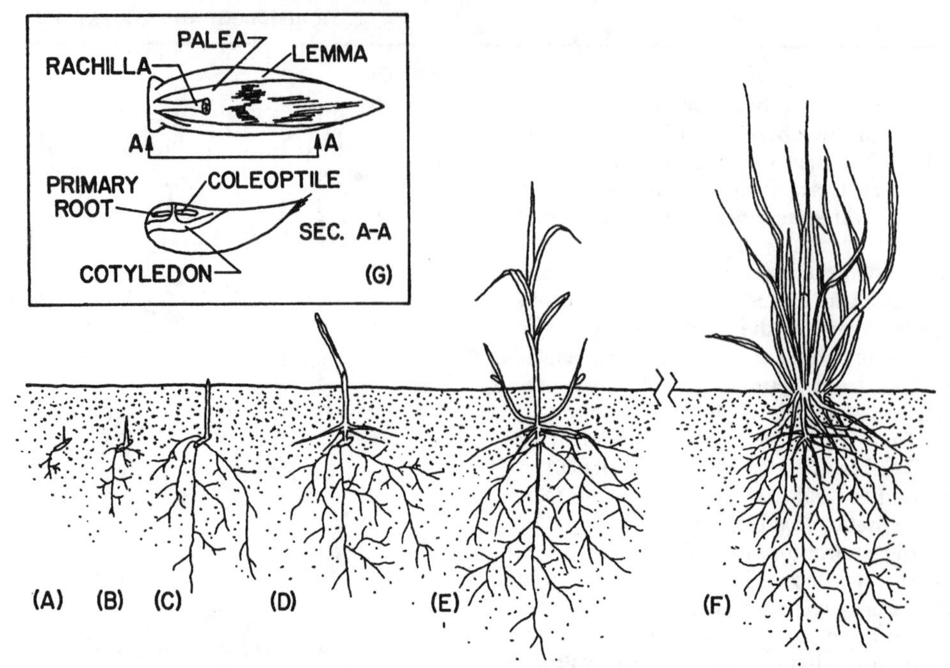

Fig. 2.3. Seedling development of grasses follows a sequence of events: (*A*) seed imbibes water and primary root emerges; (*B*) epicotyl elongates upward; seminal roots develop; (*C*) elongation of coleoptile ceases when it reaches the soil surface and the shoot (leaf blade) extends past the coleoptile; (*D*) adventitious roots begin to develop from the coleoptilar and other basal nodes; (*E*) seedling about 6 wk old; primary and seminal root systems begin to deteriorate; tillers begin to appear from axillary buds located on nodes of main stem; (*F*) plant is well established with numerous tillers, each with its own terminal bud; plant is totally dependent on the adventitious roots; (*G*) mature floret (seed unit) of tall fescue with lemma and palea intact, and with lemma and palea removed showing longitudinal section of true caryopsis (*A-A*). Note relative sizes of floret, caryopsis, and embryo.

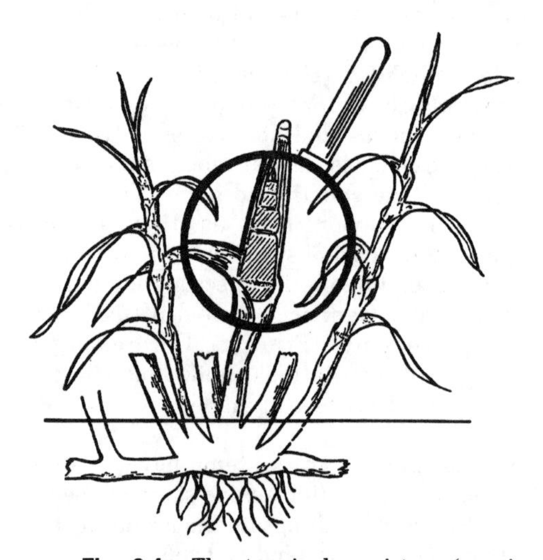

Fig. 2.4. The terminal meristem (growing point) of grasses remains near soil level, where it is protected while producing leaves and unelongated internodes. Under proper environmental conditions the terminal differentiates into the inflorescence, and the internodes elongate to elevate the inflorescence to the top of the canopy. The enlargement shows the partially elongated internodes and the developing inflorescence. Bulges on the tillers can be detected by palpation to indicate the height of the terminal meristem. (Adapted from original published by *Hoard's Dairyman*, Fort Atkinson, Wis.)

intercalary meristems located at the base of the internodes (Fig. 2.4). This separates the nodes and pushes the inflorescence upward through the whorl of sheaths to expose it at the top of the plant.

Elongated stems of a flowering grass plant are distinctly divided into nodes and internodes. The internodes of forage grasses are usually hollow but may be pithy and solid such as in big bluestem and forage sorghum. The node or joint is always solid. During vegetative regrowth, new tillers in reed canarygrass (*Phalaris arundinacea* L.), smooth bromegrass (*Bromus inermis* Leyss.), and some other grasses often elongate slightly, elevating the terminal bud to 10-20 cm above soil level, even though the terminal does not become reproductive. Close grazing in summer may remove the terminal, but both of these species have horizontal underground stems called *rhizomes* (Fig. 2.1*I*), which aid in persistence.

Leaves have vascular connections between the sheath and stem at the nodes. Lateral or axillary buds, located in the axil of each leaf, may develop into tillers. Brace roots, or adventitious roots initiated above soil level, may arise from the base of the intercalary meristem, a zone of cell division and elongation just above the node that is responsible for internode elongation. Cells of the intercalary meristem remain meristematic, capable of division and elongation, until the stem has matured. This allows lodged stems to turn upward again because of differential growth of the intercalary meristem, which is more rapid on the lower side.

In addition to vertical flowering stems, also called *culms*, grasses such as quackgrass (*Elytrigia repens* [L.] Nevski), johnsongrass (*Sorghum halepense* [L.] Pers.), Kentucky bluegrass (*Poa pratensis* L.), switchgrass (*Panicum virgatum* L.), reed canarygrass, and many others have rhizomes (Fig. 2.1*I*). The rhizome is usually the overwintering part of perennial rhizomatous grasses. It has an axillary bud located behind the scalelike leaf at each node and often contains a large reserve of stored food that supports growth of new shoots from the axillary buds. Rhizomes are desirable to allow thin stands to develop into a sodlike canopy. This makes plants such as quackgrass and johnsongrass weedlike and difficult to control. Some grasses, such as tall fescue (*Festuca arundinacea* Schreb.), have only short rhizomes, which gives the canopy a loose, bunchlike appearance.

Stolons are creeping stems located aboveground (Fig. 2.1*H*) and resemble rhizomes in that they have definite nodes and internodes. They can root at nodes and produce axillary buds from which new shoots arise. Buffalograss (*Buchloe dactyloides* [Nutt.] Engelm.) and bermudagrass (*Cynodon dactylon* [L.] Pers.) are well-known stoloniferous species.

The vertical stem of many perennial grasses has thickened lower internodes in which food accumulates, as well as lateral buds on its lower nodes from which new shoots (tillers) arise. The combination of an energy source and active meristems near ground level serves to perpetuate the plants through the winter or dormant seasons and allows regrowth to commence. Timothy (*Phleum pratense* L.) has many short, enlarged lower internodes, collectively called a *haplocorm*, which differs somewhat from a true corm. Bulbous bluegrass (*Poa bulbosa* L.) has true bulbs (Fig. 2.1*J*).

ROOTS. Established grasses have an adventitious, fibrous root system. Adventitious roots develop from lower nodes of the shoot axis of each new tiller after it has emerged aboveground and developed leaf area. The daughter tiller is then largely independent, although still attached to the mother tiller. The adventitious root system is heavily branched, especially in the upper soil horizons, making it well adapted for efficient uptake of top-dressed fertilizers (Blaser and Kimbrough 1968) and soil conservation. Depth of rooting is very dependent on soil conditions and is favored in soils that have low physical strength and good aeration (Weirsma 1959).

Species also differ in the depth and distribution of roots. For example, Gist and Smith (1948) compare root distribution of some popular cool-season grasses to a depth of 45 cm (Table 2.1). Smooth bromegrass has a root mass similar to that of Kentucky bluegrass, but a higher proportion of the roots are distributed deeper within the profile, making smooth bromegrass more drought tolerant. Timothy has a shallow root distribution, whereas orchardgrass is intermediate. Stoloniferous plants such as bermudagrass and bahiagrass (*Paspalum notatum* Flugge) form roots at the nodes of the lateral stems (Fig. 2.1*H*). These roots provide anchorage of the stolon and assume the major role of water and mineral uptake for that part of the plant. Likewise, rhizomes can form roots at their nodes to help the plant gain water and nutrients.

Warm-season grasses, especially the tall-growing prairie species such as big bluestem and switchgrass, generally have fewer roots that are larger in diameter and grow deeper into the soil than those of cool-season grasses (Weaver 1926). This added depth is important in drought-prone areas because it increases

TABLE 2.1. Total mass and percentage distribution of roots of perennial forages growing in a shallow soil

Species	Mass	Soil depth (cm) 0-7.5	7.5-15	15-22.5	22.5-30	30-45
	(kg/ha)			(%)		
Smooth bromegrass	4688	55	16	11	10	8
Orchardgrass	5234	82	8	5	3	2
Kentucky bluegrass	4292	88	7	4	<1	<1
Timothy	2771	90	5	3	2	<1

Source: Gist and Smith (1948).

the volume of soil occupied by the root system and improves access to available soil water.

INFLORESCENCE. The basic unit of the grass inflorescence is the spikelet (Fig. 2.1*A*). Spikelets usually occur in groups or clusters, collectively termed the *inflorescence*. Characteristics of the spikelets and the nature of the inflorescence offer convenient distinguishing traits for identifying grasses. With the raceme, spikelets are borne on pedicels along an unbranched axis (Fig. 2.1*F*). The spike, which is characteristic of wheat (*Triticum aestivum* [L.] emend. Thell.), western wheatgrass (*Pascopyrum smithii* [Rydb.] Löve), and perennial ryegrass (*Lolium perenne* L.), differs from the raceme in having sessile (i.e., no pedicel) spikelets (Fig. 2.1*D*). The panicle is the most common type of grass inflorescence. Here the spikelets are pedicelled in a branched inflorescence (Fig. 2.1*E*). Panicles may be open and diffuse like in smooth bromegrass, Kentucky bluegrass, and switchgrass. Alternatively, panicles can be compact, almost looking like spikes, as with timothy and pearlmillet (*Pennisetum glaucum* [L.] R. Br.).

Specialization usually takes place within the spikelet, which has a variable number of florets (flowers) ranging from one to many depending on the species (Fig. 2.1*A*). The central axis of the spikelet, the rachilla, joins the florets. In species having more than one floret per spikelet, a portion of the rachilla often remains attached to the floret as it dehisces. Two glumes (bracts) are attached at the base of the spikelet on opposite sides of the rachilla. They enclose the florets of the spikelet, giving protection, especially during early development. They also are photosynthetic and provide carbohydrate to the seed.

FLORETS. Grasses usually have small, perfect florets (Fig. 2.1*B*), although separate staminate and pistillate spikelets are characteristic of some species. Below each floret are two bracts. The larger (outer) one is the lemma, which may have an awn attached; the smaller (inner) one is the palea, which usually is partially enclosed by the lemma (Fig. 2.3*G*). There are usually three stamens. The single pistil has an ovary containing one ovule. The style commonly terminates in a feathery (plumose) two-parted stigma (Fig. 2.1*B*). In many cross-pollinated grasses two or sometimes three small structures called *lodicules* are located between the ovary and the lemma and palea at the base of the floret. These lodicules swell by absorbing water to help force open the lemma and palea at the time of anthesis to facilitate wind pollination.

Most grasses flower every year, and many temperate species have a cold temperature requirement (Chap. 5). Some perennials spread by rhizomes or stolons to cover extensive areas without flowering regularly. Most forage grasses are cross-pollinated by wind, but some forage grasses are cleistogamous (self-pollinating in the bud), similar to wheat. Some species such as Kentucky bluegrass (Carnahan and Hill 1961) and buffelgrass (*Cenchrus ciliaris* L.) (Bashaw 1974) reproduce by apomixis, a form of seed set without fertilization. Species that are cleistogamous or apomictic generally have less genetic diversity within a cultivar than do cross-pollinated species.

FRUIT OR CARYOPSIS. Almost all commonly used forage grasses have a seed unit, or caryopsis (Fig. 2.3). The *caryopsis* is a specialized form of a fruit because the ripened ovary is fused onto the outer wall of the single ovule. In most forage grasses the caryopsis remains enclosed by the lemma and palea (Figs. 2.1*C*, 2.3), even during seed harvest and processing. The lemma and palea along with the seed coat (ovary wall) protect the seed against mechanical damage, moisture loss in storage, or attack by organisms. The caryopsis consists largely of endosperm, the food used to maintain and protect the embryo during storage and to support the seedling during germination and emergence. The embryo consists of a cotyledon (scutellum), which secretes enzymes to digest the endosperm; the epicotyl, which develops into the shoot portion of the plant; and the primary root, which anchors the seedling and absorbs water (Fig. 2.3). The endosperm in most forage grass caryopses is not as massive as it is in corn or wheat. Thus, planting depth of forage grasses is critical, and relatively early on, the seedling depends on photosynthesis of emerged leaves for its food supply.

The lemma and palea may interfere with absorption of water and exchange of gases, or they may contain inhibitors that delay or slow germination until they are leached or degraded naturally (Ahring et al. 1975). For example, removing the glumes, lemma, and palea of indiangrass seed increased germination percentage in petri dishes, but field emergence from bare caryopses was decreased when compared with planting intact spikelets (Geng and Barnett 1969). Many grasses such as big bluestem and indiangrass are difficult to seed because they have large lemma and palea appendages that make the seeds light and fluffy (Chap. 32). In these grasses the actual caryopsis is only a small portion of the total weight.

Seeds of some grass species have a high level of physiological dormancy, especially some of the warm-season grasses. This is probably a protection mechanism in the embryo against early germination because dormancy is highest after seed harvest and is lost gradually with time. To obtain an accurate percentage of live seed, seed-testing laboratories routinely cold stratify seed by allowing it to imbibe water, then hold the moist seed at 4°C for 2 to 4 wk. This breaks the dormancy for the seed germination test, but a poor stand will result if the dormancy is still in the seed lot at planting time. A germination test without cold stratification just prior to planting might better predict performance under field conditions (Chap. 32).

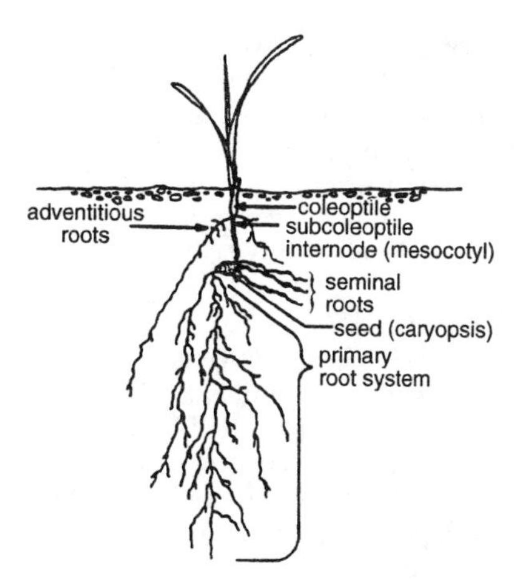

Fig. 2.5. Hypothetical seedling of a panicoid grass plant showing positions and lengths of the coleoptile, subcoleoptile internode, and the three root systems. Primary and seminal roots arising from the seed are short-lived, leaving the plant totally dependent of the adventitious roots that arise from nodes of the shoot axis near soil level. Festucoid plants have a long coleoptile and a short subcoleoptilar internode. (See Fig. 2.3.) (Adapted from Newman and Moser 1988.)

Seedling Development. All grass seedlings have hypogeal emergence, meaning the cotyledon remains below the soil (Figs. 2.3, 2.5). Following imbibition of water, the first sign of germination is the emergence of the primary root (radicle). Early in seedling development up to five seminal roots may emerge just above the cotyledon. The primary and seminal root systems are temporary but are very important for water and nutrient absorption early in the life of the grass seedling. After root growth has begun, the shoot, or epicotyl—that tissue above the cotyledon—emerges and grows toward the soil surface. These emerging leaves are protected within the tubelike coleoptile, which elongates through the soil. Once the coleoptile tip reaches light above the soil surface, it ceases growth, and the true leaves extend through it to emerge aboveground (Figs. 2.3, 2.5). Young grass leaves are very fragile and could not be forced through the soil.

Since the primary and seminal root systems of seedlings are only temporary, a permanent adventitious root system must develop before grass seedlings can become established (Ries and Svejcar 1991). These adventitious roots form at the coleoptilar node and at nodes immediately above it (Fig. 2.5). Hoshikawa (1969) evaluated over 200 species of grasses and classified them into groups based on features of underground development. Most cool-season grasses have the "festucoid" and most warm-season grasses the "panicoid" type of seedling development.

Festucoid seedlings have a long coleoptile that elongates from near the seed to reach the soil surface when the seed is planted at normal planting depth. However, many warm-season grasses have panicoid development and a very short coleoptile. In these plants, the subcoleoptilar internode (the internode between the coleoptilar and cotyledonary nodes, sometimes called a *mesocotyl*) also elongates, pushing the coleoptilar node and the short coleoptile upward to the soil surface (Fig. 2.5). Thus, the coleoptile still protects the emerging shoot even though it is very short, but it reaches the surface due to elongation of the subcoleoptile internode. This method of emergence places the coleoptilar and other nodes just below the soil surface, which may remain dry and delay or limit initiation of adventitious roots. Roots are not initiated into dry soil (Boatwright and Ferguson 1967).

Cool-season grasses generally are easier to establish under sporadic rainfall conditions than warm-season grasses. For example, blue grama (*Bouteloua gracilis* [H.B.K.] Lag. ex Steud.), a panicoid-type grass, emerges quickly, but without water the plants die when 6-10 wk old (Hyder et al. 1971). Adventitious roots do not develop because the soil surface remains dry, and seedlings die when the temporary root system is no longer functional. In contrast, crested wheatgrass (*Agropyron desertorum* [Fisch. ex Link] Schult), with festucoid seedling development, establishes under these conditions because the coleoptilar node remains at the seeding depth, where the improved water status facilitates adventitious root development (Table 2.2). The soil surface must remain moist for 2 to 4 d before adventitious roots develop on blue grama seedlings (Wilson and Briske 1979).

THE LEGUMES

There are nearly 600 genera and some 12,000 species of legumes (Willis 1973), with almost 4000 species in the US (Fernald 1950). Legumes have a narrower range of adaptation than grasses and usually require a higher management level to persist and remain productive.

Description. Legumes are dicotyledons and may be annual, biennial, or perennial. They are so named because the term *legume* indicates the type of fruit (pod) that is characteristic of plants of this family (Fig. 2.6). A legume is a monocarpellary (one-chamber) fruit that contains a single row of seed and dehisces (splits) along both sutures or ribs. Dehiscence is a good mechanism for natural seed dispersal but is a problem for seed producers as mature pods open or shatter easily and seed is lost.

Most forage legumes grow symbiotically with nitrogen-fixing bacteria that form nodules on the roots. These bacteria use plant carbohydrates to reduce atmospheric nitrogen (N_2) to forms that are available to the plant (Chap. 4). Thus, legumes are valuable components in forage mixtures with grasses and in crop rotations with corn and other cereal crops to decrease dependence on fertilizer N. Also, legumes generally are higher in cellular N (protein) than grasses.

Morphology. The morphology of forage legumes differs greatly from that of grasses. Distinct morphological differences exist between genera and between species of legumes that allow identification (Bailey 1949), and they explain differences in adaptation for spe-

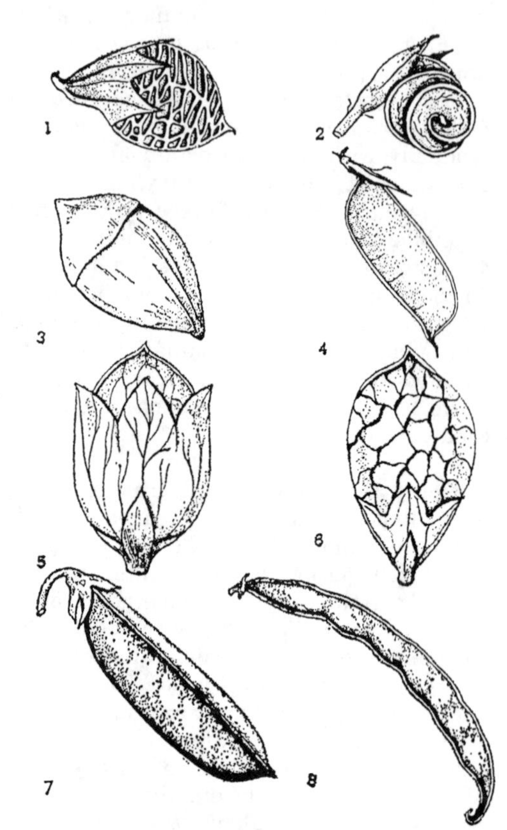

Fig. 2.6. Different types of legume seedpods showing dried sepals that remain attached and residual of the stigma: (*1*) yellow sweetclover, (*2*) alfalfa, (*3*) red clover, (*4*) hairy vetch, (*5*) common lespedeza, (*6*) korean lespedeza, (*7*) fieldpea, (*8*) cowpea. Drawings (*1*), (*3*), (*5*), and (*6*) show one seed per pod, those of the lespedezas often remaining intact during harvest so that the agronomic seed retains the pod coat.

TABLE 2.2. Morphological features of seedlings 22 days after planting at three depths in a greenhouse

Species and item	Planting depth (mm) 20	40	60
Blue grama			
Emergence (%)	100	58	0
Coleoptile length (mm)	6 ± 1	7 ± 2	...
Mesocotyl length (mm)	18 ± 2	30 ± 2	...
Depth to adventitious roots (mm)	2-6	7-10	...
Crested wheatgrass			
Emergence (%)	91	88	20
Coleoptile length (mm)	24 ± 3	34 ± 6	44 ± 9
Mesocotyl length (mm)	0	0	0
Depth to adventitious roots (mm)	20	34-40	48-60

Source: Hyder et al. (1971).

cific environmental or management conditions.

LEAVES. Leaves of legumes are arranged alternately on the stem and have characteristically large stipules, the leaflike appendages attached directly to the stem near the junction of the stem and petiole. Leaf blades are connected to the stem by a petiole. A single leaf blade can be attached directly to the petiole, in which case the leaf is termed unifoliolate, or the leaf can be compound, in which case three (trifoliolate) or more leaf blades are individually connected to the petiole by short stalks called *petiolules* (Fig. 2.7). Compound leaves can be pinnate (Fig. 2.7, drawings 1, 2, 3, 6), where the central leaf blade is connected by a longer petiolule than are the lateral leaflets, as the compound leaves are on alfalfa and sweetclover (*Melilotus officinalis* Lam.). On hairy vetch (*Vicia villosa* Roth), a tendril replaces the terminal leaflet. On palmate leaves (Fig. 2.7, drawings 4 and 5), all blades have equally short petiolules.

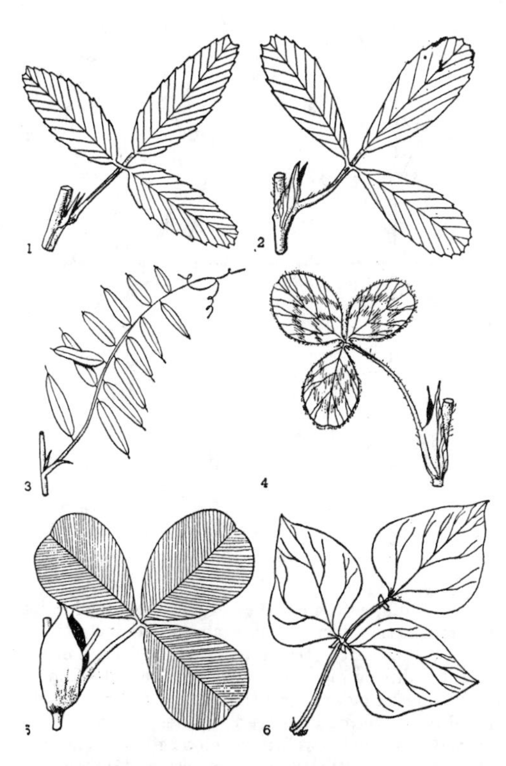

Fig. 2.7. Legume leaves differ in relative length of petioles and petiolules, presence or absence of pubescence (hair), degree of serration of leaf blades, and size and shape of stipules: (*1*) sweetclover, (*2*) alfalfa, (*3*) hairy vetch, (*4*) red clover, (*5*) korean lespedeza, (*6*) cowpea. (*1* and *2* adapted from Isely 1951.)

STEMS. Stems of legumes vary greatly among species in length, diameter, amount of branching, and woodiness. The terminal bud of the stem is always located at the stem tip, which is near the top of the canopy in legumes such as alfalfa, sweetclover, and korean lespedeza (*Kummerowia stipulacea* [Maxim.] Makino). The terminal bud of these species is removed by harvest, even with low grazing pressure, so plants must regrow from axillary buds at lower nodes, especially those near the soil surface. However, the terminal bud remains lower in the canopy with red (*Trifolium pratense* L.) and alsike (*T. hybridum* L.) clovers because the long petioles display the leaf blades well above the terminal to form the upper canopy. At early stages, animals graze mostly leaf blades and petioles and may not remove the terminal bud, allowing it to continue producing new growth. Species like white clover (*T. repens* L.) have prostrate stolons with terminal buds near ground level (Fig. 2.8). Normal cutting or grazing removes only leaf blades and petioles. The terminal bud may be damaged by hooves, but it is rarely removed by grazing.

In the axil of each leaf, located between the stipules, is an axillary bud. These buds can

Fig. 2.8. White clover spreads by means of stolons that can develop adventitious roots near each node. Leaf blades are supported by long petioles; flower heads are supported by long peduncles arising directly from nodes. Axillary buds at nodes can develop stoloniferous branches. (Adapted from Isely 1951.)

break dormancy to contribute to vegetative growth (branching) or to form flower buds. If the terminal is removed or damaged by cutting or grazing, these buds, especially the lower ones, can grow to provide new meristems.

ROOTS. Most legumes, especially the herbaceous ones, have prominent taproots (Fig. 2.9). Fine secondary roots or large branches may arise from the primary root. For example, most alfalfa and sweetclover plants have only a few fine secondary roots, the primary root surviving as a taproot until the entire plant dies. Red clover, a short-lived perennial, survives normally for only about 2 yr because the primary root becomes infected with diseases, causing the plant to die. In contrast, white clover has a taproot as a seedling, but it becomes diseased and dies about 1 to 2 yr after planting (Westbrooks and Tesar 1955). Prior to that time this stoloniferous species is able to develop adventitious roots near the nodes of the stolons (Fig. 2.8) to form a fibrous-rooted plant. Each year the cycle of producing new stolons and adventitious roots continues while old stolons and roots die. Thus, white clover can be quite drought resistant during the first year or two because the taproot can penetrate up to 1 m or more. Later, however, the plant is less drought resistant as fibrous roots from stolons proliferate mainly in the upper soil horizons.

Roots of most legumes also serve as storage organs for reserve substances, primarily carbohydrate and N-containing compounds

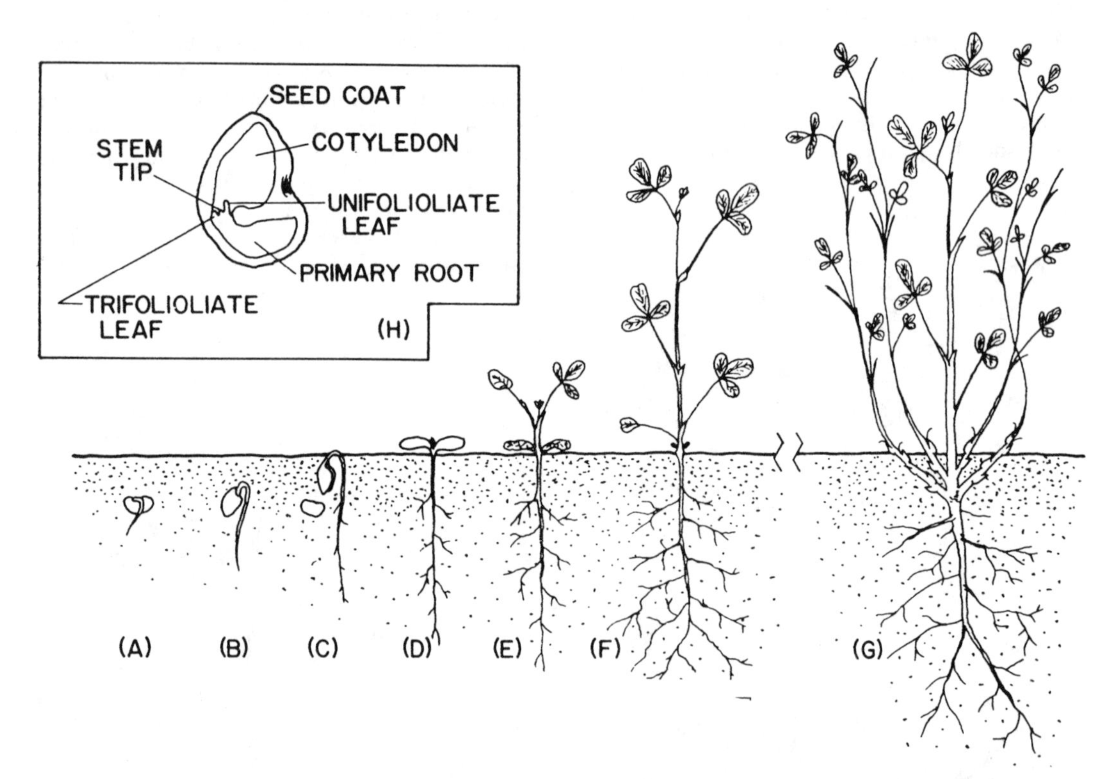

Fig. 2.9. Seedling development of a legume, such as alfalfa, with epigeal emergence: (*A*) seed imbibes water and primary root emerges; (*B*) hypocotyl becomes active, forming an arch, to penetrate the soil; (*C*) elongation of hypocotyl stops when arch reaches light; (*D*) arch straightens and cotyledons open for photosynthesis, exposing the epicotyl that was protected during movement through the soil; (*E*) primary root continues to elongate and enlarge, developing some secondary (branch) roots; unifoliolate leaf develops, followed by the first trifoliolate leaf; terminal bud is located between stipules of last developed leaf; (*F*) cotyledons fall off and axillary buds at cotyledonary node swell to develop into branches; stem continues to elongate, producing a leaf at each node; (*G*) contractile growth has occurred, taproot morphology is developing; crown is forming, with branches clearly evident from buds at the cotyledonary node and in axils of the unifoliolate and first trifoliolate leaves; crown will continue to enlarge because each new branch has unelongated internodes near or below soil level that have incomplete leaves and axillary buds to provide sites for branching or regrowth following cutting; (*H*) alfalfa seed dissected longitudinally to show one cotyledon, primary root, stem tip, and embryonic leaves.

(Chap. 3). These substances are accumulated from photosynthesis and other metabolism, then are used later to support regrowth of buds following removal of leaf area by cutting or grazing.

INFLORESCENCE. Flowers usually are arranged either in racemes as in the pea (*Pisum sativum* L.), in spikelike (compact) racemes as in alfalfa and sweetclover, in very compact racemes (heads) as in the true clovers (*Trifolium* spp.), or in umbels as in birdsfoot trefoil (*Lotus corniculatus* L.) and crownvetch (*Coronilla varia* L.). Number of flowers per inflorescence varies greatly from species to species; e.g., the short raceme of alfalfa may contain up to 20 flowers with 8-10 seeds per pod, while the longer raceme of sweetclover may contain over 60 flowers, usually with only 1 seed per pod. The clovers may have 60-120 flowers producing 1 seed in each pod. Birdsfoot trefoil usually has 4-7 flowers per umbel, with each developing pod having 5-15 seeds (Miller et al. 1975).

FLOWERS. Flowers of common legume species have floral petals (corollas) that are characteristically papilionaceous, or "butterfly-like" (Fig. 2.10). These irregular flowers have five petals: a standard, two wings, and a keel consisting of two petals that are more or less united. The calyx, or group of sepals that surrounds the corolla in the bud stage, normally is five toothed. The keel, so named for its boatlike shape, encloses the stigma and the stamens. There are usually 10 stamens, one free, the filaments of the other nine being joined at the base to form an envelope enclosing the long, slender style and ovary.

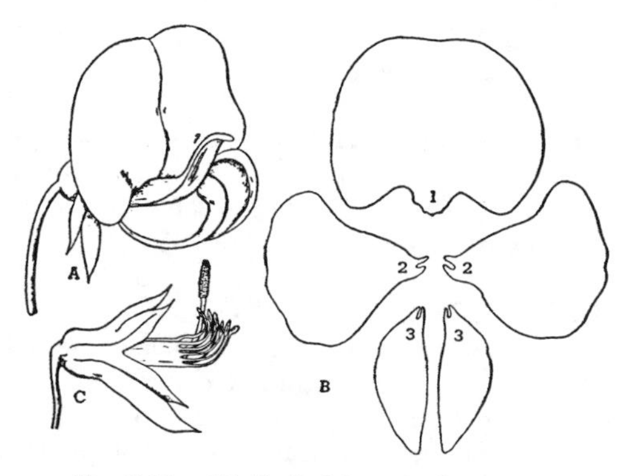

Fig. 2.10. (*A*) Typical legume flower. (*B*) The five petals separated: 1—standard; 2—two wings; 3—keel, two united petals. (*C*) The five sepals and 10 exposed stamens (1 free and 9 partially united) that surround the pistil with stigma.

The corolla tube, formed by partial joining of the five petals near their bases, varies in length among species. For example, the corolla tube of red clover, sometimes 12 mm or more in length, is relatively long for the size of the flower (see Chap. 17, Fig. 17.2). In alsike clover, white clover, sweetclover, and alfalfa the tube is much shorter. Since nectar is secreted by glands located at the corolla tube's base, the length of the corolla tube is a determining factor in the ability of bees and other insects to reach the nectar and thus is a factor in cross-pollination.

Some legumes, including many beans and peas, are ordinarily autogamous (self-pollinating) and rather completely self-compatible. They pollinate their own stigmas and need no tripping for the pollen to reach the stigma; pollen contacts the stigma as it rolls out of the opening anthers. Many forage legumes have short stamens, however, presenting the anthers too far below the stigma for pollen contact when the anthers open. These flowers must be tripped; i.e., the keel must be pressed down until the anthers and stigma spring out of the keel and the pollen is flipped into the air and given an opportunity to fall back on the stigma. Several legumes are self-sterile (e.g., red clover) or mostly self-sterile (e.g., alfalfa); therefore, little or no seed will be set unless the stigma is fertilized with pollen transported from another plant by a bee or other pollinator. Usually these flowers must be tripped by a pollinator for pollen to reach the stigma and seed to be produced.

FRUIT AND SEED. The fruit is a pod containing one to several seeds (Fig. 2.6). Legume seeds usually do not have an endosperm at maturity. Reserve food is stored in the two cotyledons (Fig. 2.9). The hilum is the scar on the seed where it was attached to the pod. Near one end and between the edges of the cotyledons is the embryo axis, which consists of the epicotyl and primary root. Each seed is enclosed in the testa (seed coat).

Many forage legumes have hard seeds, in which case the seed coat is impervious to water, thus limiting germination. With time in the soil, these coats become softened by microorganisms, wetting and drying, and freezing and thawing, which allows the seeds to germinate. Thus, volunteer stands of legumes are not uncommon, even when the area has not been seeded for several years. For example, over a period of several years in Colorado it was found that about 22% of the commercial seed of alfalfa was hard. Hard seededness of seed lots usually is decreased by storage, by blending with lots having low frequency of hard seed, or by scarification (Bass 1988). Scarification is a mechanical process of scratching the seed coat to allow imbibition (uptake) of water, but rigorous treatment can injure seed and cause abnormal seedlings and loss of vigor. Conversely, annual legumes need to reestablish each year in permanent pastures or hayfields. Hard seed can be an asset in these legumes because the seed crop dropped to the soil in 1 yr has its germination spread over several years.

Seedling Development. Seedling emergence of legumes is either epigeal or hypogeal. Most forage legumes have epigeal emergence by which the cotyledons are pushed above the ground (*epi* = above) (Fig. 2.9). In contrast, the cotyledons of hypogeal-emerging legumes remain belowground (*hypo* = below) at planting depth. The first sign of germination is the emergence of the primary root through the seed coat. If emergence is epigeal, the hypocotyl (that part of the embryo axis between the primary root and cotyledonary node) lengthens and forms an arch that penetrates the soil (Fig. 2.9*C*). The arch straightens when it reaches the soil surface and is exposed to light, the cotyledons open, and photosynthesis of the cotyledons begins. Until then the growth and respiration of the seedling are dependent on reserves stored in the cotyledons. The first true leaf can be unifoliolate, as in alfalfa and clovers, or trifoliolate, as in birdsfoot trefoil and crownvetch. In lespedezas, the first two true leaves are unifoliolate. Subsequent leaves are consistent with the characteristics of the species.

A few forage legumes, such as hairy vetch, have hypogeal emergence whereby the hypocotyl is not active. Instead, the epicotyl forms an arch that is pushed through the soil by internode elongation until it reaches light. Hypogeal emergence is probably not as effective as the hypocotyl arch emergence of epigeal legumes, but it is of advantage if the young seedling is cut or frosted back to soil level. Legumes with hypogeal emergence have one or more nodes belowground with axillary buds; thus, they have both a meristem and an energy source (cotyledons) to regrow. Epigeal seedlings in the same condition do not have axillary buds or an energy source belowground (Fig. 2.9).

The generally accepted advantages of epigeal emergence include protection of the tender epicotyl within the cotyledons when pushing through the soil and, especially, the added role of the cotyledons in photosynthesis (Shibles and MacDonald 1962). Cotyledons of cultivated forage legumes usually expand their area during and after emergence by up to 10-fold and stay active photosynthetically for 3 to 4 wk. Axillary buds at the cotyledonary node and one or two other lower nodes expand into shoots to begin formation of the crown (Teuber and Brick 1988; Fig. 2.9*G*).

Contractile Growth. About 6 to 8 wk after emergence, many perennial legumes with epigeal emergence begin contractile growth. In this unique process the first node, where the cotyledons were attached, is gradually pulled back below the soil surface (Fig. 2.9). The phenomenon is believed caused by a lateral growth of cells of the hypocotyl and upper primary root, causing these structures to thicken and shorten. Contractile growth draws the first node and the developing crown about 0.5 cm below the surface for white clover, about 1 cm for red clover and birdsfoot trefoil, and about 2 cm for alfalfa. Contractile growth of sweetclover may exceed 4 cm. Winterhardiness of these forages tends to be related to their crown depth. Deeper crown placement may allow better cold protection of the axillary buds needed for spring growth.

As each new shoot of alfalfa emerges from an axillary bud on the crown, it develops several nodes and short internodes below or near ground level (Fig. 2.9*G*). Each node has an axillary bud to provide a site for the next regrowth when the terminal is removed (Nelson and Smith 1968). In contrast, summer regrowths of birdsfoot trefoil arise from axillary buds well above the crown, but plants also develop axillary buds belowground, which survive the winter and provide spring growth. Thus, contractile growth of epigeal-emerging species allows axillary buds to develop and be protected below soil level, which adds to their persistence.

DEVELOPMENTAL STAGES

Determining the stage of development of forage crops is important to characterize maturity, predict quality, and communicate about management practices. Forage maturity may be qualitatively characterized according to height or stage of floral development by observing the most-advanced shoots. Although this may help in visualizing or communicating about the forage, this approach does not accurately characterize the total population of shoots that comprise the hayfield or pasture. Quantitative approaches have been developed for some species that are relatively easy to use, are descriptive, and make predicting forage quality more reliable.

Alfalfa. Morphological stages in alfalfa development and methods to calculate mean stage of development based on number or weight of stems in different stages are described by Fick and Mueller (1989). The 10 stages of development fit into four categories: vegetative, flower bud development, flowering, and seed production (Table 2.3).

Collection of a representative sample of the standing crop is critical. Generally all stems (shoots) more than 3 cm long should be sampled from at least four separate areas (0.1 m^2 each) or from a specified distance along four random rows within a field. Each collected shoot is assigned a stage number according to the criteria in Table 2.3. The numerical index for each stage (0 to 9) is then multiplied by the number of shoots assigned to each stage. The resulting numbers are added, then divided by the total number of shoots sampled. Hence, the mean developmental stage by count (MSC) is simply the weighted mean of the stages of individual shoots in the sample, i.e.,

$$MSC = \Sigma(S \bullet N)/C,$$

where S = stage index, N = number of shoots in stage S, and C = the total number of shoots in the sample. The mean developmental stage can be based on shoot weight (MSW) in a similar manner except total dry weight (D) of shoots in each stage is used in place of N and total weight of the sample (W) is used in place of C.

Both methods give a good estimate of the growth stage for hay-making decisions, but the counting procedure is easier to use in the field. Sometimes new alfalfa shoots (with low stage number) may emerge into older canopies (with high stage number), causing the stage to "decrease." For predicting or communicating about quality of the forage, determining the stage according to stem weight is better (Fick and Mueller 1989).

Grasses. A similar staging system has been developed that is applicable to most grasses (Moore et al. 1991): the life cycle of individual tillers is divided into five primary growth stages—germination, vegetative, elongation, reproductive, and seed development and ripening. Each stage is associated with a mnemonic code and a numerical index (Table 2.4). The germination stage is applicable only to emerging seedlings. A detailed classification system for stages of root and shoot development for grass seedlings is proposed in Moser et al. 1993. It is more helpful for determining success of seedings, timing of pesticide applications, and evaluating other establishment practices.

Classification of established grasses begins with the vegetative stage, which is based on the number of emerged leaf blades (Table 2.4). Each stage can be indexed between 1.0 and 1.9, in more detail, based on the actual number of leaves present divided by N, the number of leaves that will appear on the tiller prior to elongation. In practice, N is equal to the number of visible leaves in the vegetative stage or the number of palpable nodes at the elongation stage. Just prior to internode elongation, all leaves are visible, $n/N = 1$ and the index is 1.9. Similarly, elongation stages are based on the number of nodes that are palpable or visible. Again, if N is known, an index from 2.0 to 2.9 can be constructed based on past experience with fully developed stems of the same species or by calculating the index after seed heads begin to emerge (R_1). Reproductive and seed development stages are based on specific morphological events.

Mean developmental stage based on either count or weight is calculated for grasses just as for alfalfa. This universal system quantitatively characterizes most bunch and sod-forming grass swards. Simon and Park (1983) developed a more complex system for perennial grasses, but it is more difficult to apply in the field. Haun (1973) described a system for wheat based on leaf development that can be adapted to perennial grasses.

LOCATION AND ROLE OF MERISTEMS. The location and function of the terminal meristem, or "growing point," is a major factor in the management of grasses. The terminal meristem initiates new leaves, lays down the cells to develop into axillary buds, and develops the cells for nodes. As such it regulates the rate at which the grass plant grows or develops. When environmental conditions are met, and the plant is of sufficient size, the terminal meristem can differentiate, literally develop itself, into the inflorescence. Thereafter, the terminal will be pushed to the top of the plant by the stem intercalary meristem (Fig. 2.4), where it will produce seed and die. The plant continues to live only if at least one axillary bud at the base has differentiated into a new tiller that has developed its own root system. Therefore, any one tiller in a grass stand generally lives for only 12 to 18 mo; i.e., it is initiated, grows vegetatively until stimulated to flower, and then becomes reproductive and dies. The root system for a reproductive tiller also dies, contributing organic matter and open channels in the soil to give the characteristic soil structure of grasslands.

Good managers of grassland are always aware of the growth stage of the grass plant and especially the location of the terminal meristems in the canopy. Grazing early, before the terminal differentiates into the reproductive structure and before stem elongation occurs, gives a high-quality leafy forage and high rates of animal performance. Cutting for hay is usually delayed until the terminal has been elevated and new tillers have been initiated to replace it when cut. Delaying cutting until flowering increases the yield markedly due to stem growth, but it reduces forage quality (see Chap. 6, Vol. 2).

Axillary buds may develop into vertical tillers that grow upward within the leaf sheath of the main shoot (Fig. 2.3*F*). This tillering habit is characteristic of bunchgrasses such as orchardgrass, big bluestem, and wheatgrasses. In addition, some bunchgrasses produce few new tillers, causing the plant unit to appear as a distinct clump or bunch with open spaces between individual plants. Conversely, sod-forming grasses such as Kentucky bluegrass, smooth bromegrass, and bermudagrass often have rhizomes or stolons and tiller more profusely. The tillers grow more laterally and may emerge a short distance from the main stem. These grasses gradually spread to fill in open areas and form dense sods. Most sod-forming grasses have abundant terminal meristems and leaf area near ground level, and they tolerate frequent and close cutting or grazing better than do bunchgrasses. In contrast, bunchgrasses allow other seed to germinate in open spaces, may be more compatible with legumes, and provide better habitat for wildlife (see Chap. 18, Vol. 2).

Management Implications for Mixtures. Legumes and grasses differ markedly in morphology and do not go through growth stages at the same rate. Thus, species selected for a mixture should be matched for maturity. Further, compromises must be made in management of forage mixtures to favor the weaker competitor, encourage the desired balance of species, and maintain quality of forage. For example, red clover is commonly grown with timothy in many northern regions for hay production, whereas alfalfa is commonly grown with orchardgrass. Orchardgrass reaches each development stage earlier than timothy, and alfalfa reaches each developmental stage earlier than red clover. During summer regrowth, however, timothy commonly produces reproductive stems again, making it less well suited for grazing compared with orchardgrass, tall fescue, or Kentucky bluegrass, which remain vegetative during summer and fall with the terminal bud protected near soil level (Fig. 2.3).

Animals grazing summer regrowth of an or-

TABLE 2.3. Developmental stages, their numerical indices, and descriptions for alfalfa

Index	Stage	Description
Vegetative stages		
0	Early vegetative	Stem length ≤ 15 cm, no visible buds, flowers, or seedpods
1	Mid-vegetative	Stem length 16-30 cm, no visible buds, flowers, or seedpods
2	Late vegetative	Stem length ≥ 31 cm, no visible buds, flowers, or seedpods
Flower bud development		
3	Early bud	1-2 nodes with visible buds, no flowers or seedpods
4	Late bud	≥ 3 nodes with visible buds, no flowers or seedpods
Flowering		
5	Early flower	1 node with one open flower, no seedpods
6	Late flower	≥ 2 nodes with open flowers, no seedpods
Seed production		
7	Early seedpod	1-3 nodes with green seed pods
8	Late seedpod	≥ 4 nodes with green seed pods
9	Ripe seedpod	Nodes with mostly brown mature seedpods

Source: Fick and Mueller (1989).

TABLE 2.4. Developmental stages, their numerical indices, and descriptions for perennial grasses

Stage	Index	Description
Germination		
G_0	0.0	Dry seed
G_1	0.1	Imbibition
G_2	0.1	Primary root emergence
G_3	0.5	Coleoptile emergence
G_4	0.7	Mesocotyl and/or coleoptile elongation
G_5	0.9	Coleoptile emergence from soil
Vegetative—Leaf development		
VE or V_0	1.0	Emergence of first leaf
V_1	$(1/N) + 0.9$[a]	First leaf collared
V_2	$(2/N) + 0.9$	Second leaf collared
V_n	$(n/N) + 0.9$	*N*th leaf collared
Elongation—Stem elongation		
E_0	2.0	Onset of stem elongation
E_1	$(1/N) + 1.9$	First node palpable/visible
E_2	$(2/N) + 1.9$	Second node palpable/visible
E_n	$(n/N) + 1.9$	*N*th node palpable/visible
Reproductive—Floral development		
R_0	3.0	Boot stage
R_1	3.1	Inflorescence emergence/first spikelet visible
R_2	3.3	Spikelets fully emerged/peduncle not emerged
R_3	3.5	Inflorescence emerged/peduncle fully elongated
R_4	3.7	Anther emergence/anthesis
R_5	3.9	Post-anthesis/fertilization
Seed development and ripening		
S_0	4.0	Caryopsis visible
S_1	4.1	Milk
S_2	4.3	Soft dough
S_3	4.5	Hard dough
S_4	4.7	Endosperm hard/physiological maturity
S_5	4.9	Endosperm dry/seed ripe

Source: Moore et al. (1991).

[a]Where n equals the event number (number of leaves or nodes) and N equals the number of events within the primary stage (total number of leaves or nodes developed). General formula is $P + (n/N) - 0.1$, where P equals primary stage number (1 or 2 for vegetative and elongation, respectively) and n equals the event number. When $N > 9$, the formula $P + 0.9\ (n/N)$ should be used.

chardgrass-alfalfa mixture remove only leaf blades of orchardgrass but repeatedly remove the terminal buds from alfalfa, causing alfalfa to begin regrowth again from the crown (Fig. 2.9). If summer regrowth of white clover and tall fescue is grazed, animals remove mainly leaf blades from tall fescue and leaf blades and petioles from white clover (Fig. 2.8), leaving the stem and terminal buds of both species near soil level so they can grow and continue to initiate new leaves. Several other examples are discussed in this book. Understanding how each forage plant grows will help in understanding how it can be best managed to meet the needs of the producer.

QUESTIONS

1. Why are legumes so named? Why are they of importance to the national economy?
2. List major morphological characteristics that can be used to identify grasses and legumes.
3. How are forage grasses better adapted than forage legumes for emergence from deep soil depths?
4. Draw a grass flower and label its parts. Do the same for a legume flower.
5. Of what value are cotyledons to young legume seedlings?
6. Why are grass or legume fields used for seed production generally isolated some distance from other fields of the same species?
7. Distinguish between a stolon and a rhizome; a hypocotyl and an epicotyl.
8. Describe a raceme, corm, panicle, lodicule, seminal root, rachilla, petiole, sheath, and axillary bud.
9. Explain the role of axillary buds in persistence of grasses and legumes.
10. Why is it important to be able to identify growth stages of grasses and legumes?
11. What is the value of the binomial system for classifying plants? Who is given credit for developing this systematic plan?
12. Briefly explain why it may be more difficult to manage a grass-legume mixture than either species grown alone.

REFERENCES

Ahring, RM, JD Eastin, and CS Garrison. 1975. Seed appendages and germination of two Asiatic bluestems. Agron. J. 67:321-25.

Bailey, LH. 1949. Manual of Cultivated Plants. New York: Macmillan.

Bashaw, EC. 1974. The potential of apomictic mechanisms in grass breeding. In VG Iglovikov and AP Movsisyants, Proc. 12th Int. Grassl. Congr., Moscow, 3:698-704.

Bass, LN, CR Gunn, OB Hesterman, and EE Roos. 1988. Seed physiology, seedling performance, and seed sprouting. In AA Hanson, DK Barnes, and RR Hill, Jr. (eds.), Alfalfa and Alfalfa Improvement, Am. Soc. Agron. Monogr. 29. Madison, Wis., 961-83.

Blaser, RE, and EL Kimbrough. 1968. Potassium nutrition of forage crops with perennials. In VJ Kilmer, SE Younts, and NC Brady (eds.), The Role of Potassium in Agriculture. Madison, Wis.: American Society of Agronomy, 423-45.

Boatwright, GO, and H Ferguson. 1967. Influence of primary and/or adventitious root systems on wheat production and nutrient uptake. Agron. J. 59:299-302.

Carnahan, HL, and HD Hill. 1961. Cytology and genetics of forage grasses. Bot. Rev. 27:1-162.

Dayton, WA. 1948. The family tree of Gramineae. In USDA Yearbook of Agriculture. Washington, D.C.: US Gov. Print. Off., 637-39.

Fernald, ML. 1950. Gray's Manual of Botany. New York: American Book.

Fick, GW, and SC Mueller. 1989. Alfalfa: Quality, Maturity, and Mean Stage of Development. Cornell Univ. Coop. Ext. Inf. Bull. 217.

Geng, S, and FL Barnett. 1969. Effects of various dormancy-reducing treatments on seed germination and establishment of indiangrass, *Sorghastrum nutans* (L.). Nash. Crop Sci. 9:800-803.

Gist, GR, and RM Smith. 1948. Root development of several common forage grasses to a depth of eighteen inches. J. Am. Soc. Agron. 40:1036-42.

Gould, FW. 1968. Grass Systematics. New York: McGraw-Hill.

Haun, JR. 1973. Visual quantification of wheat development. Agron. J. 65:116-19.

Hitchcock, AS, rev. by A Chase. 1951. Manual of the Grasses of the United States. 2d ed., USDA Misc. Publ. 200. New York: Dover.

Hoshikawa, K. 1969. Underground organs of the seedlings and the systematics of Gramineae. Bot. Gaz. 130:192-203.

Hyder, DN, AC Everson, and RE Bement. 1971. Seedling morphology and seeding failures with blue grama. J. Range Manage. 24:287-92.

Isely, D. 1951. The Leguminosae of the north central United States: Loteae and Trifolieae. Iowa State Coll. J. Sci. 25:439-82.

Leithead, HL, LL Yarlett, and TN Shiflet. 1971. USDA Agric. Handb. 389. Washington, D.C.: US Gov. Print. Off.

Miller, DA, LJ Elling, JD Baldridge, PC Sandal, SG Carmer, and CP Wilsie. 1975. Predicting Seed Yields of Birdsfoot Trefoil Clones. North Central Res. Publ. 227 and Ill. Agric. Exp. Stn. Bull. 753.

Moore, KJ, LE Moser, KP Vogel, SS Waller, BE Johnson, and JF Pederson. 1991. Describing and quantifying growth stages of perennial grasses. Agron. J. 83:1073-77.

Moser, LE, KJ Moore, MS Miller, SS Waller, KP Vogel, JR Hendrickson, and LA Maddux. 1993. A quantitative system for describing the developmental morphology of grass seedling population. In Proc. 17th Int. Grassl. Congr., Palmerston North, New Zealand, 317-18.

Nelson, CJ, and D Smith. 1968. Growth of birdsfoot trefoil and alfalfa. II. Morphological development and dry matter distribution. Crop Sci. 8:21-25.

Newman, PR, and LE Moser. 1988. Grass seedling emergence, morphology, and establishment as affected by planting depth. Agron. J. 80:383-87.

Ries, RE, and TJ Svejcar. 1991. The grass seedling: When is it established? J. Range Manage. 44:574-76.

Shibles, RM, and HA MacDonald. 1962. Photosynthetic area and rate in relation to seedling vigor of birdsfoot trefoil (*Lotus corniculatus* L.). Crop Sci. 2:299-302.

Simon, U, and BH Park. 1983. A descriptive scheme for stages of development in perennial forage grasses. In JA Smith and VW Hays (eds.), Proc. 14th Int. Grassl. Congr. Lexington, Ky. Boulder, Colo.: Westview, 416-19.

Teuber, LR, and MA Brick. 1988. Morphology and anatomy. In AA Hanson, DK Barnes, and RR Hill, Jr. (eds.), Alfalfa and Alfalfa Improvement, Am. Soc. Agron. Monogr. 29. Madison, Wis., 125-62.

Watson, L, and MJ Dallwitz. 1992. The Grass Genera of the World. Wallingford, UK: CAB International.

Weaver, JE. 1926. Root Development of Field Crops. New York: McGraw-Hill.

Weirsma, D. 1959. IV. The soil environment and root development. Adv. Agron. 11:43-51.

Westbrooks, FE, and MB Tesar. 1955. Taproot survival of ladino clover. Agron. J. 47:403-10.

Willis, JC, rev. by HKA Shaw. 1973. A Dictionary of the Flowering Plants and Ferns. 8th ed. London: Cambridge Univ. Press.

Wilson, AM, and DD Briske. 1979. Seminal and adventitious root growth of blue grama seedlings on the Central Plains. J. Range Manage. 32:209-13.

3 Photosynthesis and Carbon Metabolism

C. JERRY NELSON
University of Missouri

PHOTOSYNTHESIS is the primary mechanism for increasing usable energy on this planet. With forage plants, we have a manageable process for capturing solar energy and converting it to useful products, mainly through ruminants, ungulates, and other herbivores that depend on forages to meet nutrient needs. The earth has a wide range in climates in terms of rainfall and temperatures (Chap. 5), and several land areas are not hospitable for production of cereal, oilseed, or fiber crops. Often these areas are steep, erosive, or droughty, but they receive abundant solar energy, which can be used effectively by adapted grasses, legumes, other forbs, and sedges.

Compared with cereals or tuber crops, the complex forage-livestock system has lower conversion efficiency of solar energy to useful products. Efficiency of conversion on a daily basis is low because the process sequentially involves plants that convert solar energy to plant tissue, millions of microbes that live in the rumen or cecum of the animal and convert the fibrous herbage into available nutrients, and the host animal, which converts the available nutrients into meat, milk, or fiber. The amount of retained energy is reduced during each biological step. Conversely, many forages, especially perennials, harvest solar energy during a high proportion of the year, thereby increasing the relative efficiency on an annual basis.

All green plants can convert solar energy into reduced compounds such as sugars that are useful in the plant. Some plants concentrate the sugars into reproductive parts such as the seed of cereals or into vegetative parts such as tubers of potato (*Solanum tuberosum* L.) or roots of sugarbeet (*Beta vulgaris* L.). In contrast, the goal in forage crop management is to encourage use of the sugars, amino acids, and other metabolic products to develop leaf, stem, and other plant parts, primarily in the form of digestible cell walls and cell solubles such as protein and minerals. The value of a forage plant also depends on its ability to grow and persist within a given environment, both of which are associated with the ability to carry on photosynthesis.

C. JERRY NELSON. *See Chapter 2.*

While photosynthesis is critical, it is often not the most important factor in determining yield, quality, or persistence of a grass or legume (Nelson 1988). The plant is a highly integrated entity with a wide range of interconnected activities. Thus, photosynthesis should not be considered alone but as one component in carbon (C) metabolism, which also includes photorespiration, respiration, partitioning, storage, and growth rates (Fig. 3.1). For example, most carbohydrate is used to synthesize cell wall, but some is translocated to the roots where it is respired to drive mineral uptake, and some is used to form C skeletons for synthesis of amino acids and other compounds.

THE PHOTOSYNTHETIC PROCESS

Photosynthesis is a photoreduction process that converts radiant energy to chemical energy by a photosystem complex located in the chloroplast. The chemical energy is subsequently used to reduce carbon dioxide (CO_2)

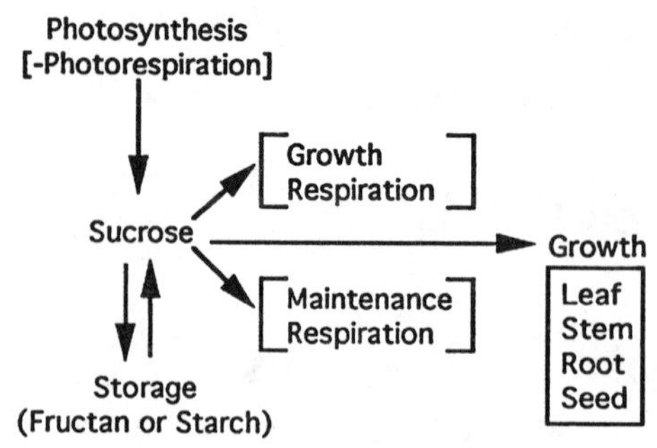

Fig. 3.1. Sugar formed in leaves by photosynthesis is translocated in the phloem to growing areas. Growth respiration is closely coupled to growth rate, whereas maintenance respiration occurs in all tissues. When photosynthesis exceeds growth and respiration needs, the excess carbohydrate is stored as starch or fructan.

from its normal oxidized state to a more reduced state (i.e., gain of electrons) to form compounds such as carbohydrates $(HCOH)_n$. The basic reactions occur in the chloroplasts, which are located mainly in the leaf blades.

Light Harvesting. Chlorophyll is the major pigment involved in capturing light for photosynthesis. Chlorophyll molecules, acting as a part of the photosystem complex of the grana, absorb light, which causes an electron to be elevated to a higher energy state (Fig. 3.2). Through coupling with a series of proteins and enzymes in the grana portion of the chloroplast, the energy in the electron is used to form stable chemical compounds that have high energy such as adenosine triphosphate (ATP) and reducing power such as reduced diphosphopyridine nucleotide ($NADPH_2$). Rates of ATP and $NADPH_2$ formation depend largely on radiation density and respond in a near-linear manner up to high densities. Beginning in the stroma part of the chloroplast and continuing in the cytoplasm, the ATP and $NADPH_2$ are oxidized through a series of linked enzymatic reactions to reduce CO_2 to form other products.

Carbon Dioxide Fixation Pathways. Two major biochemical pathways are used by grass plants to reduce CO_2 in the photosynthetic process, and these pathways are associated closely with differences in leaf anatomy (Waller and Lewis 1979). Temperate or cool-season grasses such as timothy (*Phleum pratense* L.), orchardgrass (*Dactylis glomerata* L.), and tall fescue (*Festuca arundinacea* Schreb.) have a carbon pathway in chloroplasts of mesophyll cells by which CO_2 is fixed directly by the enzyme ribulose bisphosphate carboxylase (RuBP carboxylase or Rubisco) (Table 3.1). At the enzyme site, CO_2 reacts with ribulose bisphosphate, a phosphorylated 5-carbon sugar, to form two molecules of a 3-carbon acid (hence C_3 plants) called *3-phosphoglyceric acid* (3-PGA). The 3-PGA moves out of the chloroplast to the cytoplasm, where it is metabolized into hexose, sucrose, and other compounds (Fig. 3.2). The sucrose is translocated from cell to cell, through the bundle sheath that surrounds the vascular tissue, and into the phloem for translocation out of the leaf.

The same active site on Rubisco also catalyzes a reaction whereby oxygen (O_2) reacts to cleave the ribulose bisphosphate into a 2-carbon acid (phosphoglycolate) and one 3-PGA, which results in no net increase in photosynthate. The phosphoglycolate is quickly respired to CO_2, basically as a loss. The wasteful reaction with O_2 and the subsequent oxidation of phosphoglycolate is called *photorespiration* and likely serves very little pur-

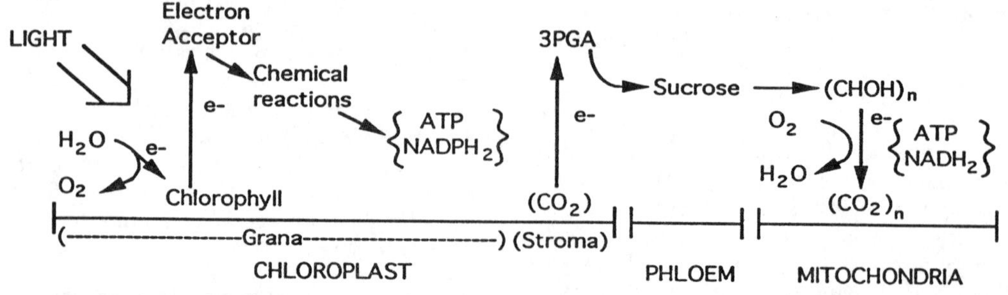

Fig. 3.2. Photosynthesis begins when light activates an electron (e^-) from chlorophyll to a higher energy state, which is used to form ATP and to reduce NADP to $NADPH_2$ in the grana. These are used to reduce CO_2 to 3-phosphoglyceric acid (3-PGA) in the stroma, and into sucrose in the cytoplasm. Sucrose is transported in the phloem to nonphotosynthesis areas where some is oxidized to form ATP and to reduce NAD to $NADH_2$, which can do work in the cell. Water hydrolysis to form carbon dioxide (CO_2) releases an e^- to replace the one lost by chlorophyll.

TABLE 3.1. Functions associated with photosynthesis in mesophyll and bundle sheath cells of leaves of C_3 and C_4 plants

	Mesophyll cells	Bundle sheath cells
C_3 species	-CO_2 fixed by Rubisco -Photorespiration occurs -Sugar formed in cytoplasm -Sucrose transported out -Have high protein content	-No chloroplasts formed -Sugars pass through enroute to phloem -Have low protein content
C_4 species	-CO_2 captured by PEP carboxylase -C_4 acid formed -C_4 acid transported out -No photorespiration occurs -Have medium protein content	-CO_2 released by C_4 acid -CO_2 fixed by Rubisco -Sugar formed in cytoplasm -No photorespiration occurs -Have medium protein content

pose. Estimates suggest 15%-40% of the light energy captured by the photosystems in C_3 plants gets wasted in photorespiration, the proportion being higher at high temperatures. Thus, C_3 plants are better adapted and more efficient in cool environments than in warm environments.

With C_3 plants, the CO_2 moves by diffusion from the atmosphere through the stomata to the Rubisco site in the stroma, where its concentration is drawn down by fixation. The low CO_2 allows O_2 to compete for the reaction site, giving rise to the wasteful photorespiration process (Ogren 1984). In tropical or warm-season species such as dallisgrass (*Paspalum dilatatum* Poir.), sudangrass (*Sorghum bicolor* [L.] Moench), and indiangrass (*Sorghastrum nutans* [L.] Nash), CO_2 is first fixed into oxaloacetate, a 4-carbon acid (hence the species are called *C_4* plants). A CO_2 molecule is added to phosphoenolpyruvate, a 3-carbon acid, in mesophyll cells by the enzyme phosphoenolpyruvate carboxylase (PEPc) (Table 3.1).

The C_4 acids formed in the mesophyll cells are rapidly translocated to the bundle sheath cells, which surround the xylem and phloem tissue and, in C_4 plants, also contain chloroplasts and Rubisco. Here the CO_2 is removed from phosphoenolpyruvate and refixed by Rubisco into 3-PGA and eventually sugars in the same manner as for mesophyll cells of C_3 plants. The phosphoenolpyruvate recycles back to the mesophyll cells to react with another CO_2. The active transfer of CO_2 in C_4 plants, via the "pumping system" involving 4-carbon acids, increases the CO_2 concentration near the reaction site of Rubisco, up to 10-fold higher than near the active site in C_3 species. The high CO_2 is very competitive with O_2 for the enzyme site, and photorespiration of C_4 species is reduced to near zero.

Due to negligible photorespiration of C_4 species, net CO_2 uptake is up to 40% higher than for C_3 species, especially at high radiation densities, where CO_2 would normally be limited. Sucrose is synthesized very near the vascular bundle in C_4 species, where it is transferred rapidly to the phloem for transport to other parts (Table 3.1).

Photosynthesis, Leaf Anatomy, and Forage Quality. As described above, leaf anatomy is different for C_3 and C_4 plants. In general, there is a higher proportion of vascular bundle tissue in C_4 plants and less mesophyll tissue (Table 3.2). While this anatomy is efficient for accommodating the CO_2 pumping system of C_4 plants, it provides a leaf tissue that is generally lower in forage quality (Akin and Chesson 1989). Mesophyll tissue has thin cell walls and digests rapidly and nearly completely in the rumen. Conversely, the vascular bundle and epidermis degrade much slower. Thus, due to differences in leaf anatomy, rumen degradation of C_3 grasses is often faster than for C_4 grasses.

Up to 50% of the soluble protein in C_3 leaves is Rubisco, and plants with high photosynthetic rates generally have high protein content. Since mesophyll cells are degraded quickly, most of their protein is released in the rumen where it is not used efficiently by the rumen microbes or the animal. Herbage of C_4 species is generally much lower in protein than that of C_3 species because less enzyme is needed in mesophyll cells due to the efficiency of PEP carboxylase, and Rubisco is located in the bundle sheath cells, which digest slowly. This may allow some bundle sheath protein of C_4 species to escape the rumen before being digested in lower stomachs, where it can be used more efficiently by the animal. The increased efficiency of protein use in C_4 species partially offsets the low concentration of protein compared with that in C_3 species.

TABLE 3.2. Percentage of cross-sectional area made up of different tissue types in leaf blades and stems of C_3 and C_4 grasses

Tissue	C_3 grasses	C_4 grasses
	% Area	
Leaf blade	(n = 6)	(n = 9)
Total vascular	15 ± 5	22 ± 8
Lignified vascular	7 ± 2	4 ± 2
Parenchyma bundle sheath	6 ± 2	15 ± 7
Epidermis	23 ± 6	35 ± 10
Sclerenchyma	6 ± 2	6 ± 3
Mesophyll	57 ± 5	38 ± 9
Stem	(n = 2)	(n = 6)
Epidermis, xylem, and sclerenchyma ring	28 ± 4	34 ± 4
Parenchyma	52 ± 3	55 ± 6

Source: Adapted from Akin and Chesson (1989).
Note: n = number of plant species sampled.

ENVIRONMENTAL EFFECTS ON PHOTOSYNTHESIS

Solar Radiation. Radiation density in the field at full sunlight during summer that is active in photosynthesis may be up to 2000 μmol photons $m^{-2}s^{-1}$. A photon is a measure of the energy of the radiation and is more meaningful than illumination or intensity units. In favorable conditions growth rate is generally a function of radiation density, which directly influences photosynthesis rate and subsequent productivity (Black 1957).

Light or radiation affects photosynthesis in a near-linear manner at low levels (Fig. 3.3), and leaf photosynthesis of C_3 and C_4 species responds similarly because electron flow is the major limitation. Photosynthesis of C_3 grasses and legumes reaches a maximum rate at 800-1200 μmol $m^{-2}s^{-1}$, but photosynthesis of C_4 grasses continues to increase up to 2000 μmol $m^{-2}s^{-1}$. Efficiency of conversion of radiation energy to dry matter at full sun may be below 3% for C_3 grasses compared with 5%-6% for C_4 grasses (Cooper 1970). Thus, at high radiation, C_4 grasses have a higher photosynthetic rate and efficiency, providing temperatures are adequate for both types.

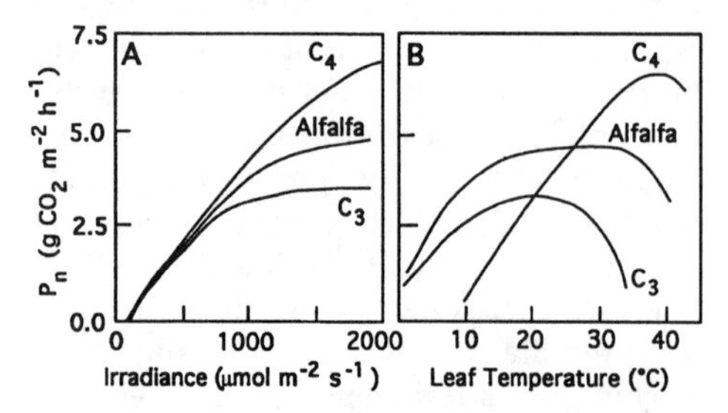

Fig. 3.3. At low radiation, net photosynthesis (corrected for photorespiration) increases in a near-linear manner for C_3 and C_4 species, but C_3 species reach a maximum at about 800-1200 μmol $m^{-2}s^{-1}$ (*A*). Net photosynthesis of C_3 species is maximal at lower temperatures than for C_4 species (*B*). Alfalfa has C_3 photosynthesis but is intermediate in both responses.

This does not mean, however, that C_4 species are always higher yielding than C_3 species (Snaydon 1991). Only leaves in the upper canopy receive high radiation, and most leaves in the lower canopy are partially shaded and function at radiation levels in which photosynthetic rates of C_3 and C_4 plants are similar (Nelson 1988). It is generally agreed that C_4 plants evolved from C_3 plants to overcome limitations due to photorespiration, which gradually increased in C_3 plants as CO_2 concentration in the air decreased (Ehleringer and Monson 1993). The main advantage of C_4 photosynthesis is climatic adaptation. The change in enzymes and their location in C_4 plants resulted in better heat tolerance (Snaydon 1991), improved water use efficiency (Brown and Simmons 1979), and increased nitrogen (N) use efficiency (Brown 1978).

Several summer annual and perennial species have the C_4 photosynthesis system (Waller and Lewis 1979). Natural abundance of C_4 grasses is higher in environments that have high temperatures, especially night temperatures in July (Fig. 3.4). Many of the most troublesome annual grass weeds have C_4 photosynthesis. Some dicots have C_4 photosynthesis, but no legumes are known to have the C_4 system.

Carbon Dioxide. The main reason C_3 plants tend to "light saturate" or plateau at about half full sun is that the CO_2 diffusion rate becomes the limiting factor allowing Rubisco to react more with O_2 and increase photorespiration. The C_4 plants have PEP carboxylase to capture CO_2, which has a much higher affini-

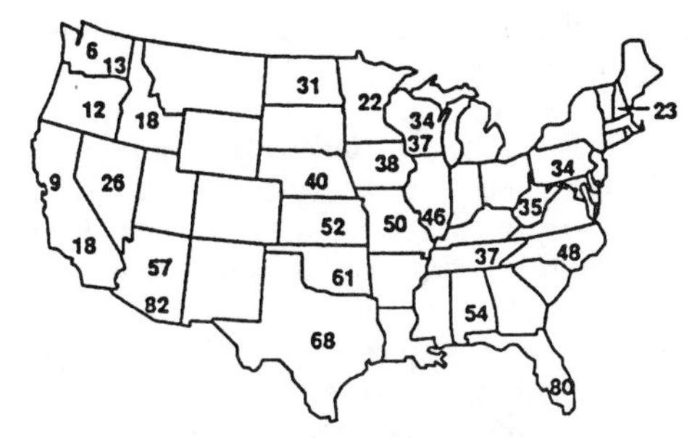

Fig. 3.4. Percentage of C_4 species among naturally occurring grass floras in several locations in the US. Species include annuals and perennials. No C_4 species were found on the arctic slopes of Alaska and in northern Manitoba, Canada. (Adapted from Teeri and Stowe 1976.)

ty for CO_2 than does Rubisco and does not react with O_2 (Ogren 1984). The low concentration of CO_2 near PEP carboxylase causes CO_2 to diffuse into the leaf faster. Since C_4 plants have the C_4-acid system to "pump" CO_2 to the bundle sheath, there is less limitation due to CO_2 than with C_3 grasses, and photorespiration is overcome.

The rate of photorespiration of C_3 species can be decreased by exposure to high CO_2 concentration in the air because diffusion through the stomata and to the chloroplast is more rapid. As CO_2 concentration in the atmosphere continues to increase from the current 350 to the expected level near 600 µl L^{-1}, photosynthesis and productivity of C_3 plants will likely increase by about 30% (Kimball 1983). Most likely, C_4 plants will increase little, if at all, and will lose some of their ecological advantage (Ehleringer and Monson 1993).

Alfalfa (*Medicago sativa* L.) tends to be intermediate in photosynthesis rate between C_3 and C_4 grasses even though its photosynthetic biochemistry is clearly C_3 (Fig. 3.3). The protein content of alfalfa leaves is very high, and a high proportion of the protein is Rubisco. Thus, along with high stomatal conductance, alfalfa may partially overcome the CO_2 limitation and the leaves will respond to higher radiation. Most forage legumes tested have rates similar to cool-season grasses.

Temperature. In general, most cool-season grasses and legumes can fix CO_2 at temperatures near freezing, have maximum rates at 20°-25°C, and have markedly reduced rates above 30°C (Fig. 3.3). The sharp reduction above 30°C is largely due to higher photorespiration and higher maintenance respiration. In contrast, photosynthesis of C_4 plants is very low at 10°C because PEP carboxylase is not very active, increases to a maximum at about 35°-40°C, and then decreases.

Alfalfa, a C_3, tends to span both types. Similar to other C_3 species it conducts photosynthesis at near 0°C, has a broad temperature optimum between 5° and 30°C, and (more similar to C_4 types) decreases in photosynthesis above 35°C. Growth temperatures show a similar range, allowing alfalfa to be grown in all 50 states of the US, which indicates its very broad adaptation among perennial forage crops.

Some forage legumes such as korean lespedeza (*Kummerowia stipulacea* [Maxim.] Makino) and sericea lespedeza (*Lespedeza cuneata* [Dum.-Cours.] G. Don) are warm-season legumes based on growth temperatures but have a C_3 photosynthetic system that is not uniquely high or efficient (Brown and Radcliffe 1986). Even so, lespedezas do not grow as well as cool-season legumes like white clover (*Trifolium repens* L.) at cool temperatures. Reasons are unknown for the differences in temperature effects on growth.

The temperature range for growth of forage plants is generally slightly narrower than for photosynthesis. This offers a conservative strategy for adaptation in that at low or high temperatures growth is slowed more than is photosynthesis. Therefore, photosynthate supply exceeds growth needs allowing the excess to be stored (Fig. 3.1) and protecting the plant from "outgrowing" its photosynthate supply.

Water Stress. Photosynthesis is slowed during drought periods due to stomatal closure, but the stress level required for closure is greater than that needed to slow leaf or stem growth (Chap. 5). During periods of high atmospheric demand or as a soil drys, growth regulator "signals" generated by the roots are transmitted to the guard cells, which cause stomata to close (Davies and Zhang 1991). Stomatal closure effectively conserves water, allowing the plant to remain somewhat turgid and to persist, but it also increases the stomatal resistance to CO_2 movement to the mesophyll cells, which reduces photosynthesis (Brown 1977). Many forage species close the stomata at stress levels near −1.5 to −1.8 megapascals, but species differ markedly in root distribution. Deep-rooted plants may be able to encounter enough water to maintain higher transpiration rates and keep the stomata open longer. In general, C_4 grasses root deeper than do C_3 grasses, which contributes to the C_4 grasses' climatic adaptation. Similarly, alfalfa has deeper roots than white clover, giving it an advantage during drought.

Growth, especially leaf growth, is very sensitive to even mild water stress (Hsiao 1973). Again, this protective mechanism allows the plant to slow growth and transpiration while maintaining photosynthesis. Interestingly, root growth continues at stress levels that slow leaf growth (Sharp and Davies 1979), effectively changing the balance between potential water uptake and water loss. The reduction in growth rate before stomata close allows storage carbohydrate to accumulate and support respiration if the water stress is severe enough to slow photosynthesis (Fig. 3.1).

Mineral Nutrition. Photosynthesis depends largely on enzymes that have N as a major constituent and often require other minerals as activators or cofactors. Fertilization of grasses with N increases the growth rate, the leaf protein content, and the photosynthetic rate per unit leaf area (Nelson et al. 1992). The major response to high N, however, is a two- to three-fold increase in leaf area and growth rate, with only a 30%-50% increase in photosynthetic rate per unit leaf area. Generally, top growth is enhanced more by N than is root growth.

Magnesium (Mg) is a constituent of the chlorophyll molecule, phosphorus (P) is intimately involved with movement of 3-PGA out of the chloroplast, and potassium (K) is a major factor in regulating stomatal opening and closing. Many minerals are moved preferentially to young leaves, and if the younger leaves are in short supply, the older ones in the lower canopy will die prematurely, releasing the minerals to be recycled to the growing tissue. The loss of lower leaves to death and decay reduces yield and decreases quality of the forage (Buxton and Fales 1994).

DARK RESPIRATION

Although called *dark respiration* to distinguish it from photorespiration, this process occurs 24 h/d in mitochondria of all living cells (Fig. 3.2). In the process, sugars are oxidized to CO_2 and water, with the energy released being used to form ATP and $NADH_2$, a molecule very similar in structure and function to the $NADPH_2$ formed in chloroplasts. Thus, sugars serve as convenient molecules to store or translocate the energy and reducing power of ATP and $NADPH_2$ generated in photosynthesis to nonphotosynthetic areas of plants where they can be formed to support plant functions.

Growth Respiration. Respiration required to drive synthesis of new dry matter is termed *growth respiration* (Fig. 3.1). Its rate is directly linked to growth rate, as some sugar transported to the meristematic region or growth zone is respired to provide energy to assemble other sugars and metabolites into cell walls and other constituents (Amthor 1989). Some sugar molecules are broken down partially (respired) to form carbon backbones, after which growth respiration is used to add an amino group ($-NH_2$) to form amino acids or to assemble lipids, nucleic acids, and other constituents.

Respiratory costs for synthesis of new tissue are strongly dependent on tissue composition. For example, it takes about 1.24 g of glucose for substrate and growth respiration to synthesize 1.00 g of cellulose (cell wall) or starch (Penning de Vries et al. 1983). Protein synthesis requires considerable respiration for uptake and reduction of NO_3^-. Thus, it takes about 2.85 g of glucose for synthesis of 1.00 g of protein or nucleic acid beginning with NO_3^-. Because they are highly reduced, it takes about 2.17 g of glucose for 1.00 g of lignin and 3.11 g for 1.00 g of lipid. Organic acids are more oxidized than carbohydrate and have synthesis costs near 0.93 g glucose per gram. Uptake of minerals such as K, Mg, Ca, and P costs about 0.05 g glucose per gram.

Thus, a leaf blade of smooth bromegrass (*Bromus inermis* Leyss) that is 16% protein, 6.5% mineral, 2.5% lipid, 4.0% organic acids, and 71% cell wall (68% carbohydrate and 3% lignin) requires about 1.49 g of glucose for synthesis of 1.00 g of tissue. If N is recycled from old protein in lower leaves, the $-NH_2$ is already formed and biosynthesis costs only 2.20 g glucose per gram of new protein. This reduces total synthesis costs to 1.38 g of glucose per gram of leaf.

Alfalfa leaves are high in protein and are more costly to synthesize than grass leaves. Similarly, due to composition, it costs more to synthesize leaves than stems, and more for stems than roots. Growth respiration occurs mostly in meristematic tissues, where cells are dividing and elongating, and in maturing tissues that are synthesizing cell wall and lignin.

Maintenance Respiration. Once formed, tissue needs to be maintained and repaired, which also requires respiration to provide energy. The major factors are repair and turnover of protein and maintenance of ion gradients. Ions must be pumped back into the cell as they gradually leak out through the membranes. Proteins gradually break down and need to be reassembled or replaced. Again, there is a respiratory cost, but no net increase in dry weight, so the process is termed *maintenance respiration*. Rates of maintenance respiration are related closely to protein content of the tissue (Amthor 1989).

Estimates are that dark respiration uses 45%-50% of each day's photosynthesis during the following 24 h, largely as maintenance respiration (Robson 1982). Maintenance respiration generally begins as a small component of total respiration in young seedlings, increases as a higher proportion of the tissue is older and non-growing, then decreases as protein content of the mature tissue decreases (McCree 1983). Wilson (1982) genetically selected for a low dark respiration rate in leaf blades of perennial ryegrass (*Lolium perenne* L.), and obtained a yield increase. Presumably, maintenance respiration was reduced, and the extra carbohydrate was available for growth.

Factors Affecting Respiration. Temperature is the major environmental factor affecting dark respiration, largely through its effect on enzymes. Temperature affects growth rate and growth respiration, as it is generally assumed the efficiency of growth respiration (i.e., respiration per unit of growth) remains similar over a range of temperatures (Amthor 1989). However, growth temperature influences the composition of the tissue formed, which affects growth respiration. For example, at temperatures above optimum for growth a higher proportion of the cell wall is lignin, and the slower growth leads to a higher proportion of protein (Buxton and Fales 1994). Both lignin and protein are expensive to synthesize, thus decreasing the respiratory efficiency of the growth process.

The rate of maintenance respiration is very temperature sensitive. At high temperatures proteins turn over faster, and membranes are more subject to leakage. Both processes require energy that is made available by increased maintenance respiration. In this manner, maintenance respiration is like photorespiration: neither is associated with net growth, and both increase rapidly with temperature. A major reason the temperature optimum for C_3 grasses is lower than for C_4 grasses (Fig. 3.3) is that the latter do not have photorespiration and are lower in protein and its associated synthesis and maintenance cost.

TRANSLOCATION OF CARBOHYDRATES

Translocation is the movement of materials from a "source" to a "sink" down a concentration gradient (Moser 1977). The source may be an assimilating organ (a green leaf) and the sink a metabolizing one (the meristem located at the tip of a root or stem, or at the base of an expanding grass leaf) or a storage organ (an alfalfa root). Later, when stored carbohydrate is used as a substrate for new growth, the former sink becomes a source as carbohydrate is exported to support regrowth (Fig. 3.1).

Expanding young leaves use most of the carbohydrate they produce, and they also import additional carbohydrate from older leaves, until the leaf is nearly fully expanded. First exports from a newly expanded leaf usually move upward to the terminal meristem, or growing point, and to younger developing leaves. As the leaf ages, some of its export moves downward, perhaps because new leaves above it are also supporting the needs of the growing point. With time, a greater proportion moves downward until finally it all does. As the leaf dies, the nonstructural carbohydrate and protein are broken down to sucrose and amino acids for translocation to new tissue for respiration and resynthesis of cell materials. Many minerals are also moved from the older leaves.

Not all sinks have the same demand for sucrose and other metabolites. In C_3 grasses, the leaf growth zone is generally supplied abundantly with carbohydrate (Nelson and Spollen 1987), storage and tiller development have a lower priority, and root tips are lowest in priority. Thus, vigorous plants in high-light environments store more energy and tiller more profusely. When over half the top growth of a grass plant is removed, photosynthesis supply is reduced dramatically because mostly young leaves are removed, and root growth ceases very rapidly and does not recover until the canopy is partially regrown (Davidson and Milthorpe 1965). Similarly, if grass is growing in shade, leaf growth is similar or perhaps enhanced, tillering and storage are reduced, and root growth is relatively slow compared with plants in full sun (Allard et al. 1991).

FOOD RESERVES

When photosynthesis exceeds the needs of respiration and growth (Fig. 3.1), legumes and grasses store carbohydrates in a readily available form in various plant parts (McIlroy 1967; Trlica 1977). The principal storage organ may be the root, as in alfalfa, red clover (*T. pratense* L.), and kudzu (*Pueraria lobata* [Willd.] Ohwi); the stolons, as in white clover, bahiagrass (*Paspalum notatum* Flugge), and buffalograss (*Buchloe dactyloides* [Nutt.] Engelm.); the rhizomes, as in smooth bromegrass, reed canarygrass (*Phalaris arundinacea* L.), and western wheatgrass (*Pascopyrum smithii* [Rybd.] A. Löve); or the stem bases, as in tall fescue, orchardgrass, dallisgrass, and big bluestem (*Andropogon gerardii* Vitman).

Reserve carbohydrates are used to support respiration and growth when leaf area and photosynthesis are low as in the spring and after each cutting. The reserves are reconverted to sucrose and translocated to meristematic areas for growth. They also are used to develop heat and cold resistance when photosynthesis is reduced, to support respiration and metabolism during periods of dormancy, to enhance flower and seed formation, and for many processes that go on within the plant during its life. In general, reserve carbohydrates provide a buffer to the carbohydrate supply from photosynthesis; thus, they are essential to the life of perennial and biennial forage species and provide a useful tool on which to base management decisions.

In addition to carbohydrates there is growing recognition the plants also have N reserves (see Chap. 1, Vol. 2). Nitrogen is critical for regulating growth processes such as cell production in meristematic regions (MacAdam et al. 1989), is stored along with carbohydrates in storage organs (Graber et al. 1927; Owensby et al. 1977), and is involved in regrowth processes, especially since root growth of grasses and N_2 fixation by legumes are reduced by defoliation. We are just learning about the roles of N reserves as part of plant adaptation and plant management.

Starch and Fructan Accumulators. Starch, a polymer of glucose, is the primary storage form of nonstructural carbohydrate accumulated in the Leguminosae. Grasses of tropical and subtropical origin also accumulate starch, but many grasses of temperate origin (species in the Hordeae, Aveneae, and Festuceae tribes) accumulate fructan (Table 3.3), basically a polymer of fructose, as the storage carbohydrate in their vegetative tissues (Smith 1968; Chatterton et al. 1989). The carbohydrate storage form is not consistent with photosynthesis types because some C_3 grasses, especially those adapted to warm temperatures such as rice (*Oryza sativa* L.), store starch rather than fructan (Bender and Smith 1973). Species in both the Leguminoseae and Gramineae families accumulate starch in their chloroplasts and seed.

Although the enzymes and cellular locations differ, the basic principle of storage and reuse is similar for starch and fructan. Starch is stored in chloroplasts of leaf tissue and amyloplasts (similar to chloroplasts but without photosynthesis enzymes) in nongreen tissues. Fructan is water soluble and accumulates in the vacuole. Short-chain forms (i.e., 3-6 hexose units) may also have an osmotic role (Nelson and Spollen 1987). Both storage carbohydrate forms are hydrolyzed to their hexose monomer, and sucrose is formed for transport when sucrose from photosynthesis is low. In general, the enzymes for fructan metabolism are functional at a lower temperature than those for starch metabolism (Pollock and Cairns 1991), which may contribute to adaptation of C_3 grasses to cool temperatures. Under cool growth temperatures plants accumulate higher concentrations of carbohydrates in their leaves (Table 3.3) and storage organs (Sullivan and Sprague 1949).

Seasonal Cycles. Maintenance of a minimal

TABLE 3.3. Total nonstructural carbohydrate (TNC) in leaves of 128 cool-season (C_3) and 57 warm-season (C_4) grasses grown at 10°/5°C (light/dark) or 25°/15°C

	TNC	Ftn	Suc	Glu	Fru	Str
			mg kg⁻¹			
C_3 grasses						
10°/5°C	312	115	58	29	24	86
25°/15°C	107	12	23	18	14	41
C_4 Grasses						
10°/5°C	166	3	66	22	14	64
25°/15°C	92	4	20	13	8	47

Source: Components of TNC include fructan (Ftn), sucrose (Suc), glucose (Glu), fructose (Fru), and starch (Str).

Note: Data from Chatterton et al. (1989).

level of food reserves in the storage organs is necessary to keep a plant vigorous and productive. Plants go through periods when carbohydrate foods are used and when they are stored (Figs. 3.5 and 3.6), and a cyclic pattern occurs between early growth and maturity. The magnitude and pattern of seasonal carbohydrate trends are similar in the principal storage organs of perennial legumes and cool-season grasses. Warm-season grasses have a similar pattern but generally store less carbohydrate than cool-season grasses, often reaching only 6%-10% of the dry weight of the storage organ for tall species like switchgrass (*Panicum virgatum* L.) (Balasko and Smith 1971; Anderson et al. 1989) and big bluestem (Owensby et al. 1977). Concentrations are even lower for caucasian bluestem (*Bothriochloa caucasica* [Trin.] C.E. Hubb.) (Forwood et al. 1988). It has a decumbent growth habit and retains active leaf area to persist when grazed closely or frequently.

With initiation of growth in spring or regrowth after cutting, carbohydrates stored in alfalfa roots are used to support new top growth (Fig. 3.5). Depletion continues until the top growth is 15-20 cm tall and there is enough leaf area to produce photosynthate to fully support respiration and growth. Thus, plants go through a transition from nearly total dependence on storage early in regrowth to total dependence on photosynthesis for later growth.

The minimum level of food reserves in other upright-growing legumes, such as red clover, also occurs about 2-3 wk after cutting, when plants are still growing vegetatively. It occurs in most temperate grasses at about the beginning of stem elongation (Fig. 3.6). However, few legumes and grasses have been studied, and differences no doubt exist among species.

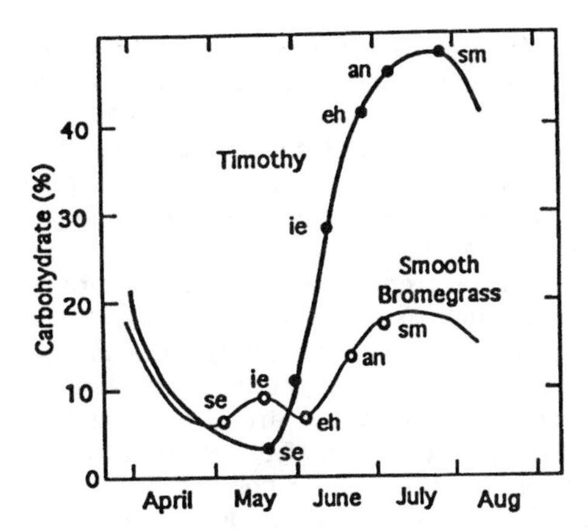

Fig. 3.6. Total nonstructural carbohydrates in the stem bases of timothy and bromegrass at successive stages of development in the field in Wisconsin: (*se*) beginning of stem elongation, (*ie*) inflorescence emergence, (*eh*) early heading, (*an*) early anthesis, and (*sm*) seed mature. (Adapted from Smith 1981.)

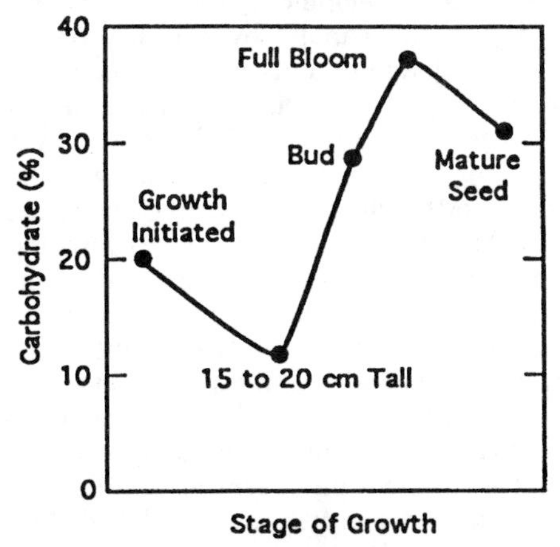

Fig. 3.5. Trend of total nonstructural carbohydrates in roots of alfalfa between initiation of growth in the spring and the mature seed stage. (Field data in Wisconsin are adapted from Graber et al. 1927.)

When alfalfa top growth exceeds 15-20 cm, photosynthate production is in excess of needs for respiration and growth (Fig. 3.1), and this excess is translocated to the roots for storage (Fig. 3.5). Accumulation in the roots continues as the tops grow, reaching the highest level of storage near the stage of full bloom. There is some removal of carbohydrates from alfalfa roots between the full bloom and mature seed stage because leaves are aging, seed is developing, and new shoots are initiated from the crown. In most forage species the maximum level of food reserves in the storage organ is reached after stem elongation has ceased or at maturity.

The cyclic pattern of use and storage of carbohydrate reserves is influenced by prevailing environmental conditions, mostly through their influence on photosynthesis, respiration, and growth (Fig. 3.1). Each process responds independently, but to different degrees, with the net effect overall being reflected in the amount of carbohydrate stored. For example, shading generally decreases photosynthesis more than growth, so storage is decreased (Paulsen and Smith 1968). High temperatures often increase respiration more than photosynthesis and hasten maturity, so storage is decreased (Sullivan and Sprague 1949; Nelson and Smith 1969). Conversely, low temperatures, low N fertilization, and limited soil moisture often cause greater reductions in growth and respiration than in photosynthesis, so storage is enhanced. When stored carbohydrate is low at time of cutting, it is often desirable to leave some leaf area for photosynthesis to help provide sugars to support respiration and regrowth.

Seasonal trends of carbohydrate storage in roots help to explain why birdsfoot trefoil (*Lotus corniculatus* L.) can be cut or grazed frequently but not closely. Carbohydrates stored over winter in roots are used to support spring growth of birdsfoot trefoil but are not restored in the roots at flowering as they are in alfalfa and red clover (Fig. 3.7). Instead, birdsfoot trefoil continues active growth, even while flowering, so storage does not occur readily. Unlike cutting alfalfa or red clover at bloom stage, one must leave a tall stubble with leaves each time birdsfoot trefoil is cut or grazed. The green leaves furnish carbohydrate needed for regrowth since little energy is available from the roots. Storage remains at a low level until growth slows down in autumn.

Stage of Cutting. Cutting or grazing when carbohydrates are at a low level may leave very little energy available to support new growth; e.g., continued cutting at immature stages of growth will eventually exhaust the plant and weaken it to the extent of death. Plants weakened by too early, too close, or too frequent cutting usually are more susceptible to drought stress, heat stress, winter injury, and invading diseases.

Usually, the closer to maturity that cutting or grazing occurs, the higher the stored food reserves will be and the easier it will be to maintain plant vigor for high productivity. However, delaying cutting to maturity usually is not compatible with the need to harvest herbage with high protein and high digestibility, both of which decrease with maturity. Thus, a compromise must be made between delaying harvest for maximum vigor and harvesting early for maximum quality of the herbage.

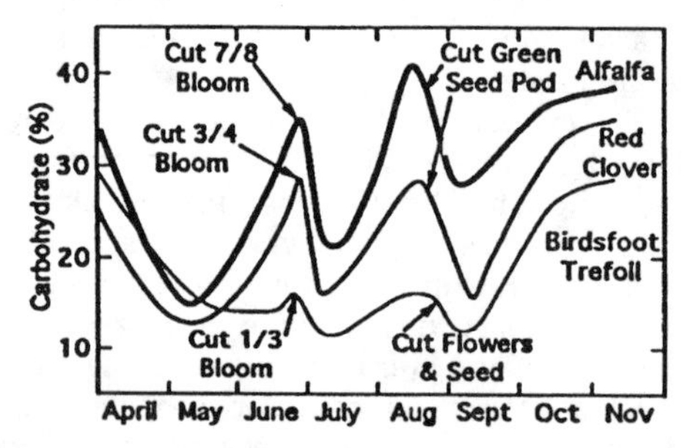

Fig. 3.7. Carbohydrate levels are high in spring in alfalfa, red clover, and birdsfoot trefoil roots, decrease as regrowth begins, then increase again in alfalfa and red clover as plants mature. The cycle is repeated after each cutting. The level in birdsfoot trefoil remains low through the summer but increases to a high level during fall. (Adapted from Smith 1962.)

MANAGING THE CANOPY

A full canopy of leaves is needed to intercept the maximum amount of radiation. Once the leaf canopy has developed to intercept more than 95% of the radiation, the density of the radiation sets a production limit, which can be achieved only when no other factor is limiting growth. To intercept 95% of the radiation, a leaf area index (LAI, or ratio of leaf blade area to land area) of 3-5 is required for flat leaf plants like clovers, 5-6 for alfalfa, and up to 7-11 for grasses like orchardgrass and perennial ryegrass (*Lolium perenne* L.) that have vertically oriented leaves (Fig. 3.8). Until these LAIs are achieved by the canopy following cutting or grazing, the growth rate is more related to amount of radiation intercepted than to photosynthetic activity per unit of leaf area (Rhodes 1973; Horst et al. 1978).

Canopy Structure. Leaf angle markedly affects light penetration into a grass or legume canopy. Leaves emerge more or less vertically at the top of grass canopies, giving a favorable leaf arrangement for light penetration (Fig. 3.8). In contrast, leaf blades of clover are folded early in development, then are moved to the top of the canopy by petiole extension, where the blade unfolds to be displayed more horizontally. This flat leaf arrangement causes most radiation to be intercepted by the young upper blades in the canopy, with little getting to the bottom. This effectively shades weeds but also causes older leaves with shorter petioles to senesce or die in the dark environment. In that case, the plant has expended energy to develop the tissue, but it is not effectively used for animal production.

Tall-growing species with most of the leaf area high on the plant often have better light penetration and photosynthesis but are almost totally dependent on stored foods for recovery since all or most of the photosynthetic area is removed with close cutting. Short-growing species with leaf area near the soil surface are not totally defoliated with close cutting or grazing and are less depen-

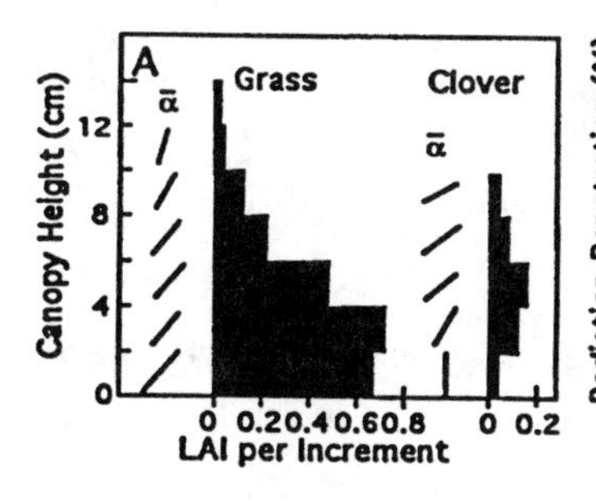

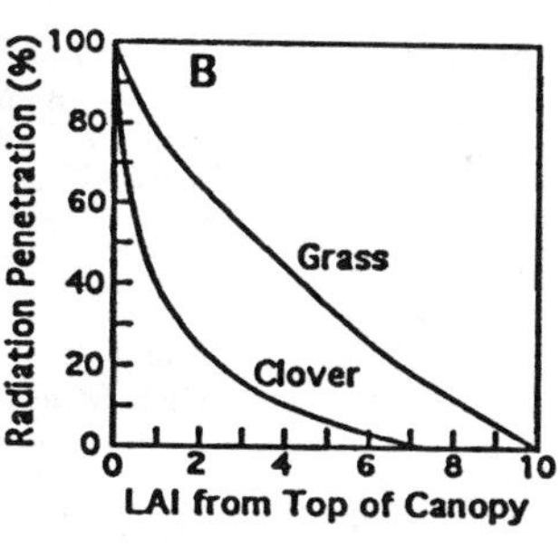

Fig. 3.8. Leaf area is distributed differently in canopies of perennial ryegrass and white clover (*A*), and the leaf angle (α) is also different. Light is absorbed mainly by upper layers of the clover canopy whereas the more vertical leaves in grasses allow penetration (*B*). (Adapted from Loomis and Williams 1969.)

dent on stored foods for support of recovery growth because part of the needed energy comes from the photosynthetic activity of intact leaves.

In a Michigan study (Harrison and Hodgson 1939), Kentucky bluegrass (*Poa pratensis* L.) was least injured by close and continuous clipping, followed, in order, by quackgrass (*Elytrigia repens* [L.] Nevski), smooth bromegrass, and timothy and orchardgrass, which were about equal. Dallisgrass, a tall bunch-type grass, was more injured in a North Carolina study by close cutting than was carpetgrass (*Axonopus affinis* Chase) or bermudagrass (*Cynodon dactylon* [L.] Pers.), both of which form canopies with most of their leaves near the soil surface (Lovvorn 1945). Forages such as Kentucky bluegrass, buffalograss, and bermudagrass depend on stored food reserves for only a short time following cutting and begin to replenish food reserves after initiation of only a few new leaves.

Location of Growing Points. Managing a canopy involves more than controlling light interception for photosynthesis and food reserves in the storage organ. Plants need meristematic areas or growing points that are driven by growth respiration to provide new growth or regrowth after cutting or grazing. Thus, location of the meristematic areas in the canopy influences how the canopy needs to be managed to optimize relations between photosynthesis, storage, growth, and respiration (Dahl and Hyder 1977).

Legumes have terminal meristems at the end of the stem and an axillary bud at each node (Chap. 2), whereas grasses have the terminal at the top of the stem, but the stem is short until just prior to flowering. The critical stage occurs when internode elongation of grasses elevates the shoot apices to a height where they are removed by cutting or grazing. At this stage food reserves are low, leaf area is removed, and subsequent growth must come from new basal axillary tillers, which may or may not be present. Grasses with many nodes and short internodes below or near the soil surface have many axillary buds for development of new tillers, which in some species may emerge as rhizomes or stolons (Dahl and Hyder 1977).

When a mixture of grasses was grazed, species with a high ratio of flowering shoots (long internodes) to vegetative tillers (unelongated internodes) decreased in frequency in favor of those with many vegetative tillers (Branson 1953). The critical period when shoot apices move above mowing or grazing height differs among grass species; mowing or grazing a mixture of grasses will damage some species more than others (Booysen et al. 1963). Timothy and smooth bromegrass often fail to persist in alfalfa mixtures because they are at the critical period when the spring crop is harvested at the early flower stage of alfalfa.

Decreased production is caused when timothy and smooth bromegrass are cut during the period between the beginning of internode elongation and inflorescence emergence (Sheard and Winch 1966; Knievel et al. 1971). During this period of growth, shoot apices are above cutting height, and carbohydrate reserves are at a low level (Fig. 3.6). There are no basal axillary tillers present for regrowth since they are not formed until about the anthesis stage, especially in timothy. Basal tiller development is apparently delayed or inhibited by the growing shoot until anthesis. Thus, regrowth may be delayed by 2 wk or more, and many cut shoots may die. Cutting or grazing before initiation of stem elongation removes only leaf blades, and the intact shoot can elongate to produce a crop. After anthesis, recovery is satisfactory because basal axillary tillers are present to produce regrowth.

Orchardgrass, in contrast, recovers rapidly when cut at almost any growth stage, even when cut at the early stages of stem elongation. The flowering shoots appear to exert less apical dominance than in timothy or smooth bromegrass. New basal tillers are produced throughout the spring period, so shoots in different stages of development are present at any given time. When orchardgrass is harvested, the cut leaves on the stubble continue to elongate, and new leaves develop rapidly on the axillary tillers, so photosynthesis of the canopy is only temporarily interrupted.

QUESTIONS

1. What are the most important environmental factors influencing photosynthesis of grasses and legumes?
2. How would change in season of the year influence photosynthetic production of a warm-season and a cool-season grass species? Would similar effects occur in legumes?
3. What is the effect of increasing temperatures on concentrations of nonstructural carbohydrates, protein, and minerals? On the digestibility of grass and legume herbage?
4. What are some of the biochemical differences between photosynthetic pathways of temperate and tropical grasses? How will these grasses likely respond to global climate change?
5. Why is a knowledge of the periods of minimum and maximum carbohydrate reserves important to the management of perennial or biennial species?
6. What is the principal vegetative storage organ for (reserve) carbohydrates in alfalfa? In ladino clover? In smooth bromegrass? In bermudagrass?
7. How do alfalfa and birdsfoot trefoil differ in their "cyclic pattern" of use and storage of carbohydrate in their roots? How does this influence decisions about the grazing management of each legume?
8. At what stage of growth are grasses most vulnerable to damage by cutting? Why?
9. How does the canopy structure influence photosynthesis activity?
10. What soil fertility conditions favor photosynthesis? The accumulation of carbohydrate reserves?

REFERENCES

Akin, DE, and A Chesson. 1989. Lignification as the major factor limiting forage feeding value especially in warm conditions. In D. Descroches (ed.), Proc. 16th Int. Grassl. Congr., Nice, France. Montrouge: Dauer, 1753-60.

Allard, G, CJ Nelson, and SG Pallardy. 1991. Shade effects on growth of tall fescue. I. Leaf anatomy and dry matter partitioning. Crop Sci. 31:163-67.

Amthor, JS. 1989. Respiration and Crop Productivity. New York: Springer-Verlag.

Anderson, BA, AG Matches, and CJ Nelson. 1989. Carbohydrate reserves and tillering of switchgrass following clipping. Agron. J. 81:13-16.

Balasko, JA, and D Smith. 1971. Influence of temperature and nitrogen fertilization on the growth and composition of switchgrass (*Panicum virgatum* L.) and timothy (*Phleum pratense* L.) at anthesis. Agron. J. 63:853-57.

Bender, MM, and D Smith. 1973. Classification of starch and fructosan accumulating grasses as C_3 or C_4 by carbon isotope analysis. J. Brit. Grassl. Soc. 28:97-100.

Black, JN. 1957. The influence of varying light intensity on the growth of herbage plants. Herb. Abstr. 27:89-98.

Booysen, P de V, NM Tainton, and JC Scott. 1963. Shoot-apex development in grasses and its importance in grassland management. Herb. Abstr. 33:209-13.

Branson, FA. 1953. Two new factors affecting resistance of grasses to grazing. J. Range Manage. 6:165-71.

Brown, RH. 1978. A difference in N use efficiency in C_3 and C_4 plants and its implications in adaptation and evolution. Crop Sci. 18:93-98.

Brown, RH, and DE Radcliffe. 1986. A comparison of apparent photosynthesis in sericea lespedeza and alfalfa. Crop Sci. 26:1208-11.

Brown, RH, and RE Simmons. 1979. Photosynthesis of grass species differing in CO_2 fixation pathways. I. Water-use efficiency. Crop Sci. 19:375-79.

Brown, RW. 1977. Water relations of range plants. In RE Sosebee (ed.), Rangeland Plant Physiology, Range Sci. Ser. 4. Denver, Colo.: Society of Range Management, 97-140.

Buxton, DR, and SL Fales. 1994. Plant environment and quality. In GC Fahey et al. (eds.), Forage Quality, Evaluation, and Utilization. Madison Wis.: American Society of Agronomy, 155-99.

Chatterton, NJ, PA Harrison, JH Bennett, and KH Asay. 1989. Carbohydrate partitioning in 185 accessions of Gramineae grown under warm and cool temperatures. J. Plant Physiol. 134:169-79.

Cooper, JP. 1970. Potential production and energy conversion in temperate and tropical grasses. Herb. Abstr. 40:1-15.

Dahl, BE, and DN Hyder. 1977. Developmental morphology and management implications. In RE Sosebee (ed.), Rangeland Plant Physiology, Range Sci. Ser. 4. Denver, Colo.: Society of Range Management, 256-90.

Davidson, JL, and FL Milthorpe. 1965. The effect of temperature on the growth of cocksfoot (*Dactylis glomerata* L.). Ann. Bot. 29:407-17.

Davies WJ, and J Zhang. 1991. Root signals and the regulation of growth and development of plants in drying soil. Annu. Rev. Plant Physiol. and Plant Mol. Biol. 42:55-76.

Ehleringer, JR, and RK Monson. 1993. Evolutionary and ecological aspects of photosynthetic pathway variation. Annu. Rev. Ecol. Syst. 24:411-39.

Forwood JR, AG Matches, and CJ Nelson. 1988. Forage yield, nonstructural carbohydrate levels, and quality trends of caucasian bluestem. Agron. J. 80:135-39.

Graber, LF, NT Nelson, WA Luekel, and WB Albert. 1927. Organic Food Reserves in Relation to the Growth of Alfalfa and Other Perennial Herbaceous Plants. Wis. Agric. Exp. Stn. Bull. 80.

Harrison, CM, and CW Hodgson. 1939. Response of certain perennial grasses to cutting treatments. J. Am. Soc. Agron. 31:418-30.

Horst, GL, CJ Nelson, and KH Asay. 1978. Relationship of leaf elongation to forage yield of tall fescue genotypes. Crop Sci. 18:715-19.

Hsiao, TC. 1973. Plant response to water stress. Annu. Rev. Plant Physiol. 24:519-70.

Kimball, BA. 1983. Carbon dioxide and agricultural yield: An assemblage and analysis of 430 prior observations. Agron. J. 75:779-88.

Knievel, DP, AVA Jacques, and D Smith. 1971. Influence of growth stage and stubble height on herbage yields and persistence of smooth bromegrass and timothy. Agron. J. 63:430-34.

Loomis, RS, and WA Williams. 1969. Productivity and the morphology of crop stands: Patterns with leaves. In JD Eastin et al. (eds.), Physiological Aspects of Crop Yield. Madison, Wis.: American Society of Agronomy, 27-47.

Lovvorn, RL. 1945. The effect of defoliation, soil fertility, temperature, and length of day on the growth of some perennial grasses. J. Am. Soc. Agron. 37:570-82.

MacAdam, JW, JJ Volenec, and CJ Nelson. 1989. Effects of nitrogen on mesophyll cell division and epidermal cell elongation in tall fescue leaf blades. Plant Physiol. 89:549-56.

McCree, KJ. 1983. Carbon balance as a function of plant size in sorghum plants. Crop Sci. 23:1173-77.

McIlroy, RJ. 1967. Carbohydrates of grassland herbage. Herb. Abstr. 37:79-87.

Moser, LE. 1977. Carbohydrate translocation in range plants. In RE Sosebee (ed.), Rangeland Plant Physiology, Range Sci. Ser. 4. Denver, Colo.: Society of Range Management, 47-71.

Nelson, CJ. 1988. Genetic associations between photosynthetic characteristics and yield: Review of the evidence. Plant. Physiol. Biochem. 26:543-54.

Nelson, CJ, and D Smith. 1969. Growth of birdsfoot trefoil and alfalfa. IV. Carbohydrate reserve levels and growth analysis under two temperature regimes. Crop Sci. 9:589-91.

Nelson, CJ, and WG Spollen. 1987. Fructans. Physiol. Plant 71:512-16.

Nelson, CJ, SY Choi, F Gastal, and JH Coutts. 1992. Nitrogen effects on relationships between leaf growth and leaf photosynthesis. In N Murata (ed.), Research in Photosynthesis, vol. 4. Boston, Mass.: Kluwer, 789-92.

Ogren, WL. 1984. Photorespiration: Pathways, regulation, and modification. Annu. Rev. Plant Physiol. 35:415-42.

Owensby, CE, EF Smith, and JR Rains. 1977. Carbohydrate and nitrogen reserve cycles for continuous, season-long and intensive-early stocked Flint Hills bluestem range. J. Range Manage. 30:258-60.

Paulsen, GM, and D Smith. 1968. Influences of several management practices on growth characteristics and available carbohydrate content of smooth bromegrass. Agron. J. 60:375-79.

Penning de Vries, FWT, HH VanLaar, and MCM Chardon. 1983. Bioenergetics of growth of seeds, fruits, and storage organs. In Potential Production of Field Crops under Different Environments. Los Banos, Philippines: International Rice Research Institute, 37-59.

Pollock, CJ, and AJ Cairns. 1991. Fructan metabolism in grasses and cereals. Annu. Rev. Plant Physiol. and Plant Mol. Biol. 42:77-101.

Rhodes, I. 1973. Relationship between canopy structure and productivity in herbage grasses and its implications for plant breeding. Herb. Abstr. 43:129-33.

Robson, MJ. 1982. The growth and carbon economy of selection lines of *Lolium perenne* cv. S23 with differing rates of dark respiration. 1. Grown as simulated swards during a regrowth period. Ann. Bot. 49:321-29.

Sharp, RE, and WJ Davies. 1979. Solute regulation and growth by roots and shoots of water stressed maize plants. Planta. 147:43-49.

Sheard, RW, and JE Winch. 1966. The use of light interception, gross morphology and time as criteria for the harvesting of timothy, smooth bromegrass and cocksfoot. J. Brit. Grassl. Soc. 21:231-37.

Smith, D. 1962. Carbohydrate root reserves in alfalfa, red clover, and birdsfoot trefoil under several management schedules. Crop Sci. 2:75-78.

______. 1968. Classification of several native North American grasses as starch or fructosan accumulators in relation to taxonomy. J. Brit. Grassl. Soc. 23:306-9.

______. 1981. Forage Management in the North. 4th ed. Dubuque, Iowa: Kendall/Hunt.

Snaydon, RW. 1991. The productivity of C_3 and C_4 plants: A reassessment. Funct. Ecol. 5:321-30.

Sullivan, JT, and VG Sprague. 1949. The effect of temperature on the growth and composition of the stubble and roots of perennial ryegrass. Plant Physiol. 24:706-19.

Teeri, JA, and LG Stowe. 1976. Climatic patterns and the distribution of C_4 grasses in North America. Oecol. 23:1-12.

Trlica, MJ. 1977. Distribution and utilization of carbohydrate reserves in range plants. In RE Sosebee (ed.), Rangeland Plant Physiology, Range Sci. Ser. 4. Denver, Colo.: Society of Range Management, 73-96.

Waller, SS, and JK Lewis. 1979. Occurrence of C_3 and C_4 photosynthetic pathways in North American grasses. J. Range Manage. 32:12-28.

Wilson, D. 1982. Response to selection for dark respiration rate in mature leaves in *Lolium perenne* and its effects on growth of young plants and simulated swards. Ann. Bot. 49:303-12.

4 Nutrient Metabolism and Nitrogen Fixation

DARRELL A. MILLER
University of Illinois

GARY H. HEICHEL
University of Illinois

ALL elements absorbed by plants are not necessarily essential for plant growth. *Nutrient metabolism* includes any element that has a certain physiological function or that cannot be substituted for a specific reaction or that is essential for the life cycle of the plant. Keeping this term in mind might prevent the confusion that sometimes results when identifying essential plant nutrients. Over 20 elements are essential or beneficial to plant growth. All are not required by all plants, but all are necessary for some plants (Taiz and Zeiger 1991).

Nutrients required by plants are displayed in Table 4.1, along with their role in animals. Elements other than carbon (C), hydrogen (H), and oxygen (O) are called *mineral nutrients.*

FACTORS AFFECTING ABSORPTION

External Factors. Although essential nutrients are present in the soil, many external factors influence whether or not they will be absorbed. (1) One factor is the concentration of the element. Although accumulation of an element is an active process, it does not function extensively on a concentration gradient or a diffusion pressure gradient. But the relative concentration does influence the possibility that the nutrient will be present at the absorption site. Also, although the movement of an ion to the xylem is passive, it will be affected by the concentration. (2) Another factor is the oxidation-reduction state of the element. Most of the elements are favored for absorption in aerated soils and in their most oxidized state. (3) A third factor is the water content of the soil. Water interacts heavily with the aeration and temperature of the soil. Water must be adequate to keep the elements in solution but at the same time dry enough to allow the soil to remain aerated and at temperatures favoring plant growth. (4) Proper aeration is required so that adequate respiration can occur and allow energy release. (5) A temperature that is favorable for both root growth and microbial activity is also needed. (6) A final factor is the effect of soil pH on root growth. Proper soil pH is essential because it influences the availability of essential nutrients. As an example, if the soil pH is below 5.0, there might be a toxic level of aluminum released.

Internal Factors. Internal factors affect absorption of elements. (1) The cell wall is a differentially permeable and selective structure that allows various cations and anions to enter the cell. The cell wall has a particular cation exchange capacity. Cations usually have some competitive advantage over anions in uptake. Three essential nutrients in the plant that are absorbed as anions are NO_3^-, $H_2PO_4^-$, and SO_4^{--}. These nutrients are need-

DARRELL A. MILLER is Professor of Plant Breeding and Genetics, University of Illinois. He received the MS degree from the University of Illinois and the PhD from Purdue University. His research has focused on allelopathy and the branched root characteristics of alfalfa.

GARY H. HEICHEL is Department Head of Agronomy, University of Illinois. He holds MS and PhD degrees from Cornell University. He has concentrated his research on the limitations of symbiotic nitrogen fixation and carbon assimilation in forage production.

TABLE 4.1. Essential nutrients other than carbon, hydrogen, and oxygen absorbed from soil and their role in plants and animals

Element	Absorption Form	Role in Plants	Role in Animals
Nitrogen	$N0_3^-$ NH_4^+	Animo acids, protein synthesis, nucleic acids	Protein synthesis
Phosphorus	$H_2P0_4^-$ $HP0_4^{--}$	Utilizing energy from food reserves, used early in life cycle, root formation	Skeleton component, energy metabolism
Potassium	K^+	Enzyme activation, winterhardiness, water relation, N uptake and protein synthesis, disease resistance, translocation, starch synthesis	Maintains acid-base balance, enzyme reactions, CHO metabolism
Sulfur	$S0_4^{--}$	Sulfhydryl groups, amino acids	Present in amino acids, acid-base balance, constituent intracellular, CHO metabolism
Calcium	Ca^{++}	Calcium pectate, cell regulation	Structural component (skeleton), blood coagulation, cell regulation
Magnesium	Mg^{++}	Chlorophyll, respiration	Skeleton development, phosphorylation, enzyme activation
Iron	Fe^{++}, Fe^{+++}	Cytochromes, enzymes	Oxygen transport in blood and muscles, cytochromes
Manganese	Mn^{++}	Formation of amino acids, chloroplast membrane, enzyme systems	Needed in bone matrix formation
Boron	$H_3\ BO_3$	Amino acids and proteins synthesis, nodule formation	Possibly required?
Copper	Cu^{++}, Cu^+	Nitrate reduction, photosynthetic electron transfer	Needed in enzymes and Fe metabolism, immune system
Zinc	Zn^{++}	Enzymatic activities	Activates enzymes and constituent in metalloenzymes
Molybdenum	$Mo0_4^{--}$	Nitrate reductase	Component of metalloenzyme, tantline oxidase
Chlorine	Cl^-	Photosynthetic phosphorylation, charge balance, osmotic pressure	Regulate extracellular osmotic pressure, maintain acid-base balance
Sodium	Na^+	Osmotic pressure, charge balance	Acts with K and Mg in extracellular components, maintaining osmotic pressure, nerve function
Iodine	I^-	Perhaps needed in tissue culture	Thyroid gland function
Cobalt	Co^{++}	N fixation in alfalfa	Constituent of vitamin B_{12}
Selenium	$Se0_3$	Not needed?	Component enzyme GSH-P_x, cellular membrane, immune system
Nickel	Ni^{++}	Part of urease; legumes need it in N fixation.	Not needed?
Silicon	$Si(OH)_4$	Drought resistance, mechanical strength	Mineralization of bones

ed and absorbed in great quantities. (2) Respiration is another internal factor affecting absorption. Movement across the plasmalemma and eventually into the vacuole involves several energy-requiring reactions. Ion uptake and growth are associated with warm, moist, well-aerated soils, and ion accumulation is linked to aerobic metabolism. Compact soils may lead to potassium (K) deficiency even if adequate potassium is present. (3) The type of cell and its stage of development affect how fast an element is absorbed. Most of the inorganic materials enter the xylem from the soil through the meristematic area, or root tips. The absorption occurs through a root hair cell. Each root hair cell may be effective in absorption for just a few days. (4) Transpiration contributes to the transport of nutrients across the cell membrane. However, the dominant factor in nutrient absorption is respiratory energy.

FUNCTION AND USE OF ESSENTIAL ELEMENTS

Most of the essential elements are involved in the structural components of the plant as well as being co-factors in many enzymatic reactions (Taiz and Zeiger 1991). A deficiency in any one element may cause structural inadequacy or block an enzymatic process; the result may be the rapid disruption of a wide array of metabolic processes. These processes may be divided into basic plant structure functions, energy storage and transfer, charge balance, enzyme activation and electron transport, and other beneficial functions (see Chap. 5, Vol. 2).

Basic Plant Structure Functions. Carbon is the backbone of organic compounds. Its four valence charges serve as points of attachment for many other elements. Carbon represents approximately 45% of the dry weight of a plant. It is found in carbohydrates (sugars and starches), proteins, lipids, and unique configurations such as carotene and nucleic acids. Carbon, in the form of carbon dioxide, is absorbed into the plant through the stomata.

Hydrogen attached to C forms a hydrocarbon. When H is transported along with an electron during photosynthesis, it represents one of the energy-generating processes in a plant system. Hydrogen is supplied to the plant from water. It composes about 6% of the plant's dry weight.

When O is inserted between C and H, it enhances the reactivity of the compound. Oxygen is primarily supplied by carbon dioxide. Plants contain approximately 43% oxygen.

These three elements—C, H, and O—serve as the backbone, fillers, and regulators of the plant's chemical structure.

Energy Storage and Transfer. Next to H, nitrogen (N) is the most important element absorbed by plants. Nitrogen is present in the air as dinitrogen gas (N_2) but must be bonded to H or O by lightning, bacteria, or some industrial process before plants can use it. Nitrogen is a key component of protoplasm, primarily as part of the protein molecule (Taiz and Zeiger 1991). Protein in the vegetative cells of plants is largely structural. Many proteins are enzymes, while others are nucleoproteins, and some are present in the chromosomes. The first effect of N deficiency is a reduction in the rate of meristematic activity, which limits cell expansion and cell division. In the early stages of N deficiency, carbohydrates may tend to accumulate. Young plants may contain 6% N in the dry matter, while older plants range from 0.5% to 2%.

Nitrogen is absorbed as NO_3^- and NH_4^+ by the roots. Most of the available N in the soil is converted to NO_3^- via bacterial action (see Chap. 5, Vol. 2). Dinitrogen [N_2] fixation is discussed later in this chapter.) Once in the plant, NO_3^- is rapidly converted to NH_4^+ via nitrate reductase. Drought or low light intensity reduces the nitrate reductase activity, resulting in an accumulation of NO_3^- in the plant.

Sulfur (S) serves as a basis for low-energy bonding in protein synthesis. Sulfur also functions in energy transfer in a manner similar to phosphorus (P). It may also be important in the resistance of protoplasm to cold and drought. Sulfur is also needed in the synthesis of sulfur-bearing amino acids. Deficiency of sulfur may result in an accumulation of nitrates as well as amides.

Phosphorus is involved in energy release and storage as well as structurally in phospholipids. Phosphorylation results in a reduction of the activation energy needed for enzymatic reactions. Phosphorus is the key constituent in the electron transport mechanism: phosphorus ions are very mobile and are translocated to cells of high metabolic activity, where the higher concentrations occur.

Charge Balance. Potassium (K) does not enter into structural components. Potassium, along with magnesium (Mg) and calcium (Ca), maintains cellular organization and provides electrical charge balance, hydration, and permeability. It also serves as an enzyme activator. Potassium is highly mobile and is present in high concentration in meristematic tissue but in low concentration in the seed. It is also involved with iron activity in chlorophyll synthesis. If the plant is deficient in K, free amino acids may accumulate. A deficiency is associated with a decrease in resistance to cold and certain diseases. Winterhardiness of both alfalfa (*Medicago sativa* L.) and bermudagrass (*Cynodon dactylon* [L.] Pers.) is greatly dependent upon adequate K.

Calcium is especially important in maintaining the organization of the protoplasm and providing the cement of cell walls as calcium pectate. Calcium is related to protein synthesis by enhancing the uptake of nitrate N, and it is associated with enzyme systems. It is rather immobile and is present more in the older leaves than in the younger ones. A deficiency results in the death of the terminal buds and prevents development of apical tips of the roots.

Magnesium plays a key role in the chlorophyll molecule. It is also involved with several enzymes and especially with phosphorus and carbohydrate metabolism. Magnesium is a mobile element and is translocated to the younger plant parts. Deficiency results in interveinal chlorosis of the leaf while the veins remain green.

Enzyme Activation and Electron Transport. Many of the micronutrients are involved with electron transfer. There is a small range of concentration levels allowed for these elements under field conditions. At relatively low concentrations boron (B), zinc (Zn), molybdenum (Mo), manganese (Mn), and copper (Cu) can be toxic to plants (see Chap. 5, Vol. 2). Soil pH is especially important in influencing the availability of molybdenum, manganese, and copper to plants.

Iron (Fe) is present in the cytochrome system and is necessary for chlorophyll synthesis. Boron is involved with amino acids and protein synthesis. It also plays a role in root nodule formation. Zinc is involved with enzymatic activities associated with H transfer. Manganese is involved with nitrate assimilation, amino acid formation, chloroplast membrane formation, and enzymatic activities associated with carbohydrate metabolism. Manganese is relatively immobile. Molybdenum is part of the nitrate reductase molecule, which is involved with nitrate reduction. A deficiency will result in a nitrate accumulation in the plant. Copper is present in proteins that act as enzymes and bring about some oxidation reactions. Phosphate accentuates a Cu deficiency. Chlorine (Cl) is involved with photosynthetic phosphorylation and osmotic pressure.

Beneficial Functions. Cobalt (Co) plays a definite role in bacterial nutrition. It is essential in vitamin B_{12} formation, which in turn is essential in the formation of leghemoglobin needed for N fixation. Sodium (Na) and silicon (Si) are needed in relation to osmotic pressure, charge balance, and mechanical strength. Silicon is probably also involved with drought resistance. Nickel (Ni) is part of the urease system, and legumes might need it for cell growth and N fixation. Iodine (I) has been reported as needed for cell growth in tissue culture.

DINITROGEN FIXATION

All plants use soil N in the form of nitrate (NO_3^-) or ammonium (NH_4^+) to meet their nutritional needs for growth and reproduction. Because mechanical harvest or grazing of forages by animals removes N in the plant dry matter of fields or range, soil amendment with manufactured fertilizers is often necessary to sustain forage or animal productivity at a profitable level. Many forage, pasture, or range systems require supplementation with purchased N fertilizers, which influences profitability and may affect environmental quality. Dinitrogen (N_2) fixation by legumes provides an opportunity for forage plants to obtain N from the earth's atmosphere rather than from commercial fertilizer. This may have economic benefits to the producer or rancher and environmental benefits to society.

Symbiosis: A Process for Nitrogen Self-sufficiency. Several forage, pasture, and rangeland species have the genetic capacity for N self-sufficiency; legumes such as alfalfa, clovers, medics, trefoils, vetches, and lespedezas are examples. This self-sufficiency, which may free the plants from the need for amendment with manufactured N fertilizer, is achieved by symbiotic fixation of inert gaseous N_2 from the atmosphere into NH_4^+. Ammonium produced from symbiosis is identical to the NH_4^+ in the soil that is used in amino acid and protein synthesis. The symbiotic N_2 fixation process occurs within tumorlike nodules on the root systems of legumes (Fig. 4.1).

Fig. 4.1. Symbiotic nitrogen fixation is a dynamic partnership between the leguminous host plant and specialized bacteria that inhabit nodules on its roots.

After infection of root hairs by nonpathogenic soil bacteria belonging to the genera *Rhizobium* or *Bradyrhizobium,* the root nodules develop, and fixation of atmospheric N_2 occurs. Forage legumes nodulate only after infection by a strain of nodule-forming bacteria specific for the plant species. For example, the bacterium that elicits nodules on alfalfa differs from the one forming nodules on clover. When nodule-forming bacteria are absent in the soil, treatment of seed before planting with a preparation of commercial inoculum containing nodule-forming bacteria, adhesive, peat, and occasionally limestone or mineral nutrients is necessary. Alternately, seed of many forage legumes that is encased in a coating containing all of these materials is available commercially. Additional information on the specificity between nodule-forming bacteria and forage species, on the infection process, or on the criteria for inoculation of seed is given by Heichel (1985) and Vance et al. (1988).

A fully developed N_2-fixing nodule of alfalfa has a growing point, or apical meristem, cells in which N_2-fixation occurs, a transport system to provide amino acids containing the fixed N_2 to other plant organs, and mature cells that no longer fix N_2 (Fig. 4.2). Some nodules, e.g., those of alfalfa and red clover (*Trifolium pratense* L.), overwinter and fix N_2 for more than one season. Other nodules, e.g., those of birdsfoot trefoil (*Lotus corniculatus* L.), newly form on the roots each year.

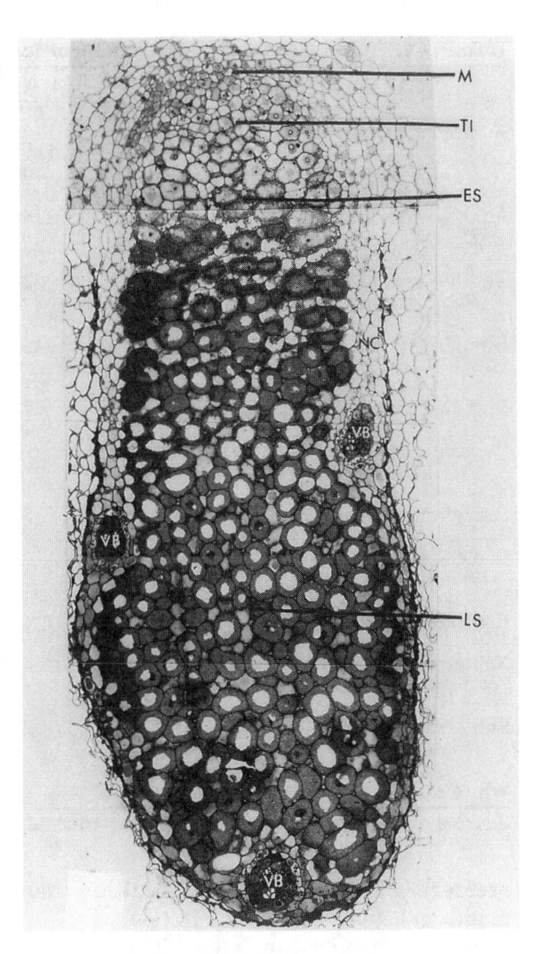

Fig. 4.2. Longitudinal section through alfalfa nodule showing four structurally distinct zones of development: (*M*) the indeterminant apical meristem, (*TI*) the thread invasion region, (*ES*) the region of early symbiotic development, and (*LS*) the region of late symbiotic development. The vascular bundles (*VB*) form a net within the perpherical cortex of the nodule.

The total amount of N_2 fixed by a forage legume species, or by a legume-grass community, varies widely (Table 4.2). Nitrogen fixation for a particular species (e.g., red clover or alfalfa) varies with geographic location because of differences in soil fertility, crop adaptation, length of growing season, and crop productivity. Even at the same location, legume species differ in N_2 fixation capacity because of adaptation, winterhardiness, crop productivity, nodule persistence, and pest pressure. In Table 4.2, compare alfalfa and white clover grown at Lexington, Kentucky, and alfalfa, red clover, and birdsfoot trefoil grown at Rosemount, Minnesota. Mixtures of forage grasses with legumes may result in

TABLE 4.2. Seasonal total of N fixation of forage legumes and forage legume-grass communities

Species	Total N_2 fixation (kg N/ha/growing season)	Location
Alfalfa (*Medicago sativa* L.)	212	Lexington, Ky., USA
	114-224	Rosemount, Minn., USA
	79-104	Sweden
Alfalfa—orchardgrass (*Dactylis glomerata* L.) sward	15-136	Lucas Co., Iowa USA
Alfalfa—reed canarygrass (*Phalaris arundinacea* L.) sward	82-254	Rosemount, Minn., USA
Birdsfoot trefoil (*Lotus corniculatus* L.)	49-112	Rosemount, Minn., USA
Birdsfoot trefoil—reed canarygrass sward	30-130	Rosemount, Minn., USA
Berseem clover (*Trifolium alexandrinum* L.)	62-235	Egypt
Hairy vetch (*Vicia villosa* Roth)	111	New Jersey, USA
Red clover (*Trifolium pratense* L.)	22-61	Rothamsted, Great Britain
	69-113	Rosemount, Minn., USA
Red clover—reed canarygrass sward	5-152	Rosemount, Minn., USA
Subterranean clover (*Trifolium subterraneum* L.)	58-183	Hopland, Calif., USA
Subterranean clover—soft chess (*Bromus mollis* L.) sward	21-103	Hopland, Calif., USA
White clover (*Trifolium repens* L.)	128	Lexington, Ky., USA

Sources: Adapted from Heichel (1987) and Heichel and Henjum (1991).

greater N_2 fixation than that resulting from stands of legumes alone.

Nitrogen fixation can, but seldom does, provide the total N requirements of legume nutrition. Legumes preferentially use NO_3^- or NH_4^+ present in the soil before nodules begin fixing N_2. Thus, the amounts of N_2 fixation in Table 4.2 may represent as little as 30% or as much as 95% of the total N in the crop; the rest is derived from soil organic matter, residual fertilizer, or animal manures. The highest rates of N_2 fixation are exhibited by long-term stands of legumes that have depleted soil N or by stands of legumes grown on soils impoverished in organic matter or other N sources. For additional information on management factors that affect N_2 fixation and how fixation changes during the growing season, see Heichel (1985) and (1987) and Vance et al. (1988).

Role of N_2 Fixation in Crop Rotations. In a sequence or rotation of different crops, the yield of a grain crop such as corn or wheat is often improved if it is preceded by a legume such as alfalfa or red clover. This *rotation effect,* the stimulation of a grain crop yield by incorporation of a preceding forage legume as a green manure crop, is often largely attributed to the N made available to the grain by legume decomposition. However, factors in addition to legume N are important in understanding the rotation effect. These are the influence of tillage practice on growth of the subsequent grain crop, the soil water depletion by the preceding crop, the release of stimulatory or inhibitory phytochemicals during crop decomposition, the soil nutrient depletion by the preceding crop, the disruption of disease and insect cycles, the legume enhancement of soil structure, and the incorporation of legume N into nonavailable forms in soil clay minerals or organic matter (Heichel 1987).

Nitrogen fertilizer application to corn can often be reduced by 100 kg N/ha to 150 kg N/ha if it is preceded by alfalfa or red clover. The reduction in recommended N fertilizer application that is attributed to the legume is usually termed the *fertilizer N equivalent* or *fertilizer N replacement value* of the legume.

Many management factors influence the amount of fertilizer N that can be replaced by a preceding legume. For alfalfa, the timing of tillage in relation to the amount of herbage regrowth is an important consideration (Table

4.3). Alfalfa herbage yields decrease successively throughout the season; the first cutting is greatest, and the last is least. Early in the season, alfalfa relies less on N_2 fixation than it does later in the season. Thus, at the time of the first hay harvest, 52 kg/ha of fixed N and 54 kg/ha of soil N are removed in the herbage. If tillage occurred immediately after the first harvest, the loss of 54 kg/ha of soil N in the herbage is not replenished by the addition of 5 kg/ha of fixed N in the roots and crowns. Thus, destruction of the stand by tillage immediately after the first harvest would cause a net deficit of 49 kg N/ha to the soil-crop system.

By the time of the second harvest, N_2 fixation has increased, and more of the fixed N is distributed to roots and crowns than occurred earlier in the season. In addition, the crop has removed less soil N than earlier in the season. If the herbage is removed on August 30, the loss of 18 kg/ha of soil N would be more than replenished by the 28 kg/ha fixed N in the roots and crowns. Although the net gain between first and second harvest is 10 kg N/ha, the soil-crop system would still be 39 kg N/ha in deficit if the stand was destroyed by tillage immediately after the second harvest.

Forgoing the second harvest on August 30 and incorporating the lush vegetative regrowth by moldboard plowing would provide 102 kg/ha of fixed N, a net contribution of 53 kg N/ha beyond the 49 kg N/ha deficit incurred after the first harvest (Table 4.3). From this detailed N budget of an alfalfa crop, plus an understanding of how N from the soil and from symbiotic fixation is distributed among herbage, roots, and crown, a general principle is apparent. Forage legumes will contribute the most fixed N to the soil, and potentially replace the most fertilizer N, if lush vegetative regrowth occurs before tillage in the fall or next spring. This is because N_2 fixation provides a greater proportion of the N in the forage later in the season than earlier, and more fixed N accumulates in herbage dry matter than in roots plus crown. If only sparse stubble remains for plow-down or no-till establishment of corn, the N contribution of legumes in rotation will be greatly reduced.

The amount of symbiotically fixed N_2 actually recovered by a succeeding crop from an incorporated legume varies with many factors. These are amount of N_2 fixed, mass of plant material incorporated, rate of decomposition of legume material, immobilization of legume N in the soil, and match of legume N mineralization with the growth needs of the succeeding crop. The emerging evidence indicates that only 25% to 60% of the yield enhancement of a grain crop following a legume is attributable to the amount of N_2 initially fixed by the legume. The rest of the yield boost is caused by incorporation and release of the soil N in the legume (not incorporated "free" from the atmosphere but merely shunted from the soils to the grain crop through the legume) and by the other components of the rotation effect (Heichel 1987).

Nitrogen Transfer in Legume-Grass Communities. Nitrogen fixation by forage legumes contributes to N self-sufficiency in mixtures of legumes and grasses by providing a source of nonfertilizer N for transfer to the grass. The occurrence of legume to grass N transfer has

TABLE 4.3. Nitrogen budget for seeding year alfalfa illustrating the distribution of symbiotically fixed N among crop components and the net incorporation of N into the soil with mid- or late-season tillage options

	Herbage Harvest		
	First (July 12)	Second (August 30)	Third (October 20)
Herbage yield (mt DM/ha)	3.5	3.0	1.2
Total N yield (herbage + roots + crown) (kg N/ha)	118	127	59
Total N_2 fixed (kg N/ha)	57	102	34
Herbage	52	74	22
Roots + crown	5	28	12
Soil N uptake (kg N/ha)	61	25	25
Herbage	54	18	16
Roots + crown	7	7	9
Tillage management options			
Tillage on October 20 (late-season)			
N incorporation/harvest (kg/ha)	−49	+10	+34
Cumulative N incorporation (kg/ha)	−49	−39	−5
Tillage on August 30 (midseason)			
N incorporation/harvest (kg/ha)	−49	+102	...
Cumulative N incorporation (kg/ha)	−49	+53	...

Source: Adapted from Heichel and Barnes (1984).

often been inferred by the greater vigor and greener color of grass growing adjacent to legume plants than the vigor and color of grass growing farther away. Only recently has information become available on the amounts of N transferred from legume to grass and the significance of the transferred N to the needs of the grass (Table 4.4).

As presented in Table 4.2, N_2 fixation of legume-grass mixtures varies with species and age of stand. Averaged over years 2 to 4 of the stand, alfalfa-grass mixtures fix more than twice the N_2 of birdsfoot trefoil or red clover-grass mixtures (Table 4.4). Nitrogen transfer from legume to grass range from insignificant amounts in the year of establishment (year 1, not shown) to as much as 28 kg N/ha/growing season in second-year red clover. Averaged over 3 yr, 7 to 14 kg N/ha/growing season are transferred from a legume to an associated grass, which represents 0.3% to 1.7% of the total N that the legume fixes annually.

Although the actual amounts of N transferred from legume to grass may seem small, they are quite significant in comparison to the annual N content of the associated grass. Excluding the year of establishment in which little N is transferred, as little as 11% to as much as 47% of the N in the grass can be derived from the legume (Table 4.4). Averaged over the 3 yr of the stand, about 30% of the N in the grass is obtained from the N_2 initially fixed by the legume (Heichel and Henjum 1991).

The mechanisms of legume-grass N transfer are not completely understood. There is some evidence of transfer of N from living roots of legumes to living roots of associated grasses. Nodule turnover may be more important in some species, e.g., birdsfoot trefoil, than in other species such as alfalfa because of species differences in the longevity of nodules. However, the mass of nodules present on the plants at any time during the season is insufficient by 30% to 60% to account for the observed amounts of N transfer (Heichel and Barnes 1984).

Current evidence points to indirect N transfer from legume to grass through death and turnover of organs of living plants in the legume-grass community or through the death of individual plants in the sward. For example, loss of diseased lower leaves of alfalfa to the soil surface, death of roots of clover or trefoil after mowing or grazing, or death of lower branches of trefoil that escape mechanical harvest may contribute substantial N.

TABLE 4.4. Nitrogen fixation and N transfer from legume to grass over 3 yr in stands of alfalfa, birdsfoot trefoil, and red clover with reed canarygrass

	Year 2			Year 3			Year 4			Average		
	Alfalfa/ Grass	Trefoil/ Grass	Red Clover/ Grass	Alfalfa/ Grass	Trefoil/ Grass	Red Clover/ Grass	Alfalfa/ Grass	Trefoil/ Grass	Red Clover/ Grass	Alfalfa/ Grass	Trefoil/ Grass	Red Clover/ Grass
Dinitrogen fixed (kg N/ha)	254	90	141	210	129	83	153	58	21	206	92	82
Season N transfer (kg N/ha)	8	28	22	11	6	18	7	8	3	7	14	14
N transferred from legume (%)[a]	29	47	33	28	11	21	36	26	25	31	28	26

Source: Adapted from Heichel and Henjum (1991).
[a]Proportion of N in the grass (whole plant basis) that was transferred from the legume.

Whatever the mechanisms of transfer, the evidence now shows that N_2 originally symbiotically fixed by the legume can cycle by various transfer mechanisms to an adjacent grass and provide up to 50% of the N requirement of the grass. This results in a significant saving of commercial fertilizer and illustrates a strategy for crafting legume-grass communities for hay or pasture that are significantly self-sufficient for N nutrition.

QUESTIONS

1. What is meant by nutrient metabolism?
2. What are the factors that affect nutrient absorption?
3. List the elements that are involved in the structural features of a plant.
4. Name several elements that are involved in energy storage and transfer within a plant.
5. Enzyme activation involves which nutrients?
6. What does symbiosis mean?
7. Name the parts of a nodule and the function of each.
8. Would you expect more N_2 fixation from legumes growing in soil high, rather than low, in available nitrates?
9. Do all legumes fix the same amount of N_4? Compare geographic locations.
10. What are the factors affecting the recovery of legume N by a succeeding crop?

REFERENCES

Ensminger, ME, and GC Odentine, Jr. 1978. Feeding beef cattle. In Feeds and Nutrition—Complete. Danville, Ill.: Ensminger Publishing, 583-700.

Heichel, GH. 1985. Symbiosis: Nodule bacteria and leguminous plants. In ME Heath, RF Barnes, and DS Metcalfe (eds.), Forages: The Science of Grassland Agriculture, 4th ed. Ames: Iowa State Univ. Press, 64-71.

______. 1987. Legume nitrogen: Symbiotic fixation and recovery by subsequent crops. In Z Helzel (ed.), Energy and World Agriculture Handbook, Vol. 2, Energy in Plant Nutrition and Pest Control. Amsterdam, Netherlands: Elsevier Science Publishing, 63-80.

Heichel, GH, and DK Barnes. 1984. Opportunities for meeting crop nitrogen needs from symbiotic nitrogen fixation. In D Bezdicek and J Power (eds.), Organic Farming: Current Technology and Its Role in a Sustainable Agriculture, Spec. Publ. 46. Madison, Wis.: American Society of Agronomy, 49-59.

Heichel, GH, and KI Henjum. 1991. Dinitrogen fixation, nitrogen transfer, and productivity of forage legume-grass communities. Crop Sci. 31:202-8.

Taiz, L, and E Zeiger. 1991. Plant Physiology. Redwood City, Calif.: Benjamin/Cummings Publishing, 107-15.

Tisdale, SL, WL Nelson, and JD Beaton (eds.). 1985. Elements required in plant nutrition. In Soil Fertility and Fertilizers, 4th ed. New York: MacMillan, 59-94.

Vance, CP, GH Heichel, and DA Phillips. 1988. Nodulation and nitrogen fixation. In AA Hanson, DK Barnes, and RR Hill, Jr. (eds.), Alfalfa and Alfalfa Improvement, Am. Soc. Agron. Monogr. 29, Madison, Wis., 229-57.

5 Environmental and Physiological Aspects of Forage Management

C. JERRY NELSON
University of Missouri

JEFFREY J. VOLENEC
Purdue University

THE environment plants experience is a combination of climate and weather. *Climate* refers to the long-term history of temperature and rainfall for a given region, whereas *weather* refers to short-term or day-to-day variations in temperature, precipitation, relative humidity, and solar radiation at a given site. Climate is the principal factor affecting adaptation of forage plants to a given location, whereas weather influences the year-to-year variation in their productivity. Agriculturalists interested in producing forages need to be aware of the climate when selecting species and cultivars to plant and need to understand how the plants can be managed to maximize the potential within the climate. Weather will influence management as well, especially in the short-term.

The growth pattern of a forage legume or grass is influenced by the genetic makeup of the plant and the environmental conditions to which it is exposed. The genetic makeup is determined by the species and cultivar selected. Growth will vary within a season and among seasons, depending on the weather complex within the climatic region. Management of the crop modifies the effects of climate and weather or alters the ability of the plant to respond to the environment. Optimum management strategies should be based on the physiological condition and stage of development of the plant. Seldom can management decisions be based on the calendar date or other fixed factors.

C. JERRY NELSON. *See Chapter 2.*

JEFFREY J. VOLENEC is Professor of Agronomy at Purdue University. He received the MS and PhD degrees from the University of Missouri. He reserches carbohydrate and nitrogen metabolism associated with regrowth and persistence of forage plants.

Special appreciation is expressed to Drs. Dale Smith, R. J. Bula, and Darell E. McCloud for use of material from the fourth edition of *Forages: The Science of Grassland Agriculture.*

CLIMATE CLASSIFICATION

Plant geographers have recognized the relationship between climate, primarily precipitation and temperature, and natural vegetation at a given site. This led Thornthwaite (1933), an American geographer, to devise a systematic classification of world climates based on precipitation effectiveness and temperature efficiency. Precipitation effectiveness depends on both rainfall and evaporation and is divided into five provinces (classes) related to relative humidity and the associated natural vegetation (Table 5.1). Similarly there are six temperature-efficiency provinces: A′ = tropical, B′ = mesothermal, C′ = microthermal, D′ = taiga, E′ = tundra, and F′ = perpetual frost.

Rain forests and forests are distinguished largely by differences in humidity. When precipitation is not limiting, the first three temperature provinces are easily distinguished. Palms or tall-buttressed tropical trees characterize A′, deciduous hardwoods characterize B′, and spruce (*Picea* spp.) and fir (*Psuedotsuga* spp.) are good indicators of C′ provinces. Sparse, stunted, somewhat open forest vegetation characterizes D′ (taiga), while E′ (tun-

TABLE 5.1. Climates classified according to precipitation effectiveness (humidity) and temperature provinces

	Temperature					
Humidity	A′	B′	C′	D′	E′	F′
A. Wet	(Rain forest)					
B. Humid		(Forest)				
C. Subhumid	(Savannas)		(Tall-grass prairie)			
D. Semiarid			(Short-grass prairie)			
E. Arid	(----------------------Desert------------------------)					

Source: Adapted from Thornthwaite (1933).
Note: Temperature A′ is tropical and F′ is perpetual frost.

dra) is treeless. There is virtually no vegetation in F′ (perpetual frost) because the soil remains frozen year-round.

Grasslands consisting of grasses, legumes, forbs, and shrubs compete naturally with trees and predominate areas where trees are less adapted. Well-managed grasslands can restrict invasion because they compete effectively with tree seedlings to limit their establishment. Conversely, if trees and other woody species become established, they shade the grasslands, are rarely grazed by domestic ruminants, and therefore become dominant in the ecosystem. Invasions by woody species such as mesquite (*Prosopis* spp.) in southwestern grasslands and eastern red cedar (*Juniperus virginiana* L.) in the central Great Plains cause severe problems. Likewise, many deciduous trees encroach pastures in eastern areas of the US.

Similar to trees, grasses exist naturally in distinct geographic regions. For example, grasses with C_4 photosynthesis (see Chap. 3) will predominate in the tall-grass prairie of the Central Great Plains, but short-grass prairies become dominant as one moves farther west to more arid areas. Grasses in short-grass prairies are mainly C_4 types in southern latitudes of the US but are C_3 types in northern latitudes and at high altitudes in southern regions (Teeri and Stowe 1976). Grassland ecosystems are discussed in more detail in Chapter 11.

In addition to humidity and temperature, the seasonal distribution of rainfall provides a basis for classifying climates. It may be subdivided into (1) areas with abundant precipitation in all seasons, (2) those sparse in summer, (3) those sparse in winter, and (4) those deficient in all seasons. Collectively, the three factors of humidity, temperature, and rainfall distribution describe 120 combinations of climate types. However, only 32 are recognized as actual climates; of these, 18 represent the major climatic regions of the world.

Forage Adaptation. Natural grasslands of the world usually developed under restricted precipitation effectiveness (Fig. 5.1) and often coevolved with grazing animals. The tall-grass prairies are located in the C humidity type and either B′ or C′ temperature provinces, as exemplified by prairies of Iowa and Illinois, eastern South Africa, the lower Danube and Po valleys of Europe, and the Pampas of Argentina. The drier grasslands, or steppes, are found in the D humidity type. The Russian and Australian steppes, the drier portion of the US Great Plains, and the African veldts are examples of short-grass vegetation.

The vast tropical savannas (CA′, Table 5.1) of South America, Asia, Australia, and Africa are transitional grasslands with characteristic scattered, low-growing, often round-topped trees such as *Acacia* spp. Savannas are characterized by pronounced wet and dry seasons.

In the D humidities, blue grama (*Bouteloua gracilis* [H.B.K.] Lag. ex Steud.), a C_4 species, is found in both the northern (C′) and southern plains (B′) regions. In contrast, blue panic (*Panicum antidotale* Retz.) and buffalograss (*Buchloe dactyloides* [Nutt.] Engelm.), also C_4 species, require a higher temperature for optimum growth. Consequently, they are mostly restricted to the B′ climates. Range grasses particularly well adapted to the drier desert (E) are Boer, Lehmann, and Wilman lovegrasses (*Eragrostis* spp.).

A few forage species such as alfalfa (*Medicago sativa* L.) contain genetic materials that are well adapted to a wide range of temperature conditions, but most grasses and legumes tend to be more restricted in temperature adaptation. In the eastern US, where moisture is adequate, temperature is the principal factor affecting distribution of grass species. Kentucky bluegrass (*Poa pratensis* L.), timothy (*Phleum pratense* L.), smooth bromegrass (*Bromus inermis* Leyss.), orchardgrass (*Dactylis glomerata* L.), and tall fescue (*Festuca arundinacea* Schreb.) are best adapted to the C′ climate. If they are grown in a B′ climate, it is either in the cooler regions or for use during winter. Forage legumes such

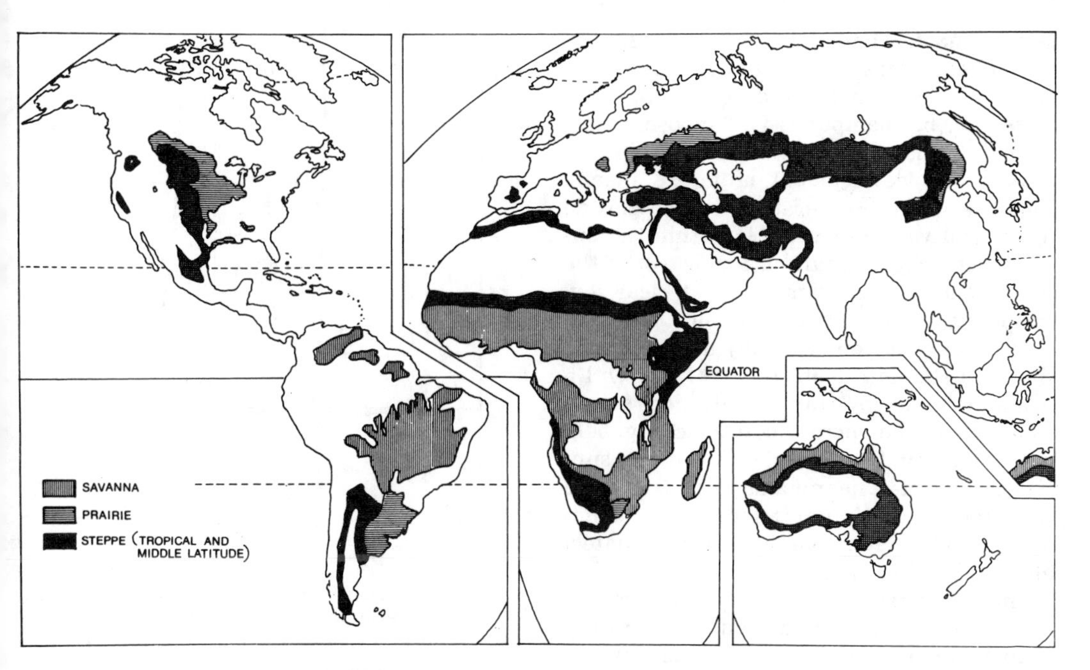

Fig. 5.1. World distribution of native grasslands. (Adapted from Trewartha et al. 1967. Used with permission of McGraw-Hill Book Co.)

as red clover (*Trifolium pratense* L.), birdsfoot trefoil (*Lotus corniculatus* L.), and alsike clover (*T. hybridum* L.) have a climatic adaptation similar to these grasses.

Characteristic grasses of B′ climates of the humid portions of the US are bermudagrass (*Cynodon dactylon* [L.] Pers.), dallisgrass (*Paspalum dilatatum* Poir.), and johnsongrass (*Sorghum halepense* [L.] Pers.). Only a small A′ area exists in the US. Adapted grasses include 'Pangola' digitgrass (*Digitaria eriantha* Steud.), bahiagrass (*Paspalum notatum* Flugge), and limpograss (*Hemarthria altissima* [Poir.] Stapf & C.E. Hubb.). Tropical legumes such as kudzu (*Pueraria phaseolides* [Roxb.] Benth.), stylo (*Stylosanthes guianensis* [Aubl.] Sw.), and phasey bean (*Macroptilium lathyroides* [L.] Urb.) are adapted to A′ climates.

Grasses and other forage species occur naturally in areas subjected to wide temperature and moisture extremes; however, productivity under these stresses may be low, and the ecosystem very fragile. Thus, economic production of a given species is largely confined to much narrower ranges of climate. As species are moved away from their center of adaptation, they must be managed even more carefully to persist and be productive. For example, inconsistent winter survival of perennial cool-season forages often restricts their geographic distribution, whereas distribution of warm-season grasses is restricted by low minimum temperatures in July (Teeri and Stowe 1976). Ability to flower and produce viable seed also restricts adaptation and perennation of annuals and short-lived perennials (Beuselinck et al. 1994).

Pasture and forage species are often grown extensively in environments that are similar but that are considerably away from their natural center of origin. In fact, nearly all forage species cultivated in the eastern US are introduced from other parts of the world. A knowledge of climates helps in understanding where species may be adapted. A convenient system of classification (Child and Byington 1981) identifies temperate zone cultivated forages, humid and subhumid rangelands, tropical zone cultivated forages, semiarid rangelands, and temperate and tropical forests. Snaydon (1981) produced a similar classification of both natural and seeded grasslands. A systematic approach developed in Sweden to estimate forage yields involves an optimum growth function, which is the product of indices of radiation, temperature, and water (Angus et al. 1983). Thus, as would be anticipated, the climatic factors that determine the suitability of a species for any region are also those that are major determinants of its productivity (Snaydon 1991).

Microclimate. The condition in and around the soil and plant canopy is called the *microclimate*. Microclimate is different from climate, can be manipulated by crop management, and is of major importance in the growth and development of forage plants. Light, temperature, moisture, carbon dioxide (CO_2), and wind are variable within the microclimate (Fig. 5.2) and are affected by the plant canopy. For example, on a clear day temperature and humidity within the forage crop canopy may be markedly different from that in a standard meteorological shelter 1.5 m above the canopy. Diurnal variations are also largest in the microclimate zone (Fig. 5.3). Near the ground, air is coolest just before sunrise and warmest near midday. At 1.5 m above the ground, maximum temperature occurs about 1 to 2 h later than it does immediately above the soil surface.

Since solar radiation strongly affects microclimate temperatures, diurnal temperature change is much greater on clear days, when incoming radiation from the sun greatly exceeds that reradiated from the earth, with the result that temperatures increase rapidly. At night the reverse occurs, and surfaces cool rapidly. On cloudy days both incoming and outgoing radiation are reduced, resulting in reduced diurnal temperature change and a smaller difference between the micro- and macroclimate. Relationships between incoming and outgoing radiation largely determine temperatures at plant and soil surfaces that affect plant growth and development and, ultimately, agronomic performance. Management can markedly alter the microclimate. For example, on clear days after a forage is cut or grazed, the soil temperature increases, especially when nearly all the canopy is removed (Fig. 5.3). This increases the temperature of buds developing near soil level that ultimately form shoots and tillers, and it can influence the regrowth rate.

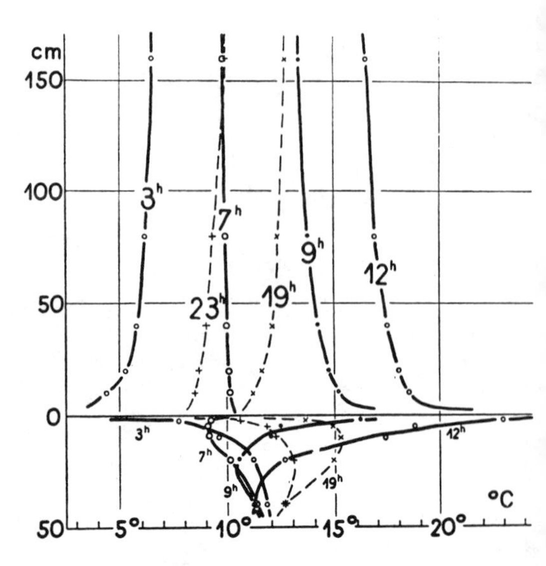

Fig. 5.3. Microclimate temperatures above and below the ground surface at different times of the day. (Geiger 1965)

SOLAR RADIATION

Light, or radiation, temperature, and soil moisture are the three cardinal environmental factors that affect vegetative development and flowering of forage species. Plant growth responses to radiation can be separated into those due to quality, density, or duration of radiation. Under field conditions these factors are often interrelated; e.g., density of radiation is usually highest during the same season that duration is longest.

Quality. Quality refers to the wavelength of the rays contributing to the radiation spectrum. Plant development is better under the full spectrum of sunlight than under any portion. For example, plants grown under only the long infrared wavelengths usually grow tall and thin and are often fragile, as in dense shade. Plants grown under only the short ultraviolet wavelengths may be retarded in

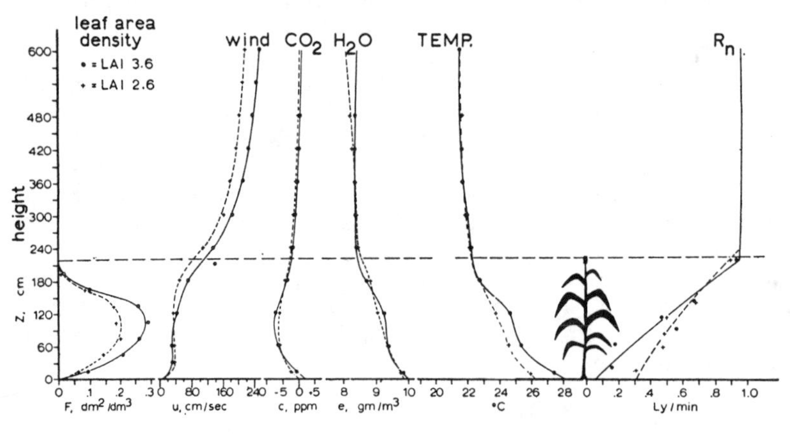

Fig. 5.2. Distribution of leaf area, wind speed, carbon dioxide, water vapor, air temperature, and net radiation in two stands of corn. *Courtesy E. R. Lemon and Wageningen Centre for Agricultural Publishing and Documentation.*

growth or have their tissues injured or even killed. Winter radiation tends to be proportionately higher in infrared radiation than summer. Plants often appear dwarfed or stunted in alpine sites, partly because less ultraviolet radiation is screened by the atmosphere than in lowland sites. Radiation in the visible range (400-700 nanometers [nm]) is most active in photosynthesis, especially that in the blue and red wavelengths. Flowering of many species is controlled by radiation in the red and far-red regions of the spectrum.

Van der Veen and Meijer (1959) divide the visible and near-visible spectrum into eight wavelength bands based on the effects on plants.

1. Longer than 800 nm, no specific effect on plants except as heat, which increases water loss.
2. Between 800 and 700 nm, elongation growth and far-red effects on the phytochrome system, which regulates flowering and other processes.
3. Between 700 and 610 nm, peak chlorophyll absorption and maximum photosynthetic activity; red effects on the phytochrome system.
4. Between 610 and 510 nm, minimal photosynthesis and little formative influences. Plants appear green because they absorb little of this radiation.
5. Between 510 and 400 nm, absorption by yellow pigments and chlorophyll resulting in photosynthetic activity; region of phototropism response.
6. Between 400 and 320 nm (UV-A radiation), some growth processes affected such as that of leaf shape; plants become shorter and leaves thicker.
7. Between 320 and 280 nm (UV-B radiation), detrimental to most plants; damage caused to DNA; mutations.
8. Shorter than 280 nm, highly deleterious; rapid death of plants.

Fortunately, the global atmosphere is capable of screening out much of the radiation above 700 nm by water vapor. Carbon dioxide also absorbs radiation above 700 nm and may contribute to global warming (see Chap. 1, Vol. 2). Ozone absorbs radiation below 320 nm, but its breakdown in the upper atmosphere leads to concern about increased transmission of UV-B radiation to the earth's surface.

Density. Radiation density in the field at full sunlight during summer may be up to 2000 μmol photons $m^{-2}s^{-1}$. Expressing radiation density in energy units (photons) is preferred over units of intensity or illumination (Chap. 3). When nutrient and water supplies are adequate, growth rate is a direct function of radiation density via the influence on photosynthesis (Black 1957). A full canopy of leaf blades is needed to intercept the maximum amount of radiation, but the amount of leaf area required depends on the leaf angle (Chap. 3). Until high leaf area is achieved by the canopy following cutting or grazing, the growth rate is more related to radiation interception than to photosynthetic activity per unit of leaf area (Horst et al. 1978; Nelson 1988).

Competition among plants for light is especially important if forages are established with a grain companion crop (Klebesadel and Smith 1959) or when forage species are grown together in mixtures. Species respond differently to variations in radiation density. For example, red clover produces more top growth under low radiation densities than alfalfa, and alfalfa more than birdsfoot trefoil (Gist and Mott 1958). This may be a major reason red clover is popular for interseeding into established grasses that compete with the young seedlings for light. Similarly, orchardgrass will grow better in low radiation than will smooth bromegrass or timothy.

Duration (Photoperiod). In addition to the photosynthetic role of light, the lengths of daily light and dark periods influence plant growth and development (Ludlow 1978). Duration of the photoperiod changes with latitude and season because of the tilt of the earth relative to its orbital path around the sun. Minimal seasonal changes occur at the equator, and the photoperiod remains near 12 h year-round. Conversely, large changes in photoperiod occur at the poles. For example, in Alaska and central Canada the daylength on June 21 can exceed 22 h but it may be only a few hours on December 21. The daylength range for a given location depends on its latitude.

Most forage species are sensitive to photoperiod, although the responses may be further modified by other climatic factors, notably temperature and water stress. Most temperate grasses and legumes flower during long photoperiods (Fig. 5.4). Flowering of tropical species is less affected by photoperiod than flowering of temperate and arctic species.

Photoperiod also affects the vegetative growth form of many forage species. Leaf and

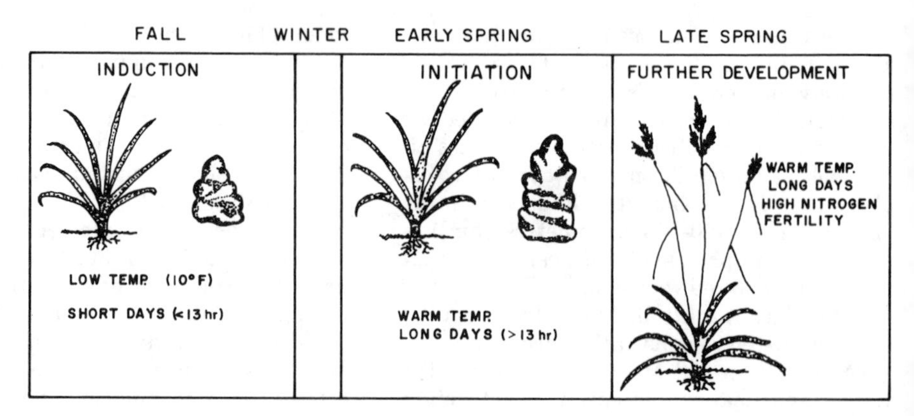

Fig. 5.4. Many long-day species, such as orchardgrass, require low temperatures and short days during autumn to induce plants to flower. The apical meristems remain vegetative during winter but gain the ability to initiate a floral structure when warmer temperatures and longer days occur in early spring. Development occurs in late spring to allow the floral structure to differentiate into inflorescence, to be elevated by internode elongation from near soil level upward through the whorl of leaves, and to be exposed at the top of the canopy. (From Gardner and Loomis 1953.)

stem growth are often erect under long photoperiods, but growth tends to be prostrate and more branched under short photoperiods. Knowledge of how photoperiod influences flowering and vegetative growth of forages facilitates proper design of pasture and hay management systems that are best adapted to the different climatic regions.

Early studies emphasize the role of daylength, but now it is known the photoperiod responses are actually regulated by length of the dark period. Even so, the daylength terminology has remained. The effect of daylength on flowering has been studied intensively since this influences both seed production and natural geographical distribution of species. *Short-day plants* flower only within a range of relatively short daylengths (actually long nights), while *long-day plants* flower only within a range of relatively long daylengths (actually short nights). For example korean lespedeza (*Kummerowia stipulacea* [Maxim.] Makino), a short-day plant, does not produce viable seed in northern states because the late flowering does not allow seed to be produced before frost. Alternatively, birdsfoot trefoil produces highest seed yields in northern states because the long daylengths easily exceed the minimum required for profuse flowering.

Other plants, referred to as *day-neutral,* are capable of flowering under either long or short daylengths. Flowering response has been found to be more complicated than this classification, commonly known as *photoperiodism,* would indicate. Many plant species require exposure to specific temperatures for flowering in addition to proper photoperiodic conditions (Fig. 5.4). Most cool-season forage grasses require a cold temperature treatment called *vernalization* in order to flower.

TEMPERATURE

Plant metabolic pathways, such as photosynthesis, respiration, and growth processes, are catalyzed by enzymes. Temperature is an important environmental factor affecting rates of these enzyme-controlled processes. Respiration, which provides the metabolic energy needed to drive growth and other metabolic processes, is related to temperature (Volenec et al. 1984). Rates of growth and other processes will vary depending on the temperature pattern to which a plant is exposed, including variation between day and night (Cooper and Tainton 1968). Temperatures during the day should be optimum for photosynthesis and growth, whereas lower temperatures at night conserve energy by reducing respiration. Optimum temperature depends on several factors including species, stage of development, and specific plant tissue. For example, optimum air temperature for vegetative growth is usually lower than that for flowering and fruit growth, and the optimum is generally 5°-8°C lower for root than for top growth.

Temperate or C_3 grasses have optimum temperatures for growth around 20°C but still grow actively at lower temperatures down to near 0°C. Tropical or C_4 grasses have a growth optimum of around 30°-35°C but grow little below 15°C (Cooper and Tainton 1968; Waller and Lewis 1979). For example, rates of photosynthesis and dry matter accumulation for seven C_3 grasses were highest in the range of

10°-20°C (Murata and Iyama 1963). However, rates for the C_4 species, bermudagrass and bahiagrass, were highest at 35°C and decreased rapidly at lower temperatures. In another study, relative growth rates of several C_4 grasses (Panicoideae) were highest at 36°C day/31°C night temperatures and were decreased about 75% at 15°/10°C (Table 5.2). In contrast, relative growth rates of C_3 grass species (Festucoideae) were highest between 21°/16° and 30°/25°C and were decreased about 40% at 36°/31°C. Switchgrass (*Panicum virgatum* L.), a C_4 species, produced the most growth and responded the most to fertilizer nitrogen (N) at the highest day/night temperature used (32°/26°C), while this occurred at lower temperatures (15°/10° and 21°/15°C) for timothy, a C_3 species (Balasko and Smith 1971).

Growth rates are a function of the rates of cell division and cell expansion. Rates of cell division are closely related to the temperature of the meristems, because duration of the mitotic cycle is temperature dependent. Temperature also affects the rate of respiration, which provides the energy needed for cell division. Rates of cell expansion likewise depend on temperature.

The influence of temperature on rates of cell division and enlargement varies with species. For example, shoot growth rates of dicots, which generally have the terminal meristem at the top of the canopy, are related to air temperature. In contrast, the terminal meristem of grasses is often near soil level, especially during vegetative stages (Chap. 2), causing shoot or tiller growth to be closely related to soil temperature. In addition, adaptation depends on temperature relationships and represents the major physiological basis for separating forage species into cool-season, warm-season, and tropical types. Interestingly, responses of growth and photosynthesis to temperature are correlated within a species, illustrating the close association between source (photosynthesis) and sink (growth) activities. Growth responses are generally more sensitive to low temperature than photosynthesis, allowing excess photosynthate to accumulate in storage organs when growth is slowed. Respiration is very responsive to temperature and at high temperatures can reduce carbohydrate storage, growth rates, and survival of forage species (Nelson and Smith 1969; Volenec et al. 1984).

TABLE 5.2. Relative seedling growth rates of several warm- and cool-season grass species grown in three day/night temperature regimes

Subfamily and species	Temperature regime (C)		
	15/10	27/22	36/31
Festucoideae			
Agropyron trichophorum (Link) Richt.	70	98	61
Bromus inermis Leyss.	75	98	65
Festuca arundinacea Schreb.	76	99	55
Poa pratensis L.	68	98	62
Trisetum spicatum (L.) Richt.	58	100	64
Phalaris arundinacea L.	66	93	61
Stipa hyalina Nees.	51	95	67
Eragrostoideae			
Chloris gayana Kunth	22	91	100
Panicoideae			
Setaria sphacelata Stapf & C. E. Hubb.	5	88	100
Cenchrus ciliaris L.	11	86	100
Paspalum dilatatum Poir.	23	90	100
Panicum coloratum L.	35	82	100
Panicum maximum Jacq.	23	89	100
Digitaria argyograpta (Nees) Stapf	32	86	100
Sorghum × almum Parodi.	26	77	100

Source: Kawanabe (1968).

Rate of plant development increases and time from seeding to flowering decreases with increasing temperatures (Fig. 5.5). In warm environments, vegetative plant development is accelerated: plants tend to be shorter and bloom earlier than in cool environments (Vough and Marten 1971). This response is part of the reason forage yields of cool-season species such as alfalfa and red clover decline during hot summer periods. Earlier onset of flowering means more frequent harvests may be needed in warm environments. Conversely, tropical species are best adapted and most productive under hot humid conditions (McWilliam 1978). When grown at high temperatures, most grasses and legumes store less nonstructural carbohydrate and produce herbage of lower digestibility but with higher protein and mineral percentages (Buxton and Casler 1993). Thus, forage digestibility decreases during spring growth due, in part, to increasing temperature as well as the decrease in the leaf:stem ratio. Conversely, forage quality of cool-season grasses decreases with age during fall at a much slower rate because temperatures are decreasing and nonstructural carbohydrates accumulate in leaves (Brown and Blaser 1965).

High-temperature Stress. Plants are subjected to continually changing temperature conditions, and stress occurs when temperatures are outside the optimal range. Severity of stress depends on stage of plant development and on its duration and intensity. High-temperature stress frequently occurs concurrent-

Fig. 5.5. Temperature affects morphological characteristics of alfalfa plants. *USDA photo.*

ly with moisture stress, making it difficult to separate the two effects (Turner and Kramer 1980). High temperatures lead to a number of metabolic disorders in plants, including enzyme inactivation, imbalance of reaction rates, membrane dysfunction, and reduced metabolic synthesis. Excessively high temperatures can induce flower sterility, especially pollen abortion, leading to poor seed production. High temperatures during late seed development can reduce subsequent germination and vigor of seedlings.

Chilling Temperature Stress. Low-temperature stress (temperatures slightly above freezing) can cause chilling injury of some plants. Translocation of photosynthate from the chloroplasts and mesophyll cells (sources) to meristems (sinks) requires metabolic energy for loading of the phloem, the major tissue involved in transport of sugars. In some tropical species, chilling temperatures may reduce respiration or alter membranes, thereby reducing translocation rates and ultimately reducing photosynthesis. For example, following night exposure to temperatures less than 15°C, Pangola digitgrass has reduced translocation of photosynthate from leaves at night (Hilliard and West 1970). Starch was retained in the chloroplasts and reduced rates of photosynthesis the following day even though day temperatures were near 30°C. Similar responses occur in other warm-season grasses and in some legumes. However, most cool-season legumes and grasses do not have this low-temperature sensitivity. Reasons for lack of sensitivity with cool-season legumes are unknown, but many cool-season grasses accumulate fructan in vacuoles in contrast with legumes and warm-season grasses, which store starch in the chloroplasts, especially at low temperatures (Chatterton et al. 1989). At low temperatures synthesis and breakdown occurs more readily for fructan than for starch (Pollock and Cairns 1991).

Freezing Temperature Stress. Winter injury can be caused by excessively cold temperatures, ice sheets, or frost heaving. Availability of new cultivars capable of withstanding adverse winters has extended forage plant adaptation to more severe winter climates (Klebesadel 1971). However, little progress has been made toward breeding cultivars to reduce heaving or ice sheet damage of forage species. Therefore, management practices are the major control mechanisms.

Frost heaving is a serious problem in regions where temperatures fluctuate around 0°C and forages are grown on imperfectly drained, fine-textured soils (Portz 1967). Especially during late winter, the alternate freezing and thawing of the surface soil causes, respectively, vertical expansion and relaxing. Plants are gripped by the freezing soil and lifted upward, but plants with taproots such as alfalfa and red clover may not settle back when the soil thaws. The next freezing event moves the plant up even more, and with repeated cycles, the taproot breaks. The plant dies because the meristems on crowns are exposed to freezing temperatures and desiccation when elevated above the protective soil.

Plants with branched root systems survive heaving conditions better than plants with a taproot system because they tend to move up and down with the soil. Prostrate plants tend to have tissue produced near soil level that acts like a mulch to reduce freezing and thawing. Similarly, a tall stubble retained over winter on upright plants helps shade the soil and trap snow. Both the stubble and snow serve as a mulch to minimize temperature changes and reduce the frequency and severity of freezing-thawing events.

Ice sheets can cause serious damage to overwintering forage plants. When the crown is completely encased in water or ice, gas exchange is inhibited, the plants accumulate respiratory by-products, and the plants die. Winter-hardy species have low rates of respiration during winter, allowing them to survive ice encasement longer (Smith 1964). Leaving a tall stubble over winter may reduce ice sheet damage since the protruding stubble provides a passage for gas diffusion from the encased plants (Freyman 1969).

Ability to survive freezing temperatures is a major factor in winter survival and plant distribution. Forage species differ widely in their ability to withstand cold. For example, freezing tests have shown biennial sweetclover (*Melilotus officinalis* [L.] Lam.) is more cold-hardy than 'Ranger' alfalfa, and Ranger is more cold-hardy than northern common red clover (Smith and Nelson 1985). Common white clover (*T. repens* L.) is more cold-hardy than ladino clover and survives longer than ladino clover when both are covered with ice (Smith 1964). Seedlings of crested wheatgrass (*Agropyron desertorum* [Fisch. ex Link] Schult) are more cold-hardy than those of western wheatgrass (*Pascopyrum smithii* [Rydb.] Löve), which are more cold-hardy than those of smooth bromegrass (Rogler 1943).

Development and Maintenance of Coldhardiness. Winter survival of a forage species depends upon the ability of the plant to make certain metabolic changes (Laude 1964; Smith 1968). Within their inherent genetic capacity, overwintering forage species develop cold resistance with the onset of the shorter days and colder temperatures of autumn. The necessity for plants to develop cold resistance to survive winter is not limited to northern areas. Forages of tropical origin are sometimes injured or killed in southern states by short periods of subfreezing temperatures (Adams and Twersky 1960).

Cold-hardening of alfalfa under artificial conditions is favored by short daylengths of around 7-8 h, generally lowering temperatures, an alternation of temperatures between warm (20°C) during the day and cold (0°-5°C) at night, and adequate radiation for good photosynthetic activity (Smith 1964). Alfalfa cold-hardens poorly in a warm temperature regime (24°/16°C) regardless of daylength (Shih et al. 1967). In a cold regime (7°/2°C), hardening is better with a short (8-h) than a long (16-h) daylength. A cold temperature is of primary importance, but metabolic processes associated with an increase in coldhardiness are also modified by photoperiod.

Development of cold resistance during autumn and its loss during spring roughly follows a cyclic pattern. In Wisconsin, perennial legumes begin to develop cold resistance in early to mid-September (Fig. 5.6) and continue until late November or early December when the soil freezes. Resistance is near maximum shortly after permanent freezing of the soil surface and after weekly air temperatures remain below freezing. A high level of cold resistance is generally maintained from early December to mid-February when snow cover helps provide protection from very low air temperatures. Cold resistance in some legume species begins to decrease slowly in mid-February with the onset of warmer temperatures and reduced snow cover and then decreases rapidly after the snow disappears and the soil surface thaws. Cold resistance is lost much faster than it is achieved.

Winter killing in the northern areas usually occurs during late winter and early spring when snow cover has disappeared and plants are exposed to severe temperature fluctuations above and below freezing. By this time plants have lost some cold resistance in response to warm temperatures, and when exposed to a sudden temperature drop, they may not have time to reharden.

Conditions resulting in active growth during hardening in autumn can hinder development of cold resistance. Cold resistance may be reduced or its development retarded by removal of top growth, especially with legumes, during the autumn hardening period (Sheaffer et al. 1988). Warm, wet weather during autumn reduces carbohydrate storage and cold resistance. Stimulating growth by fertilizing grasses with N during the hardening period reduces cold resistance (Fig. 5.7).

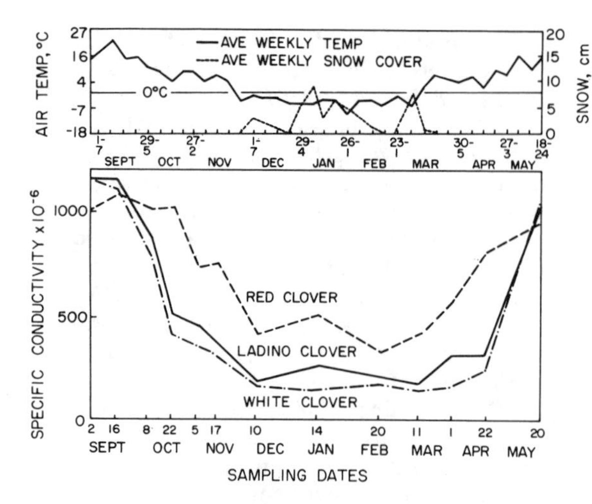

Fig. 5.6. Trend of cold resistance in red clover roots and crowns and ladino and common white clover stolons (Madison, Wis.). On each sampling date, tissue was placed in test tubes and frozen at –8°C for 14 h. Tissue then was immersed for 20 h in 50 mL water at 2°C. Electrolytes from damaged cells diffused into the water and increased electrical conductivity of the solution. Specific conductance readings decreased as plants developed cold resistance and increased as plants lost resistance. (After Ruelke and Smith 1956.)

Management to Improve Coldhardiness. Soil fertility is as important to overwintering of plants as it is to their growth during summer. Two nutrients of considerable importance are potassium (K) and N (Jung and Smith 1959; Adams and Twersky 1960). The exact role of K is not clear, but a high level of soil K is essential for development of maximum cold resistance. In contrast with K, high levels of soil N prevent maximum development of coldhardiness, particularly in grasses (Fig. 5.7) by stimulating plant growth. Cold resistance in grasses appears to be favored by a high K:N ratio (Adams and Twersky 1960).

The period prior to the first killing frost is usually a critical time in the management of forage species, particularly legumes. Plants need leaf area during autumn to synthesize carbohydrates through photosynthesis and to accumulate organic reserves before winter. Accumulated carbohydrate and N reserves are needed for developing cold resistance, maintaining dormant tissues during winter, and supporting new growth the following spring (see Chap. 1, Vol. 2). The colder and longer the winter, the more imperative it becomes that plants enter it with a high level of organic reserves. This can be ensured best by not cutting upright legumes during the cold-hardening period in autumn (Table 5.3). In northern areas it is generally recommended that overwintering legumes should not be cut or grazed during autumn, beginning 4 to 5 wk before the average date of the first killing frost (Sheaffer et al. 1988). Modern cultivars with improved winterhardiness and disease resistance are more tolerant of fall-cutting stress, especially when K is adequate (Sheaffer 1989). Cutting or grazing late in autumn, 2 to 3 wk after frost has killed the top growth, is less hazardous because organic reserves have accumulated and the plants are generally dormant and will not regrow. However, removal of herbage at this time decreases aboveground cover that helps reduce plant heaving. Cool-season grasses are less sensitive to fall grazing or cutting than are legumes.

Management of Weakened Stands. Stands weakened or injured by a severe winter usually need careful management to regain productivity (Fig. 5.8). Legumes may appear weakened and yellow, with only a few stems per plant, but can recover with proper weed control and altered cutting management. Control of weed growth in spring is important to reduce competition for light and nutrients, especially K. Delaying the first cutting to full bloom gives plants more time to heal injured tissue and develop a high level of organic re-

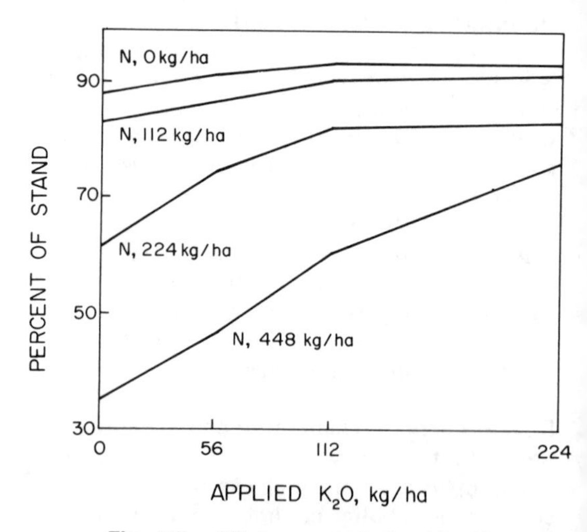

Fig. 5.7. Winter survival of 'Coastal' bermudagrass decreased with increasing levels of applied nitrogen at any given level of potassium, but at high levels of nitrogen, winter survival increased with increasing levels of applied potassium. (Adapted from Adams and Twersky 1960.)

TABLE 5.3. Plant population, height, and dry matter yield of alfalfa in year following application of postharvest autumn cuttings in Ontario, Canada

Cutting date[a]	Plants/0.09 m² at location[b]			Plant height in mid-May at location (cm)			Hay yield at location[c] (kg/ha)		
	1	2	3	1	2	3	1	2	3
Uncut	18	27	19	13	13	10	7329	4949	6814
Sept. 3	16	24	20	9	12	6	6919	4914	5764
10	13	19	17	9	11	5	6584	4317	5406
17	15	7	11	9	6	6	6974	2533	5323
24	16	12	14	10	7	7	7084	2860	5677
Oct. 1	20	15	19	10	8	8	6946	3523	6516
LSD 0.05	4	5	5	1	1	1	d	426	498

Source: Fulkerson (1970).
[a]Five and 10 d later at locations 2 and 3 respectively.
[b]Locations 1, 2, and 3 were in northern, central, and southern Ontario respectively.
[c]Total yield from two harvests during season after autumn cutting.
[d]Not significant.

serves in the storage organs. The herbage harvested may be light and weedy, but with delayed cutting the subsequent crop is often back to near-normal productivity. Cutting weakened and injured stands at immature stages of growth may kill the plants or keep them in a weakened condition. Live plants that are frost heaved less than 2.5 cm above the soil will often recover, but if they have heaved farther, the crowns will be damaged by mowing or grazing, and the weakened plants will likely die.

WATER RELATIONS

Seasonal distribution pattern, total quantity of precipitation, and evapotranspiration demands influence the adaptation of forage species (Turner and Begg 1978). Also, the quantity of water available to plants in the soil reservoir depends on the soil texture and the rooting depth of the forage plants. Water loss by plants is a physical evaporation process, which is driven largely by solar radiation. Thus, the amount of water lost from a field through soil evaporation and transpiration through plants depends largely on soil water availability and incident radiation. Air temperature, relative humidity, and wind speed have less effect than radiation on plant water use.

Drought Stress. Adequate soil moisture is essential for normal growth (Hsiao 1973; Brown 1977). The critical factor, however, is the water status within the plant tissues, which is commonly called the *water potential* and is measured in megapascals (MPa). Water po-

Fig. 5.8. Summer recovery of winter-injured alfalfa (Madison, Wis.). First crop was cut at an immature stage (*right*), allowing a heavy weed invasion, in marked contrast to the freedom from weeds (*left*) where first crop was not cut until past full bloom.

tential is influenced by the balance between availability of soil moisture and amount of transpiration from plant surfaces. Thus, plants in dry soil on cloudy days may be less stressed than plants in well-watered soil on high-transpiration days.

A decrease in water potential will affect certain plant processes more than others. Two of the first to be affected by only a moderate deficiency of water are cell division and cell enlargement, especially enlargement that may be retarded or stopped (Hsiao 1973). For example, leaf growth slows well before water stress becomes severe enough to cause stomatal closure and a decrease in photosynthesis (Fig. 5.9). Note that leaf growth, largely cell enlargement, recovers quickly when plants are rewatered after a short stress, while photosynthesis recovers slowly and not to the expected rate. In Missouri, leaves of tall fescue elongated at 15 mm/d in the field when irrigated but only 10 mm/d when not irrigated (Horst and Nelson 1979). During severe summer drought, leaf elongation of cool-season grasses nearly ceases.

A deficiency of internal water may retard growth more at one stage of development than at another. With tall-growing grasses, such as timothy or big bluestem (*Andropogon gerardii* Vitman), internal water deficiency during the period of rapid stem elongation can markedly reduce cell expansion, stem growth, and yield. In general, water deficiencies cause a reduction in vegetative growth and promote early flowering. Stress levels that limit growth differ among species and tissues within a plant because osmotically active compounds may accumulate, thereby increasing osmotic potential and enhancing water uptake, or root growth may be favored over shoot growth (Davies and Zhang 1991).

Even with adequate soil moisture, the atmospheric evaporative demand is often high enough during sunny days to cause a gradual decrease each day in water potentials of leaf and stem tissue to −1.2 MPa or lower, followed by gradual recovery at night. Consequently, when drought stressed, most cell expansion of leaves and stems occurs during the night when stomata are closed, transpiration rates are very low, and plants are rehydrating as they slowly absorb water from the soil.

If the stress has not been too severe or has not lasted an extended period, cessation of cell enlargement and similar plant responses are not permanent, and removal of the stress conditions by a reduced transpiration rate or by rewatering allows cell enlargement and related development processes to proceed (Fig. 5.9). As soil moisture levels decrease and daylight atmospheric demands impose greater water stress on plants, however, a longer period of recovery may be required before growth can resume.

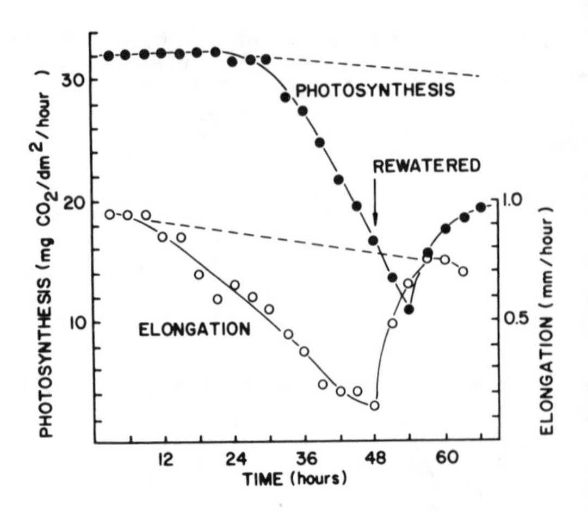

Fig. 5.9. Photosynthesis of last fully developed leaf and elongation rate of leaf emerging from whorl were measured simultaneously on *Lolium temulentum*. Experimental plants growing in pots had not received water for 2 d when data collection began at time 0. Dashed lines show responses of control plants that were not stressed. Note that leaf elongation slowed earlier than did photosynthesis. Elongation recovered quickly after rewatering at 48 h, but photosynthesis was delayed and did not fully recover. (Adapted from Wardlaw 1969.)

Under field conditions drought usually coincides with high temperatures; both climatic conditions depress plant growth. However, carbohydrate utilization for growth is generally reduced to a greater extent than is carbohydrate production via photosynthesis. Thus, drought conditions often result in increased concentrations of nonstructural carbohydrate in the plant tissue (Brown and Blaser 1970). These carbohydrates, especially hexoses, can contribute osmotically to sustain growth. Some cool-season grasses that experience drought stress during summer may have enhanced growth during fall and the following spring (Horst and Nelson 1979). Compensatory growth is due to faster rate of leaf elongation and higher weight per tiller compared with that of plants irrigated through the summer.

Excessive Moisture Conditions. Poorly drained soils in high-rainfall regions provide an unfavorable environment for growth of many forage species, especially legumes.

Some forages such as reed canarygrass (*Phalaris arundinacea* L.) and paragrass (*Brachiaria mutica* [Forsk.] Stapf) have considerable tolerance to waterlogged conditions and often are the only species that can be grown in such areas. Tolerance may be related to morphology of the root system and stem tissues or to an ability to withstand such adverse conditions as low oxygen (O), low pH, or toxic levels of aluminum (Al) or manganese (Mn) in the soil. On poorly drained soils, serious soil compaction and plant damage from hooves of livestock, especially after excessive rainfall, contribute further to poor plant growth and persistence.

Root and crown diseases are generally most severe in wet environments. Stand losses of alfalfa growing in poorly drained soils during periods of high precipitation or excessive irrigation often have been attributed to *Phytophthora* root rot (Leath et al. 1988). Damping-off of seedlings, caused by such soil fungi as *Phythium* spp., *Rhizoctonia* spp., or *Fusarium* spp., can be a serious factor in stand establishment, particularly if periods of high rainfall follow seeding. Invariably, fungal diseases are most serious during periods of excessive rainfall and high humidity.

Flooding also affects forage plant productivity and persistence, and the response depends on plant and environmental factors. Dormant plants are more resistant to being covered with water than are plants that are actively growing. A major factor is anoxia, or lack of oxygen, so plant survival is better if the water is cool to reduce respiration and flowing to aid oxygenation. Newly defoliated plants and those in early regrowth stages are more susceptible to flooding than are plants at mature stages. Siltation is also a factor as it smothers the plants. Even short periods of flooding reduce dinitrogen (N_2) fixation of legumes.

Forage species differ markedly in their ability to survive flooding. For example, in one study, most cool-season grasses survived while inundated with cold water for up to 35 d, and timothy and reed canarygrass survived up to 49 d (McKenzie 1951). Legumes survived only 14-21 d. In warm environments, hardinggrass (*Phalaris stenoptera* Hack.) and bermudagrass have very good flooding tolerance, but dallisgrass is very sensitive to inundation when it is actively growing (Colman and Wilson 1960).

QUESTIONS

1. What climatic factor determines whether the natural vegetation will be forest or grassland?
2. What region of the solar spectrum is the most important for plant growth? Why?
3. Why is plant growth maximized when about 95% of incoming solar energy is intercepted by plant leaves?
4 How would change in latitude influence maturity and potential seed production of a long-day grass species? Would responses be similar for a day-neutral species?
5. What are some physiological differences between temperate and tropical grasses? How do these grasses differ in seasons of productivity?
6. Differentiate between climatic factors involved in chilling injury and those involved in plant survival over winter.
7. Why is a knowledge of the periods of minimum and maximum organic reserves important to the management of perennial or biennial species?
8. What climatic and soil conditions favor the development of cold resistance? The loss of cold resistance?
9. How can climatic factors be used in predicting seasonal forage yields?
10. What is the diurnal pattern of growth of forage plants? How is it affected by drought stress?

REFERENCES

Adams, WE, and M Twersky. 1960. Effect of soil fertility on winter-killing of Coastal bermudagrass. Agron. J. 52:325-26.

Angus, JP, A Kornher, and BWR Torssell. 1983. A systems approach to Swedish ley production. In JA Smith and VW Hays (eds.), Proc. 14th Int. Grassl. Congr. Boulder, Colo.: Westview, 478-81.

Balasko, JA, and D Smith. 1971. Influence of temperature and nitrogen fertilization on the growth and composition of switchgrass (*Panicum virgatum* L.) and timothy (*Phleum pratense* L.) at anthesis. Agron. J. 63:853-57.

Beuselinck, PR, et al. 1994. Improving legume persistence in forage crop systems. J. Prod. Agric. 7:311-22.

Black, JN. 1957. The influence of varying light intensity on the growth of herbage plants. Herb. Abstr. 27:89-98.

Brown, RH, and RE Blaser. 1965. Relationship between reserve carbohydrate accumulation and growth rate in orchardgrass and tall fescue. Crop Sci. 5:577-82.

______. 1970. Soil moisture and temperature effects on growth and soluble carbohydrates of orchardgrass (*Dactylis glomerata*). Crop Sci. 10:213-16.

Brown, RW. 1977. Water relations in range plants. In RE Sosebee (ed.), Rangeland Plant Physiology, Range Sci. Ser. 4. Denver, Colo.: Society of Range Management, 97-140.

Buxton, DR, and MD Casler. 1993. Environmental and genetic effects on cell wall composition and degradability. In HG Jung, DR Buxton, RD Hatfield, and J Ralph (eds.), Forage Cell Wall Structure and Digestibility. Madison, Wis.: American Society of Agronomy, Crop Science Society of America, and Soil Science Society of America, 685-714.

Chatterton, NW, PA Harrison, JH Bennett, and KH Asay. 1989. Carbohydrate partitioning in 185 accessions of Gramineae grown under warm and cool temperatures. J. Plant Physiol. 134:169-79.
Child, DR, and EK Byington. 1981. Potential of the World's Forages for Ruminant Animal Production. 2d ed. Morrilton, Ark.: Winrock International Livestock-Research Training Center.
Colman, RL, and GPM Wilson. 1960. The effect of floods on pasture plants. Agric. Gaz., July, 337-47.
Cooper, JP, and NM Tainton. 1968. Light and temperature requirements for the growth of tropical and temperate grasses. Herb. Abstr. 38:167-76.
Davies, WJ, and J Zhang. 1991. Root signals and the regulation of growth and development of plants in drying soil. Annu. Rev. Plant Physiol. and Plant Mol. Biol. 42:55-76.
Freyman, S. 1969. Role of stubble in the survival of certain ice-covered forages. Agron. J. 61:105-7.
Fulkerson, RS. 1970. Location and fall harvest effects in Ontario on food reserve storage in alfalfa (*Medicago sativa* L.). In Proc. 11th Int. Grassl. Congr., Surfers Paradise, Queensland, Australia.
Gardner, FP, and WE Loomis. 1953. Floral induction and development in orchardgrass. Plant Physiol. 28:201-17.
Geiger, R. 1965. The Climate near the Ground. Rev. ed. Cambridge: Harvard Univ. Press.
Gist, GR, and GO Mott. 1958. Growth of alfalfa, red clover, and birdsfoot trefoil seedlings under various quantities of light. Agron. J. 50:583-86.
Hilliard, JH, and SH West. 1970. Starch accumulation associated with growth reduction at low temperatures in a tropical plant. Sci. 168:494-96.
Horst, GL, and CJ Nelson. 1979. Compensatory growth of tall fescue following drought. Agron. J. 71:559-63.
Horst, GL, CJ Nelson, and KH Asay. 1978. Relationship of leaf elongation to forage yield of tall fescue genotypes. Crop Sci. 18:715-19.
Hsiao, TC. 1973. Plant response to water stress. Annu. Rev. Plant Physiol. 24:519-70.
Jung, GA, and D Smith. 1959. Influence of soil potassium and phosphorus content on the cold resistance of alfalfa. Agron. J. 51:585-87.
Kawanabe, S. 1968. Temperature responses and systematics of the Gramineae. In Proc. Jap. Soc. Plant Taxon. 2:17-20.
Klebesadel, LJ. 1971. Selective modification of alfalfa towards acclimatization in a subarctic area of severe winter stress. Crop Sci. 11:609-14.
Klebesadel, LJ, and D Smith. 1959. Light and soil moisture beneath several companion crops as related to the establishment of alfalfa and red clover. Bot. Gaz. 121:39-46.
Laude, HM. 1964. Plant response to high temperatures. In Forage Plant Physiology and Soil-Range Relationships, Spec. Publ. 5. Madison, Wis.: American Society of Agronomy, 15-31.
Leath, KT, DC Erwin, and GD Griffen. 1988. Diseases and nematodes. In AA Hanson, DK Barnes, and RR Hill, Jr. (eds.), Alfalfa and Alfalfa Improvement, Am. Soc. Agron. Monogr. 29. Madison, Wis., 621-7.
Ludlow, MM. 1978. Light relations of pasture plants. In JR Wilson (ed.), Plant Relations in Pastures. Australia: CSIRO, 35-49.
McKenzie, RE. 1951. The ability of forage plants to survive early spring flooding. Sci. Agric. 31:358-67.
McWilliam, JR. 1978. Response of pasture plants to temperature. In JR Wilson (ed.), Plant Relations in Pastures. Australia: CSIRO, 17-34.
Murata, Y, and J Iyama. 1963. Studies on the photosynthesis of forage crops. II. Influence of air-temperature upon the photosynthesis of some forage and grain crops. In Proc. Crop Sci. Soc. Jap. 31:315-22.
Nelson, CJ. 1988. Genetic associations between photosynthetic characteristics and yield: Review of the evidence. Plant Physiol. Biochem. 26:543-54.
Nelson, CJ, and D Smith. 1969. Growth of birdsfoot trefoil and alfalfa. IV. Carbohydrate reserve levels and growth analysis under two temperature regimes. Crop Sci. 9:589-91.
Pollock, CJ, and AJ Cairns. 1991. Fructan metabolism in grasses and cereals. Annu. Rev. Plant Physiol. and Plant Mol. Biol. 42:77-101.
Portz, HL. 1967. Frost heaving of soil and plants. I. Influence of frost heaving of forage plants and meteorological relationships. Agron. J. 59:341-44.
Rogler, GA. 1943. Response of geographical strains of grasses to low temperatures. J. Am. Soc. Agron. 35:547-59.
Ruelke, OC, and D Smith. 1956. Overwintering trends of cold resistance and carbohydrates in medium, ladino, and common white clover. Plant Physiol. 31: 364-68.
Sheaffer, CC. 1989. Legume establishment and harvest management in the USA. In GC Marten et al. (eds.), Persistence of Forage Legumes, Proc. Trilateral Workshop, Madison, Wis.: American Society of Agronomy, 277-91.
Sheaffer, CC, GD Lacefield, and VL Marble. 1988. Cutting schedules and stands. In AA Hanson, DK Barnes, and RR Hill, Jr. (eds.), Alfalfa and Alfalfa Improvement, Am. Soc. Agron. Monogr. 29. Madison, Wis., 411-37.
Shih, SC, GA Jung, and DC Shelton. 1967. Effects of temperature and photoperiod on metabolic changes in alfalfa in relation to cold hardiness. Crop Sci. 7:385-89.
Smith, D. 1964. Winter injury and the survival of forage plants. Herb. Abstr. 34:203-9.
______. 1968. Varietal chemical differences associated with freezing resistance in forage plants. Cryobiol. 5:148-59.
Smith, D, and CJ Nelson. 1985. Physiological considerations in forage management. In ME Heath, RF Barnes, and DS Metcalfe (eds.), Forages: The Science of Grassland Agriculture, 4th ed. Ames: Iowa State Univ. Press, 326-37.
Snaydon, RW. 1991. The productivity of C_3 and C_4 plants: A reassessment. Funct. Ecol. 5:321-30.
Teeri, JA, and LG Stowe. 1976. Climatic patterns and the distribution of C_4 grasses in North America. Oecol. 23:1-12.
Trewartha, GT, AH Robinson, and GH Hammond. 1967. Elements of Geography. 5th ed. New York: McGraw-Hill.
Thornthwaite, CW. The Climates of the earth. 1933. Geogr. Rev. 23:433-40.
Turner, NC, and JE Begg. 1978. Response of pasture plants to water deficits. In JR Wilson (ed.),

Plant Relations in Pastures. Australia: CSIRO, 50-66.

Turner, NC, and PJ Kramer. 1980. Adaptation of Plants to Water and High Temperature Stress. New York: Wiley.

Van der Veen, R, and G Meijer. 1959. Light and Plant Growth. New York: MacMillan.

Volenec, JJ, CJ Nelson, and DA Sleper. 1984. Influence of temperature on leaf dark respiration of diverse tall fescue genotypes. Crop Sci. 24:907-12.

Vough, LR, and GC Marten. 1971. Influence of soil moisture and ambient temperature on yield and quality of alfalfa forage. Agron. J. 63:40-43.

Waller, SS, and JK Lewis. 1979. Occurrence of C_3 and C_4 photosynthetic pathways in North American grasses. J. Range Manage. 32:12-28.

Wardlaw, IF. 1969. The effect of water stress on translocation in relation to photosynthesis and growth. II. Effect during leaf development in *Lolium temulentum* L. Aust. J. Biol. Sci. 22:1-16.

6 Forage Fertilization

DARRELL A. MILLER
University of Illinois

HAROLD F. REETZ, JR.
Potash and Phosphate Institute

ONE of the most important components of forage production is proper soil fertility. Successful forage producers understand the importance of the basic properties and functions of soil in a forage program. Often the missing link in forage production is proper soil management and soil fertility.

Soil provides mechanical support and most of the nutrients that are needed for plant growth. Twenty essential elements are known to be necessary for plant growth. Carbon (C), hydrogen (H), and oxygen (O) come from the air. Nitrogen (N) comes from both air and soil. But the majority of these elements—phosphorus (P), potassium (K), calcium (Ca), sulfur (S), iron (Fe), magnesium (Mg), manganese (Mn), zinc (Zn), copper (Cu), boron (B), molybdenum (Mo), chlorine (Cl), sodium (Na), cobalt (Co), nickel (Ni), and silicon (Si)—come from the soil. The nutrients taken up from the soil by plants become the basis of meeting the nutritional needs for animals and humans. (See Chap. 4; see also Chap. 5, Vol. 2.)

Soil is the residue of weathering parent material and the decomposing plant and animal materials associated with it. As the mineral and biological components of the soil are broken down, nutrient elements are released and become gradually available to plants growing in the soil. This is a very slow process and generally is not sufficient to meet the needs of a growing forage crop, so supplemental fertilizers are usually required for optimum production and quality. The characteristics of a particular soil at a given site are the result of parent material, climate, living organisms, topography, and time. Soil formation is a complex mechanical, physical, chemical, and biological process.

One may describe many physical properties of the soil. *Soil texture* refers to the size of the soil particles. *Soil structure* refers to the arrangement of these particles. The main soil separates are very coarse sand, coarse sand, medium sand, fine sand, very fine sand, silt, and clay. The texture of the soil is determined predominantly by the mixture of sand, silt, and clay. In addition to these components, organic matter may or may not be present in the soil. Organic substances such as animal manure, crop residue, and leaf mulches greatly benefit the soil structure. Many producers will grow green manure crops purely for the purpose of adding organic matter to improve the soil structure.

In addition to proper soil structure and texture, soil temperature influences plant growth and soil characteristics. Proper soil temperature must be present before germination can occur, and this temperature varies with different species. When soil temperatures are below 10°C (50°F), nitrification, or the breakdown of ammonia to nitrate, is reduced (Allen et al. 1978; Christians et al. 1979). Alternating freezing and thawing im-

DARRELL A. MILLER. *See Chapter 4.*

HAROLD F. REETZ, JR., is Midwest Director of the Potash and Phosphate Institute (PPI), Monticello, Illinois. He received his MS and PhD degrees from Purdue University. He is responsible for PPI research and education programs in several Midwest states.

proves the structure of cloddy soils. Cold temperatures also reduce plants' absorption of some elements, particularly phosphorus.

SOIL TESTING

Soil testing is the most important single step in determining the rate of application of fertilizer and lime. One must know the level of fertility and especially the soil pH before one can profitably apply fertilizer. Soil test results, combined with the knowledge of what each nutrient does for plant growth, comprise the basis for planning a producer's soil fertility program for each field tested.

It is very easy to collect a soil sample for every 1 to 2 ha (2.5 to 5 A) in areas having similar cropping history, topography, and soil type. One should develop a regular pattern in a field and make a very accurate map so that the results can be interpreted correctly and the proper amount of fertilizer can be applied to each area. The most common mistake in this process is taking too few samples. Collecting samples on a specific grid pattern is recommended so that fertilizer applications will meet specific nutrient needs of different parts of the field. Modern technology uses signals from satellites to specifically identify the exact location of soil samples. Soil fertility maps generated from these samples are entered into a computer on the fertilizer applicator. The computer is also equipped to receive satellite navigation signals, which are used to guide variable-rate application of fertilizer matched to the variability in soil test levels. This leads to fertilizer rates that are agronomically sound, economically efficient, and environmentally responsible.

One should soil test every 3 to 4 yr. Late summer and fall are the best seasons for collecting soil samples. Potassium results are most reliable at these times. Since this nutrient is rather water soluble and readily absorbed by growing plants, it will be at the lower level of its availability in the late summer or early fall. One of the important goals in fertility management is to keep this minimum soil test K level above the minimum required for optimum plant growth. When the plants die, the K in the crop residue is released back to the soil by fall and winter rainfall. In the spring—following freezing and thawing (if there is sufficient soil moisture)—a higher test value will be obtained, which will be false or inflated.

Most soil tests report soil acidity, the Bray P_1 (plant-available phosphorus) level, and K (the plant-available potassium level). Some laboratories report Mehlich-3 phosphorus, a soil test that agrees well with the Bray P_1 soil test. On calcareous or high-pH soils the Olsen phosphorus soil test is commonly used. Soil tests for N are not as reliable but may be used as a general guide for N fertilization. Some soil tests may provide the availability of secondary and micronutrients, but their interpretation is less reliable than that for the test of pH, P, and K.

Organic matter is sometimes reported in soil tests. Color is related to the organic matter content of the soil. Light-colored soils usually have less than 2.5% organic matter, whereas medium-colored soils have from 2.5% to 4.5% organic matter. Dark-colored soils may have more than 4.5% organic matter. Sands are usually excluded from the determination of organic matter content.

LIME

One of the most serious limitations to soil productivity is soil acidity. On most cropland in the US, additional N has been used without a corresponding addition of limestone, and since N fertilizers are acid-forming, most US soils are consequently increasing in soil acidity. A soil test every 4 yr is the best way to keep a check on soil acidity levels. It requires about 4 units of lime to reduce the acidity resulting from 1 unit of N applied as ammonia or urea, and as much as 9 units of lime to neutralize the acidity resulting from 1 unit of N applied as ammonium sulfate.

Soil acidity affects plant growth in several ways. As soil pH becomes lower and acidity increases, several situations may exist: (1) The concentration of soluble metals may become toxic. An excess of solubility of aluminum (Al) and Mn has been established experimentally. (2) Populations and activity of organisms responding to transformations involving N, S, and P may be altered. (3) Calcium may become deficient as the pH decreases. This usually occurs when the soil's cation exchange capacity is extremely low. (4) Symbiotic N fixation is reduced greatly on acid soils. The symbiotic relationship requires a very narrow range of soil acidity compared to that necessary for the growth of plants that do not rely on N fixation. (5) Many acid soils are poorly aggregated and have poor tilth. This is particularly true for soils low in organic matter. (6) Other minerals' availability to plants may be greatly reduced, as shown in Figure 6.1, which indicates the relationship between the soil pH and nutrient availability. For example, P availability is greatest when the pH is between 6.5 and 7.5. Potassium availability drops off very rapidly below pH 6.0. Molybdenum availability increases greatly as soil acidity decreases, so Mo deficiencies are usually corrected by proper liming. Molybdenum is important in the formation and function of N-fixing nodules on legume roots. When soil pH exceeds 7.0, Fe, Mn, B, Cu, and Zn decrease in availability. When pH exceeds 8.5, leaching treatments of gypsum or S may help to lower the pH.

Various forage crops may be classified according to their sensitivity to pH, as shown in Table 6.1. For example, alfalfa (*Medicago sativa* L.) and smooth bromegrass (*Bromus inermis* Leyss.) are very sensitive to pH levels, whereas other plants, including alsike clover (*Trifolium hybridum* L.), annual lespedezas (*Kummerowia* spp.), tall fescue (*Festuca arundinacea* Schreb), and reed canarygrass (*Phalaris arundinacea* L.), may tolerate moderate acidity. It is highly recommended that when alfalfa and clover, or forage crops in general, are grown, the pH be at least 6.5 to 7.0. When a monoculture of alfalfa is grown, the pH should be 6.9 to 7.0.

One should purchase limestone on the basis of quality; premium limestone will sell for a premium price. Quality is measured by the fineness and the neutralizing power of the limestone. The value of lime in correcting soil acidity problems can be calculated using the various efficiency factors given in Table 6.2.

Depending upon the initial cost, the entire amount of limestone may be applied at one time. If cost is a factor, as when 6 mt/ha are needed, one should add the limestone in split applications—about two-thirds in the first application and the remaining one-third 3 to 4 yr later. But if alfalfa, other legumes, or especially smooth bromegrass are to be grown, one should apply all of the lime at once.

Limestone does not react with the acid in soil very far from the lime particle. When the limestone is applied to the soil surface, the limestone should be disked in, the material plowed down, and the soil worked. This distributes the limestone throughout the plow depth much more evenly than if it is just plowed directly under. If the lime is plowed down without disking it in first, the resulting plant growth may appear as light and dark green stripes corresponding to the lime distribution. Conservation requirements may re-

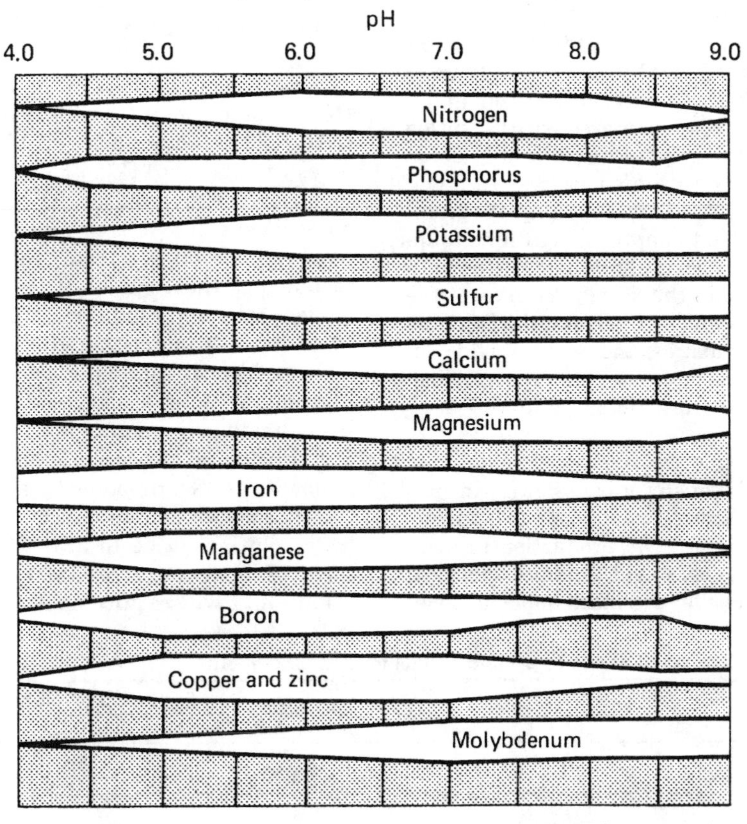

Fig. 6.1. Available nutrients in relation to soil pH. (Miller 1984, 124)

TABLE 6.1. Forage crops classified according to sensitivity to soil pH

Very sensitive to acidity (optimum pH 6.5-7.0)	Slight tolerance to acidity (optimum pH 6.0-6.5)	Moderate tolerance to acidity (optimum pH 5.5-6.0)
Alfalfa	Corn	Alsike clover
Barley	Crimson clover	Bahiagrass
Smooth bromegrass	Dallisgrass	Bermudagrass
Sweet clover	Kentucky bluegrass	Birdsfoot trefoil
	Ladino clover	Lespedeza
	Orchardgrass	Meadow fescue
	Red clover	Oats
	Ryegrass	Pearlmillet
	Timothy	Redtop
	Wheat	Reed canarygrass
		Rye
		Soybeans
		Sudangrass
		Tall fescue
		Vetch

Source: Miller (1984).

TABLE 6.2. Comparative value of limestone particle sizes

Size fraction	Efficiency factor
Through 60 mesh	100
30 to 60 mesh	50
8 to 30 mesh	20
Under 8 mesh	5

Source: Miller (1984).

strict tillage on some fields. Where necessary, lime may be left on the soil surface, but it will probably be less effective in correcting pH in the first cropping season.

Alfalfa responds to the proper pH (Table 6.3). Comparing 5-yr averages, more than 11 mt/ha was obtained with no fertilizer and a pH of 6.5 or more, as compared with 7.2 mt/ha at a pH of 5.0 to 5.5. Adding P and K boron fertilizer along with liming yielded 16 mt. If legumes are grown in combination with grasses, it is advisable to apply enough lime to increase the pH for several reasons: (1) to maintain the legume in the stand; (2) to meet the bacteria's needs and so maintain the legume's proper N-fixing ability; and (3) to provide optimum availability of other nutrients, which occurs when the pH is increased up to 7.0.

NITROGEN

The use of N fertilizer is essential in present-day agriculture if people are to continue to provide adequate crop production to meet the ever-increasing world demand for food. Nitrogen is probably the most important element needed for forage grasses. With the world's shortage of N fertilizer and energy, N fertilizer should be used in the most efficient manner possible. Nitrogen is essential for the formation of protein and must be obtained from either commercial fertilizer, N-fixing bacteria, organic matter, or rainfall. Organic matter is provided primarily by plant residues and animal manure. The organic matter not only contributes N but also provides for the improvement of soil structure, water-holding capacity, and nutrient-exchange capacity.

Most plants absorb N in the form of nitrate (NO_3). Some plants will use ammonium (NH_4^+) if NO_3^- is not present. When temperatures are above 10°C (50°F), the ammonia in fertilizer is converted to the nitrate form by bacteria before it becomes available for plants. Ammonium nitrate contains both ammonia and the readily available nitrate N. Under cooler soil conditions, this form of N is much more available than the ammonia form, which makes it especially good for early application to cool-season grasses.

The main sources of N fertilizers and their percentages of N are as follows: anhydrous ammonia, 82%; urea, 46%; ammonium nitrate, 33.5%; ammonium sulfate, 21%; and nitrate solutions, 28% to 32%. Heavy applications of N fertilizer will increase the soil acidity and will require an increase in lime-

TABLE 6.3. Response of alfalfa to pH and annual fertilizer treatment

	5-yr average, mt/ha	
Annual fertilizer treatment	pH 5.0-5.5	pH ≥ 6.5
0-0-0	7.21	11.22
0-150-300 (3)[a]	14.47	16.04

Source: Adapted from Smith (1975)
[a](3) plus boron treatment

stone applications, provided the pH is below 7.0.

Another source of N is plowing down a stand of alfalfa. (See Chapter 4 for the amount of N that alfalfa would contribute to a following crop.) From a complete stand of alfalfa, there is some carryover of N into the second year. The amount of N removed is also related to the amount of residue plowed under. Most of the N is removed with forage removal. In addition to releasing N, alfalfa also improves the tilth of the soil and adds organic matter.

A general rule is that in temperate regions forage grass yields will increase proportionally to the amount of N applied, up to the rates of 336 to 404 kg/ha annually. It has been shown that when an application exceeds 500 kg/ha, yields may decrease because of the thinning of the stand and increased competition between plants (Wedin 1974).

Most N fertilizers are top-dressed in split applications during the growing season. Grasses will consume N luxuriously if it is applied in excess, resulting in the presence of nonprotein N in the plant. As the amount of N given to forage grasses is increased, this will also increase the amount of other elements absorbed from the soil, especially K (Smith 1972). Therefore, when one is fertilizing grasses with higher rates of N, P and K should be applied properly or they will become limiting elements.

The forage grass species to be grown, the period of use, and the yield goal determine the optimum level of N fertilization (Table 6.4). A lower rate of N is recommended for fields where there is an inadequate stand, or where moisture may limit production. Kentucky bluegrass (*Poa pratensis* L.), for example, is a shallow-rooted species that is also susceptible to drought. Consequently, the most efficient use of N by Kentucky bluegrass is from an early spring application. September applications may be a second choice if there is sufficient moisture available and fall pasture is needed. Timothy (*Phleum pratense* L.) should receive N applications at approximately the same rate and time as Kentucky bluegrass.

Orchardgrass (*Dactylis glomerata* L.), smooth bromegrass, tall fescue, and reed canarygrass are more drought tolerant than are bluegrass and timothy and can use higher rates of N more efficiently than the shallow-rooted species. In these species, more uniform pasture production is obtained by splitting high rates of N into two or more applications.

The early spring application of N depends on the geographic location where the forage species is grown. The farther south in the US, N application should be earlier in the spring. Depending on organic matter and other fertility levels, spring growth may be adequate without extra N. Therefore, the first application may be delayed until after the first harvest or grazing period, so one could distribute production more uniformly throughout the summer. Total production would likely be reduced if N is applied after the first harvest rather than in the early spring. The second application is generally made immediately after the first harvest or the first grazing period. When deferring summer grazing of tall fescue, one should make a late summer application or apply additional N after the second harvest.

Legume-grass mixtures should not receive N if the legume content makes up 30% or more of the mixture. The main objective is to maintain the legume, so the emphasis should be on proper P and K fertilization rather than on N. If the legume is at least 30% of the sward, sufficient N may be fixed by the legume for the growth of the grass component of the mixture (Hojjati et al. 1978; Nuttall et al. 1980; Wagner 1954; Wedin 1974). After the legume has declined to less than 30% of the mixture, the objective of the fertility program is to increase the yield of grass. The suggested rate of N application is about 56 kg/ha when the legume makes up 20% to 30% of the

TABLE 6.4. Nitrogen application rates and time of the year for forage grasses (kg/ha; lb/A in parentheses)

Species	Time of application			
	Early spring	After 1st harvest	After 2d harvest	Early September
Kentucky bluegrass	67-90 (60-90)	. . .	. . .	
Orchardgrass	84-140 (75-125)	84-140	. . .	56 (50)[a]
Smooth bromegrass	84-140 (75-125)	84-140	. . .	56 (50)[a]
Reed canarygrass	84-140 (75-125)	84-140	. . .	56 (50)[a]
Tall fescue—deferred use	. . .	112-140 (100-125)	112-140	

Source: Miller (1984).
[a]Optional if extra fall growth is needed.

mixture, and 112 kg/ha or more when the legume is less than 20% of the mixture (Wedin 1974).

Manures generally are considered to contain 4 kg N, 2 kg P_2O_5, and 4 kg K_2O per metric ton. There is some variation in the content, depending upon the source and method of handling (Table 6.5). Regardless of the source, only 50% of the total N will be available to the crop during the first year after application because of organic N and volatilization.

Research has indicated that when good stands of alfalfa or clovers are grown with grasses, they will fix approximately 90 to 112 kg/ha of N (Wagner 1954). It was found in Maryland that on pure stands of orchardgrass and tall fescue, an excess of 180 kg/ha of N must be applied to equal the contribution of ladino clover (*T. repens* L.) grown with either of the grasses (Wagner 1954). In Wisconsin it was shown that it takes more than 270 kg/ha of N applied to grass stands to equal the contribution of alfalfa to the mixture (Smith 1972).

Effects of Nitrogen on Forage. Nitrogen fertilization will increase the concentration of P and K in the forage, provided their availability is sufficiently high in the soil. NH_4^+ will reduce the concentration of Ca, K, and Mg in the forage. Therefore, if NH_4^+ is used on grasses, grass tetany may result if the level of Mg is already on the low side (George and Thill 1979; Dougherty and Rhykerd 1985).

When the amount of N applied to forage grasses is increased, the nitrate N level will correspondingly increase. Therefore, one should be alert for the possibility of nitrate toxicity if excessive amounts of N are applied (Dougherty and Rhykerd 1985; Stritzke and McMurphy 1982; Wedin 1974). If harvest is delayed or the grasses are allowed to become more mature, the nitrate N level will decrease in the plant tissue. A further concern is that N application to the legume stand will reduce the rhizobia activity and N fixation within the nodules.

Nitrogen fertilization will decrease the plants' level of water soluble carbohydrates, primarily polysaccharides. Consequently, this may have a slight effect upon silage fermentation, since carbohydrates are required to make good-quality silage. Nitrogen application will also increase the level of total organic acids and alkaloids in grasses. Carotene levels are also increased as N is applied (Dougherty and Rhykerd 1985).

Since N fertilization tends to decrease the available carbohydrates in plants, it is more difficult to make high-quality silage from heavily N-fertilized grasses. This is especially true for cool-season grasses, as they already tend to be very low in available carbohydrates. But N application to corn (*Zea mays* L.) does not appear to have any great effect on silage quality since there is generally a high level of carbohydrates readily available (Dougherty and Rhykerd 1985). In the western corn belt and areas where drought damaged growth, high levels of NO_3 may be found in the stalk, so 30 to 45 cm stubble heights are recommended.

There are conflicting reports concerning the effect of N fertilization on the palatability and intake of forages. There are probably equal positive and negative results (Dougherty and Rhykerd 1985). Generally, comparing no added N and moderate levels of N applied to forage grasses, the added N will consistently increase animal preference. However, there is no such difference between moderate N and high levels of N. One must also keep in mind that as the N level is increased there is also a greater demand for water uptake, so moisture may become deficient with higher N fertilization. Some researchers have suggested that

TABLE 6.5. Average composition of various manures in dry and liquid form

	Dry form					
	Nitrogen (N)		Phosphorus (P_2O_5)		Potassium (K_2O)	
	kg/mt	lb/t	kg/mt	lb/t	kg/mt	lb/t
Dairy	5	11	2	5	5	11
Beef	6	14	4	9	5	11
Hogs	4	10	3	7	3	8
Chickens	8	20	7	16	3	8
	Liquid form (per 1000 gal)					
	kg	lb	kg	lb	kg	lb
Dairy	12	26	5	11	10	23
Beef	10	21	3	7	8	18
Hogs	25	56	14	30	10	22
Chickens	34	74	31	68	12	27

Source: Miller (1984)

when reed canarygrass and tall fescue are fertilized with N, the alkaloid concentration will increase, thus reducing the forage's intake and palatability (Odom et al. 1980; Wedin 1974). There appears to be no consistent effect of N fertilization on the digestibility of grasses (Dougherty and Rhykerd 1985).

Many studies have indicated that N fertilization of forages grasses will increase the beef gains. Much of this work has been done with warm-season grasses, such as 'Coastal' bermudagrass (*Cynodon dactylon* [L.] Pers.), dallisgrass (*Paspalum dilatatum* Poir.), and bahiagrass (*P. notatum* Flugge). Most warm-season grasses are grown where sufficient amounts of moisture are present. Thus, a direct correlation has been found between increased rates of N and beef gains in kg/ha (Fribourg et al. 1979; Perry and Baltensperger 1979; Rehm et al. 1977; Taliaferro et al. 1975). Increasing N fertilization on most grasses will increase the carrying capacity, thereby yielding a greater production of red meat per hectare than from unfertilized plots (Wedin 1974). Cool-season grasses also respond to N fertilization but probably not to the same extent as warm-season grasses do.

PHOSPHORUS

Phosphorus plays a very important role in the growth of plants and animals. Many of the soils of the southeast are deficient in P, whereas those in the Midwest and far West vary in their levels of P (Tisdale et al. 1985). Phosphorus is especially needed for good grass and legume growth and for N fixation by bacteria. It is crucial in the establishment of legumes and grass seedlings because it is utilized very early in the plant's life cycle and because it is very important for good root growth. Without an adequate supply of P, the plant will not reach its maximum growth or yield potential, nor will it reproduce normally. The functions of P range from the primary mechanism of energy transfer to the coding of genes, which cannot be performed by any other nutrient (Tisdale et al. 1985). In the photosynthesis process, P is essential for the transfer of carbohydrates. Phosphorus is considered the essential nutrient in the plant's electron-transport system. It is probably the main energy source within the plant because it is tied in with the phosphorylation that is required in many reactions. It provides the energy used in the synthesis of sucrose, starch, and proteins.

Adequate P fertilization will enhance root development. When root development is increased, added P decreases the amount of water that a plant needs (Duell 1974). When adequate P is applied, foliage increases, thereby shading the soil and reducing water evaporation.

Phosphorus does not readily leach out of the soil and, with continued annual applications, will gradually accumulate. Many soils fix or hold this element. When P is fixed or tightly held in the soil, the amount of leaching and availability to plants is minimized. This varies with soil type. If the soil pH exceeds 7.5, P becomes less available. Usually no more than 10% to 20% of the P that is applied is recovered in the first year of growth. Therefore, it is very important to retest the soil every 3 to 4 yr. Plant growth is highly dependent upon the amount of P that has been applied or built up over the years from previous applications.

Soil tests will generally report the available P or the total P. High-P-supplying soils are generally very favorable for good root penetration and branching throughout the soil profile. Some soils supply low amounts of P for several reasons: there may be (1) a low supply of available P in the soil profile because the parent material was low in P, or P was lost in the soil-forming process, or P is made unavailable due to high pH or calcareous material; (2) poor internal drainage, which restricts plants growth and so creates a low-P-supplying soil; (3) a dense, compact layer that inhibits root penetration or branching; (4) only a shallow layer above the bedrock, sand, or gravel; and (5) droughtiness, strong acidity, or other conditions that may restrict crop growth or reduce root depth.

Since P will be lost from the soil system through crop removal or soil erosion, and since there are minimum values required for optimum forage yields, it is recommended that soil test levels be built up to 35, 40, and 45 kg/ha in high-, medium-, and low-P-supplying regions, respectively. These values will ensure that the soil's P availability will not limit crop yield.

Research has shown that on average, it takes about 9 kg/ha of P_2O_5 to increase the P_1 soil test by 1 kg/ha. Therefore, the recommended rate of buildup of P is equal to nine times the difference between the soil test goal and the actual soil test value. The amount of P_2O_5 recommended over a 4-yr period for various soil test levels is shown in Table 6.6. Some soils will fail to reach the desired goal in 4 yr with P_2O_5 applied at the suggested rate, whereas others may exceed this goal. Therefore, it is recommended that every field be retested every 4 yr.

Most plants use phosphorus in the H_2PO_4 form, but the P content of commercial fertilizer is usually measured in the P_2O_5 form. Several sources of P in commercial fertilizers and their normal percentages of P_2O_5 are superphosphate, 20%; triple superphosphate, 45%; and rock phosphate, 41%. There are some additional sources of P that may also contain N, such as diammonium phosphate, 53% P_2O_5 and 21% N; monoammonium phosphate, 48% P_2O_5 and 11% N; and ammonium phosphate sulfate, 20% P_2O_5 and 16% N.

Prior to seeding perennial forages crops, one should broadcast and incorporate all of the buildup P needed, plus all of the maintenance P that is economically feasible. On low-fertility soils, one should apply approximately 34 kg/ha of P_2O_5, using a band seeder. If a band seeder is used, one may safely apply a maximum of 34 to 45 kg of potash (K_2O) in the band, along with the P. Usually, up to 336 kg/ha or K_2O can be safely broadcast in the seedbed without damaging the seedlings (Duell 1974).

Top-dress applications of P_2O_5 and K_2O on perennial forage crops can be applied at any convenient time, usually after the last harvest or in early fall. It is preferred that all of the P be applied after the fall harvest so that the plant has sufficient P going into the winter to maintain good root structure and early spring vigor. Potash is usually applied in split applications, half of the required amount of maintenance fertilizer after the last harvest in the fall, and the remaining half after the first harvest the following spring (Brown 1957).

Most of the P in commercial fertilizers is highly water soluble. Fertilizers with a P_2O_5 water solubility greater than 75% to 80% will not increase dry matter yields more than will those with water solubility levels of 50% to 80%. As long as the recommended rates of application and broadcast placements are used, the water solubility of P_2O_5 is of little importance under most field conditions on soils that have medium to high levels of available P. There are two exceptions, however, when water solubility is important: (1) When band placing small amounts of fertilizer to stimulate early growth, at least 40% of the applied P_2O_5 should be water soluble for acid soils, and preferably 80% for calcareous soils. (2) For calcareous soils, a high degree of water solubility is desirable, especially for soils that are low in available P.

Numerous researchers have reported the effect of P on various forage crops (Brown and Graham 1978; Christians et al. 1979; Duell 1974; Jackobs et al. 1970; Ludwick and Rumberg 1976; Nuttall 1980; Nuttall et al. 1980; Sheard 1980; Vickers and Zak 1978;

TABLE 6.6. Amount of phosphorus (P_2O_5) required annually to build up the soil (based on buildup occurring over a 4-yr period; 9 kg/ha of P_2O_5 required to change P_1 test 1 (kg/ha)

P_1 test	Soil supplying power		
	Low	Medium	High
		kg/ha	
4	103	92	81
6	99	88	76
8	94	83	72
10	90	79	68
12	86	74	63
14	81	70	58
16	76	65	54
18	72	61	50
20	68	56	45
22	63	52	40
24	58	47	36
26	54	43	32
28	50	38	27
30	45	34	22
32	40	29	18
34	36	25	14
36	32	20	9
38	27	16	4
40	22	11	0
42	18	7	0
44	14	2	0
45	11	0	0
46	9	0	0
48	4	0	0
50	0	0	0

Source: Miller (1984)

Watschke et al. 1977). In Virginia, based on a 3-yr average, alfalfa yields were found to increase from 7.2 to 11.7 mt/ha when 100 kg of P were applied per hectare. At the same location, results indicated a significant increase in orchardgrass yields with P applied at rates of 25 and 100 kg/ha (Lutz 1973). In another trial, using a 4-yr average, yields increased significantly when P was applied to an irrigated mixture of smooth bromegrass, timothy, orchardgrass, and bluegrass, with occasional red clover (*T. pratense* L.) plants (Table 6.7) (Rehm et al. 1975). Native warm-season grasses and various other grasses also respond to P application (Table 6.8) (Taliaferro et al. 1975).

Research has shown that when smooth bromegrass and alfalfa were grown together and the yields were expressed as a percentage of the control and were related to the application of N and P fertilizers and the available soil P, yields were increased by 251% over the controls. It was also noted that the proportion of alfalfa in the sward was significantly reduced by 29% when N and P fertilizer was supplied. When N was applied at 90 kg/ha and P at 20 kg/ha, the average increase in forage yield was 74%. These rates gave the most economical return for the fertilizer invested and did not significantly increase the nitrate N in the soil profile (Nutall 1980; Nutall et al. 1980).

Phosphorus increases the nodulation on legumes primarily because of increased root and nodule mass. Since a larger root system is developed, the potential for nodulation is further increased. The root mass develops earlier in the life cycle of plants that are fertilized with P, so nodules develop earlier, resulting in earlier N fixation and faster growth (Tisdale et al. 1985).

Effect of Phosphorus on Animal Health. Phosphorus is present in every living cell, in the nucleic acid fraction, and in membranes as a part of phospholipids. It is important in all phases of reproduction. In ruminants, it is vital for the proper metabolism and health of the rumen microorganisms. Phosphorus is essential in proper utilization of energy and in the process of building muscle tissue in animals. In combination with Ca, P may make up more than 70% of the ash in an animal's body. Thus, these two elements are closely related to animal health and metabolism. It is very important to keep a proper balance of Ca and P in relation to vitamin D. A desirable ratio of Ca:P is between 2:1 and 1:1. It is very important to keep this balance; even though one element may be at the minimum, the other element may be in excess of the balance, consequently creating an imbalance within the animal's body. When vitamin D is insufficiently provided in the ration, the ratio becomes less important, and more efficient utilization is made of these elements (Tisdale et al. 1985).

TABLE 6.7. Response to applied P of smooth bromegrass, timothy, orchardgrass, bluegrass, and occasional red clover plants under irrigation

Applied P, kg/ha (lb/A)	DM yield, mt/ha
0 (0)	12.2
20 (18)	15.8
40 (36)	16.7
60 (54)	19.1

Source: Adapted from Rehm et al. (1977).

TABLE 6.8. Response of various grasses to 90 kg/ha of P

Grass	Increase over check (%)
'Midland' bermudagrass	31
'Morpa' weeping lovegrass	12
'Plains' bluestem	16
Native range grass	20

Source: Adapted from Taliaferro et al. (1975).

About 99% of the Ca and 80% of the P in the body are present in bones and teeth. In dairy animals, Ca and P make up 50% of the ash of milk. Therefore, a very liberal supply of these two elements is required. The Ca and P requirements for a dairy animal, especially one that is lactating, are important. Ca:P ratios are also important for the animal's further health and longevity and for the health of the calf.

POTASSIUM

Potassium is another of the major elements for plant growth. It is used in larger amounts than is P. Within live plant tissue, the average percentage of K_2O is approximately 8 to 10 times that of P; hay or dry matter contains up to 4 times as much K_2O as P. Potassium aids plants in resisting disease, insects, cold weather, and drought, and it functions in stomatal opening and closure and sugar transport. Since plants take up a large amount of K, the soil must be replenished with K_2O, or deficiencies will show up readily, particularly when the forage is harvested as hay or silage rather than grazed.

Both legumes and grasses are very heavy users of K, especially grasses, because of their extensive fibrous root system. Generally, the first species within a grass-legume mixture to show a deficiency or die will be the legume because of its lack of an efficient uptake of K when in competition with a grass. Most soils in the US vary in K content from area to area, particularly the sandy soils, which are low in K and must be fertilized frequently with high rates of K_2O.

There are three forms of K in the soil. (1) The soluble K is the portion that is water soluble; this constitutes only a small portion of the total K in the soil. (2) There is also an exchangeable K, which is held on the soil colloids and is easily available. This also makes up a small percentage of the total K in the soil. (3) In addition, the clay fraction of the soil contains a nonexchangeable K, which is neither soluble nor readily available to plants. The nonexchangeable K makes up the largest portion of the total K in the soil, except in highly acid, sandy soils or soils that are high in organic matter, where the nonexchangeable K is relatively low.

The soluble portion of the K in the soil is subject to leaching in some soils. It is important to maintain a relatively high level of soluble K over the entire life cycle of a plant and throughout a growing season. Following droughty periods, in which the soluble K becomes less available on sandy soils or highly organic soils, a K deficiency may show up late in the growing season. Therefore, it is important to apply K to the crop annually. Some of the nonexchangeable K is made available each year from the soil, but only in limited amounts.

It has been shown that it takes approximately 4 kg/ha of K_2O to increase a soil test by 1 kg/ha (8 lb of K_2O to increase the soil test by one part per million [ppm]). As a result the recommended rate for applying K_2O to increase the soil test value to the desired goal is equal to four times the difference between the soil test goal and the actual soil test value (Table 6.9).

The major loss of K applied to the soil is through crop removal or soil erosion. It is recommended that the soil test level of K be built up to a minimum of 340 to 450 kg/ha (300 to 400 lb/A) of exchangeable K for soils in regions with low and high cation exchange capacity, respectively. These values are higher than required for maximum yield, but as in the recommendations for P, this will ensure that K availability will not limit the crop or the forage yield. Local extension offices or fertilizer dealers can provide information on the P- and K-supplying power and recommendations for specific soils.

One should apply one-half of the recommended amount of K to a forage crop after the last harvest in the fall, and the remaining amount after the first harvest the following spring. Top-dressing in this manner will ensure the availability of sufficient amounts of K throughout the growing season and will provide sufficient K in the plant tissue for winter survival. For example, for every metric ton of alfalfa DM removed from the field, approximately 7.5 kg of P (15 lb/English ton) and 30 kg of K (60 lb/English ton) are removed. To maintain an adequate fertility level, the amount of nutrient removed by the previous forage crop should be applied to the next as a top dressing. See Table 6.10 for the fertilizer rates that one should supply to maintain a potential yield goal for a forage crop (also see Fig. 6.2). The main sources of K_2O fertilizers and their percentages of K_2O

TABLE 6.9. Amount of potassium (K_2O) required annually to build up the soil (based on buildup occurring over a 4-yr period; 4 kg/ha K_2O required to change the K test 1 kg/ha)

	Soil cation exchange capacity	
K test[a]	Low[a]	High[b]
	kg/ha	
50	210	250
60	200	240
70	190	230
80	180	220
90	170	210
100	160	200
110	150	190
120	140	180
130	130	170
140	120	160
150	110	150
160	100	140
170	90	130
180	80	120
190	70	110
200	60	100
210	50	90
220	40	80
230	30	70
240	20	60
250	10	50
260	0	40
270	0	30
280	0	20
290	0	10
300	0	0

Source: Miller (1984).

[a]Soil tests taken before May or after September 30 should be adjusted downward as follows: subtract 34 kg/ha for dark-colored soils; 50 kg/ha for light-colored and fine-textured lowland soils; and 67 kg/ha for medium- and light-colored soils in southern areas.

[b]Low cation-exchange-capacity soils are those with CEC less than 12 meg/100 g soil; high are those with CEC equal to or greater than 12 meg/100 g soil.

are muriate of potash, 60%; sulfate of potash, 51%; and sulfate of potash (magnesia), 22% K_2O and 18% MgO.

Workers in Wisconsin found that most of the applied K was absorbed by roots near the soil surface when potassium sulfate (K_2SO_4) was placed at varying depths below a 2-yr stand of alfalfa (Peterson et al. 1983). It was also found that when the next three harvests were taken, 41% and 29% of the K was removed from the upper 7.6 cm and 22.9 cm from the surface, respectively. Therefore, it appears that sufficient amounts of K and P can be removed from the upper portion of the soil and that top-dressing with these elements would suffice (Doll et al. 1959).

Throughout the world the main source of K used as a fertilizer in agriculture is potassium chloride (KCl). It has been shown that grasses are less sensitive to Cl than are legumes (Blaser and Kimbrough 1968). There appears to be a Cl toxicity level at which legumes are very sensitive. From numerous research studies, investigators have set the upper limit as 336 kg/ha of KCl. This should be the maximum level at the time of seeding, due to the so-called salt effect. These writers believe that this effect results from Cl toxicity rather than from the K salt. For instance, research indicates that Cl will depress the yield of red clover (Laughlin et al. 1971). In a study using KCl and K_2SO_4, the highest yield was obtained at 200 kg/ha of K using K_2SO_4; there was severe yield depression when KCl was used (Smith 1975). There is also a possibility that some response to S was shown. In Wisconsin, it was found that when KCl was used, the level of Cl found in young alfalfa tillers was as high as 12% to 14% in the plant tissue at the time of death (Smith 1975).

Potassium has a reputation for being consumed luxuriously by plants, especially grasses. Plants have a tendency to absorb amounts of K far in excess of what is actually required for proper growth and development (Lutz 1973). The chemical composition for the critical level of K is between 2.3% and 2.5%, which is adequate for growth (Melsted et al. 1969) (Fig. 6.3). It is not unusual to find plants that are grown under high levels of K to range from 3.5% to 4.5%. This excessive percentage has not been shown to offer any advantage for either plant life or the animals consuming the forage. But no negative effects have been identified either.

Fig. 6.2. The effect of phosphorus and potassium on alfalfa top growth. (Miller 1984, 147)

The proper balance of P and K is important in maintaining productive stands of alfalfa. At the end of the fifth year, there is usually twice the production when P and K are applied compared to when no P and K are added. In addition, the number of years that a fertilized stand maintained at least 50% alfalfa was at least twice that for the control.

Potassium has a direct effect on winter survival of alfalfa. In Canada, it was found that a significant increase in the number of plants that survived was observed when K was applied beyond the control level, at a rate of 112

TABLE 6.10. Maintenance fertilizer required for various forage yields of alfalfa, grass, or alfalfa-grass mixtures

Forage yield (t/A)	P_2O_5 kg/ha (lb/A)	K_2O kg/ha (lb/A)
2	26 (24)	112 (100)
3	40 (36)	168 (150)
4	52 (48)	224 (200)
5	67 (60)	280 (250)
6	81 (72	336 (300)
7	94 (84)	392 (350)
8	108 (96)	448 (400)
9	121 (108)	504 (450)
10	135 (120)	560 (500)

Source: Miller (1984).

Fig. 6.3. Potash-deficiency symptoms on alfalfa leaflets. (Miller 1984, 148)

to 224 kg/ha, whether the temperature was –4°C or –9°C. The number of stems per plant increased in a linear manner as the rate of K was increased (Table 6.11) (Blaser and Kimbrough 1968).

Many researchers have studied the amount of K removed in various forages. It is generally concluded that as the yield of hay increases the amount of K removal increases proportionally. Yields above 10,000 kg/ha contain from 2.5% to 3.5% K; thus, it is very important to maintain a high level of K when expecting high yields.

High rates of K fertilization increase the number of nodules (Table 6.12) (Duke et al. 1980). The increased K fixed per plant is a result of the nodule number per plant and not of the acetylene reduction rates (there is a linear correlation between the number of nodules and the acetylene reduction rate). Therefore, it appears that K fertilization will increase alfalfa yields by increasing the number of nodules and so the extent of N fixation.

CALCIUM

Calcium, along with P, makes up about 70% of the mineral matter in the bodies of livestock and about 90% of their skeletons. The

TABLE 6.11. Effect of K on winter survival of alfalfa and living stems per plant, Ontario

	Winter survival, %		Living stems/plant	
K_2O rate, kg/ha (lb/A)	–4°C (25°F)	–9°C (15°F)	–4°C (25°F)	–9°C (15°F)
0 (0)	73	56	2.8	1.9
112 (100)	97	60	3.4	2.6
224 (200)	90	80	3.8	3.0
336 (300)	97	80	4.3	3.8

Source: Adapted from Blaser and Kimbrough (1968).

TABLE 6.12. Effect of K on nodulation of alfalfa

K application, kg/ha (lb/A)	Nodules/plant	Increase over control, %
0	42	. . .
448 (400) as K_2SO_4	69	166
673 (600) as K_2SO_4	111	271
673 (600) as KCl	93	223

Source: Adapted from Duke et al. (1980).

main source of Ca fertilizer is limestone. Calcium helps reduce soil acidity, it adds strength to plants, and it is needed in the formation of the bones and teeth of animals. Calcium is also needed in particularly large amounts by growing animals, pregnant animals, and those that are producing milk. Most of the Ca need by livestock can be furnished by forages, but limestone must be added to acid soils to supply the plants with amounts of Ca adequate for proper growth.

In chemical composition, green tissues of forages range from 0.9% to 1.2% Ca. Alfalfa plants prior to flowering, approximately 7.5 cm in height, will range from 1.76% to 3% Ca, while red clover at the same stage will range from 1.2% to 2% (Brown and Graham 1978).

MAGNESIUM

Magnesium needed in forage production may be supplied by dolomitic limestone or by fertilizers containing Mg, such as sulfate of potash-magnesia. Magnesium is a necessary component of chlorophyll, which plays an important role in photosynthesis and carbohydrate production. Most legumes contain amounts of Mg sufficient to meet the needs of animals. Soils that are high in K and low in Mg may produce an increased incidence of grass tetany (George and Thill 1979). This disease of cattle can be controlled by feeding the animals extra Mg or by growing legumes in the pasture mixture.

The chemical composition of alfalfa tissue ranges from 0.25% to 0.31% Mg. The percentage of Mg in alfalfa at the bud stage may exceed this amount, ranging from 0.31% to 1%, and it is 0.2% in red clover (Brown and Graham 1978; Melsted et al. 1969).

Legumes are considered very efficient users of soil Mg. They contain more Mg than do grasses; for instance, red clover contains 0.27% Mg whereas grasses have 0.19%. Magnesium deficiencies are generally associated with acid or sandy soils. Ca and Mg concentrations in legumes may be depressed by K fertilization (Blaser and Kimbrough 1968).

SULFUR

Sulfur has gained more attention recently as a fertilizer element for forage legumes. Sulfur increases root growth and helps maintain a dark green color. It is needed for protein formation and N fixation by legumes. Most of the S used by plants comes from the air or from organic matter. Sulfur is also present in many low-analysis fertilizers. There are a few soils in the southeastern US that need S fertilization. Because of the Clean Air Act, S is being removed from burning coal, so there is now an increased concern for adding S to some fertilizer recommendations. Sulfur is removed from the soil at approximately the same rate as is Mg.

Sulfur deficiencies are found generally on highly leached or sandy soils and soils that are low in organic matter. Sulfur may also become deficient under continuous alfalfa production. Most S soil test readings are reported as parts per million, and for alfalfa production, if the soil test level falls below 7 ppm, one should apply S. For readings between 7 to 12 ppm of S, it is suggested that S be applied on a trial basis, but when S readings are above 10 to 12 ppm, there is very little chance for a positive response. Since the S soil test is not considered to be as reliable as soil tests for P and K, it is recommended that tissue analysis be used as further confirmation of the need for S fertilizer. Limited data are available to show the response of alfalfa to S application. In Minnesota, on soils that were considered sufficient in S, the yield was doubled when elemental S was applied as gypsum (Table 6.13) (Lanyon and Griffith 1988). Where S was deficient, the alfalfa stand was much less winter-hardy. Alfalfa yields continued to increase up to a maximum of 80 kg/ha when S was applied. Alfalfa yields over 17,000 kg/ha dry matter (DM) contained 0.33% S and removed 55 kg/ha S (Table 6.14).

It is important to keep the proper balance between N and S. It is suggested that a ratio of 10:1 N to S be considered optimum for maximum forage production and animal utilization. If the ration exceeds 15:1, yield and protein production may be depressed.

MICRONUTRIENTS

The micronutrients needed for forage pro-

TABLE 6.13. Response of alfalfa to S application

Treatment	S application rate, kg/ha (lb/A)	DM yield, kg/ha (lb/A)
Control	0 (0)	4020 (3586)
Sulfur (elemental)	56 (50)	9410 (8394)
Gypsum	56 (50)	9690 (8643)

Source: Adapted from Lanyon and Griffith (1988).

TABLE 6.14. Effect of S on alfalfa yield, concentration of S in tissue, and amount removed when harvested at early bloom

S application rate, kg/ha (lb/A)	DM yield, kg/ha (lb/A)	S concentration, %	S removed kg/ha (lb/A)
0 (0)	7,409 (6,609)	0.16	12 (11)
40 (35)	15,114 (13,482)	0.25	38 (34)
80 (71)	17,287 (15,420)	0.33	55 (49)

Source: Adapted from Lanyon and Griffith (1988).

duction are B, Mo, Zn, Cu, Fe, Mn, Cl, Na, Co, Ni, and Si. All of these nutrients are essential for plant growth and are just as important for the proper life cycle of a forage crop as are any of the major elements (see Chap. 4; see also Chap. 5, Vol. 2). They are required in very small amounts, usually reported in ppm.

Boron. Legumes, especially alfalfa, are sensitive to low B levels. Adding B to alfalfa fertilizer is an accepted practice where deficiencies may occur. Boron deficiency is associated with low moisture levels or droughty conditions, sandy soils, and light-colored soils. Boron is relatively immobile in the plant, and the youngest tissue may show B-deficiency symptoms first. When there is a B deficiency, alfalfa may appear stunted, with the internodes shortened, and the plant may take on a general yellowish character. This condition is sometimes confused with drought problems.

Boron has a negative charge and is easily leached unless it is utilized by the legume or retained in the organic matter fraction. Boron availability is associated with the level and the decomposition of organic matter, as well as with the soil texture and soil pH. When the organic matter decomposes, B is released for plant use. The soil pH has a direct effect on the availability of B. A low pH inhibits the activity of microorganisms and the eventual release of B. Consequently, it is very important to have a soil pH of 6.5 to 7.0 to reduce a potential B deficiency. Excessive pH will also reduce B availability.

Properly fertilized and healthy legumes will usually contain 24 to 35 ppm of boron. If the B level falls below 20 ppm, an application may be advisable. For legumes, B application is recommended at the rate of 3 to 5 kg/ha of actual B. It is recommended that light-textured or permeable, well-drained soils receive annual applications of B to avoid deficiencies. Boron can be applied with blended fertilizers or by itself, as a top-dressed material. Soils that are high in organic matter and very heavy or dark colored may need only one application every 2 or 3 yr.

Boron is available in several different types of carriers. The main form of B fertilizer is borax, which ranges from 11% to more than 21% B. If B is applied annually, B should not be applied to alfalfa located where corn will be grown the following year because corn is rather sensitive to high B concentrations.

Boron is needed for both plant growth and control of the alfalfa weevil. Data from the University of Illinois indicate that high levels of B in alfalfa plant tissue may reduce the level of hatching of alfalfa weevil eggs, thus providing a potential biological control. Under extremely high levels of B, the alfalfa weevil may become completely sterilized or incapable of complete metamorphosis (Smith et al. 1974).

In addition to providing a healthy plant stand, B is essential for proper legume nodulation (Lanyon and Griffith 1988). Other researchers have shown that B is necessary for legume seed production. Excess applications of B can induce B toxicity in some crops.

Molybdenum. Molybdenum is needed in both N assimilation, or protein production, and N fixation. A deficiency of Mo appears to resemble a N deficiency in legumes. When Mo is deficient, N metabolism is reduced, so the problem appears as a N deficiency. Research has shown that protein content may be increased by addition of Mo, which is directly related to the N concentration of the plant tissue (Lanyon and Griffith 1988).

As the pH is reduced so is the availability of Mo for plant growth. Only trace amounts of Mo are required. The chemical composition of an alfalfa plant ranges from 3.3 to 3.9 ppm of Mo; deficient plants contain less than 0.5 ppm. It is sometimes falsely claimed that a small amount of Mo will substitute for lime application. Caution should be exercised when adding Mo to inoculated legume seed as a solution, since it can kill the bacteria.

Other Micronutrients. Alfalfa can absorb zinc more efficiently from soils than can many other crops. The chemical composition of alfalfa ranges from 19 to 25 ppm of Zn. Zinc is important in the enzyme system of plants.

Generally, the Zn level within the plant tissue must be below 6 ppm before a response will be shown to Zn application. Zinc deficiency is indicated by white or striped leaves (Lanyon and Griffith 1988). It is usually found on acid, sandy soils that have been recently limed.

Copper is important in the enzyme system of plants and in the formulation of hemoglobin in animals. Copper deficiency may appear on peat- or muck-type soils and heavily weathered, sandy soils. The availability of Cu is decreased on high-pH soils, and the element may show up as a plant deficiency on extremely low-pH soils. On mineral soils, application rates of 11 to 17 kg/ha of copper sulfate are usually high enough for good legume growth. For legumes grown on muck soils, this rate should be doubled (Lanyon and Griffith 1988).

Iron is essential in the formation of cholorophyll and also is in the blood hemoglobin of animals. Most soils contain sufficient amounts of Fe, except where it is tied up by very high pH. Addition of S will increase the availability of Fe in soil (Lanyon and Griffith 1988). Excessive P fertilizer may induce a deficiency of Fe and Zn.

Manganese is required in small amounts by both plants and animals. Manganese is important in the oxidative enzyme system of plants. Soils that have an extremely high pH may be deficient in available Mn. Manganese also reduces the availability of Fe for plant growth (Lanyon and Griffith 1988) and may produce Mn toxicity in plants grown on very acid soils.

Chlorine and sodium are involved with osmotic pressure and charge balance of plants. Chlorine is also involved with photosynthetic phosphorylation. Nitrogen fixation in alfalfa involves cobalt. Nickel is part of urease and is needed by legumes in N fixation. Silicon is primarily required for mechanical strength and aids in drought resistance (Tisdale et al. 1985).

NUTRIENT INTERACTIONS

When evaluating plant nutrients, we often think in terms of individual effects of one nutrient or the impact when one nutrient is limiting. But there are important interactions among nutrients, too. An interaction occurs when the effect of a combination of nutrients is greater than the sum of their individual effects.

Researchers are learning more about the interactions among essential plant nutrients and how these interactions affect not only the plant growth but also its value as a feed source. These interactions are often the key to both crop yield and quality, and managing these interactions may be the key to profitable crop and livestock production. Many nutrient interactions have been identified. The following are examples of how important they can be.

An interaction between P and Mg nutrition in plants has been identified that may dramatically reduce the incidence of grass tetany, a disease found in cattle on grass pastures with inadequate Mg content. The Mg content of pasture and forage grasses can be increased by maintaining a high P level in the soil. Potential for grass tetany can thus be reduced by adequate P fertilization. Where supplemental Mg fertilizer was applied, Mg content of the grass may actually decline unless P fertilizer is also applied (Reinbott and Blevins 1991; Blevins 1993).

Farmers can benefit from the P effect on Mg by simply following normal fertility management recommendations. Unfortunately, many farmers still do not properly manage the fertility of their forage and pasture crops. When left to depend on carryover nutrients from grain crops in the rotation, pasture and forage crops are often not adequately supplied with essential nutrients. Pasture crops are often left out of crop nutrient management plans. When forages are undernourished, they cannot provide their optimum feed value to livestock. A complete nutrient management plan for a crop rotation should include not only credits for nutrients supplied by the forage crops but also adequate nutrient supplies for the forage crops.

Nitrogen utilization is much more efficient when proper P and K levels are present. Both P and K, along with some of the micronutrients, are essential for optimum N fixation by legumes. High levels of N, such as is often found where heavy manure applications are used, can lead to severe lodging and reduced regrowth of alfalfa, unless adequate K is supplied (Volenec 1993).

QUESTIONS

1. Why is pH so important for the overall fertility program of forages?
2. Name at least two forage crops that fall into each of the following classes: those that are very sensitive to acidity; those that can tolerate moderate acidity.
3. What proportion of a legume-grass sward should the legume portion be to ensure sufficent N is be supplied to the grass?

4. How much N can a legume provide for a companion grass in a grass-legume mixture?
5. Describe the role of N in forage production in relation to both grasses and legumes.
6. What is meant by the P-supplying power and the K-supplying power of the soil?
7. Why do soils vary in their supplying power of various nutrients?
8. How much P_2O_5 and K_2O are required to build up the soil by one unit in a soil test?
9. What is the role of P in forage production? What is the role of K?
10. Why is P important in forages' utilization of water?
11. How much P_2O_5 and K_2O are removed in an 8 mt yield of alfalfa? How much fertilizer must one add to maintain the proper level?
12. Why is it important to maintain the proper balance of P and K fertilizers when applying them to forages?
13. Describe the role of K in forage fertilization.
14. In which area of the US would the rate of forage fertilization be greater: the upper Midwest or the middle Midwest?
15. Discuss the roles of B, S, Mg, and Mn in forage fertilization.

REFERENCES

Allen, SE, GL Terman, and HG Kennedy. 1978. Nutrient uptake by grass and leaching losses from soluble and S-coated urea and KC1. Agron. J. 70:264-73.

Blaser, RE, and EL Kimbrough. 1968. Potassium nutrition of forage crops with perennials. In VJ Kilmer, SE Younts, and NC Brady (eds.), The Role of Potassium in Agriculture. Madison, Wis.: American Society of Agronomy, 423-45.

Blevins, DG. 1993. Personal communication.

Brown, BA. 1957. Potassium fertilization of ladino clover. Agron. J. 49:477-80.

Brown, JC, and JH Graham. 1978. Requirements and tolerance to elements by alfalfa. Agron. J. 70:367-73.

Christians, NE, DP Marten, and JF Wilkinson. 1979. Nitrogen, phosphorus, and potassium effects on quality and growth of Kentucky bluegrass and creeping bentgrass. Agron. J. 71:564-67.

Doll, EC, AL Hatfield, and JR Todd. 1959. Vertical distribution of top-dressing fertilizer phosphorus and potassium in relation to yield and composition of pasture herbage. Agron. J. 51:645-48.

Dougherty, CT, and CL Rhykerd. 1985. The role of nitrogen in forage-animal production. In ME Heath, RF Barnes, and DS Metcalfe (eds.), Forages: The Science of Grassland Agriculture, 4th ed. Ames: Iowa State Univ. Press, 318-26.

Duell, RW. 1974. Fertilizing forage for establishment. In Forage Fertilization. Madison, Wis.: American Society of Agronomy, 67-93.

Duke, SH, M Collins, and RM Soberalske. 1980. The effects of potassium fertilization on nitrogen fixation and nodule enzymes on nitrogen metabolism in alfalfa. Crop Sci. 20:213-18.

Follett, RF, and SR Wilkinson. 1985. Soil fertility and fertilization of forages. In ME Heath, RF Barnes, and DS Metcalfe (eds.), Forages: The Science of Grassland Agriculture, 4th ed. Ames: Iowa State Univ. Press, 304-17.

Fribourg, HA, KM Barth, JM McLaren, LA Carver, JT Connell, and JM Bryan. 1979. Season trends in in vitro dry matter digestibility of N-fertilizer bermudagrass and of orchardgrass-ladino pastures. Agron. J. 71:117-20.

George, JR, and JL Thill. 1979. Cation concentration of N- and K-fertilized smooth bromegrass during the spring grass tetany season. Agron. J. 71:431-36.

Hojjati, SM, WE Templeton, Jr., and TH Taylor. 1978. Nitrogen fertilization in establishing forage legumes. Agron. J. 70:429-33.

Jackobs, JA, TR Peck, and WM Walker. 1970. Efficiency of Fertilizer Top Dressing on Alfalfa. Ill. Agric. Exp. Stn. Bull. 738.

Lanyon, LE, and WK Griffith. 1988. Nutrition and fertilizer use. In AA Hanson, DK Barnes, and RR Hill, Jr. (eds.), Alfalfa and Alfalfa Improvement, Am. Soc. Agron. Monogr. 29. Madison, Wis., 333-72.

Laughlin, WM, M Blom, and PF Martin. 1971. Red clover yield and composition as influenced by phosphorus, potassium rate and source, and chloride. Commun. Soil Sci. Plant Anal. 2(1):1-10.

Ludwick, AE, and CB Rumberg. 1976. Grass hay production as influenced by N-P top dressing and by residual P. Agron. J. 68:933-37.

Lutz, JA, Jr. 1973. Effects of potassium fertilization on yield and K content of alfalfa and on availability of subsoil K. Commun. Soil Sci. Plant Anal. 4(1):57-65.

Melsted, SW, HL Motto, and TR Peck. 1969. Critical plant nutrient composition values useful in interpreting plant analysis data. Agron. J. 61:17-20.

Miller, DA. 1984. Forage fertilization. In Forage Crops. New York: McGraw-Hill, 121-60.

Nuttall, WF. 1980. Effect of nitrogen and phosphorus fertilizers on a bromegrass and alfalfa mixture grown under two systems of pasture management. II. Nitrogen and phosphorus uptake and concentration in herbage. Agron. J. 72:295-98.

Nuttall, WF, DA Cooke, J Waddington, and JA Robertson. 1980. Effect of nitrogen and phosphorus fertilizers on a bromegrass and alfalfa mixture grown under two systems of pasture management. I. Yield, percentage legume in sward, and soil test. Agron. J. 72:289-94.

Odom, OJ, RL Haaland, CS Hoveland, and WB Anthony. 1980. Forage quality response of tall fescue, orchardgrass, and phalaris to soil fertility level. Agron. J. 72:401-2.

Perry, LJ, Jr., and DD Baltensperger. 1979. Leaf and stem yields and forage quality of three N-fertilized warm-season grasses. Agron. J. 71:355-58.

Peterson, LA, S Smith, and A Krueger. 1983. Quantitative recovery by alfalfa with time of K placed at different soil depths for two soil types. Agron. J. 75:25-30.

Rehm, GW, JT Nichols, RC Sorensen, and WJ Moline. 1975. Yield and botanical composition of an irrigated grass-legume pasture as influenced by fertilization. Agron. J. 67:64-68.

Rehm, GW, RC Soresen, and WJ Moline. 1977. Time and rate of fertilization on seeded warm-season and bluegrass pastures. II. Quality and nutrition content. Agron. J. 69:955-61.

Reinbott, TM, and DG Blevins. 1991. Phosphate interaction with uptake and leaf concentration of magnesium, calcium, and potassium in winter wheat seedlings. Agron. J. 83:1043-46.

Sheard, RW. 1980. Nitrogen in the P band for forage establishment. Agron. J. 72:89-97.

Smith, D. 1972. Influence of Nitrogen Fertilization on the Performance of Alfalfa-Bromegrass Mixture and Bromegrass Grown Alone. Wis. Agric. Exp. Stn. Res. Rep. R2384.

______. 1975. Effects of potassium top dressing a low fertility silt loam soil on alfalfa herbage yields and composition and on soil K values. Agron. J. 67:60-64.

Smith, RK, DA Miller, and EJ Armburst. 1974. Effect of boron on alfalfa weevil oviposition. J. Econ. Entomol. 67:130.

Stritzke, JF, and WE McMurphy. 1982. Shade and N effects on tall fescue production and quality. Agron. J. 74:5-8.

Taliaferro, CM, FP Horn, BB Tucker, R Totusek, and RD Morrison. 1975. Performance of three warm-season perennial grasses and a native range mixture as influenced by N and P fertilization. Agron. J. 67:289-92.

Tisdale, SL, WL Nelson, and JD Beaton (eds.). 1985. Elements required in plant nutrition. In Soil Fertility and Fertilizers, 4th ed. MacMillan, 59-94.

Vickers, JC, and JM Zak. 1978. Effects of pH, P and Al on the growth and chemical composition of crownvetch. Agron. J. 70:748-51.

Volenec, JJ. 1993. Personal communication.

Wagner, RE. 1954. Influence of legume and fertilizer nitrogen on forage production and botanical composition. Agron. J. 46:167-71.

Watschke, TL, DV Waddington, DJ Wehner, and CL Forth. 1977. Effect of P, K, and lime on growth, composition, and P absorption by Merion Kentucky bluegrass. Agron. J. 69:825-29.

Wedin, WF. 1974. Fertilization of cool-season grasses. In Forage Fertilization. Madison, Wis.: American Society of Agronomy, 95-118.

7 Forage Establishment and Weed Management

DARRELL A. MILLER
University of Illinois

JIMMY F. STRITZKE
Oklahoma State University

FORAGE ESTABLISHMENT

ONE of the most important steps in a good, efficient forage production system is establishment of a thick stand. This cannot be overemphasized since thick, vigorous stands of grasses and legumes are essential for high yields. For annuals and perennials that exist as discrete plants (e.g., alfalfa [*Medicago sativa* L.] and lovegrass [*Eragrostis* spp.]), it is critical that the proper plant populations be obtained with the initial planting. It is not as critical to obtain a full stand with the initial planting of nondiscrete forages that vegetatively propagate by stolons or rhizomes—e.g., white clover (*Trifolium repens* L.) and bermudagrass (*Cynodon dactylon* [L.] Pers.).

A thick stand of forage is one that essentially occupies all of the area with little of the soil surface visible. In the seedling year, a thick stand of alfalfa is considered to be 295 plants/m^2 (30 plants/ft^2) (Marten et al. 1963; Van Keuren 1973). Populations of alfalfa often decrease to 109 plants/m^2 (10 plants/ft^2) by the second year and to 52 plants/m^2 (5 plants/ft^2) by the third year before stabilizing. However, plant density of a 100% stand of alfalfa would be approximately 50-55 stems/ft^2 due to increased stem numbers from individual alfalfa crowns.

DARRELL A. MILLER. *See Chapter 4.*

JIMMY F. STRITZKE is Professor of Weed Science, Oklahoma State University, Stillwater. He received his MS degree from Oklahoma State University and his PhD from the University of Missouri. His research and extension activities focus on weed and brush control in pastures and rangeland.

Selecting the correct cultivar is also critical to establishment and productiveness of forage. It is important to select adapted multiple disease- and insect-resistant cultivars with good seedling vigor and good regrowth potential after harvest. In addition, cultivars should have high yield potential in response to fertilizer and other good management practices.

Site Selection and Soil Fertilization. Successful establishment of forage crops must begin by matching the correct forage species with the proper soil environment. For example, on poorly drained, waterlogged soils, species like tall fescue (*Festuca arundinacea* Schreb.) are well adapted whereas deep-rooted legumes like alfalfa require deep well-drained loam or sandy loam soils. Conducting soil tests before establishment is essential in order that soil pH, phosphorus (P), and potassium (K) levels can be adjusted before planting. The ideal pH for alfalfa and most legumes is 7.2 with values between 6.5 and 7.5 being satisfactory. Liming would be essential for alfalfa if soil pH is below 6.2. Forage grasses are more tolerant to soil pH, so they can be established on more diverse sites. Grasses do best when pH values are between 5.5 and 7.0, but some such as tall fescue can tolerate pH values as low as 4.0, and some like bermudagrass can tolerate pH values as high as 9.0.

Since P and K are not mobile in the soil, they need to be applied and incorporated before planting time. The available P level should be at least 50, and the exchangeable K should be 300 to 400. In addition to P and K, elements such as magnesium (Mg), calcium (Ca), sulfur (S), boron (B), molybdenum (Mo), manganese (Mn), and other trace elements are needed for proper growth (see Chap. 6; see also Chap. 5, Vol. 2). If they are not available in the soil, they will need to be added. It is also recommended practice to apply some nitrogen (N) immediately before or during planting. With legumes, only 11 to 22 kg/ha (10 to 20 lb/A) at planting is needed since that is enough to support the legume seedlings until they are able to synthesize their own N, especially on low-organic matter soils. With grasses, N rates of 22 to 45 kg/ha (20 to 40 lb/A) could be applied at planting, followed by additional N topdressing as needed. A detailed discussion of soil fertility is presented in Chapter 6 (also see Chap. 5, Vol. 2).

Forage Species and Cultivar Selection. For successful establishment and production of forages, one must select the best-adapted, highest-yielding, and most persistent species and cultivar for a specific soil, climate, and intended utilization. There are many different forage species and cultivars available on the market, and it is important that the proper cultivar is purchased. It is the responsibility of plant breeders to give a description of each cultivar they develop, and it should include its adaptation and degree of pest resistance. This information can be obtained from the seed dealer, and additional information on specific cultivars is often available from state land grant universities and government agencies. Obtaining high-quality seed is also very important for good seedling vigor and stand establishment. Quality is best ensured by using certified seed or brand seed from a reputable seed company.

Inoculation of Legumes. Legume roots have the ability to convert atmospheric N into a form available to plants if properly inoculated with appropriate, effective strains of N-fixing bacteria *Rhizobium*. Some of the *Rhizobium* bacteria are effective on more than one legume species, but some are specific for a single legume. This means that it is critical that legume seed be inoculated with the *Rhizobium* specific for the legume being planted. It is generally concluded that once an inoculated legume has been grown in a particular soil, the population of *Rhizobium* remaining in the soil is adequate to inoculate that legume. However, since there is no commercial test for *Rhizobium* presence in the soil and because of differences in the effectiveness of different strains of *Rhizobium,* it is recommended that legume seed be inoculated with the proper bacteria before each planting to ensure that the most effective strain is available.

Since bacteria are very sensitive to light and heat, it is important that the inoculum be maintained in a refrigerator or under cool conditions before use. Once seed is inoculated, it should be planted as soon as possible. Some alfalfa seed currently on the market has already been inoculated; however, it is important to reinoculate if the date of effective inoculation has expired or if seed has been subjected to improper storage conditions.

Planting Rates. In general, enough seed of a species is planted so that the seedling plants can occupy the area when they emerge. This means that the planting rate is influenced by seed size and then adjusted for pure live seed. In addition to seedling vigor, soil and climatic conditions can also affect the planting rate since they affect seedling establishment. For example, planting rates for alfalfa range from 11 to 13 kg/ha (10 to 12 lb/A) in the High Plains to more than 33 kg/ha (30 lb/A) in some areas of California (Smith et al. 1986). Since there are approximately 500 alfalfa seeds/g (220,000 seeds/lb), the planting rate of 33 kg/ha would be equivalent to 1631 plants/m^2 (152 plants/ft^2) if they all emerged. However, under field conditions, only 30% to 40% of the seed will actually result in viable seedlings.

There are several reasons for low seedling survival of small-seeded legumes and grasses (Cope 1982; Dutt et al. 1983; Miller 1984). (1) Seedling growth is slow due to limited food reserve in small seed. (2) It is difficult to obtain good seed-to-soil contact with the small seed. (3) Seed is often placed too deeply in the soil. (4) Seedlings are easily damaged by stresses from weeds, insects, and disease. (5) Little seedlings are very susceptible to drought conditions following planting because of their limited root systems. (6) Germination may be inhibited due to some allelopathic effects. (See chap. 16.)

Planting the forage seed too deeply is one of the most common problems associated with forage establishment. Normally, one should plant small seed no deeper than 0.6 to 1.2 cm (Smith et al. 1986; Tesar and Marble 1988). In some cases it may be necessary to place seed deeper: for example, in sandy soils and in the more arid areas of the country. However, if seed is planted deeper, it will be necessary to increase the planting rate since it has been shown that the percentage of established seedlings is greatly reduced when planting depths are increased (Table 7.1). The planting rate for each of the forage species is discussed in more detail in its respective chapter.

Time of Planting. The time of planting is related both to the forage species and the environment. In general, the optimum time of planting relates to the time of year that one is most apt to obtain a satisfactory stand. In the Midwest, spring is the generally recommended time of year for planting all forage crops, whereas in the southern half of the US, some forage crops such as tall fescue and alfalfa are planted during September and October. Planting rates in the spring depend primarily on the location. In the southern states, summer forages such as bermudagrass, bahiagrass (*Paspalum notatum* Flugge), and lespedezas (*Lespedeza cuneata* [Dum.-Cours.] G. Don and *Kummerowia* spp.) are planted as early as one can get into the field. In the areas of Kentucky and Missouri, planting time can start in February. As one goes to the upper Midwest (progressing north into Iowa and northern Illinois), planting time is the latter part of March and early April. In the northern states, planting is often delayed until May. Spring planting with a companion small grain, such as spring oat (*Avena sativa* L.), is generally done when the oats are planted.

For alfalfa and the cool-season grasses there is a transition zone running west from the Rocky Mountains through Kansas, Missouri, and Illinois where both spring and fall plantings are practiced. For example, spring plantings in the northern quarter of Illinois are usually more successful plantings, whereas late summer planting may be more desirable than spring planting in the southern one-quarter to one-third of Illinois.

Planting from late August to early November is very popular in the US, primarily south of the 40th to 41st parallel. In this area, there is often an early fall rain that ensures good germination and establishment, and this coupled with decreasing temperatures and transpiration results in good seedling growth. Success from late summer planting in this area also results from mild winters and reduced weed and insect pressures. This means that good forage stands can be established without the use of a companion crop or herbicide, since a killing frost will eliminate any summer weeds.

Late summer seeding dates are governed by the date of the killing freeze: seeding should be at least 35 to 45 d before the average killing freeze. For example, the suggested planting date for the northern one-quarter of Illinois is from August 10 to 15, whereas in the central part of the state it is August 30 to September 5, and in the southern quarter September 15 to 20. Planting should be done as close to these dates as possible to ensure that the plants become well established before winter. It is important that the plants develop an adequate root system and sufficient food reserves to carry them through the winter.

The dates suggested for Illinois correspond to those for other states at approximately the same latitude. In the West and Southwest, seedings can be made up to October 15, which allows enough time for an excellent establishment. These dates vary somewhat with species and location. For instance in Illinois, birdsfoot trefoil (*Lotus corniculatus* L.) and reed canarygrass (*Phalaris arundinacea* L.), because of their slower seedling establishment, should not be seeded as late as alfalfa or orchardgrass (*Dactylis glomerata* L.). However, in Oklahoma alfalfa needs to be planted in late August and early September before the fall rain, since rainfall after planting is required for stand establishment, whereas tall fescue can be planted later since it can be drilled into moisture.

Planting Methods. The two basic methods of planting are drilling and broadcasting. With both planting methods, it is essential that the seedbed be firm, but it should be loose enough to allow root penetration. For conventional seedbeds, an ideal surface should have particles no larger than 1 cm in diameter (about 0.5 in.) in order that good seed-to-soil contact is possible. This is usually accomplished by harrowing the seedbed when there is suffi-

TABLE 7.1. Emergence of alfalfa sown at varying depths (based on assumption of 24 viable seeds sown per ft of row)

Planting depth		Average number of seedlings	
cm	in.	emerged per ft of row	Emergence, %
1.25	0.5	14.14	60.9
2.50	1.0	10.61	45.7
3.75	1.5	6.16	26.5

Source: Smith et al. (1986).

cient moisture in the soil so that it will crumble when worked. There is usually a short period of time after each rain when soil moisture will be just right.

Drilling involves shallow planting using a conventional or no-till drill. On tilled areas, it is a common practice to pull the downspouts out and let the seed fall to the soil surface. Phosphate fertilizer (0-45-0) is often placed about 5 cm (2 in.) deep in the soil with a grain drill to give a fertilizer band below the seed (Decker and Taylor 1985). The fertilizer should be covered with soil before the forage seed is dropped. This process occurs naturally when the soil is in good working condition. Press wheels should be rolled over the forage seed to firm it into the soil surface. This will result in seed being placed 1.25 to 2.5 cm (0.5 to 1.0 in.) deep (Fig. 7.1) (Decker and Taylor 1985).

Drilling with P banded under the seed, if done correctly, is the preferred way to plant forages on a variety of soil types for the following reasons: (1) The readily available P promotes rapid development of large healthy seedlings with excellent root systems (Fig. 7.2). (2) Fewer weeds are fertilized than there would be if the P were broadcast. (3) Because of the rapid, early, vigorous growth, seedlings are able to survive more adverse soil and climatic conditions. Research has indicated that banding P is particularly advantageous if the soil is low in available P (Brown 1959; Sheard et al. 1971; Tesar et al. 1954). Another situation where banding is performed is early planting into cold and wet soils. However, if planting is done later in the spring, under more ideal conditions—such as warmer weather and high P levels—planting forages with a drill is not more advantageous than broadcasting (Smith et al. 1986).

Broadcast seeding involves spreading seed uniformly over a properly prepared seedbed, then pressing the seed into the soil surface with a corrugated roller. The fertilizer is applied earlier and incorporated with a disk during conventional seedbed preparation. If soil conditions are too loose at planting, it is recommended that the seedbed be firmed with a corrugated roller before seeding. The seed can be broadcast over the area by air or by ground equipment. However, the best tool for broadcasted planting is the double-corrugated roller-seeder since it spreads the seed

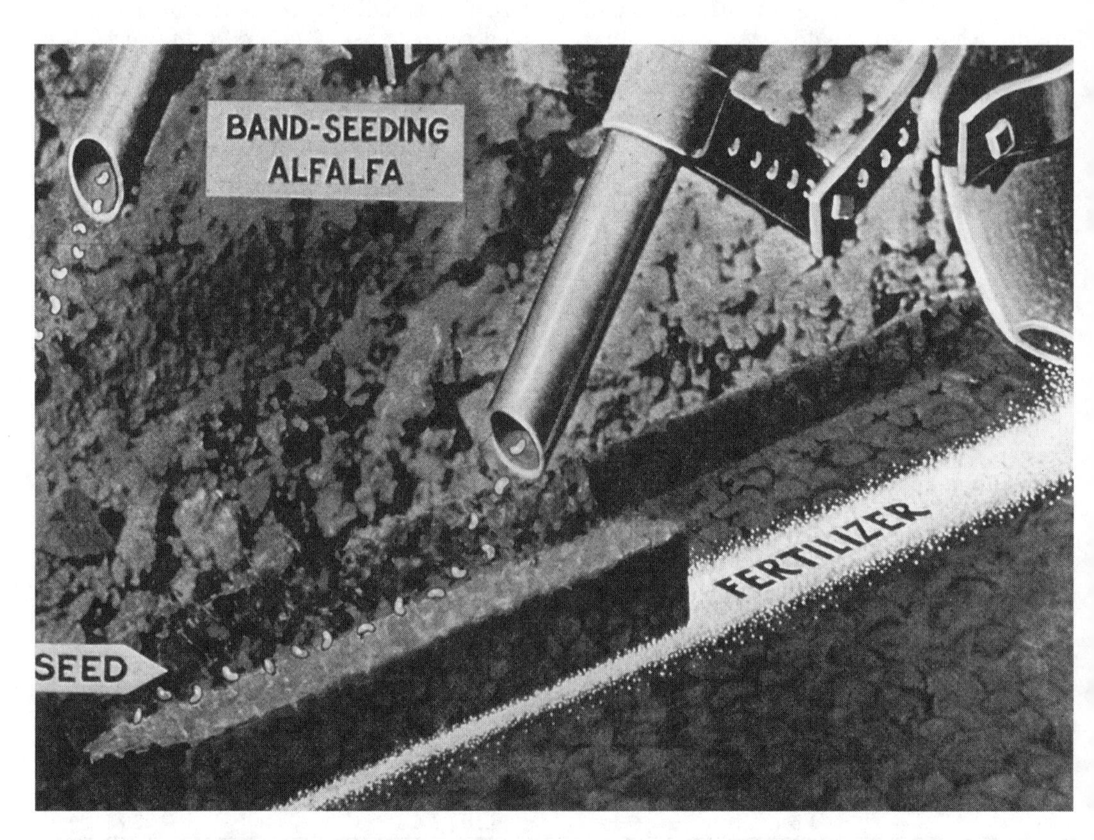

Fig. 7.1. Band-seeding of alfalfa showing the placement of fertilizer and seed. (Miller 1984)

Fig. 7.2. Rapid root development results from planting seed at the proper depth. (Miller 1984)

and packs the area (Decker and Taylor 1985; Smith et al. 1986) (Fig. 7.3). It is very important to obtain good seed-to-soil contact when seeding forage crops; this is one reason that good results are obtained when broadcasting is done with a culti-packer or corrugated roller. The culti-packer or first corrugated roller firms the seedbed and soil surface, so that the seed will not be buried when the second roller rolls over it. The seed is then placed on top of the soil and pressed into it by the second culti-packer, which forms ridges over the previous furrows and furrows over the previous ridges. This type of seeding places the seed at a uniform shallow depth (Smith et al. 1986; Van Keuren 1973).

Drilling forage into untilled areas with no-till drills can also be used for forage establishment. However, to be successful, it is critical that all of the steps of no-till establishment be followed. No-till planting is especially useful in new plantings on areas that are prone to wind erosion (sandy-textured soils) and on steep slopes where conventional tillage would result in soil erosion. With all no-till plantings, it is important that existing vegetation be effectively killed with a postemergence herbicide before planting. The vegetation should be controlled over the entire area for solid plantings of a new forage species or controlled in bands for renovation of an existing forage crop. It is also critical that P be banded with most renovations and no-till establishments, since P levels are often below recommended levels. This means that a soil test must be done to determine the rate of P to apply with the no-till drill. Phosphorus can be applied with the forage seed. However, N and K fertilizers cannot be placed with the seed since they will injure the seedling.

Companion Crops. Planting a companion crop, such as spring oats or, in some areas, winter wheat (*Triticum aestivum* L. emend. Thell.), with a forage has historically been used for forage establishment. The companion crop provides a quick ground cover, thus controlling soil erosion and reducing the invasion of weeds. A companion crop may also be planted with small-seeded forages so that something can be harvested from the area during the year of establishment. However, these companion crops compete with the young forage seedlings for nutrients, light, and moisture and can be more competition than the weeds.

To minimize competition, it may be necessary that the companion crop be planted at a reduced rate and then be removed as early as possible. It is suggested that the companion crop be harvested as a forage crop instead of a grain crop. For example, small grains should be removed as a silage or hay crop before the soft dough stage. This decreases the probability of damage to small seedlings from shading and lodging.

With the advent of herbicides for weed control, it has become an accepted practice to establish forages without a companion crop. With optimum moisture conditions, spring seeding of legumes like alfalfa without a companion crop generally produces 50% to 60% of the normal yield of an established stand during the seeding year. However, the forage yield of a spring seeding of improved grasses will only be 10% to 60% of established stands. The number of harvests taken during the seeding year may be limited by the time of planting in the spring and subsequent moisture or environmental conditions. However, there will normally be only two harvests from a new stand. With new plantings, the crowns are generally smaller and yields per cutting somewhat less than those of an established

Fig. 7.3. A brillion-type seeder utilizes two sets of corrugated rollers to ensure good seed-to-soil contact. (Miller 1984)

stand. By the second year, yields from both companion and noncompanion areas will be equal and comparable to those of older stands. In southern states, planting of alfalfa in late August to early September is usually done without a companion crop. Forage yields from these areas the next summer are essentially the same as those from 1- and 2-yr-old stands.

Management without a companion crop may require more investment. With spring planting, herbicides are essentially always needed for control of warm-season weeds and may be required for control of winter weeds with late summer-early fall seeding.

WEED MANAGEMENT

A major problem in establishing and growing quality forage is weed management. It is especially a problem with the establishment of small-seeded forage species since seedling plants are not strong competitors with weeds. Also forages are usually broadcasted or solid-seeded in narrow rows, so cultivation cannot be used to supplement chemical control as it is with corn (*Zea mays* L.) and soybeans (*Glycine max* [L.] Merr). In addition, legumes and grasses are often planted and grown as mixtures, thus making it more difficult to selectively control broadleaf or grassy weeds.

Weed management in forage crops involves integration of some of the following practices: (1) planting weed-free seed, (2) controlling weeds before planting, (3) planting forage species at the proper date, (4) maintaining the competitive nature of the crop, (5) mowing, (6) mob-grazing, (7) flaming, (8) cultivating, and (9) applying chemicals.

Weed-free Seed. One of the best cultural practices in weed management is to plant only weed-free seed. Using certified seed is a good practice since it means that seed has been produced and labeled according to established procedure, which includes having no noxious weeds and only limited amounts of other weeds. The seed of many serious problem weeds is nearly the same size, shape, and weight as some of the forage seed, so once the weed seed is mixed with the forage crop seed, removing it becomes very difficult. This means special efforts need to be practiced during seed production to prevent seed production of weeds. In most cases, this is the only reliable way you have to eliminate the weed seed problem in forage. However, there are some good mechanical seed-cleaning devices in use at present that can remove most of the noxious weed seed and decrease the weed seed to a manageable level (see Chap. 10).

Weed Control before Planting. Controlling weeds before planting involves eliminating the established weeds and reducing the weed seed in the soil. First, weeds must be controlled in cropland where forages will be planted. A crop rotation that includes a tilled crop for a couple of years is a good practice. The tillage practices will stimulate much of the weed seed to germinate, and then the weeds can be controlled with tillage or a selective herbicide. It is very important to control perennial weeds before planting forages since it is nearly impossible to selectively remove them from the forages. The perennial established weeds can be removed by tillage or postemergence herbicides prior to seedbed preparation. The seedbed preparation before forage establishment is also important. Tillage during this stage should be timely, following weed emergence after rains, and shallow enough to not bring additional weed seed to the surface. Much of the weed seed will germinate within 1 or 2 yr if environmental conditions are favorable for germination (Egley and Chandler 1983). However, seed of some weeds like jimsonweed (*Datura stramonium* L.), common lambsquarter (*Chenopodium album* L.), redroot pigweed (*Amaranthus retroflexus* L.), and velvetleaf (*Abutilon theophrasti* Medicus) may still be viable after 40 yr of burial in the soil (Ross and Lembi 1985).

Proper Date of Planting. The forage species to be established should be planted at the date that maximizes its competitiveness and probability of establishment. The seriousness of the weed problem depends somewhat upon whether the weeds are winter or summer annual; the more serious problem is that of summer annual weeds, since moisture is more apt to be limited in the summer and, as a result, forages are weakened and less able to compete with weeds. For some of the forage crops, such as alfalfa, clovers, and cool-season grasses, it may be advantageous to avoid the summer weeds by planting in the fall. However, environmental conditions can be a problem with fall planting. Some problems include fall drought, winter injury, and excessive rainfall; all contribute to poor stand establishment. Forage crops that become well established with fall planting will be competitive enough to avoid summer-annual weed problems.

If winter annual weeds are a major problem, then a spring planting may be preferred over a late summer or early fall seeding. Spring-planted forage crops will be well established by early fall and will competitively reduce the invasion of any winter annual weed. For cool-season forages that are planted in the spring, it is important that they be planted as soon as soil temperatures and moisture are favorable for germination and growth. This allows the forages to get a head start on many of the summer weeds, which minimizes the competition with many weeds. As the planting date is delayed, and the soil becomes warm, the greater is the possibility that stand failure will result due to weed competition. With availability of selective present-day herbicides, many growers are establishing forage crops when environmental conditions are most desirable and are controlling weeds with herbicides.

Competitive Nature of the Crop. The most effective way to manage weeds in forages is to maintain a thick productive stand of the forage crop. Once a stand of grasses or legumes is established, it will be competitive with weeds if it is managed correctly. This means it is very important to maintain proper fertility and practice good harvest management in order that the forage can continue to be competitive. It is also important to control disease and insects since it is known that these stresses reduces the forage's ability to compete with weeds (Berberet et al. 1987). Thus, the occurrence of weeds invading established forage crops is usually a sign of a management problem (Rohweder and Van Keuren 1985). Most often it results from the failure to keep fertility levels adequate for the forage to compete with the weeds. However, with perennials that exist as discrete plants, such as alfalfa, it may be that stand thinning due to diseases or some other stress has resulted, and the remaining forage plants just do not occupy all of the area. When this is the case, then herbicides will be needed to control the weeds, or the stand will need to be renovated.

Mowing. Mowing for weed control in forages is generally not a very effective option. Some good cases for mowing for prevention of seed production can be made, but it is usually only considered a means of minimizing the spread of weeds. For most weeds, it would require several mowings per season to prevent seed production. A major problem with mowing for weed control is that weeds occupy space and use up water and nutrients before they can be clipped. This would be of special concern during forage establishment since seeding survival could be decreased and this could lead to stand failure. A second problem is that mowing is not selective, so desirable forage plants will also be clipped. For this reason, mowing for weed control should primarily be associated with harvesting a forage crop.

Some good examples of weed management with forage harvest can be cited. Alfalfa has evolved under intensive mowing schedules, so timely harvesting of alfalfa actually kills many of the annual weeds and decreases competitiveness of perennial weeds by lowering their root carbohydrate reserves. Schreiber (1967) records that Canada thistle (*Cirsium arvense* [L.] Scop.) declines and is only of minor importance with mowing of alfalfa for hay production. Also, timely mowing and removal of weeds and forage as a silage or hay crop opens up space for desirable forages. For example, weeds like cool-season weedy grasses have some forage value if properly fertilized and harvested just before seedheads emerge. Also, fertilizing bermudagrass and removing the first harvest by mowing is an effective way to control weeds and quickly restore the competitiveness of the bermudagrass.

Mob-grazing. *Mob-grazing* is stocking of an area with a high density of animals so the available forage can be utilized in 1 or 2 wk. This method of weed control has been very successful in the establishment of bermudagrass and other forages in the South. However, there are some limitations and guidelines that need to be followed. First, it is critical that the seedling forage plants be well rooted before allowing livestock to graze. Otherwise, the seedlings will be uprooted by the grazing. Also, grazing should only be done when the soil is dry so the animals do not muddy the field.

Mob-grazing is most effective for controlling weedy grasses in desirable grasses. There are currently no herbicides registered on most forage grasses. Without some type of management, weedy grasses can cause stand failure. One positive aspect of mob-grazing is that many of the small weeds are actually fairly palatable and are readily eaten by the livestock. Thus, weeds are utilized as a forage.

Mob-grazing of established stands of alfalfa in the winter has also been shown to decrease weed density and increase forage production of alfalfa (Woodal et al. 1987). Many of the seedling weeds are essentially uprooted by the hooves of the livestock on the dry soil surface. In the winter mob-grazing of fall-planted alfalfa after it is well rooted is also effective in removing excessive growth of cool-season grasses. Allowing weeds to shade alfalfa drastically reduces its growth and competitiveness (Pike and Stritzke 1984). Removing forage early either by mob-grazing or mowing helps to minimize the effect of the weed competition.

Flaming. Prior to recent increases in the cost of petroleum products, propane and diesel burners were used to control many weeds in established alfalfa. Use was primarily just before the resumption of spring growth. Treatment at this time not only retarded or killed the weeds but it also gave some insect control and burned the overwintering residue. Some spot burning of dodder-infested areas in seed production alfalfa fields is still practiced.

Cultivation (Tillage). The main use of cultivation in forages has been in seed production, since it is essential in the production of forage seed to eliminate any potential weeds. Cultivation has also been used on summer annual forages planted in rows, such as pearlmillet (*Pennisetum americanum* [L.] Leeke) and forage sorghum (*Sorghum bicolor* [L.] Moench), and on legumes, especially alfalfa for seed production. Row planting permits weed control by cultivation and by directed and shielded application of herbicides.

Tillage has also been used in established stands. The main tillage tool used to suppress or control weeds is a spike-toothed harrow. This tool kills many annual weeds without seriously injuring the crowns of legumes. If a disk-harrow is used in legumes, considerable damage to the crowns could result, and this could provide an area of entry for diseases. However, the disk-harrow has been used to some extent, along with fertilizer, in solid grass seedings of forages. The disk-harrow damages the weeds, and the fertilizer promotes forage production of the grasses, making them more competitive with the weeds.

Herbicides. The information presented here is to be considered general information on herbicide use in forages. It is very important to consult herbicide specialists and to read annual weed control publications and the herbicide label to determine if a herbicide is currently labeled for a particular forage crop. Since forage crops are considered minor crops by commercial companies, the number of herbicides actually labeled for use is small when compared with the number for crops such as corn or soybeans. With the increased cost of registration and the reregistration requirements, the number of labeled herbicides on forages is expected to decrease.

LEGUMES. A number of herbicides have been evaluated for weed control in alfalfa (Peters and Linscott 1988). Many of them are highly effective and selective on alfalfa and other legumes. However, only a limited number of the herbicides are currently registered for application on forage legumes (Crop Protection Chemicals Reference 1993). These herbicides are listed in Table 7.2, along with some other general information.

Herbicides used in legumes can be grouped based on "time of application" into the following: preplant-incorporated (PPI), preemergence, postemergence, dormant, and between cuttings. EPTC (Eptam) is a PPI herbicide labeled for application on alfalfa, birdsfoot trefoil, clovers, and lespedeza. (EPTC is the common name, and Eptam is the trade name. Common names don't change, but trade names do with different market strategies.) Since EPTC is very volatile, it has to be applied before planting and requires immediate incorporation. Benefin (Balan) and several other dinitroanite herbicides used as PPI herbicides have excellent selectivity and give good weed control in alfalfa and clovers. The dinitroanite herbicides need to be incorporated because of their volatility and light sensitivity. Both profluralin (formerly sold as Tolban) and benefin (currently sold as Balan) were used for a number of years on forage legumes. However, profluralin is no longer manufactured, and Balan's use on forage legumes will, according to the manufacturer, be discontinued.

Ideally, PPI herbicides should be uniformly mixed into the top 2.5 cm (1 in.) of soil. Power-driven tools such as rotary tillers do the best job of incorporating herbicides, but most people don't have access to them. The tool used primarily for incorporating is the tandem disk. Best results are obtained when soil moisture is such that good mixing of herbicide with soil is possible. The disk should be set to cut about 10 cm deep with angle and speed adjusted to get good mixing. Do not apply preplant herbicides to wet soil since adequate incorporation is impossible and poor weed control will result. Use of field cultivators, spring-tooth harrows, and spike-tooth harrows is discouraged since they do not adequately mix the herbicide with the soil. Rotary hoes are ineffective for incorporation of herbicides.

There are currently no preemergence herbicides labeled for forage legumes. Propham, sold as Chem-Hoe, was once labeled and used primarily for alfalfa weed control in California.

With postemergence herbicides (bromoxynil [Buctril], 2,4-DB [Butyrac], pronamide [Kerb], and sethoxydim [Poast Plus]) plant size and growing conditions of both weeds and legume crops are important since they relate to selectivity and control. A major limitation of current labeled postemergence herbicides is that none of them control both broadleaf and weedy grasses. Buctril and Butyrac selectively control broadleaf weeds, while Kerb and Poast Plus selectively control grasses. This means that two different herbicides will need to be applied if both grass and broadleaf weeds are a problem. Most postemergence herbicides can be applied to seedling alfalfa plants after they have four trifoliolate leaves (see Fig. 7.4). Buctril, a postemergence contact herbicide, is not widely used because alfalfa seedlings not only need to have four trifoliolate leaves before spraying but also weeds need to be no larger than the four-leaf stage. In addition, Buctril is a restricted-use pesticide and has temperature and mixing restrictions with other pesticides. Sethoxydim (Poast Plus) is currently the most used herbicide for postemergence control of grass in seedling alfalfa. Poast Plus is a translocated postemergence herbicide, so no rainfall is needed for activation whereas with Kerb an inch of rainfall after application is needed to move the herbicide into the root zone where it can be taken up. However, it is very important that an oil concentrate be added to Poast Plus, and weeds need to be small and actively growing when treated.

Some herbicides are applied during the winter months when the alfalfa is dormant and not actively growing, and they are often referred to as dormant-applied herbicides. Diuron (Karmex), hexazinone (Velpar), terbacil (Sinbar), and metribuzin (Sencor, Lexone) are urea-type herbicides commonly used on dormant alfalfa. They have most of their activity through root uptake; thus, rainfall is required to move them into the soil before they can be taken up by the seedling weeds. Length of effectiveness depends on rate of application and

TABLE 7.2. Herbicides registered for weed control in legumes

Common name AI/A	Trade name (amount)	Time of application	Legume registered on	Weeds controlled	Comments
EPTC 3½ lb/A	Eptam 7-E 2 qt/A	Preplant	Alfalfa Birdsfoot trefoil Clovers Lespedezas	Annual grasses, henbit, and volunteer small grain	Thoroughly incorporate immediately after application. Eptam can be metered into irrigation water on established alfalfa stands. See instructions on the label.
Benefin 1⅛ to 1½ lb/A	Balan 3 to 4 qt/A	Preplant	Alfalfa Birdsfoot trefoil Alsike clover Ladino clover Red clover	Annual weeds and grasses	Labeled in 1993 but legumes dropped in 1994. Thoroughly incorporate into the soil soon after application. If disk is used, set it to cut 4 in. deep to get thorough mixing in the top 2 in. of soil.
Bromoxynil ¼ to ⅜ lb/A	Buctril 1 to 1½ pt/A	Early post on seedling stands (4 or more leaves)	Alfalfa	Small seedling broadleaf weeds	Contact herbicide, so good coverage important. There are some temperature restrictions in many states and unacceptable leaf burn can result if applied when temperature is above 80°F. Buctril is a restricted-use pesticide.
2,4-DB ½ to 1½ lb/A	Butyrac 200 1 to 3 qt/A	Post on seedling alfalfa (2 or more leaves) and on established stands	Alfalfa Birdsfoot trefoil Alsike clover Ladino clover Red clover	Pigsweeds, kochia, mustards, and other broadleaf weeds (except henbit)	Apply when weeds are less than 3 in. high or perennials are 6 to 8 in. tall for best results. Use on warm days when temperature is 65°F or above.
Pronamide ½ to 1½ lb/A	Kerb 50-W 1 to 3 lb/A	Post on seedling alfalfa (3 or more leaves) and on established stands	Alfalfa Birdsfoot trefoil Crown vetch Sainfoin	Winter grasses	Irrigation or rainfall is necessary to move the herbicide into the soil. Rate depends on weed problem and soil type. For control of annual bromes and volunteer wheat with the 1/2 lb/A rate, applications need to be *applied before weeds are 2 in. tall.* Kerb is a restricted-use pesticide.
Sethoxydim ⅒ to ½ lb/A	Poast Plus 12 to 60 fl oz/A	Post on seedling and established stands	Alfalfa	Annual and perennial grasses	Poast is very selective on alfalfa, so there is no size restriction on alfalfa. Poast is active on most grasses, so avoid all contact with desirable grasses. It is very important to *add an oil concentrate* and that weeds be actively growing when treated. Rate varies widely by weed species and location, so see label for additional details on rates.
Diuron 1½ to 3 lb/A	Karmex DF 2 to 3⁶⁄₁₀ lb/A	Dormant stands established 1 yr or more (Dec. to Feb.)	Alfalfa Birdsfoot trefoil in western Oregon Red clover in western Oregon	Annual grasses and weeds	Do not use on sandy soil. Do not plant treated areas to other crops for 2 yr.

TABLE 7.2. (*Continued*)

Common name AI/A	Trade name (amount)	Time of application	Legume registered on	Weeds controlled	Comments
Hexazinone ½ to 1½ lb/A	Velpar L 1 to 3 qt/A	Dormant stands established 1 yr or more	Alfalfa	Annual and some biennial weeds	Apply after alfalfa becomes dormant but before new growth begins in the spring. Not recommended in Montana, North Dakota, South Dakota, and Wyoming. Some limitations on crop rotations, so see label for details.
Terbacil ⁴⁄₁₀ to 1½ lb/A	Sinbar ½ to 1½ lb/A	Dormant stands established 1 yr or more (Dec. to Feb.)	Alfalfa	Annual grasses and weeds	Apply after alfalfa becomes dormant but before new growth begins in the spring. Use low rates for cool-season weeds and higher rates for summer weeds. Do not use on sandy soils. Best results in the South-Central states have been with the 1/2 lb/A rate applied in February to control cool-season annuals.
Metribuzin ⅜ to ¾ lb/A	Sencor Lexone DF	Dormant stands established 1 yr or more (Dec. to Feb.)	Alfalfa Sainfoin	Certain annual grasses and broadleaf weeds	Do not apply after growth begins in spring or before growth ceases in fall.
Trifluralin ¾ to 1 lb/A	Treflan E.C. 1½ to 2 pt/A	Dormant stands (Dec. to Feb.)	Alfalfa	For preemerge control of grasses and broadleaf weeds	Use incorporation equipment that will ensure some soil mxing with a minimum of damage to alfalfa. Treflan also has been registered on forage legumes used as cover crops in CRP.
Trifluralin 2 lb/A	Treflan TR-10 20 lb/A	Dormant stands (Dec. to Feb.)	Alfalfa	For preemerge control of grasses and broadleaf weeds	Has label clearance for annual control in established alfalfa where Treflan TR-10 can be activated by rainfall/overhead sprinkler irrigation or mechanical incorporation.
Paraquat	Gramoxone Extra 1½ pt/A	Dormant stands and between cuttings	Alfalfa Clovers	Dodder and seedling weeds	Contact herbicide, so coverage is important. There are many restrictions and exceptions on the Gramoxone Extra label, so it is very important to read the label before using. Gramoxone Extra is a restricted-use pesticide.

Note: Herbicides registered for application on legumes are continually changing, so it is essential to check yearly the current registered herbicides and to always read the label of the herbicide to determine its registration and use restrictions. (See current edition of Crop Protection Chemicals Reference [1993].)

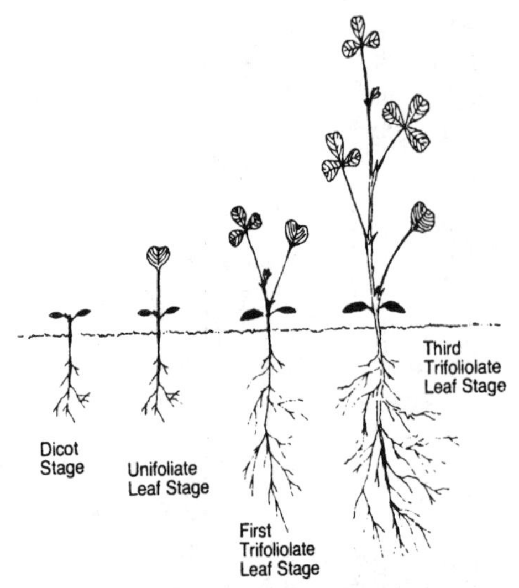

Fig. 7.4. Various seedling stages of alfalfa.

soil persistences of the individual herbicide. In general, effective weed control from these urea-type herbicides will last from 2 to 4 mo. Trifluralin (Treflan) is a dinitroanite herbicide that also has a dormant season registration on established stands. However, it has to be incorporated to be effective, and that currently is a major limitation for its use in established stands. It is best suited on sandy soils where satisfactory incorporation is usually possible. Paraquat (Gramoxone Extra) is a contact herbicide and is used both on dormant alfalfa and between cuttings. Because paraquat has no selectivity on alfalfa, there will be contact damage to alfalfa forage present at spraying, so timing has to be such that damage is minimized. This means that Gramoxone Extra is used more as a "salvage" treatment than as a conventional control.

GRASSES. Herbicides are often the best choice for controlling weeds and brush in pastures and rangelands. They are effective and selective in that weeds can be controlled without injuring desirable grasses. For example, the right formulation of 2,4-D at the correct time and rate will selectively kill most broadleaf weeds occurring on grasslands. Herbicides used on grasslands for control of weeds and brush can be grouped into foliage-applied and soil-applied (see Table 7.3).

Foliage-applied herbicides have to be applied when weeds are actively growing. This means that conditions at spraying such as soil moisture, soil temperature, and air temperature are very important and can mean the difference between no control and good control.

TABLE 7.3. Grassland herbicides by form of application

Foliage-applied	Soil-applied
Clopyralid (Reclaim)	Dicamba (Banvel)
2,4-D (several trade names)	Diuron (Karmex)
Dicamba (Banvel)	Hexazinone (Valpar)
Glyphosate (Roundup)	Picloram (Grazon PC)
Metsulfuron (Ally)	Tebuthiuron (Spike)
Paraquat (Gramoxone Extra)	
Picloram (Grazon PC)	
Triclopyr (Remedy)	

Several foliage-applied herbicides—2,4-D amine, metsulfuron (Ally), and dicamba (Banvel)—have label clearance for use on newly planted forage grasses. The amine formulation of 2,4-D at 0.45 kg/ha (0.5 lb/A) can be used to control many of the broadleaf weeds in most native and introduced grasses after they have six or more leaves. Ally currently is registered for use in the Conservation Reserve Program (CRP) in Colorado, Southern Idaho, Kansas, Montana, Nebraska, New Mexico, Oklahoma, South Dakota, Texas, Utah, and Wyoming. Bermudagrass, bluegrass (*Poa pratensis* L.), orchardgrass, smooth bromegrass (*Bromus inermis* Leyss.), timothy (*Phleum pratense* L.), and native grasses such as bluestems (*Andropogon* spp.) and grama grasses (*Bouteloua* spp.) have good tolerance. However, there are some timing restrictions on young grass seedlings, and Ally should not be used on new seedlings of tall fescue or ryegrass (*Lolium perenne* L.) since they are sensitive to it. Ally at 2.7 mL/A (0.1 oz/A) will control many of the annual broadleaf weeds but will not control the annual grasses. Banvel is also registered for broadleaf weed control on seedling grasses in CRP acres after grass seedlings exceed the three-leaf stage.

Broadleaf Weed Control in Grasses. Auxin-type foliage-applied herbicides (2,4-D, dicamba, picloram) for broadleaf weed control on established grasslands are applied when the weeds are small and easy to control and when desirable grass growth is such that it can respond to the control of the weeds. Spray time for control of summer broadleaf weeds varies according to the location and year, but generally it is done in April to May in the southern US and progressively later in northern areas, with applications as late as June or July in the northern US.

Use of herbicides on established grasslands for broadleaf weed control is as follows: 2,4-D continues to be the major herbicide used for broadleaf weed control in grasslands. It is effective on many of the weeds and still is cost-effective. Some weeds are equally susceptible to both the amine and ester formulations, but as a rule of thumb, it takes 0.34 kg (0.75 lb) of amine to equal 0.23 kg (0.5 lb) of ester. Low-volatile ester formulations are usually preferred for controlling weeds in grasslands because esters penetrate the leaves of some perennials better than amine formulations and are usually more effective on susceptible brush species like buckbrush (*Symphoricarpos orbiculatus* Moench) and sumac (*Rhus* spp.). It is critical that plants be actively growing for best results. Thus, soil moisture must be adequate, and air temperatures should be above 18°C.

1. *Formulations:* There are three general types of 2,4-D formulations. Amine formulations are actually amine salts, so they are soluble in water (they make an amber-colored solution). Amine formulations are often used since they are not volatile and are effective on small weeds and on weeds with nonwaxy leaves. HV (high-volatile) esters are not recommended for pasture spraying. They burn the leaves of weeds too quickly to control perennial weeds. Also, the vapor drift can cause problems to susceptible plants like tomatoes (*Lycopersicon esculentum* L.) and cotton (*Gossypium* spp.). LV (low-volatile) esters are the most popular formulation. They make a white, milky-looking emulsion when put into water. They are very effective on most pasture weeds and some brush species. LV esters are also the formulation to mix with fertilizer for joint applications.

2. *Time to spray:* The time to spray varies with the weed problem. Annuals, such as common broomweed (*Gutierrezia dracunculoides* [DC.] Blake), lanceleaf ragweed (*Ambrosia bidentata* Michx.), and bitter sneezeweed (*Helenium amarum* [Raf.] H. Rock), are best sprayed in early spring (late April to mid-May) when weeds are 5 to 10 cm (2 to 4 in.) tall. Some of the perennials, such as western ragweed (*A. psilostachya* DC.) and western ironweed (*Vernonia baldwinii* Torr.), also need to be sprayed in early spring. However, some perennials like horsenettle (*Solanum carolinense* L.) and Texas bullnettle (*Cnidoscolus texanus*) [Muella-Arg.] Small) need to be sprayed somewhat later (June or early July) at the time they are blooming. There are also some weeds, like musk thistle (*C. nutans* L.) and curly dock (*Rumex crispus* L.), that are best controlled by November and/or March spraying. The bottom line is that time of spraying must be tied to the weed problem.

3. *Rate to use:* The rate to use also depends on the weed problem. On annuals, we don't have to be concerned about getting herbicide movement into roots, so any rate that controls tops will kill the weeds. However, with perennial weeds, getting the herbicide down into the root system is very important. So here, rate has to be such that movement from the top of the weed into the root can occur.

Higher rates that burn the tops off too quickly will not be as effective on weeds as lower rates that allow herbicide movement into the roots. For example, only the 0.45 kg/ha (0.5 lb/A) rate of 2,4-D amine should be used on Texas bullnettle at blooming since higher rates and treating younger foliage cause browning of tops and very little herbicide movement into the roots. However, on ironweeds it is critical to use at least 1.34 kg/ha (1.5 lb/A) of LV ester on plants when spring growth is about 30 cm tall. So, again, the bottom line is that rate is tied to the weed problem.

4. *How to spray:* Since 2,4-D is a growth-regulating herbicide, it is very important to apply it to actively growing plants. That means that both temperature and moisture must be such that the weeds are growing. If it is too dry or too cold (less than 18°C), weed control may be unsatisfactory. Since 2,4-D is taken up by the leaves, it is important that application be uniform enough to get good coverage on the leaves. Best results are obtained with ground rigs with fixed booms and properly calibrated aerial applications. Good results are usually possible with boom jets, but applications with mist blowers are usually disappointing.

5. *What to expect:* Expect to see broadleaf weeds do stem bending the first week and the leaves to turn yellow and die by the second or third week. Grass release of introduced species such as bermudagrass will occur within 1 mo., and it is not uncommon to get 0.45 to 0.90 kg (1 to 2 lb) of grass released for every pound of weeds controlled. With native grasses, grass release is usually 0.45 kg of grass for each pound of weeds controlled. This can occur the first season if good summer moisture is available, or the yield response may result in the year following spraying (Powell et al. 1982).

6. *Some precautions*: All formulations of 2,4-D can drift with the wind during application, so don't spray with susceptible plants (grapes [*Vitis* spp.], tomatoes, and cotton)

downwind. Don't apply LV ester formulations near susceptible crops, no matter what direction the wind, since there is some volatility of LV ester during hot days. Also, animals tend to concentrate on sprayed areas where weeds have been controlled, so it is important to treat whole pasture units to avoid overgrazing.

Dicamba (Banvel) can be mixed with 2,4-D to get better activity on species like curly dock and thistles. Dicamba is an amine formulation and a commercial mixture; Weedmaster is a mixture of 0.45 kg of dicamba plus 1.3 kg of amine salt of 2,4-D per gallon. Refer to a Weedmaster label for a listing of rates to use with various weed species.

Picloram (Grazon PC) is another herbicide that can be mixed with 2,4-D to get better activity on some 2,4-D-resistant broadleaf weeds. Grazon P + D is a commercial mixture of 0.24 kg of picloram and 0.90 kg of 2,4-D per gallon, both as the triisopropanolamine salts. Refer to the Grazon P + D label for a listing of the rate to use on various weeds. Grazon PC is currently restricted for use only in New Mexico, Oklahoma, and Texas, but the Grazon P + D label has been expanded for use on rangelands and permanent grass pastures in Arizona, Louisiana, Alabama, Georgia, and Mississippi in addition to New Mexico, Oklahoma, and Texas.

Diuron (Karmex) has label clearance on sprigged bermudagrass. Apply 0.89 to 2.67 kg/ha (1 to 3 lb/A) of the 80% wettable powder after sprigging and before emergence of bermudagrass or weeds. Karmex can also be used postemergence at 0.45 to 0.89 kg/ha (0.5 to 1.0 lb/A) of the wettable powder on weeds less than 4 in. tall. One pint of WK surfactant is to be added to each 25 gal of spray solution with the postemergence application.

Metsulfuron (Ally) is a sulfonylurea herbicide just recently registered for selective weed control in grasses grown in pastures and rangeland. Bermudagrass, bluegrass, orchardgrass, smooth bromegrass, timothy, and native grasses have demonstrated good tolerance to Ally. Tall fescue is somewhat sensitive with reduced forage and seed production following Ally applications. Do not use on ryegrass pastures or in pastures with legumes. Ally is effective on many of the annual broadleaf weeds commonly found in cropland, plus it is also effective on some perennial and brush species like Allegheny blackberry (*Rubus allegheniensis* Porter) and multiflora rose (*Rosa multiflora* Thunb. ex Murr.). However, it is not very effective on some of the common pasture weeds like common ragweed (*Ambrosia artemisiifolia* L.), marshelder (*Iva xanthifolia* Nutt.), and horsenettle, so its future in forage weed control will most likely be as a mixture with other herbicides, such as 2,4-D. In addition to broadleaf weed control, Ally also has good activity on pensacola bahiagrass and is labeled for its control in established bermudagrass.

Brush Control in Grasses. Both foliar- and soil-applied herbicides are used for brush control. Foliar spraying for brush control is usually at least 1 mo later than spraying for broadleaf weeds since it takes longer for the soil temperature to warm in the root zone of the brush (Dahl and Sosebee 1984). Foliage-applied herbicides are usually broadcast by aerial application since brush is difficult to get over with ground rigs. Individual trees can also be sprayed with ground rigs with foliage-wetting sprays. This involves taking the herbicide broadcast rate recommended for an acre and putting it into a 100-gal mix with water. This mixture is then sprayed onto the crown of the tree until the leaves are about to drip. For brush spraying, an equal volume of diesel is often added to the herbicide before mixing with the water to aid in leaf penetration of the herbicide. This foliage-wetting type of spraying is very useful for spraying small clumps of scattered brush, but it is time-consuming. One variation of this method that has been successful has been to mix the acre recommended rate with only 50 gal instead of the 100 gal. Then the canopies are quickly sprayed (don't wet to drip since the spray is twice as concentrated). Another technique for getting good control is to return to the sprayed area as soon as browning results (usually 2 to 3 wk) and to respray the green leaves. With most of the foliage-applied herbicides, it is often necessary to retreat to obtain tree kill.

Soil-applied herbicides are taken up by the roots and quite often result in better tree kill than that obtained with foliage-applied herbicides. Since the herbicide must be moved into the soil with rainfall and then taken up by actively growing trees, the best time to apply soil herbicides to get maximum tree kill is March and April. An alternative time to get good tree kill is early in the fall before trees go dormant, providing soil moisture conditions exist and rainfall occurs to move the herbicide into the soil. With the more persistent urea-type herbicides like Spike, it is possible to get satisfactory tree kill with applications made any time of the year. However, to minimize damage to the grass with soil-applied herbicides, it is recommended that they not be applied while the grass is actively growing. Soil-applied herbicides are often used for individual tree treatment, and the rate is based on the tree size. Initially many of the soil-applied herbicides were applied as pellets or granules, but more recently, concentrated liquid formulations are also being promoted.

The response of the brush to the various herbicides can vary from no effect to total plant kill. A visual rating scale of 0 to 100 can be used with 0 being no effect, 10 to 30 being light damage, 40 to 60 being moderate damage, 70 to 90 being severe damage, and 100 being dead plants. "Brush control" would be any rating of 40 or above, which includes leaf defoliation and grass release. The length of this control might be for only one season, or if sufficient crown damage results (this would usually require a rating of 80), then brush control and grass release could be for several years. There is sometimes some confusion between brush control and eradication. Any level of brush response that releases grass is *brush control; eradication* refers to total control, which means all trees are killed. Herbicides used for brush control are as follows:

2,4-D is effective on some brush species like buckbrush, sumac, and willows (*Salix* spp.). The LV ester formulation is primarily used for brush control.

Dicamba's (Banvel) use is primarily limited to eastern persimmon (*Diospyros virginiana* L.).

Picloram (Grazon PC) is effective on several of the brush species.

Triclopyr (Remedy) is effective on many of the brush species and is generally considered to be the replacement herbicide for 2,4,5-T. A mixture of triclopyr and 2,4-D (Crossbow) is also labeled for brush control, as well as annual and perennial broadleaf weed control on rangeland, permanent grass pastures, and conservation reserve program acres.

Hexazinone (Velpar) and tebuthiuron (Spike) are urea-type herbicides that are soil-applied. Tebuthiuron has the greatest spectrum of herbicide activity and results in excellent root kill of brush on sandy sites. However, with clay soils and higher organic matter soils, its activity is poor due to soil adsorption.

Clopyralid (Reclaim) is registered for control of mesquite (*Prosopsis juliflora* [Sw.] DC.) and associated woody species and weeds on rangelands and permanent pastures in New Mexico, Oklahoma, and Texas.

RENOVATION. Paraquat (Gramoxone Extra) and glyphosate (Roundup) are labeled for use in renovation of forages. Also Gramoxone Extra is registered for use on dormant clovers in several of the northern states for desiccation of ryegrass, bluegrass, cheatgrass (*Bromus tectorum* L.), dogfennel (*Eupatorium capillifolium* [Lam.] Small), common chickweed (*Stellaria media* [L.] Vill.), and pinnate tansymustard (*Descurainia pinnata* [Walt.] Britt.). It is also registered for pasture uses that include (1) suppression of existing sod and undesirable emerged weeds prior to or at time of planting of grasses or forage legumes and (2) control of endophyte fungus-infested tall fescue. A lot of precautions and restrictions are listed on the Gramoxone Extra label, so it will be necessary to read the label before use. Roundup is registered as a broadcast spray for pasture and hay crop renovation and as a spot or wiper application for perennial weeds in pastures composed of bahiagrass, bermudagrass, bluegrass, smooth bromegrass, fescue, orchardgrass, ryegrass, timothy, wheatgrass (*Elytrigia* spp.), alfalfa, or clover. An 8-wk grazing restriction applies for the renovated areas, and a 2-wk grazing restriction is listed for spot and wiper applications.

PASTURES WITH LEGUMES. Spray legume-grass pastures only when serious weed problems exist. Some broadleaf weeds can be selectively removed from pastures containing legumes without serious damage to the legumes. 2,4-DB, sold as Butyrac, at 0.45 to 0.67 kg/ha (0.5 to 0.75 lb/A) can be used for selective removal of broadleaf weeds from some of the clovers and alfalfa. Be sure to read the herbicide label for details and restrictions.

Lespedeza and white clover are not as sensitive to 2,4-D as other legumes. This tolerance has made it possible to get selective control of some weeds such as ragweeds and bitter sneezeweed with 0.45 to 0.67 kg/ha (0.5 to 0.75 lb/A) of 2,4-D amine applied when weeds canopy over lespedeza and white clover in May.

Weeds growing in winter annual clovers such as large hop (*Trifolium campestre* Schreb.) and small hop (*T. dubium* Sibth.) and arrowleaf clovers (*T. vesiculosum* Savi) can also be controlled by delaying the spraying until the clovers have started producing seed (have brown seed heads). Seed production is not suppressed much with 0.45 kg/ha (0.5 lb/A) of 2,4-D amine even during the active vegetative stage (Conrad and Stritzke 1980). With hop clovers this will normally be in late May whereas with arrowleaf clover this will be in late June. Weeds will be harder to control at this time, so a 0.89 kg/ha (1 lb/A) application of 2,4-D LV ester will probably be required for satisfactory weed control. Banvel, Weedmaster, Grazon PC, and Grazon P + D are all damaging to legumes, so they should not be used on grass-legume pastures for weed control.

QUESTIONS

1. Outline the proper planting method for establishing a forage legume in your area. Indicate the fertility level, seeding rate, and seeding method, and explain why you have chosen this method.
2. Why is inoculation so important in crops of forage legumes?
3. Why is it so important to prepare an excellent seedbed when establishing a forage legume?
4. Why can we not seed less than 1.1 kg/ha (1 lb/A) and get a sufficient stand?
5. Compare the broadcast and band-seeding methods of establishing forages.
6. Why are weeds difficult to eliminate from forage crops?
7. What are the various methods of weed control in forages?
8. Why is it difficult to use a field cultivator to control weeds in an established stand of forages?
9. What are meant by preplant, preemergence, and postemergence treatments?
10. Outline an excellent preplant treatment for forage crops and a postemergence spray program for an established stand of legumes with a specific weed problem.
11. List two herbicides that may control a specific weed problem in grass pastures, and indicate the various precautions one must implement when using each.

REFERENCES

Berberet, RC, JF Stritzke, and AK Dowdy. 1987. Interactions of alfalfa weevil (Coleoptera: Curculionidae) and weeds in reducing yield and stand of alfalfa. J. Econ. Entomol. 80:1306-13.

Brown, BS. 1959. Band versus broadcast fertilization in alfalfa. Agron. J. 51:708-10.

Conrad, JD, and JF Stritzke. 1980. Response of arrowleaf clover to postemergence herbicides. Agron. J. 72:670-72.

Cope, WA. 1982. Inhibition of germination and seedling growth of eight forage species by leachates from seeds. Crop Sci. 22:1109-11.

Crop Protection Chemicals Reference. 1993. 9th ed. New York: Wiley.

Dahl, BE, and RE Sosebee. 1984. Timing—The Key to Herbicidal Control of Mesquite. Tex. Tech. Range and Wildl. Manage., n. 2.

Decker, AM, and TH Taylor. 1985. Establishment of new seedings and renovation of old sods. In ME Heath, RF Barnes, and DS Metcalfe (eds.), Forages: The Science of Grassland Agriculture, 4th ed. Ames: Iowa State Univ. Press, 288-97.

Dutt, TE, RG Harvey, and RS Fawcett. 1983. Influence of herbicides on yield and botanical composition of alfalfa hay. Agron. J. 75:229-33.

Egley, GH, and JM Chandler. 1983. Longevity of weed seeds after 5.5 years in the Stoneville 50-year buried-seed study. Weed Sci. 31:264-70.

Marten, GC, WF Wedin, and EF Hueg, Jr. 1963. Density of alfalfa plants as a criterion for estimating productivity of an alfalfa-bromegrass mixture on fertile soil. Agron. J. 55:343-44.

Miller, DA. 1984. Forage Crops. McGraw-Hill, 161-276.

Peters, EJ, and DL Linscott. 1988. Weeds and weed control. In AA Hanson et al. (eds.), Alfalfa and Alfalfa Management, Spec. Publ. 29. Madison, Wis.: American Society of Agronomy, 705-35.

Pike, DR, and JF Stritzke. 1984. Alfalfa (*Medicago sativa*)- cheat (*Bromus secalinus*) competition. Weed Sci. 32:751-56.

Powell, J, JF Stritzke, RW Hammond, and RD Morrison. 1982. Weather, soil, and 2,4-D effects on tallgrass prairies in Oklahoma. J. Range Manage. 35:483-88.

Rohweder, DA, and RW Van Keuren. 1985. Permanent pastures. In ME Heath, RF Barnes, and DS Metcalfe (eds.), Forages: The Science of Grassland Agriculture, 4th ed. Ames: Iowa State Univ. Press, 487-95.

Ross, MA, and CA Lembi. 1985. Applied Weed Science. Minneapolis, Minn.: Burgess Publishing.

Schreiber, MM. 1967. Effects of density and control of Canada thistle on production and utilization of alfalfa pasture. Weeds 15:138-42.

Sheard, RW, GJ Bradshaw, and DL Massey. 1971. Phosphorus placement for the establishment of alfalfa and bromegrass. Agron. J. 63:22-27.

Smith, D, RJ Bula, and RP Walgenbach. 1986. Seeding establishment and renovation of reestablished sods. In Forage Management, 5th ed. Dubuque, Iowa: Kendall/Hunt, 47-68.

Tesar, MB, and VL Marble. 1988. Alfalfa establishment. In AA Hanson, DK Barnes, and RR Hill, Jr. (eds.), Alfalfa and Alfalfa Improvement, Am. Soc. Agron. Monogr. 29. Madison, Wis., 303-22.

Tesar, MB, K Lawton, and B Kawin. 1954. Comparison of band seeding and other methods of seeding legumes. Agron. J. 46:189-94.

Van Keuren, RW. 1973. Alfalfa establishment and seeding rate studies. Ohio Rep. 58(2):52-54.

Woodal, TK, JF Stritzke, and RC Berberet. 1987. Effects of fall management and alfalfa weevil control on weed dynamics in alfalfa. In North Cent. Weed Control Conf. Proc., Kansas City, Mo., 42:70.

8 The Nutritive Evaluation of Forage

DWIGHT S. FISHER
Agricultural Research Service, USDA, and North Carolina State University

JOSEPH C. BURNS
Agricultural Research Service, USDA, and North Carolina State University

JOHN E. MOORE
University of Florida

A GENERAL background for understanding the methods of estimating forage quality and the limitations of the various techniques are presented in this chapter. The precise definitions of the terms *forage quality* and *nutritive value* have varied in scientific literature. We use *nutritive value* to refer to aspects of forage composition affecting nutrition independent of voluntary intake. *Forage quality* is used to include aspects of both nutritive value and voluntary intake. When reading scientific reports in this field of research, it is best to keep in mind that the authors may be applying different definitions to the terms (in general, the definitions of the terms are obvious from the context of their use).

The nutritive value of forages is dependent upon the animals that consume them. Forages are largely grown for, and used by, ruminant animals, and thus, an introduction to ruminant diet selection and digestion is provided in this chapter.

DWIGHT S. FISHER is Plant Physiologist, ARS, USDA, and Associate Professor of Crop Science, North Carolina State University. He holds the MS and PhD degrees from North Carolina State University. His major research is devoted to forage physiology and the modeling of forage and animal interactions.

JOSEPH C. BURNS is Plant Physiologist and Lead Scientist, ARS, USDA, and Professor of Crop Science and Animal Science, North Carolina State University. He holds the MS degree from Iowa State University and the PhD degree from Purdue University. His major research is devoted to evaluating organic fractions of grazed and stored forages as related to animal intake, digestibility, and performance.

JOHN E. MOORE is Professor of Animal Science, University of Florida. He received the MS and PhD degrees from Ohio State University. Since 1961 he has taught and conducted research in animal nutrition and forage quality evaluation.

RUMINANT DIET SELECTION AND DIGESTION

Generally, ruminant animals such as cattle *(Bos taurus* and *Bos indicus),* sheep *(Ovis aries),* and goats *(Capra hircus hircus)* collect their diets while grazing pastures of grass and legume species that vary in nutritive value and in the quantity of plant material available for grazing *(herbage mass).* In contrast to herbivores such as horses *(Equus caballus),* ruminant animals have no upper incisors and instead have a dental pad at the end of the hard palate (Hofmann 1988). They utilize the tongue, lips, lower incisors, and the dental pad to gather plant material from pasture and sever it with a ripping action. Sheep and goats have narrower muzzles than cattle and, in addition, have a split upper lip that permits greater selectivity in collecting their diet. As cattle graze, forage is collected toward the rear of the oral cavity. This process continues until enough forage is harvested to form a bolus (Fig. 8.1). The composition and size of the particles making up the bolus, along with any supplements, form the basis for the nutrition of the animal.

Ruminant species can vary in their dietary selections even within the same habitat as a result of differences in forage preference. In

Fig. 8.1. Bolus of ingesta collected from the esophagus of a cannulated steer.

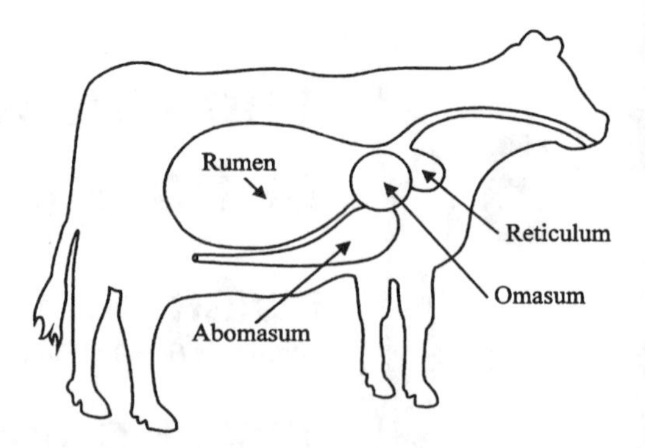

Fig. 8.2. The locations of the four compartments of the ruminant stomach.

general, cattle prefer grass, sheep prefer both grass and forbs, while goats prefer forbs and shrubbery. These preferences are a reflection of the ecological niches occupied by the wild progenitors of these domesticated livestock (Van Soest 1982). Ruminants have adapted to many ecological niches, and they may be classified based upon dietary selection within a broad range from those that select grasses to those that select fruit and seeds (Hofmann 1988). Various anatomical adaptations that facilitate herbivory are found throughout the digestive tract. The primary anatomical adaptation of the ruminant is a stomach made up of four compartments (Fig. 8.2). The compartments are the rumen, reticulum, omasum, and abomasum. Digesta typically flow from the rumen to the other compartments in that order. However, rumination and mixing between the rumen and reticulum and the rumination of regurgitated boli are important parts of the digestive process (Fig. 8.3).

Rumination consists of regurgitation of a portion, or bolus, of previously consumed feed, swallowing excess liquid regurgitated with the bolus, chewing the bolus for a few seconds to a minute, and swallowing the bolus again. Rumination results in mechanical fragmentation and creates sites for attack by rumen microorganisms. Microbial digestion aids in fragmentation by weaking the particles prior to rumination.

The rumen is the largest of the stomach compartments. In large cattle it may have a capacity of 150 L. The rumen functions as a fermentation vat with bacteria, protozoa, and fungi active in the fermentation. Many species of microbes have been identified, and often a species ferments a single substrate such as cellulose or starch. Therefore, the composition of the diet affects the type of microbial population and the rates of digestion. For example, if the diet is high in starch, starch digestion will occur rapidly, but digestion of cellulose will occur even more slowly than it would with a diet high in structural carbohydrates. The anaerobic fermentation of

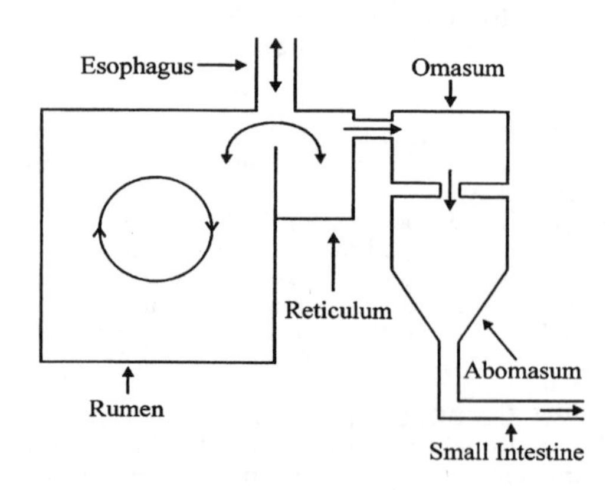

Fig. 8.3. The functional relationships of the four compartments of the ruminant stomach.

the microbes is incomplete, and energy-rich end products pass from the microbes into the rumen fluid. The major end products are the volatile fatty acids: acetic, propionic, and butyric acids. Heat and methane are also produced but cannot be used for animal production. The fermentation of structural carbohydrates produces a high acetic to propionic acid ratio, but fermentation of starch and simple sugars produces a low ratio. Saliva provides sodium bicarbonate to buffer the production of volatile fatty acids in the rumen. Rumen fluid pH is normally 6.5 to 6.8 on a forage diet but may fall to dangerous levels (<5.0) on high-starch diets.

There are muscles that divide the rumen and form what are called *pillars*. These muscles contract periodically, mixing the fluid contents. When pasture and hay are consumed, a thick mat of forage floats on top of the rumen fluid. The processes of mechanical and microbial degradation reduce the size of forage particles gradually until they settle to the bottom of the rumen and are transferred by ruminal contractions to the reticulum.

Because wires, nails, and other solid materials sometimes collect in the reticulum, it has been called the *hardware stomach*. The reticulum functions to control the movement of feed particles. For example, the reticulum forms boli prior to regurgitation and rumination. In addition, the opening between the reticulum and omasum controls the size of forage particles passing through to the omasum. Retained particles pass back and forth between the rumen and reticulum and may be ruminated.

The omasum has also been called *manyplies* because of the internal folds. The internal surface area is high compared to the volume. The omasum dehydrates the digesta passing through from the reticulum to the abomasum.

The abomasum is sometimes called the *true stomach* because it secretes hydrochloric acid and the protein-digesting enzyme pepsin. The abomasum is the only stomach compartment in domestic ruminants to synthesize a digestive enzyme. However, pepsin digests protein and has no effect on structural carbohydrates.

Predicting the end result of this complex digestive system is the key to accurate estimates of forage quality. Accurate estimates of forage quality are important to the design of efficient and economic production systems for both grazing and confinement-fed ruminants. In spite of the essential role of these estimates in animal production, most of the available assays of nutritive value and voluntary intake are expensive or difficult to obtain or are of limited use in predicting the biological response of the animal. The ultimate assay of forage quality is animal response; however, the high requirements for materials and time needed to conduct an animal response trial have stimulated the search for alternative methods of determining forage quality.

NUTRITIVE EVALUATION BY COMPONENTS

The constituents of forage can be divided into two main categories: (1) those that make up the structural components of the cell wall and (2) those existing in the cell contents (Fig. 8.4 and Table 8.1). Although most cell contents are water soluble, starch, lipids, and some proteins are examples of cell contents that are not soluble or only slightly soluble in water. Structural carbohydrates are of particular importance because their digestion is dependent on enzymes produced by gastrointestinal microorganisms. On the other hand, nonstructural carbohydrates are digested readily by enzymes of both animal and microbial origin. The physical organization of forage cells into various tissues is quite complex. The chemical constituents of forages are not distributed uniformly among different plant organs and tissues, and wide differences exist among forages in both composition and physical structure. (See Chap. 6, Vol. 2.)

Several schemes of analysis have been developed to describe the gross composition of forages and other feeds in order to predict nutritive value and voluntary intake. The two

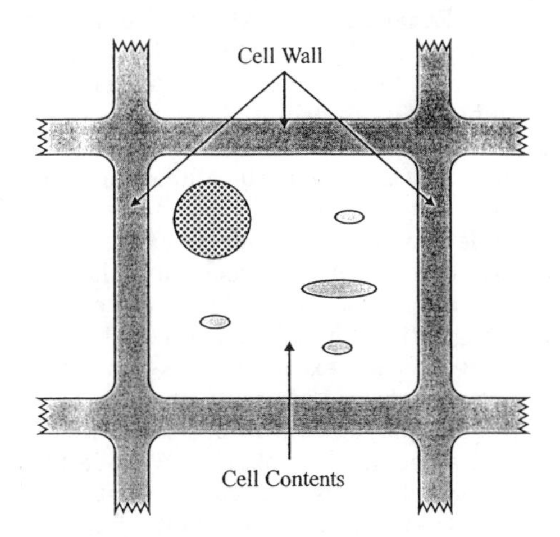

Fig. 8.4. The two primary divisions of forage are based on the cell wall component and the cell solubles.

TABLE 8.1. Nutritive constituents of forage and limitations to their utilization by ruminants

Component	Availability	Factors limiting utilization
Cellular contents		
Soluble carbohydrates	100 %	Intake
Starch	>90 %	Intake and rate of passage
Organic acids	100 %	Intake and toxicity
Protein	>90 %	Fermentation and loss as ammonia
Pectin	>98 %	Intake and rate of passage
Triglycerides and Glycolipids	>90 %	Intake and rate of passage
Plant cell wall		
Cellulose	Variable	Lignification, cutinization, and silicification
Hemicellulose	Variable	Lignification, cutinization, and silicification
Lignin, cutin, and silica	Indigestible	Not degradable
Tannins and polyphenols	Possibly limited	Generally not degraded

major systems for the analysis of forage are discussed first followed by brief discussions of vitamins, minerals, and antiquality components. The importance of voluntary intake in establishing forage quality and the difficulty of measuring intake are also discussed. Such analyses are necessary in order to formulate rations for specific animals. In addition to gross constituents, feeds and forages contain many constituents that, although in low concentration, have important effects upon nutritive value. Some of these constituents are vitamins and required minerals, but others have antiquality effects.

Proximate Analysis. Since the mid-1800s, the proximate analysis has been used widely in the evaluation of feedstuffs. In addition to the five components shown in Figure 8.5, water content may be determined by drying at 105°C. Ash is the residue remaining after burning, or combusting, at 600°C. Crude protein is determined by analyzing for nitrogen (N) by the Kjeldahl method and multiplying by 6.25. Ether extract is the total of all compounds that can be extracted with hot diethyl ether. Crude fiber is the organic matter that is insoluble in weak acid and weak alkali. Nitrogen-free extract is calculated by subtracting from 100 the sum of the other four components.

Crude protein and crude fiber have been widely used to classify feeds, and their use has made it possible to formulate diets. Roughages (including forages) have been defined, with some exceptions, as those feeds having more than 18% crude fiber in the dry matter, as opposed to concentrates, which are defined as having less than 18% crude fiber. In general, there is a negative relationship between concentrations of crude fiber and crude protein among forages and byproducts, but considerable variation exists.

Crude fiber is generally thought to be less digestible than N-free extract, but with forages there is no clear distinction. In some cases, N-free extract may actually be less digestible than crude fiber because most of the hemicellulose and part of the lignin found in the cell wall is included in N-free extract (Figs. 8.5). Hemicellulose is a structural carbohydrate, and its digestibility is similar to that of cellulose. Lignin is associated with cellulose and hemicellulose in the cell wall. Lignin is usually indigestible, but its major effect is to inhibit the digestibility of cellulose and hemicellulose. Therefore, digestibility of N-free extract is potentially high, but when hemicellulose and lignin percentages are high, the digestibility of N-free extract is depressed to values similar to or lower than

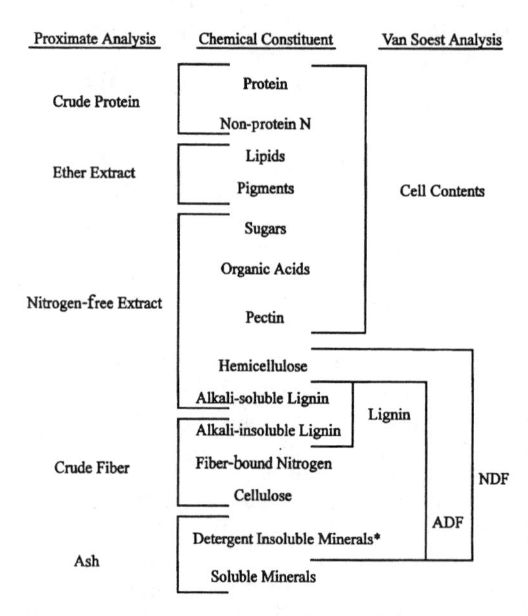

Fig. 8.5. Contrast of proximate and Van Soest methods (Van Soest et al. 1991) of forage analysis.

those for crude fiber. Proximate analysis does not always adequately characterize the nutritive value of forage carbohydrates.

Van Soest Analysis. Peter J. Van Soest developed an alternative to the proximate analysis while working at the USDA laboratories in Beltsville, Maryland, during the 1960s, and the system is widely used to analyze forages (Van Soest and Robertson 1980; Van Soest et al. 1991). This chemical method recognizes the distinction between cell walls and cell contents (Fig. 8.5). The most important procedure involves extracting a forage sample with a neutral detergent solution: the solubles are primarily the cell contents, and the insoluble residue (neutral detergent fiber [NDF]) is an excellent estimation of the total structural, or cell wall, constituents (cellulose, hemicellulose, and lignin). Neutral detergent fiber varies from roughly 10% in corn grain, which is nearly 90% digestible, to approximately 80% in straws and tropical grasses, which generally range from 20% to 50% in digestibility.

A different detergent solution is acidified with sulfuric acid and used to estimate acid detergent fiber (ADF). Acid detergent fiber is an insoluble residue, like NDF, but does not include all cell wall constituents because hemicellulose is soluble in the acid detergent solution. Acid detergent fiber ranges from approximately 3% in corn grain to 40% in mature forages and 50% in straws. Acid detergent fiber values are slightly higher than are those for crude fiber because all the lignin and some ash is included in the former.

Van Soest developed an improved analysis for lignin by treating ADF with either 72% sulfuric acid or permanganate; sulfuric acid hydrolyzes cellulose, whereas permanganate oxidizes and solubilizes lignin. Sample preparation is critical in the estimation of lignin in that a component of what is measured as lignin may actually be created by the method. For example, when plant samples are dried, sugars and amino acids may be complexed by the Maillard reaction to form artifactual lignin-like compounds. Freeze-drying minimizes this effect, and samples prepared by this method are preferred when accuracy is important. When samples are freeze-dried the moisture is removed under vacuum while the samples are still frozen.

Neutral detergent fiber is most important because it estimates that fraction of forage that, if it is to be metabolized by the animal, must first be degraded by gastrointestinal microorganisms. It is a much better measure of forage "fiber" than crude fiber; however, forages contain fiber that is not uniformly digestible. In fact, the digestibility of energy often is related closely to the digestibility of NDF in many forages, especially grasses. Microscopic studies show that the walls of some cells (such as mesophyll) undergo rapid and complete degradation by microbial and mechanical actions in the gastrointestinal tract, while lignified walls of other cells (such as xylem, bundle sheath, and epidermis) are more resistant to degradation (Fig. 8.6). Within the vascular bundle, phloem cells are generally digestible, but the bundle sheath may provide a barrier in some forages and reduce or slow down the digestion by rumen microbes. A more thorough understanding of these phenomena will lead to improvement in forage quality and forage utilization. None of the quality analyses based on chemical solubility consider rate of digestion.

The Van Soest method may also be used to predict digestibility by means of what has been termed the *Summative Equation*. The digestible dry matter is estimated to be equal to the sum of Equations (8.1) and (8.2).

$$0.98 \times (100 - \text{NDF}) \qquad (8.1)$$

$$\text{NDF} \times \qquad (8.2)$$

$$147.3 - 78.9 \times \{\text{Log}_{10}[(\text{lignin/ADF}) \times 100]\}$$
for sulfuric acid lignin

or

$$180.8 - 96.6 \times \{\text{Log}_{10}[(\text{lignin/ADF}) \times 100]\}$$
for permanganate lignin

The sum should be corrected for metabolic

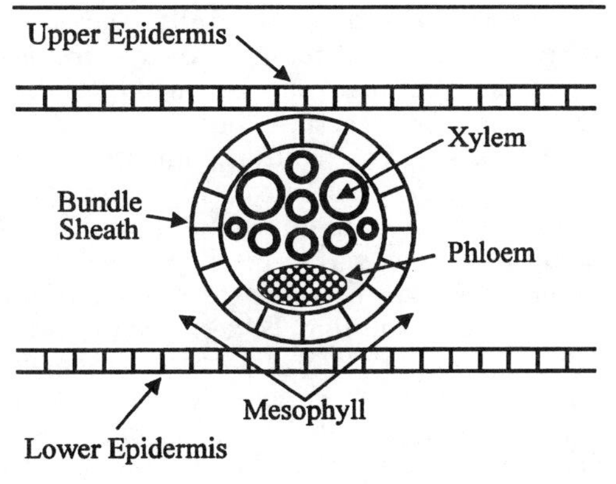

Fig. 8.6. Representation of the major cell types found in forage leaves.

fecal dry matter (subtract an average of 12.9). When necessary, the sum may be corrected for silica (subtract 3 times the silica content) and artifact lignin (subtract artifact lignin content). Generally, the summative equation does not have the accuracy and precision of some of the bioassays, but it eliminates the expense of maintaining animals in order to estimate forage quality. Equation (8.1) estimates the materials soluble in neutral detergent (100 – NDF), and these are assumed to be 98% digestible. Equation (8.2) estimates the fraction of NDF that will be digested. This is based on an empirical regression relationship between NDF digestibility and the lignin to ADF ratio. However, the regression equations may not be applicable to all forages, especially tropical grasses, because the relationship between digestibility of NDF and the lignin concentration in ADF differs among forages.

Vitamins and Minerals. Forages are important sources of vitamins A and E and the minerals essential for animal growth. (See Chap. 4 for a discussion of essential nutrients. Also see Chap. 5, Vol. 2.) There are no rapid and commonly used methods of analysis that permit the evaluation of vitamin or mineral values of feeds. Determinations of the adequacy, deficiency, or toxicity of vitamins or minerals in feeds are often based on observations of animals for specific signs or symptoms. However, subclinical toxicity or deficiency may depress intake, digestibility, or animal performance, with no signs of acute problems. Certain animal tissues and blood may be analyzed for nutrient concentration in order to determine the status of an animal with respect to a given nutrient.

Antiquality Factors. Generally speaking, some antiquality factors are always present in a forage, but many are tolerated if concentrations are below certain limits. Molds, dust, weeds, alkaloids, tannins, nitrates, and cyanides may cause toxic reactions, may cause the feed to be unpalatable, which reduces intake, or may decrease microbial activity in the rumen, which reduces digestibility. The net result may be decreased feed intake, digestibility, and animal performance. For example, there is an endophytic fungus (*Acremonium coenophialum* Morgan-Jones & Gams) that grows symbiotically between cells in the leaf sheath of tall fescue (*Festuca arundinacea* Schreb.). The plant-fungal combination produces toxins that are transported to the leaf blade and generally result in decreased animal performance. Physiological symptoms are associated with a reduced heat dissipation capacity of the animal. Accordingly, symptoms during the summer are more pronounced than during the spring or fall. The fungus has no influence, however, on estimates of nutritive value of the forage. Another example is the presence of certain alkaloids in some cultivars of reed canarygrass (*Phalaris arundinacea* L.). Chemical and bioassay estimates of nutritive value may not reveal any problem with the forage, but daily performance of steers and sheep are greatly reduced when these alkaloids reach certain thresholds and animals show physiological stress. Analyses for these factors in feeds requires a specific procedure for each class of compounds and may be time-consuming and expensive. Animal tissue analyses are used sometimes to diagnose these problems.

The presence of tannins in forages may improve disease and insect resistance of plants but also may decrease microbial degradation of dietary protein in the rumen. The latter effect may enhance animal performance by reducing the tendency for bloat and by decreasing the loss of N as ammonia due to fermentation in the rumen. Then the N is available as protein for digestion in the lower stomach and small intestine, where ammonia loss is not a factor and efficiency of utilization is higher. Excessive levels of tannin may, however, decrease rumen microbial populations and reduce animal performance by decreasing digestion rates and hence intake.

Lignin is considered an antiquality factor because it limits the extent of digestion by complexing with cellulose and hemicellulose to render lignin indigestible. The concentration of lignin is relatively easy to determine compared with some other antiquality factors and can be used to give an indication of the indigestible fraction in many forages. However, legumes and grasses differ substantially in concentrations of lignin and in the effect of lignin on digestibility. For example, the legume birdsfoot trefoil (*Lotus corniculatus* L.) may be 60% digestible while containing 10% lignin, whereas tall fescue in the same field may have only 5% lignin but is also limited to a digestibility of 60%. This occurs because of the distribution and concentration of lignin within individual cell types and because of differences in the relative proportions of various cell types (Fig. 8.6).

EVALUATION OF NUTRITIVE VALUE BY BIOASSAY

Bioassays utilize living organisms in some way to estimate forage quality. Generally these methods are based on production of milk, weight gain, or solubilizing of digestible components. Cattle, sheep, goats, rabbits (*Sylvilagus* spp.), meadow voles (*Microtus pennsylvanicus*), and crickets (*Nemobius* spp.), as well as wild ruminants such as miniature antelope (*Antilocapra* spp.) and deer (*Odocoileus* spp.) have been used as experimental animals. Ideally, the species, class, and age of the experimental animal should be the same as that of the animal to which the data will be applied. Often this is not possible because of the high cost, limited facilities, and small quantities of the experimental forage. A trial with steers as the experimental animal should be run with a minimum of six to eight animals per forage for a minimum of 90 d if animal performance is of interest. Such a trial would require from 3000 to 4000 kg of each experimental forage. This quantity of forage is often not available.

An extreme example of using alternative animals to avoid the expense of the larger animals to which the data will ultimately be applied is the use of meadow voles or crickets as the experimental animals when the application is to beef cattle. A less extreme example is the use of sheep as the animals for experiments applicable to cattle.

Digestibility may be estimated by the use of a traditional digestion trial. Four to six animals are required to test each forage. The quantity of dry matter offered as feed (OF) and the dry weight of the refused feed (RF) are determined along with the total dry weight of feces (F). The dry matter digestion coefficient may then be calculated as

$$(OF - RF - F)/(OF - RF) \quad (8.3)$$

Decisions must be made about the target level for feed refusals (0% to 20% of intake), the number of feedings per day (one or more), and the duration of the initial adjustment period (7 to 14 d).

Two additional bioassays of the nutritive value of forages are *in vitro* (in glass) and *in situ* (in position) disappearance. In these procedures, rumen microbes are utilized as the "animal" to estimate the digestibility by the disappearance of forage. Forage is incubated, either in test tubes (in vitro) or in synthetic cloth bags suspended in the rumen of cannulated animals (in situ) for a fixed period of time (usually 48 h). After incubation, samples are removed and dried, and the residual weight is determined. Disappearance is assumed to be equal to digestion. The concept of the artificial rumen or in vitro rumen fermentation technique for the evaluation of the nutritive value of forages was reported initially by Pigden and Bell in 1955. Application of the two-stage in vitro measurement with an initial microbial fermentation followed by a second acid-pepsin stage was proposed by Tilley and Terry (1963) and has become the standard for most in vitro rumen fermentation studies. It is often referred to as the "Tilley and Terry procedure," named after the two scientists who developed it, but most laboratories have modified the original procedure.

The in vitro and in situ procedures are fairly accurate, but substantial errors can result in some cases. For example, in very fibrous feeds, a 48-h incubation is not long enough to allow the extent of digestion that would occur in the rumen. This is because, with very fibrous feeds, the average length of time that a meal of forage remains in the rumen is greater than 48 h. Conversely, in feeds with very little fiber, the 48-h incubation is too long and allows for a greater extent of digestion than would occur in vivo because the average length of time that a meal of forage remains in the rumen is less than 48 h. An additional consideration is that the in vitro system is closed and the in situ system is open. This can affect the response to antiquality factors. If a very high level of tannin is present, it may reduce the in vitro disappearance and yet be diluted in the rumen in an in situ system, resulting in a higher estimate of digestibility. Both methods can be useful, but the user and interpreter of the information should be aware of discrepancies associated with the way the data were collected.

Also, enzymatic procedures have been developed that are hybrids of the chemical and bioassay procedures. Mixtures of cellulases and hemicellulases are used to digest the cell wall and are preceded usually by an acid-pepsin hydrolysis, which simulates protein digestion. This approach has the advantage of not requiring the maintenance of an animal for rumen inoculum, but hydrolysis is usually not as complete as in the in vitro procedure with a mixture of microorganisms. Regression equations are required to convert the values obtained into equivalent in vitro dry matter disappearance or in vivo digestion values.

This is not always effective for dissimilar forages but is usually satisfactory within a forage species.

Near Infrared Reflectance Spectroscopy. Most of the bioassays and chemical analyses used to predict the nutritive value of forage may be estimated using near infrared reflectance spectroscopy. Near infrared reflectance spectroscopy (NIRS) is rapidly becoming the preferred method of analysis because it is much quicker and is relatively inexpensive per analysis. Samples are scanned using near infrared light, and based on variation in the reflected light at multiple wavelengths, regression equations are developed to predict the constituent composition of interest.

INTAKE AS A COMPONENT OF FORAGE QUALITY

Because animal performance is linked strongly to intake of digestible dry matter, and intake of digestible dry matter is the product of digestibility and dry matter intake, it is essential to develop methods of predicting forage intake in order to optimize production systems (Moore 1980).

Measuring intake in confinement-fed animals is fairly straightforward. In fact, intake is often measured in the same experiments that measure digestibility. Feed intake is calculated simply as feed offered minus feed refusals. The mechanics of measuring intake are complicated by all the factors that complicate digestibility trials. A factor of primary importance is the decision about the level of feed to offer above intake. Feeding at high levels allows large quantities of refused feed and may be appropriate for measuring the potential dry matter intake of a feed, but it may not be an economic level of feeding. A range in target levels of feed refusal is an ideal approach for solving this problem.

Estimating the forage intake of free-grazing animals is more difficult. In research, intake of grazing animals is estimated using inert markers. Inert markers fall into two broad categories: internal and external markers. Internal markers such as plant pigments and lignin occur as a natural components of the feed and are assumed to be indigestible. The increase in concentration of these compounds in the forage residue as it passes through the digestive tract can be used to estimate the digestibility of the diet, but in grazing trials, internal markers are not useful for estimating intake. External markers do not occur as a natural part of the diet and are generally fed to the animal daily. The external marker used most frequently is chromic oxide (Cr_2O_3). The concentration of the marker in the feces, along with a knowledge of how much marker was fed daily, may be used to estimate fecal output. For example, if 1.0 g of external marker is fed per day and the concentration of the marker in the feces is 0.5 g kg^{-1}, then fecal output is equal to

$$1.0 \text{ g d}^{-1}/0.5 \text{ g kg}^{-1} = 2.0 \text{ kg d}^{-1} \qquad (8.4)$$

An estimate of digestibility from an internal marker or in vitro method can then be used to estimate intake by dividing fecal output by indigestibility (i.e., 100 – digestibility). For example, if the digestion coefficient was estimated to be 0.5 (50% digestible), intake would be equal to

$$2.0 \text{ kg d}^{-1}/0.5 = 4.0 \text{ kg}^{-1} \qquad (8.5)$$

Limitations to Intake. Ruminants adjust intake in response to their energy requirements, but because of the large quantities of indigestible fiber in most ruminant feeds, the quantity of feed they consume may be limited physically by the capacity of the rumen (Fig. 8.7). These observations have suggested two theoretical mechanisms that may regulate intake. *Gut fill* (distension) can limit intake before animals consume quantities of feed large enough to meet energy demands. The particle size of much of the ingested diet is too large to pass from the rumen, and the processes of digestion and rumination are instrumental in reducing particle size and making passage from the rumen possible. The rate at which digesta pass from the rumen is very closely related to feed intake. Factors such as particle size, cell wall content, degradability, and the hydration properties of the ingested feed are important in determining the rate at which digesta flow from the rumen. The composition and thickness of the cell walls interact with tissue anatomy to affect rate of degradation. The physical reduction in particle size is important because the microbes need access to the cell walls in order to digest them. Cellulolytic enzymes are synthesized by the microbes and digest the surfaces of the cell walls.

When diets with large amounts of soluble carbohydrates such as starch and sugar are fed, dietary digestible energy concentration

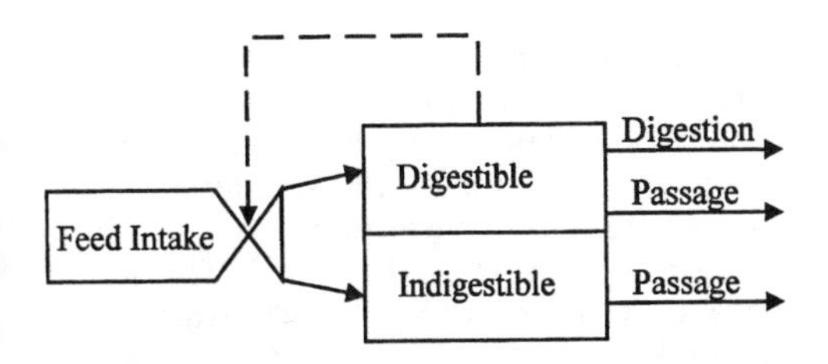

Fig. 8.7. A simple model of digestion and the regulation of intake by physical limitations.

may limit dry matter intake. The energy feedback has been termed *chemostatic* and may result in decreased dry matter intake with increased digestibility of very highly digestible diets consisting of concentrates. In forage diets, intake does not decrease with increasing digestibility. On the contrary, forage intake generally increases linearly as digestibility increases. The correlation between intake and digestibility is generally positive but may be high or low depending on the forages under consideration. When maturity within genotype is the major factor affecting quality within a small group of forages, then the correlation is high. However, with larger, more heterogeneous groups that include variation in genotypes (species), season of growth, and maturity, the correlation is low. These differing relationships show that only a portion of the variation in intake is explained by variation in digestibility. The positive correlation between intake and digestibility does not mean that chemostatic effects are irrelevant in the determination of forage intake. Distension of the gut and chemostatic effects should be viewed as interacting with the whole range of forage digestibilities in determining voluntary (ad libitum) intake, with distension being the major factor under many conditions.

GRAZING ANIMAL PERFORMANCE AND THE PLANT-ANIMAL INTERFACE

Grazing animals are affected by all the previously mentioned aspects of forage quality; however, dynamic plant-animal interactions further complicate the nutritive evaluation of pastures (see Chap. 7, Vol. 2). The behavior of grazing animals has a major effect on both the quantity and nutritive value of their diet. Their diet is determined by their selection of what portion of the landscape to graze (i.e., valley versus hillside), the plant community within the landscape, the specific portion of the plant community, and the actual bite within the plant canopy.

Intake may be estimated by the labor-intensive method of estimating bite weight, biting rate, and grazing time. The product of these three variables is equal to intake. This procedure has been useful over only short time periods. Bite weight can be measured using esophageally cannulated animals. Counting the bites taken by the animal while collecting extrusa makes it possible to calculate bite weight as the weight of forage in the bag divided by the number of bites during its collection. Mouth movements may be divided into those that collect forage from the plant canopy (gathering), those in which forage is severed from the canopy (bites), and those that aid in bolus formation for swallowing (chews). Biting severs forage and is the mouth movement of interest for calculating intake while the number of gathering and chewing mouth movements gives information on the interaction of the animal with the plant canopy. A high ratio of biting to gathering and/or chewing mouth movements indicates that the canopy is short to moderate in height or of high bulk density. A low ratio of biting to gathering and/or chewing mouth movements indicates that the canopy is tall and less dense and relatively more mouth movements are required for bolus formation prior to swallowing.

In general, animals graze from the upper portion of the plant canopy and have a strong preference for green leaves. For example, the effect of having only a small percentage of green forage in the canopy (a high dead percentage) is a low bite weight. Because leaf area is the primary source of pasture growth during most of the season, there are a number of interactions between the amount of green leaves available, plant growth rates, and amount of forage consumed by the grazing animals.

As animals remove green leaves by grazing, plant growth slows in response to the decreased leaf area. The regrowth curve of forages is sigmoidal. In other words, growth is slow when leaf area is small, it increases rapidly as leaf area expands, and then growth slows as leaf area reaches a maximum due to senescence of older leaves. Rotational defoliation schemes are designed to take advantage of the regrowth characteristics of forages by keeping the forages, as much as possible, in a rapid growth phase. This is accomplished by defoliating as growth slows near the maximum leaf area and removing the animals when most of the leaf area has been removed.

At the beginning of the defoliation period, the carbohydrate stores of forages are usually optimum, and regrowth begins rapidly once the animals are removed.

A continuous defoliation scheme may be conducted in which leaf area is managed by adjusting stocking rate to keep leaf area adequate for rapid growth. This management scheme may be successful, but the leaf area removed by grazing is generally from the younger material, and the average age of the leaf population in the pasture increases with a consequent decline in productivity per unit area. As a result of the interactions of plants and animals, even a properly managed grazing system only approaches, but does not surpass, the dry matter productivity of a hay system under the same conditions. The lower production is due partly to more efficient harvest with haying equipment and partly to losses from trampling and fouling of forage when grazed. Removal of stemmy and dead portions of a forage crop can be accomplished by hay harvest when they would be nearly impossible to remove by grazing. However, forage harvested as hay may provide a lower-quality diet and may be as much as 10 to 20 units lower in digestibility than the diet of an animal grazing selectively in the same land area. The higher quality occurs because animals select green leaves that may be much higher in nutritive value than those of the plant canopy as a whole and, also, because of the loss of digestibility that occurs as plant material is dried or ensiled. As a result, per unit land area production of marketable product (meat, milk, or fiber) may be greater from the grazing system.

Because the performance of the animal is the ultimate test of forage quality, and determination of diet quantity and nutritive value is so difficult in pasture studies, animal performance over 60 d or more is sometimes used as the sole measure of pasture productivity and nutritive value. This procedure is most often used in a research setting with *continuous stocking* (previously called *continuous grazing)* in which animal numbers are varied to maintain a preset quantity of forage on offer. This type of grazing is called *put-and-take stocking* because of the addition or removal of animals to regulate the forage quantity. A number of "tester" animals are used to determine average daily gain (ADG). These animals remain on the pasture for the entire season, and the weight gain (or loss) is assumed to apply to the put-and-take or "regulator" animals that are added and removed from time to time. The ADG is assumed to be an indication of the forage quality. However, this assumption is true only when the quantity of forage on offer is carefully regulated so that forage intake and performance are not limited by a lack of grazable forage. The animal mass that the pasture supports (testers and put-and-take animals) during the season is assumed to indicate pasture productivity, and an estimate of the effective feed unit production of the pasture can be made by adding the feed units required to maintain the animals that grazed the pasture during the year or period of interest plus the feed units required to produce gain at the level observed in the testers.

This method is best when grazing is conducted at several levels of forage on offer but is often used at a single level. The reason several levels are desirable is because the optimum level may vary for each forage being tested. The results of a study with only one level of forage on offer may be applied only to the level of forage tested. In practice, however, our knowledge of the requirements for plant growth (i.e., adequate leaf area) should prevent gross over- or undergrazing of a forage, and so the results of the trial should have wider applicability. The same relationships apply to experimental inputs that must be selected to conduct an experiment. For example, the selection of a single rate of N fertilization in a grazing study is likely to favor one species over another. It is not possible to control everything that may influence the treatment of interest, so care must be exercised in the interpretation of the results.

SUMMARY

The ultimate assay of the quality of forage is animal performance under carefully controlled conditions. Time and expense make animal evaluation difficult, and as a result, many methods of estimating the performance of animals fed forages of varying quality have been devised. Methods of estimation have been developed that range from those using chemical procedures to those using various bioassays. Bioassays include the use of smaller animals as models for the animals to which the data will ultimately be applied. Most studies of forage quality attempt to estimate digestibility, but estimates of intake are more difficult to make. Variation in intake of digestible dry matter accounts for most of the variation in projecting how nutritional a forage is. Even when estimates of digestibility are adequate, estimates of intake may be inaccurate, and as a result, accurate predictions of animal performance generally are not available. When animals are grazing, the estimation of pasture quality is complicated by the interactions among plant composition and growth, the grazing behavior of animals, and the management inputs of producers.

QUESTIONS

1. What characteristic feature in the dentition of ruminants separates them from herbivores such as horses?
2. What four steps make up the process of rumination?
3. How do proximate analysis and the Van Soest fractions differ in their determination of "fiber"?
4. How do bioassays differ from proximate analysis and the Van Soest fractions? Which is preferred?
5. How do the in vitro and in situ bioassays differ from other bioassays?
6. What are two important theoretical mechanisms proposed as limiting intake of forage by ruminants?
7. If time and money were of no concern, what assay of forage quality would be preferred?

REFERENCES

Hofmann, RR. 1988. Morphophysiological evolutionary adaptations of the ruminant digestive system. In Aspects of Digestive Physiology in Ruminants. Ithaca, N.Y.: Cornell Univ. Press, 1-20.

Moore, JE. 1980. Forage crops. In Crop Quality and Utilization. Madison, Wis.: American Society of Agronomy and Crop Science Society of America, 61-91.

Pigden, WJ, and JM Bell. 1955. The artificial rumen as a procedure for evaluating forage quality. J. Anim. Sci. 14:1239 (abstr.).

Tilley, JMA, and RA Terry. 1963. A two stage technique for in vitro digestion of forage crops. J. Brit. Grassl. Soc. 18:104-11.

Van Soest, PJ. 1982. Nutritional Ecology of the Ruminant. Corvallis, Oreg.: O and B Books.

Van Soest, PJ, and JB Robertson. 1980. Systems of analysis for evaluating fibrous feeds. In WJ Pigden et al. (eds.), Standardization of Analytical Methodology for Feeds. Ottawa, Canada: IDRC, 49-60.

Van Soest, PJ, JB Robertson, and RA Lewis. 1991. Methods for dietary fiber, neutral detergent fiber, and nonstarch polysaccharides in relation to animal nutrition. J. Dairy Sci. 74:3583-97.

9 Forage Breeding

DARRELL A. MILLER
University of Illinois

WAYNE W. HANNA
Agricultural Research Service, USDA

IMPROVED forages play a major role in providing the consumer with high-quality and economical meat, milk, and fiber products. Forages have been and continue to be important in soil conservation practices. Although many improved forages have been developed in the past, the possibilities for future improvements are almost unlimited.

Breeding forages can be more difficult than breeding grain crops. Forages represent a diverse group of plants that can be annual or perennial, cool- or warm-season, sod-forming or bunch, and grass or legume. Many forages grow in combination with other plant genera. Forages vary greatly in mode of pollination, reproductive behavior, incompatibility relationships, and chromosome numbers.

Breeding methods used to improve forages were generally patterned after those developed for grain crops. Established breeding methods are modified or new methods are developed to accomplish the breeding objective for a specific forage species. That forages are primarily perennial and often grown in combination with other forages, and that the entire plant is used, affects the way forage programs are conducted. Improvement programs with forages differ from those with grain crops in that seed production is mainly important only for propagation.

Forage improvement is complex; therefore, greater progress can be expected through a multidisciplinary team approach involving agronomists, animal scientists, cytogeneticists, molecular biologists, engineers, entomologists, pathologists, physiologists, plant breeders, and soil scientists.

DARRELL A. MILLER. *See Chapter 4.*

WAYNE W. HANNA is Research Geneticist, ARS, USDA, University of Georgia, Tifton. He holds the MS and PhD degrees from Texas A&M University. Forage cytogenetics and breeding are the main emphasis of his research.

We acknowledge the contribution of R. R. Hill from the fourth edition of *Forages: The Science of Grassland Agriculture* to the subsections "Breeding Methods" and "Forage Cultivars."

CHOOSING SPECIES

Choosing species to meet forage production requirements is one of the first major decisions of a forage breeder. The best-adapted unimproved species often may be better than the least-adapted species after genetic improvement. A breeder should ask several questions about a potential species. What are its good and poor traits? Is germplasm available to correct its faults? How good can this species be and how difficult would it be to develop the ideal forage in this species (Burton 1978)? The questions can be answered by studying the published literature, by consulting with people who have had experience with the species, and by experimentation. In the end, the breeder must select the best one or two species, assemble and study the best available germplasm, and skillfully apply breeding techniques for species improvement.

DEVELOPING BREEDING OBJECTIVES

Breeding involves the development, selection, and testing of numerous experimental lines. Forage breeding can be complex, but the process of producing improved forages can be accomplished with sound research programs and hard work. Objectives are necessary to give direction to a program and establish priorities so that problems can be solved efficiently. Objectives for a forage program will vary depending upon species, area of adaptation, and utilization (hay, pasture, silage, or another use). Some general objectives will apply to a number of species.

Dry Matter Yield. Dry matter (DM) yields are usually measured in forage-breeding programs; however, yield of digestible DM is an even more important trait. Dry matter yields can be affected by a number of genetic, cultural, and environmental factors. Tests for evaluating new cultivars should be conducted using established cultivars as controls and under conditions simulating the intended use of the forage. All experimental lines should be tested for as long and at as many locations as deemed necessary to obtain reliable data. Forages to be used in mixtures with other species should be tested under mixture conditions. Perennial species need to be evaluated over several years to determine changes in yield and vigor over time and management. Distribution of DM production over the growing season is important, and it usually adds to the usefulness and reliability of a forage cultivar.

Quality. The term *quality* is used to describe various chemical and physical constituents of a forage that affect palatability or acceptability and animal performance when the forage is grazed or harvested for feed. Forage quality can be affected by genetic, environmental, management, and physical factors (Buxton and Casler 1993). Because of large variation in forage quality parameters, the need for uniform testing procedures and good experimental techniques cannot be overemphasized.

A number of forage components such as dry matter digestibility (DMD), protein content, lignin content and type (Cherney et al. 1992), photoperiod sensitivity, and antiquality components can be improved through breeding because they are variable and under genetic control (Hanna 1993). Improvements in forage quality offer some of the greatest opportunities in future forage breeding.

Dependability. Dependability of a forage can be affected by such factors as insect and disease resistance, reseeding ability, stand persistence, and tolerance to mismanagement and adverse environmental conditions. If a plant is used for grazing, it should have equally distributed forage production; if used for hay, the plant should produce high-DM seasonal yields of forage easy to cure.

Propagation. Propagation is an important consideration in a breeding program. Regardless of how good an improved cultivar may be, it cannot be used if it is not available to the grower. Forages can be propagated either by seed or vegetatively. Annual species usually are seed propagated. Good seed production usually is necessary in annual species to make commercial production feasible. Perennial forages are either seed or vegetatively propagated. Lower seed yields and higher seed costs can more easily be tolerated with perennials than with annuals. Vegetative propagation offers the breeder the advantages (1) of using wide crosses to rapidly develop new cultivars and (2) of producing a single superior plant with many desirable traits that can be released as an improved cultivar. It eliminates the need for progeny testing. Male and female sterility in the hybrid helps to prevent the forage from becoming a weed problem. 'Coastal' bermudagrass (*Cynodon dactylon* [L.] Pers.) is an excellent example of a successful vegetatively propagated hybrid (see Chap. 33).

Poor seedling vigor is a major problem in establishing grasses and legumes. Seedlings must be able to establish quickly and survive unfavorable environmental conditions such as drought, pests, weeds, and competition from associated species.

MODE OF POLLINATION

Information on mode of pollination is useful for estimating the amount of genetic variation present in individual plants and in populations and as an aid in determining the most effective breeding method to use. This information is often available in the literature. If not, it should be determined before an intensive improvement program is initiated.

Plants may be either cross- or self-pollinated. In general, most annual species of both grasses and legumes are self-pollinated, and perennial species are cross-pollinated (Table 9.1). There are exceptions. Cross-pollination may result from pollen blown by wind or transferred by insects. Most forage species have perfect flowers (anthers and pistils in the same flower), and the cross-pollinated forages often show some degree of self-incompatibility (Figure 9.1). Self-incompatibility causes self-sterility when pollen from a plant will not function on pistils of that same plant.

Self-incompatibility, common in many for-

TABLE 9.1. Modes of pollination and growth habits of various forage grass and legume species

Species		Chromosome number	Growth habit
	Self-pollinated forage grasses		
Lovegrass, weeping	*Eragrostis curvula*	40	Perennial
Sudangrass	*Sorghum bicolor var. sudan*	20	Annual
Wheatgrass, slendor	*Elymus trachycaulus*	28	Perennial
	Self-pollinated forage legumes		
Lespedeza, common	*Kummerowia striata*	22	Annual
Lespedeza, korean	*K. stipulacea*	20	Annual
Vetch, common	*Vicia sativa*	12	Winter annual
Vetch, hairy	*V. villosa*	14	Winter annual
	Cross-pollinated forage grasses		
Bromegrass, smooth	*Bromus inermis*	42,56	Perennial
Bermudagrass	*Cynodon dactylon*	30, 36	Perennial
Fescue, meadow	*Festuca pratensis*	14, 28, 42, 70	Perennial
Fescue, tall	*F. arundinacea*	42	Perennial
Grama, blue	*Bouteloua gracilis*	Varied	Perennial
Grama, sideoats	*B. curtipendula*	Varied	Perennial
Orchardgrass	*Dactylis glomerata*	28	Perennial
Redtop	*Agrostis alba*	28, 42	Perennial
Reed canarygrass	*Phalaris arundiacea*	14, 28	Perennial
Ryegrass, perennial	*Lolium perenne*	14	Perennial
Timothy	*Phleum pratense*	14, 42	Perennial
Wheatgrass, crested	*Agropyron desertorum*	14	Perennial
Wheatgrass, western	*Pascopyrum smithii*	42, 56	Perennial
	Cross-pollinated forage legumes		
Alfalfa	*Medicago sativa*	32	Perennial
Alfalfa	*M. falcata*	16, 32	Perennial
Birdsfoot trefoil	*Lotus corniculatus*	12, 24	Perennial
Clover, alsike	*Trifolium hybridum*	16	Perennial
Clover, crimson	*T. incarnatum*	14, 16	Winter annual
Clover, red	*T. pratense*	14	Perennial
Clover, white	*T. repens*	32	Perennial
Sweet clover, white-flowered	*Melilotus alba*	16	Annual, biennial
Sweet clover, yellow-flowered	*M. officinalis*	16	Biennial
	Apomictic species		
Dallisgrass	*Paspalum dilatatum*	40, 50	Perennial
Kentucky bluegrass	*Poa pratensis*	28, 56, 70	Perennial
	Dioecious species		
Buffalograss	*Buchloe dactyloides*	56, 60	Perennial

Source: Adapted from Miller (1984, 62).

age grasses and legumes, is usually controlled by a few genes with a number of alleles at each locus (Knox et al. 1986). Selfed seed within a species may vary from 0% to nearly 100% depending on the genotype controlling the incompatibility system. The amount of self-incompatibility can be estimated easily by comparing seed set on selfed flowers to seed set on flowers that have been open-pollinated.

Advantages of self-incompatibility are that it tends to maintain highly variable populations and it simplifies making crosses in species with perfect flowers and without male sterility. Disadvantages are that homozygous

Fig. 9.1. Sudangrass (*Sorghum bicolor* [L.] Moench) floret at anthesis showing receptive stigmas and exserted anthers ready to dehisce pollen.

lines are difficult to develop, and problems may be encountered in breeding and seed production. Breeding methods based on selfing cannot be easily used with self-incompatible forages.

CYTOGENETICS

Cytogenetic information on chromosomes as well as reproductive behavior is invaluable to a breeding program. It can aid in bypassing or eliminating barriers to plant improvement, in producing new mutations and genetic systems, in tapping the germplasm in our world gene pools, and in understanding, enhancing, and fixing hybrid vigor. Plant breeders should become familiar with the published information and acquire the additional necessary cytogenetic information as the breeding program progresses.

Chromosomes. Probably the greatest contribution of cytogenetics to plant breeding has been in the areas of determining chromosome numbers, ploidy levels, chromosome behavior and abnormalities, genome relationships, fertility, and incompatibility relationships. These are basic characteristics that must be understood in every breeding program. They will continue to be important as we explore the germplasm in the world collections and wild species.

Reproductive Behavior. Forages can reproduce sexually or by apomixis. *Apomixis* is an asexual method of reproduction through the seed. It is present in many of the forage grass species (Hanna and Bashaw 1987). There are no reports of apomixis in forage legumes; however, a thorough search of diverse germplasm may reveal this method of reproduction in the forage legumes. The most obvious advantage of obligate apomixis is that it will fix heterosis in a particular genotype even though it reproduces by seed. Apomictic cultivars do not require isolation for commercial seed production, and they allow great potential for rapidly developing superior gene combinations. Superior apomictic selections from introductions can be released as new cultivars. For apomixis to be used in plant breeding, cross-compatible sexual and apomictic plants must be available. If not, a search for them within the species or in a related species should be initiated.

Wide Crosses. A breeder should first evaluate the germplasm within a species for specific genes before attempting to use germplasm from other species and genera. Crosses between species (interspecific) and between genera (intergeneric) offer exciting potential in forage improvement but can result in crossing incompatibility and fertility problems (Sanchez-Monge and Garcia-Olmedo 1977).

GERMPLASM

Improved cultivars cannot be any better than the genetic potential within the germplasm used. The presence of genetic diversity is an essential characteristic of germplasm. Without diversity, breeding success will be minimal. The breeder must also effectively and efficiently evaluate, utilize, maintain, and store germplasm.

Collection, Evaluation, and Utilization. The choice of populations from which superior individuals are selected or developed includes natural ecotypes, or landraces, older adapted cultivars, germplasm pools created to produce new gene combinations, specific crosses between selected parents, and plant introductions. This germplasm may originate from domestic or foreign sources.

The first step in using germplasm is to assess the variation from which to select or develop superior parents from the population. These parents can be used in the breeding program to develop an improved cultivar. In self-pollinated and apomictic species it may be possible to directly select superior plants without genetic improvement for commercial utilization.

Maintenance and Storage. Open-pollinated seed can be harvested from evaluation nurseries of self-pollinated and apomictic species and maintained in cold storage. Cross-pollinated crops also may be maintained by seed, but the seed should be produced by sibbing (crossing onto sister plants) or selfing under isolation to prevent contamination. The identity of germplasm of cross-pollinated species is rapidly lost under open-pollination without isolation. Valuable recessive genes also may be lost or become difficult to identify if cross-pollinated species are maintained by open-pollination.

Seed should be maintained in cold storage. Seed to be stored should be dried to the minimum moisture content possible or tolerated by the species and placed in a seed room or environmental chamber of at least 40% relative humidity and 4°C. If humidity control is not possible, seed can be placed at 4°C in airtight containers. Maintenance of seed viability is favored by low temperature and low humidity.

Perennials also may be vegetatively maintained in small plots. Planting in methyl bromide-fumigated soil helps to prevent contamination, controls weeds, and reduces maintenance costs. It may be necessary to move the nurseries every few years to maintain purity and improve growth of accessions.

CROSSING TECHNIQUES

Crossing of various parental selections is necessary to increase variability and produce desired gene combinations when breeding procedures are used to improve a species. Many crossing techniques are available, with variation among species and plant breeder preference. Some techniques for emasculating and crossing include

1. Mechanical removal of anthers from flowers (emasculation)
2. Hot water or alcohol treatment to sterilize pollen
3. Use of self-incompatibility in parental lines
4. Mutual pollination or placing of inflorescences in proximity to each other and relying on chance hybrids (this technique works best if marker genes are available)
5. Use of humidity chamber to delay anther dehiscence
6. Genetic and cytoplasmic-nuclear male sterility
7. Pollen sterilization by enclosing inflorescences in plastic bags

A helpful summary of crossing techniques used in various grasses and legumes has been published by Fehr and Hadley (1980).

BREEDING METHODS

Many different breeding methods (Allard 1960; Jensen 1988) have been used to develop improved forage cultivars. Several methods may be employed at different stages in the development of a single cultivar. Factors that influence the choice of a breeding method include (1) objectives of the breeding program, (2) mode of reproduction of the species, (3) gene action and heritability of the traits to be considered, and (4) resources available to the breeder. There are few situations, if any, in forage breeding where one method will be distinctly superior to all others, and the better method is not always apparent.

Mass Selection. With mass selection, a source population of individual plants is examined, and the most desirable plants, or seed from the most desirable plants, are selected to form the improved population. The source population may be plant introductions, an old field or pasture, a field subjected to high levels of disease or insect damage, old strain tests, or other fields or pastures in which potentially valuable plants are likely to be found. Selected individuals may be entered into other phases of the breeding program, or they may be used as parents of a cultivar. Mass selection was used extensively in the development of many of our first improved cultivars. Selection in plant introductions, in old fields, or in fields with very high levels of disease or insect damage offers a unique opportunity to find new and valuable germplasm.

Backcross Breeding. Backcross breeding is described in most plant breeding texts (Allard 1960). The procedure is used when the breeder wishes to transfer a highly heritable (easy to measure) trait into a population, but the method used in forage breeding differs from that described in most plant breeding textbooks. The donor parent (the one with the highly heritable trait) is usually a clone or a population instead of an inbred line. The recurrent parent is often a heterozygous, heterogenous (genetically and morphologically variable) population. Forage breeders seldom attempt to achieve homogeneity (genetic uniformity) in a backcrossing program because it is not needed and may be disadvantageous for most forage cultivars. The backcross procedure will often be used in combination with other breeding methods.

Recurrent Selection. Recurrent selection is a method of plant breeding in which selected plants are intercrossed to produce a new population. This is the first cycle (cycle one) of the recurrent selection. Then, individual plants are again selected for the various desired characteristics and intercrossed, producing cycle two of the recurrent selection method. Selection continues until the desired progress is achieved.

Hallauer (1992) describes the methods and uses for the different types of recurrent selection. The basic steps of recurrent selection are evaluation of a large population of individual plants and selection of the superior individual plants to form the improved population. If selection is based on observable characteristics, it is called *phenotypic recurrent selection.* It is called *genotypic* (or *progeny test* or *family*) re-

current selection if plants are selected based on the performance of the selected plants. Hill and Haag (1974) discuss the advantages and disadvantages of the various type of genotypic selection. Response to recurrent selection can be doubled if the superior plants are selected and intermated in isolation, because pollen from unselected plants would not contribute to the next cycle of selection (Burton 1992).

Molecular Techniques. In recent years genetic engineering and gene-splicing have been investigated in forages. This area of research includes moving genes or pieces of genes between species and genera. In the future, this methodology could prove to be a valuable tool for developing improved forage cultivars (Mujeeb-Kazi and Sitch 1988; Hodges et al. 1993). (See Chap. 2, Vol. 2.)

FORAGE CULTIVARS

The final step in a forage-breeding program is the utilization of germplasm selected by one or more of the preceding breeding methods. The type of cultivar produced is determined by the mode of reproduction of the species. Individual populations of plants se-

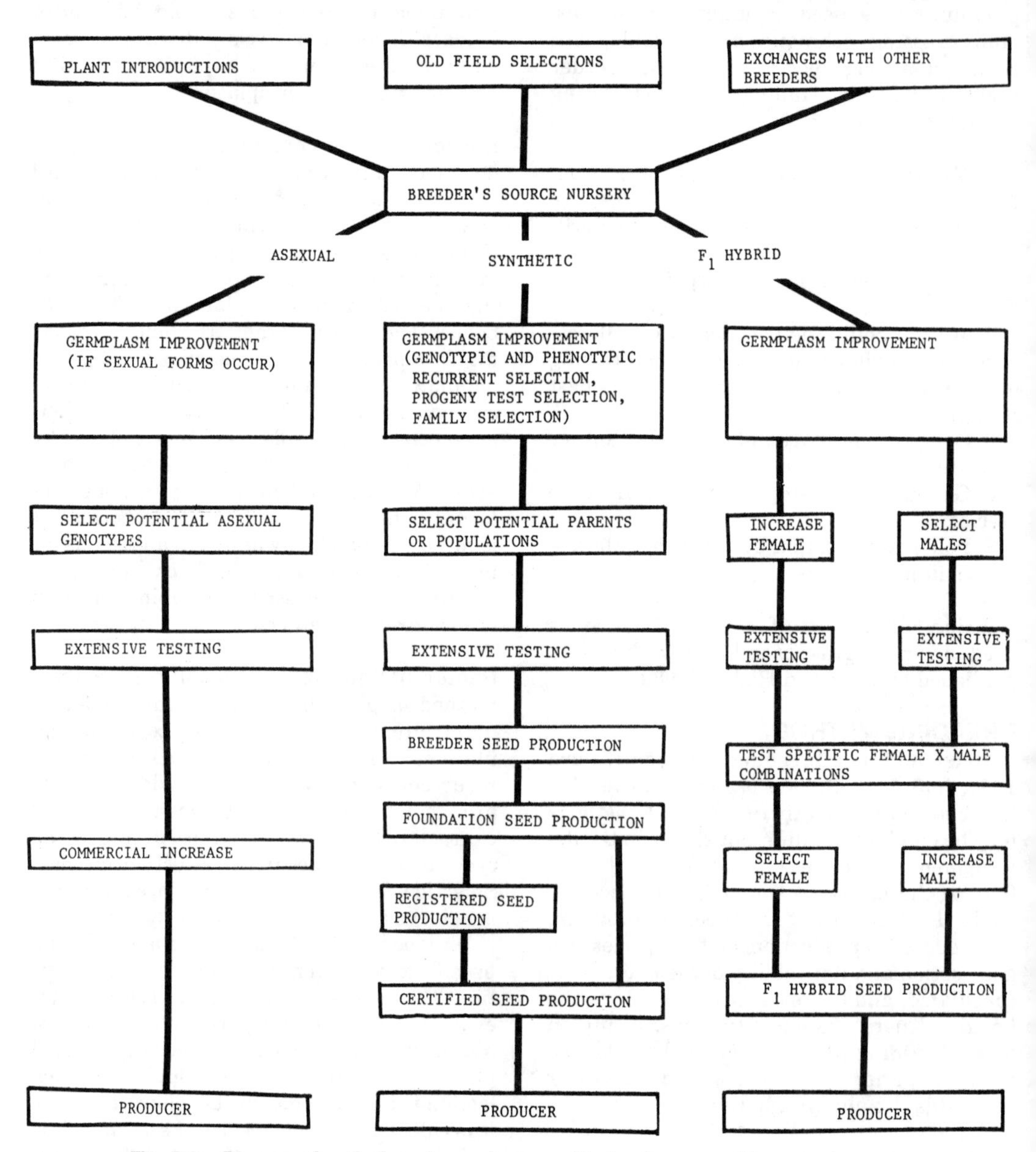

Fig. 9.2. Diagram of methods and procedures used to develop new cultivars.

lected by mass or recurrent selection are potentially new cultivars and may not need further work.

Asexual Propagation. Some forage cultivars are propagated as clones (vegetatively) or apomictically (vegetatively through the seed). Bermudagrass is probably the best-known example of a forage that is clonally reproduced. Kentucky bluegrass (*Poa pratensis* L.), dallisgrass (*Paspalum dialtatum* Poir.), and buffelgrass (*Cenchrus ciliaris* L.) are examples of forages that reproduce apomictically (Fig. 9.2).

Synthetics. Individual plants or lines selected by one of the preceding breeding methods are potential parents of a new synthetic cultivar. Synthetics are usually used to produce cultivars in cross-pollinated crops (Fig. 9.2). A synthetic is produced by recombining (in isolation) the best clones for a particular breeding objective (Fig. 9.3). The best synthetics are produced by including lines that combine well in crosses with many different lines (good general combining ability). The most vigor can be expected in the first generation because the population consists of many F_1 plants. Some loss of vigor can be expected in

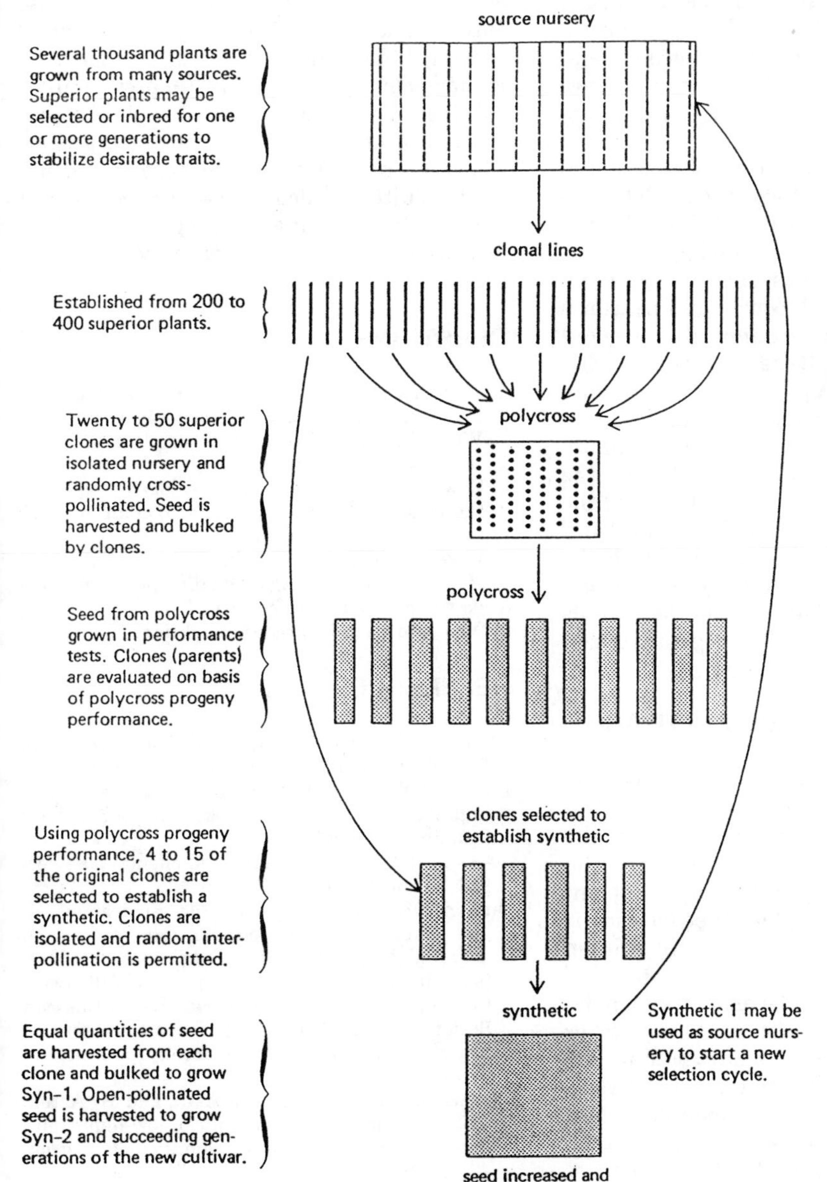

Fig. 9.3. An example of how a synthetic forage cultivar is developed. (Adapted from Miller 1984, 67.)

succeeding generations depending on such factors as number of clones used to produce the synthetic (Hill and Elgin 1981), amount of cross-pollination, and ploidy. Seed production out of the area of adaptation and unequal fertility of different genotypes in the synthetic may result in a shift of its genetic composition, thus changing its yield potential.

F_1 Hybrid. The F_1 hybrid is considered by many plant breeders to be the ultimate type of cultivar because it can maximize yield by crossing two parents with the highest yield potential (Fig. 9.2). A disadvantage of an F_1 hybrid in sexually reproducing species is that only first-generation seed can be used, so it has to be remade each time. However, an advantage of remaking the hybrid each generation is that genetic shifts that may change yield, as in synthetics, are not likely in F_1 hybrids.

A method of pollination control is essential for the production of F_1 hybrid seed. Cytoplasmic-nuclear male sterility (CMS) is the method used in forages to produce commercial F_1 hybrids in crops such as sorghum (*Sorghum bicolor* [L.] Moench) and pearlmillet (*Pennisetum americanum* [L.] Leeke) (Edwardson 1970). CMS has also been reported in other forage crops but has not to date been utilized in commercial hybrids.

F_1 hybrid production requires the maintenance of at least three lines: the CMS line, the B-line or maintainer of the CMS line, and the male pollinator of the hybrid. Fertility restoration of forage hybrids is usually not desirable (unlike required fertility in corn hybrids) because lack of seed set on the hybrids tends to keep plants more vegetative and maintains higher forage quality.

SEED AVAILABILITY AND CULTIVAR MANAGEMENT

The success of a new cultivar will depend to a large extent on the availability of seed or planting material and on the ability of the grower to use it. No one has more interest in the success of a new variety than the plant breeder. Therefore, the plant breeder's job is not finished when the improved cultivar is developed and released.

Arrangements need to be made through national or state foundation seed projects or commercial companies to rapidly produce adequate quantities of seed to meet user needs. Failure to produce sufficient amounts of seed will result in limited use of even the best genetically improved forage. The breeder must provide information on seed increase to the seed producer when this information is not known.

Proper management of the improved cultivar by the grower to maximize forage production is necessary to ensure its utilization and success. Management information should be obtained in cooperation with other disciplines during the breeding and evaluation phases. Management factors of value to the user are (1) frequencies and rates of application of fertilizer and (2) grazing intensities or cutting heights and frequencies needed to maximize production of high-quality forage. Animal performance data are important. Special management information is needed for those species that reseed naturally. In perennial species, information should be available on management practices that improve winter survival and persistence. This is only a partial list of important management practices. The grower needs a total package for the improved cultivar, adequate seed supplies at a reasonable price, and the proper cultural information to ensure maximum economic returns.

QUESTIONS

1. What are some of the characteristics that plant breeders look for when developing a cultivar? What characteristics do they wish to avoid?
2. Describe mass selection.
3. How is a synthetic cultivar developed?
4. Name several ways in which a plant breeder evaluates breeding material.
5. What information should one have about a species before starting a breeding program? How do you obtain this information?
6. What are the advantages of hybrids?

REFERENCES

Allard, RW. 1960. Principles of Plant Breeding. New York: John Wiley and Sons.

Burton, GW. 1978. The philosophy of the genetic improvement of forages grasses. Ark. Agric. Exp. Stn. Spec. Rep. 63.

———. 1992. Recurrent restricted phenotypic selection. Plant Breed. Rev. 9:101-13.

Buxton, DR, and MD Casler. 1993. Environmental and genetic effects on cell wall composition and degradability. In HG Jung, DR Buxton, RD Hatfield, and J Ralph (eds.), Forage Cell Wall Structure and Digestibility. Madison, Wis.: American Society of Agronomy, Crop Science Society of America, and Soil Science Society of America, 685-714.

Cherney JH, JR Cherney, DE Akin, and JD Axtell. 1992. Potential of brown-midrib, low-lignin mutants in improving forage quality. Adv. Agron. 46:157-98.

Edwardson, JR. 1970. Cytoplasmic male sterility. Bot. Rev. 36:341-420.

Fehr, WR, and HH Hadley (eds.). 1980. Hybridization of Crop Plants. Madison, Wis.: American Society of Agronomy, Crop Science Society of America.

Hallauer, AR. 1992. Recurrent selection in maize. Plant Breed. Rev. 9:115-74.

Hanna, WW. 1993. Improving forage quality by breeding. Int. Crop Sci. Madison, Wis.: Crop Science Society of America, 671-75.

Hanna, WW, and EC Bashaw. 1987. Apomixis: Its identification and use in plant breeding. Crop Sci. 27:1136-39.

Hill, RR, and JH Elgin, Jr. 1981. Effect of the number of parents on performance of alfalfa synthetics. Crop Sci. 21:298-300.

Hill, RR, and WL Haag. 1974. Comparison of selection methods for autotetraploids. Crop Sci. 14:587-90.

Hodges, TK, KS Rathore, and J Peng. 1993. Advances in genetic transformation of plants. In Proc. 17th Int. Grassl. Congr. Palmerston North, New Zealand, 1013-24.

Jensen, NF. 1988. Plant Breeding Methodology. New York: John Wiley and Sons.

Knox, RB, EG Williams, and C Dumas. 1986. Pollen pistil and reproductive function in plants. Plant Breed. Rev. 4:9-80.

Miller, DA. 1984. Forage Crops. New York: McGraw-Hill, 62-67.

Mujeeb-Kazi, A, and LA Sitch (eds.). 1988. Review of advances in plant biotechnology, 1985-88. In Proc. 2d Int. Symp. on Genet. Manipulation in Crops. Mexico, FD Mexico, and Manila, Philippines: CIMMYT and IRRI.

Sanchez-Monge, E, and F Gracia-Olmedo (eds.). 1977. Interspecific hybridization in crop plants. In Proc. 8th Congr. of Eucarpia, May 23-25. Madrid, Spain, 407.

10 Seed Production Principles

DARRELL A. MILLER
University of Illinois

JEFFREY J. STEINER
Agricultural Research Service, USDA

THE seed industry provides consumers with dependable quantities of high-quality seed. Special rules are used by seed producers to help ensure that cultivars remain true to the genetic composition as originally developed by plant breeders. In the past, when farmers needed seed for planting, they allowed fields to produce seed as a by-product of forage production. Presently, most seed utilized for forage production is produced as specialty seed crops in regions geographically removed from areas where the seed will be utilized but where environmental conditions are conducive to dependable seed yields.

Several factors affect the capacity of forage crops to produce seed. Forage crops are selected for yield, forage quality, and other characteristics related to livestock productivity. These desirable forage characteristics often conflict with those characteristics needed for maximum seed production. Once forage plants begin to flower and produce seed, the quality of the forage often decreases, which results in reduced animal performance. However, for high seed yields to be obtained by seed growers, the forage plants must have the capacity to flower prolifically and to produce seed.

Most forage cultivars are bred to be grazed or cut frequently for hay and silage and are rarely allowed to complete their life cycles to produce seed. Because of this, when allowed to produce seed, these plants are exposed to pressures in the field not normally experienced when grown for forage. Many of the flowers produced may fail to set viable seed as a result of disease and insect pressures during reproduction or due to a lack of pollination if cross-pollination is required. Once the seed begins to mature after pollination, there are stresses from inclement weather that lower seed quality and lower the ability of seed to produce vigorous plants after seeding. Numerous problems may also exist during seed harves and seed conditioning that can reduce the amount of seed that will be eventually sold.

Many of these problems can be reduced by growing forages for seed in enviroments that are more conducive to reproduction than those areas where forages are predominantly consumed. When developing new forage cultivars, seed yield must be evaluated in the region where seed production will occur to ensure that cultivars can produce adequate amounts of seed to make it available to forage producers (Rumbaugh et al. 1971; Miller et al. 1975).

AREAS OF SEED PRODUCTION

Most of native and introduced warm-season grass cultivars are produced for seed in the southern Great Plains where they are normally used for grazing. Most seed produc-

DARRELL A. MILLER. *See Chapter 4.*

JEFFREY J. STEINER is Research Agronomist with the USDA/ARS at the National Forage Seed Production Research Center in Corvallis, Oregon, and researches forage legume seed production systems and germplasm. He received the MS degree from California State University, Fresno, and the PhD from Oregon State University.

tion of the temperate grasses is located in the western portion of the US, with a majority of perennial (*Lolium perenne* L.) and annual ryegrass (*L. multiflorum* Lam.), tall and fine fescue (*Festuca arundinacea* Schreb.), and creeping bentgrass (*Agrostis stolonifera* L. var. *palistris* [Huds.] Farw.) grown in western Oregon and Washington. Significant acreages of Kentucky bluegrass (*Poa pratensis* L.) seed is produced in eastern Oregon and Washington and western Idaho.

As with temperate grasses, significant amounts of some forage legume species are still grown for seed in areas other than the western US. Seed production of most of the improved red clover (*Trifolium pratense* L.) cultivars now occurs in the West, but a great deal of uncertified red clover seed is produced in Illinois, Indiana, Ohio, Missouri, and central Canada. Similarly, most of the production area of alfalfa (*Medicago sativa* L.) seed shifted in the late 1940s from the Great Plains states to California and regions of the Pacific Northwest east of the Cascade Mountains, but some acreage still remains in Nebraska, Kansas, and Oklahoma.

White clover (*T. repens* L.) seed is produced in California's Sacramento Valley with limited amounts in western Oregon and Idaho. However, much of the white clover seed used in the US is imported from New Zealand. Most crimson clover (*T. incarnatum* L.) seed is produced in the Willamette Valley of western Oregon, but some seed production of this species as well as that for korean lespedeza (*Kummerowia stipulacea* [Maxim] Makino) and timothy (*Phleum pratense* L.) is still produced in the midwestern or southeastern states (Wheeler and Hill 1957; Youngberg and Buker 1985). Most birdsfoot trefoil (*Lotus corniculatus* L.) seed is produced in Minnesota, Wisconsin, Michigan, and adjacent regions of Canada. Limited acreage of birdsfoot trefoil seed is grown in western Oregon. Specialty winter annual forage legumes such as common vetch (*Vicia sativa* L.), rose clover (*T. hirtum* All.), subclover (*T. subterraneum* L.), arrowleaf clover (*T. vesiculosum* Savi), and berseem clover (*T. alexandrinum* L.) are produced in limited amounts in either California or western Oregon.

THE ECOLOGY OF SEED PRODUCTION

Consistent production of high-quality forage seed is dependent upon all the environmental and crop culture methods that affect the reproductive development, seed yield, and genetic composition of each cultivar. Since most forage seed crops are grown in regions different from where the seed will be utilized, numerous aspects of the seed production environment must be considered in relation to plant growth and reproduction. It is critical to maintain the genetic integrity of cultivars as the volume of seed is increased in each successive generation of seed production. Each crop cultivar starts as a small amount of seed produced by private and public cultivar breeding programs, and it is increased until there are large enough quantities to be sold in the marketplace. Factors such as photoperiod and seasonal temperature ranges along with rainfall patterns have traditionally determined the regions where cultivars may be successfully produced. However, economic and social concerns may also substantially influence the capacity of the seed industry to produce consistent amounts of high-quality seed in a given seed production environment.

Much of the past change in geographic distribution of forage grass and legume seed production from the midwestern to western US was due to more dependable weather conditions during seed maturation and harvest (Rincker et al. 1988). The climatic conditions of the West are best characterized by relatively high summer air temperatures with low humidity, as well as drier periods during harvest time (Youngberg and Buker 1985). Frequent and untimely rainfall in the midwestern states contributed to reduced seed yields and frequent crop failures in that region.

Even in the western states, certain specific regional climates may be more conducive to seed production than others. High ambient air temperatures during reproduction may reduce seed yields by decreasing the number of seeds that are successfully formed in each flower. The number of alfalfa florets setting pods decreases when the air temperature exceeds 38°C, with a 27°C optimum temperature for seed production. As the relative humidity decreases, seed production usually increases, but low humidity and high temperatures can make easy-to-shatter seed crops more difficult to produce than they would be in climates that reduce seed shed from pod dehiscence when the seeds are ready to harvest. High light intensity is also essential for good seed production, and clear skies and warm temperatures favor insect pollinator activity, which is essential for good seed set in crops requiring bees for cross-pollination.

Care must be taken when choosing a region

for seed production based on the reproductive physiology of the cultivar. The differential effects of environmental factors such as temperature and photoperiod length may shift the genetic composition of a cultivar during seed production. For example, the temperatures in the southern California and Arizona seed production regions are characteristically mild in winter and hot in summer compared to the cold winter weather of the Midwest or upper Great Lakes region. When cold-resistant alfalfa cultivars are grown for seed in milder climates, plants produced from that seed are taller than the original parent material, have more plant growth in the fall, and are more susceptible to winter injury than plants produced from seed grown in climates more like those where the cultivar was developed (Garrison and Bula 1961). As the number of generations of seed production increases in regions different from the regions of adaptation, the resulting genetic shift is toward plants more similar to cultivars adapted to milder climates. A reverse genetic shift may occur if southern species are increased for several generations in much colder regions.

Another type of genetic shift may occur from the different flowering responses of individual plants in a population to photoperiod (Bula et al. 1964; Garrison and Bula 1961; Taylor et al. 1990). Seed production at southern latitudes of red clover cultivars adapted to northern regions may result in a disproportionate production of non-winter hardy genotypes that are not adapted to cold continental winter climates. This phenomenon can be useful for producing cultivars that are strongly vegetative at southern latitudes. When grown at northern latitudes, where long daylengths induce more flowering, some cultivars of white clover and birdsfoot trefoil may produce more seed than when they are grown at southern latitudes.

CROP CULTURAL FACTORS

The cultural practices needed for seed production are often very different from those of conventional forage operations. These may include special planting methods and plant population configurations, soil fertility management requirements, and cultivation and forage growth management operations. To achieve economic seed yields, seed crops may have different requirements than forage crops for irrigation water amount and timing, weed control, and insect and disease pest management. Also, special practices not used in forage production such as prebloom hay harvest, pollinator management, and seed harvest are required to produce seed.

Crop Establishment. Most forages grown for seed are planted in rows 15 to 122 cm apart at very low seeding rates compared to rates for forage production systems. However, broadcast aerial seedings are primarily used in large white clover seed fields in California. With fields planted in rows, conventional grain-type drills are used to place the seed in rows as narrow as 15 to 30 cm wide. Single-row box-type planters that space rows 70 cm or more may be used when wide row spacing is required. There is some evidence that supports planting individual alfalfa seeds in a spaced fashion similar to that used for vegetables to minimize intrarow plant competition; evidence suggests this increases seed yield compared to more dense within row plantings.

Since seed yield, as opposed to optimum phytomass production in forage systems, is the desired end product of seed production, lower plant densities favor reproductive development. Additionally, row-planted configurations may facilitate seed production through (1) weed control by mechanical cultivation where appropriate; (2) better penetration of insecticides through the entire plant canopy; (3) more erect plant growth, which allows better light penetration into the canopy, reduced plant lodging, and greater accessibility to flowers by insect pollinators; (4) easier control and application of irrigation water where surface irrigation is appropriate; (5) easier roguing of individual plants that are not true to the cultivar description; and (6) reduced foliar disease problems.

Fertilization Requirements. Grass seed crops generally need supplemental nitrogen (N) fertilization (Youngberg and Buker 1985). The time of N application is dependent upon the growth habit of the crop. Some grasses primarily grow and develop their flower buds during the summer and thus respond to one N application. Other species, such as cool-season grasses, initiate their flower primordia during the fall or winter and respond to N applications that are split between the fall and early spring. However, the optimum dates and rates of N application for cool-season grasses vary for different species. The total amount of N that needs to be applied depends upon residual soil fertility from previous fertilizer applications and from

N that remains in the soil from previous crops in the rotation cycle. As grass stands become older, more N is needed to keep seed production high because there are greater amounts of total crop phytomass that require more N to support active growth.

Species-specific symbiotic N-fixing rhizobia should be applied to the legume seeds at the time of planting so that N in the soil and air will be biologically fixed by these rhizobia and made available for plant growth. Legumes need adequate levels of phosphorus (P), potassium (K), calcium (Ca), and other nutrients to ensure proper growth. The amount of each nutrient required for seed production may be less than that needed for maximum forage production. High P levels in white clover seed fields may reduce seed yield because vegetative growth is favored over reproductive development (Clifford 1987). However, low levels of the micronutrient boron (B) in the soil may reduce seed yields in legumes such as white clover (Johnson and Wear 1967).

Local soil tests can help to determine the amounts of each nutrient needed for adequate plant growth. Soil pH may greatly influence legume seed crop success. Many legume seed crops are very sensitive to acid soil conditions. Strawberry clover (*T. fragiferum* L.) is better adapted to alkaline than acid soils. Crimson clover is very sensitive to acid soils and will not adequately fix atmospheric N unless soil acidity is modified with lime.

Soil and Water Management. Legume seed production in California, Idaho, and eastern Oregon and Washington requires supplemental irrigation. Red and white clover seed in western Oregon and Washington is mostly grown without irrigation, but these crops respond well to properly timed water applications. Most grass seed crops are produced in western Oregon and Washington without irrigation except some high-value bentgrass types. Bluegrass seed produced in Idaho and eastern Oregon and Washington is mostly irrigated. Seed yields of many grass species do not respond to added irrigation water unless the crop is growing under atypical drought conditions.

Specific water requirements vary by species. Nonreproductive vegetative growth of forage legumes can reduce seed yields but can be managed by proper water application time and amount. If alfalfa, white clover, and birdsfoot trefoil are overwatered, seed yields will either be greatly reduced or not be different from plants that do not receive supplemental irrigation (Taylor et al. 1959; Steiner et al. 1992). Red clover responds well to increased water applications if grown on well-drained soils. In general, it is important that plant growth be balanced with active flower development to achieve good seed production. In addition to differences between species, crop water requirements depend upon soil water-holding capacity and soil depth, amount and pattern of natural precipitation, air temperature and wind speed (which affect evapotranspiration), and the length of the growing season. The shallow-rooted white clover needs more frequent water applications than does the deep-rooted alfalfa when grown in California. To reduce plant vegetative growth, white clover grown with supplemental irrigation is best adapted to soils that have low water-holding capacities. Alfalfa, with its long taproot, may utilize water from deep within the soil profile and survive much longer dry periods than either white or red clover.

Pollination Management. With most forages, pollen must be transferred from the stamen to the stigma within the flower for seed production to occur. After the pollen is placed on the stigma, it germinates, a pollen tube grows down the style, and a gamete unites with an ovule, completing fertilization. Legumes such as alfalfa, red clover, white clover, arrowleaf clover, birdsfoot trefoil, and common vetch require cross-pollination by insect pollinators. Even though alfalfa and crimson clover can self-pollinate, seed yields and forage yield potential are usually greater when insect pollinators are used. Grasses such as perennial ryegrass, orchardgrass (*Dactylis glomerata* L.), smooth bromegrass (*Bromus inermis* Leyss.), and tall fescue are also cross-pollinated, but this is usually accomplished by the wind. Some important grasses such as Kentucky bluegrass, dallisgrass (*Paspalum dilatatum* Poir.), and buffelgrass (*Cenchrus ciliaris* L.) reproduce by apomixis and set seed without fertilization of their ovules (Bashaw and Funk 1987).

Cross-pollinated legumes require the pollinating insect to manipulate the flower in a specific manner before fertilization can be accomplished (McGregor 1976). In alfalfa, the pistil of the flower is held under pressure within the keel, and fertilization cannot occur until the keel is opened, releasing the reproductive column and allowing it to strike the standard flower. This action is called *tripping* (Figs. 10.1 and 10.2). Other common legume flowers do not have this mechanism and therefore are not tripped. Instead, bees force the reproductive column out of the keel in a pistonlike, reversible action that brings the pollen in contact with the insect. When a pollinator visits a flower, pollen from that flower either adheres to small hairs or sticks to its body. As the insect travels from one flower to another, pollen is transfered between flowers of different plants. The amount of forage legume seed produced is directly related to pollinator activity (McGregor 1976).

The honeybee (*Apis mellifera*) is an important pollinator of many forage legume seed crops that require cross-pollination. Many of the pollinating insects used in California alfalfa and white clover seed fields are honeybees. Honeybees are also used for most clover, vetch, and birdsfoot trefoil seed production in the other seed-producing states. A general recommendation is to use an average of four to nine honeybee hives per hectare for adequate pollination. Some situations may warrant using a greater number of hives, but the cost of additional hives may limit optimum economic returns. When using honeybees, it is probably best in large fields to have the hives spread no more that 100 m apart. Each hive should have about 160 cm^2 of healthy brood in all growth stages and sufficient bees to blanket 15-20 combs to facilitate cross-pollination (Todd and Reed 1970).

In eastern Oregon and Washington, Idaho, and Nevada alfalfa seed production, the leafcutter bee (*Megachile rotundata*) is the most widely used bee. The leafcutter bee is also being used to a limited extent in California. The leafcutter bee nests in grooved, laminated boards composed of wood, particle board, polystyrene plastic, or other solid materials with drilled holes that can serve as nests (Fig. 10.3). The blocks or nesting boards are placed in shelters that can be moved from field to field as sufficient flowers develop that are ready for pollination. A unique characteristic of the leafcutter bee is that it collects pollen or nectar to provision the brood cell. The egg is laid on the mixture of nectar and pollen in an individual cell. Because the leafcutter bee forages for pollen, it is a very effective pollinator of alfalfa (Rincker et al. 1988). About one female leafcutter bee is required per 4 m^2 of alfalfa flowers (Bohart 1967).

The alkali bee (*Nomia melanderi*) is a native pollinator that has been used in alfalfa seed production in California and south-central Washington. These bees nest in the soil and have been cultured in gravel and soil beds lined with plastic so that the moisture level of the beds may be controlled (Fig. 10.4) (Frick et al. 1960). Salt is often added to the surface of the soil to maintain proper soil moisture by reducing evaporation and controlling weeds (Rincker et al. 1988). Such plastic-lined beds are difficult to move and are expensive to establish. Approximately 10 m^2 of well-populated nesting sites are required per hectare of alfalfa seed field (McGregor 1976).

The bumblebee (*Bombus* spp.) is also a very effective pollinator in alfalfa and all other forage legume seed crops. Bumblebees are

Fig. 10.1. Three alfalfa florets untripped. (Miller 1984, 79)

Fig. 10.2. Three alfalfa florets tripped. (Miller 1984, 79)

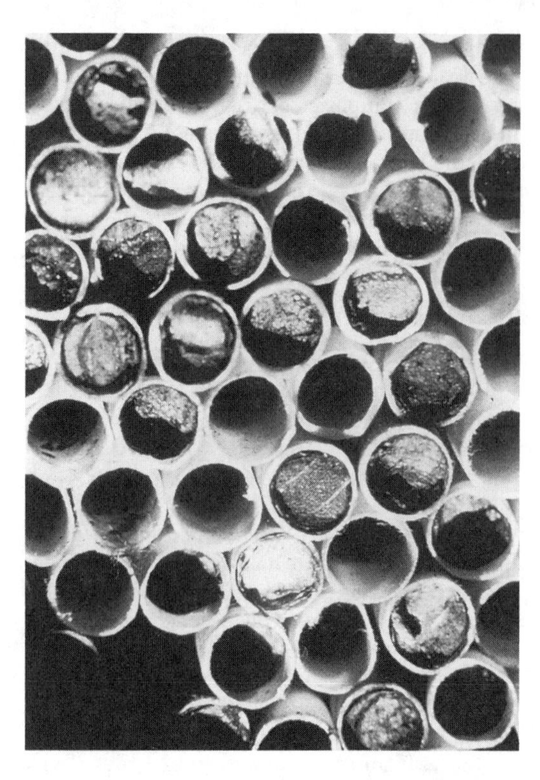

Fig. 10.3. Close-up of the leafcutter bee nest. (Miller 1984, 83)

very difficult to culture but are abundant in seed fields that are adjacent to wild areas along roadsides, ditches, and wooded areas. They nest primarily in undisturbed soil—often in abandoned rodent holes. Only impregnated female bumblebees overwinter, and they are very particular about nesting conditions. The general solitary nature of bumblebees makes them difficult to domesticate, but nesting can be encouraged (Heinrich 1979).

Fig. 10.4. Alkali bee on the ground at a nesting site. (Miller 1984, 83)

Weed Management. Weed control methods vary by seed crop and by the associated weeds. Weeds may be a serious problem when stands are thin because the competitive nature of weeds lowers crop seed yields and can delay harvest. Also, depending on weed seed shape and size, weed seed that cannot be removed from crop seed by seed conditioning after harvest may render the crop unsaleable.

Depending on planting method, appropriate strategies must be employed to control weeds. Effective crop rotations, properly prepared seed beds, and planting at a time that allows rapid and uniform establishment of the seed crop stand can increase the effectiveness of herbicides. When seed crops are planted in wide rows, mechanical cultivation methods can be employed that will control weeds or help the crop plants to gain a competitive advantage over the weeds. Many fall-planted temperate grass seed crops are seeded under a narrow band of activated charcoal sprayed over the planting row at planting time (Lee 1973). Diuron herbicide is then sprayed over the field and activated by rainfall. Weeds and volunteer crop seed between the treated rows are killed by the herbicide, but the crop emerges unharmed from the charcoal-treated row because the herbicide is bound to the charcoal and rendered inactive. Early-maturing weeds in several perennial legume seed crops can be controlled by removal of winter and early spring plant regrowth by grazing, mowing, or hay harvest. Mowing can also be an effective method to reduce weed competition during the summer or early fall in seed crops that were planted earlier in spring.

Preemergence herbicides are often a very important management tool for achieving good initial crop stands. Postemergence herbicides are effective when specific sensitive weeds must be selectively controlled in established seed crop stands. The effectiveness of herbicides for reducing weed establishment can be greatly enhanced when used in combination with other control methods (Lee 1965).

Dodder (*Cuscuta* spp.), a parasitic plant, is one of the most important weeds to control in legume seed production. Since this is a primary noxious weed, no dodder seeds may be found in seed lots that are to be sold. This weed must be aggressively eradicated from seed fields. Seed fields are monitored from the ground and air to identify infested areas, which are then sprayed locally and burned. Dodder infestations can be reduced by the herbicides chlorpropham and granular triflorilan, if properly applied. Most of the few remaining dodder seeds in seed lots can be removed during seed conditioning with special cleaning equipment. However, this reduces the net amount of clean seed after conditioning and adds to the final cost of production (Purdy et al. 1961; Youngberg and Buker 1985).

There are several problems associated with relying on herbicides as the only means of controlling weeds, particularly in perennial seed crops. Included among these are: (1) There is induced selection for herbicide-resistant weeds (e.g., resistant populations of annual ryegrass have developed in wheat fields where diclofop has been continually used). (2) Weed species populations shift toward species that are not susceptible to commonly used herbicides (e.g., downy chess [*Bromus tectorum* L.] and smooth bromegrass have become the predominate competitors in some grass seed crops). (3) Some herbicides that have been very effective in the past are no longer available or registered for use in seed crops (e.g., dinoseb was a very effective herbicide for controlling broadleaf weeds in grass and legume seed crops, but it is not available any longer because of health risks).

It is likely that few new herbicides will be developed and introduced into seed production systems. Fewer of the herbicides presently used for seed production will remain available because of increased costs to reregister these compounds. The judicious use of available registered herbicides combined with crop rotations, cultural practices that reduce weed establishment and seed populations in the soil, and biocontrol products offer the best future possibilities for effective weed control in seed fields. Specific weed control systems will need to be developed that fit local production conditions and that utilize multiple control methods.

Insect Management. Insect control is a more serious and consistent problem in legume seed fields than in grasses. In addition to those pests that may attack crop foliage, some pests specifically attack the florets of legumes and ruin or greatly reduce the yield. Most of the insect pests in grass seed production are sporadic problems and can be controlled by manipulating the insects' natural environment.

For both grass and legume seed crops, cultural practices can be employed that completely control certain pests or reduce the need for or improve the efficiency of insecticides. Billbugs (*Sphenophorus parvulus* Gyll.), a considerable pest in older stands of orchardgrass grown for seed, can be reduced by reducing the number of continuous years the crop is grown in the rotation cycle. The sod webworm (*Crambus* spp.) can attack Kentucky bluegrass and reduce stand longevity. The use of deep-rooted cultivars can reduce sod webworm crop damage. In the Pacific Northwest, burning of grass seed field crop residues after harvest controls some insect pests, but the accumulation of charcoal on the soil surface can also bind and make insecticides ineffective when applied at labeled application rates (Kamm and Montgomery 1990). Burning has also been used to control certain insect pests in alfalfa seed fields in Nevada.

Organophosphate insecticides, some of which have systemic properties, are frequently used in legume seed production. Depending on the seed production region and crop, one to as many as four applications are required to adequately control pests such as lygus bugs (*Lygus elisus* Van Duzee), aphids (*Acyrthosiphon* spp.), and leafhoppers (*Empoasca fabae* Harris). More recently, pyrethroid-type insecticides, which are more benign toward pollinators and which require fewer applications per season, have come into wider use and are replacing the organophosphates. Special care must be taken to use pyrethroid-type insecticides properly because some insect pests develop resistance to these types of chemicals and seed crop forage and harvest residues cannot be utilized for livestock feed.

Some insect pests cannot be controlled by insecticides, so other methods need to be used. There is no chemical available to control the clover seed midge (*Dasyneura leguminicola* Lintner) in red clover, but its life cycle can be disrupted by removing a hay crop during the early bloom period in spring. Seed chalcid (*Bruchophagus roddi* Gussakovaky) damage in alfalfa can be reduced by incorporating residues into the soil after seed harvest using cultivation and irrigation to rot infested seeds; by performing spring clip back of early spring regrowth to delay initial crop flowering time, which removes available host sites for females trying to oviposition; and by not extending the pollination period into late summer, which allows a buildup of chalcid populations. This insect has become a more significant pest in California since pyrethroid insecticides have replaced organophosphates. Control of alternate nesting sites in weeds can also be used to control some insect pests. A dogfennel (*Eupatorium capillifolium* [Lam.] Small) weed control program in winter and early spring both alongside and within red clover seed fields removes feeding and oviposition sites for lygus bugs and thus reduces pest pressure later in the season (Kamm 1987). This combined weed and insect control practice can probably be used for other legume seed crops where lygus bugs are a problem.

Insect control must be carefully planned and done in a manner that optimizes pest control, minimizes harm to insect pollinators and beneficial insects, and avoids chemical trespass from the field where the product is applied. A serious problem for forage seed growers is the diminishing number of insecticides that are available for insect pest management. Because forage seed crops are grown on a relatively limited area compared to major crops such as wheat, corn, and soybeans, they are considered to be minor crops. This makes the cost high for registering new products or registrating old chemicals that require new use labels. It is expected that fewer insecticides will be available for seed growers in the future.

Disease Management. Most diseases are best controlled by selecting and planting disease-resistant cultivars. However, most cultivars are genetically selected for their performance in the region where they will be grown for forage and not necessarily for the region where they will be grown for seed. These cultivars are often exposed to different disease races and disease pressures in the seed production regions compared to where they are grown for forage. Also, since the crop is grown until the reproductive phase of development is complete, it is exposed to possible disease infection for longer periods of time than when grown for forage. There appear to be more diseases that greatly reduce seed yields in grasses than in legumes.

Diseases that affect grass seed production attack both the inflorescences and foliage. The most important diseases in grass seed production are the foliar rusts (Hardison 1963). The rusts can be controlled with resistant cultivars and fungicides. However, grass seed fields must be carefully monitored since once rust symptoms are apparent the effectiveness of fungicides is greatly reduced. Grass seed crops are also susceptible to diseases that attack the flowers and seeds. The epidemiology of these diseases is becoming better understood, but there are still no completely effective chemical controls. The use of cultural controls and genetic resistance requires further research (Alderman 1991).

The diseases that affect forage legume seed crops are primarily the same as those that attack forage crops. Most important are the root-rotting diseases such as *Fusarium* and *Phytophthora*, which affect stand persistence. More recently, verticillium wilt (*Verticillium albo-atrum* Reinke & Berthier) has become an important disease problem in some alfalfa seed-producing regions because it can be seedborne. When red clover and other legume seed fields are not properly rotated, *Sclerotinia* can become a serious problem with long-term implications. Sclerotia, the fungal-resting structures, remain viable in the soil for a long time and cannot be eliminated except by long-term removal of susceptible hosts. Northern anthracnose (*Kabatiella caulivora* [Kirch] Karak.), a seedborne disease of crimson clover, can be controlled by seed treatment and by a proper planting time in the fall (Leach 1962).

SEED HARVEST

Much of the equipment used to harvest forage seed is similar to that used for grain crops, but some special equipment and practices must be used to maximize yields. Many forage species have a tendency to shatter before all seeds are mature or to produce flowers in an indeterminate fashion so that seeds in different stages of maturity are present on the same plant at the same time. Legume seeds produced in naked pods, e.g., birdsfoot trefoil and vetches, tend to dehisce when ripe, releasing seed in all directions. Almost all forage legumes and grasses are susceptible to shattering when the seed has dried and is ready for harvest.

Forage seed crops are harvested with a combine, either by combining the standing mature crop or by first windrowing the crop before shattering begins and then combining the windrowed material at a later time (Rincker et al. 1988). The windrow method allows cutting the crop when the foliage is slightly green and before shattering has begun (Fig. 10.5). Once the seed has dried adequately for proper threshing, a special continuous belt attachment mounted to the mouth of the combine gently picks up the cut windrow without shaking the plant and thus minimizes seed shatter. When direct combining is involved, a desiccant such as diquat may be used to aid crop drying and minimize shatter losses. The success of desiccants for preparing seed plants for harvest is highly dependent upon dry weather and an absence of rain, which can reinitiate plant growth. Monitoring of seed moisture content during maturation can help determine the proper time to harvest forage seed crops by maximizing seed maturity and minimizing losses to seed shattering (Klein and Harmond 1971).

Other kinds of specialty equipment have been developed to solve unique harvest problems for particular seed crops. For example, most of the alfalfa seed produced in California is direct harvested with combines. Even though the alfalfa seed pods are not as easily shattered as in some other crops, the mechanical shaking of the plant by the cutter bar mounted on the combine can result in significant seed losses. Air-jets mounted on extended lifter fingers and directed toward the mouth of the combine blow pods into the combine as they begin to fall from the plants (Fig. 10.6).

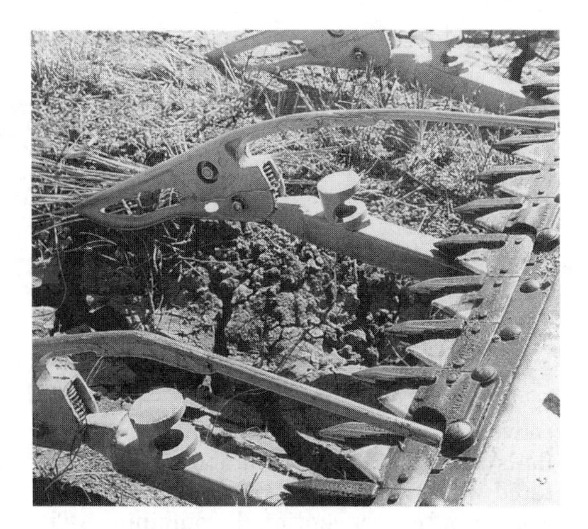

Fig. 10.6. Experimental air-jet lifter guard reduces seed losses when harvesting alfalfa seed. *R. J. Buker photo.*

Subclover seed, which is produced on the surface of the soil underneath the crop canopy, is vacuumed up with loose soil by Horwood Bagshaw harvesters developed

Fig. 10.5. Red clover mowed into windrows to dry before combining. *Photo courtesy of Jeffrey Steiner, USDA-ARS, Corvallis, Ore.*

TABLE 10.1. Methods for separating crop seeds from contaminants

Physical property of seed	Method
Size and shape	Air-screen cleaner (screen sections)
Shape	Spiral separator, vibrating separator
Length	Indent disc, cylinder machine
Thickness, width	Indent disc
Weight	Gravity table, aspirator, air-screen cleaner
Surface texture	Roll mill, magnetic separator, vibrating separator, velvet roll
Color	Color sorter

Source: Berlage and Brandenburg (1984); Purdy et al. (1961); Smith (1988).

especially for this crop in Australia. Another specialty harvest device is a Murphy pickup. It is a header attachment mounted to a conventional combine with rotating rubber flails that create a vacuum and pick up shattered white clover florets that remain on the ground after conventional combining. The Murphy pickup has also been used in New Zealand to harvest white clover from windrows.

SEED CONDITIONING

Following harvest, the seed must be conditioned and suitably prepared for sale as a product free of other seed and nonseed contaminants. Seed conditioning involves the removal of trash, weed seeds, other crop seeds, and foreign material by utilizing the physical characteristics of the seed. The most commonly utilized seed characteristics for separating crop seed from contaminants are seed size, shape, length, thickness, width, weight, and surface texture (Table 10.1).

The three most commonly used machines for conditioning forage seed are the air-screen machine, indent disc separator, and gravity deck. The air-screen machine is the basic tool of seed conditioning and separates seed from contaminants by passing the material over a series of screens with different-sized and -shaped openings. The seed is then separated from lightweight materials by aspiration. Seed cleaned in this machine is dimensionally similar and relatively free of foreign materials (Vaughan et al. 1968).

Indent disc machines can further separate seed from other materials that are slightly different in length but that are similar in width and thickness (Harmond et al. 1968). Examples of contaminants separated from seed by the indent disc are bentgrass from perennial ryegrass and Canada thistle (*Cirsium arvense* [L.] Scop.) from white clover (Vaughan et al. 1968).

The gravity deck removes inert materials, such as small stones, that are similar in size and shape as the seed by differences in the densities of the materials (Harmond et al. 1968). The gravity deck can also be used to upgrade seed quality in forage legumes by removing seed that contains seed chalcid larvae.

If harvested seed is not free from adhering inflorescence structures, a hammer mill-type machine or rubber rollers may be used to help break the seed away from various attached appendages to make the seed free flowing and ready for basic conditioning.

After basic conditioning is complete, other specialty conditioning operations may be

Fig. 10.7. A magnetic seed separator used to clean dodder seed from alfalfa seed. *Photo courtesy of Jeffrey Steiner, USDA-ARS, Corvallis, Ore.*

required to make the seed suitable for sale. Seed lots of alfalfa that contain dodder can be conditioned with magnetic separator machines (Fig. 10.7) that remove the dodder seeds after the seed lot is treated with small amounts of water and iron filings that adhere to the weed seeds (Klein and Harmond 1971). Also, legume seeds that have hard seed coats may require that the impermeable seed coat be scarified so that it can absorb water and germinate. Scarification can be done by buffing after conditioning with a special abrasive oil, treating the seed with acid or heat, or mechanical abrasion (Klein and Harmond 1971).

SEED STORAGE AND LABELING

Once a seed lot has been conditioned and is ready for sale, it is packaged and stored until sold and used. It is important that seed to be planted is stored under low-humidity and low-temperature conditions to maintain its viability and vigor. Packaging in moisture-resistant bags or containers helps to ensure that quality is maintained when the seed is shipped from the drier western states to more humid areas (Smith 1988). To maintain maximum viability, most seed is stored at a fixed moisture content and at a fixed temperature between 0° and 4°C (32° to 39°F). In this temperature range, water in the seeds does not freeze, and the activity of enzymes within the seeds needed for germination is retarded. A rule of thumb for maintaining high seed quality during storage is that relative humidity (RH) + temperature (F°) ≤ 100 (Harrington 1960).

An important requirement of seed sales is the seed label should be attached to all seed that is marketed. To regulate interstate and foreign sales of seed, the Federal Seed Act was approved by Congress in 1939. The purpose of this law is to ensure that consumers are purchasing seed of known quality. The law requires that all seed purchased be properly labeled to prevent misrepresentation. The label attached to every seed lot must contain specific information that details the kind of seed being sold and its origin, a listing of seed lot purity and whether or not there is any noxious weed seed, and the results of a recent germination test (Copeland and McDonald 1985).

SEED CERTIFICATION

Seed certification was established to assure consumers that the cultivar identity and genetic purity of seed being sold are properly represented. Seed fields are certified in each state by a seed certifying agency. The minimum uniform rules and standards for certification were developed by the Association of Official Seed Certifying Agencies (AOSCA) to facilitate seed trade (AOSCA 1971). Seed fields are inspected to ensure that the crop has the appearance of the cultivar being grown and that there are no other cultivars or undesirable weeds present. After harvest and seed conditioning, the seed lot is tested for purity and germination to ensure that minimum quality standards are met before marketing.

Rules for seed certification have been established to ensure that a cultivar remains genetically true to the original seed used to increase that cultivar for the market. Most of our forage seed is produced and certified in areas outside the cultivar's region of origin or adaptation. Through successive generations of seed increase, the genetic nature of a cultivar may change. This change may result naturally from environmental pressures placed on the crop during seed production or from contaminants that are introduced into the seed field. Rules have been established to safeguard a cultivar's genetic identity and thus maintain its productivity, winterhardiness, and disease and insect resistance. Data have shown that genetic shifts have occurred in red clover, ladino clover (*Trifolium repens* L.), and alfalfa (Bula et al. 1969; McLennan et al. 1960). These findings support the concept of limiting the number of successive generations that a seed lot can be increased and the number of years that a field can be used to produce seed for sale, especially for those cultivars or species having considerable amounts of genetic variability.

To regulate the number of successive generations of seed increase, seed certifying agencies have established four classes of seed that correspond to a generation of seed increase. Each class is identified with a specific colored seed tag.

1. Breeders' seed is the original seed produced and designated as such by the originating or sponsoring plant breeder or institution. It carries a white seed tag.
2. Foundation seed is produced by growing the breeder class seed. It also carries a white seed tag.
3. Registered seed is produced from foundation seed. With some species, or in some states, this seed class may be eliminated in order to reduce the number of generations

between breeders' and certified seed. Registered-class seed carries a purple seed tag.

4. Certified seed is produced from foundation or registered seed. This class of seed is available for consumer use. It carries a blue seed tag, signifying the seed is of high quality.

Certified seed fields are inspected during the seed production process for proper isolation; freedom from off-types; noxious weeds that would be spread to other fields when the seed is sold; and in some cases for seed-borne diseases since these would be difficult to detect when the seed is sold. These steps help to ensure that cultivar integrity is maintained before seed harvest. In addition to field inspections, harvest and transportation equipment must be inspected before certified seed is shipped to a seed-conditioning facility. At the conditioning facility, precautions are also taken to ensure that seed lots are not contaminated by other seed lots at the factility. Records are maintained to follow the progress of each seed lot as it is received, conditioned, and packaged for sale.

Certified seed labels contain information about the percentage of pure seed, germination, hard seed percentage, other crop seed, inert matter, and weed seed content. When forage growers purchase uncertified seed, there is no guarantee that the seed is of highest quality. Uncertified seed lots may be a source of a new weed problem or may perform less than optimumly when compared to a seed lot of an improved cultivar. Forage growers should never decide to buy their seed by only looking for lowest seed price. A difficulty with forage seed availability in the US is that much of the seed sold is not certified. For crops other than alfalfa, many consumers have decided not to pay a premium for named cultivars that carry the certified blue seed label and often are satisfied with purchasing a common class of seed that is not named. With the increase in international trade, seed may be imported from other countries that is cheaper than some US cultivars but that is not adapted to local growing conditions and therefore performs poorly.

Seed certification plays an important role in international seed trade. The European market requires quality seed of improved cultivars that are adapted to their local conditions. Because seed production in Europe is often poor and undependable due to unfavorable climatic conditions and limited seed-conditioning facilities, the European market depends greatly on seed imports. Recently, the Organization of Economic Cooperation and Development (OECD) was established for the export of improved cultivars and the maintenance of high-quality seed for the European market. No seed is exported to European countries unless the cultivar is on an approved list. The basis for maintaining such international trade is through seed certification. A special tag is issued for the export seed trade that certifies that the seed meets specified quality standards. This system requires close cooperation between seed companies in the US and Europe. A similar arrangement is also used for seed trade with the market in Japan.

PLANT MATERIAL AND CULTIVAR PROTECTION, EXCLUSIVE RELEASE RIGHTS, AND SEED COMMERCE

The Plant Patent Act (PPA) of 1930 (35 USC Section 161-164) allowed patenting of asexually propagated cultivars and was designed to encourage research investment in asexually reproduced plant species, other than tuber-propagated plants and plants found in an uncultivated state.

Since 1970, sexually propagated cultivars could be protected under the Plant Variety Protection Act (PVPA) (Caldwell and Schillinger 1989). To secure protection under the PVPA, the cultivar must be novel as determined by its distinctiveness, uniformity, and stability. The Plant Variety Protection Office, Agricultural Marketing Service, USDA, administers the PVPA and issues plant variety protection certificates. The owner of a certificated variety is entitled to prohibit others from selling, importing, or exporting the variety; from sexually multiplying a variety for marketing; or from producing another variety from the certificated variety (Greengrass 1993). However, there are two exemptions associated with PVPA, which are referred to as the research and farmer exemptions. The research exemption permits the use of protected seed as a parent for future breeding activities. Under the farmer exemption, farmers are allowed to save seed to plant the following year's crop and to sell surplus seed to other farmers, provided they are to primarily be engaged in producing seed for sale. These privileges have been abused, and legislation is being considered for modifying and limiting these provisions in accordance with the 1991 Convention of the International Union for the Protection of New Varieties of Plants (UPOV) (Barton 1993).

A further development of the 1991 UPOV

convention regarding the distinctiveness of a cultivar may be summarized by the concept of "essentially derived" varieties. Considerable debate is expected to ensue over this concept in the future (Baenziger et al. 1993).

With the advent of the modern era of biotechnology, and as a result of the landmark decision by the US Supreme Court in 1980 in *Diamond* vs *Chakrabarty* and the action in 1985 by the US Patent and Trademark Office action on *ex parte Hibberd,* living organisms, traits, and genes may now be protected by utility patents. As a result biological and legal issues have developed that are associated with ownership and proprietary protection of plant material.

Following the development of a new cultivar, the plant breeder, company, or institution has choices for plant protection. These choices now include PPA, PVPA, and utility patents. What option to choose will depend upon which will result in the greatest protection and benefit. Each state or company has its own regulations about giving an exclusive release or license to a company for the purpose of selling that particular cultivar. When exclusive release rights are given, only one company can participate in the sale of a cultivar to consumers. Allowing companies to bid for the rights to sell a cultivar gives the plant breeder an incentive to develop new cultivars.

The price of seed is directly related to consumer demand and the availability of a cultivar and competing cultivars in the market. When new cultivars or species are released and seed supplies are limited, the price of seed often increases. Seed supply may become limited when growing conditions are unfavorable for seed production. As more seed of a desirable cultivar becomes available, the price usually decreases. If an abundant supply of seed of a species is available due to high production or imports, the price of that seed decreases more. Weather conditions in seed production regions can still greatly affect the availability of seed supplies.

QUESTIONS

1 Why are most of the forage seed-producing areas located in the western US?
2. Why are most forage species grown in rows for seed production?
3. How does water management for seed production differ from that for forage production?
4. Describe the steps involved in alfalfa flower pollination.
5. Why are forage seed crops susceptible to different pest problems than forage crops?
6. Which element of grass seed pest management is the most important?
7. What must a seed grower be cautious of when applying insecticides to legume seed crops? Why is this not a problem for grass seed crops?
8. Name two effective insect pollen collectors. Why are they good pollinators?
9. What can be done to avoid dodder seed contamination in alfalfa seed?
10. What is the greatest harvest problem of forage seed crops?
11. What is meant by genetic shift, and how does it affect seed quality?
12. Describe the practice of limited generations of seed increase and its role in avoiding genetic shift.

REFERENCES

Alderman, SC. 1991. Assessment of ergot and blind seed diseases of grasses in the Willamette Valley of Oregon. Plant Dis. 75:1038-41.

Association of Official Seed Certifying Agencies. 1971. AOSCA Certification Handbook. Publ. 23.

Baenziger, PS, RA Kleese, and RF Barnes (eds.). 1993. Intellectual Property Rights: Protection of Plant Materials. CSSA Spec. Publ. 21. Madison, Wis.: Crop Science Society of America, American Society of Agronomy, and Soil Science Society of America.

Barton, J. 1993. Introduction: Intellectual property rights workshop. In PS Baenziger, RA Kleese, and RF Barnes (eds.), Intellectual Property Rights: Protection of Plant Materials, CSSA Spec. Publ. 21. Madison, Wis.: Crop Science Society of America, American Society of Agronomy, and Soil Science Society of America, 13-19.

Bashaw, EC, and CR Funk. 1987. Apomictic grasses. In WR Fehr (ed.), Principles of Cultivar Development. New York: Macmillan, 40-82.

Berlage, AG, and NR Brandenburg. 1984. Seeding conditioning equipment research. Seed Sci. and Technol. 12:895-908.

Bohart, GE. 1967. Management of wild bees. In Beekeeping in the United States, USDA Agric. Handb. 335. Washington, D.C.: US Gov. Print. Off, 411.

Bula, RJ, RG May, CS Garrison, and JG Dean. 1969. Floral response, winter survival, and leaf mark frequency of advanced generation seed increases of Dollard red clover. Crop Sci. 9:181-84.

Bula, RJ, RG May, CS Garrison, CM Rincker, and DR McAllister. 1964. Growth responses of white clover progenies from five diverse geographic locations. Crop Sci. 4:295-97.

Caldwell, BE, and JA Schillinger (eds.). 1989. Intellectual Property Rights Associated with Plants. ASA Spec. Publ. 52. Madison, Wis.: Crop Science Society of America, American Society of Agronomy, and Soil Science Society of America.

Clifford, PTP. 1987. Producing high seed yields from high forage producing white clover cultivars. J. Appl. Seed Prod. 5:1-9.

Copeland, LO, and MB McDonald, Jr. 1985. Principles of seed science and technology. Minneapolis, Minn.: Burgess.

Frick, KE, H Porter, and H Weaver. 1960. Development and maintenance of alkali bee nesting sites. Wash. Agric. Exp. Stn. Circ. 366.

Garrison, CS, and RJ Bula. 1961. Growing seeds of forages outside their regions of use. In Seed, USDA Yearbook of Agriculture. Washington, D.C.: US Gov. Print. Off., 401-6.

Greengrass, B. 1993. Non-US protection procedures and practices—Implications for US innovators. In PS Baenziger, RA Kleese, and RF Barnes (eds.), Intellectual Property Rights Protection of Plant Materials, CSSA Spec. Publ. 21. Madison, Wis.: Crop Science Society of America, 41-59.

Hardison, JR. 1963. Commercial control of *Puccinia striiformis* and other rusts in seed crops of *Poa pratensis* by nickel fungicides. Phytopathol. 53:209-16.

Harmond, JE, NR Brandenburg, and LM Klein. 1968. Mechanical Seed Cleaning and Handling. Washington, D.C.: US Gov. Print. Off.

Harrington, JF. 1960. Drying, storing, and packaging seed to maintain germination, vigor. Seedsmen's Dig. 11.

Heinrich, B. 1979. Bumblebee Economics. Cambridge, Mass.: Harvard Univ. Press.

Johnson, WC, and JI Wear. 1967. Effect of boron on white clover (*Trifolium repens* L.) seed production. Agron. J. 59:205-6.

Kamm, JA. 1987. Impact of feeding by *Lygus hesperus* (Heteroptera: Miridae) on red clover grown for seed. J. Econ. Entomol. 80:1018-21.

Kamm, JA, and ML Montgomery. 1990. Reduction of insecticide activity by carbon residue produced by burning grass seed fields after harvest. J. Econ. Entomol. 83:55-58.

Klein, LM, and JE Harmond. 1971. Seed moisture—A harvest timing index for maximum yields. Trans. Am. Soc. Agric. Eng. 14:124-26.

Leach, CM. 1962. *Kabatiella caulivora,* a seed-borne pathogen of *Trifolium incarnatum* in Oregon. Phytopathol. 52:1184-90.

Lee, WO. 1965. Herbicides in seed bed preparation for establishment of grass seed fields. Weeds 13:293-97.

______. 1973. Clean grass seed crops established with activated carbon bands and herbicides. Weed Sci. 21:537-41.

McGregor, SE. 1976. Insect Pollination of Cultivated Crop Plants. USDA-ARS Agric. Handb. 496. Washington, D.C.: US Gov. Print. Off.

McLennan, JE, R Greenshields, and RM MacVicar. 1960. A genetic analysis of population shifts in pedigree generations of Lasalle red clover. Can. J. Plant Sci. 40:509-15.

Miller, DA, LJ Elling, JD Baldridge, PC Sandal, SG Cramer, and CP Wilsie. 1975. Predicting seed yields of birdsfoot trefoil clones. North Cent. Reg. Res. Publ. 227 and Ill. Agric. Exp. Stn. Bull. 753.

Purdy, LH, JE Harmond, and GB Welch. 1961. Special processing and treatment of seeds. In Seed, Yearbook of Agriculture. Washington, D.C.: US Gov. Print. Off., 322-30.

Rincker, CM, VL Marble, DE Brown, and CA Johansen. 1988. Seed Production practices. In AA Hanson, DK Barnes, and RR Hill, Jr. (eds.), Alfalfa and Alfalfa Improvement, Am. Soc. Agron. Monogr. 29. Madison, Wis., 985-1021.

Rumbaugh, MD, WR Fehr, JD Axtell, LJ Elling, EL Sorensen, and CP Wilsie. 1971. Predicting Seed Yield of Alfalfa Clones. North Cent. Reg. Res. Publ. 207.

Smith, DL. 1988. The seed industry. In AA Hanson, DK Barnes, and RR Hill, Jr. (eds.), Alfalfa and Alfalfa Improvement, Am. Soc. Agron. Monogr. 29. Madison, Wis., 1023-36.

Steiner, JJ, RB Hutmacher, SD Gamble, JE Ayars, and SS Vail. 1992. Alfalfa seed water management: I. Crop reproductive development and seed yield. Crop Sci. 32:476-81.

Taylor, NL, CM Rincker, CS Garrison, RR Smith, and PL Cornelius. 1990. Effect of seed multiplication regimes on genetic stability of Kenstar red clover. J. Appl. Seed Prod. 8:21-27.

Taylor, SA, JL Haddock, and MW Pedersen. 1959. Alfalfa irrigation for maximum seed production. Agron. J. 51:357-60.

Todd, FE, and CB Reed. 1970. Brood measurement as a valid index to the value of honey bees as pollinators. J. Econ. Entomol. 63: 148-49.

Vaughan, CE, BR Gregg, and JC Delouche. 1968. Seed Processing and Handling. State College: Mississippi State Univ.

Wheeler, WA, and DD Hill. 1957. Seed Production of Grassland Seeds. Princeton, N.J.: Van Nostrand.

Youngberg, HW, and RJ Buker. 1985. Grass and legume seed production. In ME Heath, RF Barnes, and DS Metcalfe (eds.), Forages: The Science of Grassland Agriculture, 4th ed. Ames: Iowa State Univ. Press, 72-79.

11 Grassland Ecosystems and Their Improvement

KEVIN D. KEPHART
South Dakota State University

CHARLES P. WEST
University of Arkansas

DAVID A. WEDIN
University of Toronto

GRASSLANDS are plant associations that are predominantly grasses (family Poaceae), and they usually include other plant families such as legumes (Leguminosae), composites (Compositae), and sedges (Cyperaceae), sometimes mixed with shrubs and sparse trees. These areas dominate much of the earth's land surface. The once abundant natural grasslands of the US (Fig. 11.1) have been largely replaced since European settlement by high-input crops, hay meadows, introduced pastures, and urban development. These landscape changes have led to increased losses of topsoil and decreased species diversity. There is growing interest in shifting to more sustainable systems of grassland management, with the intent to protect or improve soil and plant resources and yet maintain economic viability.

This chapter presents principles of grassland ecology that provide rationale for management and improvement of grazing lands. Many physical (abiotic) and biological (biotic) factors interact to determine the species composition, productivity, and dynamics of grasslands. Abiotic factors include limitations of plant productivity by water (especially drought), light, nutrients, temperature, and fire, whereas biotic factors include herbivory by above- and belowground animals, diseases, decomposing organisms, and disturbance by digging animals. The study of grassland ecology emphasizes the interactions of these different processes. Although these interactions play at least a partial role in the composition of most grasslands, their relative importance varies widely among grassland types. Thus, only a few generalizations hold true for all grassland types. When pasture or range management practices fail in a particular location, it is usually because managers applied over-

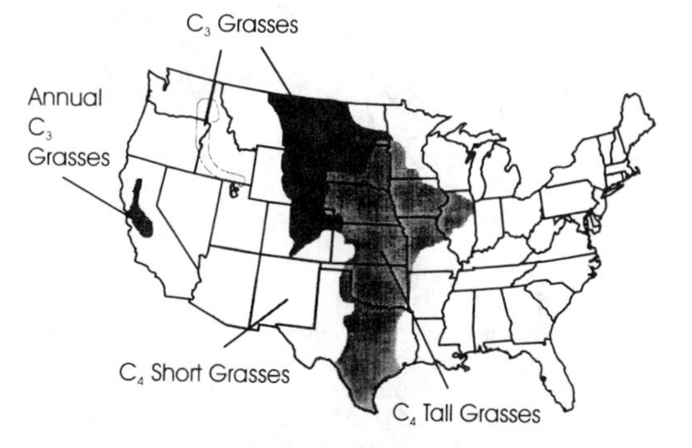

Fig. 11.1. General regions for major grasslands of the US. (Adapted from Brown 1993 and Gould and Shaw 1983).

KEVIN D. KEPHART is Associate Professor of Plant Science, South Dakota State University. He received his MS degree from the University of Wyoming and his PhD from Iowa State University. He teaches forage crop management and crop physiology and conducts research on morphology, yield components, and quality of forages.

CHARLES P. WEST is Associate Professor of Agronomy, University of Arkansas. He served as a postdoctoral research fellow in New Zealand from 1982 to 1984, conducting research on N_2 fixation in pastures. He received his MS degree from the University of Minnesota in 1978 and his PhD from Iowa State University. He specializes in the physioecology of grass-endophyte interactions.

DAVID A. WEDIN is Assistant Professor of Botany at the University of Toronto. He received his PhD degree from the University of Minnesota. He specializes in the study of plant competition and nutrient dynamics of grassland ecosystems.

generalized lessons from other grassland types and failed to first understand the ecology of their particular system.

CLIMATE AND THE DISTRIBUTION OF GRASSLANDS

Water availability is the principal climatic determinant affecting development of grasslands. Grasslands generally exist in semiarid to subhumid regions with 300-1000 mm of annual precipitation. Drier areas support deserts, and wetter areas generally support forests (Fig. 11.2). Temperature and soils interact with precipitation to determine water availability and thus are also important determinants of grassland distribution (see Chap. 5).

Evapotranspiration is the total amount of water lost from the soil: the sum of transpiration through plants plus evaporation from the soil surface. Both temperature and relative humidity influence evapotranspiration. For example, Minneapolis, Minnesota, has a mean annual precipitation of 725 mm and a mean annual temperature of 5.5°C. Manhattan, Kansas, has a mean annual precipitation of 835 mm and a mean annual temperature of 12.8°C. Although Manhattan receives greater precipitation, it has greater evapotranspiration (740 mm H_2O yr^{-1}) than Minneapolis (590 mm H_2O yr^{-1}) and is, in effect, the drier of the two sites. Both of these tall-grass prairie sites would be considered humid, however, because annual evapotranspiration is less than annual precipitation, indicating that, on average, water is not greatly limiting to plant productivity. In the semiarid shortgrass prairie of northeastern Colorado, potential evapotranspiration (605 mm H_2O yr^{-1}) is almost double actual mean precipitation (310 mm H_2O yr^{-1}). In the arid desert grasslands of southern New Mexico, potential evapotranspiration (800 mm H_2O yr^{-1}) greatly exceeds actual precipitation (230 mm H_2O yr^{-1}).

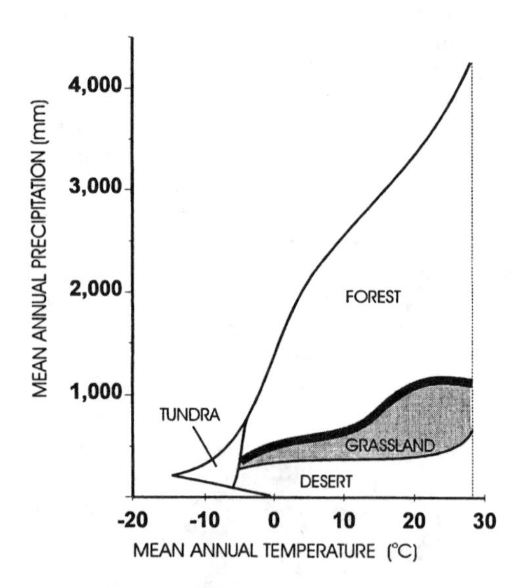

Fig. 11.2. Seasonal temperatures and precipitation are the major climatic factors that influence the class of the vegetational ecosystem. Tundra forms in cold, dry climates. Grasslands form in regions that are too wet for deserts but too dry for forests; however, there may be a broad region of mutual adaptation for forests and grasslands. (Adapted from Whittaker 1975.)

Although grasslands generally occur in areas too dry to support forests, natural grasslands also developed in more humid zones, such as the central North American Corn Belt. Grass dominance in humid environments has often resulted from periodic disturbance, such as fire or forest clearing, in combination with highly variable precipitation both within and between years (Borchert 1950). Intermittent drought disproportionately stresses woody vegetation relative to drought-adapted grasses.

Natural grasslands may also occur in regions with relatively high precipitation because of unusual soil conditions. Tall-grass formations once occurred as far east as Long Island, New York, and the coastal plain of the southeastern US. These grasslands occurred on sandy soils, which have both low-nutrient and low-water-holding capacity compared to adjacent loamy forest soils. Fire was also important in maintaining these grassland communities. Other grasslands in the eastern US, such as the Blackland Prairies of Mississippi and Alabama, occurred on poorly drained clay soils that were waterlogged from fall to late spring but dried during summer. Finally, grasslands may dominate on extremely shallow soils where forests may be water-limited because of limited rooting depth.

In the western US, topography interacts with precipitation in determining the composition of vegetation types. Semiarid conditions generally occur in rain shadow regions located on the leeward side of mountain ranges. These include valleys east of the Pacific Coast mountain ranges, the Great Basin region east of the Cascade and Sierra Nevada Mountains, and the Great Plains region east of the Rocky Mountains (Fig 11.1). A diagrammatic longitudinal transect (Fig 11.3) shows woodland associated with subhumid to humid zones and grassland associated with arid, semiarid, and subhumid zones. The intermontane Great Basin receives an average annual rainfall of 100-300 mm and is dominated by shrub-bunchgrass vegetation.

The Great Plains region extends from just east of the Rocky Mountains to the eastern deciduous forests and contains the widest diversity of grassland types on earth. This region is also called the North American grassland biome. The west-to-east gradient of increasing precipitation is associated with increasing grassland productivity and increased canopy height. Sparse, short grasses (15-30 cm) such as blue grama (*Bouteloua gracilis* [H.B.K.] Lag. ex Steud.), buffalograss (*Buchloe dactyloides* [Nutt.] Engelm.), and western wheatgrass (*Pascopyrum smithii* [Rydb.] Löve) are the dominant species of the semiarid shortgrass prairie, also called the *shortgrass steppe*.

The tall-grass prairie formed in the eastern region of the grassland biome. Here, dominant species grow up to 180 cm and include big bluestem (*Andropogon gerardii* Vitman), little bluestem (*Schizachyrium scoparium* [Michx.] Nash), switchgrass (*Panicum virgatum* L.), and indiangrass (*Sorghastrum nutans* [L.] Nash). The native grasslands of the central grassland biome are the mixed-grass prairies, dominated by species characteristic of both tallgrass and shortgrass regions together with other wheatgrasses (*Elymus, Elytrigia,* and *Pascopyrum* spp.) and needlegrasses (*Stipa* spp.).

Whereas a water availability gradient extends from east to west across the grassland biome, a temperature gradient extends from south to north. Grasses are classified into two groups according to photosynthetic pathway, C_3 and C_4 (see Chap. 3). The C_3 species have temperature optima of 15°-30°C and are often

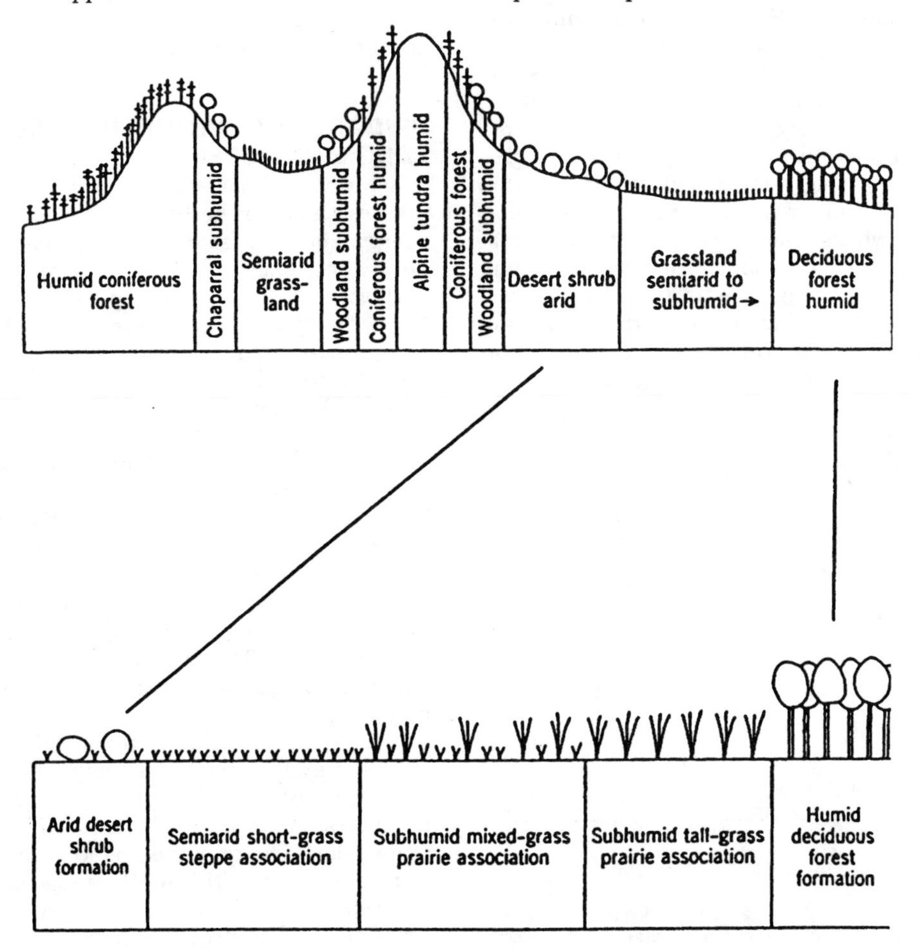

Fig. 11.3. A diagrammatic description of continental regions where grassland ecosystems develop. Major grasslands form in rain shadows on leeward sides of major mountain ranges. Such regions are too warm for tundra, too wet for deserts, and too dry for forest development. (Adapted from Harlan 1956.)

called *cool-season grasses* because of their frost tolerance. They have optimum growth during spring and autumn. The C_4 species have temperature optima of 25°-40°C and are referred to as *warm-season grasses* because they reach their maximum productivity during summer. Although most of the North American grassland biome is dominated by C_4 grasses, C_3 species become more important with increasing latitude and elevation because of the cooler climate (Fig. 11.4).

Ecologists recognize the key role climate plays in determining the distribution of grasslands relative to other vegetation types as well as the distribution of different grassland types within a larger grassland biome. Recent research, however, has led grassland ecologists to reconsider many of the classic assumptions about the distribution of grasslands and the dynamics of plant communities. An older view is that a single and predictable set of plant species, a climax community, existed for each set of climatic conditions (Clements 1915). This climax community was thought to be relatively stable in terms of both productivity and species composition. Disturbances such as overgrazing, drought, fire, or plowing were thought to change this climax community to earlier successional stages, which, following a predictable successional sequence, would return to the climax state. Range condition is frequently determined by the degree of similarity of existing vegetation to the presumed climax vegetation of the area prior to livestock grazing (Sampson 1917).

Today, grassland communities are viewed as being much more dynamic. It is now clear

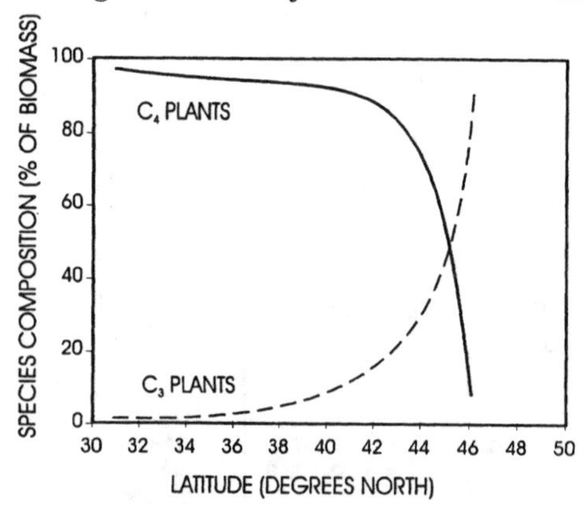

Fig. 11.4. Cool-season grasses (C_3) are better adapted to cool climates than warm-season grasses (C_4). As latitude increases, C_3 species become more dominant in the grassland ecosystem. (Adapted from Sims 1988.)

that climate varies dramatically on time scales of years, decades, centuries, or millennia. Even without human intervention, the distribution of grassland types across North America would have changed because of natural climatic changes in this century. Within the current climate, some disturbed grassland communities will not return to their previous status. For example, black grama (*Bouteloua eriopoda* [Torr.] Torr.) grasslands of the Southwest apparently originated during a global cooling period that lasted from roughly 1600 to 1900 (Nielson 1986). Climatic conditions that allow successful germination and establishment of black grama have occurred in only 7 yr since 1915. It is not realistic to expect management practices to cause these grasslands to return to climax black grama grasslands if they are overgrazed or other stresses occur. In some cases, grassland degradation is reversible, as predicted by older climax vegetation theory, but in other cases, particularly in arid landscapes, it may not be.

PRIMARY PRODUCTION IN GRASSLAND ECOSYSTEMS

Being the primary producers in grasslands, grasses and other herbaceous species provide energy and nutrients for large mammalian herbivores, such as cattle, and for smaller herbivores, such as grasshoppers and nematodes. These organisms that consume the vegetation are secondary producers and form a second trophic level in the food chain. Predators and humans that consume animal products form a third trophic level in the food chain. In a parallel food chain, plant and animal tissues not consumed by higher trophic levels pass through a decomposer food chain. Invertebrates, fungi, and bacteria ultimately break down plant and animal organic matter to carbon dioxide (CO_2) and inorganic forms of nitrogen (N), phosphorus (P), sulfur (S), and other minerals. The consequence of decomposition is a release of energy and recycling of nutrients.

If we consider the total combined biomass of plants and animals in a temperate region pasture, it is clear that plants, the primary producers, make up the bulk of the biomass (Table 11.1). The primary producers provide a tremendous foundation for the support of other forms of life. During the 19th century, the grasslands of central North America nourished 60-70 million bison (*Bison bison* L.) and about 50 million antelope (*Antilocapra americana* Ord), in addition to the smaller wildlife, insects, and other herbivores that lived on the grasslands (Chadwick and Brandenburg 1993). It is astounding to consider that the combined weight of the bison population alone was greater than that of all the human population currently living in the US and Canada. Equally astounding is that the second most abundant group of organisms in a grassland are not the mammals and birds that one generally notices but instead the fungi, bacteria, arthropods, worms, nematodes, and other fauna found below the surface. Most of these organisms are involved in decomposition. The combined weight of belowground animals (arthropods, worms, and nematodes) may be three to four times that of livestock at average stocking rates.

Energy flowing through both the consumer and decomposer food chains originates from solar radiation intercepted and used for photosynthesis. Net primary production (NPP) is the annual growth of above- and belowground plant biomass produced per unit area (kg ha^{-1} yr^{-1}).

Primary production in grasslands is usually limited by the supply of essential nutrients and water or by unfavorable temperatures. Occasionally, excess water or the accumulation of nutrients, salts, or other minerals to toxic levels may limit grassland NPP. Variation in factors that limit NPP define differences among ecosystems and the rationale used to manage them. An obvious gradient in grassland productivity is caused by water availability, with low NPP in desert or arid grasslands and relatively high NPP in humid grasslands. Even in arid grasslands, however, nutrient limitations to NPP may occur. Semiarid short-grass prairies in Colorado and desert grasslands in New Mexico show large increases in productivity with irrigation, but they also show small increases in dry land productivity with nutrient additions. In semi-

TABLE 11.1. Biomass yield of several organism classes for a nongrazed temperate grassland

Organism class	Biomass (kg fresh weight ha^{-1})
Plants	20,000
Fungi	4000
Bacteria	3000
Arthropods	1000
Annelids (worms)	1320
Protozoa	380
Algae	200
Nematodes	120
Mammals	1.2
Birds	0.3

Source: Pimentel et al. (1992).

humid and humid grasslands, nutrient limitation rather than soil water is usually the dominant constraint on NPP.

One cannot fully understand grassland ecosystems without also understanding belowground processes. Belowground NPP exceeds aboveground NPP in most grasslands by as much as 5- to 10-fold. Grasses alter photosynthate partitioning between above- and belowground structures in response to various stresses. For example, grazing and low light availability cause reduced root growth relative to top growth. Conversely, drought increases root growth relative to top growth. Even when aboveground productivity changes little, large changes usually occur in belowground productivity, which can ultimately undermine ecosystem stability.

Most soil organic matter is derived from belowground biomass. The fertile soils of the world's richest agricultural regions, such as in the midwestern US, Ukraine, and the pampas of South America, were all produced by grasslands. Grasslands are also effective in rebuilding soil organic matter after it has been depleted by crop production or erosion. Three factors contribute to this process: high rates of belowground productivity, slow decomposition of grass roots and litter, and low to moderate rainfall in grassland regions, which discourages leaching of nutrients and minerals.

NUTRIENT CYCLING IN GRASSLAND ECOSYSTEMS

The dynamics of carbon (C) and nutrients are tightly linked in grassland ecosystems. In most temperate terrestrial ecosystems, N is the major nutrient limiting plant growth. Nitrogen is not a component of parental rock materials, which are the sources for most mineral nutrients. Moreover, the large pool of dinitrogen (N_2) gas in the atmosphere is not available to most plants. Phosphorous limits plant growth in many tropical regions that have soils that are highly weathered and therefore have depleted P.

Nitrogen cycling in grassland ecosystems is an important regulator of both primary and secondary productivity (Fig. 11.5). Most natural grasslands are characterized by a relatively closed N cycle. These ecosystems have relatively low inputs and losses of N in relation to the annual exchange of N among soil, vegetation, and animal components. Managed grasslands, such as pastures or hayfields, are characterized by both increased N inputs, amount removed, and natural losses (Ball and Ryden 1984).

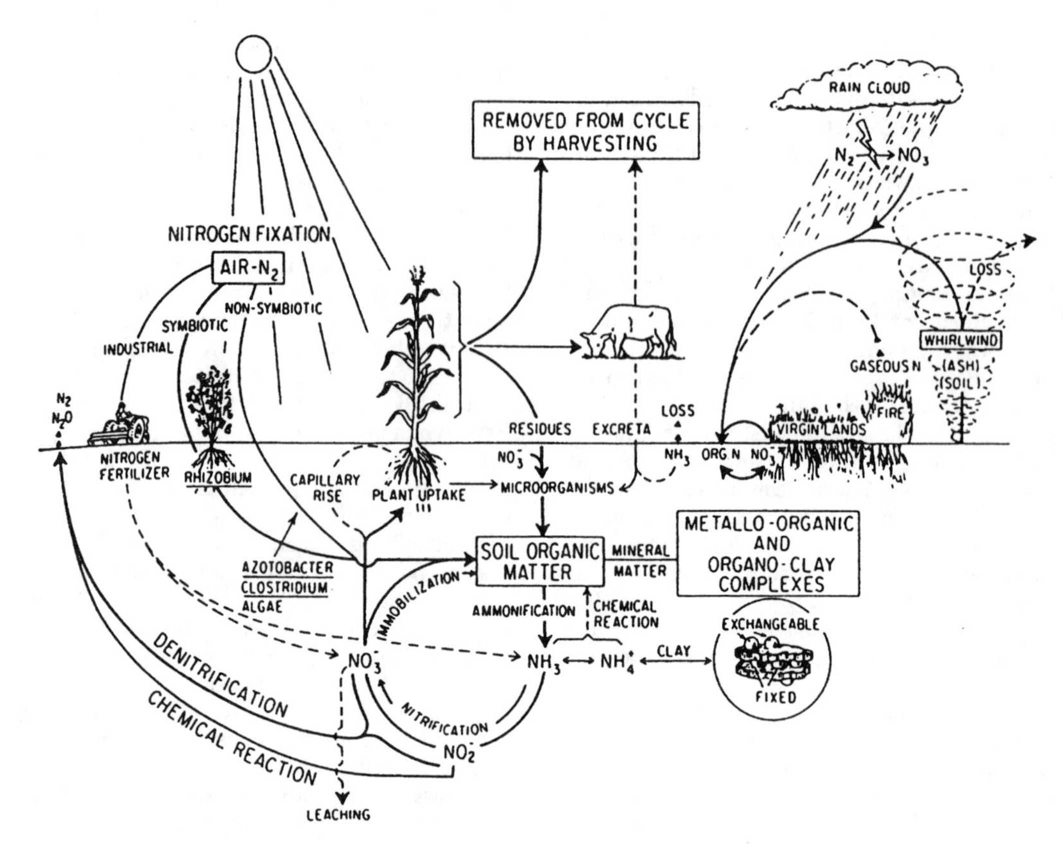

Fig. 11.5. A generalized flow of nitrogenous compounds in a grassland ecosystem. (Adapted from Tisdale and Nelson 1975.)

Primary inputs of N into natural grasslands include fixation, or reduction, of N_2 gas by free-living bacteria or by *Rhizobium* bacteria associated with legumes. Annual N inputs in natural grasslands tend to be small compared to release (mineralization) of ammonium (NH_4^+) and nitrate (NO_3^-) from plant litter and from soil organic matter by microbial decomposers. The total annual supply of available N from both mineralization and fixation for most unfertilized grasslands ranges from 30 to 80 kg N ha^{-1} yr^{-1}. This is lower than rates in most deciduous forests or plowed croplands. Historically, inputs of NH_4^+ and NO_3^- into grassland ecosystems by atmospheric deposition, often during thunderstorms, were very low (1-5 kg N ha^{-1} yr^{-1}); however, this may no longer be true because of air pollution.

Most grassland soils tend to be well aerated and have neutral to basic pH. Under these conditions, excess NH_4^+ in the soil is converted to NO_3^- by bacteria. Because of its negative charge and the lack of positively charged sites on most soil particles, NO_3^- is not tightly held in the soil and is readily leached with water moving through the soil. Although N losses are generally low in intact natural grasslands, they can be significant under high stocking density or when N inputs from manure, fertilizer, or pollution exceed the uptake capacity of plants. Under these conditions, the volatilization of ammonia (NH_3) and nitrogenous gases, such as nitrous oxide (N_2O), may also lead to significant loss of N from the ecosystem. Understanding these N losses and the efficiency with which N is cycled in grasslands is important because fertilizer N inputs are costly and N losses can be detrimental to the environment. For example, NO_3^- leaching below the rooting zone can enter groundwater, where it is a major pollutant.

Availability of soil N for plant growth is generally low in intact grasslands. Relatively low N concentrations in senescent grass tissues, together with the generally dry climates, lead to slow decomposition of grass litter and roots. Nitrogen is particularly limiting to decomposers and is effectively reincorporated into microbial biomass in the litter or soil rather than being released in mineralized forms available for plant uptake. Such soils, however, are highly fertile when plowed and cultivated for crops because of the gradual release of nutrients from a large organic pool with relatively low loss. Typically, grasslands contain 1000 kg N ha^{-1} in the top 15 cm of soil for every 1% of soil organic matter. Increases in temperature and aeration due to long-term cropping along with reduced inputs of plant biomass cause a net reduction in soil N and C. Soil organic matter levels have declined an average of 50% across the North American grassland biome since European settlement (Buyanovsky et al. 1987).

By contrast, in tropical rainforest ecosystems, most C and nutrients are in living biomass. As rainforest litter decomposes, nutrients are quickly released and made available again. The concentrations of available soil nutrients are high, supporting the high productivity of the vegetation. When living vegetation in the rainforest is harvested or burned, however, large amounts of nutrients are easily lost through leaching. Such soils rapidly become infertile for crop production.

HERBIVORY IN GRASSLAND ECOSYSTEMS

Grasslands often have high amounts of herbivory; roughly 20%-60% of NPP is consumed by vertebrate and invertebrate herbivores. In contrast, 50%-75% of aboveground NPP is consumed by grazers in intensively managed temperate pastures (Ball and Ryden 1984). In natural grasslands, invertebrate herbivory may account for roughly 10% of the consumption of NPP (Detling 1988). Belowground herbivores frequently consume 25% of belowground NPP, with nematodes alone accounting for 10%-15% (Detling 1988).

Grazing has both negative and positive impacts on plants. It affects plants negatively by (1) reducing leaf area, (2) removing active apical meristems, (3) depleting plant stores of nutrients, and (4) shifting allocation of energy and nutrients from roots to shoots in order to replace lost photosynthetic tissues. Conversely, grazing benefits plants by (1) increasing light availability through the plant canopy, (2) removing older leaves and increasing the proportion of younger leaves with higher photosynthetic rates, and (3) activating dormant meristems in crowns and rhizomes. Grasses are particularly capable of enduring frequent defoliation because they have intercalary meristems for leaf and stem growth, they often have dense tillering from a single plant, and grasses can endure greater shoot desiccation to avoid drought (Coughenour 1985). On the other hand, legumes, forbs, and shrubs usually have elevated growing points, which are removed when grazed.

Nitrogen removed from managed grasslands in the form of meat, milk, and other animal products accounts for only 5%-20% of the N consumed by the animals (Ball and Ryden 1984). In spite of this relatively small amount of removal, herbivory not only affects the physiology of the grazed plants but also resource allocation for the overall grassland ecosystem. Grazing ruminants gather forage N (largely plant protein) that is fairly uniformly distributed across the landscape, and they concentrate it into small areas as urine and feces, especially where the animals congregate.

Protein (16% N) in the forage is readily converted to NH_3 in the rumen. Excess NH_3 that is not consumed by rumen microflora is absorbed through the rumen wall and excreted in the urine. Excretion of N in urine occurs on small areas and at high frequency around certain sites in pastures, such as near water and in shaded areas. Application rates of N in urine patches are typically 300 kg N ha^{-1}, far in excess of the potential for plant absorption in a short time. The resulting high amounts of labile N compounds in the soil solution are readily volatilized as NH_3 or leached as NO_3^-. Thus, grazing alters N distribution, hastens N cycling, and increases the vulnerability of the grassland to N losses.

In addition to removing and redistributing nutrients, grazing changes the energy balance at the soil surface, creates soil disturbances, and alters plant colonization. Grazing frequently has great effects on the species composition of the plant community. Differences among plant species in palatability to herbivores and in their tolerances to heavy or lax grazing pressure (animal units per unit of available forage) can lead to competitive superiority of one species over another. Grazing generally shifts the competitive balance to favor shorter or prostrate species such as white clover (*Trifolium repens* L.), Kentucky bluegrass (*Poa pratensis* L.), buffalograss, and blue grama. Lax grazing or leaving areas ungrazed allows taller species to intercept light at the expense of shorter species.

Plant Responses to Managed Herbivory. Plants have generally evolved through one of two responses to herbivory: tolerance or avoidance. Many grasses tolerate grazing by having efficient mechanisms for regrowth following tissue loss. Other plants avoid grazing with either physical deterrents, such as spines, or chemical deterrents, such as alkaloids and tannins. Although chemical defenses against herbivory are less common in grasses than in other plants, *Acremonium* fungal endophytes produce alkaloids in tall fescue (*Festuca arundinacea* Schreb.) and perennial ryegrass (*Lolium perenne* L.) that cause animal toxicosis and lead to reduced forage intake. Endophyte presence in these grasses also inhibits insect and nematode herbivory and so confers a competitive advantage to infected individuals over noninfected grasses.

Given the diversity of ways plants respond to herbivory both physiologically and evolutionarily, it is not surprising that grassland types respond differently to grazing. For example, in a comparative study of cattle grazing across the North American grassland biome, each of the four major grassland types responded differently (Detling 1988). In tallgrass prairie, grazing resulted in a greater abundance of forbs and C_3 grasses relative to ungrazed treatments. In mixed-grass prairie, grazed plots had an increased abundance of C_4 grasses. In short-grass prairie, grazed plots had an increased abundance of succulent species (cacti), and in desert grasslands, grazed plots had an increased abundance of forbs and shrubs.

Although herbivory, especially grazing by large mammals, is important in the continental US and most grasslands, some grasslands had no history of large mammalian grazers prior to European settlement. These include Hawaii, the South American pampas, and the tussock grasslands of New Zealand and Australia. The intermontane grasslands of North America west of the Rocky Mountains had only light grazing pressure compared to the central North American grasslands east of the Rocky Mountains. Each of these grassland ecosystems has undergone dramatic changes in species composition and productivity over the last century in response to the introduction of domestic livestock or the altered use by large herbivores.

A major question in grassland ecology is "Under what conditions does moderate to heavy grazing increase grassland productivity?" (Levin 1993). If defoliation from grazing is either too intense or too frequent, negative impacts dominate and result in decreased potential NPP. Levels at which excessive grazing pressure occur differ greatly among grass species and among grassland types. Ecological studies of grasslands have been used to support arguments that grazing increases grassland productivity and is beneficial to grasses (Savory 1988). There are also arguments that grazing leads to degradation and desertification in grasslands (Schlesinger et al. 1990). It is clear there is no single prescription for grazing management that optimizes livestock productivity while sustaining grassland productivity across all grassland types.

Two insights have emerged from recent ecological studies of grazing impacts on grassland ecosystems. The first is the importance of belowground processes. Although a short-term increase in aboveground NPP may be seen in response to heavy grazing pressure, the long-term productivity of grasslands depends on belowground productivity, which often decreases sharply as grasses repartition resources aboveground. The second insight is the importance of N availability. Although short-term productivity may increase in response to increased N availability in grazed grasslands, a long-term decline in grassland productivity will occur if increased N losses associated with herbivory are not balanced by increased inputs. These effects are likely to be greatest in humid ecosystems.

FIRE IN GRASSLAND ECOSYSTEMS

Fire was critical for maintaining several types of native grasslands prior to European settlement, including the grasslands of central North America, the high velds of South Africa, and the pampas of South America (Daubenmire 1968; Huntley and Walker 1982; Collins and Wallace 1990). In just a few decades, savannah or forest can displace grasslands when defoliation by fire, mowing, or grazing is absent (Fig. 11.6). Fires restrict invasion by woody and weedy vegetation.

The value of fire as a management tool varies between arid and humid grassland ecosystems because it can also alter botanical composition. In water-limited short-grass communities, forage availability is low, but forage quality tends to be high, even in senesced plant material. Consequently, fire in arid grasslands is detrimental to livestock production because it removes valuable forage. On the contrary, in subhumid to humid grasslands, there is a buildup of C_4 grass biomass that contains low N concentrations and high concentrations of fiber, both of which limit plant litter decomposition. Fire improves forage quality by removing this low-quality dead biomass, thus allowing regrowth of nutritious young forage. Therefore, fire is a practical, low-input tool for reducing undesir-

Fig. 11.6. Periodic fires prevented establishment of forests in some regions, permitting grasses to dominate the landscape. The top photograph was taken in 1874 during an expedition of the Black Hills of Dakota Territory, led by Lt. Col. George Armstrong Custer. The lower photograph was taken at the same location 100 yr later. Note the greater forest growth in the later photograph. Extended forest growth in the Black Hills has resulted after several decades of fire prevention management. *Photographs courtesy of South Dakota State University.*

able vegetation and improving forage quality for livestock production in environments where water availability allows rapid recovery of the stand without loss of desirable perennial grasses. An alternative way to consider fire is suggested by Bell (1982): "Fire is an herbivore that does not require protein for growth."

IMPROVEMENT

Grassland improvement in an agricultural context entails the use of management to increase primary and secondary production. Common improvement strategies involve the introduction or encouragement of plant genotypes of high nutritional quality, high yield, and long stand life and the use of animal genotypes with efficient feed conversion. Inputs are commonly used to alleviate limitations to growth, such as soil fertility, and to control undesirable competitors, predators, and parasites. Additionally, grassland improvement involves controlling the intensity and timing of grazing. Finally, grassland improvement includes measures that avert environmental degradation, such as soil erosion, and that improve nutrient cycling from animal excreta back into the forage. Moreover, improvement may entail nonagricultural goals such as species diversity, preservation of natural conditions, recreation, and landscape beautification.

The goals of improvement depend upon the status of the ecosystem. For productive grasslands, improvement goals are likely aimed to maintain economic or environmental viability. More drastic or expensive improvement measures are often used to revitalize grasslands in poor condition. Milton et al. (1994) propose a five-step model of changes in status of arid and semiarid grassland vegetation, beginning with optimum secondary production and ending with denudation (Table 11.2). A key element of this model is that natural oscillations in vegetative composition occur within each step. Also, each step has a class of management practices for improvement. As a grassland ecosystem condition decreases toward denudation, the expense of improvement practices generally increases and the likelihood of restoring secondary production to economically useful levels diminishes.

Nutrient Management. Net productivity increases with the supply of a limiting resource such as water or N. Chemical or organic fer-

TABLE 11.2. Stepwise degradation model for arid and semiarid grasslands

Step Number	Description	Symptoms	Management option	Management level
0	Biomass and botanical composition varies with climatic and stochastic factors.	Perennial vegetation varies with weather.	Practice adaptive management.	Secondary producers
1	Herbivory reduces production of palatable plants, permitting increased production of less palatable species.	Demography of plant population changes.	Allow controlled grazing.	Secondary producers
2	Certain plant species fail to persist and are lost, as are their associated symbionts and predators.	Lost species diversity results in reduced secondary productivity.	Manage the vegetation; add seed, replant, interseed, apply herbicides, and improve fertility.	Primary producers
3	Primary biomass production varies as ephemeral species dominate landscape instead of perennials.	Perennial grass production decreases, and number of short-lived plants increases.	Manage soil cover and prevent erosion.	Physical environment
4	Denudation and desertification take place.	Bare ground and widespread erosion result.	Develop alternative land-use options.	

Source: Milton et al. (1994).

Note: The description and symptoms refer to the status of the grassland to economically support secondary production. Management options are the actions that a grassland manager can take to improve the production status. The management level refers to the production or trophic level at which management options should be focused.

tilizers, such as organic wastes, are frequently applied to pastures to remove a major limitation to plant growth, such as too little N, especially where precipitation is abundant. Nitrogen is the most frequently applied nutrient to grass-dominated pastures because of the high N requirement of grasses for growth. Potassium (K) and P are more likely to be limiting in legume-dominated systems.

Optimization of soil fertility and pH enhances plant growth and allows an increase in stocking density, but it often has little or no effect on forage quality. Fertilization with P and K may increase individual animal performance by increasing legume composition in a legume-grass mixture. Adding magnesium (Mg) fertilizer may reduce risk of grass tetany.

In New Zealand, applying P to perennial ryegrass/white clover pastures has an amplifying effect on nutrient acquisition and total productivity. Since P is the major limiting nutrient for legumes in New Zealand, P fertilization allows the clover to compete with ryegrass. This also promotes N_2 fixation by the legume and, in turn, is consumed by grazers, excreted, and mineralized by soil bacteria. Consequently, this N stimulates growth of the ryegrass. Regular P inputs are designed to maintain clover competitiveness and to replace P removed by livestock.

Grasses and legumes exploit different niches regarding N acquisition. Grasses depend on soil NO_3^- and NH_4^+. Legumes also take up NO_3^- and NH_4^+ from the soil, but when those forms are in short supply, they have the additional capacity to assimilate atmospheric N_2 derived from symbiosis with *Rhizobium* and *Bradyrhizobium*. Encouraging legume growth in pastures has an enhancing effect on the flow of energy from the sun to a marketable animal product. Relatively small inexpensive inputs of off-farm energy in the form of seed, inoculum, P and K fertilizer, and the manager's time can result in the capture of "free" N in legume forage.

Grazing Management Systems. The objectives of grazing management are varied but usually are aimed toward optimizing production per animal and animal production per land area. Other objectives include improved sustainability, uniformity, and convenience of animal production.

Extensive systems of grazing management are characterized by low animal densities on large areas. Such lands usually have a major climatic constraint, such as low precipitation or a cold temperature, which limits the amount and duration of forage production. Yet livestock production constitutes the best agricultural use of these lands because environmental constraints usually prohibit crop production. An example is the Great Basin region of western North America. Low carrying capacity and high climatic risk make it less economically attractive to invest in management inputs such as weed and brush control, fertilization, and reseeding.

In contrast to extensive systems, intensive systems of grazing management are practiced where precipitation and temperature allow high forage growth rates over a relatively long growing season and thus permit high stocking rates. The main limiting factors to production are supplemented with inputs such as fertilizers. Intensive grazing management may also include high input of the manager's time and the manager's decision making on stocking rate, timing of moving animals, positioning of supplemental feeds, and other management considerations. Thus, the manager can control the flow of forage into animals as opposed to simply trying to produce more forage.

Grazing methods can be used to improve the spectrum of grassland systems from extensive rangeland to intensive annual forage crops. In some cases herds are concentrated into pasture subdivisions for limited periods to enforce a more complete utilization of the forage. After the desired degree of grazing is achieved, the herd is shifted to another subdivision, and the grazed pasture is allowed to recover. The results are a greater proportion of plant energy directed through the animal and less wasted plant residue.

A rotational stocking system is the main grazing method used to control the degree of defoliation. Continuous stocking can also be controlled but requires careful monitoring of grazing pressure, grassland condition, and placement of water, feed, and shade. The desired forage species must be tolerant of close, repeated defoliation.

New Plantings. Another option for improving grassland is seeding new species or cultivars that introduce germplasm with enhanced NPP, forage quality, grazing tolerance, resistance to pests, or tolerance of climatic extremes. New plantings are also useful for noneconomic objectives, such as to build soil organic matter and fertility, to reduce soil erosion, to encourage diversity of plant and animal wildlife, and to beautify the landscape.

New plantings do not necessarily involve only introduced, nonindigenous species. Native species are sometimes seeded into semiarid and subhumid grasslands to speed recovery of desirable species composition in conjunction with brush control, controlled stocking and burning, and fertilization. Seedbed preparation and replanting on water-limited rangeland carries substantial risk of failure. Therefore, species selection must carefully match the environment and intended use, and replanting must be considered as merely one component in a package of improvement practices. Understanding climatic and edaphic conditions is essential when selecting species. Rather than sowing the same species to all terrain, appropriate species should be matched to specific soil and climatic conditions at each site.

Establishment of new seedings entails disturbance of the existing ecosystem by opening the soil cover, perhaps applying fertilizer, and minimizing competition to the seedlings. Seeding directly into an existing stand affords a means of incorporating a complementary species to diversify animal diets, to provide symbiotic N_2 fixation with legumes, or to lengthen the effective growing season. Interseeding, or sod seeding, using reduced or no-till techniques minimizes soil disturbance, thereby decreasing the risk of soil erosion. Control of competition from existing vegetation with herbicides, controlled stocking, or mowing is essential for successful interseeding.

GRASSLANDS, SOCIETY, AND GLOBAL CHANGE

The end of the 20th century brings unprecedented environmental change caused by human activity. The controversy over environmental warming has caused concern about grassland ecosystems. The rise in atmospheric CO_2 concentrations from 280 $\mu L\ L^{-1}$ prior to 1850 to over 360 $\mu L\ L^{-1}$ in the 1980s is not debated. Because CO_2 is often a limiting resource for photosynthesis, this increase may have major positive or negative impacts on both grassland productivity and forage quality. For example, evidence suggests that increased atmospheric CO_2 concentrations will favor C_3 species at the expense of C_4 species (Mooney et al. 1991).

Increases in pollutants containing S and N have occurred, particularly since 1950. Rates of atmospheric N deposition are now up to 20 kg N ha^{-1} in eastern North America and up to 50 kg N ha^{-1} in northern Europe. This atmospheric deposition is caused by increased fossil fuel combustion and increased use of N fertilizer. Consequently, changes in management may be necessary for both natural and artificial grasslands. For example, routine mowing and haying may be needed to remove excess N from the ecosystem. Grazing would be less effective for nutrient removal.

Biological diversity at the ecosystem, species, and genetic population levels is also in jeopardy. Loss of diversity is caused by environmental degradation and large-scale destruction and fragmentation of natural ecosystems. There is a growing realization that much of the biodiversity in grasslands is directly impacted by agricultural practices. In Minnesota, for example, only a few thousand hectares remain of the millions of hectares of original tall-grass prairie. Of Minnesota's 250 rare and endangered species of plants and animals, 105 are associated with native prairie.

Of what value is grassland biodiversity? For a breeder it represents a genetic resource for use in developing new cultivars adapted to particular climates or pests. Grassland biodiversity may also have a role in grassland stability during drought or other disturbances. A Minnesota study found that the effect of a severe drought differed across a series of grassland plots depending on how many plant species the plots contained. Native prairie plots with higher species diversity recovered more quickly than less diverse plots (Tilman and Downing 1994).

As yet, we have little understanding of the functional role of most species in grassland ecosystems. Even in relatively simple temperate pastures dominated by a few forage species and a single large grazer, there may be up to 1000 arthropod species per hectare (Pimentel et al. 1992). Some of these are herbivores, competing with domestic livestock for forage, but most are part of the decomposer food web or are predators on other arthropods. Although simplification of grassland ecosystems through loss of species appears to be an unavoidable consequence of intensive management, managers should minimize these losses of biodiversity until their consequences are better understood.

New approaches to land management, such as integrated resource management, focus on balancing all aspects of ecosystem management (Pimentel et al. 1992). Decisions are balanced between economic (crop, forage, and grazing) and environmental (soil, wildlife, and water quality) concerns. The goal is to develop approaches that are economically and environmentally sustainable in a multiple-use landscape. Grasslands can conserve soil organic matter and topsoil, reduce rates of nutrient leaching and runoff, and provide wildlife habitat while still maintaining economic levels of livestock production.

QUESTIONS

1. Outline the major abiotic and biotic factors that determine where grasslands form. Give particular emphasis to major climatic factors.
2. List the dominant grass species in the short-, mid-, and tall-grass prairies.
3. What is the fate of prolonged overgrazing? Will a grassland necessarily return to a productive status once grazers are removed? Elaborate on how grazing can benefit grasslands.
4. Propose a simplified N flow diagram for a grassland in your region. Take into account species and conditions that occur in your area.
5. List some situations where fire is beneficial to grassland development.
6. Discuss improvement techniques for (a) grasslands with declining populations of desirable species, (b) grasslands that have lost certain key species, (c) grasslands suffering from lack of prolonged production or extensive erosion, (d) denuded grasslands.
7. Explain the importance of grassland ecosystems to the future of agriculture.
8. List some regions where you think extensive and intensive systems of grazing management might be located.

REFERENCES

Ball, PR, and JC Ryden. 1984. Nitrogen relationships in intensively managed temperate grasslands. Plant and Soil 76:23-33.

Bell, RHV. 1982. The effect of soil nutrient availability on community structure in African ecosystems. In BJ Huntley and BH Walker (eds.), Ecology of Tropical Savannas, Ecological Studies 42. Berlin: Springer-Verlag, 193-216.

Borchert, JR. 1950. The climate of the central North American grassland. Ann. Assoc. Am. Geogr. 40:1-39

Brown, DA. 1993. Early nineteenth-century grasslands of the midcontinent plains. Ann. Assoc. Am. Geogr. 83:589-612.

Buyanovsky, GA, CL Kucera, and GH Wagner. 1987. Comparative analyses of carbon dynamics in native and cultivated ecosystems. Ecol. 68:2023-31.

Chadwick, DH, and J Brandenburg. 1993. The American prairie: Roots of the sky. Natl. Geogr. 184(4):90-119.

Clements, FE. 1915. Plant Succession. Publ. 242. Washington, D.C.: Carnegie Inst.

Collins, SL, and LL Wallace. 1990. Fire in North American tallgrass prairie. Norman: Univ. of Oklahoma Press,

Coughenour, MB. 1985. Graminoid responses to grazing by large herbivores: Adaptation, exaptations, and interacting processes. Ann. Mo. Bot. Gard. 72:852-63.

Daubenmire, R. 1968. Plant Communities: A Text-Book of Plant Synecology. New York: Harper and Row.

Detling, JK. 1988. Grasslands and savannas: Regulation of energy flow and nutrient cycling by herbivores. In LR Pomeroy and JJ Alberts (eds.), Concepts of Ecosystem Ecology: A Comparative View. New York: Springer-Verlag, 131-54.

Gould, FW, and RB Shaw. 1983. Grassland Systematics. 2d ed. College Station: Texas A&M Univ. Press.

Harlan, JR. 1956. Theory and Dynamics of Grassland Agriculture. Princeton, N.J.: D. Van Nostrand Co.

Huntley, BJ, and BH Walker (eds.). 1982. Ecology of Tropical Savannas. Ecological Studies 42. Berlin: Springer-Verlag.

Levin, SA. 1993. Grazing theory and rangeland management. Ecol. Appl. 3:1

Milton, SJ, WRJ Dean, MA du Plessis, and WR Siegfried. 1994. A conceptual model of arid rangeland degradation: The escalating cost of declining productivity. BioSci. 44:70-76.

Mooney, HA, BG Drake, RJ Luxmoorre, WC Oechel, and LF Pitelka. 1991. Predicting ecosystem responses to elevated CO_2 concentrations. BioSci. 41:96-104.

Nielson, RP. 1986. High-resolution climatic analysis and Southwest biogeography. Sci. 232:27-34.

Pimentel, D, U Stachow, DA Takacs, HW Brubaker, AR Dumas, JJ Meaney, JAS O'Neil, DE Onsi, and DB Corzilius. 1992. Conserving biological diversity in agricultural/forestry systems. BioSci. 42:354-62.

Sampson, AW. 1917. Plant succession as a factor in range management. J. For. 15:593-96.

Savory, A. 1988. Holistic Resource Management. Washington, D.C.: Island Press.

Schlesinger, WH, JF Reynolds, GL Cunningham, LF Huenneke, WM Jarrell, RA Virginia, and WG Whitford. 1990. Biological feedbacks in global desertification. Sci. 247:1043-48.

Sims, PL. 1988. Grasslands. In MG Barbour and WD Billings (eds.), North American Terrestrial Vegetation. New York: Cambridge Univ. Press, 266-86.

Tilman, D, and JA Downing. 1994 Biodiversity and stability in grasslands. Nature 367:363-65.

Tisdale, SL, and WL Nelson. 1975. Soil Fertility and Fertilizers. 3d ed. New York: Macmillan.

Whittaker, RH. 1975. Communities and Ecosystems. New York: Macmillan.

12 Hay and Silage Management

KENNETH A. ALBRECHT
University of Wisconsin

MARVIN H. HALL
Pennsylvania State University

FORAGES grown for hay and silage occupy approximately 25 million ha and result annually in production of over 130 million metric tons (mt) of dry matter in the US (Fig. 12.1). This production is valued at over 11 billion dollars to US agriculture. The additional environmental value of forages to society in the form of reduced soil erosion and nitrogen (N) fertilization is significant but unmeasured in economic or global scales. The land area occupied by forages for hay and hay silage production has remained relatively constant during the past 20 yr. However, during the same time period, annual forage production has increased by nearly 30%. This indicates that improved management practices have been adopted and/or improved genetic material is being utilized to increase production per unit area. Most likely, this has resulted from application of technology generated by research. In general, the technology base is still ahead of farmer practices, leaving opportunities for continued improvement through educational efforts.

KENNETH A. ALBRECHT is Professor of Agronomy, University of Wisconsin, Madison. He holds the MS degree from the University of Minnesota and the PhD from Iowa State University. His major research interest is in forage management and utilization.

MARVIN H. HALL is Associate Professor of Agronomy, Pennsylvania State University. He holds the MS degree from Ohio State University and the PhD from the University of Minnesota. His major extension and research interests are in management and utilization of forages.

The cost of producing, harvesting, and storing forage for livestock is rising steadily. A portion of this cost is fixed by factors that are independent of management and are not affected by hay yield or quality. Examples include land, machinery, and labor. Undersander (1994) demonstrates that there is little relationship between total cost of production per hectare and harvested yield of alfalfa (*Medicago sativa* L.). Production cost was $160/ha when yields ranged from 5.6 to 12.3 mt/ha in a 1993 Wisconsin farm survey. This is because many management inputs such as crop or cultivar selection, fertilizer, pest control, and timely harvest represent variable costs that are only a small percentage of the total production cost. In dairy forage systems in Wisconsin, which involve stored forage, each additional metric ton of alfalfa produced results in an increased return of $243/ha when the added product value of milk is included in the analysis (Fig. 12.2). Therefore, a management goal should be to optimize productivity (in terms of yield and appropriate quality for the livestock enterprise) of fields devoted to production of stored forages.

Proper management is essential to ensure and maintain profitability of the forage component of the farming enterprise. Within a set of climatic constraints that regulate potential productivity, management is the major factor controlling the actual productivity of hay and hay silage crops. In this chapter, management is considered in a holistic sense, as more than applying the proper amounts of fertilizer or harvesting at the proper time. While these

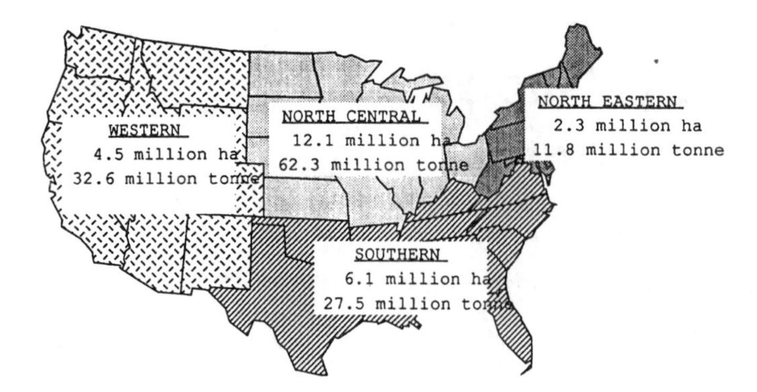

Fig. 12.1. Hay and silage production in the US by region.

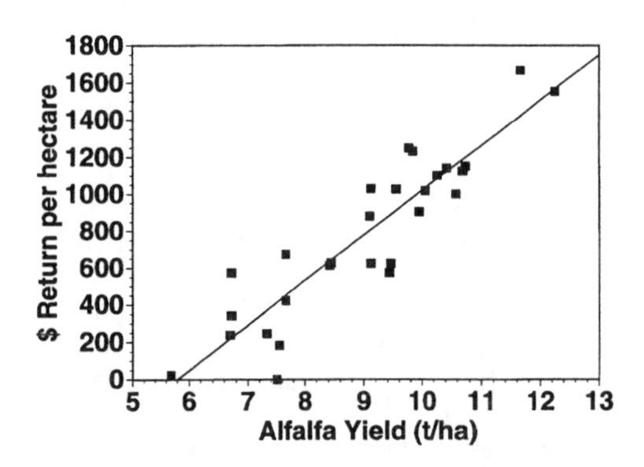

Fig. 12.2. Effect of alfalfa yield on total dollar return per hectare when fed to dairy cows in Wisconsin. (Adapted from Undersander 1994.)

practices are critical, management involves all aspects of forage crop production, beginning before seeding (assessing the soil conditions and selecting the proper species and cultivar) and continuing through harvest (obtaining hay and hay silage quality that matches the nutritional needs of the consuming animal) and concluding with the storage of the material.

SELECTING A SPECIES AND CULTIVAR

Perennial forage crops are used for hay and silage production in all parts of the US. The species, cultivar, or mixture that is most appropriate for a particular location depends on climate, soil factors, disease and insect problems, and the type of livestock enterprise that will utilize the forage. There is a wide range in quality, yield, and persistence potentials among forage crops that are adapted to a particular area, and these potentials should be considered when selecting a forage species and then a cultivar. In many conditions a mixture of grasses and legumes is more appropriate than a monoculture. For example, heaving, or the lifting of alfalfa plants from the soil by freeze-and-thaw action, can be a serious problem in some environments. If fibrous-rooted grasses are grown in mixtures with legumes, the incidence of heaving is reduced because of a change in physical conditions around the legume roots and because the insulation of snow on the soil surface through the winter reduces fluctuations in soil temperatures (Smith et al. 1986). Annual or perennial legumes are often grown with bermudagrass (*Cynodon dactylon* [L.] Pers.) and tall fescue (*Festuca arundinacea* Schreb.) to improve forage quality and to add biologically fixed N to the cropping systems (Ball et al. 1991). The goal is to match the forage crop to the land resource and livestock enterprise.

The choice of forage species is limited to those adapted to a location (Table 12.1). Adaptation is usually determined by relative ability to persist and to produce a yield. Broad climatic patterns and local edaphic conditions must be considered when selecting the appropriate forage. Further selection for quality can be made within the adapted species. In general, if the environment is conducive to persistence and production of legumes, they will be grown in preference to or in mixtures with grass for hay and haylage because they are also high in quality and can support intensive livestock production. For example, strong dairy industries developed in areas where high-quality alfalfa can consistently be produced and conserved. In the southeastern US, where conditions generally do not favor production of forage legumes, bermudagrass is the hay crop of choice because of its excellent yield and persistence and its quality meets the needs of the cow-calf industry.

MANAGING FORAGES IN THE SEEDING YEAR

Perennial forage crops are normally in production for many years; thus poor establishment may have a long-term negative impact.

TABLE 12.1. Characteristics of several hay and silage forage species

Crop	Seedling Vigor	Tolerance to soil limitations: Droughty	Wet	Low pH[a]
Alfalfa	M	H	L	L
Birdsfoot trefoil	L	M	H	H
Red clover	H	L	M	M
Sericea lespedeza	L	H	L	H
White clover	M	L	H	M
Bahiagrass	L	H	L	H
Bermudagrass	M	H	M	M
Orchardgrass	H	M	M	M
Perennial ryegrass	H	L	M	M
Reed canarygrass	L	H	H	H
Smooth bromegrass	H	H	M	M
Tall fescue	H	M	M	H
Timothy	M	L	L	M

Note: L = low, M = Moderate, and H = high.
[a]pH below 6.0.

Therefore attention should be given to several management factors that are known to influence establishment success. These include amending the soil if pH or fertility levels are not in line with crop needs (Chap. 6), proper seedbed preparation, weed and insect control (Chap. 7), and harvest management. When companion crops are used during establishment, appropriate harvest management of the companion crop is also critical.

Managing Companion Crops. Companion crops (especially oat [*Avena sativa* L.]) are often seeded with some legumes and grasses for a number of reasons. Rapid canopy development of the companion crop reduces opportunities for weed development and also reduces the potential for soil erosion. Also, companion crops can be harvested for grain and straw or for forage in the summer. However, excessive competition with the establishing forage crop can occur and has led to the recommendation that small grain companion crops be removed as forage before maturing, at the soft dough stage or earlier (Sheaffer et al. 1988a; Taylor et al. 1979).

Harvest Management of Forages Established without a Companion Crop. Harvest management strategies have been developed to maximize yield and quality of red clover (*Trifolium pratense* L.) and alfalfa without negatively affecting persistence or yield the following year (Hall and Eckert 1992). Alfalfa can be harvested 60 d after emergence and then at 30- to 35-d intervals, until 40 d before a killing frost, to obtain optimum yield and quality during the established year (Sheaffer 1983). There is evidence that red clover persistence is improved by harvesting frequently enough to keep it from flowering in the year of establishment. Keeping red clover in a vegetative condition seems to result in greater crown and root development and increased survival of plants over the first winter (Smith et al. 1986). This usually means that two harvests should be taken in the North and three in the South, beginning 60 to 80 d after seedling emergence (Hall and Eckert 1992). Intensive management of these two legumes in the seeding year can result in forage yields that are 50% to 80% of those obtained from previously established stands.

In general, grasses are less sensitive to seeding year harvest management than are legumes. Seedling vigor varies widely among grasses. For example, reed canarygrass (*Phalaris arundinacea* L.) and bermudagrass have relatively noncompetitive and nonproductive seedlings, and the ryegrasses (*Lolium* spp.) are very competitive and productive during the establishment year. Harvest management decisions for establishing grasses should be based on whether or not enough forage is present to justify harvesting and whether or not harvesting will benefit the desired grass by removing weed competition.

MANAGING ESTABLISHED FORAGE STANDS

After a good forage stand has been established and properly managed through the seeding year, continued production and persistence depends on sound management. Good management practices include maintaining optimum levels of soil fertility, controlling weed and insect pests, and implementing a proper harvest schedule. (For more detail on soil fertility see Chap. 6; also see Chap. 5, Vol. 2. For information on forage pests see Chap. 4, Vol. 2.)

The goal of most forage programs is to maximize the economic yield of high-quality forage while ensuring stand persistence. Harvesting more frequently improves forage quality because plants are harvested at younger stages, while harvesting less frequently generally results in greater yields and increased stand longevity because plants can restore reserves between cuttings to maintain vigor (Brink and Marten 1989). Therefore, harvest management of perennial forages requires a compromise between yield, quality, and persistence. The frequency at which forage crops are harvested should depend on the nutrient needs of the livestock that will be consuming the forage (Fig. 12.3), as well as the desired longevity of the stand. Because of sudden changes in weather and year-to-year variation in growing seasons, there is no simple rule to follow when making a decision to harvest, but general principles have been developed. Decisions on when to harvest have to be made based on long-term weather records and a sound understanding of how the plant grows and survives.

FORAGE QUALITY

The stage of maturity at which forages are cut has the major influence on the quality of that forage. Forage crops generally decline in nutritive value as they mature (Table 12.2). As forage plants mature, there usually is an increase in acid detergent fiber and neutral detegent fiber concentrations whereas crude protein concentration declines. Thus, relative feed value (RFV), an index that ranks forages relative to the digestible dry matter intake of full bloom alfalfa (RFV = 100), declines with maturity.

Quality changes with maturity are related to reduced stem quality and an increased proportion of stems in the forage. Leaves of both grasses and legumes contain a much greater

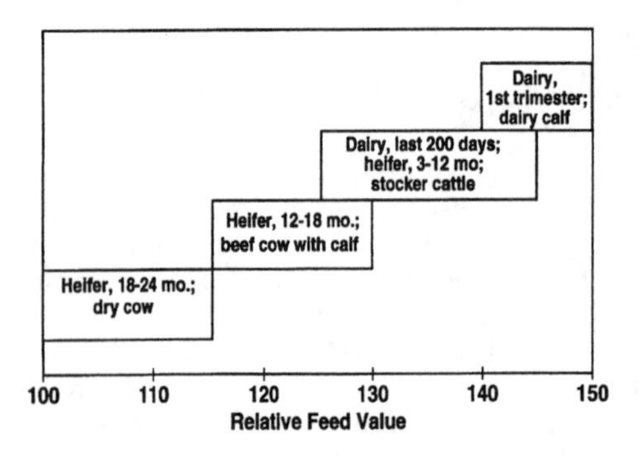

Fig. 12.3. Matching forage quality to livestock requirements. (Adapted from Undersander et al. 1991.)

concentration of digestible nutrients than do stems (Albrecht et al. 1987; Buxton 1990), and the proportion of leaves declines as lower leaves senesce or are damaged by disease. A short delay in harvest can result in forage of much lower quality. For example, it is well known that alfalfa has a high potential for quality, but cutting alfalfa late will result in forage of lower quality than other species. However, cutting when the plants are immature in order to improve quality often causes a reduction in yield and persistence (Brink and Marten 1989). If a forage stand is to be kept for only 2-3 yr, the forage may be harvested for higher quality, but if a long-lived stand is desired, the forage should be harvested later.

Energy Reserves. The initial growth of perennial forages in the spring and after every harvest depends on energy (food) reserves stored in the taproot, crown, or stem base of the plants (Chap. 3). High levels of energy reserves are important for fast regrowth, which results in higher yields. Substantial energy reserves are also needed for the development of coldhardiness, which allows the plant to persist during the winter and still have enough energy available for good spring growth (McKenzie et al. 1988). Energy reserves are usually highest when the plant is in the full bloom stage and usually lowest a short time after harvest, when the plant is growing rapidly (Smith et al. 1986).

Indicators on Which to Base Harvest. The stage of plant development is generally a reliable predictor of energy reserve status and when the plants can be harvested without negatively affecting persistence. However, when the weather is extremely cool and cloudy for an extended period, development may be delayed. Relying on the calendar date alone to make a decision to harvest may also be unwise. Light, temperature, and moisture vary from year to year and have a direct effect on maturation (Buxton and Fales 1994). The most practical method to determine when to harvest is the stage of plant development in conjunction with the calendar date, since seasonal weather variations can alter the relationship between stage of development and energy reserves (Sheaffer et al. 1988b).

Establishing a Harvest Strategy. The intensity of cutting management (i.e., the number of cuttings made per year) should be based on the desired quality and life expectancy of the

TABLE 12.2. Market hay grades for legumes, legume-grass mixtures, and grasses

Grade	Species and stage	RFV	CP	ADF	NDF	DDM	DMI
				(% dry matter)			(% of body weight)
Prime	Legume, pre-bloom	>151	>19	<31	<40	>65	>3.0
1	Legume, early bloom; 20% grass, vegetative	125-151	17-19	31-35	40-46	62-65	2.6-3.0
2	Legume, mid-bloom; 30% grass, early-head	101-124	14-16	36-40	47-53	58-61	2.3-2.5
3	Legume, full bloom; 40% grass, headed	86-100	11-13	40-42	53-60	56-57	2.0-2.2
4	Legume, full bloom; 50% grass, headed	77-85	8-10	43-45	61-65	53-55	1.8-1.9
5	Grass, headed and/or rain-damaged	<77	<8	>45	>65	<53	<1.8

Source: Adapted from Rohweder et al. (1978).

Note: RFV = relative feed value [calculated as (DDM × DMI)/1.29]; CP = crude protein; ADF = acid detergent fiber; NDF = neutral detergent fiber; DDM = digestible dry matter [calculated as 88.9 – (0.779 × ADF%)]; DMI = dry matter intake [calculated as 120/NDF%].

crop. If the goal is to have a long-lived stand, then a longer interval between cuttings should be considered. If the crop is being grown under a short rotation (3 yr or less), then more cuttings to maximize forage quality may be desirable.

The first harvest in the spring can be made as early as the bud (legume) or early-heading (grass) stage of development (Table 12.3). During the spring there is generally limited environmental stress, and the forage crop can normally tolerate early harvesting. Harvesting at an early stage provides high-quality forage and allows for more cuttings to be made during the year. However, in order to cut this early, soil fertility should be at optimum levels.

For legumes, harvests during the summer should be made when the crop is in the bud to early bloom stage of development. Grasses, which remain vegetative following spring harvest, should be harvested based on a set number of days (e.g., 35 d) after the previous harvest or when adequate forage has accumulated to justify harvesting. A cutting interval, for hay or silage, that is consistently shorter than 30 d can be extremely stressful to the stand because this is insufficient time for energy reserves to accumulate in preparation for the next regrowth. Low energy reserves lead to slow regrowth, which results in reduced yields and reduced competitive ability with weeds or associated species (Graber et al. 1927). When severe, there can be an actual loss of stand—sometimes in 1 yr.

Cutting management of a legume-grass mixture should be based on the harvest schedule of the legume. Thus, it is critical to include a grass species that is compatible with the desired legume. Especially critical is matching the reproductive growth phases of the grass and legume components. For example, if the legume is alfalfa and is subject to intense cutting (four or more harvests per year), species such as orchardgrass (*Dactylis glomerata* L.), perennial ryegrass (*L. perenne*

TABLE 12.3. Summary of when to harvest various forage species and mixtures for hay or silage to optimize quality, yield, and persistence

Species	First harvest	Regrowth
Alfalfa	Bud or first flower	Bud or first flower of alfalfa
Alfalfa—orchardgrass	When orchardgrass heads	Bud or first flower
Alfalfa—smooth brome and other grasses	Alfalfa bud or first flower	Bud or first flower of alfalfa
Red clover	First flower to 25% bloom	First flower of red clover
Red clover—grasses	When grasses at early-head	First flower of red clover
Ladino clover alone or with grasses	10 to 50% bloom of ladino	Every 30 to 35 d
Birdsfoot trefoil alone or with grasses	10 to 50% bloom of birdsfoot trefoil	10 to 50% bloom of birdsfoot trefoil
Smooth brome, orchardgrass or timothy	When heads emerge	Vegetative
Reed canarygrass or tall fescue	Flag-leaf to early heading	Every 30 to 40 d
Bermudagrass or bahiagrass	When 40 cm tall	Every 4 to 5 wk

Source: Adapted from Miller (1984).

L.), and reed canarygrass are compatible because they are tolerant of frequent cutting. A less intensive system (three cuttings per year or fewer) is compatible with timothy (*Phleum pratense* L.) or smooth bromegrass (*Bromus inermis* Leyss.), which requires longer intervals between cuttings and generally does not tolerate early spring harvest.

Fall-cutting Management. During the late summer and early fall, forage plants are preparing for winter by developing freezing tolerance and accumulating energy reserves in their storage organs (roots, crowns, stem bases, rhizomes, and stolons). Depending on the timing, fall harvest may interfere with this process (Bowley 1989). Harvesting at a time that will allow only a few weeks of regrowth before the herbage is killed by frost will greatly reduce energy reserves. Harvesting will also remove stubble, which catches snow and serves as a layer of insulation from extremely cold air temperatures. Stubble also improves survival by penetrating ice sheets and reducing heaving. Inadequate energy reserves and loss of stubble increase the risk of forage stand loss, especially for legumes. Generally, grasses tend to be less sensitive than legumes to fall harvest management (Smith et al. 1986).

Winter environmental conditions can aggravate the effects of fall harvesting. Soil and snow serve as insulation between plants and cold air temperatures. Lack of snow increases the risk of winter kill. Wet soils freeze and thaw more intensively, which increases the amount of frost heaving. Fields that have a history of frost heaving or accumulating little snow cover should not be fall harvested.

Risks to stand persistence in alfalfa can be minimized by (1) delaying at least one of the summer harvests until the early flower stage or later, (2) maintaining high soil fertility levels, especially potassium (K), (3) selecting fall-dormant alfalfa cultivars that have good disease resistance and winterhardiness.

Although fall harvesting increases the risk of stand loss or decreased productivity the following spring, the risk might have to be taken because the need for a forage or the value of a forage may be too great. Use of a scoring system to assess the risk of late summer or fall harvesting of alfalfa may be helpful (Table 12.4).

Optimum levels of K in the soil enhance the ability of forage plants to overwinter and support good spring growth. This has been well documented for alfalfa (Jung and Smith 1959) and bermudagrass (Adams and Twersky 1960). It is important that adequate K be available during the late summer and early fall since the storage of energy reserves for winter survival occurs during this time.

MANAGEMENT OF FORAGE PESTS

Pests that are of concern in forage production are weeds, diseases, and insects. Cultural, biological, and chemical control of weed pests is discussed in Chapter 7. Disease and insect pests are often difficult to control with cultural practices, and chemical control is often uneconomical in all but high-value forage legumes.

Forage diseases are mainly caused by fungi. Bacterial and viral diseases are prevalent in particular forage species and locations; however, nationally their impact on forage production is relatively small compared to the economic impact of fungi. Nematodes can be a severe limitation to forage production in some areas. Cultural practices such as proper species and cultivar selection to match the site, selecting disease-resistant cultivars, management practices that optimize forage growth and vigor, and mowing new forage stands before old stands can help minimize the infection and spread of diseases (Ball et al. 1991; Leath et al. 1988). In general, the use of chemicals to restrict disease infection and spread is not economical for forage production.

Insect damage to forage ranges from highly predictable in some areas to extremely sporadic in other areas. This wide range in levels of damage stems from the large impact weather and climatic conditions can have on insect reproduction and growth. Some forage species (e.g., birdsfoot trefoil [*Lotus corniculatus* L.]) are more resistant to insect feeding than others. This is thought to be a result of the higher tannin content in these species. Some insects can be controlled by timely harvesting. That is, harvesting the forage will remove adequate numbers of insects or disrupt their life cycle so that the population remains below the threshold population at which chemical control is justified. Biological control has been highly successful with some insects. The use of parasitic wasps to control alfalfa weevil (*Hypera postica* Gyll.) populations is one such example. Insecticides can also be used to reduce or eliminate insect pests; however, their use also reduces or eliminates the population of beneficial insects in the forage.

TABLE 12.4. Calculating the risk of alfalfa winter injury due to late summer or fall harvest in the northern temperate region

	Points	Score
1. *Stand age*		
—≥3 yr	3	
— 2-3 yr	2	
—≤1 yr	1	____
2. *Alfalfa cultivar*		
a. Winter dormancy rating (fall growth score)		
—Moderately winter-hardy (3.4-5.2)	3	
—Winter-hardy (5.3-7.0)	2	
—Very winter-hardy (7.1-8.0)	1	
	____	
b. Disease resistance		
—Moderate resistance to only bacterial wilt	4	
—Moderate resistance to bacterial wilt plus either phytophthora root rot, fusarium wilt, or anthracnose	3	
—Moderate resistance to all above-mentioned diseases	1	
Cultivar score (multiply a × b)	____	____
3. *Soil exchangeable K level*		
—Low	4	
—Medium	3	
—High	1	____
4. *Soil drainage*		
—Poor	3	
—Medium	1	
—Excellent (sandy soils)	0	____
5. *Harvest frequency*		
—Four cuts by Sept. 1	5	
—Four cuts by Sept. 15	5	
—Four cuts by Oct. 15	2	____
—Three cuts by Sept. 15	4	
—Three cuts by Oct. 15	3	
—Three cuts by Sept. 1	1	
—Two cuts by Oct. 15	1	
—Two cuts by Sept. 15	1	
—Two cuts by Sept. 1	0	____
6. *Will a 15-cm stubble remain after a mid-Sept. or October cut?*		
—No	1	
—Yes	0	____

7. Total score	*Fall-cutting risk*
3-7	low, below average
8-12	moderate, average
13-17	high, above average
18 or more points	very high, dangerous

Source: Sheaffer (1989). Harvest schedules are applicable to Minnesota and Wisconsin.

In general, the use of insecticides should be employed as a last resort and only when economically justifiable. For additional information about forage pests, refer to Chapter 4, volume 2.

QUESTIONS

1. What factors should be considered when deciding on the forage species and cultivar for a particular area?
2. Why is fall management of some crops, such as alfalfa, so important?
3. In general, how would harvest management for maximum yield differ from management for maximum quality? Why?
4. What is the best management strategy to reduce disease problems in most forages?
5. Why is the production cost per hectare similar for fields that produce 5 or 12 mt of alfalfa per hectare?

REFERENCES

Adams, WE, and M Twersky. 1960. Effect of soil fertility on winter-killing of coastal bermudagrass. Agron. J. 53:325-26.

Albrecht, KA, WF Wedin, and DR Buxton. 1987. Cell-wall composition and digestibility of alfalfa stems and leaves. Crop Sci. 27:735-41.

Albrecht, KA, SK Barnhart, CP West, and WF Wedin. 1982. Potassium fertilization and harvest schedule effects on yield and quality of alfalfa. In Proc. Am. Forage and Grassl. Counc., 79.

Ball, DM, CS Hoveland, and GD Lacefield. 1991. Southern Forages. Atlanta, Ga.: Potash and Phosphate Institute.

Bowley, SR. 1989. Debate: Why we need a fall rest period. In Proc. Am. Forage and Grassl. Counc., 17-22.

Brink GE, and GC Marten. 1989. Harvest management of alfalfa—nutrient yield vs forage quality and relationship to persistence. J. Prod. Agric. 2:32-36.

Buxton, DR. 1990. Cell wall components in divergent germplasms of four perennial forage grass species. Crop Sci. 30:402-8.

Buxton, DR, and SL Fales. 1994. Plant environment and quality. In GC Fahey et al. (eds), Forage Quality, Evaluation, and Utilization. Madison, Wis.: American Society of Agronomy, 155-99.

Graber, LF, NT Nelson, WA Luekel, and WB Albert. 1927. Organic Food Reserves in Relation to the Growth of Alfalfa and Other Perennial Herbaceous Plants. Wis. Agric. Exp. Stn. Bull. 80. Madison, Wis.

Hall, MH, and JW Eckert. 1992. Seeding year harvest management of red clover. J. Prod. Agric. 5:52-56.

Jung, GA, and D Smith. 1959. Influence of soil potassium and phosporus content on the cold resistance of alfalfa. Agron. J. 51:585-87.

Leath, KT, DC Erwin, and GD Griffin. 1988. Diseases and nematodes. In Hanson et al. (eds.), Alfalfa and Alfalfa Improvement, Am. Soc. Agron. Monogr. 29. Madison, Wis., 621-70.

McKenzie, JS, R Paquin, and SH Duke. 1988. Cold and heat tolerance. In AA Hanson, DK Barnes, and RR Hill, Jr. (eds.), Alfalfa and Alfalfa Improvement, Am. Soc. Agron. Monogr. 29. Madison, Wis., 259-302.

Miller, DA. 1984. Forage Crops. New York: McGraw-Hill.

Rohweder, DA, RF Barnes, and N Jorgensen. 1978. Proposed hay grading standards based on laboratory analyses for evaluating quality. J. Anim. Sci. 47:747-59.

Sheaffer, CC. 1983. Seeding year harvest management of alfalfa. Agron. J. 75:115-19.

______. 1989. Fall cutting is a management option in the north. In Proc. Am. Forage and Grassl. Counc., 23-29.

Sheaffer, CC, DK Barnes, and GC Marten. 1988a. Companion crop vs. solo seeding: Effect on alfalfa seeding year forage and N yields. J. Prod. Agric. 1:270-74.

Sheaffer, CC, GD Lacefield, and VL Marble. 1988b. Cutting schedules and stands. In AA Hanson, DK Barnes, and RR Hill, Jr. (eds.), Alfalfa and Alfalfa Improvement, Am. Soc. Agron. Monogr. 29. Madison, Wis., 411-37.

Smith, D, RJ Bula, and RP Walgenbach. 1986. Forage Management. 5th ed. Dubuque, Iowa: Kendall/Hunt.

Taylor, TH, WF Wedin, and WC Templeton, Jr. 1979. Stand establishment and renovation of old sods for forage. In RC Buckner and LP Bush (eds.), Tall Fescue. Am. Soc. Agron. Monogr. 20. Madison, Wis., 155-70.

Undersander, DJ. 1994. Green gold results. In Proc. Wis. Forage Counc. 18th Forage Prod. and Use Symp.:65-68.

Undersander, DJ, N Martin, D Cosgrove, K Kelling, M Schmitt, J Wedberg, R Becker, C Grau, and J Doll. 1991. Alfalfa Management Guide. Madison, Wis.: American Society of Agronomy.

13 Harvesting and Storage

DARRELL A. MILLER
University of Illinois

C. ALAN ROTZ
Agricultural Research Service, USDA

IN many areas of the world, forage cannot be grown during a major portion of the year due to cold temperatures or lack of moisture. Since animals require a continuous supply of feed, forage conservation is critical to enable animal production in these areas. Forage is conserved by harvesting and storing crops grown during favorable conditions for use during less favorable conditions. In the US, many dairy and beef operations use only conserved forages in order to obtain a more consistent, reliable, and predictable feed supply.

Although there are many variations in forage conservation systems, the primary methods involve either the harvest of dry hay or silage. To produce dry hay, the crop is mowed and dried in the field to a moisture level that allows stable storage, normally 150-200 g kg^{-1} moisture. Hay at this moisture can be stored for many months. Higher-moisture forage, 500-850 g kg^{-1}, can be stored as silage. The crop is often mowed, dried in the field to the appropriate moisture, and chopped into lengths of 0.5-2.0 cm. Corn and sometimes grass crops are mowed and chopped in a single direct-cut operation. For both silage harvest procedures, the chopped forage is placed into a silo, which limits the availability of oxygen (O_2). Fermentation occurs in this anaerobic environment, dropping the silage pH. As long as exposure to O_2 is limited, silage with a pH of 4.0-5.0 is stable for long storage periods.

Each forage conservation method offers advantages and disadvantages relative to the other. Silage systems enable greater mechanization of the handling and feeding processes. With greater mechanization, less labor is required. The chopped material is also more conveniently used in total mixed rations. On the other hand, silage systems require more power or energy for harvesting, handling, and feeding and a greater investment in machinery and storage structures, both of which lead to greater costs. Baled dry hay requires less storage volume and is easier to transport and market. Average total crop loss is 24%-28% when hay is properly stored in a shed or barn. Most of this loss occurs during harvest, with about 5% loss during storage. Average total loss in silage production is 14%-24%, with about half of this loss occurring during storage (Buckmaster et al. 1990). Since neither system offers a clear advantage over the other, both will likely always be used in animal agriculture.

DARRELL A. MILLER. *See Chapter 4.*

C. ALAN ROTZ is Agricultural Engineer, ARS, USDA, conducting research on forage harvest and storage systems at the US Dairy Forage Research Center, East Lansing, Michigan. He received MS and PhD degrees from Pennsylvania State University.

MOWING AND CONDITIONING

Mowing is the first step in most forage harvest systems. In mowing, the plant top is severed from its root 5-10 cm above the soil surface. Conditioning is a forage treatment that helps speed the field drying process. Mowing and conditioning are most often combined in one machine called a *mower-conditioner* (Fig. 13.1).

Fig. 13.1. A mower-conditioner cutting and conditioning alfalfa for field drying.

Mowing. Mowing machines use either cutterbar, rotary disk, rotary drum, or flail mowing devices. For many years, the reciprocating cutterbar or sicklebar mower has proven a reliable and relatively inexpensive mowing machine. A disadvantage is that the cutting capacity of the device limits field speed. The rotary disk mower is capable of greater speeds. In fact, the field speed of the disk mower is often limited only by the operator's ability to control the machine. A disadvantage of disk mowers is higher costs of operation. The purchase cost is 30% more per unit width. Repair costs are low initially, but as the machine ages, wear on the many gears and other moving parts may lead to high costs. Disk mowers require about four times as much power to operate (about 5.0 kW m^{-1} of width compared to 1.2 kW m^{-1} for a cutterbar mower [Rotz and Muhtar 1991]). Therefore, a larger tractor is required, and more fuel is consumed per hour of use. With faster field speeds, less labor and tractor time are required. All things considered, neither mower type provides a universal advantage.

Rotary drum and flail mowers have more serious disadvantages. These require up to twice the power of a rotary disk mower. Drum mowers also tend to leave a less uniform swath, which may cause uneven drying. Flail mowers provide faster drying, but the chopping action of the flail causes at least twice the loss incurred by other mowers (Rotz and Sprott 1984). Flail mowers are also more likely to cause a ragged cut, which retards regrowth and reduces stand persistence. For these reasons, rotary drum and flail mowers are not widely used for mowing forage crops.

Mechanical Conditioning. Many different types of mechanical conditioning devices are used to speed forage drying. Most use either rolls or flails to condition the crop. Roll devices provide suitable drying with low losses when conditioning alfalfa. Many combinations of intermeshing and nonintermeshing rubber and steel rolls are used, but intermeshing rubber rolls are most commonly used in North America. The rolls smash and/or break the plant stems, and moisture evaporates more easily from these broken areas. Most roll-conditioning designs provide similar improvements in drying when properly adjusted (Shinners et al. 1991; Rotz and Sprott 1984). Flail conditioners, developed in Europe for grass crops, use rotating metal or plastic flails to abrade the waxy outer plant layer and break plant stems. Flail conditioning can cause high losses when used on alfalfa, but it is often preferred when mowing grass crops.

In humid climates, roll conditioning is very effective on first-cutting alfalfa, but it may be less effective on later cuttings (Rotz et al. 1987). On first cutting, the conditioning treatment increases the drying rate about 80%, which reduces drying time about 2 d. On second cutting, a 35% increase reduces the average curing time about 1 d, but little benefit may be obtained in the following cuttings. The difference among cuttings is due to plant structure. The stem is normally relatively thick in first-cutting alfalfa, but it is smaller in the regrowth of later cuttings. Since thick stems are more difficult to dry, the smashing and breaking action has more potential for improving drying. The finer stems of later cuttings also tend to flow between the rolls and receive little damage.

Typical losses with well-adjusted mower-conditioners vary between 1% and 5% of the dry matter (DM) yield, with similar losses among the two major types of mowers (Shinners et al. 1991; Koegel et al. 1985; Rotz and Sprott 1984). Among roll-type conditioning machines, roll design does not have much effect on loss. Adjustment of roll pressure and clearance has more effect than the configuration of or material used in the rolls. The difference in loss between a well-adjusted mower-conditioner and a similar mower without a conditioner is 1%-2% of the crop yield. Losses are much higher with flail mowing machines, averaging between 6% and 11% of yield. Flail conditioning devices adjusted for very aggressive conditioning may also double the loss in alfalfa mowing. The mowing and conditioning loss is mostly leaves. Because leaves are higher in crude protein (CP) and lower in fiber than stem material, a typical loss causes a small decrease in CP and a small increase in fiber concentrations in the remaining forage.

Chemical Conditioning. Chemical conditioning is a newer process for increasing the drying rate of alfalfa. A chemical, referred to as a *conditioner* or *drying agent,* is sprayed on the crop at the time of mowing (Rotz and Davis 1986). The chemical affects the waxy surface of the plant to allow easier moisture removal. Potassium and sodium carbonates are the most commonly used chemicals. To apply the treatment, a solution is made by mixing about 30 g of potassium carbonate or 15 g each of potassium and sodium carbonates per liter of water. The treatment is more effective as more solution is applied, but the optimum amount is about 280 L ha^{-1} for yields under 3.5 mt ha^{-1} and 470 L ha^{-1} for greater yields. The treatment is effective on alfalfa and other forage legumes, but it has no effect on grass crops. The greatest drying improvement is obtained when the crop is dried in a relatively thin swath.

Chemical conditioning is effective on all alfalfa cuttings, but most effective on midsummer cuttings. In Michigan, the chemical provides about a 40% increase in the drying rate on first cutting (Rotz et al. 1987). On second and third cuttings, a 50%-175% increase in the drying rate can be obtained with an average increase of 85%. In fall harvests the increase is small, only about 25%. The chemical treatment's effectiveness is directly related to the drying conditions; i.e., better drying is obtained on sunny, warm days and with thin, wide swaths. Compared to mechanical conditioning alone, an average reduction in field-curing time for hay harvest is 0.5 d on first cutting and about 1.0 d on later cuttings.

SWATH MANIPULATION

As forage dries in the field, the top of the swath dries more rapidly than the bottom. Manipulation of the swath can speed the drying process by moving the wetter material to the upper surface where it dries more quickly. Swath manipulation can also improve drying by spreading the hay over more of the field surface. Spreading exposes more of the crop to the radiant solar energy and drying air. There are three operations used in haymaking that manipulate the swath: tedding, swath inversion, and raking.

Tedding. Interest in tedding has grown in recent years following the introduction of the European-style tedder. This device uses rotating tines to stir, spread, and fluff the swath. Tedding can be done anytime during field curing, but it is best to do so before the crop is too dry (above 400 g kg^{-1} moisture content). The stirring or fluffing aspect of tedding typically reduces field-curing time up to 0.5 d. Tedders are sometimes used to spread narrow swaths formed by the mower-conditioner over the entire field surface. When done soon after mowing, the average curing time is reduced up to 2 d compared with drying in a narrow swath. Tedding may also allow more uniform drying with fewer wet spots in the swath.

Losses caused by tedding are normally

between 1% and 3% of the crop yield, but much greater loss can occur. The beating action of the tedder is most damaging to legume crops due to a more delicate attachment of leaves and stems. Leaf loss is greatly influenced by crop moisture content, particularly in legume crops. As the crop dries, leaf shatter increases exponentially with a very high potential loss in forage containing less than 300 g kg^{-1} moisture (Savoie 1988). With 70% of the loss being leaf material, tedding affects the quality of the remaining forage by reducing the portion of leaves. In alfalfa, up to a 1% DM decrease in CP concentration and a 2% DM increase in neutral detergent fiber (NDF) can occur as a result of the operation.

Tedding can also cause greater raking loss. When a light crop (less than 2.5 mt DM ha^{-1}) is spread over the field surface, raking loss can be more than double that when narrower swaths are raked. In alfalfa, the loss caused by tedding is often greater than the average rain loss avoided by faster drying (Rotz and Savoie 1991). The decision to use tedding should be made by comparing the probable loss from more time laying in the field with the known loss and cost of tedding. The increased machinery, fuel, and labor costs may not justify routine use of tedding.

Swath Inversion. Swath-inverting machines provide a more gentle method for manipulating field-curing swaths. Although many machine designs are used, a pickup device normally lifts the swath onto a platform or moving belts. The swath is turned and dropped back to the soil surface and is inverted from its original position (Savoie and Beauregard 1990). Exposing the wet bottom layer of the swath speeds drying enough to reduce the average field-curing time a few hours. Although not as effective as tedding in improving drying rate, leaf shatter is reduced. Increases in field loss due to swath inversion are small, varying between 0% and 1.5% of yield in predominantly alfalfa crops. With less drying benefit though, there is less reduction in the probability of rain-induced loss. Again, the added labor, fuel, and machinery costs of the operation may be greater than the benefit received (Rotz and Savoie 1991).

Raking. Raking is another form of swath manipulation. Wetter hay from the bottom of the swath is rolled to the outer surface of the windrow, which speeds drying. Following the initial improvement, the heavy swath formed by raking can reduce the drying rate, so the moisture content of the crop when raked is important. If the crop is too wet, the wet material rolled into the center of the windrow dries slowly. For minimal loss and optimum drying, hay should be raked at a moisture content between 300 and 400 g kg^{-1}. Raking hay in the morning of the day in which it is anticipated to be ready for baling can reduce the field-curing time by 1 or 2 h, allowing an earlier start at baling. Another alternative is to create a narrow swath with the mower-conditioner and then to use no manipulation during field curing. Hay dries more slowly, requiring up to 2 d more in the field, so the probability of rain damage increases. In humid climates, narrow swath drying without raking generally is best for silage production, but when making dry hay, wide swath drying with raking is often preferred.

Raking causes a loss of 1%-20% of crop yield (Rotz and Muck 1994). The loss increases as the crop dries, particularly below a moisture content of 300 g kg^{-1}. When the crop is spread over much of the field surface, it is more difficult to gather with the rake, and the loss increases. Greater loss is generally found with a rotary windrower compared with the more standard side-delivery rake. The rotary windrower uses rotating tines to sweep hay into a windrow. The sweeping action allows more crop to become entangled with the stubble and lost. Side-delivery rakes provide a rolling and wrapping action that reduces entanglement with the stubble and increases entanglement among the plants in the swath. With similar loss of leaves and stems during raking, the quality of the remaining forage is affected little by the loss.

HAY BALING

Hay balers are mobile machines that include a pickup device to lift the hay windrow from the field surface, a compression chamber where the hay is packed and formed into a bale, and a tying mechanism that completes the bale. Many types of hay balers are available that produce bales of a variety of sizes and shapes. Advantages and disadvantages are associated with each. Small rectangular bales have been most popular in the past 50 yr, but now large round bales are becoming the predominate hay package. Various sizes of larger rectangular packages are also becoming more popular. Other packages including small round bales, low-density stacks, and cubes are sometimes used, but they constitute a relatively small portion of the hay produced.

Rectangular Bales. The most popular rectangular bale is 36 by 46 cm in cross section with lengths up to 130 cm. An advantage of the small rectangular bales is that the 25-35 kg bale can be manually handled in stacking and feeding. A disadvantage though is that the bale handling is labor-intensive. During a typical baling operation where hay is transported and stacked in storage simultaneously with baling, up to five people are required. The labor requirement for hay harvest varies between 1 and 2 work h mt^{-1} DM produced (Table 13.1). Baling is less energy intensive than most forage harvest operations; it requires 5-10 L of fuel per mt DM of hay produced. A relatively small tractor of 26-45 kW is sufficient to power small rectangular balers. The baler pickup normally extends from the right side of the machine, so the tractor pulling the baler is driven on the left side of the windrow. In a newer design, referred to as the *centerline baler,* the pickup device is centered behind the tractor, with the compression chamber located above the center of the pickup (Fig. 13.2.). The centerline design is promoted for better maneuverability and less loss during baling.

To reduce the manual labor in bale handling, several laborsaving devices have been developed. A popular option is a bale thrower mounted at the exit of the baler. The device throws the bales into a trailing wagon, reducing the need for handling and stacking on the wagon. At the unloading site, bales are sometimes dropped from an elevator to avoid stacking in storage. Automatic bale wagons are sometimes used in large operations. These wagons mechanically lift bales dropped on the field surface by the baler. Hydraulically driven mechanisms stack the bales, and the completed stack can be transferred to storage.

Fig. 13.2. A centerline baler with a bale ejector and trailing wagon.

Large, high-density bales are becoming more popular, particularly for hay transported long distances. The high-density bales are transported more efficiently than small bales. These bales have a height and width of 60-130 cm, a length of 120-250 cm, and a mass of 200-1000 kg. Special equipment is thus needed for lifting, transporting, and feeding the large bales. Balers producing these large packages offer greater baling capacity, with the ability to harvest up to twice as much hay per hour as the small package balers. The centerline design is used in these balers. More power is required, with a minimum tractor power of 90 kW required for the largest balers.

Typical DM losses during hay baling vary between 2% and 5% of the yield, with the loss equally divided between pickup and chamber losses (Shinners et al. 1992; Rotz and Abrams 1988; Koegel et al. 1985). Pickup losses are high when the machine is pulled at a faster speed than the rotating speed of the pickup device. The machine tends to overrun the swath, causing loss as high as 5%. Chamber loss is largely influenced by crop moisture content, with greater loss in drier material. When hay is baled at night, leaf moisture is higher, similar to stem moisture, and cham-

TABLE 13.1. Approximate labor and fuel requirements for forage harvesting

	Labor, work h mt^{-1}		Fuel, L mt^{-1} DM	
	Range	Typical	Range	Typical
Operations				
Mowing	0.1–0.6	0.2	1.0–5.0	1.5
Tedding	0.1–0.7	0.1	0.3–3.0	0.7
Raking	0.1–0.5	0.2	0.5–3.0	0.9
Baling (large round)	0.1–0.4	0.2	1.4–4.0	2.0
Baling (small rectangular)[a]	0.7–4.0	1.0	3.0–10.0	4.0
Chopping[a]	0.4–2.0	0.6	4.0–19.0	7.0
Harvest systems				
Round bale hay	0.5–1.4	0.7	4.0–10.0	6.0
Rectangular bale hay	0.9–1.6	1.3	5.5–13.0	7.0
Wilted silage	0.6–2.0	0.8	9.0–23.0	10.0
Direct-cut silage	0.4–1.0	0.6	5.0–13.0	7.0

[a]Includes transport and unloading.

ber loss can be cut in half. Chamber loss is mostly high-quality leaf material, so excessive chamber loss has more effect on the quality of the remaining forage than most other machine losses. By maintaining the chamber loss below 3%, the effect on forage quality is relatively small.

Large Round Bales. Large round bales have become popular primarily because they require less manual labor. In round balers, the hay windrow is lifted into the baler by a pickup and rolled into a bale in the compression chamber. The cylindrical bales are 90 to 180 cm in diameter and 120 to 160 cm in length with a mass of 200 to 900 kg. Completed bales are transported to the storage site with a tractor-mounted loader or a wagon. When bales are not transported simultaneously with harvest, one person can perform the entire operation.

Large round balers require more power than small rectangular balers. Recommended minimum tractor sizes vary with baler size from 35 to 55 kW. The power requirement is dependent on the baler design. Round balers use either variable or fixed chambers. Variable chambers usually use belts to apply pressure on the rolling hay throughout bale formation. This forms a bale of uniform density. Fixed chambers apply pressure only near the completion of the bale, which forms a low density or soft core in the center of the bale. Fixed chamber balers require about 50% more power in a typical baling operation (Rotz and Muhtar 1991). Fuel and labor requirements with round bale harvest are largely dependent on the method of transport. When bales are hauled to a storage site on wagons carrying four to eight bales, fuel use is comparable to that with small bale systems, and labor use is about half. Both labor and fuel requirements can become excessive when bales are individually transported.

Losses in large round balers vary among the baler designs. For a variable chamber baler, chamber loss is about 40% greater than that in a small rectangular baler. For a fixed chamber baler, the loss can be three times as much (Koegel et al. 1985). Chamber loss is very sensitive to the feed rate of hay. Excessive loss occurs at low feed rates because the bale is rolled in the chamber too much per unit of hay baled.

HAY STORAGE

Hay is either stored inside a shelter or outside with varying amounts of protection from the weather. When well protected, hay is relatively stable during storage with only minor respiration by microorganisms on the hay. The respiration transforms DM to heat and gases that leave the hay, causing DM loss and nutrient change. In dry hay stored under cover, little respiration occurs, and DM loss over 6 mo of storage is about 5% (Rotz and Muck 1994). Respiration reduces forage quality by removing some of the most digestible nutrients for the animal. Hay stored outside and unprotected experiences the same loss as hay stored inside plus additional loss from weathering of hay on the exposed surface. Loss often increases an additional 10%-15% DM with outside storage of large round bales.

Inside Storage. Rectangular bales are most often stored inside a shelter. Round bales can be but often are not. Many shed designs are used. A common design is a pole barn enclosed on at least three sides. A roof with no side walls is sometimes used, but this design does not provide quite as much protection. The size of shed required depends on the bale size, shape, and density and the height of the haystack. For rectangular bales stacked in a 5-m-high shed, about 1.5 m^2 of floor area is required to store each mt DM of hay. Round bales require a little more space since they cannot be stacked as tightly or perhaps as high. For bales 1.5 m in diameter stacked three high, the space requirement is about 1.75 m^2 mt^{-1} DM. The requirement is about 2.2 m^2 mt^{-1} DM for bales 1.2 m in diameter. When stacked two high, the requirement is 2.3-2.9 m^2 mt^{-1} DM.

For protected hay, the major factor influencing preservation is the moisture content as it enters storage. Hay with less than 150 g kg^{-1} moisture is relatively stable. In hay containing more moisture, microbial respiration causes the hay to heat during the first 3 to 5 wk of storage. The amount of heating and the associated loss increases with moisture content with about 1% DM loss in hay of 150 g kg^{-1} moisture to 8% DM loss in 300 g kg^{-1} moisture hay (Rotz et al. 1991b). Heating also increases with hay density (Buckmaster et al. 1989a). For hay containing more than 300 g kg^{-1} of moisture, excessive loss and even spontaneous combustion can occur. Heating during the first month of storage helps dry the hay. After the initial heating and drying period, hay is relatively stable, but a small loss of about 0.5% DM per month continues throughout storage. Similar heating and loss occur in hay of all forage species.

Quality changes are relatively small in hay entering storage with less than 200 g kg^{-1} moisture. Typical changes in alfalfa hay during 6 mo of storage are a 1%-2% decrease in digestible DM with little change in the concentration of available protein and a 1% DM increase in NDF content (Rotz and Abrams 1988). Only in higher-moisture hay do substantial changes occur that affect the diet and performance of animals consuming the forage (Buckmaster et al. 1989a). For example, in hay of 250 g kg^{-1} moisture, an 8% DM loss causes about a 3% decrease in digestible DM content with a 3.5% DM increase in NDF content. Heating of high-moisture hay also causes a Maillard reaction, which may reduce the protein available to the animal.

Outside Storage. Round bales are often stored outside to eliminate the investment in a storage structure. Rectangular bales are sometimes stacked outdoors, but they are normally covered with a tarp to protect the hay. Due to their shape, round bales shed precipitation better than rectangular bales, but some moisture is absorbed, which leads to greater loss. Various storage methods provide a range of protection from the environment. The least protected are uncovered bales set on soil. Setting bales on crushed stone, old tires, or other material to break the soil contact improves preservation of the bottom of the bale. Plastic wrap around the bale circumference provides further protection. Normally the best protection outdoors is attained by covering a stack of bales with a tarp. When placed on a well-drained surface, bales protected by this method have DM and nutrient losses similar to bales stored inside.

Loss in large round bales stored outside varies from 3% to 40% (Rotz and Muck 1994). The loss is again primarily caused by microorganisms on the hay, and the biological activity is greatest when the hay is moist and warm. Therefore, the loss is decreased in hay stored over winter periods in northern climates or in hay stored in more arid climates where the hay remains relatively dry. Losses range from 0.5% to 1.5% DM per month of storage in drier climates to 1.0% to 3.0% DM per month in wetter climates. In general the DM lost is the most digestible portion of hay, which decreases digestible DM and increases fiber concentration. Crude protein content varies from a substantial increase to a substantial decrease. Available protein decreases with the heating of hay and the loss of more soluble protein.

Storage conditions affect losses by protecting the hay from moisture. The center of bales stored outside is preserved similarly to hay stored in a shed. Much greater loss occurs in the outer 10-20 cm of the bale, where hay is exposed to rain and perhaps damp soil. Compared to elevated bales, bales on soil additionally lose about 3% DM. When rain occurs on exposed bales, a portion of the rain is absorbed in the outer layer of the bale. The moisture content in this layer is increased to between 250 and 400 g kg^{-1} (Harrigan and Rotz 1992). The moisture level inside the bale may also increase a small amount. Increased moisture leads to greater microbial activity and loss. A plastic wrap around the circumference of the bale greatly reduces the moisture accumulation, particularly when the bales are elevated. A plastic wrap on an elevated bale reduces the weathering loss in storage by about 35% (Rotz and Muck 1994).

Hay Preservatives. Harvest losses in hay making can be reduced by baling hay at a moisture content near 250 g kg^{-1}. Baling moist hay reduces baler chamber losses, providing a small improvement in harvested yield and quality. Raking and pickup losses may also be reduced a small amount. Field-curing time on the average is reduced about 1 d, which reduces the potential for rain damage. With all of these factors combined, harvested yield is increased an average of 7%. However, the moist hay deteriorates rapidly in storage, offsetting the benefit of reduced field losses unless the hay is treated to enhance preservation.

Materials used for the preservation of high-moisture hay include propionic and other organic acids, buffered acid mixtures, anhydrous ammonia, and bacterial inoculants. Propionic acid and similar organic acids normally reduce mold growth and heating of high-moisture hay when applied at rates of 1%-2% of hay weight (Rotz et al. 1991b). Acid treatments reduce storage loss in damp hay during the first couple months of storage, but losses are higher than those in dry hay. Acid-treated hay does not dry out as much during storage. Apparently the more moist environment in the hay maintains a higher level of microbial activity even with the acid treatment. Over a 6-mo storage period, the loss in acid-treated hay catches up to that in untreated high-moisture hays, the treatment providing little improvement in losses and nutrient changes.

Organic acids promote corrosion in balers and bale-handling equipment. To reduce corrosion, buffered acid products have been developed. The acid is blended with ammonia or another compatible chemical to increase the pH of the treatment. Buffered mixtures may be as effective as propionic acid when equivalent amounts of propionate are applied, but they are less effective when applied at rates below 1% of hay weight (Rotz et al. 1991b).

Anhydrous ammonia is a very effective hay preservative (Rotz et al. 1986). Storage DM loss is reduced or eliminated in hay with up to 350 g kg^{-1} moisture when wrapped in plastic and treated with ammonia at 1% or more of hay weight. The ammonia prevents heating, and it may eliminate mold development while the hay is covered. The ammonia also adds nonprotein nitrogen (N) to the forage, which in some rations may be beneficial to animals. With less storage loss, the increase in fiber content that normally occurs during storage is reduced. Increases in DM, hemicellulose and cellulose digestion, and energy content may occur. Although anhydrous ammonia provides the most effective preservation of forage, animal and human safety concerns deter its use. Ammonia treatment of forage has caused toxicity to animals when not used properly. Toxicity most often occurs when ammonia is used with high-quality hay at higher-than-recommended application rates (greater than 3% of hay weight). Direct exposure to anhydrous ammonia can cause severe burns, blindness, and death.

Bacterial inoculants are sometimes used to preserve high-moisture hay. Inoculation with a few forms of lactobacillus has shown no effect on mold, color, heating, DM loss, or quality change in high-moisture hay (Rotz et al. 1988). Bacillus inoculants were found to improve hay appearance with little effect on DM loss and quality compared with untreated hay of similar moisture (Tomes et al. 1990). Until a more tangible benefit is found, inoculant products do not appear to be viable hay preservatives.

SILAGE CHOPPING

To make silage, a forage harvester is used to chop the crop for easier handling and better packing in a silo (Fig. 13.3). The forage harvester consists of a pickup or cutting mechanism that carries the forage into the machine. Rotating knives are used to cut the forage into lengths from 6 to 60 mm. The chopped material is conveyed into a blower mechanism that throws the material through a spout into a trailing wagon or truck. The chopped forage is transported to the silo where it is dumped into a bunker silo or elevated into a tower silo with a forage blower.

Labor requirements for silage harvest are normally less than that required for hay harvest, but energy requirements are greater. Silage harvest requires a minimum of two people, one operating the forage harvester and the other conducting the transport and unloading functions. Higher-capacity systems often require two transport operators. With bunker silos, an additional person is required at the unloading site to operate a tractor used to load and pack the silo. The labor requirement for the whole silage harvest system is 0.9-2.0 work h mt^{-1} DM (Table 13.1). The low end is for direct-cut harvest of crops such as corn silage. Forage harvesters require much power. Machines are available with minimum tractor size requirements of 30-150 kW with most greater than 50 kW. Fuel requirements range from 12 to 20 L mt^{-1} DM, about double that of typical hay harvest systems.

Losses in forage harvesting vary from 2% to 6%, with similar amounts lost from the pickup and by drift (Rotz and Muck 1994). Drift losses occur as the chopped material exits the spout of the harvester and travels toward the trailing wagon or truck. Drift losses are influenced by the crop moisture content, the wind conditions, machine adjustment, and operator skill. The quality of the lost material is similar to that harvested, so the loss has little effect on the quality of the remaining forage.

SILAGE STORAGE

Chopped forage is ensiled in tower silos,

Fig. 13.3. A forage harvester chopping wilted alfalfa silage.

bunker silos, or silage bags. Tower silos include both unsealed and sealed silo designs. Unsealed silos are constructed of concrete staves or poured concrete reinforced with steel rods to contain the silage pressure. With this type of silo, silage is normally unloaded from the top surface for feeding. Sealed silos are constructed of steel plates or poured concrete. The top of the silo is sealed, and a special coating on the walls may be used to reduce O_2 infiltration into the structure. By reducing O_2 availability, better fermentation and reduced respiration are attained. Silage is normally unloaded from the bottom of sealed silos.

Horizontal or bunker silos are becoming more popular for storing all types of forage. Bunker silos are normally constructed of concrete panels used to form a rectangular structure with one or two open ends. Silage is packed within the walls to a depth of 2.5-5.0 m. Silage is not preserved as well in this type of silo, but this design offers ease in filling and emptying. Better preservation can be obtained by covering the packed silage with a sheet of plastic. This step is often avoided to reduce the labor required for installation and removal of the plastic. Silage can also be pressed into long plastic tubes or bags, but special equipment is needed. As long as a tight seal is maintained in the bag, this process results in high-quality storage of silage. Use of bags eliminates the need for permanent structures and allows segregation of higher-quality forage for more efficient allocation to meet animal feed requirements.

Silos can be built with nearly any storage capacity. Common sizes and capacities of tower silos are listed in Table 13.2. Silage is packed more tightly in larger silos, which extends their capacity. Bunker silos are commonly built in 1.5-m increments of length and width. Silage is normally packed to a density near 190 kg DM/m^3. Packed at an average depth of 3 m, about 1.9 m^2 of bunker floor area is required for each metric ton DM of stored forage. At a depth of 4 m, the size requirement is 1.4 m^2 mt^{-1} DM. For forage pressed into plastic tubes or bags, 2.7 m diameter bags hold 1.0-1.5 mt DM m^{-1} of bag length.

Forage Changes in the Silo. Regardless of the type of silo used, the ensiling process can be described in four major phases: preseal, fermentation, infiltration, and feedout (Buckmaster et al. 1989b). The preseal phase occurs during the first few days as the silo is filled. Oxygen enters the silo, with the plant enzymes and microorganisms on the silage enabling respiration. The respiration causes heating and a small DM loss. The O_2 is quickly used, leaving an anaerobic environment for fermentation. Fermentation occurs over a period of a couple weeks. The goal with good fermentation is a rapid drop in forage pH to a level that allows stable storage. During fermentation, a small DM loss occurs, and a portion of the hemicellulose is broken down, causing a decrease in the NDF concentration. A portion of the true protein is transformed to nonprotein N by proteolysis. Proteolysis increases with silage temperature, so rapid filling and sealing of the silo is important to reduce the preseal heating phase. In silage containing more than 700 g kg^{-1} moisture, pressure in the silo squeezes plant moisture from the silage, or effluent. This seepage from high-moisture silage causes nutrient losses and the problems of effluent handling and disposal.

The major silage change during the remainder of the storage period is due to aerobic respiration. Throughout storage, O_2 penetrates the silo wall (silo cover for bunker) at a slow rate, allowing microbial respiration. Forage DM (primarily available carbohydrates) is transformed to heat and gases that leave the silo, causing a loss. Oxygen infiltration is initially controlled by the permeability of the silo wall or cover. As the silage carbohydrates on the outside of the silo are respired, the O_2 must permeate farther into the silo, and this reduces the infiltration rate. Better sealing of the silo reduces this loss. The final loss occurs as the silo is emptied.

TABLE 13.2. Approximate silo capacities, mt DM

	Silo diameter, m						
Silo height, m	4.9	5.5	6.1	6.7	7.3	7.9	9.1
15.2	71	90	111	135	160	188	250
18.3	92	117	144	174	207	248	324
21.3	---	147	179	216	257	306	401
24.4	---	---	206	256	303	356	472

Source: Adapted from American Society of Agricultural Engineers (1991).
Note: Metric tons of moist silage = mt DM/(1 – moisture content).

The surface of the silo is again exposed to O_2, and respiration continues in the loose forage in the feed bunk. Loss during this phase is related to the exposed surface area in the silo and the rate at which the silo is emptied.

In silos emptied over a 1-yr period, typical forage DM losses range from 6% to 11% in bottom-unloaded sealed silos, 7% to 14% in top-unloaded stave silos, and 10% to 15% in well-managed bunker silos (Buckmaster et al. 1989b). The loss during the preseal and fermentation phases is normally about 1%. The remaining loss occurs from O_2 infiltration during storage and feedout, with up to half occurring at feedout. When effluent occurs, up to 5% additional DM is lost.

Dry matter lost from the four phases is primarily highly digestible carbohydrates, which increases the concentration of other constituents. For example, little loss of N normally occurs unless there is effluent loss. Consequently, a 1%-2% DM increase in CP may occur, depending on the CP of the crop entering the silo and the DM lost during silo storage. Neutral detergent fiber content is affected by enzymatic and acid hydrolysis of structural carbohydrates during storage. Changes in NDF levels range from a 1%-2% DM decline to a 4% DM or more increase, depending on the respiration loss relative to the amount of cell wall hydrolysis. With good silo management, acid detergent fiber has little change, so the concentration increases 2%-5% DM with the loss of other carbohydrates.

Silage Additives. A wide variety of additives are used to enhance silage fermentation and preservation. The principal additives include bacteria, enzyme, ammonia, and acid products (Rotz and Muck 1994). The most common additives in the US are bacterial inoculants that supplement the natural lactic acid bacteria on the crop. When the added bacteria dominate fermentation, the silage has more lactic acid and a lower pH than expected from the natural fermentation. The improved fermentation may reduce DM loss by an average of 2.5% DM with a small reduction in proteolysis (Muck and Bolsen 1991). Inoculant success is most likely when the number of natural lactic acid bacteria on the forage is low. Inoculants are often more effective on grasses and legumes than on corn and sorghum silage because more natural lactic acid bacteria are typically found on the grain crops.

Enzyme products consist of a mixture of cellulases, hemicellulases, pectinases, and amylases that are used to break down forage fiber. Addition of these products normally reduces the fiber concentration in grass silage and sometimes in alfalfa silage (Muck and Bolsen 1991). The smaller effect on alfalfa is likely due to differences in its cell wall structure. Typical reductions in fiber concentration are 1%-5% DM. Dry matter loss may also be reduced by 5 percentage units. This reduction in respiration loss may be due to better silage packing in the silo with the breakdown of structural plant fiber.

Ammonia and acid products are often used in direct-cut silage systems. Anhydrous ammonia and urea are common additives to corn silage. The nonprotein N of these additives boosts the forage CP content. The added ammonia also kills aerobic microorganisms, which improves silage stability. These additives may improve DM and fiber digestibilities and reduce proteolysis. In spite of these benefits, DM losses may be increased, and little improvement in animal performance may be obtained (Muck and Bolsen 1991). Formic acid and similar products are sometimes used on direct-cut grass silage. The added acid quickly drops the pH of the forage, preventing undesirable forms of fermentation that naturally occur in very high-moisture silage. Effluent still remains a deterrent to this type of silage production.

Caution in Silage Storage. Gases formed during silo filling and several weeks thereafter can accumulate to a level that causes sickness or death. The gases formed during ensiling are carbon dioxide (CO_2) and nitric oxide (NO), which in turn produce nitrogen dioxide (NO_2) (Peterson et al. 1958). Carbon dioxide is not toxic, but it can displace O_2 and thus cause suffocation. Carbon dioxide is colorless and odorless. Nitrogen dioxide has a reddish yellow color and a pungent odor. It is heavier than air, accumulating in the silo for up to 10 d after filling. Through further oxidation in combination with water, nitric acid (N_2O_5) is formed. When such oxidation occurs within the body, permanent lung damage results (Stoneberg et al. 1968). Excess nitrates (NO_3) in forage may remain in the silage at a level potentially toxic to livestock.

Several steps can be taken to minimize the danger of silage gases. The silo should not be entered for a couple weeks after it is filled. If the silo must be entered, the forage blower should be operated for 20 min to help expel the gases. Be alert for any bleachlike odor or a yellowish brown gas that might be seen near the silo. If an enclosed room exists at the base of the silo, the room should be well ventilated for at least 2 wk after filling. The heavier-than-air gases may seep out of the silo and fill the room. Doors between the silo room and barn should remain closed to prevent these gases from entering the barn and killing livestock. Enclosures connected to the silo should be barricaded to prevent accidental exposure of gases to children or others. A person exposed to silage gas should see a doctor immediately to prevent potential lung damage and the development of pneumonia.

FORAGE SYSTEMS

Several forage harvest and storage processes are combined to form the forage system for a given farm. Although some systems are best for certain situations, none are generally preferred for all applications. Many factors must be considered when developing a best system for a given farm. Factors like the investment in equipment and structures and the labor and fuel requirements are important considerations. Hay systems normally require a lower investment in equipment and structures compared to silage systems. With large round bales, a storage facility may not be used, but the greater DM and nutrient losses that occur may not justify the lower investment. Systems using small rectangular bales can require twice the labor input (Table 13.1). A labor advantage in round bale hay systems and many silage systems is that less manual labor is required. Fewer people may be required to complete the work, but more of their time may be required. In silage systems manual labor is replaced by mechanical power; thus, substantially more fuel is required (Table 13.1).

Dry matter and nutrient losses are also important considerations. Average harvest and storage losses are between 20% and 30% for most forage systems (Fig. 13.4). Losses are less in direct-cut silage systems when the silage is well preserved with little effluent loss. Harvest losses are greater in hay systems compared with silage systems, but storage losses are less if inside storage is used. The loss in all systems is primarily leaves and soluble carbohydrates extracted from the plant tissue. The total loss typically causes a substantial gain in fiber concentration, with the potential for a small loss to a small gain in CP concentration (Rotz and Muck 1994). These changes often increase the need for supplemental feeds and may limit animal intake and performance. The loss of feed value to animals consuming the forage is therefore greater than the DM loss.

The overall goal for most producers is to harvest forage with minimum nutrient loss at the lowest cost. Other factors like timeliness of harvest and labor availability may also be important constraints. Speeding the drying

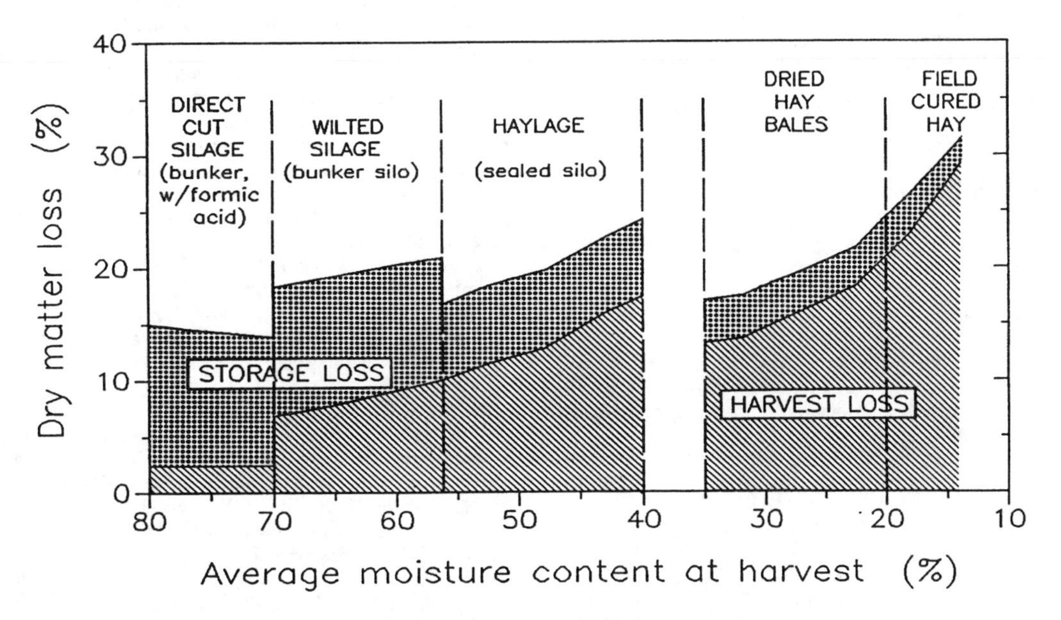

Fig. 13.4. Average dry matter losses in various alfalfa harvest and storage systems over many years of Upper Midwest weather conditions. (Rotz et al. 1991a)

process, reducing the number of machine operations, and modification and adjustment of machines can all be used to reduce losses and perhaps costs. Only by balancing the losses and costs with other benefits incurred can the best system for forage harvest be selected. Best options vary with climatic regions, crops grown, and farm management styles.

The preservation of forages is environmentally sound. Proper management of forage crops on potentially erodible soils plus harvesting, storing, and eventually feeding of the DM produced to livestock can allow a sustainable ecosystem. Livestock manure can be returned as fertilizer, completing the cycle with little soil and nutrient loss to the environment.

QUESTIONS

1. What are the various methods of harvesting forages?
2. Give the approximate moisture level for each storage condition.
3. What are the advantages of a rotary disk mower over a cutterbar mower?
4. What are the advantages of a conditioner? Are there any disadvantages?
5. Name the various methods of how one can speed up the drying of a forage in the field.
6. List the various types of balers and the advantage of each.
7. What causes heat buildup in harvested hay? List the various additives that one can use to control heat buildup in baled hay.
8. Describe the silage process.
9. What are some silage additives?
10. What are some precautions that one should follow or be aware of when making silage?
11. Consider a forage-harvesting system for your farm, and outline the various considerations in your decision.

REFERENCES

American Society of Agricultural Engineers. 1991. ASAE Standards. D2521.1. St. Joseph, Mo.: ASAE.

Buckmaster, DR, CA Rotz, and JR Black. 1990. Value of alfalfa losses on dairy farms. Trans. ASAE 33(2):351-60.

Buckmaster, DR, CA Rotz, and DR Mertens. 1989a. A model of alfalfa hay storage. Trans. ASAE 32(1):30-36.

Buckmaster, DR, CA Rotz, and RE Muck. 1989b. A comprehensive model of forage changes in the silo. Trans. ASAE 32(4):1143-52.

Harrigan, TM, and CA Rotz. 1992. Net, plastic and twine wrapped large round bale storage loss. Am. Soc. Agric. Eng. paper 921572. St. Joseph, Mich.

Koegel, RG, RJ Straub, and RP Walgenbach. 1985. Quantification of mechanical losses in forage harvesting. Trans. ASAE 28(4):1047-51.

Muck, RE, and KK Bolsen. 1991. Silage preservation and silage additive products. In KK Bolsen (ed.), Field Guide for Hay and Silage Management in North America. Des Moines, Iowa: NFIA, 105-26.

Peterson, WH, RH Burris, R Sant, and HN Little. 1958. Production of toxic gas (nitrogen oxides) in silage making. J. Agric. Food Chem. 6:121-26.

Rotz, CA, and SM Abrams. 1988. Losses and quality changes during alfalfa hay harvest and storage. Trans. ASAE 31(2):350-55.

Rotz, CA, and RJ Davis. 1986. Sprayer design for chemical conditioning of alfalfa. Trans. ASAE 29(1):26-30.

Rotz, CA, and RE Muck. 1994. Changes in forage quality during harvest and storage. In Forage Quality, Evaluation and Utilization. Madison, Wis.: American Society of Agronomy, Crop Science Society of America, and Soil Science Society of America, 828-68.

Rotz, CA, and HA Muhtar. 1991. Rotary power requirements for agricultural equipment. Am. Soc. Agric. Eng. paper 911550. St. Joseph, Mich.

Rotz, CA, and P. Savoie. 1991. Economics of swath manipulation during field curing of alfalfa. Appl. Eng. Agric. 7(3):316-23.

Rotz, CA, and DJ Sprott. 1984. Drying rates, losses and fuel requirements for mowing and conditioning alfalfa. Trans. ASAE 27(3):715-20.

Rotz, CA, SM Abrams, and RJ Davis. 1987. Alfalfa drying, loss and quality as influenced by mechanical and chemical conditioning. Trans. ASAE 30(3):630-35.

Rotz, CA, LR Borton, and JR Black. 1991a. Harvest and storage losses with alternative forage harvesting methods. In Proc. Am. Forage and Grassl. Counc. Georgetown, Tex, 210-13.

Rotz, CA, DJ Sprott, and JW Thomas. 1986. Anhydrous ammonia injection into baled forage. Appl. Eng. Agric. 2(2):64-69.

Rotz, CA, RJ Davis, DR Buckmaster, and MS Allen. 1991b. Preservation of alfalfa hay with propionic acid. Appl. Eng. Agric. 7(1):33-40.

Rotz, CA, RJ Davis, DR Buckmaster, and JW Thomas. 1988. Bacterial inoculants for preservation of alfalfa hay. J. Prod. Agric. 1(4):362-67.

Savoie, P. 1988. Hay tedding losses. Can. Agric. Eng. 30:39-42.

Savoie, P, and S Beauregard. 1990. Hay windrow inversion. Appl. Eng. Agric. 6(2):138-42.

Shinners, KJ, RG Koegel, and RJ Straub. 1991. Leaf loss and drying rate of alfalfa as affected by conditioning roll type. Appl. Eng. Agric. 7(1):46-49.

Shinners, KJ, RJ Straub, and RG Koegel. 1992. Performance of two small rectangular baler configurations. Appl. Eng. Agric. 8(3):309-13.

Stoneberg, EG, FW Schaller, DO Hull, VM Meyer, NJ Wardle, N Gay, and DE Voelker. 1968. Silage production and use. Iowa State Univ. Pam. 417.

Tomes, NJ, S Soderlund, J Lamptey, S Croak-Brossman, and G Dana. 1990. Preservation of alfalfa hay by microbial inoculation at baling. J. Prod. Agric. 3:491-97.

14 Forages in a Livestock System

JEROME H. CHERNEY
Cornell University

VIVIEN G. ALLEN
Virginia Polytechnic Institute and State University

FORAGE-livestock systems are the integrated combination of animal, plant, soil, and other environmental components managed to achieve a productive agroecosystem. Forage crops provide one half or more of the total feed requirements of dairy cattle (*Bos taurus*) and most of the feed requirements for beef cattle and sheep (*Ovis aries*). Forages also provide a physical site to keep the animals part-time or full-time. Successful forage-livestock systems must be ecologically and economically sustainable.

Preceding chapters consider the separate components of forage production and management. What are the primary interrelationships among the forage management components? How can these pieces be organized to develop forage selection and management programs that meet nutritional and other needs of an environmentally sound livestock system? Most farmers as well as many agricultural researchers, state and federal agency employees, and private consultants are uncomfortable with such organizational and management questions. It is easier to deal with discrete blocks of technical information, but interrelationships among management components provide the key to successful livestock farming.

Although integration of all forage-livestock enterprise components is a complex subject, the principles involved are relatively simple: (1) match adapted forage species with available soil resources, (2) minimize nutrient losses and prevent water quality problems, (3) match quantity and quality of forage produced with needs of the specific animal class(es), and (4) match grazing method(s) to plant requirements and the target for animal production. Integration is achieved by balancing the relative efficiencies of forage production, forage consumption, and livestock production (Hodgson 1990).

A forage-livestock enterprise is only one component of a much larger agricultural system, consisting of biophysical (i.e., physical and biological components), work, socioeconomic, and farm economic subsystems (Fick and Cherney 1994). The larger agricultural system may also include the regional organization of individual farming systems (Loomis and Conner 1992). Although ultimate success of the forage-livestock enterprise must be judged in the context of this larger agricultural system, a narrow focus is needed here to specifically assess the contribution of forages to the system. Success in the context of the livestock enterprise is measured as profitability and ecological stability. A strategic man-

JEROME H. CHERNEY is Associate Professor of Soil, Crop, and Atmospheric Sciences, Cornell University. He received the MS degree from the University of Wisconsin and the PhD degree from the University of Minnesota. His research deals with forage crop production and utilization, with particular emphasis on integrated dairy forage management systems.

VIVIEN G. ALLEN is Professor of Agronomy, Virginia Polytechnic Institute and State University. She received the MS and PhD degrees from Louisiana State University. Her research involves development and testing of forage-livestock systems and integrating forage-livestock systems into sustainable, whole-farm systems.

agement plan must be in place if profitability is to be realized or maximized. This chapter considers forage crops as one component of a dairy, beef, or sheep enterprise.

A primary goal of ruminant-feeding management is to provide forages with sufficient nutrients to support high levels of productivity, while at the same time minimizing feed costs (Miller 1979). Ideal feeding management is more of a process than it is a fixed set of conditions or rules. The process of developing an animal-feeding system integrates the soil resource, forage species, storage and other physical facilities, and animal. Matching forages to the needs of specific animal groups optimizes animal performance and minimizes nutrient waste. There are several general factors to consider when evaluating forages in a livestock system.

GENERAL CONSTRAINTS AFFECTING FORAGES IN A LIVESTOCK SYSTEM

Perennial forage crops under consideration for a livestock enterprise must be selected from the subset of forage species adapted to the environmental parameters of a particular region. Nutrient management in the broad sense, scale of farm operations, and production goals also affect forage selection and management.

Nutrient Balance and Water Quality. Forage-livestock management decisions now and even more so in the future will be affected, if not controlled, by manure management concerns. Intensive livestock management is often considered an economic necessity, particularly on dairy farms. With intensive management often comes an increased amount of crop imports onto the livestock farm. Simple nutrient balance calculations make it clear that when imports of a given nutrient into a farm ecosystem exceed exports in the form of meat, milk, or fiber the nutrient will eventually accumulate on the farm and may lead to water quality and other problems. A manure management plan can reduce the severity of the problem but may not be able to solve it. Grasses have advantages in nutrient management over legumes because they can use excess nitrogen (N), support vehicle traffic, and can better tolerate saturated soils. However, grass production for high-quality dairy forage requires a higher level of management compared with legume production.

Nutrient balance concerns can be viewed from an animal perspective. The number of animals per unit land and the capacity of land to receive waste must be in harmony. The Netherlands has enacted government regulations to control the amount of phosphorus (P) that can be applied to land. The maximum amount of P per hectare corresponds approximately to that produced by three cows per hectare. In reality, such calculations are dependent on crop rotation and the specific soil resource. Land-use regulations are being considered in the US, particularly in regions that supply water to large metropolitan centers. For example, water for New York City is transferred long distances from upstate New York through aqueducts. New York City has legal authority to restrict agricultural practices in the 0.5 million ha upstate watershed in order to control water quality.

Forages can be considered the solution to many potential environmental problems, with forage quality playing a key role in nutrient balance on the farm. Use of forages with high-intake potential can reduce the amount of concentrate in the diet, such that forage management for optimum quality results in minimizing feed imports onto the farm. Reduction in feed imports is a powerful tool to help balance nutrients in a livestock system (Koepf 1989). When a broader viewpoint than just the livestock enterprise is taken, however, minimizing feed imports may not always be in society's best interest. Livestock can be viewed as environmentally safe recyclers of many coproducts from industries such as distillers, brewers, and bakers, which would otherwise have their own disposal problems (Pell 1992). Broiler litter, a coproduct of the poultry industry, is an excellent source of N and several minerals that can be fed to ruminants. The use of intensive-grazing management also has the potential to alleviate many nutrient management problems, especially those related to manure handling and storage.

Scale of the Farm Operation and Production Goals. Generally the size of a farm operation is positively correlated with profitability. Increased mechanization is often required as herd size increases, and the move to more mechanization typically improves labor and overall production efficiency. As farm size increases, feasibility of grazing decreases, and the tendency increases to adopt systems for year-round feeding of stored forages. Options for stored feed and for forage delivery systems are defined, in large part, by the size of the operation.

Large-scale operations with maximum production efficiency are not a requisite for profitability. Smaller operations that control inputs and that benefit from grazing methods that intensify the system can be quite productive. The scale of the operation affects how forages fit into livestock systems by altering the number of potential options to be considered in the forage integration process.

Animal production goals impact all aspects of the forage-livestock system. The level of animal production is directly related to the efficiency and management of the operation. In addition, the larger the herd size, the more that high production goals translate into increased profitability. As goals for production per animal increase, the amount of concentrate fed generally increases, thus increasing feed imports and resulting in more nutrient balance problems on the farm. Production goals, as with the scale of farm operations, help to focus on the subset of options considered during the forage integration process.

ENTERPRISE PLANNING

The goal of enterprise planning is to successfully integrate forage and livestock components to increase overall efficiency and, therefore, to increase profitability of the livestock enterprise. Knowledge of the basic components of forage and livestock management and an understanding of the main linkages between these components are necessary before the integration process can be attempted. Concept maps (read from top to bottom) are useful for this purpose (Fig. 14.1). Although Figure 14.1 specifically describes perennial forage management, it can apply equally well to annual forage species if the stand life (persistence) component is removed. These basic components of forage management must be kept in mind as well as the general constraints already discussed.

Basics. Proper management begins with forage selection. Perennial forage species and cultivar selection interact with site selection to provide a forage or combination of forages that are (1) persistent and (2) capable of producing acceptable yields of a desirable quality (Fig. 14.1). This is accomplished by selecting species and cultivars that are free of antiquality components or at least with tolerable levels of antiquality components (e.g., low-alkaloid reed canarygrass [*Phalaris arundinacea* L.] or low-endophyte tall fescue [*Festuca arundinacea* Schreb.]). The chosen species must persist under the given climatic and site conditions.

After the above requirements are met, the forage should be selected to match the protein, fiber, and energy needs of the livestock. Alfalfa (*Medicago sativa* L.) and corn (*Zea mays* L.) silage have a high-quality image compared to perennial grasses that leads to their preferred use, even on land not suited to these crops. Corn often is planted on land that cannot produce silage yields that are high enough to return costs of production. Alfalfa is frequently established on sites where it cannot persist because of soil constraints. These problems are due to a lack of integrated analysis of the forage management component of the livestock enterprise.

Species, cultivar, and site selection all have a direct influence on forage yield potential (Chap. 5). Species selection, particularly when comparing legumes and grasses, directly influences forage quality, while cultivar selection within a species improves quality primarily through control of antiquality components. Site selection by itself has a minimal effect on forage quality. The interaction of species, cultivar, and site, along with harvest management and nutrient management, determine stand life. Stand life obviously directly influences, and is directly influenced by, the crop rotation plan.

The interaction of species, cultivar, and site determines the appropriate nutrient management strategy. Soil fertility status is amended by the use of commercial fertilizer and/or manure to provide for maximum economic forage yield without negatively affecting water quality or nutrient balance of a given site. Nutrient management has a positive effect on forage yield and stand life and also influences harvest management.

Harvest/grazing management is the single most critical component of the forage management scheme because it controls forage yield, botanical composition, and forage quality. The appropriate harvest/grazing management is dependent on species selection, but the time interval between harvests will be influenced by nutrient management. The primary goal of harvest and nutrient management is not necessarily to maximize stand life but to provide the necessary forage quality and yield while at the same time maintaining an acceptable stand life. As the importance of grazing increases in a livestock system, the storage component decreases in importance. Grazing adds another level of complexity to the livestock system and, as such, requires a higher level of management than stored feeding systems.

Storage options are described in the pre-

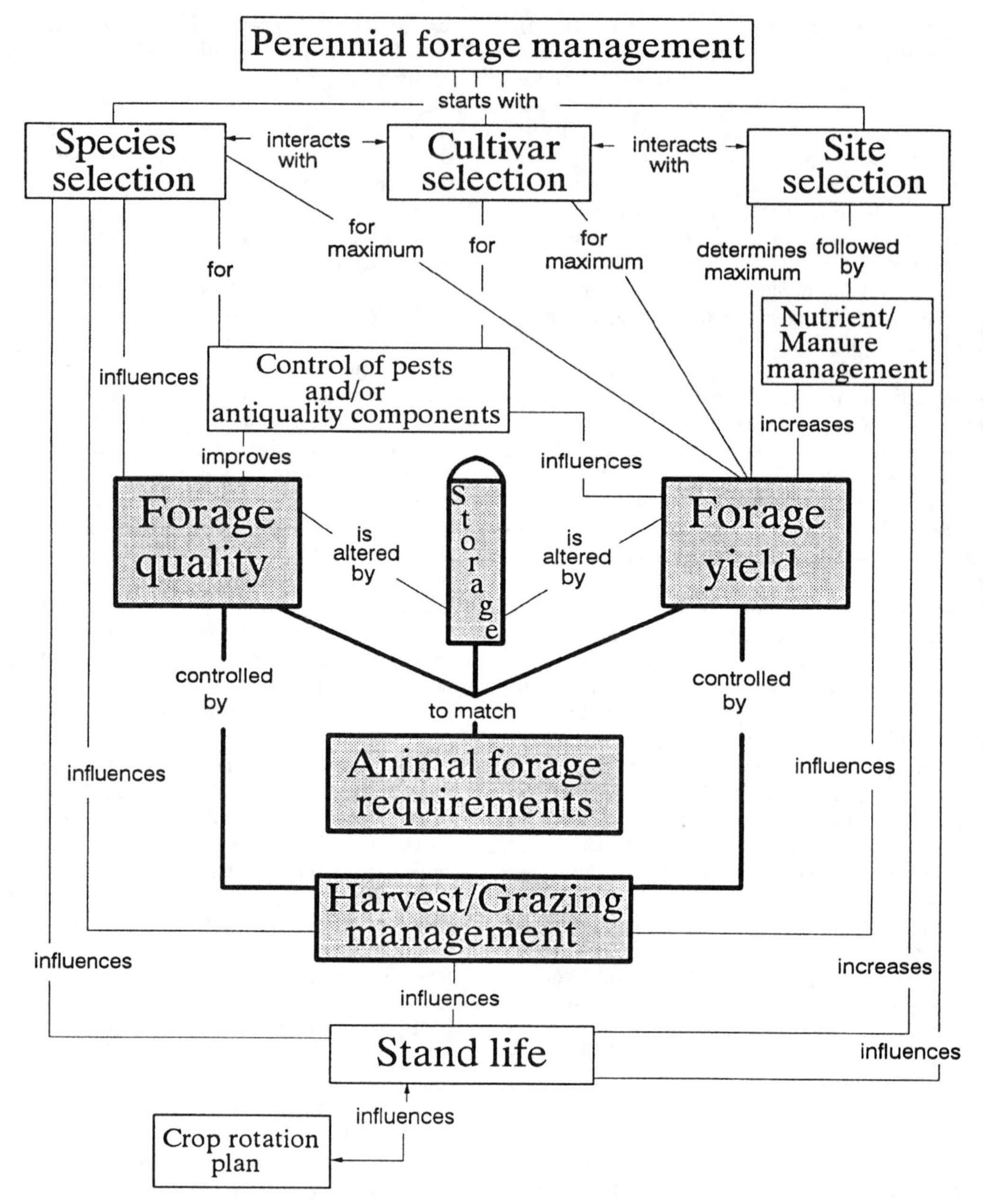

Fig. 14.1. Concept map of major linkages in a forage-livestock system. As the importance of grazing increases in a livestock system, the storage component decreases in importance. (Adapted from Fick and Cherney 1994.)

ceding chapter. Both forage quality and yield are altered by storage, and the degree of change depends on the particular storage method. Along with forage quantity and quality, storage options and access to storage must be evaluated and integrated to match animal requirements for forages.

Animal feed requirements depend on maintenance needs and production potential of each class of livestock. Maintenance and production requirements are affected by animal species, age, sex, production level, and environment. It is often desirable to group animals within a class by production level and to provide forage to match the needs of each group. Profitability is reduced if low-quality forage is fed to animals with high nutrient requirements or if high-quality forage is fed to animals with low nutrient requirements. Availability and cost of concentrate feeds also need to be considered because concentrate can be used as a substitute for some of the forage in a ration, thereby affecting forage requirements. For high-producing animal groups the overall objective is to maximize performance by maximizing feed intake, subject to feed cost restraints.

Integrating Forages with Livestock Requirements. In enterprise planning estimates of forage, storage, and animal requirements are used to provide a framework for operating the business. The process of integration for a particular livestock enterprise involves setting up a strategic plan that estimates quantity and quality of forage needed for each animal class, where forage will be produced, how forage will be harvested, and what storage and access options are available.

Strategies for dairy enterprises typically differ considerably from those of beef and sheep. The emphasis in dairy operations is on producing high-quality forage, and grazing may be employed as one of several harvest management options. Generally, beef and sheep operations are based on grazing, with much less emphasis on storage. Thus, different approaches are used in strategic planning for dairy enterprises compared with those for beef and sheep.

GRAZING SYSTEMS

Grazing systems are the integrated combination of animal, plant, soil, and other environmental components and the grazing methods by which the systems are managed to achieve specific results or goals (FGTC 1992). By comparison, a grazing method, such as creep grazing or rotational stocking, is a specific grazing management designed to achieve an objective. In reality then, a grazing method can be transferred from one place to another and used as a specific, described, intact technique. A grazing system, however, cannot be transferred because it is unique to the precise set of local conditions, combinations of resources, kinds of livestock, forage species, soil, climate, and other components that affect the system.

Principles or components of a grazing system design are transferable among locations. Thus, it is important to understand the principles of grazing management and soil-plant-animal-environmental-management interrelationships (Fig. 14.1), as well as the management objectives, in order to create appropriate systems within a specific set of resources. These principles can apply equally well to dairy cattle on pasture as they do to beef and sheep, but the adopted grazing system will be unique to the situation.

Grazing pressure is an animal-to-forage available relationship that may be modified to improve grazing systems (Fig. 14.2). When there is excess forage, grazing pressure is low, and forage is wasted. Individual animal performance may be high due to the animal's ability to selectively graze, but output from the system is likely to be decreased. Adding animals or reducing the area to be grazed such that there is less forage per animal increases grazing pressure. This can result in a more intensive grazing system and improved output per unit land area. However, going too far in this direction results in too little forage to support animal needs. Forage plant survival, individual animal performance, and output per unit land area are sacrificed. The optimum range will be determined by production goals, kind of forage, environment, opportunity, and the relative profitability of individual animal performance versus production per unit area. The optimum range should also be within a range of plant use that ensures vigorous regrowth, a dense sod that protects the soil from erosion, and an animal stocking density that must not create problems with water quality due to soil compaction and/or excessive loading of the system with manure.

Grazing systems must also allocate quality of forage appropriately to meet the various requirements of animals. Likewise, nutritional needs of animals must be matched as closely as possible to the potential for forage production. Successful systems allocate nutrition to the responsive animal class while ensuring that the varying nutritional needs are met. Seasonal variations in forage quality and quantity generally mean that more than one forage, managed separately in more than one paddock, is required to provide adequate forage over the grazing season or throughout the year (Fig. 14.3). Advantage can be taken of the differences among forage species, and even the differences among cultivars, to extend the grazing season and match forage quality and quantity with animal requirements.

Forages differ in their inherent quality potential, and there are seasonal variations in forage quality. The best estimate of the potential of a forage for animal production is the intake multiplied by its digestibility (Jorgensen and Howard 1982). Legumes generally are higher in intake, digestibility, rate of passage, and potential for animal performance than grasses. In general, cool-season annuals rank

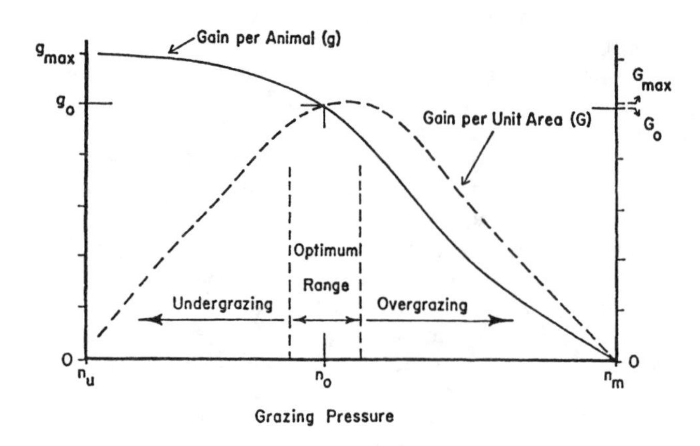

Fig. 14.2. Grazing pressure (n) effects on gain per animal (g) and gain per unit area (G). (Adapted from Mott 1973.)

second to legumes in forage quality and expected animal performance. These are followed by cool-season perennials and warm-season annuals while warm-season perennials generally present the greatest challenge to supporting high levels of animal production. With any forage, maintaining a desirable growth stage and leafy, vegetative, optimally fertilized green plant material is critical. Having forage present in sufficient quantity and having a physical structure of the plant canopy that optimizes intake per bite are also primary considerations.

Grazing systems must have sufficient flexibility or elasticity to accommodate varying environmental, market, and production conditions. Conservation of excess forage should be planned to provide for needs during expected or unexpected forage deficits, but the supply should be matched closely to the expected needs unless this excess can be sold later and is a planned part of the total system. Stockpiling forages for grazing at a later time can reduce stored forage needs and can often provide opportunities for high-quality grazed forage (Fig. 14.3).

Systems are made up of many interactive parts that must function as a whole to obtain the desired outcome. The sum of the parts of a system rarely if ever equal the whole. A change in one part of a system, such as a change of forage species or animal breed, can affect the balance among other parts of the system and may either improve the system or cause it to fail. A grazing system out of balance results in gaps of insufficient feed, periods with excess forage growth, inappropriate animal performance, degradation of the environment, and loss of economic profitability. Successful grazing systems are sustainable environmentally, economically, and in terms of labor demands as well as total productive capacity.

FORAGES IN DAIRY SYSTEMS

Among classes of ruminant livestock, dairy cattle have the highest nutrient requirements, and milk production is more sensitive to the balance of nutrients fed than is meat or wool production. Lactating dairy cows are more efficient (i.e., there is more milk per unit feed) at higher levels of milk production be-

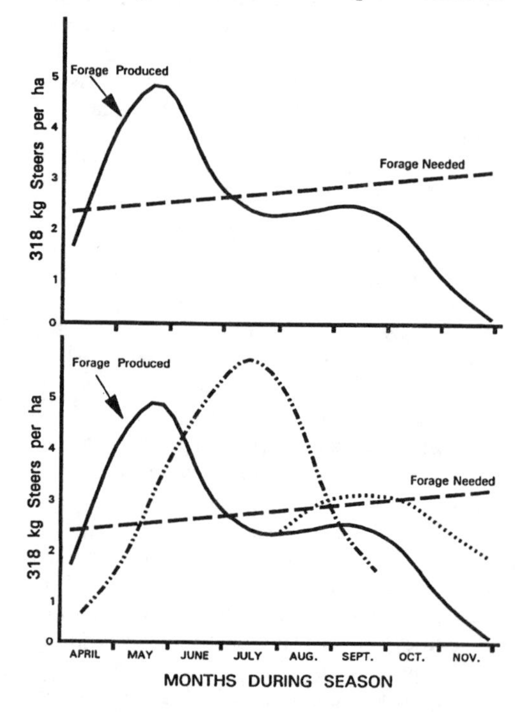

Fig. 14.3. Production rate of cool-season perennial forages does not match animal needs. Use of cool- (*solid line*) and warm-season (*dotted line*) perennials provides a better seasonal distribution. Increased forage is needed as the grazing season progresses because animal weight is increasing. Stockpiling (*dashed line*) forage for grazing during late fall and winter reduces needs for harvested feeds. (Adapted from Blaser et al. 1973.)

cause most production costs are fixed and are not affected by level of production (Fig. 14.4). Animal maintenance requirements can also be considered fixed costs. As milk production per cow is increased, the feed cost per unit of milk produced decreases, and profits increase. Relatively low concentrate feed prices in the US allow dairy managers to increase profitability by using high-energy rations to increase milk production. This is not feasible in most other countries where concentrate feeds are expensive and milk-marketing quotas may exist. These factors affect dairy-forage integration.

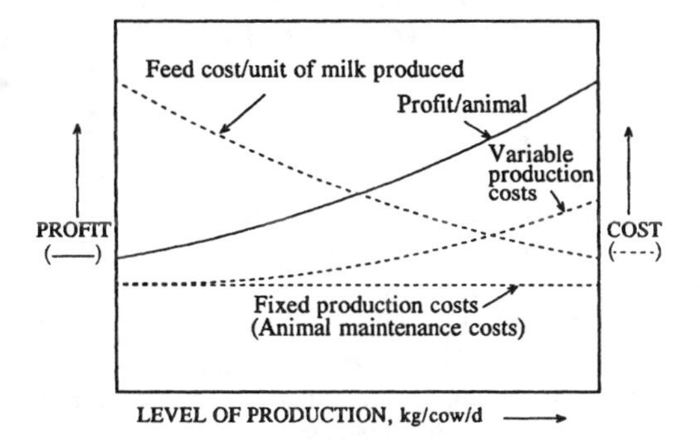

Fig. 14.4. Increased production per cow results in increased efficiency and increased profits.

Integrating Forages into a Dairy Enterprise. The following description is based on the Pitt-Conway dairy-forage integration process model (Fig. 14.5) described by Conway (1993). The model can be used to organize and to integrate the information needed by a dairy farm manager to optimize use of resources and maximize profits. This includes a setup phase prior to actual integration, where initial estimates are developed for the upcoming year for each parameter using detailed, yet relatively simple, calculations. In the model the livestock enterprise under consideration is assumed to be forage based with minimal cash cropping, and grazing is considered to be a relatively minor component of the overall enterprise.

ESTIMATION OF DAIRY ANIMAL FORAGE REQUIREMENTS. Average daily dry matter consumption per animal is estimated for each animal class. The total number of animals in each class, on average throughout the year, is then used to calculate animal forage require-

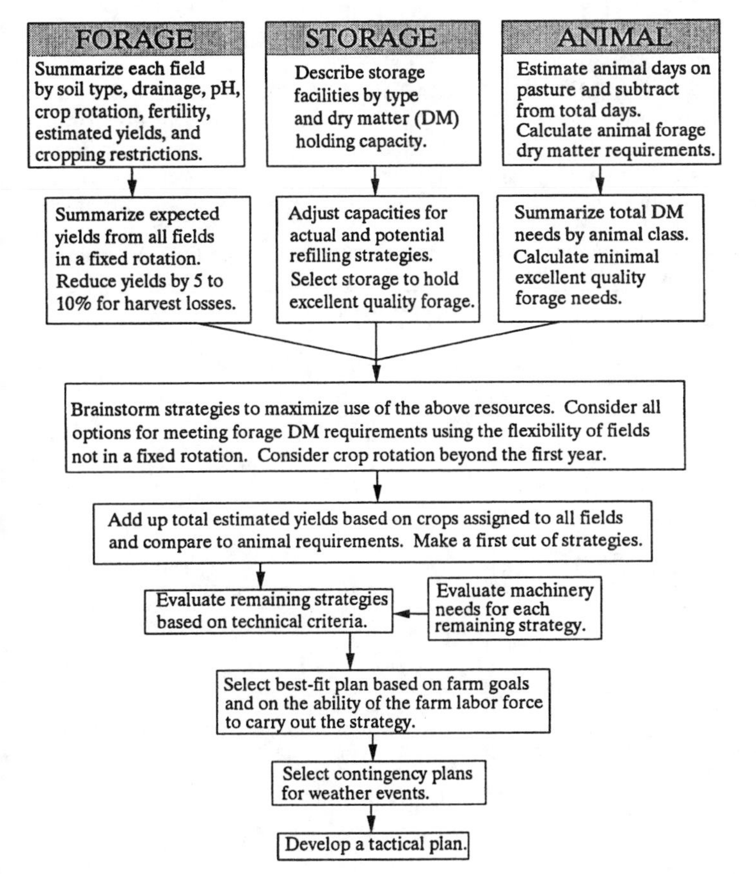

Fig. 14.5. Pitt-Conway dairy-forage integration model. The objective is to match the quantity and quality of forage available with requirements of specific livestock classes and with available storage options.

ments, i.e., the total forage dry matter needs for the year. The estimated number of animal days on pasture for each animal class is subtracted from the yearly dry matter needs of the herd, and the remaining total represents annual need for stored forage.

It is practical to consider two heifers as equivalent to one mature animal unit and to estimate animal forage needs accordingly for the strategic feeding plan. However, it is more accurate to estimate the forage needs of heifers and cows based on body weight. Tables are available to estimate dry matter intake according to body weight and forage quality (Table 14.1). Intake values need to be adjusted for the forage to corn silage ratio, if these are the primary two sources of forage in the diet. Feed requirements also differ by stage of lactation and by level of milk production, but dairy cattle will consume a relatively consistent amount of forage, depending on their body weight and forage quality, with the variable factor being concentrate supplement.

A separate category is required for the minimum amount of excellent quality forage needed for young heifers up to 5 mo in age and for all cows during the first 100 d of lactation. Excellent quality forage is defined as alfalfa with a neutral detergent fiber (NDF) concentration between 350 and 450 g kg^{-1} dry matter (DM) or grass with an NDF concentration between 450 and 550 g kg^{-1} DM. Alfalfa and grass with lower NDF concentrations than 350 or 450 g kg^{-1}, respectively, may not have sufficient fiber to be used as a primary fiber source in a ration. When developing actual tactical feeding plans, it is often advantageous to divide the cow herd into four groups: early lactation, mid-lactation, late lactation, and dry, followed by ration balancing specific to each group.

EVALUATION OF STORAGE CAPACITY IN A DAIRY SYSTEM. Upright, bunker, and/or trench silos are described in terms of silo dimensions, filling strategy, calculated holding capacity, and practical holding capacity. Practical holding capacity includes possible refills during the year and may include the filling of bunker or trench silos beyond actual silo dimensions. Dry hay storage is described in terms of dimensions and practical holding capacity. Flexible storage systems (see Chap. 13) are considered for their ability to fit in with available machinery and feeding systems currently on the farm. However, holding capacities for flexible systems are not included in the total estimate of storage capacity. Limitations of each storage option are described in terms of forage moisture content, spoilage, unloading ease, and ability to separate forage based on quality. If carryover of excess forage from the previous year is anticipated, the amount of carryover and its location should be included. Excellent quality forage must be accessible throughout the year.

ESTIMATION OF FORAGE PRODUCTION IN A DAIRY SYSTEM. Each field on the farm is described in terms of size, soil type or land-use class, drainage class, fertility status, pH, and crop rotation plan, including a listing of pos-

TABLE 14.1. Estimated dry matter intake per day of legume forage and corn silage (in a 2:1 ratio) for poor, average, and excellent quality forage

	DM intake per day (kg)					
	Heifer weight (kg)			Cow weight (kg)		
Feed	140	280	420	450	550	650
Poor quality[a]						
Legume	[b]	2.87	4.59	6.40	7.60	9.00
Corn silage	---	1.44	2.30	3.20	3.80	4.50
Average quality[c]						
Legume	1.65	3.75	6.25	7.32	8.80	10.40
Corn silage	0.83	1.88	3.12	3.66	4.40	5.20
Excellent quality[d]						
Legume	1.96	4.70	6.54	8.58	10.34	12.26
Corn silage	0.98	2.35	3.27	4.29	5.17	6.13

Note: Other forage:corn silage ratios can be estimated based on this data. Intake of grass forage can be estimated from legume intake in each forage quality category.

[a]Poor quality is forage greater than 530 g kg^{-1} NDF for legume or 600 g kg^{-1} NDF for grass. Estimated intake of poor quality grass forage is 80% of the legume intake value.

[b]Never feed young heifers poor quality forage.

[c]Average quality is forage between 450 and 530 g kg^{-1} NDF for legume or between 550 and 600 g kg^{-1} NDF for grass. Estimated intake of average quality grass forage is 90% of the legume intake value.

[d]Excellent quality is forage between 350 and 450 g kg^{-1} NDF for legume and between 450 and 550 g kg^{-1} NDF for grass. Estimated intake of excellent quality grass forage is similar to that of legume forage.

sible forage crops that are suited to the field. Field use may be restricted by drainage, topography, distance from the home farm, previous pesticide applications, or government program regulations. Estimated yield of the previous crop is compared with yield estimates for other forage crops suited to the field. Dry matter yields from all fields definitely returning to the same crop as the previous year, or otherwise in a fixed rotation, are summed.

All strategic plans for dairy-forage systems must be based on an adequate supply of high-quality forage. It is assumed that most of the remaining forage produced is of average quality, with only a small amount of low-quality forage. Some low-quality forage can be used in heifer or dry cow diets, but such forage has limited usefulness in a dairy enterprise.

PROCESS OF INTEGRATION IN A DAIRY SYSTEM. Dry matter produced on all fields in the fixed rotation is subtracted from the total animal requirement for forage dry matter. Then everyone involved in strategic planning of the enterprise must brainstorm to develop several cropping strategies to meet the remaining dry matter needs from fields not in a fixed rotation. Brainstorming involves asking a series of questions constrained only by farm size and specific storage and feeding facilities of the dairy enterprise under consideration. Selected strategies must complement the available machinery and labor resources and also fit in with the overall goals of the farm enterprise. The purpose of the questions is to develop a series of strategies for meeting the forage dry matter needs of the dairy enterprise. Example questions are as follows:

Can I use an all-grass field to aid in "disposal" of manure and at the same time provide a modest quality, low-potassium (K) forage source for dry cows?

Would an all-silage program allow me to minimize the weather problems and produce more high-quality forage?

Can I produce and feed silage in bags or tubes without equipment or labor force additions?

Can a portion of the land be rotationally stocked to avoid building additional storage structures?

Do I need to produce some high-quality hay in mid- to late summer to help ensure that high-quality forage is always available?

Can I buy and transport some of my forage from a neighbor more economically than I can produce it myself?

Should the erosive hillside field be removed from the rotation and set aside as a permanent pasture?

Can I buy corn grain for less than my cost of producing high-moisture corn grain?

Additional examples can be found in Conway 1993. The more questions asked, the more scenarios brainstormed from which a strategy can be selected. Looking ahead more than 1 yr is advantageous but much more complex. Strategies are evaluated based on overall farm goals, feasibility, and availability of necessary labor, machinery, storage structures, and feeding systems. After a "best fit" strategy is selected, a tactical plan for actually accomplishing the selected strategy is developed that should be coordinated with all other farm activities. In reality, the plan will need to be flexible and include several contingency plans to account for uncontrollable factors such as weather.

Modeling Dairy-Forage Management. The process of integrating a large number of options to arrive at one or more of the most favorable scenarios is complicated. It is also very difficult to evaluate what effect a change in one component will have on the entire dairy system. Computer models, such as DAFOSYM (Rotz et al. 1989), are being developed and constantly updated to assist in understanding the process of integration and also to help evaluate how changes in single components affect profitability of the entire system. Results from modeling are at best only as good as the data entered and assumptions made in the model. Models continue to be very effective for identifying gaps in our information base, which can then be further researched (Spedding 1984).

FORAGES IN BEEF SYSTEMS

Beef production is generally divided into three production phases: cow-calf, stockers or yearlings, and finishing cattle. Nutritional needs of beef cattle depend on the stage of production. Young calves and finishing beef cattle have the highest nutritional needs and should be provided the highest-quality forage. Cows need high nutrition from the time of calving until the cow is rebred.

Cow-Calf Production. When calves are about 4 mo old, they begin to utilize forage for more of their nutrition. Therefore, the nutritional needs of the cow are lower because milk production has declined, and it is not profitable to continue to feed the cow at a high nutritional level (Blaser et al. 1986). At this point, allocation of high-quality forage is better shifted to the calf through a grazing method such as creep grazing.

The breeding herd should generally be maintained in a forage system based on perennial forage species. Whether the perennial base is warm- or cool-season forages, or some combination of the two, will depend on geographical location and climatic conditions (Chap. 5). Grazing should be maximized with as little use of stored forages as possible. The stored forage should be hay or silage generally made during periods of excess forage supply. However, consideration should be given to purchasing harvested forages. The high cost of equipment, time, and labor required for harvesting operations may make purchased forage a more attractive option, particularly for smaller operations.

Use of annual forage species in the system increases expense and risk, unless the species are present due to other farm enterprises. For example, a coproduct of grain production can be corn stalks that can be grazed or small grain straw, perhaps treated with urea to increase quality and digestibility, that can be fed. The high quality of vegetative small grains, annual ryegrass (*Lolium* L. spp.), and annual legumes may not be justifiable as forage for cows, thus being used by stockers or yearlings. On the other hand, crop residues can often be used by the cow herd.

Furthermore, designing the system to incorporate first-last grazing (Chap. 13, Vol. 2) allows cows to graze residual forage last after another group of animals has first utilized the highest-quality forage through selective grazing. This grazing method matches forage and nutritional requirements, which can sometimes improve profitability and total forage utilization within the system.

Calving season should be adjusted such that there will be adequate forage quality and quantity for cows from calving through rebreeding. When calves are about 4 mo old, restricting forage intake of cows by increasing stocking density, using a lower-quality forage, or restricting the area where they may eat high-quality forage is more cost effective than maintaining cows on good supplies of high-quality forage. High-quality forages should be used for younger animals.

The forages used for the cow herd can also be used to meet the higher nutritional needs of calves. However, these forages must be managed to maintain high quality and quantity to maximize intake. Alternatively, allowing calves access to areas not accessible by the cow allows calves to selectively graze and maximize intake and gains on a variety of forages (Allen et al. 1992). Such access can be achieved by use of specially designed creep gates (Fig. 14.6). Spaced posts with adjustment and proper insulation of electric fences to allow calves entrance to the creep-grazing area can also be used. Hayfields, firebreaks, conservation strips, and other forage-use areas adjacent to the base pasture can be used for creep grazing. Regardless of the forage species, the system should be designed to ensure an adequate supply of grazable forage in the creep pasture that encourages selective grazing as the calf size and feed needs increase. The forage must allow selectivity or be higher in quality than the base pasture.

Stocker Production. Stocker cattle can use forage efficiently for economical gains. The objective for stocker cattle is to support growth and development, but fattening at this stage is generally not desirable. Replacement heifers may need to be managed differently from stocker cattle. Generally the objective is to breed heifers at 15 mo to calve as 2 yr olds. Successful breeding at 15 mo is more dependent on reaching the target body weights than age per se (Schultz 1969). Heifers should reach approximately 65% of their expected mature body weight by the time they are bred and about 85% by the time they calve (Chessmore 1979). Gaining weight too rapidly during growth can lead to fat deposits in mammary tissue and depressed milk production potential (Swanson 1967).

Forage systems for stocker cattle should be based on the target end weight and date, and the needed rate of gain to reach this objective should be calculated. The quality and quantity of forage required to meet this objective can then be calculated and added to the system. Since the objective is growth rather than fattening, emphasis should be placed on ensuring optimum intake of utilizable protein, energy, and minerals. Profits are generally highest when feed input from grazing is maximized and that from stored forages or supplements is minimized. Pastures based on perennial species are often most cost-effective, but overseeding of base pastures to improve quality or extend the season may be attractive options. For example, overseeding of

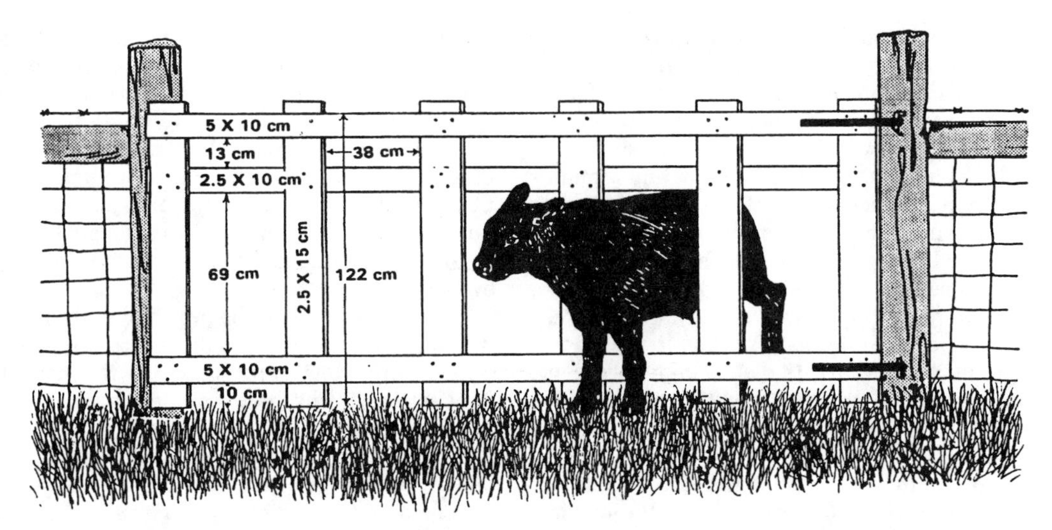

Fig. 14.6. A creep gate is designed to allow calves to graze areas that their dams cannot access at the same time. (Adapted from Blaser 1986.)

bermudagrass (*Cynodon dactylon* [L.] Pers.) with annual ryegrass (*L. multiflorum* Lam.) in autumn gives nearly year-round grazing in the South. Overseeding tall fescue with a legume such as alfalfa, ladino clover (*Trifolium repens* L.), or red clover (*T. pratense* L.) improves summer productivity and, especially, the forage quality of the pasture.

Grazing systems can be intensified by selection of grazing method(s), but the method of grazing should be tailored (1) to the requirements of the animal to achieve the desired rate of gain and (2) to the need for rapid regrowth and longevity of the forage. Thus, the appropriate grazing method will vary with specific conditions. Since a single forage base is rarely appropriate year-round, sequencing the use of several different forages and/or forage combinations growing in separate pastures generally achieves the best match of forage growth and quality with animal requirements. For example, this could be accomplished by use of tall fescue for grazing during winter, spring, and early autumn and caucasian bluestem (*Bothriochloa caucasica* [Trin.] C.E. Hubb.) or bermudagrass for grazing during midsummer. In the lower Midwest, grazing wheat (*Triticum aestivum* [L.] emend. Thell.) pastures in spring followed by old world bluestems (*Bothriochloa* spp.) provides an appropriate sequence for grazing. Differences in growth periods due to forage species characteristics and to the influence of the local environment largely determine which forage species will be appropriate for sequence grazing. The objective is to make the transition from one species or species combination to the next with minimal overlap and to avoid gaps of insufficient forage production (Fig. 14.3).

Cost-effectiveness of stocker systems should take into account the amount of animal gain produced per unit area of land and the relative value per kilogram of gain. The value (i.e., selling price) per kilogram of calf at weaning is nearly always higher than the value per kilogram at the end of a stocker phase. Therefore, it could be more profitable to increase the cow herd and sell more calves with higher weaning weights than to retain ownership of the calves beyond weaning through the stocker phase. Alternatively, purchase of low- to medium-weight calves at weaning and using all forage diets of appropriate quality through a stocker phase can be an economically attractive option. Economic success generally comes with intensification of the grazing system.

Finishing Systems. Finishing systems for beef cattle generally begin at about 350-450 kg and continue to about 550 to 650 kg, depending on breed of cattle and desired amount of finish. The feed supply can be all forage, mainly forage with some additional energy supplemented as grain while grazed forage is utilized, grazed forage followed by a short period on a high-energy silage, or a feedlot finishing phase where little forage is fed and cattle are finished mainly on a high-concentrate diet (see Chap. 19, Vol. 2). Because of the relatively low cost of grain in the US, finishing cattle on high-concentrate diets has been an attractive alternative. Shifts in the relative cost of grains, environmental and animal welfare concerns with high concentrations of animals, and increasing concerns over dietary fat and other issues of diet and human health have increased interest in finishing cattle using a higher proportion of forage in the diet. Finishing systems based exclusively on forages generally require more time for animals to reach the desired carcass weight and grade since their diet is generally lower in energy than a concentrate diet. A high degree of management skill is required to maximize intake of forage for high performance of finishing cattle.

Similar to forage systems for lactating dairy cows, those for finishing cattle should use the highest-quality forage available and incorporate grazing methods that promote the highest-possible intake of digestible dry matter. Cool-season annual grasses such as the cereal grains and annual ryegrass can be used to advantage in finishing systems. Cool-season perennial forages managed for high digestibility and intake can also support excellent gains. Corn silage supplemented with crude protein is an excellent high-energy forage for finishing beef cattle (Hammes et al. 1964). One hectare of corn will usually produce sufficient silage to fatten about 10 steers over an approximate 120-d feeding period. Generally, silages made from perennial grasses or legumes are relatively lower in energy, and cattle fed these silages require either an additional energy supplement or longer feeding periods to achieve target slaughter weights and grades.

FORAGES IN SHEEP SYSTEMS

Sheep are among the most efficient livestock in utilizing all-forage diets for profitable production. Although sheep breed very seasonally, the diversity of suitable forage species and their growth characteristics allow many interesting forage systems to be designed. Furthermore, new technology in sheep-breeding programs offers the opportunity for out-of-season breeding, allowing additional options for designing grazing systems to make maximum use of forages.

Principles of sheep systems are similar to those of cattle, but the production cycle is shorter. Ewes need maximum energy and high-quality forages during the last 4 to 6 wk of gestation. During this period, growth of the fetus restricts intake through reduced capacity of the digestive tract. Thus, forage must be high in digestible dry matter. High-quality forage is also needed during the first 4 wk of lactation to support high milk production. During this period, about 75% of the variation in lamb growth is attributed directly to milk production of the ewe. Need of high forage quality for the ewe declines after this period. Between 4 and 8 wk postpartum, milk production declines to about 25% of the peak level, and the need for high-quality forage for the ewe declines.

Dry ewes should not be provided high-quality forage as this can result in fattening and difficult breeding. They are in an ideal period to effectively utilize lower-quality forage such as crop residues. Dry ewes can also be used as last grazers, following selective grazing by lambs as first grazers. This grazing method can also aid in reducing internal parasite problems in lambs since they always graze a paddock prior to its being occupied by the mature animal.

Two weeks prior to breeding, ewes again need access to high-quality forage and continue to need it from about 2 to 4 wk into the breeding season. This "flushing" helps to increase ovulation and embryo survival. However, ewes should not be continued on high-quality forages too long since this is costly and can result in fat ewes that have difficulty during late pregnancy and parturition.

Lambs need access to high-quality forages from shortly after birth. After the first month, lamb gains are nearly totally dependent upon the feed they consume, rather than the ewes' milk. Forages that are high in quality and palatability, and are readily accessible to the lambs, can provide this needed feed. Since this is a time when the nutritional needs of the ewe are declining, a method to partition this forage to the lambs is needed. This can be through creep grazing, forward creep grazing (a form of first-last grazing), or lambs being given access to paddocks during times when the ewes are restricted to dry lot. Lambs can generally be weaned at about 8 wk. Lambs should be wormed at weaning and moved to a "clean" paddock to aid in controlling internal parasites.

Lambs born in late winter or early spring can graze small grains, other cool-season annuals, and early growth of cool-season perennial forages. Such forages are generally very high in quality and are growing rapidly. As lambs are weaned at about 8 wk, they can take advantage of these forages. Systems in which lambs are born later in the spring have been successful in Virginia, where lambs remain with the ewes and eat cool-season perennial forages through the summer until ready for slaughter by early autumn (Notter et al. 1991). First-last grazing or creep grazing should be used with such systems to provide high-quality forage to lambs while restricting intake of ewes. If these lambs have not reached a finished condition by autumn, successful lamb finishing can be accomplished on annual brassica forages (Koch et al. 1987; Chap. 36). In areas where tall fescue is adapted, autumn lambing can utilize stockpiled herbage to good advantage. Lambs born in September or October are ready to graze high-quality stockpiled fescue by early November and can continue to graze through much of the winter. If small grain forages are available for early spring grazing, these high-quality forages could provide forage for finishing lambs by April or May.

In the far West, winter grazing of alfalfa provides opportunities for lamb-finishing systems. Alfalfa is managed for hay production during spring and summer. Alfalfa grows slower during autumn and winter and can be rotationally stocked to give lamb gains of 0.11 to 0.3 kg d^{-1} and to finish to choice grades.

CLASS PLANNING EXERCISE

The principles of forage-livestock integration are best understood by working through an abbreviated version of the planning process for a real or hypothetical farm. The set of agronomic inputs needed for this exercise are similar for a dairy, beef, or sheep enterprise on a given farm, but actual numbers will vary considerably by region of the country. Agronomic inputs for a particular region include size of the farm, number and size of fields, past crop rotation, land-use and drainage classification for each field, map of field layout, yield potential for each land-use class, and cropping restrictions. Tables for estimated dry matter intake of various livestock classes are used to estimate animal forage requirements. Class exercises for planning dairy, beef, and sheep forage production programs are available (Fick 1993; Stringer et al. 1987).

SUMMARY

Successful forage-livestock systems are profitable, sustainable enterprises that exist in harmony with nature. Although specific goals of individual livestock enterprises vary considerably, all livestock enterprises must include a goal for environmental stewardship of the land. Environmental problems resulting from poor nutrient/manure management can occur regardless of the size of the livestock operation. Therefore, a nutrient/manure management plan to achieve nutrient balance and maintain excellent water quality is essential for every livestock farm. Successful integration of forage crops into the livestock system can provide a solution for many of the real and perceived environmental problems associated with livestock farming.

Forage selection and establishment are based on the land resource and climate, while harvest management is based on animal forage requirements. Forages grown on the farm can be an economical feed source for livestock, but their true value is dependent on the cost and availability of on-farm and off-farm supplemental feeds. Dairy enterprises have a relatively constant demand for forage throughout the year. There is a strong emphasis placed on forage quality, with a heavy reliance on stored feeding systems, unless grazing is preferred. Beef and sheep enterprises rely heavily on grazing and must be able to match the annual production cycles of crops and animals. Pasture-based systems that maximize forage use and minimize stored feed are more complicated in terms of their feeding system but less complicated in terms of nutrient/manure management and potential environmental problems.

Although good forage management may mean the difference between profit and loss in a livestock enterprise, forage management by itself will not guarantee success. Strategic planning, including alternative plans, will optimize the use of resources and maximize the probability of providing the correct forage for each class of livestock. A balance between the efficiencies of forage production, forage consumption, and livestock production should be the overall goal of a livestock enterprise.

QUESTIONS

1. In order to achieve a sustainable forage-livestock system, what factors must be considered?
2. What general forage types would be expected to support the highest level of animal performance in your region?
3. What plant and animal characteristics should be considered when planning a forage-livestock system?
4. What are the ways a livestock farmer can alter forage quality and forage yield?
5. Describe three different options available to a livestock farmer who requires more storage capacity for forage. How would each option affect the rest of the enterprise?
6. What is the difference between a grazing system and a grazing method?
7. Describe three different options available to a dairy farmer who anticipates an inadequate supply of high-quality forage after completing the spring harvest of forages.

8. What factors must a dairy farmer consider when estimating animal forage requirements?
9. Why do some classes of beef cattle have higher nutritional needs than others and require the highest quality of forage?
10. Evaluate alternative approaches to finishing beef cattle in terms of environmental quality.
11. Describe the change in nutritional needs for ewes and lambs, and design a forage system to meet these needs in your region.
12. What grazing methods can be employed to meet the simultaneous, yet different, nutritional needs of ewes and lambs?

REFERENCES

Allen, VG, JP Fontenot, DR Notter, and RC Hammes, Jr. 1992. Forage systems for beef production from conception to slaughter: I. Cow-calf production. J. Anim. Sci. 70:576-87.

Blaser, RE, DD Wolf, and HT Bryant. 1973. Systems of grazing management. In ME Heath, DS Metcalfe, and RF Barnes (eds.) Forages: The Science of Grassland Agriculture. 3d ed. Ames: Iowa State Univ. Press, 581-95.

Blaser, RE, RC Hammes, Jr., JP Fontenot, HT Bryant, CE Polan, DD Wolf, RS McClaugherty, RG Kline, and JS Moore. 1986. Forage-Animal Management Systems. Va. Agric. Exp. Stn. Bull. 86-87.

Chessmore, RA. 1979. Profitable Pasture Management. Danville, Ill.: Interstate Printers and Publishers.

Conway, JF. 1993. Dynamic forage resource management. In Silage Production: From Seed to Animal, Proc. Natl. Silage Prod. Conf., NRAES-67. Ithaca, N.Y.: Northeast Regional Agricultural Engineering Service.

Fick, GW. 1993. Planning a forage production program. Dept. of Soil, Crop, and Atmos. Sci. Teach. Ser. no. T93-3. Ithaca, N.Y.: Cornell Univ.

Fick, GW, and JH Cherney. 1994. Forage crop production. In WO Lamp and G Armbrust (eds.), Protection of Perennial Forage Crops. New York: J Wiley and Sons.

Forage and Grazing Terminology Committee. 1992. Terminology for grazing lands and grazing animals. J. Prod. Agric. 5:191-201.

Hammes, RC, Jr., JP Fontenot, HT Bryant, RE Blaser and RW Engel. 1964. Value of high-silage ration for fattening beef cattle. J. Anim. Sci. 23:795-801.

Hodgson, JG. 1990. Grazing Management: Science into Practice. New York: John Wiley and Sons.

Jorgensen, NA, and WT Howard. 1982. Reaching 20,000 pounds of milk with forage. In Proc. Am. Forage and Grassl. Counc., 103-8.

Koch, DW, FC Ernst, Jr., NR Leonard, RR Hedberg, TJ Blenk, and JR Mitchell. 1987. Lamb performance on extended-season grazing of tyfon. J. Anim. Sci. 64:1275-79.

Koepf, HH. 1989. Nutrient recycling in balanced livestock-crop systems. In Sustaining the Smaller Dairy Farm in the Northeast. New Milford, Conn.: Sunny Valley Foundation.

Loomis, RS, and DJ Conner. 1992. Crop ecology: Productivity and management in agricultural systems. Cambridge, Australia: Cambridge Univ. Press

Miller, WJ. 1979. Dairy Cattle Feeding and Nutrition. New York: Academic Press.

Mott, GO. 1973. Evaluating forage production. In ME Heath, DS Metcalfe, and RF Barnes (eds.), Forages: The Science of Grassland Agriculture. 3d ed. Ames: Iowa State Univ. Press, 126-35.

Notter, DR, RF Kelly, and FS McClaugherty. 1991. Effects of ewe breed and management system on efficiency of lamb production: II. Lamb growth, survival and carcass characteristics. J. Anim. Sci. 69:22-33.

Pell, AN. 1992. Does ration balancing affect nutrient management? In Cornell Nutrition Conference for Feed Manufacturers. Ithaca, N.Y.: Cornell Univ.

Rotz, CA, DR Buckmaster, DR Mertens, and JR Black. 1989. DAFOSYM: A dairy forage system model for evaluating alternatives in forage conservation. J. Dairy Sci. 72:3050-63.

Schultz, LH. 1969. Relationship of rearing rate of dairy heifers to mature performance. J. Dairy Sci. 52:1321-29.

Spedding, CRW. 1984. Agricultural systems and the role of modeling. In Agricultural Ecosystems. New York: John Wiley and Sons.

Stringer, WC, NS Hill, and BW Pinkerton. 1987. FORBEEF: A forage-livestock system computer model used as a teaching aid for decision making. J. Agron. Educ. 16:85-87.

Swanson, EW. 1967. Optimum growth patterns for dairy cattle. J. Dairy Sci. 50:244-52.

15 Economics of Forage Production and Utilization

KEVIN C. MOORE
University of Missouri

C. JERRY NELSON
University of Missouri

FORAGES are an important part of American agriculture. Of the 427 million ha (1.054 billion A) in the US devoted to producing crops or forages, 266 million ha (656 million A) or 62% are classified as either cropland pasture or grassland pasture (USDA 1991). Many economic decisions related to production and utilization of this vast resource face the forage producer and the forage industry. Our intent is to raise awareness of a range of economic issues and to illustrate economic analyses within the context of forage management decisions.

Important *macroeconomic*, or more general, questions for the forage industry include, Where should forage production take place? How can forages compete economically with row crops and other land uses? What factors influence the choice of forage species in a given location of forage production? And how do governmental policies affect production and utilization of forages? *Microeconomic* questions involve issues facing the individual producer or buyer of forages, i.e., how does economics enter the decision-making process for site-specific problems?

Farmers or ranchers need to be aware of macroeconomic issues and their implications, but they generally deal with microeconomic questions. For example: "Will adding a legume to my grass pasture increase my net income?" "What fertility program should I use for my rented hayfield?" "Is it more profitable for me to mechanically harvest my forage and transport it to my livestock or to have my animals harvest the forage directly?" "Should I buy hay or produce it myself?" "How long should I try to make my alfalfa (*Medicago sativa* L.) stand last?" And the list goes on and on.

KEVIN C. MOORE is Associate Professor of Agricultural Economics at the University of Missouri. He earned the BS degree from Illinois State University and MS and PhD degrees in agricultural economics from Iowa State University. He conducts research and extension programs on forage-livestock economics and farm management.

C. JERRY NELSON. *See Chapter 2.*

MACROECONOMIC QUESTIONS

Location of Forage Production. Forages are produced in particular areas within the country in large part because they have a comparative advantage relative to other land uses. The advantage of a given species is generally linked with its adaptation to and relative productivity in a given environment. The economic principle of comparative advantage asserts that a country, region, individual, or any other defined economic entity tends to specialize in production of commodities in which that entity has the greatest *relative* advantage. In the law of *comparative advantage,* the ratios of the physical production of one good to that of another are used to determine which good(s) should be produced where, with trade between the producers resulting in a situation where everyone is at least as well off as before trading and with at least one party being better off.

Comparative advantage is based solely on physical opportunity costs. An *opportunity cost* is the value of a foregone opportunity, thus it measures the amount of one product given up in order to produce the alternative good. The law of comparative advantage does not account for other cost and return tradeoffs that ultimately determine patterns of production and trade. For example, alfalfa hay is typically of higher quality and more valuable than grass hay. Further, alfalfa must compete with corn (*Zea mays* L.) production in some states, whereas it must compete with wheat (*Triticum aestivum* [L.] emend. Thell.) production in others because comparative advantage is also a factor in grain production. Land values are generally based on potential for crop production, and transportation costs may offset trade opportunities. Thus, several factors work together with comparative advantage to affect the competitiveness of an enterprise and to determine what specific forages are grown in a given area.

Economic Competitiveness of Forages. Producers and consumers look to price as the guide to determine the value of an item, but pricing forages is not as accurate as desired. To determine the ultimate competitiveness of forages as an agricultural enterprise, factors affecting market prices and production costs must be examined.

VALUE OF FORAGES. Supply and demand determine market prices for forages, but the value of forages depends on many factors. Forages provide low-cost feed for ruminants, which convert these plants into desirable products such as meat, milk, and wool. For forage as feed, forage quality is the most important factor determining value per unit weight, as forage must compete as a feed nutrient source with grains and other feedstuffs. When used in crop rotations, forages can also provide benefits such as fixed nitrogen (N), reduced soil loss, improved water quality, and breaking of pest cycles, but these are more difficult to assign a value.

Many forage crops are used as ground covers for both recreational and aesthetic purposes, and long-term stands of forages and grasslands provide food, nesting sites, and winter cover for a variety of wildlife. Products produced from forages have many human and industrial uses and find their way into the market structure. Each use gives forages an economic value, but in many cases it cannot be determined accurately. For example, what is the value of wildlife or the value of clean water?

Currently, most forage is marketed based on visual appearance, making it difficult to price and market if the buyer cannot examine the product. Further, there is a gradual increase in understanding that the fiber component (i.e., cell wall), rather than the protein, is the major quality criterion in forage. Also, the nature of the livestock enterprise that will use the forage will determine the quality characters that are important. For example, high-quality alfalfa hay has a much higher relative value than grass hay for high-producing dairy cows, but the value differential is much smaller for beef cows.

The US is attempting to develop market standards for forages that include a quality measurement (Rohweder et al. 1981; Petritz 1989). It is broadly recognized that odor and visual features such as leafiness, green color, forage and weed species present, and stage of maturity are indicators, but not quantifiers, of forage quality (see Chap. 8). Several forage quality indices have been developed and tested, but none has been accepted universally (Moore 1994).

The relative feeding value (RFV) is based on fiber solubility and tends to work as an index for high-quality hays in dairy rations. However, large differences in leaf anatomy between cool- and warm-season grasses (Chap. 3) and differences in fiber-lignin complexes between grasses and legumes make it difficult to use the same index for comparative analyses among a wide range of forage species. Near infrared reflectance technology offers rapid analysis, but coefficients based on animal performance need to be developed for different types of forages. Accurate field techniques are needed to estimate the value of forage quality for the pricing of forages.

FORAGE PRODUCTION COSTS. Many factors influence the cost of producing forages, land costs being one of the largest. These costs vary geographically, based on productive potential of the region, the competition from other crops and nonagricultural uses, and change with time. Fertilizer and application costs can be high for some forages and are related to land quality. Harvest costs increase with yield and depend on method of harvest. Storage costs are also a major factor and vary with climate. Establishment costs can be high, as can costs of fencing, water, and other essentials of a grazing system.

TRANSPORTATION COSTS. Transportation costs are a major factor causing forages to be grown over a wide range of environments and localities. Harvested forage is bulky, subject to spoilage, and has a relatively low value per unit volume. Much hay is bought and sold, but the majority of sales go to neighbors. Still, some hay is transported from production areas in the northern plains or intermountain areas to consumption areas several states away, e.g., in the Southeast or Southwest. Most often this is high-quality legume hay with a high value per unit of mass. Some hay in compressed bales or in pelleted form gets traded internationally. But usually the high cost of handling and transportation causes forages to be produced near the livestock that consume it.

Role of Government Policies. Governmental policies influence the geographic distribution of forage production and the management practices of farmers. Two notable areas of farm legislation that profoundly influence agricultural landscapes are the Conservation Reserve Program (CRP), which began in 1985, and the conservation compliance provisions of the 1990 Farm Bill.

The Conservation Reserve Program is a 10-yr contract in which the land owner promises, in return for an annual payment, to cease row crop production on highly erosive lands and establish and maintain long-term vegetative cover on land enrolled in the program. The encouraged shift in land use to forages should enhance water quality, improve wildlife habitat on such lands, reduce the large stocks of feed grains and wheat being stored, and reduce government expenditures to agriculture through smaller deficiency payments for grain crops.

After 11 sign-up periods through 1993, 14.3 million ha (35.4 million A) of highly erodible land had been enrolled in the CRP, with the majority of the land being seeded to perennial grasses. At the end of the contract the land will be available for haying and grazing. It may again be used for row crops if approved conservation measures are used, but the emphasis is on forages continuing to occupy much of this land.

Deficiency payments encourage producers to participate in federal farm commodity programs that influence the planting of crops such as corn, grain sorghum (*Sorghum bicolor* [L.] Moench), wheat, and cotton (*Gossypium hirsutum* L.). The conservation compliance provisions of the 1990 Farm Bill place restrictions on soil erosion occurring in crop rotations if the producer is in the government program. To remain eligible, erosion on designated fields must be limited to *T* (the soil tolerance level) or that amount of soil loss that does not diminish long-term productivity.

In some cases, conservation tillage alone will not bring a rotation based solely on row crops into compliance. Thus, a solid-seeded forage crop, which results in little soil loss, may be included in the rotation to reduce the *average* annual erosion of the entire rotation to *T*. The forage crop can be pastured or cut for hay but provides added value by allowing other high-income, but more erosive, row crops to be grown. Each landowner must evaluate the costs of compliance against the benefits. Comparisons include costs for adding a forage in the rotation, using different tillage equipment, constructing and managing terraces, and adopting different cultural practices such as planting row crops without tillage (no-till) or in contour strips. Each field or farm will have a unique set of alternatives and reasons for meeting conservation compliance. Some producers may use forages in the rotation but not maintain livestock; this depends on the existence of outside markets to sell the forages. Others may use forages to alter disease, insect, and weed problems on subsequent crops. Forage legumes fix N to benefit the rotation.

International trade agreements such as the North American Free Trade Agreement (NAFTA) affect forage production. NAFTA influences the trade of beef, pork, and milk between the United States, Mexico, and Canada, and the economic situation facing these industries affects the amount and geographic distribution of forage production. Groundwater legislation may limit fertilizer and pesticide use for row crops and cause a shift to perennial meadows and hayfields. Monetary and fiscal policies (i.e., how the government decides to reduce the federal budget deficit and the resulting impact on interest rates) may influence the economic competitiveness of forages with more intensive cropping strategies. Unfortunately, few macroeconomic studies have focused on forages to give insight for directing these governmental policy decisions.

The preceding discussion illustrates the complexity of government influence and the individual task of farmers in deciding what, how many, and what kind of forages and livestock opportunities should be in their individual farm plans and when and where to include them. Answers to these more specific and local questions involve microeconomics.

MICROECONOMIC QUESTIONS

Using economics in farm management decisions requires translating technical information from many associated disciplines into financial terms to examine the potential profitability of an alteration in a management practice. This complex task may be simplified by recognizing three general types of analyses used in production economics: factor-product analysis, factor-factor analysis, and product-product analysis.

Factor-Product Analyses. Factor-product analyses deal with relationships between input (factors of production) costs and output (product) prices. The essential question is how much of an input to use in order to maximize net returns from production. Some sample factor-product problems are "How much N should I topdress on my hayfield?" "Should I overseed more red clover (*Trifolium pratense* L.) into my grass pasture to improve animal performance?" Production functions, constructed to show how the level of one input affects the amount of output (holding all other inputs constant), are used as the bases for these analyses.

To illustrate factor-product analysis, we consider a case dealing with optimizing N fertilizer rates for subirrigated meadow hay production in Nebraska (Clark et al. 1991). These meadows in the Nebraska Sandhills consist mainly of cool-season grasses, several species of sedges (*Carex* spp.) and rushes (*Juncus* and *Eleocharis* spp.), a few warm-season grasses, and a few legumes. Various rates of N were applied each year, and one crop of hay was harvested between late June and mid-July each year, the exact date depending on soil wetness. Data presented are the 4-yr averages (Fig. 15.1). These graphic relationships are often referred to as *production functions.*

The mathematical function for the response of hay production to applied N was statistically estimated using data from several years to be able to understand the typical response because N must be applied each year before yield is known. Therefore, confidence limits, based on variation in response among the 4 yr of data, are also shown (Fig. 15.1). One can be 95% confident the yield response will fall between the limits. Thus, confidence limits give an indication of reliability or risk.

Yields responded most at low rates of applied N, increasing over 25 kg per kg N added. But at higher N rates, yield change per kg N lessened; i.e., the response curve decreased in slope and became flat. This shape is typical

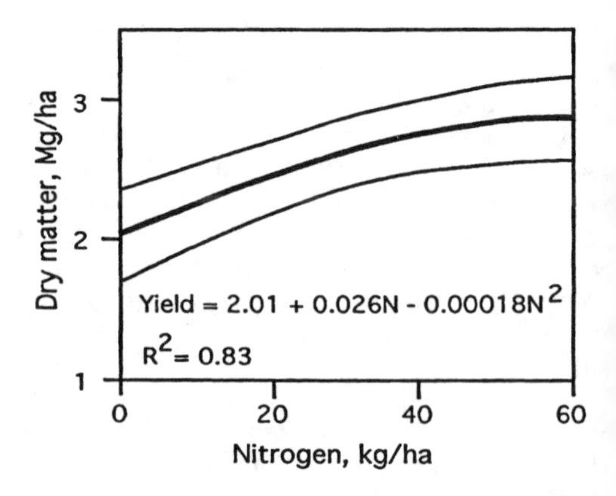

Fig. 15.1. Dry matter yield of subirrigated meadow hay in Nebraska increases in response to applied nitrogen fertilizer, but the rate of response decreases. Upper and lower lines indicate the 95% confidence limits based on 4 yr of data. (Adapted from Clark et al. 1991.)

for a production function. *Marginal* (meaning additional) *physical product* (MPP) represents the change in output generated from a unit of added input. Graphically, MPP is the slope of the production function at any level of input use. Maximum yields are usually higher than the yield level that provides the highest economic return.

The economically optimal level of input use is the N rate at which net return or profit is maximized. This occurs where the value of the additional output produced is equal to the cost of the last unit of input (i.e., there is no net return for that input). To do the economic analysis one must calculate the costs of N (including application) and estimate or determine the anticipated hay value. Then, a matrix is set up considering input costs (N) and output value (cost to purchase the same hay).

The economically optimum rate of N depends on the cost of N and price of hay (Table 15.1). Again, the annual decision is difficult because the costs of N and its application are known at time of application, but the value of the hay is unknown. Here one must use long-term averages or other means to estimate the value of the hay. At a given price of N the analysis shows it pays to use more fertilizer N as the value of the hay increases. For any given hay value, however, as cost of N increases, optimum rates of applied N decrease. This table illustrates how the profit-maximizing rule under factor-product analysis is often applied.

Another example involves response of tall fescue (*Festuca arundinacea* Schreb.) to N fer-

TABLE 15.1. Hay value and N price (including application costs) affect rate of N fertilizer to apply for profit maximization

Hay value	N price, $/kg			
	0.44	0.66	0.88	1.10
$/Mg	N rate, kg/ha			
30	23	6	0	0
40	38	25	10	0
50	48	36	25	13
70	56	48	40	34
90	59	54	48	42
125	63	59	55	51

Source: Adapted from Clark et al. (1991).

tilization in Georgia (Fig. 15.2). In this case, N was applied four times annually, and the forage was harvested several times during the growing season. Note the annual rates of N were much higher, and the response function had a shape different from that in Figure 15.1, especially at low rates of N. The yield response to N was fitted best by a logistic function. This likely happened because tall fescue first responds to N by increasing tiller density to fill in the canopy (Nelson and Zarrough 1981), but the tillers are small and contribute little to yield. At higher N rates, there is an increase in vertical growth and an increase in tiller weight, so yield responds.

One of the research goals was to determine the efficiency of N use (i.e., change in yield per unit change in N), which was highest at about 125 kg/ha and 80% N utilization (i.e., 100 kg of N was removed per ha in the forage). Yield responded to N up to 400 kg/ha, but at that rate the forage removed only 230 kg/ha. Thus, utilization was 58%, and this raises environmental questions about the fate of the unused N. The biological understanding of N use is important and, independent from economics, environmental risk (ethics) or regulations (laws) may influence how much N is actually used.

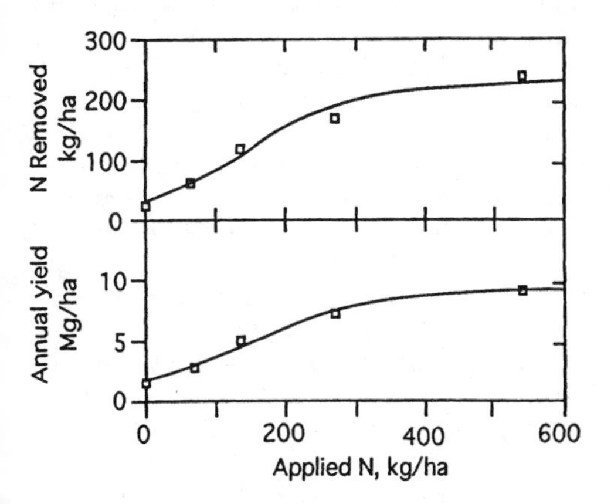

Fig. 15.2. Yield of tall fescue in Georgia increases as rate of nitrogen increases (*lower panel*). The nitrogen was split into four applications per year. Upper panel shows the amount of nitrogen removed. (Adapted from Overman and Wilkerson 1993.)

Factor-Factor Analyses. Factor-factor problems involve determining the best combination of inputs to meet a defined objective. Common types of factor-factor questions include, "Should I apply manure or purchase fertilizer to increase production from my pastures?" "To reduce the influence of tall fescue endophyte (*Acremonium coenophialum* Morgan-Jones & Gams) on performance of my cattle, should I introduce a legume in the pasture or use feed supplements?" "Which of several forage species would work best for raising sheep on my farm?" Each of these questions includes a goal related to output, but the real question is what the appropriate level is for each input to meet the production goal most economically.

The production function for factor-factor analyses requires fixing output at the target level, while varying the level of two or more inputs, to determine the substitution value for one input relative to the other. The goal is to find the least-cost combination of inputs required. In practice, scientists find it difficult to fix output at a target level. Thus, most often the research goal is to determine which input is most cost-effective in meeting a similar production target.

An example involves storage of large round bales of hay. Several storage alternatives are available. The main purpose is to decrease spoilage and loss of forage quality. The operator who needs a given amount of quality forage for livestock has options: more land can be used to produce enough extra forage to offset the losses due to on-ground field storage, which are often as high as 30%-40% of the dry matter, or a method can be invested in that retains a larger proportion of the dry matter and quality.

The value of protection depends on the value lost during storage, which varies with location, time in storage, and quality of the hay. Buckmaster (1993) studied alternative methods for cost-effective storage of round hay bales based largely on conditions in Pennsylvania. All loss cannot be eliminated, even with barn storage, so the gain from storage is the value of the avoidable loss. For example, losses based on several studies averaged 6% for barn storage, 20% for outdoor storage under plastic, and 27% for open storage on the ground. Buckmaster developed a computer spreadsheet to analyze the cost and storage losses associated with several storage methods.

In Figure 15.3 breakeven costs are presented for barn construction for various levels of hay values and outside storage losses. If hay is more valuable and/or outside storage losses become greater, the value of a hay barn increases; i.e., one can pay more for the barn based on the value of hay losses if hay is stored outside. For each storage alternative, a value of savings per unit weight of hay stored was calculated. In addition, the time needed to recover the cost of the initial investment was calculated.

Each cover method was cost-effective. On an annualized basis, savings were \$7.17/Mg (Mg = Megagrams = 1000 kg) for barn storage, \$9.61/Mg for group storage, and \$6.89/Mg for individually wrapped bales. The value of the barn was underestimated if it could be used to store other items as the hay

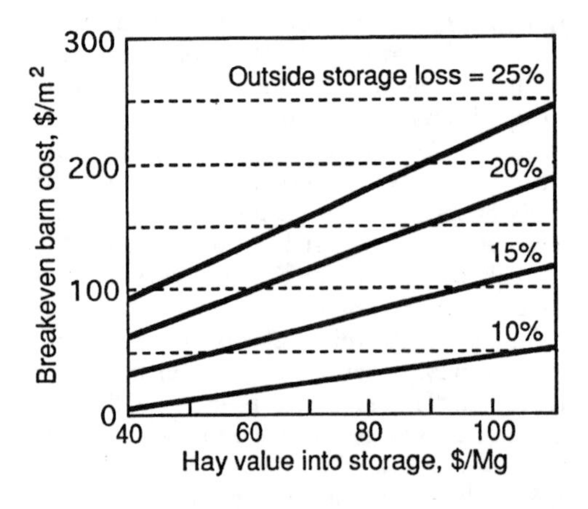

Fig. 15.3. Hay value and amount of outside storage loss influence the amount one can invest in barn storage. Costs include machinery and labor associated with the additional operations to store the hay. (Adapted from Buckmaster 1993.)

was fed out. Although not calculated, the farmer also gains the use of the "saved" land equivalent for producing another crop. These values can guide producer decisions regarding alternatives for hay storage.

Another example is more complex but uses the same basic analysis. Knowlton et al. (1993) examined the profitability of using an enzyme treatment on alfalfa silage fed to dairy cattle. It is important to harvest alfalfa early, before the neutral detergent fiber (NDF) increases above a given level, say 40%. But it is difficult to harvest large amounts of alfalfa at the proper stage, and the last fields harvested may have NDF values that are higher than desired. In that case more corn grain must be fed to maintain a high-energy ration. But, certain enzyme treatments can digest some of the cellulose to release sugars and reduce NDF, allowing silage made from alfalfa at later stages to be "brought back" to acceptable quality.

The question is whether or not enzyme treatment of alfalfa can economically offset late harvest, thereby substituting for grain. Also, the enzyme treatment hydrolyzes some cell wall materials to release sugars and lower NDF. It has been hypothesized the sugars released by treatment and the remaining cell wall (NDF) will be easier to digest in the rumen, which will increase the rate of passage and feed intake. The end result is increased milk production.

Untreated silage was compared with enzyme-treated silage in a model that predicted both feed intake and milk output. Knowlton et al. held output (milk) constant and determined the savings in corn grain due to silage treatment (Table 15.2, scenario 1). The treatment was not economical (cost was \$1.36/Mg silage), even when silage NDF was reduced by 6 percentage units; i.e., it was cheaper to feed grain than treat the silage. Rather than hold output constant, they assumed an increase in sugars released (energy) and milk output (scenario 2). The increased milk production was enough to offset the enzyme cost. Note the value of the enzyme treatment depended largely on the amount of decrease in NDF.

In scenario 3 they assumed NDF reduction would increase intake and milk production, a factor needing to be tested biologically. In an independent test, the model indicated a silage of 57% NDF treated to reduce the NDF to 51% was not as good as an untreated silage of 51% when both were fed at the same level, perhaps because lignin and other factors were not reduced. They concluded the enzyme must reduce NDF by a large amount and increase intake to be cost competitive with a feeding supplement.

In another example, Heitschmidt et al. (1990) evaluated returns from cow-calf production on semiarid grasslands in Texas using four alternative grazing systems for 4 yr. They compiled data on animal performance such as cow weights, conception rates, and calf-weaning weights and compared economic returns from (1) a heavily stocked continuous grazing system, (2) a moderately stocked continuous grazing system, (3) a three-herd, four-pasture deferred rotation grazing system, and (4) a one-herd intensive rotational grazing system (Fig. 15.4). Residual return to land, management, and profit was greatest under the intensive rotational grazing system ($16.38/ha), followed by the deferred system ($15.98/ha), the heavily stocked system ($12.90/ha), and the moderately stocked system ($11.01/ha). This basically followed the stocking density response; i.e., return increased with less land area per cow. However, if calculations were made on a cow basis, performance was best at low stocking densities. This is not uncommon as cows and calves can be more selective in grazing, and it illustrates that economic evaluations of forage-livestock studies depend on whether the objective is increasing production per land area or per animal. One goal for intensive grazing systems is to increase output per land area while maintaining individual animal performance.

Obtaining data to answer factor-factor questions is more difficult than for factor-product questions. In real life it is difficult to keep output levels constant while varying inputs. The first step to answering factor-factor questions is to evaluate how output is enhanced due to increasing each input independently. Then the interrelationship between the two inputs must be determined based on the degree of substitution possible. In some cases complementarity exists between sets of inputs; e.g., feeding supplements can increase gains of beef calves with minimal effects on the pasture, but adding a legume to improve gains may change seasonal productivity

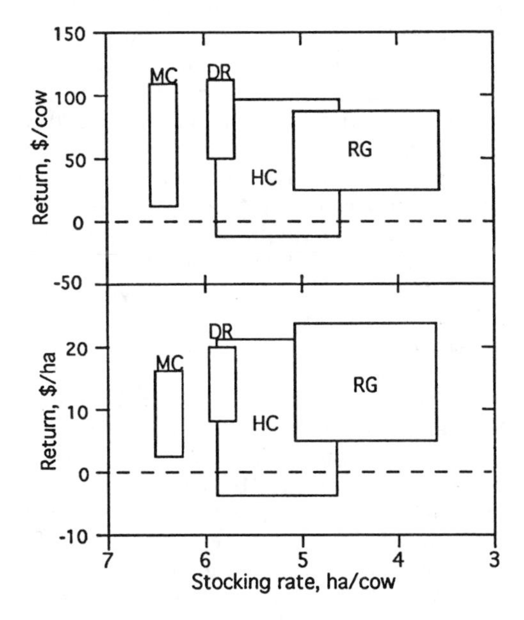

Fig. 15.4. Relationships between stocking rate and return to land, management, and profit for heavy continuous (*HC*), moderate continuous (*MC*), deferred rotation (*DR*), and rotational grazing (*RG*) systems in Texas. Rectangle dimensions indicate the range in responses over the 4-yr study. Average response is near the center of each rectangle. (Adapted from Heitschmidt et al. 1990.)

TABLE 15.2. The effects of reducing NDF of alfalfa silage for milk cows by 2, 4, or 6 percentage points due to enzyme treatment

	NDF reduction, percentage points		
	2	4	6
Scenario 1			
Savings in diet costs, $/Mg silage	$0.23	$0.41	$0.61
Scenario 2			
Increase in milk prod, kg/d	0.05	0.18	0.27
Value at $0.26/kg milk, $/Mg silage	$1.83	$3.71	$5.55
Scenario 3			
Increase in intake, kg/d	0.34	0.72	1.10
Increase in milk prod, kg/d	0.86	1.80	2.74
Value at $0.26/kg milk, $/Mg silage	$14.05	$28.56	$42.94

Source: Adapted from Knowlton et al. (1993).
Note: Alfalfa was 46% NDF before treatment. Treatment cost was $1.36/Mg silage.

and/or carrying capacity of the pasture. Labor and equipment costs will also vary.

Least-cost ration formulation is a good example of a multiple input factor-factor problem. First, the goal for the particular enterprise is determined: perhaps a ration is needed for 544-kg (1200-lb) dairy cows to produce, on average, 9524 kg (21,000 lb) of milk per lactation. Nutritional requirements of the animals are specified in terms of energy, protein, and other factors that are derived from data based upon factor-product studies relating animal performance (milk output) with nutritive components of feedstuffs in the diet. Various feedstuffs are then characterized by the specific levels of energy and nutrients they provide. Finally, using techniques such as linear programming, the problem is solved for the least-cost combination of inputs that will meet the stated requirements. Limits on amounts of specific components should be set if total substitutability between some inputs is not possible.

Product-Product Analyses. Many farm management considerations are product-product questions such as, "Should I raise corn and soybeans (*Glycine max* [L.] Merr.) on my farm, or would I be better off with forages and cattle?" "Is my land resource better suited for pure alfalfa or a grass-legume mixture?" "Should I sell calves directly from my cow-calf operation, or should I carry the calves to yearlings?" Product-product analysis deals with the question of which output(s) or product(s) should be produced with a given set of limited resources. In the simplest two-product case, the production function actually represents the set of available resources and how they can be used to produce a finite combination of the two final products.

A question may be, for example, how much corn and/or alfalfa should be produced on a given farm with fixed amounts of land of varying quality. Profits are maximized based on factors such as specified amounts of tractor horsepower, hours of available labor, and money available for operating expenses. Based upon the available resources, a "production possibilities frontier" can be constructed; this represents the maximum output of both products possible for various allocations of fully employed inputs. The production possibilities curve can depict substitutability between inputs (e.g., the value of corn residue for feed) or complementarity (e.g., the value of N fixed by alfalfa) (see Calkins and DiPietre 1983).

Foltz et al. (1993) compared profitability of alfalfa with corn and soybeans on representative Corn Belt farms. They developed budgets detailing the costs and returns for the alternative products. At a yield of 7.8 Mg/ha (3.5 ton/A), alfalfa was profitable on farms with low-productivity soils but was not as profitable as corn and soybeans and therefore did not enter the optimal farm plan. A yield of 10 Mg/ha (4.5 ton/A) was required before alfalfa would replace corn and soybeans on some of the land area. Required yields for alfalfa on farms with high-productivity soil were about 40%-45% higher. In addition to soil, results depended on climate and specific enterprise costs being used in the study.

Product-product analyses are more difficult than factor-product or factor-factor analyses because they necessarily involve components based on the other two types. To determine which product mix to produce with a given set of resources, one must first know (1) how each resource can be used individually in the production of each output (factor-product) and (2) how resources interact in production of any given product (factor-factor). To these direct questions, product-product analyses add the final dimension of how each potential product competes with the other for the limited resources.

Linear Programming. Linear programming using a computer is a technique commonly used to answer product-product questions based on the best use of limited resources within a complex system. The problem is solved by formulation of a mathematical function that must be linear in form and will be either maximized (profits) or minimized (costs) depending on the objective. A set of crop and livestock activities is specified that draws upon a set of limited resources (inputs) including land, labor, capital, and equipment that are available on the farm. The final questions to be answered are which and how much of each production activity to undertake in order to maximize net farm income. Coefficients, developed from simpler analyses, are assigned to the inputs in the linear formula. Solving the linear program results in a solution stating which (and how much of each) crop and livestock activity will be produced, how much of the farm's resources are being used, and the resulting total net farm income. The program can be used to answer what if questions such as, "What happens to profit if I reduce corn silage production and expand my alfalfa production?" "Would I be better off raising sheep than dairy heifers on my permanent pastures?"

ENTERPRISE BUDGETING

Enterprise budgets represent the costs and returns for a given crop or livestock production activity within a total system, and they contain data that represent factor-product, factor-factor, and sometimes product-product relationships. As such they are the cornerstones of many farm management decisions. Enterprise budgets for pasture and hay usually use the land unit as the basis for production. Livestock budgets are often calculated on a per head basis, but they can also be useful when based on land area. Any budget can be converted to a different base unit by dividing or multiplying by the appropriate conversion.

Enterprise Returns. Measuring income per hectare would at first seem to be a routine computation of yield times price, the usual case for annual crop budgets such as for corn silage. A common mistake in enterprises such as backgrounding steers is to value income as the selling price per unit weight multiplied by the weight gained at the farm. In this case the net returns are overstated. The correct way to compute value of gain is to subtract the beginning value of the calf from the sales value and to divide this difference by the weight gained. Many factors contribute to returns for grazing livestock enterprises.

For breeding livestock programs like cow-calf, most income comes from the sale of calves (Table 15.3). Unless twinning becomes the norm, there will be less than one salable calf per cow annually. The sales income from bulls and cull cows must be added, again calculated on a per cow basis. For breeding enterprises, sources of income other than young stock sales (i.e., milk from dairy animals, wool from the ewe flock, etc.) are also added to calculate total gross income.

Enterprise Costs. Economists like to use enterprise budgets to calculate total costs of production. Therefore, each factor of production (land, labor, capital, and management) is allocated a return even if a cash cost is not involved. *Opportunity costs,* the value the resource would have earned in the next best alternative use, are used to represent production costs when no cash expense is incurred. For example, if pasture used for grazing is owned debt free, no cash expense is reflected in costs of production. But the land could have

TABLE 15.3. Enterprise budget for a Missouri beef cow-calf operation with spring calving, 87% calf crop, 238 kg calves sold in fall, 15% cow replacement rate, and 1% cow death loss

Income per cow:		
Calf sales: (0.87 – 0.15) =		
0.72 calf × 238 kg @ $1.35/kg		$321.30
Cull cows: 0.14 × 454 kg @ $1.15/kg		$ 72.80
Total gross income per cow	$394.10	
Costs per cow:		
Variable costs: (feed value based on purchase price)		
Corn grain (272 kg × $0.08/kg)		$ 22.50
Supplements, salt (68 kg × $0.33/kg)		$ 22.50
Hay (1,450 kg × $0.06/kg)		$ 88.00
Total feed costs	$133.00	
Machinery operation, feed storage and preparation, etc.		$ 26.00
Veterinary, and medicine		$ 15.00
Commissions, hauling, other materials, & services		$ 14.00
Utilities, insurance, repairs, taxes, miscellaneous		$ 24.00
Interest on variable costs		$ 7.00
Total other variable costs	$ 86.00	
Total feed and other variable costs		$219.00
Income over variable costs	$175.10	
Fixed costs:		
Interest on breeding herd ($600 @ 8%)		$ 48.00
Interest and deprec. on machinery, equip.		$ 16.00
Taxes and deprec. on real estate		$ 10.00
Total fixed costs	$ 74.00	
Return to land, labor, and management		$101.10
Labor (8 h × $6.00/h) = $48.00		
Return to land and management		$ 53.10
Pasture (1.2 ha/cow × $62.50/ha) = $75.00		
Return to management		–$ 21.90

Note: Data are presented on a per cow basis.

been rented. Thus, the foregone cash rental value becomes the opportunity cost used in the budget.

Variable costs involve cash expenditure and depend on the level of production. Examples include feed, veterinary, seed, fertilizer, and purchased livestock expenses. *Fixed costs* occur whether or not anything is produced and do not vary with the level of production. They include depreciation, interest, repairs, taxes, and insurance. Interest can be an opportunity cost, or it can be a cash cost if resources are acquired by renting or debt financing.

Enterprise Net Returns. The net return calculated from an enterprise budget computed in the preceding manner cannot be interpreted as profit in the usual accounting sense. Net income over fixed and variable costs, or profit as calculated on an income statement, represents the return to unpaid labor, management, and equity capital. In the enterprise budget, *return to land, labor, and management* closely represents net farm income or what is typically thought of as profit. This assumes interest actually paid is included in variable costs along with hired labor and rental payments on land, machinery, and other components. The return to operator labor is generally used to pay for family living expenses. *Return to land and management* represents the income from production that is available to pay for land. Variable costs and operator labor have been covered. If each of these resources is allocated a return in an enterprise budget, however, the result is an economic profit, or *return to management,* as shown in Table 15.3. All factors of production, including a return to capital (both debt and equity), labor (both hired and operator), and land (rented and owned), have been covered.

Return to management is negative in Table 15.3, but this does not mean cow-calf enterprises in Missouri are not profitable. A return to operator labor, all land, and capital have been taken out. For many producers these costs do not involve cash costs and are overlooked when calculating "profit." Some capital would very likely be equity capital, and in most cases some land would be owned (at least partially) by the operator.

ROLE OF TIME IN FORAGE MANAGEMENT DECISIONS

Time plays an especially important role in forage budgeting and economic analysis because of the perennial nature of many species. Time is a minimal factor for annual crops such as corn silage because costs and returns occur within a single year. Costs for money invested in inputs is reflected in operating interest charges or an opportunity cost for equity capital investments. Conversely, investments in seeding perennial grasses or legumes generate returns that accrue over an extended period of time. Dollars returned 2 yr, 3 yr, or more after the initial investment are not as valuable as dollars received during the first year, and waiting for return of future dollars involves more risk. Thus, explicit accounting for time is needed in perennial enterprise budgeting and for management decisions.

Consider the decision of seeding alfalfa and deciding on the potassium (K) fertilization strategy as part of the long-term management program (Fig. 15.5). The budget for this enterprise is most appropriately a series of annual budgets representing the stream of costs and returns during the life of the stand (Table 15.4). Establishment costs in year 1 include seed, chemicals, tillage, and seeding expenses. Yields are low in the establishment year, are higher in subsequent years as the stand reaches its full potential, and decrease again as the stand eventually thins. Budgets for each year include returns that year along with costs of harvesting and material and application costs for fertilizer, herbicides, and insecticides. For example, costs increase in years 5 and 6 because herbicides were needed to control weeds in the thinning stand. Net return is the difference between annual costs and annual return and represents return to land, labor, and management.

Net Present Value Analysis. Net present value (NPV) is commonly used to evaluate investment decisions like the alfalfa example when time is a factor. In NPV analysis future costs and returns, obtained from annual enterprise budgets, are discounted to represent their values today and then summed over the life of the investment. The discount rate is often thought of as an interest rate. If $1.00 is invested today at 6% interest, it will be worth $1.06 in 1 yr. Likewise, the present value of $1.06 to be received in 1 yr, discounted at 6%, is worth $1.00 today. By converting future income to today's value, the producer can make a direct comparison with the initial investment and can decide, before investing, if the investment will give the needed rate of return.

As an example, $1000 is invested with an expected net return of $500 per year for 3 yr.

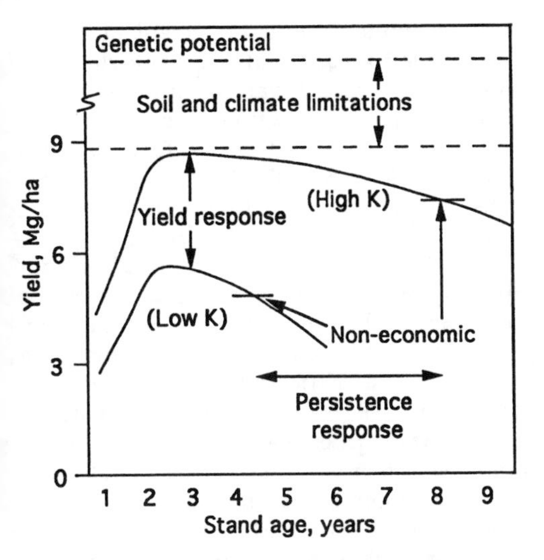

Fig. 15.5. Alfalfa yields are lower the establishment year than in subsequent years, remain stable for several years depending on management, and gradually decrease as the stand thins. Fertilization with potassium increases yield and persistence. (Missouri data adapted from Moore et al. 1991 and Nelson et al. 1992.

The alternative is a secure investment that will return 8%, so a 10% discount rate is selected to offset risk. The initial cost of the investment occurs in year 0 and is not discounted. Calculation of NPV gives

$$\text{NPV} = -\$1000 + \$500/(1 + 0.10) + \$500/(1 + 0.10)^2 + \$500/(1 + 0.10)^3$$

$$-\$1000 + \$454.54 + \$413.22 + \$375.66$$

$243.43 Net Present Value.

Thus, the total of $1500 returned over 3 yr is equivalent to $1243.43 today, so the return is positive (i.e., it exceeded a 10% return by $243.43 and is a better investment than the one at 8%). The discount rate selected indicates the needed rate of return. The interest rate on borrowed money is often used, especially when debt is the primary source of funds. If cash reserves are being used, the opportunity cost of the equity (i.e., the current rate of return on equity capital) is useful. A weighted average of the two is appropriate when both are used.

Regardless, the discount rate must provide a return that embodies (1) a risk-free rate, which reflects the fact that having the return sooner than later is always preferred, (2) an inflation premium that estimates and accounts for any declining purchasing power of money through inflation, and (3) a risk premium that adjusts for the differences in risk among alternatives. The higher the risk, the greater the risk premium, since it represents a charge for accepting more risk. Short- and long-term risks must be evaluated. A lower-yielding forage may be preferred to a higher-yielding forage if it is more winter-hardy.

A similar analysis was conducted in the case of alfalfa (Table 15.4). The net returns for each year were discounted to present values and then added. Initial investment costs (i.e., plow-down fertilizer, lime, and other establishment costs) were subtracted. In this case the investment was positive if the stand lasted at least 5 yr and was not kept more than 10 yr.

Annualized Net Present Value. It is also critical in NPV analysis to make an adjustment

TABLE 15.4. Calculation of annuity equivalent net present values (AENPV) for varying alfalfa stand lives

Stand age	Est. yield	Annual return	Annual costs	Net return	NPV	PVIFA	AENPV
	(Mg/ha)	($/ha)	($/ha)	($/ha)	($/ha)		($/ha)
0	---	---	---	0	0	0.91	0
1	5.29	423	847	–424	–350	1.74	–202
2	9.59	767	624	143	–243	2.49	–98
3	9.74	779	629	150	–140	3.17	–44
4	9.74	779	629	150	–48	3.79	–12
5	9.58	767	658	109	14	4.36	3
6	9.28	742	699	43	36	4.87	7
7	8.81	705	682	23	46	5.34	9
8	8.20	656	660	–4	45	5.76	8
9	7.43	595	632	–37	30	6.14	5
10	6.51	521	598	–77	3	6.50	0
11	5.44	435	559	–124	–36	6.81	–5
12	4.21	337	515	–178	–88	7.10	–12

Note: Hay price is $80/Mg; discount rate is 10%. Example is for 112 kg K plow-down and 202 kg /ha topdressed annually. Net present value (NPV) is divided by the present value interest factor of an annuity (PVIFA) to obtain the AENPV.

when comparing investments of unequal lifespan. If the 3-yr return on the $1000 investment was extended to year 4, but had a net return of only $10 in year 4, the NPV would be

$$NPV = -\$1000 + \$500/(1 + 0.10) + \$500/(1 + 0.10)^2 + \$500/(1 + 0.10)^3 + \$10/(1 + 0.10)^4$$

$$= \$250.26 \text{ Net Present Value.}$$

The net return of $10.00 in year 4 is worth only $6.83 ($250.26 – $243.43) at time 0, but for the same $1000 outlay, the 4-yr investment is chosen because it has a larger NPV. However, there is some question about maintaining the investment into year 4 since the return drops off dramatically. If possible, it would be better to repeat the 3-yr investment. This leads to comparing unequal investment lives using annuity equivalents.

An annuity is an equal series of receipts or payments over time. The NPVs are annualized by calculating annuity equivalents over the life of the investment by using factors representing the present value of a uniform series. These factors are found in most farm management or agricultural finance textbooks. For example, the NPV of $1.00 received at the end of each year for 3 yr, discounted at 10%, is $2.49. This factor (actually 2.4869) can be used to calculate the amount of a uniform series received for 3 yr, which is equivalent to the present value of the 3-yr example:

$$NPV/factor = \text{Annuity Equivalent}$$

$$\$243.43/2.4869 = \$97.88.$$

An investor would be equally satisfied to receive an annual annuity payment of $97.88 each year for 3 sequential years or $243.43 today. The "annualized" value or annuity equivalent for the 4-yr investment uses the factor for a uniform series lasting 4 yr (3.1699), thus $250.26/3.1699 = $78.95. The annuity value is higher for the 3-yr investment, and it is preferred.

Table 15.5 provides an example of how annuity equivalents can be used to select the most profitable fertility management program for yield and persistence of alfalfa in southern Missouri (for additional details see McDonald 1989; Nelson et al. 1992). Potassium at various rates was plowed down before alfalfa was planted, and K was top-dressed annually. Other plant nutrients were maintained at high levels. Plants were cut three times the establishment year, then five times annually. Enterprise budgets were constructed detailing annual costs for harvest, herbicides, labor, and other inputs.

Analysis of NPV and calculated annuity equivalent net present values (AENPV) were based on discount rates of 5% or 10% and hay prices of $80 or $100/Mg. The stand life for maximum profit was determined by selecting the highest AENPV among years within each treatment (e.g., $9.00 and 7 yr in Table 15.4). Note the plow-down treatment had little economic effect on persistence compared with the topdressing treatment (Table 15.5), partly because costs of plow-down K occurred early in the stand life and gave a slow return on investment. Cost of topdressing at 202 kg/ha occurred annually and was very cost-effective for stand life.

The dollar value of the highest AENPV for each fertility treatment can be compared directly so that the most economic fertility regime and stand life combination can be cho-

TABLE 15.5. Summary of greatest annuity equivalent net present values (AENPV) and optimal stand life for various potassium K treatments on alfalfa

		10%, $80		5%, $80		10%, $100	
PD	TD	AENPV	Yr	AENPV	Yr	AENPV	Yr
(kg/ha)		($/ha)	(no.)	($/ha)	(no.)	($/ha)	(no.)
0	0	–56	4	–51	4	34	5
0	202	–30	8	–22	8	94	9
0	404	–32	8	–23	8	113	9
112	0	–51	4	–47	4	43	4
112	202	9	7	16	7	154	8
112	404	–26	8	–17	8	122	9
224	0	–2	4	4	4	117	4
224	202	17	6	26	6	161	6
224	404	–25	8	–16	7	127	8
448	0	–40	4	–34	4	69	4
448	202	3	6	11	6	150	7
448	404	–14	8	–4	7	145	8

Note: The K was either plowed down before establishment (PD) or top-dressed annually (TD). Discount rates are 5% or 10% and hay price is $80 or $100/Mg.

sen (Table 15.5). Because of high establishment costs, several treatments were not economical unless K was added or hay prices were high. The optimal alfalfa management strategy at a 10% discount rate and hay value of $80 is 224 kg/ha plow-down, 202 kg/ha annual topdress, and a 6-yr stand life. The AENPV of $17.00/ha represents an economic profit (as defined earlier) since the enterprise budgets included all costs (establishment, labor, land, machinery, etc.). Since the AENPV decreases after year 6, it would be more economical to repeat the same 6-yr investment than to continue the stand.

Other columns show how AENPV and optimal stand age change as discount rate or hay price change. While the optimal choice is not altered, a lower discount rate results in future profits (positive AENPV) being worth more and future losses (negative AENPV) being less costly. And as expected, higher hay prices result in greater returns, as each strategy meets the 10% required rate of return when alfalfa brings $100/Mg. More comparisons could be made for soil types, yield potentials, and other management practices.

Net present value analysis is a very useful technique for evaluating many different types of investments. It can be applied to asset purchase problems, loan-refinancing alternatives, lease versus buy decisions, and a host of other situations. Forage examples include deciding to buy versus to lease pastureland, determining a bid price for buying land, choosing between forage species for renovation of haylands or grazing lands, and many other decisions involving multiyear consequences. If returns and/or costs of an investment are spread out over a number of years, NPV analysis should probably be evaluated. Because results of many decisions regarding forages involve more than 1 yr (in contrast with annual crops), NPV analysis is a valuable tool for the forage producer.

QUESTIONS

1. What factors beyond comparative advantage determine the economic competitiveness of forages in a geographic area?
2. How do government commodity programs for feed grains affect forage and forage-livestock producers?
3. Discuss the choice of the discount rate used for a net present value analysis. What factors should be represented in the value used?
4. What inputs can forages substitute for? In what instances might there be some complementarity between forages and other inputs?
5. What are the difficulties in doing product-product research?
6. Discuss the differences between factor-product, factor-factor, and product-product analyses. What information for these analyses is provided by the agronomist, the agricultural economist, the agricultural engineer, and the animal scientist?
7. When is it important to consider an economic analysis on a land area basis? When is it better to consider it on an animal basis?
8. Discuss the importance of time in forage management and decision making. How does it influence enterprise budgeting and investment analyses?

REFERENCES

Barry, PJ, JA Hopkin, and CB Baker. 1988. Financial Management in Agriculture. Danville, Ill.: Interstate.

Beneke, RR, and R Winterboer. 1973. Linear Programming Applications to Agriculture. Ames: Iowa State Univ. Press.

Boehlje, MD, and VR Eidman. 1984. Farm Management. New York: John Wiley.

Buckmaster, DR. 1993. Evaluator for round hay bale storage. J. Prod. Agric. 6:378-85.

Calkins, PH, and DD DiPietre. 1983. Farm Business Management. New York: Macmillan.

Clark, RT, JT Nichols, and KM Eskridge. 1991. Economic optimum fertilization rates for subirrigated meadow hay production, including values for hay quality. J. Prod. Agric. 4:233-40.

Foltz, JC, JG Lee, and MA Martin. 1993. Farm-level economic and environmental impacts of eastern Corn Belt cropping systems. J. Prod. Agric. 6:290-96.

Heady, EO. 1962. Economic aspects of forage production. In ME Heath et al. (eds.), Forages: The Science of Grassland Agriculture, 2d ed. Ames: Iowa State Univ. Press, 22-30.

Heitschmidt, RK, JR Conner, SK Canon, WE Pinchak, JW Walker, and SL Dowhower. 1990. Cow/calf production and economic returns from yearlong continuous, deferred rotation and rotational grazing treatments. J. Prod. Agric. 3:92-99.

Jacobs, VE. 1973. Forage production economics. In ME Heath et al. (eds.), Forages: The Science of Grassland Agriculture, 3d ed. Ames: Iowa State Univ. Press, 21-29.

Knowlton, KF, RE Pitt, and DG Fox. 1993. Model-predicted value of enzyme-treated alfalfa silage for lactating dairy cows. J. Prod. Agric. 6:280-86.

Lee, WF, MD Boehlje, AG Nelson, and WG Murray. 1980. Agricultural Finance. 7th ed. Ames: Iowa State Univ. Press.

McDonald, MVH. 1989. Economics of alfalfa persistence and replacement under varying potassium and phosphorous fertility levels. MS thesis, Department of Agricultural Economics, Univ. of Missouri, Columbia.

Moore, JE. 1994. Forage quality indices: Development and application. In GC Fahey et al. (eds.), Forage Quality, Evaluation, and Utilization. Madison, Wis.: American Society of Agronomy, 967-98.

Moore, KC. 1992. Farm Management Newsletter FM 92-3. Univ. Ext., Univ. of Missouri, Columbia.

Moore, KC, MV McDonald, JH Coutts, DD Buchholz, and CJ Nelson. 1991. Economics of alfalfa persistence and replacement. In Proc. Forage and Grassl. Conf., Columbia, Mo. Georgetown, Tex.: American Forage and Grassland Council, 233-35.

Nelson, CJ, and KM Zarrough. 1981. Tiller density and tiller weight as yield determinants of vegetative swards. In CE Wright (ed.), Plant Physiology and Herbage Production, Occup. Symp. 13. Hurley, UK: British Grassland Society, 25-29.

Nelson, J, D Buchholz, K Moore, and J Jennings. 1992. Alfalfa persistence with P and K fertility. Better Crops 76(3):18-21.

Overman, AR, and SR Wilkinson. 1993. Modeling tall fescue cultivar response to applied nitrogen. Agron. J. 85:1156-58.

Petritz, D. 1989. Confusing terms don't make the grade. Hay and Forage Grow., March, 38.

Rohweder DA, N Jorgenson, and RF Barnes. 1983. Proposed hay-grading standards based on laboratory analyses of evaluating forage quality. In JA Smith and VW Hays (eds.), Proc. 14th Int. Grassl. Congr., Lexington, Ky. Boulder, Colo.: Westview, 534-38.

Tyner, FH, and JC Purcell. 1985. Forage production economics. In ME Heath et al. (eds.), Forages: The Science of Grassland Agriculture, 4th ed. Ames: Iowa State Univ. Press, 43-50.

USDA. 1991. Agricultural Statistics. Washington, D.C.: US Gov. Print. Off.

PART 2

Forage Legumes and Grasses

16 Alfalfa

DONALD K. BARNES
Agricultural Research Service, USDA, and University of Minnesota

CRAIG C. SHEAFFER
University of Minnesota

ALFALFA, *Medicago sativa* L., originated near Iran, but related forms and species are found as wild plants scattered over central Asia and into Siberia. Its value as feed for horses and other animals is described as early as 490 B.C. by the Roman writers Pliny and Strabo. Alfalfa was carried into Europe and later South America by invading armies, explorers, and missionaries.

Alfalfa was first introduced into the eastern US by the colonists in 1736. The crop only persisted in a few calcareous soils. Spanish sources of alfalfa were introduced into the southwestern US about 1850 from South America. Those types spread into northern California and as far east as Kansas, where they became parents of the first cultivars used in the southeastern US and southern Great Plains. Between 1858 and 1910 two winter-hardy germplasm sources from Europe and Russia were introduced into the upper Midwest and Canada (Rumbaugh 1979). Three types that were not winter-hardy were introduced into the southwestern US from Peru (1899), India (1913 and 1956), and Africa (1924). A fourth type that was not winter-hardy was introduced from the Arabian peninsula for evaluation by scientists (1980s). Intermediate, winter-hardy types were introduced from northern France (1947) and the southern Russia, Iran, Afghanistan, and Turkey area (between 1898 and about 1925). In total, nine sources represent the basic alfalfa germplasm used in present US cultivars (Barnes et al. 1977).

DONALD K. BARNES is a Plant Geneticist, USDA-ARS, and Adjunct Professor of Agronomy, University of Minnesota, St. Paul. He received the MS degree from the University of Minnesota and the PhD from Pennsylvania State University. Since 1960 his major area of research has been alfalfa breeding and genetics with special emphasis on disease and insect resistance, pollination control, nitrogen fixation, and germplasm enhancement.

CRAIG C. SHEAFFER is Professor of Agronomy, University of Minnesota, St. Paul. He received the MS and PhD from the University of Maryland. Since 1977 his major research has been in forage management with emphasis on alfalfa management and cropping systems.

DISTRIBUTION AND ADAPTATION

Alfalfa, or *lucerne,* as it is called in many European countries, is worldwide in distribution. In the US it is one of only a few crops grown in every state. During the last decade between 10 and 11 million ha of alfalfa have been grown annually. There was a marked increase in alfalfa usage in the 1950s from about 6 million ha to about 12 million ha. About 57% of the production is located in the north central states, with Wisconsin, North Dakota, South Dakota, and Minnesota each having 800,000 or more ha. The western states account for another 25% of the production followed by about 10% in the northeastern states and about 8% in the mid-Atlantic and southeastern states (Melton et. al. 1988).

Alfalfa can survive temperatures below –25°C in Alaska and above 50°C in California. Alfalfa is highly drought tolerant. In Minnesota, yields of alfalfa were reduced less by drought than yields of birdsfoot trefoil (*Lotus corniculatus* L.), red clover (*Trifolium pratense* L.), or cicer milkvetch (*Astragalus cicer* L.) (Peterson et al. 1992). Average yield of drought-stressed alfalfa was 120% greater than yields of drought-stressed birdsfoot trefoil and cicer milkvetch, and 165% greater than the yield of similarly stressed red clover. Alfalfa becomes dormant during periods of severe drought and resumes growth when moisture conditions become favorable (Hall et al. 1988). These periods of drought-induced dormancy can last for 1 or 2 yr.

Alfalfa grows well on irrigated, fertile soils in the dry climates of the western US. Irrigation of alfalfa is less effective on most soils in the eastern US. This is because there is usually sufficient natural rainfall, and the frequent periods of wet soil and high humidity can cause increased losses from root and foliar disease. Water use varies with climate, cultivar, and soil fertility, but an average of 5.6 to 8.3 cm/ha of water are required per metric ton of dry forage (Sheaffer et al. 1988c). Maximum daily water use rates are from 1.3 to 15 mm per day and from 400 to 1900 mm per season.

PLANT DESCRIPTION

Alfalfa is a herbaceous perennial legume (Teuber and Brick 1988). Leaves are normally pinnately trifoliolate and arranged alternately on the stem (Fig. 16.1). Plants with compound multifoliolate leaves (having more than three leaflets) are being used in cultivar development. A mature alfalfa plant may have from 5 to 25 stems, which usually reach a height of 60-90 cm.

The crown is first formed at the cotyledon node, at or beneath the soil surface, through contractile growth of the hypocotyl. Sec-

Fig. 16.1. Alfalfa plant: (*A*) a crown having many shoots and a branched taproot; (*B*) stems with flowers; (*C*) three views of a single flower—*left,* face view; *center,* flower before tripping, showing the closed keel; and *right,* flower after tripping, showing the sexual column; (*D*) seedpods and a seed. *USDA drawing.*

ondary and tertiary bud and stem development also occur at this node and other basal nodes. Extensive bud development in combination with additional contractile growth and secondary stem thickenings results in a perennial site of meristem activity.

Cultivars differ in crown type and stem numbers. Some winter-hardy cultivars develop broad crowns with many upright stems developing from rhizomes originating at the crown. A few very winter-hardy cultivars can produce adventitious stems from primordia arising from roots (Teuber and Brick 1988). These plant types are referred to as being *creeping rooted*. Winter-hardy cultivars have broad crowns that can reach 15-30 cm in 3- to 4-yr-old plants. Crowns of plants that are not winter-hardy and are of the same age will be much smaller.

Regrowth following harvest can occur from either crown or axillary stem buds depending on height of cutting and plant type. In winter-hardy alfalfas, most new shoots elongate from the crown when the prior growth is harvested at first flower. In types that are not winter-hardy new shoots elongate from the bases of previously cut stems. If alfalfa is uncut during the growing season, one or more regrowths will occur. The accumulated regrowth can smother the plant.

The alfalfa root system is usually a distinct taproot, which penetrates the soil 7-9 m or more. However, it is not unusual for the root system to be extremely branched. Root mass decreases logarithmically with depth, with about 60%-70% of the total root mass in the upper 15 cm of the soil profile (Heichel 1982). Fibrous roots proliferate in the upper 20 cm of the profile and bear most of the nodules.

IMPORTANCE AND USE

Alfalfa, often called "Queen of the Forages," is one of the most important forage plants in the US. It has the highest feeding value for farm animals of all commonly grown hay crops (Marten et al. 1988). It produces more protein per hectare than grain or oil seed crops. The most efficient use of alfalfa for dairy animals is often in combination with corn silage, where the protein from alfalfa complements the energy from corn (*Zea mays* L.). Alfalfa has a high mineral content and contains at least 10 different vitamins. It is an important source of vitamin A. These characteristics make alfalfa used as hay, meal, or silage a desirable ration component for most farm animals.

Alfalfa in combination with the bacteria *Rhizobium meliloti* Dangead is a highly effective symbiosis for biological nitrogen (N) fixation. The bacteria infect plant roots, and finger-shaped nodules are formed. Bacteria in the nodules convert atmospheric N into a form readily used by the plant. Selecting plants for increased nodulation has increased N fixation (Heichel et al. 1989). The level of N fixation also can be influenced by the strain of bacteria. Alfalfa can increase subsequent crop productivity when it is used in crop rotations (Baldock et al. 1981; Hesterman et al. 1986). The rotation effect after alfalfa is caused by improved water-holding capacity, increased soil organic matter, and reduction of some soil pathogens in addition to the N residues supplied. Alfalfa also helps minimize pollution by reducing water runoff and soil erosion and by removing soil N from greater depths than can be reached by annual crops. The nondormant alfalfa cultivar 'Nitro' was the first alfalfa cultivar selected specifically for providing forage and N for plow-down in cropping systems (Barnes et al. 1988).

Alfalfa is the primary honey crop in the US. It accounts for about one-third of the annual honey production by honeybees. Some honey production results as a by-product of seed production. Research to select alfalfa plants with increased nectar production has been successful, thereby suggesting the possibility of someday producing cultivars specifically for use as bee pasture (Teuber and Barnes 1979). Breeding for increased nectar production has resulted in increased seed production (Teuber et al. 1990).

Alfalfa is primarily harvested and stored as either hay or silage and used on the farm. A substantial domestic and international market also exists for hay and dehydration products. At harvest (the first flower stage of maturity) slightly more than half of the forage consists of leaves. Leaves contain more protein, digestible nutrients, and vitamins than stems (Marten et al. 1988; Kalu and Fick 1981). Alfalfa leaf loss increases during harvesting operations as hay moisture concentration decreases. Forage-conditioning equipment, which crushes, crimps, or breaks alfalfa stems, increases rate of drying and reduces leaf loss. It also preserves quality by reducing losses due to weather. Dry matter (DM) losses from field operations including raking and baling range from 5% to 50% (Pitt 1990). All of these factors can influence the feeding value of alfalfa hay.

Preservation of alfalfa as silage decreases field losses because less drying time is required and leaf shattering is minimized (Pitt 1990). Storage losses generally are greater for silage than for hay. Harvesting alfalfa as haylage (40%-60% moisture) or as wilted silage (60%-75% moisture) will increase forage quality and decrease storage losses as compared with direct cutting alfalfa. Haylage often is stored in oxygen-limiting tower silo structures to minimize fermentation losses.

Some alfalfa is dehydrated or fractionated, either wet or dry process, by commercial enterprises (Jorgenson and Koegel 1988). Dehydration involves the artificial drying at high temperatures of partially air-dried chopped forage. Dehydrated alfalfa can be pelleted, cubed, or sold as meal. Meal can be separated into leaf and stem fractions. Wet fractionation consists of crushing and expression of the plant juice. A high-fiber pressed forage suitable for ensiling and a high-protein juice is produced. A protein curd can be separated from the juice by heat coagulation.

Alfalfa is high-quality pasture for all classes of livestock. One study reported that cattle grazing alfalfa gained 0.67 kg/ha per day, which was less than that for cattle grazing birdsfoot trefoil or sainfoin (*Onobrychis viciifolia* Scop.) (an average of 0.80 kg/ha), but alfalfa had greater carrying capacity and persistence than either legume (Marten et al. 1987). Mixtures of alfalfa with grasses reduce the risk of bloat for grazing ruminants and decrease weed infestation. Reed canarygrass (*Phalaris arundinacea* L.), orchardgrass (*Dactylis glomerata* L.), and smooth bromegrass (*Bromus inermis* Leyss) are grasses most frequently used in mixtures with alfalfa in the northern US (Sheaffer et al. 1990). Orchardgrass tends to be more competitive in mixtures than reed canarygrass and smooth bromegrass. Excellent milk production can be obtained from grazing dairy animals. However, supplemental feeding of grain is required for maximizing milk production and fattening beef and weaned lambs.

CULTIVARS AND BREEDING

Many alfalfa cultivars have been released during the 1900s in the US. The rate of cultivar release increased from about 0.3/yr between 1901 and 1940 to about 1/yr between 1941 and 1960, to about 12/yr between 1976 and 1980, and to about 34/yr between 1988 and 1992 (see Fig. 16.2). The increase in numbers has been due to a need for cultivars with increased multiple pest resistance, to an increase in numbers of private alfalfa breeding programs, and to a change in marketing philosophy so that more small seed dealers own their own cultivars. Before 1925 most US alfalfa-breeding research was concerned with selecting germplasms with increased winterhardiness. 'Grimm' (1900), 'Cossack' (1907), and 'Ladak' (1910) were very winter-hardy cultivars that were released during that period (Rumbaugh 1979). During the next 25 to 30 yr the primary objective was to develop cultivars that were winter-hardy and resistant to bacterial wilt caused by *Clavibacter michiganense* subsp. *insidiosum* (McCull) Davis et al. 'Ranger' (1942) and 'Vernal' (1953) were examples of the first cultivars that resulted from this effort (see Table 16.1).

During the last 25 yr many diseases, insects, and nematodes were recognized as important problems for alfalfa (Leath et al. 1988; Fox et. al. 1991). Prior to 1982 public breeding programs developed research methodologies and bred cultivars with new types of host plant pest resistance. A chronological listing of cultivars bred for pest resistances is presented in Table 16.2. Following the release of a public cultivar with a new type of resistance, numerous cultivars with new combinations of pest resistances were

TABLE 16.1 Forage yields associated with differences in winterhardiness and bacterial wilt resistance for seven alfalfa cultivars established in 1967 at Rosemount, Minnesota

	Forage yield (mt/ha)						% Stand	Level	
Cultivar	1969	1971	1973	1975	1977	10-yr total	after 10 yr	Winter-hardiness[a]	Bacterial wilt resistance[b]
Ramsey	10.5	9.5	7.2	11.3	8.8	89.6	41	VW	R
Iroquois	10.5	9.6	7.3	10.3	7.5	87.5	23	W	VR
Vernal 10.0	8.9	6.3	10.5	7.6	84.1	20.0	W	R	
Saranac	10.6	9.4	6.5	10.0	5.6	81.2	10	MW	R
Ranger	9.4	8.5	6.0	9.6	6.6	77.2	10	W	MR
Arnim	10.2	7.6	0.0	0.0	0.0	37.6	1	MW	S
Alt Franken-Schmidt	9.8	6.8	0.0	0.0	0.0	35.9	1	W	S
LSD (0.5)	0.9	1.0	0.9	1.0	1.1				

Sources: University of Minnesota and USDA test.
[a]VW = very winter-hardy; W = winter-hardy; MW = moderately winter-hardy.
[b]VR = very resistant; R = resistant; MR = moderately resistant: S = susceptible.

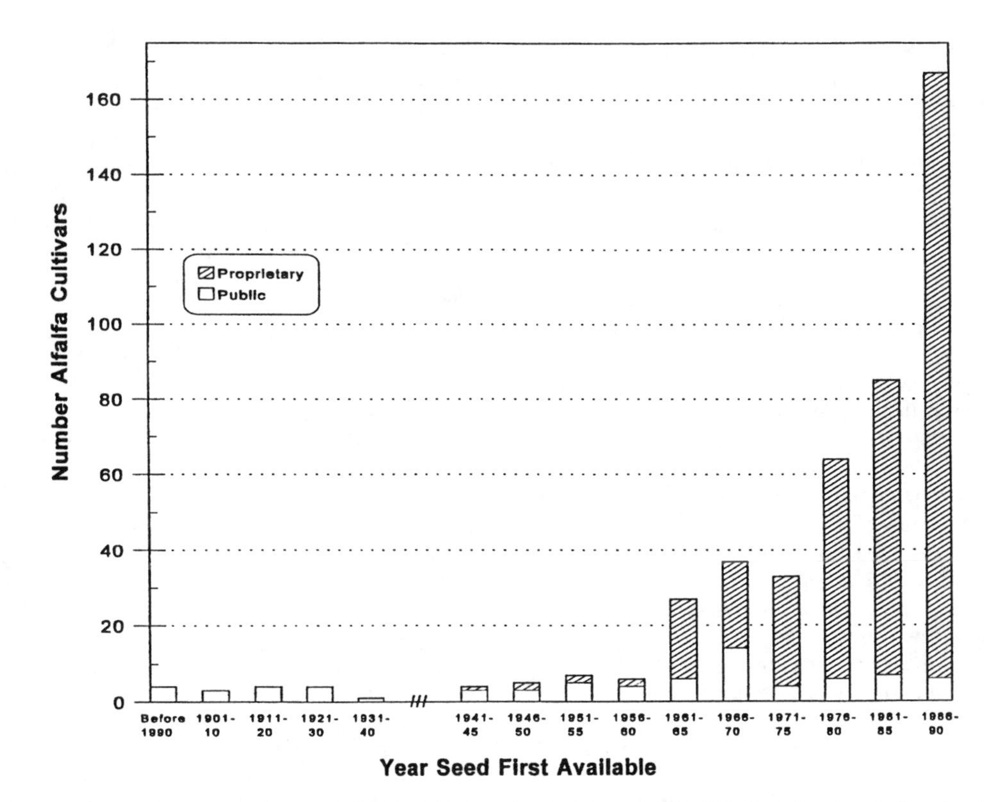

Fig. 16.2. The number of alfalfa cultivars developed by public (USDA and university) and private plant breeding programs in the US during the 1900s.

usually developed by private breeding programs. In 1982 the first cultivars with resistance to verticillium wilt (caused by *Verticillium albo-atrum*, Reinke and Berthier) were simultaneously released by public and private programs. The historical role of the public alfalfa breeders being responsible for cultivar development then changed. Public programs began to emphasize "basic, high-risk, and multidiscipline research," the development of new breeding procedures, and germplasm development. Industry programs concentrated on cultivar development. Development of procedures to select for verticillium wilt resistance by a public program (Christen and Peaden 1981) and the immediate use of this information by industry was the first example of the present public-private industry research cooperation. Basic studies of aphanomyces root rot (caused by *Aphanomyces euteiches* Drechs.) by a public research program (Delwiche et al. 1987) fol-

TABLE 16.2. Developments in breeding for alfalfa pest resistance in the US, 1942-1982

Type of resistance	First cultivar	Year released	Developing agency
Bacterial wilt	Ranger	1942	USDA, Nebraska AES
Stem nematode	Lahontan	1954	USDA, Nevada AES
Spotted alfalfa aphid	Moapa	1957	USDA, Nevada AES
	Zia	1957	New Mexico AES
Pea aphid	Washoe	1966	USDA, Nevada AES
Alfalfa weevil	Team	1967	USDA, North Carolina AES
Phytophthora root rot	Agate	1972	USDA, Minnesota AES
Anthracnose	Arc	1973	USDA, North Carolina, Maryland, Virginia, and Pennsylvania AES
Blue alfalfa aphid	CUF 101	1976	USDA, California AES
Lygus sp.	Rincon	1978	New Mexico AES
Verticillium wilt	Vernema	1982	USDA, Washington AES
	Apollo II	1982	North American Plant Breeders
	C/W 8015	1982	Cal/West Seeds
	Trumpetor	1982	Northrup King Co.
	WL 316	1982	W-L Research, Inc.

lowed by the release of resistant cultivars by industry programs further illustrated the effectiveness of the present system. The development of cultivars with multifoliolate leaves (1987) and improved forage qualities (1991) are the most recent examples of new product development.

Types of Cultivars. Alfalfa is a naturally outcrossing, tetraploid species that depends upon bees for pollination. All three of these factors contribute to its genetic heterogeneity. Part of the variability can be traced to the introgression of yellow-flowered *M. falcata* L. with purple flowered *M. sativa* (Rumbaugh et. al. 1988). The wide range of flower colors in many winter-hardy cultivars illustrates the combination of these two related species. It also illustrates the potential variability that exists for all characteristics. An alfalfa cultivar is a population of plants, and each plant is genetically different from the other. Cultivars can be described according to the mean response of all plants, such as average yield, and as a frequency of certain types of plants, such as the percentage of plants resistant to some pest.

Five general types of plant-breeding procedures or germplasm sources have been used to develop alfalfa cultivars. These are plant introductions, improved populations, synthetics, hybrids, and strain crosses. Most of the US cultivars between 1900 and 1940 were plant introductions from other countries (Barnes et al. 1977). Between 1940 and 1950 most new cultivars were improved populations that resulted from intercrossing a large number of plants that were selected by natural selection or by a form of mass selection such as recurrent phenotypic selection for one or more agronomic or pest resistance traits (Hanson et al. 1972). During the period of 1950 to 1975 synthetics were the most common cultivar types. A synthetic cultivar is produced by intercrossing a set, usually 4-16 selected parent plants (clones), and then increasing the seed through three or four generations without selection. Parent plants in synthetics are usually selected for good general combining ability for yield in addition to agronomic and pest resistance traits. Yields of synthetics usually decline with each generation following synthesis (Rumbaugh et al. 1988).

Maximum heterozygosity among genes is necessary to achieve maximum yield. This occurs in either the second generation (Syn 2) of a synthetic cultivar or in a double-cross hybrid (Dunbier and Bingham 1975). Hybrids result from crossing two individuals (plants or lines) of unlike genetic constitution. Some type of pollen control is necessary for hybrid seed production. Cytoplasmic male sterility systems are available in alfalfa (Viands et al. 1988). However, unlike corn, which is monoecious and wind pollinated, the alfalfa flower is bisexual and cross-pollinated by bees. This makes pollination control difficult and seed production of hybrids low. A few high-yielding hybrids have been produced, but they are not available as cultivars because of the high cost of seed production.

Strain crossing between two alfalfa strains with different types of pest resistances frequently was used in cultivar development during 1975 to about 1985 (Elgin et al. 1983). This combined the advantages of the synthetic and hybrid cultivars without requiring absolute pollen control. The primary advantage of strain crossing is that it enables plant breeders to quickly incorporate multiple pest-resistant traits, several from each strain, into one cultivar without losing yield from inbreeding (Rumbaugh et al. 1988). Currently many breeding procedures are being used in cultivar development, but synthetic varieties are most frequently released. (Future breeding techniques and cultivar development is discussed in Chap. 2, Vol. 2.)

Choosing a Cultivar. More than 440 cultivars were approved for certified seed production in the US between 1962 and 1992. They have included both public and private cultivars. Besides these recognized cultivars there are also brands and mixtures (blends) being marketed. Even though the performance of some brands and blends may be satisfactory, there often is no guarantee that they will have the same constitution from year to year. Some brands and blends contain seed of the newer cultivars; others may contain seed from old cultivars and common seed.

Before selecting a cultivar, growers need to outline the characteristics that best fit their operations. Selected cultivars should produce high yields of quality forage, have sufficient winterhardiness, but not more than needed to escape injury, and be resistant to pests that are problems in the production area.

CULTURE AND MANAGEMENT

Alfalfa can be a highly productive crop, but it requires a fertile soil, adequate water, and a good seedbed at establishment. Yield and persistence potentials of new cultivars cannot be realized without good management practice.

Fertilizer Requirements. Alfalfa is sensitive to soil acidity; pH values of 6.5-7.0 are required for maximizing forage production (Lanyon and Griffith 1988). Soil pH influences symbiotic N_2 fixation and the availability of essential and toxic elements. Potassium (K), phosphorus (P), sulfur (S), and boron (B) are the most common nutrients limiting alfalfa production although deficiencies of other nutrients may occur on certain soils. Soil fertilization programs should be based on local soil test results. Potassium concentration in alfalfa herbage exceeds that for every other element except N; thus high rates of removal occur with high yields. In addition to increasing yields, K fertilization has increased alfalfa tolerance to severe harvest management and winter injury (Smith 1975; Goplen et al. 1980; Sheaffer et al. 1986). Potassium fertilization increased nodule number, nodule mass, and N_2 fixation (acetylene reduction) rates (Duke et al. 1980).

Phosphorus concentration in alfalfa herbage is less than for K. However, it is a limiting nutrient for alfalfa production in many portions of the US. It is important for seedling development and is frequently banded beneath alfalfa seed at establishment.

Alfalfa obtains 40%-80% of its N from N_2 fixation, depending on plant age and soil N status (Vance et al. 1988). When the soil N status is high, nodulation will be reduced or prevented until the plants use the available N. Therefore, it is not recommended that N be applied to alfalfa. The only exception is that N is sometimes recommended when seeding alfalfa on low-N soils.

Seedings. Most alfalfa seedings in the US occur either in the early spring or in late summer and fall. Time of alfalfa seeding is influenced by precipitation patterns, temperature, and cropping patterns. Spring seedings allow for harvest during the seeding year, but weed control is usually required. Late summer and fall seedings usually avoid weed competition and unfavorable summer temperatures and moisture conditions but must allow for adequate seedling development prior to the onset of winter.

With spring seedings into conventionally prepared seedbeds, alfalfa is frequently established with a companion crop such as oat (*Avena sativa* L.) or barley (*Hordeum vulgare* L.) on soils with erosion potential or where grain and straw are desired (Simmons et al. 1992). Seeding rates for small grain companion crops should be reduced 25%-35%, and the small grain should be removed at boot or soft dough stage for forage to reduce competition with the alfalfa seedlings for moisture and light. Despite these precautions, alfalfa yields in the establishment and subsequent year sometimes are reduced. Spring establishment of pure stands of alfalfa with preplant-incorporated herbicides allows for greatest alfalfa yields and stands in the establishment and subsequent years. With an early seeding date and an initial harvest 60 d following alfalfa emergence, two and often three harvests can be made by early September in the north central US (Sheaffer 1983). Selection of herbicides for legume establishment should be based on state recommendations. Alfalfa can also be successfully established without a small grain companion crop or herbicides, and weeds can be removed during routine alfalfa harvesting (Sheaffer et al. 1988a). Often the forage quality of weeds invading seedling alfalfa stands is equal to or greater than that of an oat companion crop, and alfalfa regrowth following harvest will crowd out annual weeds (Andersen and Marten 1975).

Permanent pastures and rangeland can be renovated by introducing alfalfa. For successful interseeding in northern regions, grass competition must be reduced by use of chemicals or close grazing. Spring is the most successful time for interseeding because soil moisture levels are favorable (Groya and Sheaffer 1981). In frost seeding of pastures where less than optimum establishment conditions exist, alfalfa is inferior to red clover and white clover (*T. repens* L.) (Schmid and Martin 1976).

Attempts to reestablish alfalfa immediately following a previous alfalfa crop or to thicken old alfalfa stands have sometimes resulted in establishment failures. These failures have been related to autotoxicity (Hegde and Miller 1990).

Most alfalfa cultivars have approximately 100,000 seeds/kg. A goal of 135-270 plants/m^2 at establishment is desirable for maximizing seeding year yields and to ensure good stands for extended periods. Since only 20%-50% of the planted seed produces plants with present establishment practices, high seeding rates frequently are recommended. Increasing seeding rates above 10-13 kg/ha is not economically justifiable in northern regions when well-prepared firm seedbeds are used (Sund and Barrington 1976). In southern and western regions where seedling diseases and inadequate soil moisture reduce seedling survival, seeding rates of 17-34 kg/ha are not uncommon (Tesar and Marble 1988). In the years following establishment, alfalfa yield is not increased with stands greater than 45-54 plants/m^2 (Sund and Barrington 1976; Bolger and Meyer 1983). Decreased plant numbers are compensated by greater tiller production by existing plants. Higher seeding rates do not affect stem numbers per unit area or leaf:stem ratios and have no affect on forage quality.

Alfalfa should be seeded in a firm seedbed to a depth of 1.3 cm. This may be somewhat shallower or deeper on fine and coarse-textured soils, respectively. Use of culti-packer seeders or press wheels on conventional drills is desirable in obtaining good soil-seed contact. Band-seeding (the placement of a band of alfalfa seed over a band of fertilizer placed 2.5 to 5.0 cm deep) using a modified grain drill increases alfalfa stands and seedling development on soils low in fertility or when unfavorable environmental conditions exist.

Inoculation of alfalfa seed with *Rhizobium* is not required for most high-organic matter soils previously planted to alfalfa. In soils where inoculation is required, bacteria can be supplied via preinoculated seed or with commercial cultures at planting (Vance et al. 1988).

Harvest Schedules. The number of harvests or grazing cycles per year varies from 1 in northern and arid regions to a maximum of 10 in irrigated regions of the Southwest. The number of harvests per year depends on the stage of maturity at harvest and favorable environmental conditions. Average hay yields in the US are greatest in California and Arizona at about 14.6 mt/ha (USDA 1981). They are about 7.6 mt/ha in the major alfalfa-producing states of the upper Midwest. These average yields are frequently doubled by producers using improved cultivars and improved soil and crop management practices.

In response to cool weather and short daylengths in the fall, winter-hardy cultivars become dormant. The dormancy reaction involves complex physiological changes by the plant in preparation for winter and results in decreased herbage growth and increased carbohydrate storage (McKenzie et al. 1988). Root and crown carbohydrate reserves are utilized to initiate regrowth in the spring and after each cutting. The N_2 fixation system also is affected by carbohydrate reserves and dormancy responses. Nodule growth and N_2 fixation follow a cyclic pattern where activity is low during dormancy and after harvest. New growth increases N_2 fixation, which reaches maximum levels as alfalfa reaches full-bloom (Vance et al. 1979).

During a regrowth cycle following dormancy or harvesting, herbage DM yield increases until flowering and then declines due to leaf loss. Yield increases are associated with increases in stem mass and decreases in leaf:stem ratio (Sheaffer et al. 1988b). Nutrient concentrations are greater in leaves than in stems. With increased plant maturity, stem quality declines at a greater rate than leaf quality. Alfalfa maturity is described by estimating morphological development of individual alfalfa stems (Kalu and Fick 1981). *Vegetative, bud, flowering,* and *seedpod* are terms describing stages of alfalfa development.

Recommended harvest schedules must consider forage yield, forage quality, stand persistence, and morphological development of the alfalfa plant. As alfalfa matures, apical dominance is broken, and new stems elongate from buds at crowns or stem bases, provided sufficient root carbohydrate levels exist (Fick 1977). Cutting by calendar date or time interval cannot take all of the necessary factors into consideration. For hay and silage harvesting systems, harvesting at early-flowering (first flower) is the best compromise between having acceptable forage and nutrient yields and stand persistence. Producers who value alfalfa as a protein and energy supplement often sacrifice yield and persistence to obtain greater forage quality by harvesting at bud stages. In pastures, grazing should begin when alfalfa is in vegetative stages and continued for a maximum of 7-10 d. Normally alfalfa should be rotationally grazed with a 30-40 d recovery period because continuous grazing will deplete root carbohydrate reserves and cause stand loss. However, the recently released cultivar 'Alfagraze' was selected to tolerate and yield well under continuous grazing (Brummer and Bouton 1991).

In the northern portions of the US and in Canada where winter injury to alfalfa frequently occurs, the traditional recommendation has been to allow 4-6 wk to elapse between the time of the last harvest in August or September and the first killing frost (–3°C). This management allows for adequate vegetative regrowth and the accumulation of root reserves that are essential for winter survival. While this harvest strategy based on a "critical fall period" poses the least risk to stand persistence, recent research (Tesar and Yager 1985; Sheaffer and Marten 1990) indicates that this recommendation can be liberalized to afford greater management flexibility. A modern theory of fall harvest decisions is based on risk assessment (Sheaffer et al. 1988b). Risks associated with fall cutting can be minimized for well-drained, high-fertility soils, improved cultivars, and stands of less than 3 yr. In addition, it appears that the interval between the final harvest and seasonal harvest is more important than date of final harvest (Edminsten et al. 1988).

Final harvests taken following the first killing frost should leave a 14-cm stubble in the field. Stubble can reduce freezing and thawing of the soil and can catch snow to provide insulation and reduce damage from ice sheeting.

Highest seasonal yields are obtained from cutting as low as equipment limitations permit. This normally results in a 7- to 10-cm stubble. Higher stubble heights, which leave more leaf area for photosynthesis and sites for regrowth, are beneficial for persistence under frequent harvest management or when alfalfa is winter injured.

SEED PRODUCTION

Alfalfa seed production is a specialized industry. Areas with low relative humidities and moderate to high temperatures are preferred for alfalfa seed production because of a low incidence of leaf disease, long periods of pollinator activity, and favorable harvest conditions. About two-thirds of US alfalfa seed is produced under irrigation in five western states: California, Idaho, Nevada, Oregon, and Washington (USDA 1981). South Dakota and Kansas produce significant quantities of seed under dryland conditions.

The alfalfa flower is morphologically complex and has a unique tripping mechanism (Barnes 1980; Teuber and Brick 1988). The flower has five petals: a large standard petal, two lateral wing petals, and two fused keel petals (Fig. 16.1). Each flower has a stigma and an awl-shaped style on a cylindrical ovary containing 10 to 12 ovules. The ovary is surrounded by a staminal column consisting of one free and nine fused stamens. The nectary is located at the base of the ovary. Before pollination can occur, the sexual column (ovary and staminal column) must be tripped (released). This is usually done by bees foraging for nectar and pollen. When tripping occurs, the stigma strikes either the bee or the standard petal. If the stigma fails to contact foreign pollen on the bee, its own pollen, which surrounds the stigma, will germinate and often fertilize ovules. Flowers that are self-pollinated will form fewer seedpods and will have fewer seeds per pod than cross-pollinated flowers. About 50% of field-produced alfalfa seed results from cross-pollination.

The honeybee, *Apis mellifera* L.; leaf-cutter bee, *Megachile rotundata* (F.); and alkali bee, *Nomia melanderi* Ckll., are propagated for alfalfa seed pollination. Bumblebees, *Bombus* spp., and other wild bees are very effective trippers and often contribute to seed production in the Midwest. Leaf-cutter and alkali bees are the most effective pollinators in the Northwest, and honeybees are most effective in California. Good pollination management is essential to successful seed production. This includes timely pesticide application and pollinator release to coincide with flowering of the crop. Consideration must also be given to control of pollinator predators and diseases and increasing parasites and predators of crop pests (Rincker et al. 1988). The most serious insect pests for alfalfa seed production are two species of *Lygus* bugs, *Lygus elisus* Van Duzee and *L. hesperus* Knight; spotted alfalfa aphid, *Therioaphis maculata* (Buckton); and alfalfa seed chalcid, *Bruchophagus raddi* (Guss.).

Alfalfa for seed production usually is planted in rows at seeding rates of 1-3 kg/ha. Rows are generally superior to solid planting because they fit well with irrigation and they make it easy to control weeds and volunteer alfalfa seedlings. Fertility programs and seedbed preparation are similar to those for hay production. Alfalfa seed production is increased by a minimal water stress, which suppresses vegetative growth and increases rate of maturity. In areas where multiple regrowths and seed sets occur, alfalfa is irrigated following each seed set. All seed does not mature at the same time, and seed can shatter easily. To prevent seed losses, it is necessary to harvest when about two-thirds of the pods have turned dark brown. Nearly all alfalfa seed is direct combined following the spraying of a defoliant (Rincker et al. 1988). Seed yields in excess of 2100 kg/ha have been reported, but average seed yields vary from about 750 kg/ha in the western US to about 50 kg/ha in the midwestern states.

DISEASES

More than 20 diseases are serious problems for alfalfa in the US (Leath et al. 1988; Stuteville and Erwin 1990). These include fungal and bacterial wilts, leaf spots, crown and root rots, viruses, and nematodes. Important wilts are bacterial wilt, fusarium wilt caused by *Fusarium oxysporum* Schlecht. f. sp. *medicaginis* (Weimer) Snyd. & Hans., and verticillium wilt. The most serious leaf spots are common leaf spot caused by *Pseudopeziza medicaginis* (Lib.) Sacc., lepto leaf spot caused by *Leptosphaerulina briosianna* (Pollacci) J. H. Graham & Luttrell, stemphylium leaf spot caused by *Stemphylium botryosum* Wallr., and summer blackstem caused by *Cercospora medicaginis* Ellis & Everh.

Important crown and root rots include anthracnose caused by *Colletotrichum trifolii* Bain & Essary, aphanomyces root rot, spring blackstem caused by *Phoma medicaginis* Malbr. & Roum. var. *medicaginis*, phytophthora root rot caused by *Phytophthora megasperma* Drechs., rhizoctonia diseases caused by *Rhizoctonia solani* Kuehn., and sclerotina crown and stem rot caused by *Sclerotina trifoliorium* sensu Kohn. Alfalfa mosaic is the primary virus disease. Alfalfa stem nematode *Ditylenchus dipsaci* (Kuhn) Filpjev, root-knot nematodes (*Meloidogyne* spp.), and root-lesion nematodes (*Pratylenchus* spp.) are the most prevalent nematode species on alfalfa. Resistant cultivars are available for most of the diseases and nematodes listed.

PESTS

There are a number of insect pests on alfalfa in the US (Manglitz and Ratcliffe 1988). The insect pests most likely to be problems during seed production are already listed. The insect pests that interfere with forage production include the potato leafhopper, *Empoasca fabae* (Harris); the alfalfa weevil, *Hypera postica* (Gyll.); the spotted alfalfa aphid; the pea aphid, *Acyrthosiphon pisum* (Harris); the blue alfalfa aphid, *A. kondoi* Shinji; the alfalfa plant bug, *Adelphocoris lineolatus* (Goeze); and the meadow spittlebug, *Philaenus spumarius* (L.). The potato leafhopper is the most problematic pest and causes damage throughout most alfalfa-producing areas in the eastern and central US. It causes yellowing of the foliage and stunting of stems. The damage results in significant losses in yield and forage quality, especially loss in carotene.

OTHER SPECIES

There are about 20 annual species of *Medicago;* these are called *medics*. Annual medics evolved in the Mediterranean climate, which typically has hot, dry summers and mild, wet winters with variable amounts of rainfall. In regions where they are adapted, medics have been traditionally used as winter annuals (Crawford et al. 1989). They germinate rapidly with the onset of moisture and initiate flowering and seed production soon after emerging. In Australia, the forage is extensively used for pasture and green manure in ley-cropping systems, where cereals and pasture crops are alternated in the same field. Only 3 medics are widely distributed and used in the US: California burclover (*M. polymorpha* L.) and spotted burclover (*M. arabica* Huds.) are found in the Pacific states and in the South, respectively. Black medic (*M. lupulina* L.) is widely distributed in waste areas and permanent pastures. 'Georges' black medic, a selected ecotype, is grown as a replacement for summer fallow in Montana and surrounding states (Sims and Slinkard 1991). There is potential in the US for annual medics to be used as a cover crop and/or a smother plant intercropped with corn, soybeans, and small grains and/or a N source grown in rotational systems.

QUESTIONS

1. Alfalfa has a diversity of uses. Describe important uses of alfalfa.
2. What are the soil fertility needs for good establishment and persistence of alfalfa?
3. List and discuss factors that are important for successful alfalfa establishment in prepared seedbeds.
4. What criteria would you use to schedule alfalfa harvests that would provide for maximum nutrient yields per hectare while providing reasonable stand persistence? How would this criteria change for producers whose goal is to produce forage with high nutrient concentrations?
5. What is the average yield of alfalfa forage in your state? How many harvests usually occur per year?
6. How is management of alfalfa for seed production different from management for forage production? What factors limit seed production in your area?
7. What breeding methods are currently used in producing most alfalfa cultivars? Describe the historical and present roles of public and private industry plant breeders in cultivar development.
8. List criteria useful in selecting an alfalfa cultivar for your locality. Identify several adapted cultivars.
9. Describe the fall dormancy reaction of alfalfa. What factors affect fall dormancy, and why is it important for plant persistence?
10. What disease and insect pests have an important influence on profitable alfalfa production in

your state? What are the control measures for these pests?

REFERENCES

Andersen, RN, and GC Marten. 1975. Forage nutritive value and palatability of 12 common annual weeds. Crop Sci. 15:821-29.

Baldock, JO, RE Higgs, WH Paulson, JA Jackobs, and WD Shrader. 1981. Legume and mineral N effects on crop yields in several crop sequences in the upper Mississippi valley. Agron. J. 73:885-90.

Barnes, DK. 1980. Alfalfa. In WR Fehr and HH Hadley (eds.), Hybridization of Crop Plants. Madison, Wis.: American Society of Agronomy and Crop Science Society of America, 177-87.

Barnes, DK, ET Bingham, RP Murphy, OJ Hunt, DF Beard, WH Skrdla, and LR Teuber. 1977. Alfalfa Germplasm in the United States: Genetic Vulnerability, Use, Improvement, and Maintenance. USDA Tech. Bull. 1571.

Barnes, DK, CC Sheaffer, GH Heichel, DM Smith, and RN Peaden. 1988. Registration of 'Nitro' alfalfa. Crop Sci. 28:718.

Bolger, TP, and DW Meyer. 1983. Influence of plant density on alfalfa yield and quality. In Proc. Am. Forage and Grassl. Counc. Eau Claire, Wis., 37-41.

Brummer, EC, and JH Bouton. 1991. Plant traits associated with grazing-tolerant alfalfa. Agron. J. 83:996-1000.

Christen, AA, and RN Peaden. 1981. Verticillium wilt in alfalfa. Plant Dis. 65:319-21.

Crawford EJ, AWH Lake, and KG Boyce. 1989. Breeding annual Medicago species for semiarid conditions in southern Australia. Adv. Agron. 42:399-435.

Delwiche, PA, CR Grau, EB Holub, and JB Perry. 1987. Characterization of *Aphanomyces euteiches* isolates recovered from alfalfa in Wisconsin. Plant Dis. 71:155-61.

Duke, SH, M Collins, and RM Soberalske. 1980. The effects of potassium fertilization on nitrogen fixation and nodule enzymes of nitrogen metabolism in alfalfa. Crop Sci. 20:213-19.

Dunbier, MW, and ET Bingham. 1975. Using haploid-derived autotetraploids. Crop Sci. 15:527-31.

Edminsten, KL, DD Wolf, and M Lentner. 1988. Fall harvest management of alfalfa. I. Date of fall harvest and length of growing period prior to fall harvest. Agron. J. 80:688-93.

Elgin, Jr., JH, JE McMurtrey III, BJ Hartman, BD Thyr, EL Sorensen, DK Barnes, FI Frosheiser, RN Peaden, RR Hill, Jr., and KT Leath. 1983. Use of strain crosses in the development of multiple pest resistant alfalfa with improved field performance. Crop Sci. 23:57-64.

Fick, GW. 1977. The mechanism of alfalfa regrowth: A computer simulation approach. Cornell Agric. Exp. Stn. Search Agric. and Agron. 7:1-27.

Fox, CC, R Berberet, FA Gray, CR Grau, DL Jessen, and MA Petersen. 1991. Standard Tests to Characterize Alfalfa Cultivars. 3d ed., North Am. Alfalfa Improv. Conf., Beltsville, Md.

Goplen, GP, H Baenziger, LB Bailey, ATH Gross, MR Hanna, R Michauc, KW Richards, and J Waddington. 1980. Growing and Managing Alfalfa. Can. Publ. 1705.

Groya, FL, and CC Sheaffer. 1981. Establishment of sod-seeded alfalfa at various levels of soil moisture and grass competition. Agron. J. 73:560-65.

Hall, MH, CC Sheaffer, and GH Heichel. 1988. Partitioning and mobilization of photoassimilate by alfalfa subjected to water deficits. Crop Sci. 28:964-69.

Hanson, CH, TH Busbice, RR Hill, Jr., OJ Hunt, and AJ Oakes. 1972. Directed mass selection for developing multiple pest resistance and conserving germplasm in alfalfa. J. Environ. Qual. 1:106-11.

Hegde, RS, and DA Miller. 1990. Allelopathy and autotoxicity in alfalfa: Characterization and effects of preceding crops and residue incorporation. Crop Sci. 30:1255-59.

Heichel, GH. 1982. Alfalfa. In ID Teare and MM Peet (eds.), Crop-Water Relations. New York: Wiley, 127-55.

Heichel, GH, DK Barnes, CP Vance, and CC Sheaffer. 1989. Dinitrogen fixation technology for alfalfa improvement. J. Prod. Agric. 2:24-32.

Hesterman, OB, CC Sheaffer, DK Barnes, WE Lueschen, and JH Ford. 1986. Alfalfa dry matter and N response in legume-corn rotations. Agron. J. 78:19-23.

Jorgenson, NA, and RG Koegel. 1988. Wet fractionation processes and products. In AA Hanson, DK Barnes, and RR Hill, Jr. (eds.), Alfalfa and Alfalfa Improvement, Am. Soc. Agron. Monogr. 29. Madison, Wis., 553-66.

Kalu, BA, and GW Fick. 1981. Quantifying morphological development of alfalfa for studies of herbage quality. Crop Sci. 21:267-71.

Lanyon, LE, and WK Griffith. 1988. Nutrition and fertilizer use. In AA Hanson, DK Barnes, and RR Hill, Jr. (eds.), Alfalfa and Alfalfa Improvement, Am. Soc. Agron. Monogr. 29. Madison, Wis., 333-72.

Leath, KT, DC Erwin, and GD Griffin. 1988. Diseases and nematodes. In AA Hanson, DK Barnes, and RR Hill, Jr. (eds.), Alfalfa and Alfalfa Improvement, Am. Soc. Agron. Monogr. 29. Madison, Wis., 621-70.

McKenzie, JS, R Paquin, and SH Duke. 1988. Cold and heat tolerance. In AA Hanson, DK Barnes, and RR Hill, Jr. (eds.), Alfalfa and Alfalfa Improvement. Am. Soc. Agron. Monogr. 29. Madison, Wis., 259-302.

Manglitz, GR, and RH Ratcliffe. 1988. Insects and mites. In AA Hanson, DK Barnes, and RR Hill, Jr. (eds.), Alfalfa and Alfalfa Improvement, Am. Soc. Agron. Monogr. 29. Madison, Wis., 671-704.

Marten, GC, DR Buxton, and RF Barnes. 1988. Feeding value (forage quality). In AA Hanson, DK Barnes, and RR Hill, Jr. (eds.), Alfalfa and Alfalfa Improvement, Am. Soc. Agron. Monogr. 29. Madison, Wis., 463-91.

Marten, GC, FR Ehle, and EA Ristau. 1987. Performance and photosensitization of cattle related to forage quality of four legumes. Crop Sci. 27:138-45.

Melton, B, JB Moutray, and JH Bouton. 1988. Geographic adaptation and cultivar selection. In AA

Hanson, DK Barnes, and RR Hill, Jr. (eds.), Alfalfa and Alfalfa Improvement, Am. Soc. Agron. Monogr. 29. Madison, Wis., 545-620.

Peterson, PR, CC Sheaffer, and MH Hall. 1992. Drought effects on perennial forage legume yield and quality. Agron. J. 84:774-79.

Pitt, RE. 1990. Silage and Hay Preservation. NRAES-5. Ithaca, N.Y.: Northeast Regional Agricultural Engineering Service, Cooperative Extension.

Rincker, CM, VL Marble, DE Brown, and CA Johansen. 1988. Seed production practices. In AA Hanson, DK Barnes, and RR Hill, Jr. (eds.), Alfalfa and Alfalfa Improvement. Am. Soc. Agron. Monogr. 29. Madison, Wis., 985-1021.

Rumbaugh, MD. 1979. NE Hanson's Contributions to Alfalfa Breeding in North America. S.Dak. State Univ. Publ. 3538A.

Rumbaugh, MD, JL Caddel, and DE Rowe. 1988. Breeding and quantitative genetics. In AA Hanson, DK Barnes, and RR Hill, Jr. (eds.), Alfalfa and Alfalfa Improvement, Am. Soc. Agron. Monogr. 29. Madison, Wis., 777-808.

Schmid, AR, and NP Martin. 1976. Pasture renovation opportunities in Minnesota. In Proc. 2d Annu. Symp. Minn. Forage and Grassl. Counc., Bloomington, Minn., 30-37.

Sheaffer, CC. 1983. Seeding year harvest management of alfalfa. Agron. J. 75:115-19.

Sheaffer, CC, and GC Marten. 1990. Alfalfa cutting frequency and date of fall cutting. J. Prod. Agric. 3:486-91.

Sheaffer, CC, MP Russelle, OB Hesterman, and RE Stucker. 1986. Alfalfa response to potassium, irrigation, and harvest management. Agron. J. 78:464-68.

Sheaffer, CC, DK Barnes, and GC Marten. 1988a. Companion crop vs. solo seeding: Effect on alfalfa seeding year forage and N yields. J. Prod. Agric. 1:270-74.

Sheaffer, CC, GD Lacefield, and VL Marble. 1988b. Cutting schedules and stands. In AA Hanson, DK Barnes, and RR Hill, Jr. (eds.), Alfalfa and Alfalfa Improvement, Am. Soc. Agron. Monogr. 29. Madison, Wis., 411-37.

Sheaffer, CC, CB Tanner, and MB Kirkham. 1988c. Alfalfa water relation and irrigation. In AA Hanson, DK Barnes, and RR Hill, Jr. (eds.), Alfalfa and Alfalfa Improvement, Am. Soc. Agron. Monogr. 29. Madison, Wis., 373-409.

Sheaffer, CC, DW Miller, and GC Marten. 1990. Perennial grass-alfalfa mixtures: Grass dominance and mixture yield and quality. J. Prod. Agric. 3:480-85.

Simmons, SR, NP Martin, CC Sheaffer, DD Stuthman, EL Schiefelbein, and T Haugen. 1992. Companion crop forage establishment: Producer practices and perceptions. J. Prod. Agric. 5:67-72.

Sims, JR, and AE Slinkard. 1991. Development and evaluation of germplasm and cultivars of cover crops. In WL Hargrove (ed.), Cover Crops for Clean Water, Proc. Soil Conserv. Soc. Am., 9-11 Apr., Jackson, Tenn. Ankeny, Iowa, 121-29.

Smith, D. 1975. Effects of potassium topdressing a low fertility silt loam soil on alfalfa herbage yields and composition on soil K values. Agron. J. 67:60-64.

Stuteville, DL, and DC Erwin. 1990. Compendium of Alfalfa Diseases. 2d ed. St. Paul, Minn.: American Phytopathological Society Press.

Sund, JM, and GP Barrington. 1976. Alfalfa Seeding Rates: Their Influence on Dry Matter Yield, Stand Density and Survival, Root Size and Forage Quality. Wis. Res. Bull. R2786.

Tesar, MB, and VL Marble. 1988. Alfalfa establishment. In AA Hanson, DK Barnes, and RR Hill, Jr. (eds.), Alfalfa and Alfalfa Improvement. Am. Soc. Agron. Monogr. 29. Madison, Wis., 303-32.

Tesar, MB, and JL Yager. 1985. Fall cutting of alfalfa in the north central USA. Agron. J. 77:774-78.

Teuber, LR, and DK Barnes. 1979. Breeding alfalfa for increased nectar production. Md. Agric. Exp. Stn. Misc. Publ. 1:109-16.

Teuber, LR, and MA Brick. 1988. Morphology and anatomy. In AA Hanson, DK Barnes, and RR Hill, Jr. (eds.), Alfalfa and Alfalfa Improvement, Am. Soc. Agron. Monogr. 29. Madison, Wis., 125-62.

Teuber, LR, CM Rinker, and DK Barnes. 1990. Seed yield characteristics of alfalfa populations selected from receptacle diameter and nectar volume. Crop Sci. 30:579-83.

US Department of Agriculture. 1981. Agricultural Statistics. Washington, D.C.: US Gov. Print. Off.

Vance, CP, GH Heichel, and DA Phillips. 1988. Nodulation and nitrogen fixation. In AA Hanson, DK Barnes, RR Hill, Jr. (eds.), Alfalfa and Alfalfa Improvement, Am. Soc. Agron. Monogr. 29. Madison, Wis., 229-57.

Vance, CP, GH Heichel, DK Barnes, JW Bryan, and LE Johnson. 1979. Nitrogen fixation, nodule development, vegetative regrowth of alfalfa (*Medicago sativa* L.) following harvest. Plant Physiol. 64:1-8.

Viands, DR, P Sun, and DK Barnes. 1988. Pollination control: Mechanical and sterility. In AA Hanson, DK Barnes, RR Hill, Jr. (eds.), Alfalfa and Alfalfa Improvement, Am. Soc. Agron. Monogr. 29. Madison, Wis., 931-60.

17 Red Clover

NORMAN L. TAYLOR
University of Kentucky

RICHARD R. SMITH
Agricultural Research Service, USDA, and University of Wisconsin

RED CLOVER, *Trifolium pratense* L., is the most widely grown of all the true clovers. Grown alone and with grasses, it is a very important legume hay crop in the northeastern US. It is used for hay, pasture, and soil improvement, and it fits well into 3-and 4-yr rotations. Nationwide, red clover and grass mixtures declined from approximately 6 million to 5 million ha during the period of 1950 to 1970, and based on seed disappearance they probably declined to about 4.5 million ha by 1987.

Red clover is thought to have originated in Asia Minor and southeastern Europe. An early-flowering type known in Spain by 1500 spread to Holland and Lombardy by 1550 and then to the German Rhineland. Red clover was introduced to England from Germany about 1650 and was carried to America by English colonists (Merkenschlager 1934).

DISTRIBUTION AND ADAPTATION

All red clovers may be grouped into three divisions: early-flowering, late-flowering, and wild red. The wild red clover found in England is unknown in North America. Most American red clovers are of the early-flowering type, known collectively as *medium red clover*. This type is characterized by two or three hay crops per year and a biennial or short-lived perennial growth habit. American mammoth red clover is the principal late-flowering type grown in the US. The late-flowering or single-cut type usually produces one crop plus an aftermath.

In addition to early and late types there are many intermediate forms in Europe. The major difference among early-and late-flowering types is related to daylength response, the single-cut type requiring a longer photoperiod. Tetraploid forms of red clover have been developed that are of agricultural significance primarily in Europe.

Red clover is best adapted where summer temperatures are moderately cool to warm and moisture is sufficient throughout the growing season. Extensively grown in the humid region of the US, it extends into eastern North and South Dakota, Nebraska, and Kansas. It extends north into Ontario and Quebec and south into Tennessee and North Carolina and is used as a winter annual in the southeastern US. It is also grown in much of the Pacific Northwest, primarily under irrigation (Fig. 17.1).

PLANT DESCRIPTION

Red clover is a herbaceous plant made up of numerous leafy stems rising from a crown. The flowers are borne on heads (compact clusters) at the tips of the branches (Fig. 17.2). Heads usually consist of up to 125 flowers.

Most American strains when grown under favorable conditions have corolla tubes 9.0-

NORMAN L. TAYLOR is Professor in Charge of Legume Breeding, University of Kentucky. He received his MS degree from the University of Kentucky and PhD from Cornell University. His area of special interest is breeding and genetics of red clover and evolutionary genetics of *Trifolium*.

RICHARD R. SMITH is Research Geneticist on Clover Breeding for the US Department of Agriculture, Agricultural Research Service, at the US Dairy Forage Research Center at the University of Wisconsin, Madison, and he is also Professor of Agronomy at the university. He received his MS from the University of Illinois and PhD from Iowa State University. His special areas of interest are breeding and genetics of disease resistance and persistence and quality in red clover.

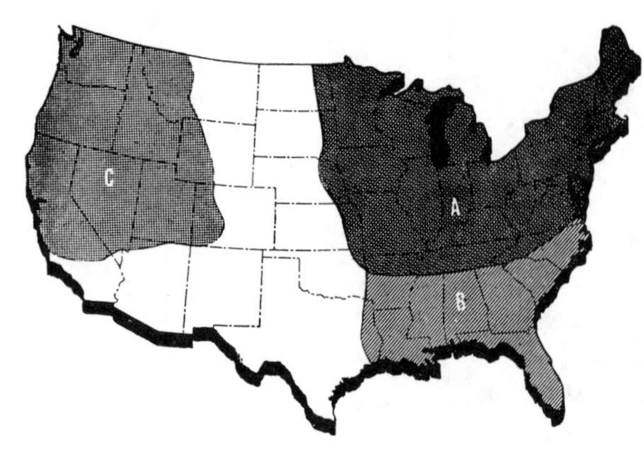

Fig. 17.1. The three major red clover-producing areas of the US are (*A*) general region, extensively grown; (*B*) used as a winter annual except at high elevations; and (*C*) grown under irrigation except at high elevations.

10.5 mm in length. Flower color usually is rose purple or magenta. The shape is somewhat similar to pea flowers but is more elongated and much smaller (Fig. 17.2). Seeds are short and mitten shaped, 2-3 mm long, varying from pure yellow to purple. The stems and leaves of American strains are generally hairy, but those of European strains are smooth. Each leaf usually is divided into three oblong leaflets; however, much variation exists, and leaves with from one to eight leaflets have been found (Fig. 17.3).

Red clover follows the typical leguminous growth pattern in germination, producing two cotyledonary leaves, followed by a unifoliolate leaf, then a maximum of four or five trifoliolate leaves on the primary shoot. However, in contrast to alfalfa (*Medicago sativa* L.), the primary shoot does not elongate but instead produces shoots at the axils of the primary leaf that exhibits internode elongation (Fig.17.4). Red clover has a taproot system with many secondary branches.

Normally the red clover taproot disintegrates in the second year, and plants that survive do so by developing secondary roots. Plants of the single-cut type form a leafy rosette growth the first year, producing no flowering stems. In the double-cut type, many plants flower in the seedling year. If these plants are mowed in late summer (September 15-20 in Kentucky and September 1 in the northern zone) and are allowed to develop strong rosettes before the onset of freezing temperatures, they will escape serious winter injury.

Fig. 17.2. *Left* (about 2×): flowers are borne on heads at tips of branches; *right* (about 10×): corolla tubes are 9.0-10.5 mm in length. *Univ. of Kentucky photo.*

Fig. 17.3. Leaves usually consist of three leaflets, but variations from one to eight occur. *Univ. of Kentucky photo.*

Fig. 17.4. The primary shoot does not elongate but instead produces shoots at the axils of the primary leaf that elongate. *Univ. of Kentucky photo*

CULTIVARS AND STRAINS

Red clover is highly self-incompatible and sets seed almost entirely by cross-fertilization. Distinct plant types have evolved through natural selection. In Europe these types are distributed largely according to latitude. Early-blooming forms predominate south of 50°N, while late, single-cut types are favored north of 60°N. Between these latitudes a full range is found, with medium late and late types predominating. Numerous European cultivars have been listed and described (Arseniuk 1989a). European workers have developed tetraploid cultivars that are reported to be more disease resistant, higher yielding, and more winter-hardy than diploid types (Jonsson 1985; Drobets and Krasnaya 1985).

In North America many regional and local strains have evolved. Both single-cut and double-cut cultivars are grown in Canada, whereas in the US the double-cut type predominates. Practically all American cultivars are of the diploid type.

Seed from foreign sources often does not perform well in the US. In Iowa six English cultivars produced only 65% as much forage as standard US cultivars. Although Canadian cultivars have performed well in the northern US, most red clover cultivars are not adapted to areas far removed from where they were developed. In addition, seed of cultivars increased outside the area of origin may be subjected to genetic shifts. Genetic shift in 'Kenland' and 'Kenstar' red clover has been greater at Shafter, California, than at Prosser, Washington; this appears to be related to higher seed yield of early plants than of late plants. The earlier types produce the most seed but are less persistent (Taylor et al. 1990). More flowering and fewer winter-hardy types in 'Dollard' red clover plants have been found from seed increased at Shafter than from seed increased at Prosser. Multiplication of seed of Finnish 'Tepa' tetraploid red clover cultivar in Israel has resulted in an increase in the number of early-flowering plants as compared with the original seed. Practices that maintain trueness to type in cultivars under seed multiplication include limitations of number of generations of certification outside the area of origin, limitation to more northern areas where genetic shifts are less, and prevention of seed production on first-year stands. Shifts have been shown to be as great with the narrowly bred Kenstar cultivar as compared with the broadly based Kenland (Taylor et al. 1990). Also recommended is strict adherence to Association of Official Seed Certifying Agencies (AOSCA) requirements for land history, isolation, control of volunteering, and seed mixtures.

In North America cultivars of red clover are produced primarily by agricultural experiment stations, the USDA and Agricultural Research Service, and private plant-breeding companies. Cultivars registered with the Crop Science Society of America published in *Crop Science* and approved for certification are listed in Table 17.1 (Jones 1992). The cultivars 'Acclaim', 'Atlas', 'Cinnamon', 'Scarlett', 'Florex', 'Hamidori', 'Hayakita', 'Kuhn', 'Marino', 'Persist', 'Red Star', 'Concorde', 'Renegade', 'Ruby Red', 'SBR-R-8603', 'Sapporo', 'Tristan', 'WD-1419', and 'Walter' have also been approved for certification in 1992 but have not been registered with the Crop Science Society of America, and little breeding information is available for them (Jones 1992).

No tetraploid cultivars have been registered by the Crop Science Society of America, but several are listed by the AOSCA as eligible for certification (Jones 1992). However, the cultivar 'Bytown' has been released in Canada, as has a Hungarian tetraploid cultivar, 'Hungaropoly.' It is possible that seed of both of these cultivars may be marketed in the US.

Single-Cut Cultivars. Mammoth red clover produces no flowering stems the first year but forms a rosette type of growth instead. During the second year this type comes into bloom 10-14 d later than the double-cut. Usually only one crop of hay may be harvested, but if moisture is favorable, some aftermath develops.

TABLE 17.1. Medium red clover cultivars registered with the Crop Science Society of America for which certified seed was produced in 1992

Cultivar	Reference	Developer	Characteristic	Adaptation
Arlington	Smith et al. 1973	USDA, Univ. of Wis.	Resistance to northern anthracnose, powdery mildew, and bean yellow mosaic virus (BYMV)	North central and northeastern US
Kenland	Hollowell 1951	USDA, Univ. of Ky.	Resistance to southern anthracnose	South central and southeastern US
Kenstar	Taylor and Anderson 1973	USDA, Univ. of Ky.	Resistance to southern anthracnose and BYMV; persistence; later than Kenland	South central US
Marathon	Smith 1994	USDA, Univ. of Wis.	Resistance to northern anthracnose; persistence	North central and northeastern US
Pennscott	Hollowell 1953	USDA, Penn. State Univ.	Naturalized from farm fields	Northeastern US
Reddy	Stratton et al. 1986	Farmers Forage Research Cooperative	Resistance to northern anthracnose and moderate resistance to southern anthracnose and powdery mildew	North central and northeastern US
Redland II	Moutray and Mansfield 1985	North American Plant Breeders, Inc.	Resistance to northern anthracnose and powdery mildew	North central and northeastern US
Redman	Buker et al. 1979	Farmers Forage Research Cooperative	Resistance to northern anthracnose and powdery mildew	Central and south central US

'Altaswede' was developed by the University of Alberta, Edmonton, Canada, from Swedish red clover. It is extremely hardy and well adapted to conditions in Alberta. 'Norlac' was developed by the Agriculture Canada Research Station, Lacombe, Alberta, by selection for resistance to northern anthracnose. Three to 6 d earlier than Altaswede and slightly shorter, it is adapted to Canada and possibly to the northern US.

Variability can be expected in noncertified seed lots. In one University of Kentucky study, noncertified lots labeled Kenland, variety unknown, and medium red clover in the second year of production yielded 6%-18% of the yield of certified Kenstar. Seed lots from locations distant to the area of use can be expected to be particularly low yielding and lacking in longevity. The only way to be sure of performance is to purchase certified seed of adapted cultivars.

Hybrid Red Clover. The possibility of developing double-cross hybrid medium red clover has been investigated in the US. The breeding procedure utilizes inbreds homozygous for S-alleles of the gametophytic self-incompatibility system. Double-cross hybrids that have been produced have not been superior to the Kenstar synthetic cultivar. Although control of crossing by the S-allele system was effective, isolation of superior combining ability among inbred clones to produce high yield and persistence may require greater funding than is usual in most red clover breeding programs.

Interspecific Hybrids. Because genetic variability for persistence and resistance to certain diseases is limited in red clover, efforts have been made to hybridize it with other species of the genus. Hybridization has been possible with two annual species, *T. diffusum* Ehrh. and *T. pallidum* Waldst. & Kit. With in vitro embryo rescue it has been possible to obtain crosses with *T. sarosiense* Hazsl. (Phillips et al. 1982), *T. medium* L. (Merker 1984), and *T. alpestre* L. (Phillips et al. 1992; Merker 1988), but most if not all F_1 plants have been sterile and much less vigorous than their parents. If the hybrid infertility barriers can be overcome through backcrossing, great potential exists to improve persistence and disease resistance of red clover.

CULTURE

Fertile, well-drained soils of high moisture-holding capacity are best for red clover. Loams, silt loams, and even fairly heavy-textured soils are preferred to light sandy or gravelly soils. The red clover taproot often is much branched, and a large part of the root system is concentrated in the top 30 cm of soil. Red clover will grow on moderately acid soils, but maximum yields are obtained only when calcium (Ca) is adequate and the pH is 6.0 or higher. Phosphorus (P) is used in large quantities by red clover. The concentration of P in the dry matter (DM) of red clover should be 0.2%-0.4% for maximum yields. An annual 10,000 kg/ha yield removes about 20-40 kg/ha that must be replaced. Potassium (K) may also be limiting in some areas. The critical content of K below which red clover suffers from K deficiency is about 1.8% of the DM. Therefore, for a yield of 10,000 kg/ha, 180 kg K/ha is necessary. Many soils can supply this much K without additions, but in any case soil tests will indicate the amount to apply.

Seeding. Red clover usually is sown with a small grain companion crop such as oats, barley, flax, or winter wheat. Small grain seeding rates are preferably reduced 50%-75% from normal. One of the primary reasons for the success of red clover is its tolerance of shading. The light compensation point for red clover seedlings is approximately 6% of daylight. Other forage legumes will not tolerate this low light intensity.

Early spring seeding is favored for red clover establishment, especially in the northern part of the Red Clover Belt, because of more favorable moisture conditions. Some clover is seeded still earlier, broadcast on winter wheat (*Triticum* L. [Poaceae]) or winter rye (*Secale* L. [Poaceae]) in February or March.

Farther south, as in Kentucky and Tennessee, seeding between February 15 and March 15 usually is preferred. Red clover may be seeded in the late summer in the southern section of the Red Clover Belt and prior to August 15 in the northern section if moisture conditions are favorable. South of Tennessee, where clover is used as a winter annual, seedings are usually made from October 15 to December 15. In the western states seedings may be made in the winter through January.

When seeded alone in the Midwest, the usual rate is 9-11 kg/ha. In the eastern states a little more seed is often used. Red clover usually is seeded in mixtures with a grass, in which case 4.5-7.0 kg/ha is usual.

Pasture Renovation. Red clover is used extensively in pasture mixtures and for renovating old pastures. It is the easiest legume to establish in closely grazed or renovated sods. Experience in the upper South has shown that it is usually best to disturb the existing sod by disking or other means of mulch tillage in the late fall or winter and to sow the clover in February. Grass should be kept clipped or grazed until the clover is well established. Rotational rather than continuous grazing will result in longer life of the stand. The inclusion of grass in clover pastures is desirable to control soil erosion, and cattle are less likely to bloat on mixtures than on clover alone. Reproductive disturbances of livestock also have been reported from grazing of predominantly red clover pastures. These are apparently caused by the estrogenic activities of the isoflavones-formononetin, biochanin A, and to a lesser extent daidzein and genistein.

MANAGEMENT

Many stands of red clover are lost by improper management in the first year. It is desirable to remove the small grain companion crop and straw before it can smother the clover. Clover should be clipped in August in the Corn Belt and probably not later than September 15 in the southern section of the Red Clover Belt. If the small grain is harvested for hay or silage, it is usually possible to harvest at least one hay crop or two rotational grazings. After renovation seeding in February two crops of hay may be harvested.

In the second year, first-crop red clover should be harvested at approximately 20% bloom (Ohlsson and Wedin 1989) for the best compromise between forage quality and yield. In one study, protein content declined from 28% to 14% and total digestible nutrients, from 88% to 65%, whereas DM yields increased from 933 to 7105 kg/ha, with advance in maturity from the vegetative to the full-bloom stages. Buxton et al. (1985) reports that red clover stems had greater digestibility than did the leaves at each stage of maturity. Red clover stems and leaves were more digestible than stems and leaves of either alfalfa or birdsfoot trefoil (*Lotus corniculatus* L.). Correspondingly, the rate of decline in digestibility with the advance of maturity was less in red clover than it was in either alfalfa or birdsfoot trefoil. At each respective growth stage, protein content in red clover was slightly lower than in other forage legumes (Buxton et al. 1985). However, degradation of red clover protein by ruminant animals was significantly less than for other forage legumes (Albrecht and Broderick 1990). Thus, more protein escaped ruminant digestion and was available for optimum rumen microbial growth and digestion by the animal in the small intestine (Broderick and Buxton 1991; Klopfenstein 1991). Rumen protein escape values for red clover ranged from 30% for fresh herbage to 46% for forced air dried herbage at the bud stage of growth.

The second crop is used for additional hay, silage, or pasture or is left for a seed crop. In many seasons it is difficult to determine the date of first bloom because of lack of flowering as a result of attack by the lesser clover leaf weevil, *Hypera nigrirostis* (F.). Usually the first harvest will be made in early May in the southern part of the Red Clover Belt and in early to mid-June in the northern part. Subsequent harvests as hay or pasture may be made at 6- to 7-wk intervals, depending upon soil moisture availability. Generally, this will be two harvests in the northern part of the Red Clover Belt and three in the southern part.

Second-cutting hay also should be harvested at 20% bloom for optimum quality and yield. In years when moisture is adequate, regrowth will occur after the second hay or seed crop. This growth may be removed by pasturing or for hay if flowering occurs early. Three hay harvests in the second year usually will result in lower yields in the third year than will two harvests, but quality is higher in the second year for the three-cutting system. A management of one hay harvest and one seed harvest usually is no more detrimental to the stand than are two hay harvests. As in the first year, it is best not to cut or graze late in the season. The growth present after a killing frost may be grazed or removed as hay. In the third year, management should be similar to that of the second year.

SEED PRODUCTION

Production of red clover seed in the humid region of the US usually is secondary to forage production. The first crop generally is harvested for hay and the second crop for seed. In the northwestern states where seed production is the primary enterprise, yields have been much higher and more dependable.

Red clover seed production depends upon pollination by insects. Bumblebees, *Bombus* spp., are especially effective in pollinating red clover but are sometimes inadequate to ensure a good seed crop (Fig. 17.5). Honeybees, *Apis mellifera* L., pollinate red clover, particularly when they are collecting pollen. The presence of more attractive nectar-producing plants in the vicinity is a factor. The availability of sweetclover (*Melilotus* Mill. [Fabaceae]), white clover (*T. repens* L.), and alsike clover (*T. hybridum* L.), plus the flowers of many weed plants growing in the vicinity of the red clover field, often results in inadequate red clover pollination.

Regardless of whether the second crop is to be used for seed or hay, early harvesting (10-15 d after first bloom) of the first crop for hay is recommended. Red clover is harvested for seed with combines when the heads have turned brown and the stems are yellow-brown. Windrow curing is a common method. Cutting may be with a mower equipped with a curler attachment or with a swather of the type used for windrowing small grain. Direct combining is used by some growers, with little attempt being made to obtain clean seed. The "rough" seed is dried to prevent heating, is rethreshed, and then is cleaned by use of a stationary combine.

DISEASES AND PESTS

Red clover is subject to a number of diseases, most of which are either controlled by breeding or left uncontrolled. Ascospores and mycelium of the crown rot fungus, *Sclerotinia trifoliorum* Eriks., attack red clover between late August and mid-October at relatively low temperatures and contiue to colonize the red clover crowns thoughout the winter and early spring. The crown rot fungus affects fall-sown more that spring-sown stands and is one of the more destructive diseases in the southern part of the Red Clover Belt. Induction of tetraploidy in red clover has been reported by researchers in Poland to increase resistance to crown rot (Arseniuk 1989b).

In the northern states, root rot may seriously deplete or even eliminate the stand after the first cutting. Organisms commonly associated with root rot are *Fusarium oxysporum* Schlect., *F. solani* (Mart.) Sacc., and *F. asetivum* Link. These are weakly pathogenic organisms that may severely injure red clover when it is in a poor state of nutrition or under stress. Rots also are associated with an internal breakdown disorder for which no causual agent has been found.

Northern anthracnose is a major disease in the cooler parts of the Red Clover Belt. It occurs from Massachusetts to Minnesota and south to Delaware and Missouri. The disease is caused by the fungus *Kabatiella caulivora* (Kirchn.) Karak. and is most destructive in wet weather at temperatures of 20°-25°C (Fig. 17.6).

Southern anthracnose caused by *Colletotrichum trifolii* Bain has been reported as far north as southern Canada, but its destructive force is felt mainly in the southern part of the Red Clover Belt. In some seasons, however, it has been problematic as far north as central Iowa. Powdery mildew caused by *Erysiphe polygoni* DC. is a disease manifested primarily on the leaflets. The fungus forms a cobweb of mycelia that gives leaflet surfaces a white, dusty appearance. Usually the disease is more prevalent during dry, cool, cloudy weather of early fall. The disease may weaken plants, but it seldom kills them (Fig. 17.7).

Virus diseases have been shown to be destructive to red clover stands. Red clover is susceptible to alfalfa mosaic virus, bean yellow mosaic virus (BYMV), clover yellow vein virus, peanut stunt virus, red clover vein mo-

Fig. 17.5. Bumblebees are especially effective in pollinating red clover. *USDA, ARS photo.*

Fig. 17.6. Northern anthracnose is most destructive in wet weather at temperatures of 20°-25°C. *USDA, ARS photo.*

saic virus (RCVMV), and white clover mosaic virus (McLaughlin and Boykin 1988), but BYMV and RCVMV were the most prevalent in Minnesota from 1957 to 1960. BYMV strain 204-1 was the most destructive virus in Kentucky. More recently, peanut stunt virus has become prevalent in Kentucky. A new strain of BYMV designated "RC" was discovered recently, but it is apparently not as serious a disease as strain 204-1 (Taylor et al. 1986). Stem-blackening diseases include spring black stem, *Phoma trifolii* Johnson & Valleau, and summer black stem, *Cercospora zebrina* Pass. Other diseases often causing serious losses include target spot, *Stemphylium sarcinaeforme* (Cav.) Wiltshire, and pepper spot (common leaf spot), *Pseudoplea trifolii* (Biv.-Bern.) Fckl.

Second-cutting red clover hay occasionally has been reported to be unpalatable or to cause slobbering by livestock. This condition is associated with black patch disease caused by *Rhizoctonia leguminicola* Gough & Elliott. An alkaloid salivation factor, "slaframine," has been isolated from the fungus.

Black patch may be avoided by timely and frequent harvests, and the alkaloid may be diluted by the use of grasses with clover and by mixing first and second cuttings during feeding. The alkaloid also tends to disappear after about 1 yr of storage. The most-promising approach toward control of the important red clover diseases would appear to be development of resistant cultivars. Notable progress has been made in developing cultivars resistant to northern and southern anthracnose, powdery mildew, and virus diseases.

Red clover is also attacked by many insects. The clover root borer, *Hylastinus obscurus* (Marsham), often kills plants by the end of

Fig. 17.7. Powdery mildew gives leaflets a dusty white appearance. *Univ. of Kentucky photo.*

the first crop year. Studies in the Midwest indicate it is a major factor in reduction of second-year stands.

Another insect that causes considerable damage is the clover root curculio, *Sitona hispidula* (F.), which is closely related to the sweetclover weevil *S. cyclindricollis* Fåhr. The clover seed chalcid, *Bruchophagus platyptera* (Walker), is one of the most-damaging insects to clover and other legume seed. Eggs are laid in the soft green seedpods when the pods are about half-grown. The larvae feed on the young developing seed. Several generations may develop in one season. Late-developing larvae overwinter in dry seed, either harvested or on plants growing along field borders.

As noted earlier, in the South the lesser clover leaf weevil usually attacks developing clover heads of the first growth and prevents flowering. The potato leafhopper, *Empoasca fabae* (Harris), often causes great damage to strains of European origin, which lack the heavy pubescence of stems and petioles characteristic of most plants in strains of American red clover.

Other insects injurious to red clover include the yellow clover aphid, *Therioaphis trifolii* (Monell); the meadow spittlebug, *Philaenus spumarius* (L.); the clover seed midge, *Dasyneura leguminicola* (Linter); the clover leafhopper, *Aceratagallia sanguinolenta* (Prov.); and the pea aphid, *Acyrthosiphon pisum* (Harris). Red clover germplasms have been released with resistance to the pea aphid and the yellow clover aphid, but no cultivars have been released. Other control measures including biological, cultural, and chemical are given by Manglitz (1985).

RELATED SPECIES

The species most closely related to red clover as shown by taxonomic treatments and interspecific hybridization are the annual species *T. diffusum* and *T. pallidum,* which have been placed in subsection I (Trifolium) of the Trifolium section along with *T. pratense* (Zohary and Heller 1984). Somewhat less closely related are the species *T. medium* (zigzag clover) and its subspecies, *T. sarosiense,* which are in subsection II Intermedia along with *T. pignatii* (Brogn. & Bory), *T. heldreichianum* (Hausskn. ex Gib. & Belli), *T. patulum* (Tausch), *T. velebiticum* (Deg.), and *T. wettstenii* (Dorfl. & Hay) (Zohary and Heller 1984). Also related as shown by interspecific hybridization is *T. alpestre* (Phillips et al. 1992; Merker 1988), which has been hybridized with *T. rubens* L. and was placed in subsection III Alpestria by Zohary and Heller (1984). With the exception of *T. medium,* none have any agronomic importance in North America.

Zigzag clover is a high-polyploid perennial species that resembles red clover. It is apparently native to central and southern Europe and western Asia but is now naturalized in North America in eastern Canada and New England. Zigzag clover possesses a strong perennial root system and spreads vigorously by rhizomes. Stems and leaves are smooth. Heads are somewhat larger than those of red clover. Florets contain one pale brown seed. Many different forms exist, but no cultivars have been developed. It is highly self- incompatible and is cross-pollinated primarily by bumblebees inasmuch as the corolla tube is too long for pollination by honeybees. Its usage has been limited by low seed yields. Seed failure has been reported to be due to genetic factors, to clover seed chalcid damage, and to preference of pollinators for other plants. Breeding to increase forage and seed yield of zigzag clover has been conducted, and several germplasms have been released (Taylor 1991). However, forage and seed yield remain too low for widespread use of zigzag clover.

QUESTIONS

1. What is the relationship between latitude and adaptation of the principal types of red clover?
2. Suggest reasons why red clover cultivars from one region are not likely to be adapted to another region.
3. How is red clover normally pollinated? Why are honeybees less effective pollinators in Illinois or Iowa than in California?
4. What are two of the newer developments in the breeding of red clover? Why were they undertaken?
5. In the management of red clover, why is it important not to harvest or graze late in the growing season?
6. What is the optimum time for harvesting the first crop of red clover for hay? What are the benefits of harvesting at this time?
7. Name several important diseases of red clover. What is the most promising method of attacking these disease problems?
8. What are the principal factors that influence seed production? Where in the US is seed production best? Why?

REFERENCES

Albrecht, KA, and GA Broderick. 1990. Degradation

of forage legume protein by rumen microorganisms. Am. Soc. Agron. Abstr., Madison, Wis., 185.

Arseniuk, E. 1989a. Effect of induced autoploidy on response to *Sclerotinia* clover rot in *Trifolium pratense* L. Plant Breed. 103:310-18.

_____. 1989b. Effect of polyploidization of red clover (*Trifolium pratense* L.) on winter hardiness and resistance to some diseases. Howowla-Roslin,-Aklimatyzacja-i-Nasiennictwo 33:3-5.

Broderick, GA, and DR Buxton. 1991. Genetic variation in alfalfa for ruminal protein degradability. Can. J. Plant Sci. 71:755-60.

Buker, RJ, SJ Baluch, and PA Sellers. 1979. Registration of Redman red clover. Crop Sci. 19:928.

Buxton, DR, JS Horstein, WF Wedin, and GC Marten. 1985. Forage quality in stratified canopies of alfalfa, birdsfoot trefoil and red clover. Crop Sci. 25:273-79.

Droblets, PT, and TS Krasnaya. 1985. The Possibility of Breeding for Increased Fodder Value in Red Clover. Selektsiya-i-Semenovodstvo,-USSR 16.

Hollowell, EA. 1951. Registration of varieties and strains of red clover, II. Agron. J. 43:242.

———. 1953. Registration of varieties and strains of red clover, III. Agron. J. 45:574.

Jones, LL. 1992. Report of acreage applied for certification in 1992 by seed certification agencies. Assoc. Off. Seed Certif. Agencies Prod. Publ. 46:139-40.

Jonsson, HA. 1985. Red clover (*Trifolium pratense*) 'Sara.' Agri-Hortique-Genetica 43:43-51.

Klopfenstein, T. 1991. Utilization of alfalfa protein by ruminent livestock. In Proc. 21st Natl. Alfalfa Symp., Rochester, Minn., 47-55.

Mclaughlin, MR, and DL Boykin. 1988. Virus diseases of seven species of forage legumes in the southeastern United States. Plant Dis. 72:539-42.

Manglitz, GR. 1985. Insects and related pests. In NL Taylor (ed.), Clover Science and Technology, Am. Soc. Agron. Monogr. 25. Madison, Wis., 269-94.

Merkenschlager, F. 1934. Migration and distribution of red clover in Europe. Herb Rev. 2:88-92.

Merker, A. 1984. Hybrids between *Trifolium medium* and *Trifolium pratense*. Hereditas 101:267-68.

———. 1988. Amphiploids betweem *Trifolium alpestre* and *Trifolium pratense*. Hereditas 108:267.

Moutray, JB, and JL Mansfield. 1985. Registration of Redland II red clover. Crop Sci. 25:708.

Ohlsson, C, and WF Wedin. 1989. Phenological staging schemes for predicting red clover quality. Crop Sci. 29:416-20.

Phillips, GC, GB Collins, and NL Taylor. 1982. Interspecific hybridization of red clover (*Trifolium pratense* L.) with T. sarosiense Hazsl. using in vitro embryo rescue. Theor. Appl. Genet. 62:17-24.

Phillips, GC, JW Grosser, S Berger, NL Taylor, and GB Collins. 1992. Interspecific hybridization between red clover and *Trifolium alpestre* using in vitro embryo rescue. Crop Sci. 32:1113-15.

Smith, RR. 1994. Registration of Marathon red clover. Crop Sci. 34:1125.

Smith, RR, DP Maxwell, EW Hanson, and WK Smith. 1973. Registration of Arlington red clover. Crop Sci. 13:771.

Stratton, SD, CW Edminster, RR Ronnenkamp, and PA Sellers. 1986. Registration of Reddy red clover. Crop Sci. 26:196.

Taylor, NL. 1991. Registration of KY-M-2 zigzag clover germplasm. Crop Sci. 31:1395-96.

Taylor, NL, and MK Anderson. 1973. Registration of Kenstar red clover. Crop Sci. 13:772.

Taylor, NL, SA Ghabrial, S Diachun, and PL Cornelius. 1986. Inheritance and backcross breeding of the hypersensitive reaction to bean yellow mosaic virus in red clover. Crop Sci. 26:68-74.

Taylor, NL, CM Rincker, CS Garrison, RR Smith, and PL Cornelius. 1990. Effect of seed multiplication regimes on genetic stability of Kenstar red clover. J. Appl. Seed Prod. 8:21-27.

Zohary, M, and D Heller. 1984. The Genus *Trifolium*. Jerusalem: Israel Academy of Sciences and Humanities.

18 White Clover and Other Perennial Clovers

GARY A. PEDERSON

Agricultural Research Service, USDA, and Mississippi State University

WHITE clover, *Trifolium repens* L., is one of the most important and widely distributed forage legumes in the world. It has a wide range of climatic adaptation, high nutritional quality, and the ability to fix atmospheric nitrogen. The Mediterranean region is the center of origin of 110 of the 237 *Trifolium* spp. including white clover (Zohary and Heller 1984).White clover may have evolved in the Mediterranean region from primitive clovers that had originated in North America and migrated through Asia to the Mediterranean. Early European settlers introduced white clover into the US. Benjamin Franklin in 1749 and others note that white clover would appear whenever land was disturbed or cleared. By 1794, Strickland reports that white clover was found throughout the US (Carrier and Bort 1916). White clover spread so rapidly that some Indian tribes called it "white man's foot grass" from the idea that it grew wherever white men walked. In fact, white clover spread rapidly through the Ohio River valley and the Midwest prior to extensive settlement by colonists.

DISTRIBUTION AND ADAPTATION

White clover will grow almost anywhere throughout the humid, temperate regions of the world. Its climatic area of adaptation ranges from the Arctic regions of Russia and Canada to the subtropical areas of Australia and South America. White clover grows in locations with annual precipitation ranging from 31 to 191 cm, mean temperature ranging from 4.3° to 21.8°C, and soil pH ranging from 4.5 to 8.2 (Duke 1981). White clover is considered to be indigenous to the entire European continent, eastern Asia, and northern Africa. Reports of white clover grown under cultivation in Europe date to as early as 1608 in the Netherlands and Belgium and 1653 in England (Fussel 1964; Zeven 1991). By animal, human, or other means, white clover spread throughout most of Asia. Early European settlers introduced white clover into New Zealand and temperate regions of Australia. White clover is found in the North African countries of Morocco and Tunisia as well as in eastern and southern Africa along rivers and in cooler mountain areas. It occurs throughout the temperate regions of Central and South America.

In North America, white clover has been grown in pastures in every state of the US and every province of Canada wherever adequate soil moisture from rainfall or irrigation is available. About half of the 45 million ha of humid or irrigated pastures in the US are estimated to contain some white clover (Duke 1981). The prostrate growth and volunteering nature of white clover make it difficult to obtain an accurate measure of its value in livestock production. The area of white clover utilization and perceived value in the US has changed over time. In the early part of the 1900s, white clover was thought to be well

GARY A. PEDERSON is Research Geneticist for the Forage Research Unit, ARS, USDA, Mississippi State, and Adjunct Associate Professor of Agronomy at Mississippi State University. He received the MS degree and the PhD from Pennsylvania State University. He has conducted research on the genetics, breeding, and pest resistance of white clover.

adapted and an important forage mainly in the Northeast and Midwest. By 1949, ladino white clover was extensively grown in the West and Southeast and was the foundation of intensive pasture systems in the Northeast. With increased interest in alfalfa hay production in the early 1950s, the use of grazed forages such as white clover decreased in the North. Presently, white clover is predominantly grown in the Southeast due to a strong reliance on grazed forages rather than spring hay production in this region. With the advent of sustainable agriculture, white clover is again being considered as a component in pastures of the Northeast and Midwest. In the western US, white clover is located mainly near sources of soil moisture in river valleys and irrigated fields (Fig 18.1).

PLANT DESCRIPTION

White clover is a tetraploid species with a somatic chromosome number of 32. Since chromosome pairing is regular with 16 bivalents and no polyvalents, white clover is an amphidiploid with disomic inheritance. Plants are glabrous with prostrate stolons spreading indeterminately along the ground and rooting from nodes. Leaves are palmately trifoliate with long petioles.

White clover seeds are very small, with approximately 1,374,000-1,764,000/kg. Seeds are round to ovate, about 1.1-1.2 mm long, and 0.9-1.0 mm wide. The seeds are smooth and dull and range from yellow to red-brown, depending on age and environment. Most white clover seeds will be hard seeded or impermeable to moisture when ripened under dry conditions. When ripened under moist conditions, most white clover seeds can immediately germinate. Hard-seeded white clover

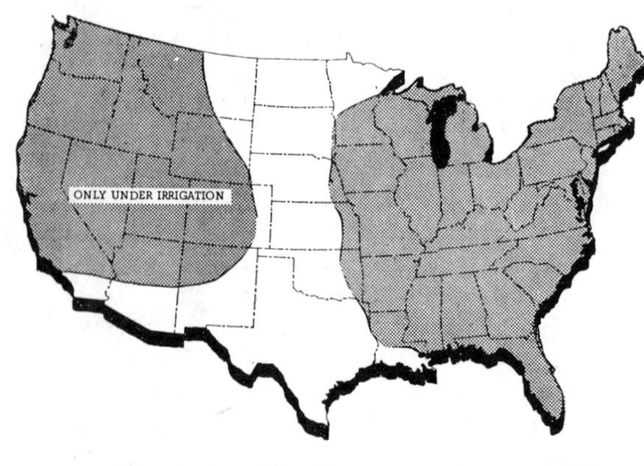

Fig. 18.1. White clover producing areas of the US.

seed may remain viable in the soil for up to 30 yr. Under subfreezing storage, germination of white clover seed may remain above 90% for at least 20 yr (Rincker 1983). Seed will generally germinate under relatively cool conditions but will remain dormant under high temperatures. Scarification or alternating temperatures will stimulate germination of dormant hard seeds.

During germination, the seed imbibes moisture, and the embryo swells greatly in volume. The growing radicle splits open the seed coat and extends into the soil. In the typical epigeal mode of germination, the hypocotyl elongates, forming a hook shape as it pulls the cotyledons above the soil surface. The first leaf formed is a unifoliolate. All subsequent leaves are trifoliolates. Successive leaves are borne on progressively longer petioles. The primary seedling root elongates and branches, forming a typical taproot system. During the first 4 wk, the seedling grows in a rosette shape with the primary stem showing little internode elongation. By 6-8 wk, up to five primary stolons begin to grow from axillary buds of the primary stem, and growth of the primary stem ceases.

Stolons are the mode of vegetative growth, spread, and survival of white clover. Each stolon is indeterminant in growth with nodes, internodes, and an apical meristem. Each node produces a trifoliate leaf, up to two adventitious roots, and either a lateral stolon or a flower, depending on environmental conditions and genotype. Under moist conditions, a stolon node will commonly produce one adventitious root on the side of the stolon closest to the soil.

The seedling taproot system generally persists for the first 1-2 yr of growth. Basal portions of primary stolons die, but stolon growth continues at the tip. Death of the taproot, primary stem, and basal portions of primary stolons usually results in a doughnut-shaped area of younger growth about the site of the original plant. Individual stolons may be genetically identical but are no longer connected. As each stolon grows forward at the tip, forming new stolon branches and spreading, the oldest portion dies. In this manner, white clover plants theoretically could live forever since new stolons can be formed indefinitely. In practice, this is improbable though Harberd (1963) observed a white clover clone estimated by amount of spread to be at least 100 yr old in Scotland. Parent clones of 'Regal' and 'Tillman' have been maintained for over 40 yr under greenhouse conditions.

Each individual stolon is totally dependent on its adventitious root system for growth and survival. Stolon adventitious root systems vary between genotypes in many root characters (Pederson 1989; Caradus 1990). White clover growth may be effectively supported by either taproot or adventitious root systems. Over 80% of white clover roots are usually located in the top 20 cm of soil though soil type, genotype, and management will affect root distribution (Caradus 1990).

As with other legumes, white clover is able to fix atmospheric nitrogen (N) for its own growth and for growth of associated grasses. White clover plants produce nodules following infection of root hairs by inoculated or indigenous strains of *Rhizobium leguminosarum* var. *trifolii* Jordan. Estimates of the amount of N fixed in a year by white clover range from 42 to 390 kg N/ha (Frame and Newbould 1986; Caradus 1990). The amount of N fixed depends predominantly on soil N levels and leaf area along with temperature, soil pH, cultivar/*Rhizobium* strain interactions, geographic location, management, shade, drought, diseases, and other stresses (Frame and Newbould 1986; Caradus 1990).

White clover leaves are formed alternately on stolon nodes at a rate of one every 3-4 d in favorable growing conditions. Leaf size may be dependent on genotype, environment, and management. White clover adapts to continual defoliation by phenotypic plasticity, which is a change in the size of the plant leaves and other organs in response to stress (Forde et al. 1989). Leaves are trifoliolate though leaves with four or more leaflets (i.e., four-leaf clover) are commonly observed. Leaves live for about 40 d from bud formation to senescence, depending on the environment. Most white clover leaves have inherited white or red leaf marks on the upper leaf surface with numerous variations of white leaf marks contained in an allelic series (Quesenberry et al. 1991).

Leaves of white clover may contain cyanogenic glucosides that are released as hydrocyanic acid (HCN) by enzymes when leaves are damaged. Cyanogenesis is controlled by two genes, Ac and Li, both of which must be present in the dominant form to produce the glucosides and enzymes. Cyanogenesis is a defense mechanism against herbivore feeding as slugs, snails, weevils, grasshoppers, locusts, and other insects prefer acyanogenic to cyanogenic white clover plants (Kakes 1990; Ellsbury et al. 1992). White clover leaves also contain an estrogen, coumestrol; however, neither HCN nor coumestrol has been shown to produce any adverse effects on animal production (Gibson and Cope 1985).

Flowers are formed from axillary buds on stolons. Each node may form either a flower or a lateral stolon but not both. Profuse flowering in white clover occurs at the cost of vegetative spread and ultimately persistence. White clover is considered to be a long-day plant though temperature, genotype, nutrition, plant age, soil moisture, and management may influence floral initiation, intensity, and duration. A minimum photoperiod of 13-16 h has been reported for white clover (Turkington and Burdon 1983; Thomas 1987). Flower heads containing 20-150 florets are borne on peduncles slightly longer than leaf petioles. Florets mature progressively from bottom to top of the flower head. Each floret is a perfect flower with a white corolla of five petals (standard petal, two wing petals, and two keel petals). The ovary contains three to seven ovules. White clover is cross-pollinated and self-incompatible though a few self-fertile plants have been found. Honeybees (*Apis mellifera* L.) or bumblebees (*Bombus* spp.) land on the wing and keel petals to obtain pollen or nectar near the base of the stamen and ovary. The weight of the bee opens the keel petals, exposing the stigma to pollen from the bee and depositing pollen from the anthers onto the bee. The keel closes following each visit. Fertilization occurs about 8 h after pollination, followed by floret reflexing within 24 h. Up to 3-4 seeds per floret mature about 28 d after fertilization.

TYPES AND CULTIVARS

Types. Type classifications are used to arbitrarily divide white clover into groups with similar characteristics. These types are all cross-compatible, with the F_1 plants being intermediate to the parents. In the US, white clover is commonly classified as one of three different types: small (*T. repens* L. f. *repens*), intermediate (*T. repens* f. *hollandicum* Erith ex Jav. & Soo), and large (*T. repens* var. *giganteum*) (Fick and Luckow 1991). Other countries use slightly different classifications of type in white clover (Szabo 1988).

Small white clover has short petioles, small leaflets, short peduncles, and small flowers. It is found throughout the world in lawns, pastures, parks, playgrounds, and areas that are mowed or grazed. Though it produces little forage yield, small white clover persists well due to a prostrate growth habit. Small white clover cultivars, such as 'Kent Wild White', have rarely been seeded in improved pastures in the US.

Intermediate white clover is intermediate in size between the small and large types for petiole and peduncle length, leaflet and flower size, and stolon diameter. Most ecotypes described as 'Common' or 'White Dutch' are intermediate types (Zeven 1991). Intermediate white clovers flower more profusely than large types. This enables intermediate white clovers to be grown as reseeding winter annuals in southeastern US pastures.

Large white clover has the largest petioles, peduncles, leaflets, flowers, and stolons of all three types and the greatest forage yields, especially under rotational grazing or clipping. Large white clover does not flower as readily as intermediate types and does not persist as well as small types. The large type was originally represented by an ecotype from irrigated pastures in northern Italy called 'Ladino Gigante Lodigiano', first grown in the US in the late 1800s (Tabor 1957). It was from this ecotype name that the term *ladino clover* was derived: first to describe the ecotype and later to describe all large-type white clovers. As presently used, the term *ladino clover* is archaic and incorrect (Williams 1987) since large-type white clovers have been bred and selected that are quite different from the original Italian ecotype.

Cultivars and Breeding. Over 230 white clover cultivars have been developed in the world during the last 60 yr (Caradus 1986). Only 27 of these cultivars have originated in the US, with few presently available. In the southeastern US, cultivars certified are the large-type 'Osceola' and Regal and the intermediate 'Louisiana S-1'. In the western and northern US, cultivars certified are the large-type 'California Ladino', 'Canopy', 'Sacramento', and 'Titan' and the small-intermediate 'Star' (AOSCA 1982-91).

Breeding principles and procedures for white clover are similar to other cross-pollinated forages. Numerous techniques have been utilized in white clover breeding programs including ecotype selection, phenotypic selection, clonal selection, progeny testing, recurrent selection, inbreeding, developing hybrids, mutation breeding, and polyploidy. Breeding programs in the US have attempted to improve white clover persistence through the use of multiple pest resistance and improved agronomic characters (Gibson and Cope 1985; Pederson et al. 1993).

Interspecific Hybrids, Tissue Culture, and Biotechnology. White clover has been successfully hybridized with two annual *Trifolium* species (*T. nigrescens* Viv. and *T. isthmocarpum* Brot.) and four perennial species (*T. uniflorum* L., *T. occidentale* Coombe, *T. ambiguum* Bieb., and *T. hybridum* L.) (Przywara et al. 1989). Potential benefits of interspecific hybridization include disease and insect resistance, increased persistence, improved root systems, cold and drought tolerance, and improved seedling vigor. No improved cultivars have yet been developed through interspecific hybridization of white clover with other species. Tetraploid and octoploid clones of *T. ambiguum* × *T. repens* have recently been released that could incorporate the virus resistance and/or rhizomatous root system of *T. ambiguum* into white clover (Williams et al. 1990; Taylor et al. 1991).

Novel ways of introducing new characters into white clover include somaclonal variation, protoplast fusion, and genetic transformation. These techniques require efficient regeneration procedures. White clover plants have been regenerated from callus cultures, protoplasts, cell suspensions, immature embryos, ovule culture, and meristem tips using various protocols (Williams 1987). In most cases, regeneration has been successful with only a few genotypes. Selections are underway to develop highly regenerative white clover populations for future studies (Oxtoby and Hughes 1989). Both root and stolon cultures of white clover have been transformed using *Agrobacterium* vectors; however, no transformed plants have been regenerated (Webb et al. 1990; White and Greenwood 1987).

CULTURE

Growth of white clover is best under cool, moist conditions in fertile soils. White clover is more suited to well-drained clay and loam soils than sandy soils, which are often low in soil moisture and fertility. Although it will grow in slightly acid soils, it yields and persists better in soils of pH 6.0 to 6.5. Aluminum (Al) toxicity may be a problem below pH 5.5 (Frame and Newbould 1986).

Performance of white clover is often dependent on ample supplies of phosphorus (P) and potassium (K). The fine, highly branched roots of grasses are more efficient at obtaining nutrients than the coarser roots of white clover. Except for initial starter fertilization in areas of poor fertility, white clover does not require N fertilizer since symbiotic N fixation provides the plant with all the N it requires.

Except when planted for seed production, white clover is usually established in association with grass. The choice of the grass is dependent on geographic location and compatibility with white clover. Grasses that have been grown with white clover in the US include tall fescue (*Festuca arundinacea* Schreb.), orchardgrass (*Dactylis glomerata* L.), Kentucky bluegrass (*Poa pratensis* L.), ryegrasses (*Lolium* spp.), centipedegrass (*Eremochloa ophiuroides* [Munro] Hack.), carpetgrass (*Axonopus affinis* Chase), dallisgrass *Paspalum dilatatum* Poir.), bahiagrass (*Paspalum notatum* Flugge), bermudagrass (*Cynodon dactylon* [L.] Pers.), smooth bromegrass (*Bromus inermis* Leyss.), and timothy (*Phleum pratense* L.) (Gibson and Cope 1985). Generally white clover is more compatible with a bunch-type grass than a sod-forming species (Frame 1990) though there are exceptions (Brink and Fairbrother 1991).

White clover may be planted into a prepared seedbed or seeded directly into an established grass sod. A prepared seedbed should be firm, smooth, and free of weeds. Lime, P, K, and minor elements should be incorporated prior to planting at rates determined by a soil test. The seed should be inoculated with the appropriate *Rhizobium* culture immediately prior to planting, or alternately, a commercial coated seed may be used. Depending on the mixture used, the rate of seeding should be 2-5 kg/ha (Frame and Newbould 1986; Ball et al. 1991). To reduce competition with white clover seedlings, a grass should be seeded at half of the recommended rate for a pure grass stand. A useful practice that establishes a good competitive balance is to drill the grass in wide-spaced rows and broadcast white clover over the entire area. White clover should be seeded in the fall in the southern US to allow maximum growth prior to the hot, dry summer. In the North, spring or summer seedings are best to allow for maximum growth prior to winter. Due to its small seed size, white clover should be seeded at the surface or no deeper than 5-10 mm. Seed may be pressed into the soil by a roller or culti-packer.

It is more difficult to establish white clover into a grass sod than a prepared seedbed. Competition from the grass sod for light and moisture should be minimized prior to seeding by mowing, grazing closely, tilling lightly, or spraying bands of a herbicide. White clover may be broadcast or seeded with a no-till drill. The seed must be placed in contact with the soil for successful establishment and soil compacted about the seed. Insect damage to emerging clover seedlings may be extensive in grass sods, so insecticides may be needed. Grass competition and weed invasion during the first few weeks of establishment may be reduced by grazing or mowing.

MANAGEMENT

The key to maintaining good stands of white clover in a mixed pasture is to manage the pasture for the clover rather than for the grass. This is an art that was commonly practiced throughout the US prior to the late 1950s. Since the 1950s, farmers have made extensive use of N fertilizer and herbicides for broadleaf weed control in improved pastures. These reduce the clover-to-grass ratio by stimulating the grass to outcompete the clover for light and moisture (N fertilizer) or by killing the clover directly (herbicides). With increased public concerns about environmental hazards such as nitrates and pesticides in farming, management of pastures for legume components is again being practiced.

The ideal proportion of white clover to maintain in a mixed pasture is 20%-40%. The pasture should be closely grazed or mowed to reduce shading of white clover by tall grass or weeds. Animal treading effects, excreta, and selective grazing may reduce clover-to-grass ratios under grazing. Alternate and controlled continuous stocking have both been successfully used on white clover-grass pastures. Under close continuous stocking, leaf size of white clover is often reduced, and more stolon branching may occur. An open canopy produced under continuous stocking is conducive to seedling growth in environments where white clover acts as a reseeding annual. Alternate stocking provides white clover with a rest period when leaves and petioles can expand to full size. Rest periods may be based on forage height or time intervals. Paddocks should be grazed to about 5 cm and then allowed to regrow to 20-25 cm. Rest periods based on time intervals may be as short as 4 wk during periods of active growth. Alternate stocking requires more management skills and greater flexibility due to dependence on climatic changes.

The nutritive and feeding value of white clover makes it a valuable component of a pasture. Nutritive value differences between white clover and erect forages are mainly due to the stoloniferous growth habit of white clover. Animals, especially cattle, will graze highly nutritious leaves, petioles, flowers, and peduncles of white clover. Prostrate stolons are rarely grazed. In upright forages, lower-quality stems are grazed, and the leaf-to-stem ratio declines with advancing maturity (Brink and Fairbrother 1992). White clover is lower than grasses in cellulose, hemicellulose, water soluble carbohydrates, lignin, sodium (Na), and manganese (Mn) and higher in total N, organic acids, calcium (Ca), magnesium (Mg), iron (Fe), copper (Cu), cobalt (Co), molybdenum (Mo), boron (B), and selenium (Se) (Frame and Newbould 1986). Digestibility of white clover remains relatively high throughout the growing season as old leaves and petioles are continually replaced. Organic matter, cell walls, and protein are rapidly degraded in the rumen, which increases voluntary intake of white clover by 20% over grasses (Thomson 1984; Frame and Newbould 1986). This rapid degradation of cell walls and protein may be related to bloat incidence in cattle. Due to the risk of bloat, cattle should not be allowed to graze pastures of 60%-100% white clover. Producers can greatly reduce the risk of bloat by providing surfactant blocks, which reduce stable foam development in the rumen (Ball et al. 1991).

Cattle and sheep production has been improved in pastures containing 20%-40% white clover compared with that in pure grass stands. White clover and clover mixtures are also being promoted and utilized as food for deer and other wildlife (Ball et al. 1991). Studies have shown improved sheep and cattle weight gains (up to 0.42 kg/d), milk production (up to 3 kg/d), milk quality (higher protein), and conception rates for white clover-grass pastures compared with rates for grass with N fertilizer (Thomson 1984; Burns and Standaert 1985; Gibson and Cope 1985). White clover-grass pastures may also be more cost-effective because, of 37 pasture treatments, white clover-tall fescue had the lowest pasture costs per amount of gain by steers (Ball and Crews 1992). Burns and Standaert (1985) estimate that white clover-tall fescue would be cheaper than N-tall fescue when N fertilizer costs $0.44/kg or more and white clover stand life is 3 yr or more.

SEED PRODUCTION

Most of the commercial white clover seed production occurs under irrigation in the western US. During 1982-91, 84% of the certified white clover seed production was in California, 14% was in Oregon, and 2% was in Washington, Idaho, Nevada, and Mississippi (AOSCA 1982-91). An average of 2800 ha/yr were in certified white clover seed production. With a seed yield of 420 kg/ha (US Department Commerce 1987), total certified white clover seed production averaged about 1.2 million kg seed/yr during 1982-91.

Floral density is the most significant component of white clover seed yield. Proper management and environmental conditions can greatly affect the final harvested seed yield. Seed production fields of white clover should be grazed or clipped until May 15 in California prior to flowering. Vegetative growth is removed, and light penetrates into the canopy, encouraging formation and active growth of more stolon tips per unit area. More stolon tips increases the number of potential sites for flowers. Once seed production begins, white clover growth should be slowed through controlled irrigation to prevent excessive vegetative growth, which can cause lodging and germination of seed in the heads. Honeybees are routinely used for pollination with 3-10 hives/ha. Usually degree of bloom and seed head maturity are the guides used to stop irrigation prior to harvest (Marble et al. 1970). Seed should be harvested when 90%-95% of the heads are brown and peduncles have started to dry.

DISEASES

White clover is susceptible to various foliar diseases, stolon and root rots, viruses, and nematodes. Diseases may reduce white clover forage yields, persistence, nodulation, N fixation, quality, and stress resistance. Stand losses may often be due to combinations of these diseases with various environmental and management constraints. Since white clover is usually grown in mixed stands under grazing, these diseases are rarely dramatic enough to be observed by producers as direct causes of poor legume performance.

Foliar Diseases. A number of fungal organisms can cause leaf diseases of white clover. Though these diseases may be severe in localized areas, they are rarely of major concern. Effects of foliar diseases can often be reduced by grazing or harvesting foliage to reduce the amount of inoculum present or by using resistant cultivars. Foliar diseases of white clover and their causal organisms are pepper spot (*Leptosphaerulina trifolii* [Rostr.] Petr.), sooty blotch (*Mycosphaerella killianii* Petr. [formerly *Cymadothea trifolii*]), rust (*Uromyces* spp.), and various leaf spots (*Cercospora zebrina* Pass., *Curvularia trifolii* [Kauff.] Boed., *Stem-*

phylium spp., *Pseudopeziza trifolii* [Biv.-Bern.] Fckl., and *Stagonospora meliloti* [Lasch] Petr.) (Latch and Skipp 1987).

Stolon and Root Rots. Fungal stolon and root rots often cause a greater reduction in white clover stands than foliar diseases. By reducing effects of these fungal pathogens, fungicide treatments have resulted in greater white clover yields and summer stolon survival (Pederson et al. 1991; Pratt and Pederson 1992). Numerous soil fungi such as *Sclerotinia trifoliorum* Eriks., *Sclerotium rolfsii* Sacc., and others can weaken or kill mature plants. Species of *Fusarium, Rhizoctonia, Curvularia, Colletotrichum, Macrophomina,* and *Leptodiscus* alone and in combination cause stolon and root rots of white clover (Gibson and Cope 1985; Latch and Skipp 1987). Stolon and root rot diseases can be minimized by avoiding highly susceptible varieties or by grazing or harvesting excess foliage to improve soil drying.

Viruses and Phyllody. White clover is susceptible to numerous viruses that have spread throughout the US (Barnett and Gibson 1975; McLaughlin and Boykin 1988). Common viruses infecting white clover include peanut stunt (PSV), alfalfa mosaic (AMV), clover yellow vein (CYVV), white clover mosaic (WCMV), red clover vein mosaic, clover yellow mosaic (CYMV), and tobacco ringspot virus (Barnett and Diachun 1985). All of these viruses are transmitted by aphid species or other insects, except for WCMV and CYMV, which are mechanically transmitted by machinery.

White clover plants infected with either PSV, CYVV, or AMV have reduced stolon growth, leaf dry weight, root dry weight, vigor, and nodulation as compared with uninfected plants (Cope et al. 1978; Gibson et al. 1981; Barnett and Diachun 1985). Infection commonly occurs with more than one virus, increasing the susceptibility of plants to damage or death due to other stresses (McLaughlin et al. 1992). Resistant cultivars are the only practical means of control. A multiple-virus-resistant white clover germplasm, SRVR, was recently released (Gibson et al. 1989).

Clover phyllody is a disease of white clover that causes flowers to be modified into leafy structures. This disease is caused by a mycoplasma-like organism that is transmitted by leafhoppers or dodder. Infected plants are often stunted and have poor vigor, nodulation, and winterhardiness. Phyllody can reduce white clover seed yields since the leafy structures replacing the flowers do not produce seed.

Nematodes. Root-knot nematodes (*Meloidogyne incognita* [Kofoid & White] Chitwood, *M. hapla* Chitwood, *M. arenaria* [Neal] Chitwood, and *M. javanica* Chitwood) are the most important nematodes affecting white clover growth and persistence in the US. White clover roots infected by root-knot nematodes may be stunted with numerous galls. Nodulation is greatly decreased, and often the root system deteriorates, causing plant death. The southern root-knot nematode, *M. incognita,* can cause yield reductions of 6% to 17% and stand losses of 12% to 62% in white clover (Pederson et al. 1991). White clover cultivars and germplasms have little resistance to root-knot nematodes (Quesenberry et al. 1986; Windham and Pederson 1991) though progress has been made in selecting white clover for tolerance and resistance to *M. incognita* (Gibson 1973; Pederson and Windham 1991). White clover is also susceptible to the stem nematode (*Ditylenchus dipsaci* [Kuhn] Filpjev.), clover cyst nematode (*Heterodera trifolii* Goffart), and root lesion nematode (*Pratylenchus* spp.) though these nematodes are not of major importance in the US.

PESTS

A number of insect species, spider mites, snails, and slugs may adversely affect white clover establishment, growth, or seed production. Pest damage to white clover may not be as severe as on other forage legumes. White clover is usually grown in mixtures with grasses rather than in pure stands and relies on stolons with adventitious roots for persistence rather than a single taproot.

Pests may damage foliage, root systems, or flowers and seed of white clover. Pests that feed on leaves or suck sap from leaves and petioles include the clover leaf weevil (*Hypera punctata* [F.]), alfalfa weevil (*H. postica* [Gyll.]), potato leafhopper (*Empoasca fabae* [Harris]), meadow spittlebug (*Philaenus spumarius* [L.]), and various species of aphids, spider mites, crickets, grasshoppers, snails, and slugs. Insects that feed on root systems include clover root curculio (*Sitona hispidula* [F.]) and clover root borer (*Hylastinus obscurus* [Marsham]). Insects that feed on flower parts or developing seed of white clover include clover head weevil (*Hypera meles* [F.]), lesser clover leaf weevil (*H. ni-*

grirostris [F.]), clover seed weevil (*Tychius picirostris* F.), ladino clover seed midge (*Dasyneura gentneri* Pritchard), and clover head caterpillar (*Grapholita interstinctana* Clemens).

Insects also transmit various diseases to white clover plants. Most viruses affecting white clover in the US are spread in a nonpersistent manner by aphid vectors. Leafhoppers spread the mycoplasma-like organism that causes clover phyllody in white clover. Clover root curculio larvae cause damage to white clover roots that exposes them to fungal root rots.

Insects can be controlled by insecticides, resistant varieties, biological control, and management. Insecticides have improved white clover yields by 52% to 85%, stolon and root weight by 70%, and stand density by 27% to 39% (James et al. 1980; Pederson et al. 1991). Yet due to economics and environmental concerns, insecticides are rarely used on pastures to control white clover insect damage. Little research has been done on developing insect-resistant varieties of white clover, and few predators or parasites have been released to control white clover insects. Management practices, such as close grazing, are often utilized for insect control.

OTHER PERENNIAL CLOVERS

Kura Clover. Kura clover (*Trifolium ambiguum* Bieb.) is a long-lived rhizomatous perennial species that is also called *caucasian* or *pellett clover*. Kura clover is a self-incompatible, cross-pollinated species with diploid (2x = 16), tetraploid, and hexaploid types. 'Rhizo', the first US cultivar, was released in 1988 following selection of hexaploid plants for persistence, rhizome growth, and disease and insect resistance (Henry and Taylor 1989). A number of kura clover germplasms (ARS-2678, KY-1, MS-2X, MS-4X, MS-6X) have also been released since 1988.

Kura clover has poor seedling vigor and grows slowly during the establishment year (Speer and Allinson 1985; Townsend 1985). Once established, kura clover is able to survive extreme environmental conditions with its deep rhizomatous root system. It tolerates water-logged and poor fertility soils, goes dormant during summer droughts, and is very winter-hardy. Kura clover is resistant to most viruses that affect white clover and has resistance to the southern root-knot nematode (Pederson and McLaughlin 1989; Pederson and Windham 1989). It is valuable as a honey crop, producing nectar with a high sugar content. The area of adaptation for kura clover is similar to that of white clover though it is more productive in the northern US, where summer moisture is available. Kura clover has good stand persistence and forage quality under frequent defoliation or under grazing by sheep (Sheaffer and Marten 1991; Peterson et al. 1992; Sheaffer et al. 1992). Bloat may occur in pastures with high concentrations of kura clover as it does with white clover or alfalfa (Sheaffer et al. 1992).

Alsike Clover. Alsike clover (*Trifolium hybridum* L.) is a short-lived perennial species utilized for forage and hay production in mixtures with red clover and grasses. Alsike is a self-incompatible, cross-pollinated, diploid species with 16 chromosomes. No cultivars have been developed in the US. Two diploid cultivars, 'Aurora' and 'Dawn', have been developed in Canada and several induced tetraploid cultivars have been developed in Europe. Alsike clover is resistant to northern and southern anthracnose but is susceptible to most other diseases and insects that attack red clover (Duke 1981; Townsend 1985).

Alsike clover prefers a slightly colder and wetter climate and is better suited to poor fertility soils than red clover. It tolerates acid and alkaline soils better than most clovers. Alsike clover has fine stems, making it susceptible to lodging during hay production. Other species with strong stems are usually grown with alsike clover to reduce lodging incidence. Alsike does especially well in high-altitude pastures of the western US. Alsike clover should be harvested for hay when in full bloom, and so it usually produces only one harvest per year.

Strawberry Clover. Strawberry clover (*Trifolim fragiferum* L.) is a stoloniferous perennial legume similar to white clover in growth habit. The initial taproot of strawberry clover survives longer than white clover's, but the stolons do not spread as extensively. The flowers and seed heads are round and mostly pink and white, resembling a strawberry. Two cultivars, 'Salina' and 'Fresa', have been released in the US. Strawberry clover is a self-incompatible, cross-pollinated, diploid species with 16 chromosomes.

Strawberry clover is widely used on wet, saline soils. It is very valuable for grazing or as a green manure corp in irrigated areas where poor drainage and saline soils limit production of other crops. It will tolerate flooding better than most clovers because it elevates stolon tips above the surface of the water (Townsend 1985). Strawberry clover is quite susceptible to root-knot nematodes (Quesenberry et al. 1986). Strawberry clover is utilized in the western US in pastures under close continuous grazing and in lawns as a ground cover.

QUESTIONS

1. Why is white clover valuable as the legume component of pastures?
2. Why is white clover more widely utilized in the southern than in the northern US?
3. Compare the primary stem and taproot system of white clover and alfalfa.
4. Why does the quality of white clover remain relatively high while the quality of many forages declines with advancing maturity?
5. Give the management practices that change the botanical composition of a white clover-grass pasture, causing a shift toward (a) more white clover and (b) more grass.
6. How can bloat be controlled when grazing white clover pastures?
7. Why are diseases and insects rarely noted as primary causes of white clover stand loss?

REFERENCES

Association of Official Seed Certifying Agencies (AOSCA). 1982-91. Report of Acres Applied for Certification by Seed Certification Agencies. Prod. Publ. 36-45.

Ball, DM, and JR Crews. 1992. Comparison of Selected Alabama Forage Crops as Pasture for Stocker Steers. Ala. Coop. Ext. Serv. Circ. ANR-764.

Ball, DM, CS Hoveland, and GD Lacefield. 1991. Southern Forages. Atlanta, Ga: Potash and Phosphate Institute.

Barnett, OW, and S Diachun. 1985. Virus diseases of clovers. In Clover Science and Technology, Am. Soc. Agron. Monogr. 25, Madison, Wis., 235-68.

Barnett, OW, and PB Gibson. 1975. Identification and prevalence of white clover viruses and the resistance of *Trifolium* species to these viruses. Crop Sci. 15:32-37.

Brink, GE, and TE Fairbrother. 1991. Yield and quality of subterranean and white clover-bermudagrass and tall fescue associations. J. Prod. Agric. 4:500-504.

______. 1992. Forage quality and morphological components of diverse clovers during primary spring growth. Crop Sci. 32:1043-48.

Burns, JC, and JE Standaert. 1985. Productivity and economics of legume-based vs. nitrogen-fertilized grass-based pasture in the United States. In Forage Legumes for Energy-efficient Animal Production. Washington, D.C.: USDA and ARS.

Caradus, JR. 1986. World checklist of white clover varieties. N.Z. J. Exp. Agric. 14:119-64.

______. 1990. The structure and function of white clover root systems. Adv. Agron. 43:1-46.

Carrier, L, and KS Bort. 1916. The history of Kentucky bluegrass and white clover in the United States. J. Am. Soc. Agron. 8:256-66.

Cope, WA, SK Walker, and LT Lucas. 1978. Evaluation of selected white clover clones for resistance to viruses in the field. Plant Dis. Rep. 62:267-70.

Duke, JA. 1981. Handbook of Legumes of World Economic Importance. New York: Plenum.

Ellsbury, MM, GA Pederson, and TE Fairbrother. 1992. Resistance to foliar-feeding hyperine weevils (Coleoptera: Curculionidae) in cyanogenic white clover. J. Econ. Entomol. 85:2467-72.

Fick, GW, and MA Luckow. 1991. What we need to know about scientific names: An example with white clover. J. Agron. Educ. 20:141-47.

Forde, MB, MJM Hay, and JL Brock. 1989. Development and growth characteristics of temperate perennial legumes. In Persistence of Forage Legumes. Madison, Wis.: American Society of Agronomy, 91-109.

Frame, J. 1990. Herbage productivity of a range of grass species in association with white clover. Grass and Forage Sci. 45:57-64.

Frame, J, and P Newbould. 1986. Agronomy of white clover. Adv. Agron. 40:1-88.

Fussel, GE. 1964. The grasses and grassland cultivation of Britain. I. Before 1700. J. Br. Grassl. Soc. 19:49-54.

Gibson, PB. 1973. Registration of SC-1 white clover germplasm. Crop Sci. 13:131.

Gibson, PB, and WA Cope. 1985. White clover. In Clover Science and Technology, Am. Soc. Agron. Monogr. 25, Madison, Wis., 471-90.

Gibson, PB, OW Barnett, HD Skipper, and MR McLaughlin. 1981. Effects of three viruses on growth of white clover. Plant Dis. 65:50-51.

Gibson, PB, OW Barnett, GA Pederson, MR McLaughlin, WE Knight, JD Miller, WA Cope, and SA Tolin. 1989. Registration of southern regional virus resistant white clover germplasm. Crop Sci. 29:241-42.

Harberd, DJ. 1963. Observations on natural clones of *Trifolium repens* L. New Phytol. 62:198-204.

Henry, DS, and NL Taylor. 1989. Registration of 'Rhizo' kura clover. Crop Sci. 29:1572.

James, JR, LT Lucas, DS Chamblee, and WV Campbell. 1980. Influence of fungicide and insecticide applications on persistence of ladino clover. Agron. J. 72:781-84.

Kakes, P. 1990. Properties and functions of the cyanogenic system in higher plants. Euphytica 48:25-43.

Latch, GCM, and RA Skipp. 1987. Diseases. In White Clover. Wallingford, UK: CAB International, 421-60.

McLaughlin, MR, and DL Boykin. 1988. Virus diseases of seven species of forage legumes in the southeastern United States. Plant Dis. 72:539-42.

McLaughlin, MR, GA Pederson, RR Evans, and RL Ivy. 1992. Virus diseases and stand decline in a white clover pasture. Plant Dis. 76:158-62.

Marble, VL, LG Jones, JR Goss, RB Jeter, VE Burton, and DH Hall. 1970. Ladino Clover Seed Production in California. Calif. Agric. Exp. Stn. Circ. 554.

Oxtoby, E, and MA Hughes. 1989. Selection of rapidly regenerating genotypes of white clover for use in *Agrobacterium* transformation. In Vitro 25:61A.

Pederson, GA. 1989. Taproot and adventitious root growth of white clover as influenced by nitrogen nutrition. Crop Sci. 29:764-68.

Pederson, GA, and MR McLaughlin. 1989. Resistance to viruses in *Trifolium* interspecific hybrids related to white clover. Plant Dis. 73:997-99.

Pederson, GA, and GL Windham. 1989. Resistance to *Meloidogyne incognita* in *Trifolium* interspecific hybrids and species related to white clover. Plant Dis. 73:567-69.

______. 1991. Breeding strategies for developing *Meloidogyne incognita* resistance in white clover. In Symposium on Plant Breeding in the 1990s, N.C. State Univ. Crop Sci. Dept. Res. Rep. 130, 39.

Pederson, GA, GL Windham, MM Ellsbury, MR McLaughlin, RG Pratt, and GE Brink. 1991. White clover yield and persistence as influenced by cypermethrin, benomyl, and root-knot nematode. Crop Sci. 31:1297-1302.

Pederson, GA, GL Windham, MR McLaughlin, RG Pratt, and GE Brink. 1993. Breeding for multiple pest resistance as a strategy to improve white clover persistence. In Proc. 17th Int. Grassl. Congr., Palmerston North, New Zealand, and Australia, 926-27.

Peterson, PR, CC Sheaffer, and RM Jordan. 1992. Kura clover yield and quality under sheep grazing and clipping. In Proc. Am. Forage and Grassl. Counc. 1:185-89.

Pratt, RG, and GA Pederson. 1992. Enhanced survival of white clover stolons over summer by applications of benomyl. Phytopathol. 82:1159-60.

Przywara, L, DWR White, PM Sanders, and D Maher. 1989. Interspecific hybridization of *Trifolium repens* with *T. hybridum* using in-ovulo embryo and embryo culture. Annu. Bot. 64:613-24.

Quesenberry, KH, DD Baltensperger, and RA Dunn. 1986. Screening *Trifolium* spp. for response to *Meloidogyne* spp. Crop Sci. 26:61-64.

Quesenberry, KH, RR Smith, NL Taylor, DD Baltensperger, and WA Parrott. 1991. Genetic nomenclature in clovers and special-purpose legumes: I. Red and white clover. Crop Sci. 31:861-67.

Rincker, CM. 1983. Germination of forage crop seeds after 20 years of subfreezing storage. Crop Sci. 23:229-31.

Sheaffer, CC, and GC Marten. 1991. Kura clover forage yield, forage quality, and stand dynamics. Can. J. Plant Sci. 71:1169-72.

Sheaffer, CC, GC Marten, RM Jordan, and EA Ristau. 1992. Forage potential of kura clover and birdsfoot trefoil when grazed by sheep. Agron. J. 84:176-80.

Speer, GS, and DW Allinson. 1985. Kura clover (*Trifolium ambiguum*): Legume for forage and soil conservation. Econ. Bot. 39:165-76.

Szabo, AT. 1988. The white clover (*Trifolium repens* L.) gene pool. I. Taxonomical review and proposals. Acta Bot. Hung. 34:225-41.

Tabor, P. 1957. Early references to giant white clover. Agron. J. 49:520-21.

Taylor, NL, JA Anderson, EG Williams, and WM Williams. 1991. Registration of octoploid hybrid clover germplasm from the cross of *Trifolium ambiguum* × *T. repens*. Crop Sci. 31:1395.

Thomas, RG. 1987. Reproductive development. In White Clover. Wallingford, UK: CAB International, 63-123.

Thomson, DJ. 1984. The nutritive value of white clover. In Forage Legumes, Proc. Br. Grassl. Soc. Occup. Symp. 16:78-92.

Townsend, CE. 1985. Miscellaneous perennial clovers. In Clover Science and Technology, Am. Soc. Agron. Monogr. 25, Madison, Wis., 563-78.

Turkington, R, and JJ Burdon. 1983. The biology of Canadian weeds. 57. *Trifolium repens* L. Can. J. Plant Sci. 63:243-66.

US Department of Commerce. 1987. Census of Agriculture. Vol. 1, Geogr. Area Ser. Washington, D.C.: Bureau of the Census.

Webb, KJ, S Jones, MP Robbins, and FR Minchin. 1990. Characterization of transgenic root cultures of *Trifolium repens, Trifolium pratense* and *Lotus corniculatus* and transgenic plants of *Lotus corniculatus*. Plant Sci. 70:243-54.

White, DWR, and D Greenwood. 1987. Transformation of the forage legume *Trifolium repens* L. using binary *Agrobacterium* vectors. Plant Mol. Biol. 8:461-69.

Williams, EG, NL Taylor, J van den Bosch, and WM Williams. 1990. Registration of tetraploid hybrid clover germplasm from the cross of *Trifolium ambiguum* × *T. repens*. Crop Sci. 30:427.

Williams, WM. 1987. Genetics and breeding. White clover taxonomy and biosystematics. In White Clover. Wallingford, UK: CAB International, 323-419.

Windham, GL, and GA Pederson. 1991. Reaction of *Trifolium repens* cultivars and germplasms to *Meloidogyne incognita*. J. Nematol. 23:593-97.

Zeven, AC. 1991. Four hundred years of cultivation of dutch white clover landraces. Euphytica 54:93-99.

Zohary, M, and D Heller. 1984. The Genus *Trifolium*. Jerusalem: Israel Academy of Sciences and Humanities

19 Birdsfoot Trefoil

PAUL R. BEUSELINCK
Agricultural Research Service, USDA, and University of Missouri

WILLIAM F. GRANT
McGill University

BIRDSFOOT trefoil, *Lotus corniculatus* L., is a cross-pollinated perennial legume used for pasture or for hay and silage production. It does not cause bloat and can be managed to reseed so that stands may be maintained. Two distinct types, Empire and European, are grown in the US and Canada. Empire types are related to the cultivar 'Empire', a naturalized ecotype discovered in Albany County, New York. European-type cultivars such as 'Cascade', 'Viking', and 'Leo' were developed from European introductions. Empire-type cultivars generally are finer stemmed, more prostrate in growth habit, 10-14 d later in flowering, more indeterminate in growth and flowering habit, more winter-hardy, and slower in seedling growth and recovery growth rate after harvest than are European types.

Species of *Lotus* are widely distributed throughout the world, but birdsfoot trefoil is native to Europe, North Africa, and parts of Asia. The greatest genetic diversity of *L. corniculatus* occurs in the Mediterranean basin, considered the regional center of origin. Nineteenth-century reports show it grew naturally in many pastures of Europe and was a good feed for cattle and horses. However, it was not until after 1900 that birdsfoot trefoil was cultivated in Europe (MacDonald 1946). It is not certain when or how birdsfoot trefoil was introduced into the US. The crop first received recognition in 1934 when a naturalized stand was found growing in New York (Johnstone-Wallace 1938).

DISTRIBUTION AND ADAPTATION

Birdsfoot trefoil is grown in the British Isles; throughout much of Europe; in several South American countries (Argentina, Brazil, Chile, and Uruguay); and in areas of India, Australia, and New Zealand. In the US it first became naturalized in New York, Washington, Oregon, and California. Major hectarage now occurs in the area shown in Figure 19.1. In Canada considerable birdsfoot trefoil is grown, primarily in the eastern provinces. Birdsfoot trefoil does not persist well in most parts of the southeastern US where reseeding cannot offset serious stand losses caused by crown and root diseases. The southern limit of adaptation includes higher elevations in North Carolina, Alabama, Georgia, and Arkansas.

Birdsfoot trefoil grows on many different types of soils, from clays to sandy loams. It will grow on droughty, infertile, acid, or mildly alkaline soils, on mine spoils, and under saline and waterlogged conditions. Birdsfoot trefoil is more resistant than alfalfa (*Medicago sativa* L.) to wet or waterlogged soil conditions because it is not susceptible to *Phytophthora megasperma* Drechs. (Chi and Sabo

PAUL R. BEUSELINCK is Research Geneticist, Agricultural Research Service, US Department of Agriculture, and Associate Professor, University of Missouri. He received his PhD from Oregon State University. His research focuses on the improved persistence of birdsfoot trefoil. His studies have included disease resistance, selection for rhizomes, reproduction, and the collection and evaluation of new *Lotus* germplasm.

WILLIAM F. GRANT is Professor of Genetics, Department of Plant Science, Macdonald Campus of McGill University. He received his PhD degree from the University of Virginia. His research focuses on the cytogenetic relationships of birdsfoot trefoil and closely related diploid species and the development of indehiscence. His studies have included anther culture, production of amphidiploids, karyotype analyses, herbicide tolerance, and biochemical tehniques.

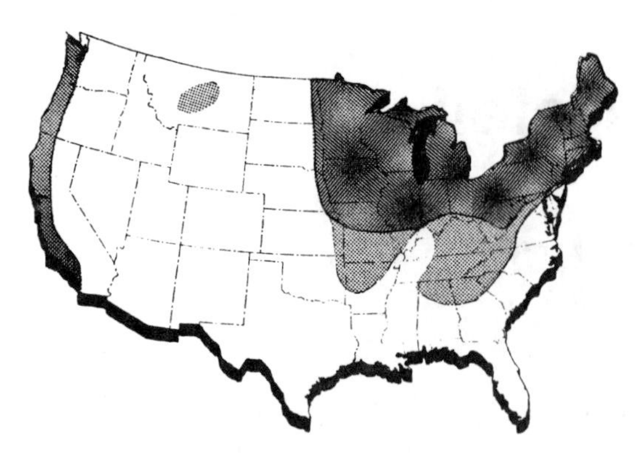

Fig. 19.1. Areas of the US in which birdsfoot trefoil is adapted.

1978) and tolerates low oxygen (O_2) levels by effectively removing the ethanol produced from its roots (Barta 1980). Although it tolerates adverse soil conditions, it is most productive on fertile, moderately to well-drained soils having a pH of 6.2-6.5 or higher. Most North American and European cultivars (European and Empire types) are winter-hardy in the northern US and southern Canada, but winter-killing of European-type cultivars from South America makes them unadapted north of 45°N latitude. Severe winter conditions, or winters with variable temperatures, can be detrimental to all birdsfoot trefoil cultivars. Winter snow cover is critical to birdsfoot trefoil survival above 40°N latitude.

Fig. 19.2. Birdsfoot trefoil with many well-branched stems. *N.Y. Agric. Exp. Stn. photo.*

PLANT DESCRIPTION

Mature plants of birdsfoot trefoil have many well-branched stems arising from a single crown. Under favorable growing conditions the main stems attain a length of 60-90 cm. Stems generally are smaller in diameter and less rigid than those of alfalfa (Fig. 19.2). Plants can be propagated by stem cuttings because roots develop from callous tissue of stem internodes and both roots and shoots develop from axillary buds of the node. Compound leaves, attached alternately on opposite sides of the stem, are pentafoliolate, with three leaflets attached to the terminal end of the petiole and two smaller leaflets attached at the base. During darkness the leaflets close around the petiole and stem. When cut and dried for hay, the leaves fold around the stem and become inconspicuous, often giving the impression of excessive leaf loss.

Birdsfoot trefoil has a well-developed taproot with numerous lateral branches in the upper 30-60 cm of soil. The taproot does not penetrate as deeply as alfalfa, and the distribution of branch roots in the upper soil is more extensive. Roots have the ability to produce new shoots, and segments taken below the crown will develop shoots and roots and may aid plant survival. Plants generally live for about 2-4 yr, making reseeding or production of new shoots from plant pieces necessary for persistence. Birdsfoot trefoil that persists by producing rhizomes has been found, and plant breeders have been attempting to transfer this trait into commercial cultivars (Pedersen et al. 1986; Beuselinck 1992).

Birdsfoot trefoil is a long-day plant, with most North American cultivars requiring a 16- to 18-h day for full flowering. The inflorescence is a typical umbel having four to eight florets attached at the end of a relatively long peduncle (Fig. 19.3). Flower color varies from light to dark yellow and may be tinged with orange or red stripes. A shorter daylength restricts flowering and results in plants having a more prostrate, rosette growth habit (Seaney and Henson 1970). Flower buds may abort even under optimum photoperiod. Factors causing bud abortion or flower drop include adverse temperature, low nutrient level, insects, and diseases.

Fertilization and seed set depend on pollination of flowers by insects, primarily species of *Hymenoptera* or *Megachile*. Although flowers are perfect, having both stamens and pistils, most seed results from cross-pollination. Even though self-pollination occurs, a self-incompatibility mechanism limits the development of self-seed (Dobrofsky and Grant 1980). Ten to 15 seeds are borne in long cylindrical pods, which turn brown to almost black at maturity. The pods readily dehisce when mature. Three to five pods, 2.5-4.0 cm long, attached at right angles to the end of the peduncle give the appearance of a bird's foot, thus the common name *birdsfoot trefoil*. Seeds are small, 600,000-925,000/kg, and vary from olive-green to brown and almost black. Seed coats often are mottled with black spots that vary in size from small dots to large patches.

Nutritive Value. Birdsfoot trefoil has nutritive value equal to or greater than that of alfalfa (Langille and Calder 1971; Marten and Jordan 1979) and has high feeding value as pasture, hay, or silage (Seaney and Henson 1970). Birdsfoot trefoil does not cause bloat, perhaps because it contains tannin, which precipitates the soluble proteins and renders them incapable of producing stable foams in the rumen. Also, tannins are hypothesized to aid rumen bypass of highly digestible proteins for later digestion and absorption in the lower gut. The cell walls of birdsfoot trefoil rupture more slowly than those of alfalfa or clovers (*Trifolium* spp.); this may allow slower release of bloat-causing substances (Lees et al. 1981; Howarth et al. 1982).

Fig. 19.3. Young florets of birdsfoot trefoil, seedpods, and mature dehisced pods showing spiral twisting.

Most birdsfoot trefoil plants contain a cyanogenic glucoside, which can be hydrolyzed by enzymatic action to produce free hydrocyanic acid (HCN). About 4% of the plants of the cultivar Viking are acyanogenic (Seaney and Henson 1970). Concentration of HCN in some species of *Lotus* is poisonous to animals (Jones et al. 1978), but there have been no authenticated reports of *L. corniculatus* causing HCN poisoning when grazed. Some have speculated that presence of HCN may contribute to low insect injury to birdsfoot trefoil.

Importance and Use. The hectarage of birdsfoot trefoil in the US has steadily increased from about 260,000 in 1957 to the present estimate of close to 1 million ha. Over 200,000 ha are grown in Canada.

Because of wide adaptation, persistence, and relatively high forage yield, birdsfoot trefoil has replaced much of the red (*Trifolium pratense* L.) and white (*T. repens* L.) clovers previously grown with grasses in the Northeast. In this area it also is used on the poorly drained, shallow soils previously planted to timothy (*Phleum pratense* L.). In the Midwest it has gained importance as a persistent and productive permanent pasture legume, usually in combination with various grasses (Fig. 19.4). The range of adaptation of birdsfoot trefoil has expanded into the southeastern US with the development of cultivars selected from Mediterranean germplasm (Hoveland et al. 1990). In addition to forage production, its ability to reseed, perennial life-cycle, and prodigious attractive flowers make birdsfoot trefoil popular for plantings on highway slopes and medians for beautification, erosion control, and soil improvement.

Although pure stands of birdsfoot trefoil provide the best quality forage, companion grasses contribute to higher forage yields by utilizing the nitrogen (N) fixed by the birds-

Fig. 19.4. Birdsfoot trefoil, an excellent pasture legume. *N.Y. Agric. Exp. Stn. photo.*

foot trefoil and filling in vacant areas that frequently develop in stands and allow weed invasion (Marten and Jordan 1979; Sheaffer et al. 1984). Declining plant density in pure stands can be attributed to competition, winter-killing, and disease (Beuselinck et al. 1984). Grasses in mixture with birdsfoot trefoil reduce lodging, decrease severity of frost heaving in the southern Corn Belt (not usually a problem farther north where soils remain frozen), and improve the rate of hay drying and curing.

Birdsfoot trefoil flowers later and is more compatible with native grasses, like big bluestem (*Andropogon gerardi* Vitman), than are legumes like red clover and alfalfa, which are more active in growth earlier in spring and smother the warm-season species. Kentucky bluegrass (*Poa pratensis* L.) and red fescue (*Festuca rubra* L.) are less competitive with birdsfoot trefoil than are orchardgrass (*Dactylis glomerata* L.), tall fescue (*F. arundinaceae* Schreb.), smooth bromegrass (*Bromus inermis* Leyss.), or perennial ryegrass (*Lolium perenne* L.). Less competitive or low-yielding grasses in the mixture initially will permit the birdsfoot trefoil to dominate. However, competitive or high-yielding grasses can be used in mixtures if management favoring regrowth and reseeding of birdsfoot trefoil is practiced (Sheaffer et al. 1984; Beuselinck et al. 1992).

Carbohydrate root reserves of birdsfoot trefoil decrease during spring growth and, due to the warm summer temperatures, remain relatively low until the end of the growing season when they are replenished (Nelson and Smith 1969). Regrowth after grazing, or cutting for hay, is dependent on carbohydrates provided by photosynthesis or stored in roots. Since carbohydrate root reserves are low during the growing season, birdsfoot trefoil can be grazed frequently but should not be completely defoliated. Cutting low reduces the number of axillary buds that produce regrowth and removes leaf area needed to produce carbohydrates (Greub and Wedin 1971).

Pasture. Prostrate and intermediate Empire-type cultivars persist better than erect European-type cultivars in pastures that are closely grazed, but erect cultivars can be used in pasture (Seaney and Henson 1970; Hoveland et al. 1990). As already discussed, more leaves remain on prostrate types after grazing, ensuring vigorous regrowth, and this condition also favors the production of flowers to aid reseeding.

Persistence of birdsfoot trefoil plants and yield of animal product/ha will be greater if rotational, rather than continuous, stocking is practiced (Seaney and Henson 1970). Rotational stocking promotes greater photosynthetic area after grazing than does continous stocking, so there is a reduced demand during regrowth for carbohydrate stored in the roots. Because root reserves are lower under warm growth conditions than under cool ones, rotational stocking is needed most during the summer months (Nelson and Smith 1969). Rotational stocking also favors stand persistence through reseeding, since flowers on low-growing stems have a greater opportunity to set and mature seed.

Stockpiling, the practice of accumulating forage in a pasture during the spring and early summer until it is needed for grazing, can be used with birdsfoot trefoil to provide late summer forage in some locations because it retains its high-quality leaves on mature growth better than does alfalfa. Success with stockpiling birdsfoot trefoil is favored by dry, cool environments (Marten and Jordan 1979). Hot, humid climates impair stockpiling because *Rhizoctonia* foliar blights damage leaves and stems, lowering forage productivity and quality (Beuselinck et al. 1984).

Hay. European-type cultivars with erect growth have been developed for hay production, but prostrate Empire-type cultivars persist better when clipped frequently (Nelson and Smith 1969). Birdsfoot trefoil will provide two to three hay crops, depending on length of the growing season (Seaney and Henson 1970), but yields usually are 50%-80% of those of alfalfa. Quality of hay is equal to or better than alfalfa hay. In northern areas, hay harvests after early September may retard the storage of root carbohydrates required for winter survival and spring growth. Susceptibility to lodging, slow establishment, and high seed costs are factors against using birdsfoot trefoil as a hay crop in short rotations.

CULTIVARS

At least 25 cultivars have been developed in the US and Canada. European and Empire types are distinct (Fig. 19.5). Some cultivars in use are listed in Table 19.1 with a brief description of their characteristics.

Empire, 'Norcen', 'Carroll', 'Dawn', and Leo are grown extensively in the north central area. In the upper South, Dawn, 'Fergus', and 'AU-Dewey' have performed well. In eastern Canada, Empire, Leo, and Viking account for most of the hectarage.

Considerable variability exists within birdsfoot trefoil germplasm for further selection of new cultivars. Various breeding methods are available, or in development, for improving certain traits. These methods include interspecific hybridization, tissue culture, protoplast fusion, and genetic transformation technology. *Lotus corniculatus* is a tetraploid species with 24 chromosomes. Interspecific hybridization with closely related diploid *Lotus* species offers a means of improving certain traits (Williams 1987). The most successful method has been to produce interspecific diploid hybrids, double the chromosome number with colchicine, and cross the resulting amphidiploids directly to *L. corniculatus* (Somaroo and Grant 1972). Tissue culture techniques, including protoplast fusion and somaclonal variants, are being used to develop

Fig. 19.5. 'Empire' (*left*) illustrates the more decumbent growth habit of the Empire-type cultivars; 'Maitland' (*right*) exhibits the more erect growth habit of the European-type cultivars.

TABLE 19.1. Birdsfoot trefoil and big trefoil cultivars

Cultivar	Characteristics
AU-Dewey	Prostrate growth habit, good vigor and adaptation; excellent stand persistence; high yield potential in southeast US; Alabama selection.
Bull	Leo characteristics; higher yielding, high survival under soil compaction; eastern Canadian selection; Ontario-developed from synthetics by University of Guelph, Canada.
Carroll	A winter-hardy pasture type; more upright and maturing slightly earlier than Empire; Iowa-developed.
Cascade	Early maturing, but not as winter-hardy as Viking; developed from French selections.
Cree	Seed yield superior to Leo; winterhardiness comparable to Leo and Empire; western Canadian, selected at Saskatoon.
Dawn	Semierect Empire type; selected for increased resistance to root rots and leaf and stem diseases; wide adaptation within the north central US; Missouri-selected.
Empire	Semierect, flowers 10-14 d later than European types; growth and flowering indeterminate; seedling, spring, and recovery growth rates slower than European types; moderately winter-hardy; selected for persistence; the first birdsfoot trefoil cultivar, derived from a selected ecotype found in Albany, NY
Fergus	Naturalized ecotype from Kentucky, semiprostrate strain, good seedling vigor, disease tolerant; adapted to southeast US.
Grasslands Maku	Big trefoil (*L. uliginosus*); tetraploid (2n = 4× = 24); developed in New Zealand.
Kaiser	Big trefoil (*L. uliginosus*); suited as a pasture legume in mixed pasture on the fragipansoils of the lower Ohio River valley; developed at Purdue.
Kalo	Rarely exceeds 0.5 m in height; Oregon-selected from Dwarf English variety *L. corniculatus* var. *arvensis* Pers. for grazing, ground cover, and soil erosion.
Leo	Early spring vigor, winter-hardy, and good seed yield; developed from USSR selection (Morshansk) at Macdonald Campus, McGill University.
Maitland	Seedling vigor and forage yield greater than Viking; developed at University of Guelph, Canada, from an 11-clone synthetic.
Marshfield	Big trefoil (*L. uliginosus*); survives frequent flooding in winter and is tolerant of brackish overflows.
Norcen	Broad-leaved, intermediate type; diverse genetic background provides resiliency to different environments within north central US.
Viking	Erect, early maturing, 15%-20% greater forage yield than Empire in New York; persists where Empire types may be lost due to competition from other crops in a mixture; derived from Danish and New York ecotypes.
Upstart	Leo characteristics; higher seedling vigor; eastern Canadian selection; Ontario-developed from synthetics by University of Guelph, Canada.

new genotypes (Aziz et al. 1990; Damiani et al. 1990; Niizeki and Saito 1989; Yang et al. 1990) and herbicide-resistant cultivars (Boerboom et al. 1991; MacLean and Grant 1987; Pofelis et al. 1992).

CULTURE AND MANAGEMENT

Inoculation. Strains of *Rhizobium* and *Bradyrhizobium* bacteria specific for birdsfoot trefoil are required for effective nodulation (Seaney and Henson 1970). Unlike alfalfa nodules, those on birdsfoot trefoil senesce after each forage harvest, and a new population of nodules must be formed as plants regrow (Vance et al. 1982). Birdsfoot trefoil has a greater nodule mass than does alfalfa, but a lower rate of N fixation and lower total seasonal N fixation (Vance et al. 1982). The level of N fixed by pure stands of birdsfoot trefoil varies with cultural conditions but averages about 90 kg/ha annually (Heichel et al. 1985).

Seeding. Seedling vigor is positively associated with seed size or weight (Carleton and Cooper 1972; Beuselinck and McGraw 1983). High-quality seed of commercial cultivars is the best ensurance of vigorous seedlings. Since seed size is approximately 60% that of alfalfa and 75% that of red clover, the growth rate of birdsfoot trefoil seedlings usually is slower, but cultivars may vary. For example, Empire has less seedling vigor than does Viking, which in turn has less vigor than 'Maitland'.

Seed coats of mature birdsfoot trefoil seed often are impermeable to water and are termed *hard seed*. Seed lots can be as high as 90% hard seed, but most seed is sold at or below 40% hard seed. Hard seed typically gets scarified during the harvest and processing of seed. Scarification of hard seed encourages quick germination.

Availability of ample soil moisture is the key to successful seedings. Depending upon location of plantings, either spring or late July to early August seedings can be successful (Fribourg and Strand 1973; Marten and Jordan 1979). Autumn seedings may not succeed in northern areas because climatic conditions do not allow sufficient time for plants to develop winterhardiness (Laskey and Wakefield 1978), but in areas with moderate winters, autumn seeding is common. If irrigation is available, an autumn planting is usually favored because temperatures are typically cooler and less stressful on birdsfoot trefoil seedlings during establishment. Broadcast, band, and no-tillage seedings are used successfully to establish birdsfoot trefoil in new plantings or to interseed into established grass pasture.

Recommended seeding rates range from 5 to 12 kg/ha. In new plantings, seedbeds should be smooth and firm and seed not deeply buried because seed is small and seedlings grow slowly. Maximum desired planting depth is usually less than 1.3 cm. Packing of the soil before and after seeding improves contact of the seed with soil moisture, which promotes better seedling establishment. Since birdsfoot trefoil seedlings are small and stands are slow to establish, patience is a vital but inexpensive input.

Birdsfoot trefoil can be successfully seeded with a spring small grain companion crop to reduce erosion on slopes and suppress weed growth during the establishment year. A reduced seeding rate of the small grain and early removal of the grain companion crop for forage reduces competition for light and increases the probability of successful birdsfoot trefoil establishment. Spring-seeded winter grains as a companion crop sometimes do not allow successful establishment of birdsfoot trefoil. Companion crops are too competitive for summer seedings. Herbicides for weed control are an alternative to a companion crop (Seaney and Henson 1970).

Perennial grass pastures can be renovated by interseeding birdsfoot trefoil into the grass sod. Animal production usually is increased, and the economic return is greater than that from grasses fertilized with N (Langille and Calder 1971). The grass should be heavily stocked or mowed short through late summer and autumn prior to seeding that autumn or subsequent spring. Various types of planters or sod drills are useful in establishing birdsfoot trefoil in grass sods. These machines essentially prepare a small partially tilled strip in the sod and place seed and fertilizer into the soil. Herbicides are often applied before interseeding to suppress the grass sod. Good stands have also been obtained on high-pH soils by broadcasting a mixture of seed and fertilizer on pastures that cannot be tilled (Winch et al. 1969). Pasture yields have been increased four- to fivefold, and the stocking season extended 2 to 5 mo. This method of establishment often is combined with light stocking later in the season to neutralize competition from spring growth of grasses and to control broad-leaved weeds.

Weed Control. Weeds can be controlled during establishment by clipping or mowing. However, success depends upon the particular weed problem and the relative growth stage of weeds and birdsfoot trefoil. Under some conditions mowing broad-leaved weed species allows weed grasses to grow and become more competitive than the original broad-leaved weeds. Herbicides can be used for control of weed competition in birdsfoot trefoil. Local recommendations should be reviewed each year to determine available herbicides and to ensure their proper use.

The development of cultivars resistant to herbicides may simplify weed control in birdsfoot trefoil. Recurrent selection was used to develop 'T-68', a birdsfoot trefoil germplasm that is up to five times more tolerant to 2,4-D (2,4-dichlorophenoxyacetic acid) than Viking (Devine et al. 1975). Recurrent selection has also been used to develop glyphosate (N-[phosphonomethyl]-glycine) tolerant birdsfoot trefoil (Boerboom et al. 1991). In vitro selection for herbicide tolerance appears practical, and progress has been made for resistance to sulfonylurea herbicides (Pofelis et al. 1992; MacLean and Grant 1987; Vessabutr 1992). Cultivars with resistance to herbicides have not, as yet, been developed for commercial use. It will be necessary to develop this technology in concert with chemical companies, who will need to obtain clearances for use of herbicides on birdsfoot trefoil.

Soil and Soil Fertility. Birdsfoot trefoil has greater tolerance than alfalfa to acid, infertile, or poorly drained soils. Even though it will grow and persist on such soils, forage yields are significantly increased by proper applications of lime and fertilizer. At a pH lower than 6.2-6.5, seedling growth and establishment may be slow, and nodulation and N fixation may be retarded or completely inhibited. The soil pH required for maximum forage production is only slightly lower than for alfalfa.

Adequate amounts of phosphorus (P) promote early, vigorous seedling growth, thereby increasing the probability of successful establishment. Dry matter production of birdsfoot trefoil seedlings generally increases with increased availability of P, but not potassium (K) (Russelle et al. 1989). Birdsfoot trefoil responds to applications of P and K on deficient soils, but applications to fertile soils do not appear to benefit forage and seed yield or winterhardiness (Russelle et al. 1991).

Birdsfoot trefoil is recognized for its ability to tolerate problem soils, including saline conditions. Birdsfoot trefoil shows high tolerance to salt and moderate tolerance to pH 4.5 (Ayers 1948; Schachtman and Kelman 1991). The cultivar AU-Dewey is more tolerant of soil acidity than other cultivars, with good root and forage growth to pH 4.8 (Hoveland et al. 1990). Poorly drained soils or soils subject to high water tables are candidates for plantings of birdsfoot trefoil.

SEED PRODUCTION

Pollination and Growth. As mentioned previously, birdsfoot trefoil is a qualitative, long-day plant requiring at least 16 h of daylight for full flowering; a shorter daylength restricts flower production. Birdsfoot trefoil flowers indeterminately and sets seed over an extended period. These characteristics of the crop necessitate special climatic requirements for commercial production of seed. Long days at latitudes above 40° favor a contracted period of intense blooming, which is advantageous for seed production (Beuselinck and McGraw 1988). The majority of birdsfoot trefoil seed is produced commercially in Minnesota, Wisconsin, and Michigan, but Oregon, New York, and Canada also produce significant quantities. Although seed yields can approach 600 kg/ha, average yields range from about 50 to 175 kg/ha and typically are 100 kg/ha or less (McGraw et al. 1986).

To ensure maximum pollination and seed set, native pollinating insects are supplemented by hives of domesticated honeybees (*Apis mellifera* L.). Populations of 11 bees/m^2 are needed to pollinate all flowers (Seaney and Henson 1970). Multiple visitations by bees to flowers increase pod and seed set, so a high population of pollinators is recommended. Insects force their way toward the base of flowers, extruding the pistil through the keel tip, which facilitates the transfer of pollen to the stigma. Ovaries contain an average of 45 ovules, but the number of ovaries that develop into mature seed varies from few to many. Fertilization occurs within 24-48 h after pollination.

Pods develop rapidly after pollination and reach maximum size within 3 wk. As pods mature, they change from green to light brown then become almost black. Seeds become physiologically mature slightly before the pods turn light brown. Pod and seed maturity occur 25-50 d after pollination, being accelerated by warm temperatures (Seaney and Henson 1970). As the long, narrow seedpod matures, the loss of moisture from the tissue results in increased tension between fiber layers, separation at the sutures, and a twisting and splitting of the two valves of the pod (Fig. 19.3.) Pod dehiscence and subsequent seed loss due to shattering can be high when relative humidity drops below 40%.

Seeding rates to establish fields for seed production can be lower than required for forage. Rates of 0.9 to 1.8 kg/ha for wide row spacings and 3.6 to 8.8 kg/ha for broadcast or narrow rows were found to be most favorable for seed production (Pankiw et al. 1977; McGraw et al. 1986). Narrower row spacings are used in Canada, probably because of a contracted growing season at higher latitudes. Most seedings are established without irrigation, so planting when sufficient moisture is available will allow rapid germination and adequate plant growth to successfully overwinter.

Clipping seed fields during spring to reduce vegetative growth can delay flowering and seed harvest and decrease seed yields (McGraw and Beuselinck 1983). Early clipping to delay flowering and seed harvest is not generally recommended or used by seed growers. However, in Oregon seed production fields, spring growth is earlier than in other commercial locations, and a hay harvest in May is required because flowering is reduced in stands with dense growth.

Harvest. Under optimum conditions, in Minnesota, Michigan, Wisconsin, Oregon, and New York seed yields can exceed 600 kg/ha. However, indeterminate flowering, limited distribution of the products of photosynthesis to reproductive growth, flower and pod abortion, and pod dehiscence lead to low seed yield (McGraw and Beuselinck 1983).

Pods within a field mature unevenly. Any delay in cutting after pods are completely mature results in significant seed loss from pod dehiscence and seed shattering. Frequent observation and critical timing of harvest are necessary to obtain maximum seed yields. Recommendations for exact timing of harvest vary from when pods are light green to when they are brown.

The most common method of seed harvest is to mow, windrow, then combine directly from the partially dried windrow. The combine should be fitted with a special attachment to the mouth of the combine that gently lifts the windrow to avoid seed shattering. Direct combining has been used to some extent, but the passage of large quantities of green forage through the combine often clogs the cleaning mechanism, slowing harvest and increasing seed loss.

DISEASES

Crown and root rots are the most important diseases of birdsfoot trefoil. The presence of crown and root rots has been reported in most regions where birdsfoot trefoil is grown, although no single organism has been identified as the primary cause (Seaney and Henson 1970). Losses of 68% to 88% of stands by the end of the second year have been reported by Henson (1962), who found over 80% of surviving plants badly diseased by root rot; 90% of stands were lost in 2 yr in Missouri regardless of management (Beuselinck et al. 1984). Severe loss from these diseases is usually associated with warm weather and high humidity, and thus the diseases are of greater importance in the southern than in the northeastern or north central US.

Crown and root rots are not caused by a single organism but are the result of a parasite-saprophyte complex that may vary in different environments. Organisms that have been found in this complex are species of *Fusarium, Verticillium, Macrophomina, Mycoleptodiscus, Rhizoctonia,* and *Sclerotinia* (Ostazeski 1967). The predominance of the pathogens involved with the complex and the gross appearance of the symptoms change with the geographic environment in which the birdsfoot trefoil is grown (Ostazeski 1967). Plants severely infected with root rot have extensive decay in the central portion of the upper taproot and crown and often fail to regrow after harvest. Entire fields may be decimated after harvest when environmental conditions are optimal for disease development.

Differences in susceptibility and tolerance to root rots have allowed selection and breeding to increase resistance (Henson 1962), but the interaction of several pathogens with the growing environment complicates genetic improvement. Compared with the cultivars Viking and Empire, Dawn shows significant tolerance to crown and root rot in Missouri.

Several disease organisms attack stems and leaves. *Sclerotinia trifoliorum* Eriks. causes a rot of lower stems and crown, usually under heavy snow cover in late winter or early spring (Barr and Callen 1963). Under warm, humid conditions, *Sclerotium rolfsii* Sacc. and *Rhizoctonia solani* Kuehn attack the crown and lower foliage, causing leaf blight, which may lead to death of affected plants. *Rhizoctonia solani* is partially controlled by harvesting or mowing foliage when plants first show symptoms of infection, as these managements eliminate the humid conditions that this fungus requires to spread within the canopy. The most widespread foliar disease is *Stemphylium loti* Graham. This organism causes reddish brown stem and leaf lesions and results in premature leaf drop or death of stems. Immature seedpods may also be attacked, resulting in shriveling and discoloration of seed (Graham 1953). Foliar and stem blights caused by *Phomopsis loti* and *Diaporthe phaseolorum* are less common but can be severe in some locations. Occasional epidemics of rust (*Uromyces striatus* var. loti) can cause considerable damage to leaves and stems, but plant death is uncommon (Zeiders 1985).

Birdsfoot trefoil is suseptible to parasitism by root-knot (*Meloidogyne hapla, M. javanica, M. arenaria,* and *M. incognita*) and root-lesion (*Pratylenchus penetrans* Cobb) nematodes. Synergism between *Fusarium* spp. and *R. solani* with root-lesion nematodes lowers birdsfoot trefoil productivity and increases plant mortality (Thompson and Willis 1975).

PESTS

A number of insects cause losses of forage and seed (Seaney and Henson 1970; Mackun and Baker 1990). The meadow spittlebug, *Philaenus spumarius* L., feeds by sucking plant sap and causes plant stunting and abortion of flower buds. This insect produces a characteristic white foamy mass on stems and leaves. Three plant bugs, *Adelphocoris lineolaris* Goeze, *Lygus lineolaris* Palisot de Beavois, and *Plagiognathus chrysanthemi* Wolff, and the potato leafhopper, *Empoasca fabae* Harris, also cause injury (Wipfli et al. 1990). The plant bugs destroy stem terminals and flowers. The potato leafhopper causes a characteristic yellowing and reddening of leaves and general stunting; heavy infestations can reduce forage yield and quality.

The trefoil seed chalcid, *Bruchophagus platypterus* Walker, is a small, black, wasp-like, host-specific insect that can greatly reduce seed yields by parasitizing seed. Fertilized females are attracted to seedpods by the presence of volatile compounds that elicit landing and egg-laying behavior (Kamm 1989). Eggs are laid in developing seedpods, and the larvae feed inside the maturing ovule, leaving only a hollow, inviable seed. The trefoil seed chalcid differs from other seed chalcids in that it will not seek its host unless it sees yellow-colored flowers (Kamm 1992). It is estimated that commercial seed yields may be reduced as much as 40% or more by this pest. Pesticides do not effectively control the seed chalcid, so growers must rely on cultural practices to reduce seed losses. Large amounts of seed left on the ground after harvest may increase infestations the following year. Chalcid populations can be reduced by avoiding delayed, late-season seed harvest and by burning or burying combine trash after harvest to destroy infested seed.

OTHER SPECIES

In addition to *L. corniculatus*, two other species of *Lotus* are important as pasture legumes in the US. Narrowleaf trefoil, *L. tenuis* Waldst. & Kit. ex Willd., has narrow linear-lanceolate leaflets on slender weak stems with comparatively long internodes. The flowers are slightly smaller and fewer in number than *L. corniculatus* and usually change from yellow to orange-red with age. Plants are less erect than Empire-type cultivars, have very shallow root systems, are adapted to wet soil, and are tolerant of saline soils (Ayers 1948). *Lotus tenuis* is a diploid species with a somatic chromosome number of 12. It is grown in New York, California, Oregon, and Washington. Naturalized ecotypes are produced, but currently there are no named commercial cultivars. Narrowleaf trefoil has been grown on wet soils where *L. corniculatus* is adapted, but it is not as winter-hardy as *L. corniculatus* and has less seedling vigor (Beuselinck and McGraw 1983). Like birdsfoot trefoil, narrowleaf trefoil has a long-day flowering response. It requires the same specific inoculum as *L. corniculatus* for N fixation. The nutrient value of narrowleaf trefoil is similar to that of birdsfoot trefoil. Because of its close resemblance with *L. corniculatus,* some investigators suggest that *L. tenuis* may be a progenitor species (Dawson 1941), although other diploid species have been implicated (Raelson and Grant 1988).

Big trefoil, *L. uliginosus* Schkuhr. (sometimes called *L. pedunculatus* Cav.), is a nonbloating perennial legume with fine stems, relatively large leaves, and vigorous rhizomes. It is a diploid species, like *L. tenuis,* with a somatic chromosome number of 12, but tetraploids of *L. uliginosus* are found. Big trefoil is a long-lived legume in grass pastures in which the plants become more vigorous and increase their stand via rhizomes (Kaiser and Heath 1988). Some cultivars are listed in Table 19.1 with a brief description of their characteristics. Big trefoil shows good adaptation on wet, poorly drained soils and upland areas with high rainfall and is naturalized in the Pacific Northwest (northern California, Oregon, and Washington). Big trefoil is adapted to wet sites where *L. corniculatus* is adapted, but it is not as winter-hardy as *L. corniculatus*. Big trefoil has smaller seed and less seedling vigor than *L. tenuis* (Beuselinck and McGraw 1983). Big trefoil is another long-day species requiring a 16-h photoperiod for complete floral induction. Unlike *L. tenuis,* it requires a specific inoculum that is different from *L. corniculatus* for N fixation. The nutrient value of big trefoil compares favorably with that of birdsfoot trefoil, but it has a higher level of tannin, which can affect palatability.

There are native *Lotus* species that are of value for conservation purposes. An induced tetraploid of the native Spanish clover (*L. purshianus* [Benth.] Clem. & Clem.) may have value for California rangelands as dryland pasture for wildlife (MacLean and Grant 1986). This species can colonize copper-enriched soil by the development of copper tolerance in both the legume plant and its symbiont (Wu and Lin 1990). The species is a soil N enricher and can enhance wildlife habitat in forestry operations (Schoeneberger et al. 1989). Selections of big deervetch (*L. crassifolius* [Benth.] Greene) have been made by the Soil Conservation Service for use in conservation uses, erosion control in crop- and pastureland, and reforestation projects in the Pacific Northwest (Corning 1987).

QUESTIONS

1. What are the desirable and undesirable characteristics of birdsfoot trefoil?
2. What different soil conditions will birdsfoot trefoil tolerate that other legumes will not?
3. How would you manage birdsfoot trefoil to ensure a persistent pasture?
4. What factors determine the successful establishment of birdsfoot trefoil?
5. Describe ways to use birdsfoot trefoil to improve unproductive pastures.
6. List characteristics of birdsfoot trefoil that make seed production difficult.
7. The seedpod of birdsfoot trefoil dehisces when mature. Is this a favorable or unfavorable trait?
8. Give reasons why birdsfoot trefoil does not cause bloat in cattle.

REFERENCES

Ayers, AD. 1948. Salt tolerance of birdsfoot trefoil. J. Am. Soc. Agron. 40:331-34.

Aziz, MA, PK Chand, JB Power, and MR Davey. 1990. Somatic hybrids between the forage legumes *L. corniculatus* and *L. tenuis* Waldst. et Kit. J. Exp. Bot. 41:471-79.

Barr, DJS, and EO Callen. 1963. *Sclerotinia trifoliorum* Erikss., a destructive pathogen of birdsfoot trefoil (*Lotus corniculatus* L.). Phytoprotection 44:18-24.

Barta, AL. 1980. Regrowth and alcohol dehydrogenase activity in waterlogged alfalfa and birdsfoot trefoil. Agron. J. 72:1017-20.

Beuselinck, PR. 1992. Rhizomes of Birdsfoot Trefoil: A Progress Report. 12th *Trifolium* Conf., 25-27 March, Gainesville, Fla.

Beuselinck, PR, and RL McGraw. 1983. Seedling vigor of three *Lotus* species. Crop Sci. 23:390-91.

———. 1988. Indeterminate flowering and reproductive successs in birdsfoot trefoil. Crop Sci. 28:842-44.

Beuselinck, PR, EJ Peters, and RL McGraw. 1984. Cultivar and management effects on stand persistence of birdsfoot trefoil. Agron. J. 76:490-92.

Beuselinck, PR, DA Sleper, SS Bughrara, and CA Roberts. 1992. Effects of harvest frequency on mono and mixed cultures of tall fescue and birdsfoot trefoil. Agron. J. 84:133-37.

Boerboom, CM, NJ Ehlke, DL Wyse, and DA Somers. 1991. Recurrent selection for glyphosate tolerance in birdsfoot trefoil. Crop Sci. 31:1124-29.

Carleton, AE, and CS Cooper. 1972. Seed size effects upon seedling vigor of three forage legumes. Crop Sci. 12:183-86.

Chi, CC, and FE Sabo. 1978. Chemotaxis of zoospores of *Phytopththora megasperma* to primary roots of alfalfa seedlings. Can. J. Bot. 56:795-800.

Corning, MA. 1987. Special purpose legumes for the Pacific Northwest. Prog. Rep. Clovers Spec. Legume Res. 20:38-39.

Damiani, F, M Pezzotti, and S Arcioni. 1990. Somaclonal variation in *L. corniculatus* L. in relation to plant breeding purposes. Euphytica 46:35-41.

Dawson, CDR. 1941. Tetrasomic inheritance in *L. corniculatus* L. J. Genet. 42:49-72.

Devine, TE, RR Seaney, DL Linscott, RD Hagin, and N Brace. 1975. Results of breeding for tolerance to 2,4-D in birdsfoot trefoil. Crop Sci. 15:721-24.

Dobrofsky, S, and WF Grant. 1980. An investigation into the mechanism for reduced seed yield in *Lotus corniculatus*. Theor. Appl. Genet. 57:157-60.

Fribourg, HA, and RH Strand. 1973. Influence of seeding dates and methods on establishment of small-seeded legumes. Agron. J. 65:804-7.

Graham, JH. 1953. A disease of birdsfoot trefoil caused by a new species of *Stemphylium*. Phytopathol. 43:577-79.

Greub, LJ, and WF Wedin. 1971. Leaf area, dry-matter production, and carbohydrate reserve levels of birdsfoot trefoil as influenced by cutting height. Crop Sci. 11:734-38.

Heichel, GH, CP Vance, DK Barnes, and KI Henjum. 1985. Dinitrogen fixation, and N and dry matter distribution during 4 year stands of birdsfoot trefoil and red clover. Crop Sci. 25:101-5.

Henson, PR. 1962. Breeding for resistance to crown and root rots in birdsfoot trefoil. Crop Sci. 2:429-32.

Hoveland, CS, MW Alison, Jr., NS Hill, RS Lowrey, Jr., SL Fales, RG Durham, JW Dobson, Jr., EE Worley, PC Worley, VH Calvert II, and JF Newsome. 1990. Birdsfoot Trefoil Research in Georgia. Ga. Agric. Exp. Stn. Res. Bull. 396.

Howarth, RE, BP Goplen, SA Brandt, and KJ Cheng. 1982. Disruption of leaf tissues by rumen microorganisms: An approach to breeding bloat-safe forage legumes. Crop Sci. 22:564-68.

Johnstone-Wallace, DB. 1938. Pasture Improvement and Management. N.Y. State Coll. Agric. Ext. Bull. 393.

Jones, D, RJ Keymer, and WM Ellis. 1978. Cyanogenesis in plants and animal feeding. In JB Harborne (ed.), Biochemical Aspects of Plant and Animal Coevolution. New York: Academic Press, 27-32.

Kaiser, CJ, and ME Heath. 1988. Big trefoil, a new legume for pastures on fragipan soils. In J Janick and JE Simon (eds.), Advances in New Crops, Proc. 1st Natl. Symp. New Crops: Res., Dev., and Econ. Portland, Oreg.: Timber Press, 191-94.

Kamm, JA. 1989. In-flight assessment of host and nonhost odors by alfalfa seed chalcid (Hymenoptera: Eurytomidae). Environ. Entomol. 18:56-60.

———. 1992. Influence of celestial light on visual and olfactory behavior of seed chalcids (Hymenoptera: Eurytomidae). J. Insect Behav. 5:273-87.

Langille, JE, and FW Calder. 1971. Effect of harvesting practices on foliage and root development, digestibility, cold hardiness and nodulation of birdsfoot trefoil. Can. J. Plant Sci. 51:499-504.

Laskey, BC, and RC Wakefield. 1978. Competitive effects of several grass species and weeds on the establishment of birdsfoot trefoil. Agron. J. 70:146-48.

Lees, GL, RE Howarth, BP Goplen, and AC Fesser. 1981. Mechanical disruption of leaf tissues and cells in some bloat-causing and bloat-safe legumes. Crop Sci. 21:444-48.

MacDonald, HA. 1946. Birdsfoot Trefoil (*Lotus corniculatus* L.): Its Characteristics and Potentialities as a Forage Legume. Cornell Univ. Agric. Exp. Stn. Memo 261.

McGraw, RL, and PR Beuselinck. 1983. Growth and yield characteristics of birdsfoot trefoil. Agron. J. 75:443-46.

McGraw, RL, PR Beuselinck, and KT Ingram. 1986. Plant population density effects on seed yield of birdsfoot trefoil. Agron. J. 78:201-5.

Mackun, IR, and BS Baker. 1990. Insect populations and feeding damage among birdsfoot trefoil-grass mixtures under different cutting schedules. J. Econ. Entomol. 83:260-67.

MacLean, NL, and WF Grant. 1986. The induction of tetraploid *Lotus purshianus* as a potential range plant in California. Field Crops Res. 14:193-96.

———. 1987. Evaluation of birdsfoot trefoil (*Lotus corniculatus*) regenerated plants following in vitro selection for herbicide tolerance. Can. J. Bot. 65:1275-80.

Marten, GC, and RM Jordan. 1979. Substitution value of birdsfoot trefoil for alfalfa-grass in pasture systems. Agron. J. 71:55-59.

Nelson, CJ, and D Smith. 1969. Growth of birdsfoot trefoil and alfalfa. IV. Carbohydrate reserve levels and growth analysis under two temperature regimes. Crop Sci. 9:589-91.

Niizeki, M, and K Saito. 1989. Callus formation from protoplast fusion between leguminous species of *Medicago sativa* and *Lotus corniculatus*. Jap. J. Breed. 39:373-77.

Ostazeski, SA. 1967. An undescribed fungus associated with a root and crown rot of birdsfoot trefoil (*Lotus corniculatus*). Mycologia 59:970-75.

Pankiw, P, SG Bonin, and JAC Lieverse. 1977. Effects of row spacing and seeding rates on seed yield in red clover, alsike clover and birdsfoot trefoil. Can. J. Plant Sci. 57:413-18.

Pedersen, JF, RL Haaland, and CS Hoveland. 1986. Registration of 'AU-Dewey' birdsfoot trefoil. Crop Sci. 26:1081.

Pofelis, S, H Le, and WF Grant. 1992. The development of sulfonylurea herbicide resistant birdsfoot trefoil (*L. corniculatus*) plants from in vitro selection. Theor. Appl. Genet. 83:480-88.

Raelson, JV, and WF Grant. 1988. An isoenzyme study in the genus *Lotus* (Fabaceae). Evaluation of hypotheses concerning the origin of *L. corniculatus* using isoenzyme data. Theor. Appl. Genet. 76:267-76.

Russelle, MP, RL McGraw, and RH Leep. 1991. Birdsfoot trefoil response to phosphorus and potassium. J. Prod. Agric. 4:114-20.

Russelle, MP, LL Meyers, and RL McGraw. 1989. Birdsfoot trefoil seedling response to soil phosphorus and potassium availability indexes. Soil Sci. 53:828-36.

Schachtman, DP, and WM Kelman. 1991. Potential of *Lotus* germplasm for the development of salt, aluminium and manganese tolerant pasture plants. Aust. J. Agric. Res. 42:139-49.

Schoeneberger, MM, RJ Volk, and CB Davey 1989. Factors influencing early performance of leguminous plants in forest soils. Soil Sci. Soc. Am. J. 53:1429-34.

Seaney, RR, and PR Henson. 1970. Birdsfoot trefoil. Adv. Agron. 22:119-57.

Sheaffer, CC, GC Marten, and DL Rabas. 1984. Influence of grass species on composition, yield, and quality of birdsfoot trefoil mixtures. Agron. J. 76:627-32.

Somaroo, BH, and WF Grant. 1972. Crossing relationships between synthetic *Lotus* amphidiploids and *L. corniculatus*. Crop Sci. 12:103-5.

Thompson, LS, and CB Willis. 1975. Influence of fensulfothion and fenamiphos on root lesion nematode numbers and yield of forage legumes. Can. J. Plant Sci. 55:727-35.

Vance, CP, LEB Johnson, S Stade, and RG Groat. 1982. Birdsfoot trefoil (*L. corniculatus*) root nodules: Morphogenesis and the effect of forage harvest on structure and function. Can. J. Bot. 60:505-18.

Vessabutr, S. 1992. Transfer of chlorsulfuron resistance from tobacco to birdsfoot trefoil (*Lotus corniculatus*) by asymetric somatic hybridization. PhD thesis, McGill Univ., Montreal.

Williams, EG. 1987. Interspecific hybridization in pasture legumes. Plant Breed. Rev. 5:237-305.

Winch, JE, EM Watkin, GW Anderson, and TL Collins. 1969. The use of mixtures of granular dalapon, birdsfoot trefoil seed and fertilizer for roughland pasture renovation. J. Br. Grassl. Soc. 24:302-7.

Wipfli, MS, JL Wedberg, and DB Hogg. 1990. Damage potentials of three plant bug (Hemiptera: Heteroptera: Miridae) species to birdsfoot trefoil grown for seed in Wisconsin, USA. J. Econ. Entomol. 83:580-84.

Wu, L, and SL Lin. 1990. Copper tolerance and copper uptake of *L. purshianus* (Benth.) Clem. & Clem. and its symbiotic rhizobium *Loti* derived from a copper mine waste population. New Phytol. 116:531-39.

Yang, JB, DA Somers, RL Wright, and RL McGraw. 1990. Seed pod dehiscence in birdsfoot trefoil, *Lotus conimbricensis,* and their interspecific somatic hybrid. Can. J. Plant Sci. 70:279-84.

Zeiders, KE. 1985. First report of rust caused by *Uromyces* species on birdsfoot trefoil in the United States. Plant Dis. 69:727.

20 Arrowleaf, Crimson, and Other Annual Clovers

CARL S. HOVELAND
University of Georgia

GERALD W. EVERS
Texas A&M University

ANNUAL clovers are used where perennial clovers do not persist. They are used extensively in the southeastern US for overseeding warm-season perennial grasses and with winter annual grasses to reduce costs and improve animal performance. The benefits are high-quality forage (Brink and Fairbrother 1992), extended grazing season (Hoveland et al. 1978), reduction or elimination of nitrogen (N) fertilizer (Knight 1970), and spring weed control (Evers 1983). On the West Coast, where summers are dry, annual clovers are grown with cool-season annual grasses (Murphy et al. 1973). Public concern about nitrate pollution of groundwater has revived the use of annual clovers in summer crop rotations (Hargrove and Frye 1987). Minor uses include roadside beautification and stabilization, a winter cover crop in vegetable production, and attraction of beneficial insects in orchards.

There are some management practices common to all clover species discussed in this chapter. Soil pH should be between 5.5 and 6.2 for adequate uptake of plant nutrients by most annual clovers. Boron (B) may be required on sandy soils to enhance root growth and seed production. Fertilizer rates should be based on a soil test. When perennial grass sods are overseeded, better clover stands occur if the grass competition is reduced by grazing or close mowing, by application of a chemical desiccant, or by light disking.

Arrowleaf Clover

Arrowleaf clover, *Trifolium vesiculosum* Savi, is a winter annual legume native to the Mediterranean region, Balkan Peninsula, and areas west and north of the Black Sea. It was introduced into the US in 1956 (Miller and Wells 1985). Desirable characteristics are late maturity with high forage yield and quality, high percentage of hard seed, good reseeding ability, and low incidence of bloat. Susceptibility to viruses is the greatest limitation to arrowleaf clover.

Arrowleaf clover is grown successfully from eastern Texas and Oklahoma to South Carolina and southward to the Gulf of Mexico (Fig. 20.1). Arrowleaf clover will thrive on well-drained sandy and clay soils but is less tolerant of acidity and low fertility than other clovers. It does not tolerate alkaline soils or

CARL S. HOVELAND is Terrell Distinguished Professor of Crop and Soil Science at the University of Georgia. A native of Wisconsin, he earned the MS degree from the University of Wisconsin and the PhD from the University of Florida. He has conducted forage management and utilization research in Texas, Florida, Alabama, Georgia, and New Zealand.

GERALD W. EVERS is Professor of Forage Physiology, Texas A&M University, and Texas Agricultural Experiment Station, Overton. He earned the MS and PhD degrees from Texas A&M University. His special interests are forage establishment, overseeding cool-season annuals on warm-season perennial grasses, and management of legume-grass pastures.

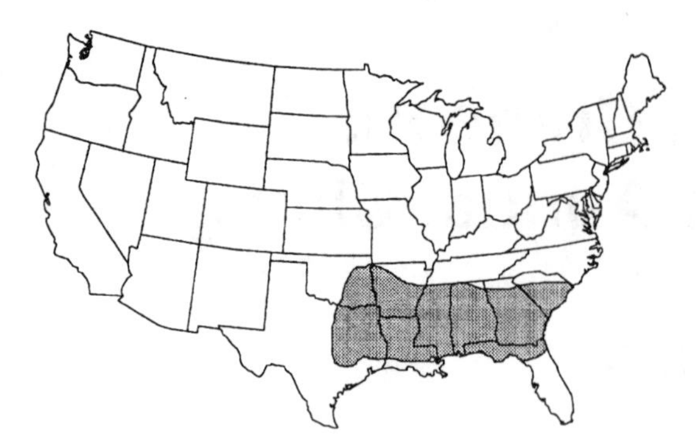

Fig. 20.1. Primary arrowleaf clover-growing area in the US.

those with poor drainage (Hoveland and Webster 1965).

PLANT DESCRIPTION

Arrowleaf clover forms a leafy rosette during early growth and then produces branching stems that curve upward to a length of 60-150 cm. The smooth, thick, hollow stems are often purple, becoming hard near maturity. Long, white, pointed stipules are at the base of each leaf petiole. Each arrow-shaped, glabrous leaflet is 3.5-7.0 cm long and has pronounced veins. Leaf marking ranges from none to a large, white, V-shaped mark (Fig. 20.2).

Flower heads are conical, often exceeding 6 cm in length and 3 cm in diameter. From 150 to 170 individual florets make up the heads; the corolla is white initially and later turns pink to purple. Flowering occurs over a 4-7 wk period, the later cultivars blooming from late May to July. New blooms continue to develop, while other heads have mature seed. Each floret produces two to three rough, brown seeds. Seeds are intermediate in size with about 880,000/kg.

IMPORTANCE AND USE

Arrowleaf clover is grown on more hectares in the US than any other annual clover species, with the largest concentration in eastern Oklahoma and Texas (Table 20.1). It is the latest-maturing annual clover and therefore usually the highest yielding. Forage digestibility of arrowleaf is slightly higher than that of crimson clover at all stages of maturity (Hoveland et al. 1972). Digestibility declines with maturity, primarily because of decreased stem digestibility and an increase in the percentage of stems (Beaty et al. 1977).

Arrowleaf clover extends the grazing season, reduces N fertilizer input, and improves animal gain when mixed with cool-season annual grasses for stocker cattle (Anthony et al. 1971). It is also used successfully in overseeding warm-season perennial grasses in the autumn to furnish grazing in late winter and spring for cow-calf operations (Hoveland et al. 1978). Arrowleaf is not recommended for overseeding warm-season perennial grass hay meadows because the late clover growth restricts grass production during May-June.

CULTIVARS

Three cultivars of arrowleaf clover have been released in the US. 'Amclo', the earliest maturing cultivar, reaches full bloom in central Georgia by mid-May (Miller and Wells 1985). 'Yuchi' blooms 3 to 4 wk later while 'Meechee' is about 5 wk later than Amclo. When soil moisture is sufficient, the later-maturing cultivars are the most productive. Meechee is the most winter-hardy. Seed of the three cultivars cannot be distinguished, so the use of certified seed is recommended.

MANAGEMENT

Recommended seeding rates for arrowleaf clover range from 7 to 11 kg ha^{-1} of scarified seed. The autumn planting date should be delayed until the daily minimum temperature is ≤15°C for good germination (Evers 1980). Planting 1-2 cm deep in a prepared seedbed results in the best seedling emergence and nodulation (Rich et al. 1983). Arrowleaf clover

TABLE 20.1. Hectarage of annual clovers in the US, 1992

Clover species	Pasture, hay, green manure	Roadside beautification	Seed production
Arrowleaf	545,600	2,000	6,500
Ball	28,000	1,200	40
Berseem	11,000	0	800
Crimson	205,200	13,200	7,400
Persian	47,400	0	0
Rose	20,400	0	200
Subterranean	137,400	0	0

Note: Estimates from forage extension specialists and/or seed companies in each state where these clovers are grown.

Fig. 20.2. Leaves and flower heads of (*A*) arrowleaf, (*B*) berseem, (*C*) crimson, (*D*) persian, (*E*) rose, and (*F*) subterranean clovers.

has only fair seedling vigor and nodulation and therefore is sensitive to shading from taller grass sods (Evers 1982). Cutting the grass sward to 2.5 cm and then disking lightly can enhance early-season clover production. Problems with aluminum (Al) toxicity occur below pH 5.0 and with iron-deficiency (Fe-deficiency) chlorosis above pH 7.5.

Arrowleaf clover will still produce some seed if grazed continually at a moderate stocking rate until maturity in June or July. A common practice in east Texas is to plant half the normal seeding rate the second autumn since most of the seed produced is hard. However, if a hay or seed crop is desired, grazing of arrowleaf should be terminated by early April (Hoveland et al. 1970, 1972). Regrowth is reduced by cutting hay in mid-April, and stands are virtually eliminated by cutting in May. Under hay conditions, arrowleaf clover has virtually no buds at the base of plants in late April and May, which accounts for the lack of regrowth after cutting. Maximum forage yields of arrowleaf have been obtained by frequent cutting or grazing until early April, followed by a hay harvest at early bloom in late May. Bloat potential of arrowleaf clover is low.

SEED PRODUCTION

Yuchi and Meechee arrowleaf begin blooming in late May or early June and produce mature seed in late June or July in the southeastern US. In western Oregon, seed is ready to harvest in August. Seed is borne in clustered pods produced at the tips of stems that remain erect if the plants have not made too much vegetative growth. Shattering is not a major problem. Arrowleaf clover is cross-pollinated, and bees, *Apis mellifera* L., are essential for pollination. Unless enough native bees are present, two to three colonies per hectare of honeybees should be placed in the seed field. Seed yields of 100-400 kg ha^{-1} are obtained in the southeastern US and 670-900 kg ha^{-1} in western Oregon. Dodder (*Cuscuta* sp.) is a serious weed problem because dodder seed and arrowleaf clover seed are extremely difficult to separate. Herbicides are available that control dodder.

Potential seed fields should be grazed until April. Failure to graze will result in extremely heavy vegetative growth that mats down, rots, and produces only a small number of seed heads. Thin stands or late-planted clover may give little or no grazing but often are the most productive seed fields. Seed of arrowleaf clover does not thresh readily because it is protected by the inflated calyx and will remain in the head if not dislodged in the threshing cylinder. If the combine is not properly adjusted, seed heads will be carried over the straw rack and out of the machine.

Diseases and Pests. The major disease problems with arrowleaf clover are bean yellow mosaic and peanut stunt viruses, which are transmitted by aphids (*Aphidae* spp.). Leaves and petioles of infected plants turn bright red and may result in plant senescence before flower production. Genetic resistance to these viruses has been identified and is being used in cultivar development (Pemberton et al. 1991). Virus infections also predispose plants to *Phytophthora* root rot. Crown and stem rot (*Sclerotinia trifoliorum* Eriks.) may also be a problem if a heavy accumulation of forage occurs during warm wet weather. The striped field cricket (*Nemobius fasciatus* Deg.) can destroy small arrowleaf clover seedlings in a warm-season perennial grass sod. Miller and Wells (1985) provide a more extensive discussion on disease and insect problems.

Ball Clover

Ball clover, *Trifolium nigrescens* Viv., introduced from Turkey in 1953, was first planted by the Soil Conservation Service near Selma, Alabama (Hoveland and Johnson 1967). Ball clover hectarage is largely centered in east Texas, Arkansas, Louisiana, Mississippi, and Alabama. It has survived and grown well from Arkansas and northern Alabama to the Gulf Coast in areas where wet clay or loam soils predominate (Hoveland and Webster 1965). It is especially adapted to soils that are wet during winter but are too dry in summer for survival of white clover.

Ball clover has smooth, egg-shaped leaflets and small, white to yellowish white fragrant florets. Stems are succulent and prostrate to partially erect and may reach a length of 75 cm. Flowering is about 1 wk later than that of crimson clover. The yellow to light brown egg-shaped seeds are slightly smaller than white clover seeds, with about 2.2 million seeds/kg.

Ball clover occupies a unique ecological niche on loam or clay soils in the lower South that are wet in winter but may be droughty in summer or have severe competition from warm-season perennial grasses that reduce persistence of white clover. Although the productive season of this clover is short, it is a dependable reseeder and so is a valuable component in many pastures.

Only common seed is available because there are no cultivars of ball clover. The large genetic diversity of ball clover has resulted in local ecotypes being developed from old reseeding fields; those originating farther south grow more in winter and bloom slightly earlier than ecotypes from fields farther north (Hoveland and Johnson 1967). Because of the variation within ecotypes, it is doubtful if they will remain stable after several generations of reseeding in a different environment.

MANAGEMENT

Seed of ball clover can be surface broadcast at 1-2 kg ha^{-1} after warm-season perennial grasses have ceased growth in autumn. Ball clover is more tolerant of soil acidity than other clovers. Forage yields are somewhat less than for crimson clover, with most of the production occurring in late spring. Ball clover tolerates hard grazing and produces seedheads close to the ground, ensuring adequate seed for reseeding. It is a good reseeder, a result of the high hard seed percentage and ability to germinate at low temperatures during late autumn and winter when moisture conditions are more favorable (Hoveland and Elkins 1965). Bloat can be a problem.

Ball clover is an excellent seed producer, with yields ranging from 100 to over 300 kg ha^{-1}. It requires insect pollination for seed production (Weaver and Weihing 1960) and is very attractive to bees and yields nectar profusely for honey production. Combine-harvested seed will have more than 60% hard seed.

No major disease problems have been encountered with ball clover. It is relatively resistant to alfalfa weevil (*Hypera postica* Gyll.), but clover head weevil (*H. meles* F.) may seriously reduce seed yields. Striped field crickets can be destructive to ball clover seedlings in grass sods during autumn.

Berseem Clover

Berseem clover, *Trifolium alexandrinum* L., also called *Egyptian clover,* is believed to have originated in Syria and was first introduced into Egypt about the sixth century (Knight 1985b). It was introduced into the US in 1896 and grown successfully in California in 1918. Berseem clover is very productive on alkaline soils but is limited to the southern areas of the US because of poor cold tolerance.

Berseem clover is grown in the southeastern US and under irrigation in southern California, Arizona, and northern Mexico. The species is not cold-hardy and should not be planted where temperatures repeatedly reach –6°C or lower (Knight 1985b). Berseem clover does best on loam to clay soils with a pH of 6.5-8.0. If inundated with water for 3-4 d, the topgrowth will senesce, but new growth will initiate from the crown.

Berseem clover has an erect growth habit, hollow stems and oblong leaflets, and self-sterile yellowish white florets (Fig. 20.2). A short taproot limits its use on sandy soils with low water-holding capacity. Flowering occurs in May. Seeds range in color from yellow to purple; there are approximately 440,000/kg. Production of hard seed is very low, which limits natural reseeding.

Production of berseem clover in the US is small (Table 20.1). There is some use of berseem under irrigation in southern California for green chop and silage. Its greatest potential in the southeastern US is on alkaline upland and riverbottom soils with good internal drainage. Berseem has good early forage production equal to crimson clover under mild winter conditions in the Gulf Coast area. However, berseem outyields crimson because of its later maturity.

CULTIVARS

Early cultivars lacked cold tolerance and were limited to southern California and Arizona and the immediate Gulf Coast area. 'Bigbee' berseem was selected for improved cold tolerance in north Mississippi out of the Italian cultivar 'Sacromonte' and released in 1985 (Knight 1985b). 'Multicut' berseem, flowering 10 d later than Bigbee, was released in 1988 for use in the Central Valley of California and in the irrigated desert valleys of southern California and northern Mexico (Graves et al. 1989). In those areas, it can be harvested five to six times, beginning from mid-December to June.

MANAGEMENT

Berseem clover should be seeded at 11-18 kg ha^{-1} during September or October. Germination was found to exceed 90% at day/night temperatures of 25°/15°C but dropped to 64% at 30°/20°C and to 10% at 35°/25°C (Evers 1991). Cutting or grazing should be initiated when the clover reaches a height of 20-25 cm. Allowing berseem to get taller than 40 cm before the initial defoliation will increase the chance of freeze damage (Joost and Mooso 1989). Berseem clover always had a lower leaf and higher stem percentage than red (*T. pratense* L.), white (*T. repens* L.), and subterranean (*T. subterraneum* L.) clovers when undefoliated growth was harvested at 10-d intervals during the spring (Brink and Fairbrother 1992). Digestible dry matter and crude protein were also usually lower than in the other species. With irrigation in southern California, dry forage yields may average 12-16 Mg ha^{-1}, and N_2-fixation was estimated to be 260-350 kg N ha^{-1} (Williams et al. 1990). In the southeastern US, dry matter yields of berseem clover range from 5 to 7 Mg ha^{-1} (Joost and Mooso 1989). Berseem clover has moderate bloat potential.

Berseem has poor reseeding potential because of low hard seed production. Seed yields of irrigated berseem clover in the southwestern US may exceed 1000 kg ha^{-1} (Knight 1985b). Seed fields should be planted in 60-cm rows and not defoliated after February. There are no known disease or insect problems specific for berseem clover.

Crimson Clover

Crimson clover, *Trifolium incarnatum* L., is native to southern Europe, where it was cultivated as a forage and green manure crop during the 18th century. This clover was introduced into the US in 1819, and by 1855 seed was widely distributed (Knight 1985a). Advantages of crimson clover are good seedling vigor and ease of establishment, early forage production, and adaptability to a wide range of soils. The primary limitation is poor reseeding because of low production of hard seed and rapid softening of that hard seed.

Crimson clover is grown throughout the southeastern US as far north as Kentucky and as far south as the Gulf Coast except for peninsular Florida (Knight 1985a) (Fig. 20.3). It is also grown in the Pacific Coast states, with the primary seed production area in the Willamette River valley south of Portland, Oregon. Crimson clover is used to a limited extent as a summer annual green manure crop in the northern US. It is adapted to a wider range of soil types than other annual clovers but does not tolerate poor drainage (Hoveland and Webster 1965). It can grow on soils with pH 5.0 to 8.0, but optimum growth and nodulation occur between pH 5.5 and 7.5. Aluminum may limit nutrient uptake on soils below pH 5.0, and Fe chlorosis becomes a problem on alkaline soils near pH 8.0.

PLANT DESCRIPTION

Crimson clover seeds are yellow and spherical, with approximately 330,000/kg. Under favorable soil moisture conditions, seedlings make rapid growth and form a dense crown or rosette type of leaf development. Leaves are broadly obovate at the tip, narrow at the base, and densely covered with hairs, with plants reaching a height of 30-60 cm (Fig. 20.2). Development of flower stems is initiated when daylength exceeds 12 h in the spring (Knight and Hollowell 1958). Conical bright crimson flower heads are composed of 75-125 florets. Flowers are mostly self-fertile but not self-pollinating. Bees seeking nectar or pollen trip the flowers and bring about pollination. Seed mature in approximately 24-30 d.

IMPORTANCE AND USE

Superior seedling vigor and relatively large seed make crimson one of the easiest annual clovers to establish. The primary use of crimson clover is for overseeding warm-season perennial grasses for winter and spring grazing (Knight 1970; Dunavin 1982). Grazing of crimson clover pastures can begin earlier than that of most other annual clovers because of its greater autumn and winter growth. Crimson clover is preferred for overseeding bermudagrass (*Cynodon dactylon* [L.] Pers.) hay meadows because it is early maturing and therefore less competitive to bermudagrass in late spring.

Yields of 3-6 Mg ha^{-1} of good quality hay may be obtained by harvesting in April when

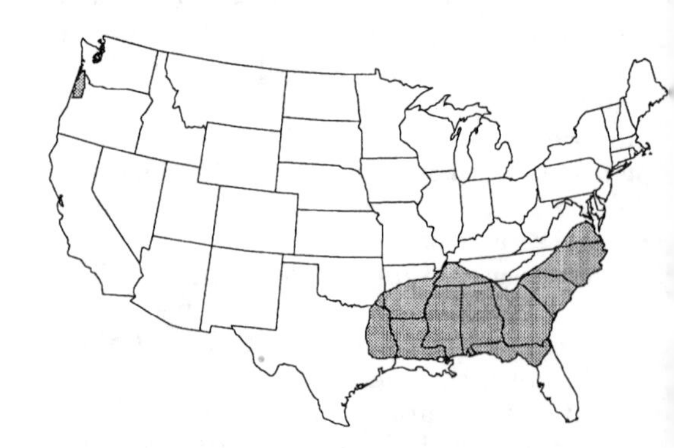

Fig. 20.3. Primary crimson clover-growing area in the US. Crimson is more winter-hardy than arrowleaf clover, so it can be grown farther north, but this is rarely done because of well-adapted perennial clovers.

the majority of plants are not yet beyond the half-bloom stage. However, poor drying conditions at that time often makes hay harvest difficult.

The early maturity of crimson clover also makes it one of the best cool-season annual forage legumes for no-tillage rotations with corn (Holderbaum et al. 1990) and grain sorghum (Touchton et al. 1982) in the southeastern US. Maximum clover top growth N content and subsequent N release of about 70 kg N ha^{-1} occurs when crimson clover is allowed to reach the late bloom stage (Ranells and Wagger 1992). For over 40 yr crimson clover has been used for roadside stabilization and beautification (Table 20.1). Poor reseeding is the main disadvantage of crimson clover.

CULTIVARS

Common crimson clover was the only type grown until 'Dixie', selected for higher hard seed percentage, was released in 1946 (Knight 1985a). It is still the dominant cultivar planted in the US today. Of other cultivars released, 'Frontier', 'Tibbee', 'AU Robin', and 'Flame' flower 1 to 2 wk earlier than Dixie, 'Auburn', and 'Autauga' while 'Chief', 'Kentucky', and 'Talladega' flower about 1 wk later. Currently, Chief, Tibbee, Talladega, AU Robin, and Flame are the only certified cultivars in the US.

MANAGEMENT

Crimson clover is generally planted between mid-August and November. Early plantings are limited to prepared seedbeds while warm-season perennial grasses are overseeded during late September through November. Crimson clover will germinate at temperatures over 30°C (Evers 1980), so earlier plantings can be made, but they are frequently lost to drought and virus infections. In northern Mississippi, crimson clover planted August 15 produced the highest yields over a 6-yr period (Knight 1985a). Planting November 15 produced only 25% as much forage as planting August 15 because of cooler temperatures during establishment.

Planting rates for crimson clover range from 12 to 28 kg ha^{-1}, depending on intended use of the clover, planting method, and whether planted alone or in mixtures. Higher seeding rates are recommended when broadcasting the seed on a grass sod rather than drilling because a smaller percentage of the germinating seed becomes established. Two-thirds of the normal seeding rate is used when planting with small grains and/or annual ryegrass.

Forage production of a bermudagrass and crimson clover mixture in Mississippi was 63% greater than that of grass with no N and 27% greater than that of grass alone with 179 kg N ha^{-1} (Knight 1967). In northern Florida a crimson clover-bahiagrass (*Paspalum notatum* Flugge) mixture produced twice as much forage as grass alone with no N fertilizer (Dunavin 1982). Annual N_2 fixation of Tibbee crimson clover was 155 kg ha^{-1} when planted in a pure stand on a prepared seedbed (Brink 1990).

SEED PRODUCTION

Seed yields of crimson clover are highest from lower seeding rates of 6 to 9 kg ha^{-1} (Rampton 1969). Seed yields average 600-900 kg ha^{-1} under ideal conditions in the Willamette River valley of Oregon. Yields in the southeastern US are lower and less dependable because of higher humidity and rainfall during seed maturation. Placing colonies of honeybees in or adjacent to seed fields is recommended since the tripping of florets is important in setting seed.

DISEASES AND PESTS

The most widespread and serious crimson clover disease is crown and stem rot caused by *Sclerotinia*. This soil-borne pathogen attacks the crown under a thick canopy of foliage during cool, wet weather in winter and causes crown rot. Crimson clover is also susceptible to a *Phytophthora* root rot (Knight 1985a). To avoid stand losses, crimson clover should not be grown on low-lying, poorly drained soils subject to frequent flooding or prolonged high moisture. Other disease problems are fusarium wilt disease caused by a form of *Fusarium oxysporum* Schlect, sooty blotch (*Mycosphaerella killianii* Petr.), leaf and stem spot (*Cercospora zebrina* Pass.), and viruses.

Problem insects for young crimson clover seedlings are the fall armyworm (*Spodoptera frugiperda*), several cutworms, the yellow-striped armyworm (*S. ornithogalli*), the Hawaiian beet webworm (*Hymenia recurvalis*), and several bean beetles. The clover head weevil and the lesser clover weevil (*H. nigrirostris* Fab.) destroy seed in the head, resulting in low seed yields and poor natural reseeding.

Rose Clover

Rose clover (*Trifolium hirtum* All.) is native to the Mediterranean region and Asia Minor at elevations above 200 m. The original seed was brought to the US from Turkey in 1936 (Love 1985). Although not high yielding, rose clover can grow and persist in droughty, infertile soils and drier climates unsuitable for other clovers. It is well adapted on the slightly alkaline soils of the 60-90 cm rainfall area of central Texas and Oklahoma. Rose clover is grown extensively on rangeland in California. It will grow on a wide range of soil types from pH 5.5 to nearly 8.0. Iron deficiency chlorosis becomes a problem as soil pH nears 8.0, and aluminum toxicity may be a problem at pH below 5.0. Rose clover is not adapted to poorly drained soils.

PLANT DESCRIPTION

Stems, leaves, and flower heads are pubescent and pale green. Leaflets are wedge shaped with rounded tops and may have a pale crescent with a dark line above it (Fig. 20.2). Growth habit is semiprostrate to erect, reaching a height of 40-50 cm. Rooting depth reaches 2 m, making rose clover more drought tolerant than shallow-rooted species such as subterranean clover (Love 1985). Flowers are pale pink, about 2 cm in diameter, and contain 20-40 sessile florets. Seeds are straw colored and tinged with green and brown, with approximately 375,000 seeds/kg. About 90% of the seeds have hard seed coats, contributing to excellent reseeding (Williams and Elliott 1960).

IMPORTANCE AND USE

Presently, rose clover is used almost exclusively on California rangelands that have wet winters and dry summers. Rose clover is usually planted in mixtures with subterranean and crimson clovers but may be planted alone, especially if the soil is extremely infertile. High production of hard seed, coupled with a slow rate of seed softening (Smith 1993), gives rose clover better reseeding potential than other that of annual clover species. Poor seedling vigor is the major disadvantage of rose clover.

CULTIVARS

'Wilton', the first rose clover cultivar, was derived from a plant introduction from Turkey (Love 1985). 'Kondinin', 'Hykon', 'Sirint', and 'Olympus' are earlier-maturing cultivars than Wilton that were selected in Australia. 'Overton R-18' rose clover, increased from a Spanish introduction, is later maturing than Wilton (Smith et al. 1992).

MANAGEMENT

Seeding rates of 4-8 kg ha^{-1} are sufficient for seed production and for range seeding when rose clover is drilled in California rangeland (Love 1985), but rates of 10-15 kg ha^{-1} are recommended if seed is broadcast on rangeland or planted in the southeastern US. Minimum daily temperatures must be $\leq$15°C for good germination (Young et al. 1973; Evers 1991). Grazing can be initiated on California rangelands when rose clover is 5-15 cm high to utilize weedy grasses before they head out and are avoided by livestock (Love 1985). Livestock should be removed before the last spring rains to allow natural reseeding. After rose clover has produced seed, grazing may continue during the dry summer. Isoflavone, which can cause infertility in sheep, is very low in rose clover (Gildersleeve et al. 1991). Bloat potential of rose clover is low.

In California, seed production comes from range seedings with yields of 300-500 kg ha^{-1}. There is some seed production of Overton R-18 in Texas and Oregon. A rasp-bar combine should be used to harvest rose clover seed because the seed is held tightly by the calyx. To date there are no major disease or insect problems affecting rose clover.

Subterranean Clover

Subterranean clover, or subclover, is the common name for three *Trifolium* species that mature seed in burrs below or near the soil surface (McGuire 1985). The three species differ genetically, in morphological features, and in edaphic adaptation. *Trifolium subterra-*

neum L. has black seed, has some pubescence on leaves and stems, and originated in the Mediterranean basin and southern England. Seed of *T. yanninicum* Katzn. & Morley is cream colored; this clover has glabrous plant parts except for the upper leaf surface and is adapted to areas of water-logged soils. All plant parts are glabrous in *T. brachycalcycinum* Katzn. & Morley, but the calyx only covers the base to lower third of the pod rather than most of the pod as in the other species. *Trifolium brachycalcycinum* is adapted to alkaline soils but is not as cold-hardy as the other two species. The prostrate growth habit of the subclovers furnishes tolerance to close grazing and dependable seed production.

Subclover, primarily *T. subterraneum,* is used most extensively in Australia, where it was inadvertently introduced over 100 yr ago. Subclover is grown in New Zealand, South Africa, Argentina, Chile, Uruguay, and the US as well as the Mediterranean area (McGuire 1985). The major subclover area in the US is west of the Cascade and Sierra Nevada ranges in California and Oregon where winters are mild and moist and summers are dry. The primary subclover species grown are *T. subterraneum* with some *T. brachycalcycinum* in southern Texas and California because of the alkaline soils and mild winters. The *T. subterranum* species does well on soils with pH up to about 7.3 and then suffers from iron-deficiency chlorosis (Gildersleeve and Ocumpaugh 1989). Some subclover is grown in the southeastern US on sandy loam to clay soils. Subclover does poorly on deep sandy soils because it does not develop a deep root system to survive dry periods. Reseeding of subclover is not as good in the southeastern US where rain falls throughout the year as in Mediterranean climates such as on the US West Coast and southern Australia (Evers et al. 1988). Subclover seed maturing under moist soil conditions has a lower initial hard seed level (Smith 1988), and summer rainfall causes germination.

PLANT DESCRIPTION

Subclover produces dense, prostrate growth from nonrooting stems and stolons. Undefoliated stands only reach a height of 20-25 cm. This growth habit allows it to tolerate close grazing by sheep, horses, and wildlife. The inflorescence of subclover is inconspicuous and contains three to seven (usually four) white or pinkish florets (Fig. 20.2). After flowering and fertilization, a seed-bearing burr is formed and the peduncles elongate toward the ground. The stiff forked bristles on the seed burr serve as mechanisms that bury a portion of the burrs in the soil. There are approximately 150,000 seeds/kg.

IMPORTANCE AND USE

Subclover is the base of much of southern Australia's 20 million ha of improved grassland where there is little or no summer rainfall and sheep are the main livestock enterprise. All available subclover cultivars were selected under Mediterranean climatic conditions in Australia, which may explain their inconsistent performance in the southeastern US where rainfall occurs throughout the year.

Subclover has good seedling vigor and early nodulation, enhancing establishment and making it competitive with warm-season perennial grasses (Evers 1982). Forage production of subclover grown with bermudagrass or bahiagrass without N fertilizer was found to be equivalent to adding 160-254 kg N ha^{-1} to the grasses alone (Evers 1985). Brink (1990) reports subclover fixed 104-206 kg N ha^{-1} when grown in a pure stand on a prepared seedbed in northern Mississippi. Subclover was found to maintain a higher leaf and lower stem percentage during spring than berseem clover, which has an upright growth habit (Brink and Fairbrother 1992). Because leaves and petioles are the primary subclover components available to grazing livestock, crude protein, in vitro digestible dry matter, and neutral detergent fiber concentrations were similar when subclover-bermudagrass pastures were grazed at three levels of forage availability (Fairbrother et al. 1992).

CULTIVARS

There are many commercial cultivars of the three species of subclover; all cultivars were selected in Australia (McGuire 1985). Length of growing season before seed maturity is the major difference among cultivars. Cultivars differ in level of hardseededness; high levels are desirable for reseeding and estrogenic activity, and low levels are desirable because estrogens reduce reproductive fertility in sheep. Cultivars used in the US have low estrogenic activity. 'Mt. Barker', 'Woogenellup', and 'Nangeela' are the most frequently used cultivars of *T. subterraneum* in the US. 'Clare', 'Koala', and 'Rosedale' are available *T. brachycalcycinum* cultivars, with Rosedale being the most cold tolerant.

MANAGEMENT

All subclover species have high germination at day/night temperatures of 25°/15°C or lower, but germination rate decreases as temperature decreases (Evers 1991). Cultivars of *T. brachycalcycinum* germinate well in cooler temperatures, approximately 60% at 30°/20°C, which indicates they could be planted earlier in autumn than other species. Early planting of subclover is critical. Jones et al. (1978) found that healthy subclover plants developed when the mean ambient air temperature was between 10° and 17°C for 6 wk following germination. Poor clover stands developed when the mean temperature for the first 6 wk was about 7°C. The most rapid nodulation of subclover occurs at 30°C and begins to decrease as temperature drops below 22°C. Seeding rates range from 10 to 22 kg ha^{-1}.

In the southeastern US, subclover's greatest potential is in mixtures with warm-season perennial grasses (Evers 1985; Brink 1990). Yield increases were not as great when subclover was grown with a temperate perennial grass because of similar growing seasons (Brink and Fairbrother 1991). The decumbent growth habit of subclover permits closer defoliation during the growing season than does the growth habits of other annual clovers, with an optimum grazing height of 3-6 cm (Fairbrother et al. 1992). Volunteer reseeding of subclover can be enhanced by limiting N fertilization of the grass sod during summer and grazing the warm-season grass short until low temperatures limit grass growth even though subclover seedlings may have emerged earlier (Evers et al. 1988). Bloat potential is moderate.

SEED PRODUCTION

Seed yields of subclover generally range from 500 to 1000 kg ha^{-1} in Australia. Grazing of seed fields opens up the canopy and improves seed yields. Maximum seed yields in Oregon occurred following grazing from just prior to the start of flowering until the time of early burr fill (Steiner and Grabe 1986). Harvesting of the burrs buried in soil requires use of specialized suction combines developed in Australia (McGuire 1985). Subclover reseeding in pasture systems in the southeastern US can be enhanced by grazing to 3-6 cm by early April.

DISEASES AND PESTS

There are no major diseases restricting subclover acreage in the US. Powdery mildew (*Erisiphe polygoni* DC.), *Sclerotinia,* pepper spot (*Leptosphaerulina trifolii* [Rostr.] Petr.), and root rot caused by species of *Pythium, Fusarium,* and *Rhizoctonia* can occur on subclover (McGuire 1985). Virus diseases that are transmitted by aphids include clover stunt, bean yellow mosaic, and red leaf.

Other Clovers

PERSIAN CLOVER

Persian clover, *Trifolium resupinatum* L., is native to southern Asia Minor and the Mediterranean countries. Its exact date of introduction in the US is not known, but it was found in Wilcox County, Alabama, in 1923 and near Hamburg, Louisiana, in 1928 (Knight 1985b). Persian clover is found on loam and clay soils in low-lying areas in the southeastern US. Although seedling growth and nodulation are best at pH 6.4 or higher, it has grown well on wet clay soils with pH 5.7 in southeast Texas.

Persian clover is glabrous and has a decumbent growth habit under grazing but does not root at the nodes. Small, light purple flowers are produced on axillary peduncles 1-5 cm long, which are self-pollinating and self-fertile (Fig. 20.2). There are approximately 2 million seeds/kg, which range in color from yellowish green to reddish brown. Stems are hollow and reach 60 cm if not defoliated. About 95% of the seed is hard, which contributes to its good natural reseeding and stand persistence (Weihing 1962).

Persian clover is planted at 5-8 kg ha^{-1} in autumn in the southeastern US. It can be grown for grazing, hay, and green manure (Knight 1985b). Annual ryegrass or another cool-season grass should be grown in a pasture with persian clover because of the high bloat potential. Blister beetles, *Epicauta vittata* Fabricius, can be a major insect problem in persian clover hay as consumption of four to five dead blister beetles is fatal to horses.

'Abon' persian clover, selected in southeast Texas, can be grazed 4-8 wk earlier and 4 wk later than common persian. Abon grows to 1 m tall and has larger flowers and leaves than common. Abon has survived low temperatures of −11°C. Poor seedling growth and high bloat potential are the main limitations of persian clover.

HOP CLOVERS

Hop clover, *Trifolium agrarium* L.; large hop clover, *T. campestre* Schreb.; and small hop clover, *T. dubium* Sibth., are similar in appearance. All have small, round heads of yellow flowers that turn brown upon ripening of seed, resembling the flower of a hop (*Humulus lupulus* L.) plant. Hop clovers are adapted to droughty, infertile, and eroded soils of the southern US. All three species are highly palatable pasture plants, but they produce low yields of forage for only a short period in the spring (Knight 1985b).

MISCELLANEOUS ANNUALS

Additional species of annual clovers include cluster clover, *T. glomeratum* L., adapted to the Gulf Coast and especially to southern Mississippi; Mike's, or bigflower, clover, *T. michelianum* Savi, a tall coarse species also adapted to the Gulf Coast; lappa clover, *T. lappaceum* L., adapted to low-lying wet calcareous clay soils of the lower southeastern US; striate clover, *T. striatum* L., adapted to clay soils of the lower Southeast; and rabbitfoot clover, *T. arvense* L., adapted to infertile dry sandy soils (Knight 1985b). Whitetip clover, *T. variegatum* Nutt. ex Torr. & Gray; cup clover, *T. cyathiferum* Lindl.; tomcat clover, *T. tridentatum* Lindl.; maiden clover, *T. microcephalum* Pursh.; squarehead clover, *T. microdon* Hook & Arnott; pinlole clover, *T. bifidum* A. Gray; tree clover, *T. cilolatum* Benth.; and pin-point clover, *T. gracilentum* Torr. & Gray, are native annuals in rangelands extending from the Pacific Coast states of California and Oregon to British Columbia (Crampton 1985).

QUESTIONS

1. Why are winter annual clovers used in many areas rather than perennial clovers?
2. What is the most widely grown winter annual clover in the US, and what characteristics account for this distribution?
3. Which annual clovers require well-drained soils, and which tolerate flooding and poor drainage?
4. What accounts for the natural reseeding properties of certain annual clovers?
5. How do the annual clovers differ in bloat potential in cattle?
6. What characteristics of crimson clover make it so successful as a green manure crop in rotation with summer crops?
7. What are the soil pH requirements of the different annual clovers?
8. What are the advantages of growing a winter annual clover in a bermudagrass sod?
9. Why is rose clover widely grown on California rangeland but little used in the southeastern US?
10. How do temperature requirements for seed germination differ among annual clovers, and why is this important in determining planting date?

REFERENCES

Anthony, WB, CS Hoveland, EL Mayton, and HE Burgess. 1971. Rye-Ryegrass-Yuchi Arrowleaf Clover for Production of Slaughter Cattle. Ala. Agric. Exp. Stn. Circ. 182.

Beaty, ER, AE Smith, and JD Powell. 1977. Leaf, petiole, and stem accumulation and digestibility in 'Amclo' clover. Agron. J. 69:682-84.

Brink, GE. 1990. Seasonal dry matter, nitrogen, and dinitrogen fixation patterns of crimson and subterranean clovers. Crop Sci. 30:1115-18.

Brink, GE, and TE Fairbrother. 1991. Yield and quality of subterranean and white clover-bermudagrass and tall fescue associations. J. Prod. Agric. 4:500-504.

___. 1992. Forage quality and morphological components of diverse clovers during primary spring growth. Crop Sci. 32:1043-48.

Crampton, B. 1985. Native range clovers. In NL Taylor (ed.), Clover Science and Technology, Am. Soc. Agron. Monogr. 25. Madison, Wis., 579-90.

Dunavin, LS. 1982. Vetch and clover overseeded on a bahiagrass sod. Agron. J. 74:793-96.

Evers, GW. 1980. Germination of cool season annual clovers. Agron. J. 72:537-40.

___. 1982. Seedling growth and nodulation of arrowleaf, crimson, and subterranean clovers. Agron. J. 74:629-32.

___. 1983. Weed control on warm season perennial grass pastures with clovers. Crop Sci. 23:170-71.

___. 1985. Forage and nitrogen contributions of arrowleaf and subterranean clovers overseeded on bermudagrass and bahiagrass. Agron. J. 77:960-63.

___. 1991. Germination response of subterranean, berseem, and rose clovers to alternating temperatures. Agron. J. 83:1000-1004.

Evers, GW, GR Smith, and PE Beal. 1988. Subterranean clover reseeding. Agron. J. 80:855-59.

Fairbrother, TE, GE Brink, and RL Ivy. 1992. Effect of forage availability on steer performance of a subterranean clover-bermudagrass forage system. J. Prod. Agric. 5:28-33.

Gildersleeve, RR, and WR Ocumpaugh. 1989. Greenhouse evaluation of subterranean clover species for susceptibility to iron-deficiency chlorosis. Crop Sci. 29:949-51.

Gildersleeve, RR, GR Smith, IJ Pemberton, and CL

Gilbert. 1991. Screening rose clover and subterranean clover germplasm for isoflavones. Crop Sci. 31:1374-76.

Graves, WL, WA Williams, VA Wegrzyn, and D Calderon. 1989. Registration of 'Multicut' berseem clover. Crop Sci. 29:235-36.

Hargrove, WL, and WW Frye. 1987. The need for legume cover crops in conservation tillage production. In JF Power (ed.), The Role of Legumes in Conservation Tillage Systems, Proc. Natl. Conf. Soil Conserv. Soc. Am., 27-29 Apr., Athens, Ga., 1-5.

Holderbaum, JF, AM Decker, JJ Meisinger, FR Mulford, and LR Vough. 1990. Full-seeded legume cover crops for no-tillage corn in the humid East. Agron. J. 82:117-24.

Hoveland, CS, and DM Elkins. 1965. Germination response of arrowleaf, ball, and crimson clover varieties to temperature. Crop Sci. 5:244-46.

Hoveland, CS, and WC Johnson. 1967. Natural selection of ball clover ecotypes in reseeding pastures. Crop Sci. 7:544.

Hoveland, CS, and HL Webster. 1965. Tolerance of five clovers to flooding. Agron. J. 57:3-4.

Hoveland, CS, EL Carden, WB Anthony, and JP Cunningham. 1970. Management effects on forage production and digestibility of Yuchi arrowleaf clover (*Trifolium vesiculosum* Savi). Agron. J. 62:115-16.

Hoveland, CS, RF McCormick, and WB Anthony. 1972. Productivity and forage quality of Yuchi arrowleaf clover. Agron. J. 64:552-55.

Hoveland, CS, WB Anthony, JA McGuire, and JC Starling. 1978. Beef cow-calf performance on Coastal bermudagrass overseeded with winter annual clovers and grasses. Agron. J. 70:418-20.

Jones, MB, JC Burton, and CE Vaughn. 1978. Role of inoculation in establishing subclover on California annual grasslands. Agron. J. 70:1081-85.

Joost, RE, and JD Mooso. 1989. Harvest management and regrowth of *Trifolium alexandrinum*. In D Desroches (ed.), Proc. 16th Int. Grassl. Congr., Montrouge: Dauer, 523-24.

Knight, WE. 1967. Effect of seeding rate, fall disking, and nitrogen level on stand establishment of crimson clover in a grass sod. Agron. J. 59:33-36.

___. 1970. Productivity of crimson and arrowleaf clovers grown in Coastal bermudagrass sod. Agron. J. 62:773-75.

___. 1985a. Crimson clover. In NL Taylor (ed.), Clover Science and Technology, Am. Soc. Agron. Monogr. 25. Madison, Wis., 491-502.

___. 1985b. Miscellaneous annual clovers. In NL Taylor (ed.), Clover Science and Technology, Am. Soc. Agron. Monogr. 25. Madison, Wis., 547-62.

Knight, WE, and EA Hollowell. 1958. The influence of temperature and photoperiod on growth and flowering of crimson clover (*Trifolium incarnatum* L.). Agron. J. 50:295-98.

Love, RM. 1985. Rose clover. In NL Taylor (ed.), Clover Science and Technology, Am. Soc. Agron. Monogr. 25. Madison, Wis., 535-46.

McGuire, WS. 1985. Subterranean clover. In NL Taylor (ed.), Clover Science and Technology, Am. Soc. Agron. Monogr. 25. Madison, Wis., 515-34.

Miller, JD, and HD Wells. 1985. Arrowleaf clover. In NL Taylor (ed.), Clover Science and Technology, Am. Soc. Agron. Monogr. 25. Madison, Wis., 503-14.

Murphy AH, MB Jones, JW Clawson, and JE Street. 1973. Management of Clovers on California Annual Grasslands. Calif. Agric. Exp. Stn. Circ. 564.

Pemberton, IJ, MR McLaughlin, and GR Smith. 1991. Inheritance of resistance to virus-induced lethal wilt in arrowleaf clover. Phytopathol. 81:1001-5.

Rampton, HH. 1969. Influence of planting rates and mowing on yield and quality of crimson clover seed. Agron. J. 61:92-95.

Ranells, NN, and MG Wagger. 1992. Nitrogen release from crimson clover in relation to plant growth stage and composition. Agron. J. 84:424-30.

Rich, PA, EC Holt, and RW Weaver. 1983. Establishment and nodulation of arrowleaf clover. Agron. J. 75:83-86.

Smith, GR. 1988. Screening subterranean clover for persistent hard seed. Crop Sci. 28:998-1000.

___. 1993. Improvement of reseeding in annual clovers. In Proc. 17th Int. Grassl. Congr., Palmerston North, New Zealand, and Australia, 415-16.

Smith, GR, FM Rouquette, Jr., GW Evers, MA Hussey, WR Ocumpaugh, JC Read, and AM Schubert. 1992. Registration of 'Overton R-18' rose clover. Crop Sci. 32:1507.

Steiner, JJ, and DF Grabe. 1986. Sheep grazing effects on subterranean clover development and seed production in western Oregon. Crop Sci. 26:367-72.

Touchton, JT, WA Gardner, WL Hargrove, and RR Duncan. 1982. Reseeding crimson clover as an N source for no-tillage grain sorghum production. Agron. J. 74:283-87.

Weaver, N, and RM Weihing. 1960. Pollination of several clovers by honeybees. Agron. J. 52:183-85.

Weihing, RM. 1962. Selecting persian clover for hard seed. Crop Sci. 2:381-82.

Williams, WA, and JR Elliott. 1960. Ecological significance of seed coat impermeability to moisture in crimson, subterranean, and rose clovers in a Mediterranean-type climate. Ecol. 41:785-90.

Williams, WA, WL Graves, and KG Cassman. 1990. Nitrogen fixation by irrigated berseem clover versus soil nitrogen supply. J. Agron. and Crop Sci. 164:202-7.

Young, JA, RA Evans, and BL Kay. 1973. Temperature requirements for seed germination in an annual-type rangeland community. Agron. J. 65:656-59

21 Lespedezas

ROBERT L. McGRAW
University of Missouri

CARL S. HOVELAND
University of Georgia

Annual Lespedezas

ANNUAL lespedeza species, commonly called "striate" (*Kummerowia striata* [Thunb.] Schindler) and "korean" (*K. stipulacea* [Maxim.] Makino), are warm-season legumes used for pasture and hay. Many references, including earlier editions of this book, list the genus as *Lespedeza*. The name was changed based on the general acceptance of *Kummerowia* as the appropriate genus by the community of systematic botanists. The first International Legume Conference, held at Kew, England, in 1978, was directed principally toward an improved classification of the legume tribes and genera (Polhill and Raven 1981). *Kummerowia* was accepted as the genus for these annual lespedezas at the conference (Ohashi et al. 1981). The USDA Germplasm Resource Information Network (GRIN) accepts *Kummerowia* as the genus for both species (Wiersema et al. 1990).

Both species are introductions from eastern Asia (Henson 1957). The exact introduction of striate lespedeza into the US is unknown; however, it was first identified in Georgia in 1846 and spread quickly over much of the southeastern US under the names *common lespedeza* or *Japan clover* (McNair and Mercier 1911). Korean lespedeza was introduced into the US in 1919 when Adrian Pieters of the USDA received a package of lespedeza seed from Ralph Mill, a medical missionary in Seoul, Korea (Pieters 1934). Seed was increased at the USDA experimental farm in Arlington, Virginia and released as 'Korean'. Korean became the common name in the US for *K. stipulacea*.

ROBERT L. MCGRAW is Associate Professor of Agronomy at the University of Missouri. He recieved the MS and PhD degrees from the University of Florida and has conducted research on forage management, physiology, and seed production in Minnesota and Missouri.

CARL S. HOVELAND. *See Chapter 20.*

DISTRIBUTION AND ADAPTATION

Annual lespedezas are grown in a wide belt extending from eastern Texas, Oklahoma, and Kansas into southern Iowa and eastward to the Atlantic Coast (Fig. 21.1). The western boundary is limited by lack of moisture. The northern boundary is limited by the plant's inability to produce seed before a killing frost. Photoperiod and temperature influence growth and floral initiation in the annual lespedezas (Nakata 1952; Offutt 1968). Plants flower when the daylength is less than the critical photoperiod. In northern latitudes, floral initiation of the annual lespedezas is delayed because the critical photoperiod occurs later in the summer. Seed maturation also is delayed in northern latitudes because cooler temperatures slow seed development. Thus, annual lespedeza, when grown too far north, will not persist naturally because frost kills the plants before viable seed is produced to reestablish the stand.

The critical photoperiod is longer for early-maturing cultivars than for those that mature later (Smith 1941). Korean cultivars tend to mature earlier than striate cultivars and are grown in the upper two-thirds of the lespedeza region. Striate matures later and is more important in the southern part of the region; however, a newly released early-maturing striate cultivar, 'Marion', may move its range farther north.

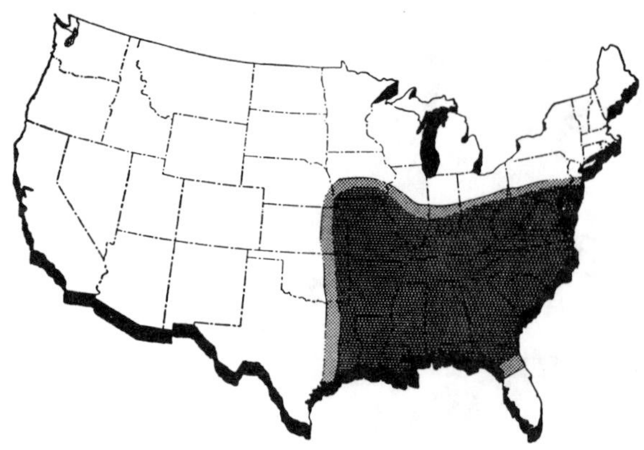

Fig. 21.1. Area of the US adapted to the lespedezas.

PLANT DESCRIPTION

Both annual lespedeza species are fine stemmed, leafy, herbaceous legumes with shallow taproot systems (Fig. 21.2). When not cut during the season, plants will grow to a height of 40-50 cm. They are considered to be self-pollinated although plants may have both chasmogamous and cleistogamous flowers. Outcrossing is thought to be rare in the chasmogamous flowers. Cleistogamous flowers are apetalous and chasmogamous flowers have pink to purple corollas. The pod contains a single seed that is blue-black and may or may not be mottled.

Striate cultivars are easily distinguished from korean cultivars (Fig. 21.3). Striate cotyledons have an indentation on one edge near the outer end, whereas those of korean are nearly elliptical. Leaves of korean are broader and the stipules larger than those of striate. Pods initiate in leaf axils along the entire stem of striate, whereas pods of korean are borne in clusters at the tips of branches developing from leaf axils. Leaves of korean turn forward around the developing pods, giving protection against shattering when the seeds are mature. Leaves of striate do not turn forward after flowering, so this species has a greater potential for seed shattering. The calyx covers less than one-half the pod in korean and more than one-half the pod in striate (see Chap. 2, Fig. 2.6). Pubescence on stems of korean is appressed upward, whereas pubescence is appressed downward in striate.

Fig. 21.2. 'Marion' striate annual lespedeza growing in a seed production field in Boone County, Missouri, in early August. *Mo. Agric. Exp. Stn. photo.*

IMPORTANCE AND USE

The total area of annual lespedeza in the US reached a maximum of about 20 million ha in 1949 (Henson 1957). Annual lespedeza was often grown in rotations with small grains. It provided pasture, hay, or seed after the grain crop was harvested, fixed nitrogen (N) for the rotation, and then reseeded itself for the next year (Baldridge 1957). Increased fertilizer use after World War II and emphasis on higher-yielding crops resulted in a shift away from annual lespedeza. Today it is used mainly as a pasture legume with limited amounts being cut for hay. Annual lespedeza has naturalized over a substantial portion of its range and is common in many pastures, especially where little or no N fertilizer is applied.

CULTIVARS

Four cultivars of striate lespedeza have been produced commercially in the US. The first, known as *common lespedeza,* was small, fine stemmed, and almost prostrate (Cope 1957). It was replaced by 'Tennessee 76', which resulted from a selection program begun in 1912 at the Tennessee Agricultural Experiment Station. Neither cultivar is grown commercially today. They were replaced by 'Kobe', which was introduced into the US from Japan in 1919 by J. B. Norton. Plants of Kobe are larger with coarser stems and bigger leaves and seeds than those of common lespedeza. Most striate lespedeza seed sold today is Kobe.

Marion was released in 1989 by the Missouri and Arkansas Agricultural Experiment Stations and the USDA-ARS (Beuselinck and McGraw 1990) (Table 21.1). At Columbia, Missouri, Marion flowered about 3 wk earlier than 'Summit' (a korean type) and 4 wk earlier than Kobe. Thus, Marion can produce seed and persist in the northern part of the lespedeza region. Marion is also resistant to many foliar diseases that affect korean cultivars, resulting in greater leaf retention and better forage quality.

Several korean lespedeza cultivars have been released; Korean, 'Harbin', 'Iowa 6', 'Iowa 39', 'Iowa 48', 'Climax', 'Rowan', 'Sum-

Fig. 21.3. Striate lespedeza is easily distinguished from korean lespedeza. (*A*) Cotyledons of striate (*right*) have distinct indentations. (*B*) Striate plants set their seed in leaf axils along the entire length of stems (*left*), whereas seed of korean is borne on clusters at the tips of branches developing from leaf axils. Note the turning forward of the leaves around the flowers and pods of korean (*C*). Striate stipules (*left*) are not as wide as those of korean. (*D*) Striate leaflets (*left*) are not as wide as those of korean. *Univ. of Arkansas photo.*

TABLE 21.1. Lespedeza cultivars for which seed is currently available, 1993

Cultivar	Reference	Release agencies	Characteristics
Annual lespedeza			
Marion	Beuselinck and McGraw 1990	Univ. of Missouri, Univ. of Arkansas, and USDA	Early-maturing striate type, resistant to foliar diseases
Summit	Offutt 1963	Univ. of Missouri, Univ. of Arkansas and USDA	Korean type with improved resistance to bacterial wilt, tar spot, and southern blight
Sericea lespedeza			
Appalow	Henry and Taylor 1981	SCS, USDA, and Univ. of Kentucky	Prostrate, leafy, very high tannin
AU Donnelly	Mosjidis and Donnelly 1989	Auburn Univ.	Low tannin, high yield
AU Lotan	Donnelly 1981	Auburn Univ.	Low tannin, lower yield than Serala
Interstate	Donnelly 1971	Auburn Univ.	Profuse branching, fine stem, short height, high tannin
Interstate 76	Donnelly and Minton 1979	Auburn Univ. and USDA	Root-knot nematode resistance, intermediate height, dense growth, high tannin
Serala	Donnelly 1965	Auburn Univ.	Fine, soft stems, high tannin
Serala 76	Donnelly and Minton 1979	Auburn Univ. and USDA	Root-knot resistance, tall, fine stem, high tannin

mit', and 'Yadki'. Few were grown widely, and the seed of most are no longer available commercially. Most *K. stipulacea* seed is sold under the common name "korean," although it can be from any cultivar. Seed of Summit is still being produced (Table 21.1). Summit was released in 1963 by the Missouri and Arkansas Agricultural Experiment Stations and the USDA-ARS (Offutt 1963). Summit has greater forage and seed yields and better disease resistance than the cultivar Korean.

MANAGEMENT

Soil Fertility. Annual lespedezas grow relatively well on eroded, acidic soils low in phosphorus (P) that will not support many other forage legumes. Their ability to grow on low-fertility soils makes them valuable legumes for low-input pasture systems. However, annual lespedezas do respond to lime and fertilization and grow best on productive, well-drained soils (Baldridge 1957). Korean is less tolerant of acid soil and more tolerant of alkaline soil than is striate.

Establishment. Annual lespedezas are among the easiest of the pasture legumes to establish. They can be sown from midwinter to early spring. Broadcasting without covering in late winter, and allowing frost heaving to bury the seed, will normally provide good stands in grass pastures. Korean seed that has not been hulled should be seeded at 20-30 kg/ha, whereas the seeding rate for striate should be somewhat higher. Seed should be inoculated with *Bradyrhizobium* spp. (cowpea miscellany group) the first time it is planted in a new area.

If pastures are managed properly, annual lespedeza should reseed itself. In Missouri, plants should not be grazed or cut from early September until the killing frost to allow for adequate seed production (Beuselinck and McGraw 1991). Reseeding or establishment in cool-season grass pastures is improved by grazing or cutting the grass in early spring to reduce competition. High rates of N fertilizer on grass sods will reduce the potential for successful establishment.

Soon after maturity, annual lespedeza seed may have poor germination. The germination percentage of a seed lot increases with time (Kenney 1932), a characteristic important for stand persistence in pastures that reseed naturally. Seed with little or no dormancy may germinate during a warm period in fall or winter and be killed by frost. In early spring, germinating seed may be killed by late frosts. Because much seed remains dormant, there is usually sufficient seed left to make a solid stand after the potential for spring frost is past.

Physiology and Harvest Management. Hay yields of annual lespedeza usually range from 2000 to 4000 kg/ha; however, on soils with good fertility and irrigation, yields can exceed 6000 kg/ha (McGraw et al. 1990). Hay can be harvested either once or twice per year. If harvested once, it should be cut in the early bloom stage for optimum yield and quality. If harvested twice, the first cut should be in mid- to late July when the lower leaves are just beginning to senesce and again at first bloom (Gray and Reynolds 1970)

Annual lespedezas do not rely on root nonstructural carbohydrates for regrowth. Only 5 to 45 g/kg of total nonstructural carbohydrate

(TNC) accumulate in roots during the growing season (Davis et al. 1991). Little change in root TNC occurs due to harvest or regrowth. Some leaf area should be left on the stubble after cutting to allow photosynthesis for regrowth.

Forage Quality. Annual lespedezas are excellent pasture legumes for summer grazing. Although they do not produce much forage, the forage that is produced is free of bloat hazard and develops when it is most needed. They are most productive during late summer, when the growth rate and quality of most perennial pastures, especially cool-season grasses, are low. Annual lespedezas maintain good quality during summer (McGraw et al. 1989). In a 2-yr study conducted at Columbia, Missouri, Marion and Summit lespedeza produced their maximum growth during July and August. During that period Marion maintained an average crude protein concentration of 138 g/kg, an acid detergent fiber of 335 g/kg, and an NDF of 549 g/kg, and Summit maintained concentrations of 139 g/kg, 394 g/kg, and 555 g/kg, respectively.

SEED PRODUCTION

Commercial seed yields of annual lespedeza usually range from 200 to 300 kg/ha; however, greater yields are often achieved. Seed is normally marketed unhulled (seed still enclosed by the pod). Seed is usually harvested directly with a combine when the leaves become dry and the pods have turned brown or after a killing frost. Harvesting before leaves become completely dry is desirable to reduce seed shattering, especially for striate cultivars. Under these conditions, seed moisture is too high for storage, and seed must be dried to maintain seed viability. Seed shattering is sufficient to ensure dense volunteer stands the next season.

Diseases and Pests. Diseases can cause great losses, with korean cultivars being the most susceptible. Bacterial wilt (*Xanthomonas campestris* pv. *lespedezae* [Ayers et al.] Dye) is a serious problem on korean cultivars in the northern part of the lespedeza region. The entire plant may wilt and die, but usually only a part of the plant is affected, causing stunting and leaf loss (Offutt and Baldridge 1956). Tar spot (*Phyllachora lespedeza* [Schw.] Sacc.) causes heavy leaf spotting, defoliation, and a reduction of yield and quality (Hanson et al. 1956). *Rhizoctonia solani* Kuehn, powdery mildew (*Microsphaera diffusa* C. & P.), and southern blight (*Sclerotium rolfsii* Sacc.) are sometimes serious in the southern part of the lespedeza-growing region.

Root-knot nematodes (*Meloidogyne incognita* [Kofoid & White] Chitwood) can damage annual lespedeza in the southeastern coastal plain. Lespedeza is also a favorable host for the soybean cyst nematode (*Heterodera glycines* Ichinhoe), tobacco stunt nematode (*Tylenchorhynchus claytoni* Steiner), and sting nematode (*Belonolaimus gracilus* Steiner) (Webster 1972).

Insect damage to annual lespedeza is usually slight, but damage can be caused by grasshoppers, armyworms (*Pseudaletia* spp.), lespedeza webworms (*Tetralopha scortealis* [Lederer]), and the three-cornered alfalfa hopper (*Spissistilus festinus* [Say]).

Sericea Lespedeza

Sericea lespedeza (*Lespedeza cuneata* [Dum.-Cours.] G. Don) is native to eastern China, Korea, and Japan (Pieters 1939). This warm-season perennial legume was widely planted in the southeastern US during the 1930s and 1940s for forage and soil conservation. However, it generally was considered to be a low-quality forage plant that had to be managed carefully to produce forage of acceptable quality for livestock. Newer cultivars are much improved and provide high-quality forage for summer grazing or hay.

DISTRIBUTION AND ADAPTATION

Sericea is well adapted to the humid eastern US from Missouri and Ohio to the Gulf Coast (Fig 21.1). This legume is not grown to any extent in similar climatic zones elsewhere in the world, although it has shown excellent promise in southern Africa and southern Brazil.

PLANT DESCRIPTION

Sericea is a leafy, herbaceous, long-lived perennial legume. It has an erect growth habit and coarse to fine stems (Fig. 21.4). When not cut during the season, it will grow to a height of 0.5-1.0 m. Sericea produces trifoliolate leaves distributed along the entire stem. Leaflets are long and narrow, and the tips are truncate or slightly indented and end in a sharp point (Fig. 21.5). Flowers are borne on short pedicels in leaf axils along the stem and in colors varying from cream to purple. Flowers are cleistogamous and chasmogamous. The former are always self-fertilized, while the latter often are cross-fertilized (Donnelly 1979). As a summer-growing legume, the forage growth is killed back to the ground each autumn and renewed from crown buds the following spring. This drought tolerant legume develops a deep branched taproot.

IMPORTANCE AND USE

Sericea was planted widely for forage and soil conservation during the 1930s, reaching a peak in the 1950s, after which higher-quality forages, N-fertilized grasses, and row crops replaced much of the hectarage. Since then, most new sericea plantings have been on eroded land, on mine spoils, and along highways. The advent of new higher-quality sericea cultivars offers an opportunity for low-input forage production satisfactory for beef cattle.

CULTIVARS

Old plantings of sericea consist of the common thick stem types. However, the release of 'Serala' by Auburn University in 1962 resulted in substantial plantings of this high-tannin cultivar with fine, soft stems (Table 21.1). Subsequent releases from the Alabama breeding program include 'Interstate', a profuse shorter branching cultivar for highway planting, and two cultivars with resistance to root-knot nematode, 'Serala 76' and 'Interstate 76'. 'Appalow', released in 1978 by the SCS, USDA, and University of Kentucky, is a prostrate, leafy cultivar for highway use.

Since high tannin content is less desirable for forage use, the release of the tall fine-stemmed, low-tannin cultivar 'AU Lotan' in 1980 by Auburn University was a breakthrough in sericea breeding. It contains about

Fig. 21.4. Sericea perennial lespedeza growing along a roadside in Boone County, Missouri, in early August. *Mo. Agric. Exp. Stn. photo.*

Fig. 21.5. Single leaf (*A*) and stem (*B*) of sericea perennial lespedeza. *Mo. Agric. Exp. Stn. photo.*

one-half the tannin and averages about 27% higher digestibility than other cultivars. However, because the low yield of AU Lotan was undesirable, it has been improved in the low-tannin cultivar 'AU Donnelly', released in 1989, which has yields similar to high-tannin cultivars.

MANAGEMENT

Soil Fertility. Sericea is well adapted to clay or loam soils and grows well under conditions of poor soil fertility where many other legumes do not thrive. Yield response to P and potassium (K) fertilization has been poor (Wilson 1955; Joost et al. 1989). Phosphorus, K, and lime have little effect on tannin content of sericea except where plant growth is severely inhibited (Wilson 1955).

An attractive feature of sericea is its tolerance to acid soils with aluminum (Al) levels that are toxic to many species (Joost and Hoveland 1986). In addition, sericea genotypes differ in tolerance to Al (Joost et al. 1986), offering further potential for use on these sites. Sericea has been valuable in revegetating mine spoils where it will grow on soils below pH 5 and where Al toxicity is a problem (Powell et al. 1983). The mechanism of Al tolerance in acid soils is related to the exclusion of Al by roots (Joost and Hoveland 1985). Sericea does not respond to lime on agricultural soils in Alabama (Wilson 1955).

Establishment. Sericea produces a high percentage of hard seed; therefore, scarified seed should be used to facilitate prompt germination. Inoculation with *Bradyrhizobium* spp. (cowpea miscellany group) is generally unnecessary but is inexpensive insurance, especially when planting on mine spoils or on fields where warm-season legumes were not previously grown (Hoveland et al. 1990). A well-prepared seedbed is essential, and seeding into grass sod has generally been unsuccessful.

Sericea has a reputation for extremely slow establishment and being a poor competitor with weeds (Wright et al. 1978). Slow germination of sericea seed may be caused partially by a water soluble inhibitor in the seed coat that must degrade or dissolve naturally (Logan et al. 1969). Seed germination and seedling emergence of sericea increase linearly with increasing temperature, and seedling growth is greatly reduced at lower temperatures and shorter daylengths (Mosjidis 1990). Thus, the traditional early planting of sericea in March or April to avoid weeds and drought results in slow establishment. Later planting dates may be more desirable if weeds can be controlled. Herbicide application at planting permits a lower seeding rate and should result in a hay harvest the first year (Hoveland et al. 1971). Once established, sericea is quite competitive with weeds, but herbicides may be needed for control of annual winter weeds (Smith and Calvert 1987).

Physiology and Harvest Management. Sericea can be harvested for hay two to three times each year. Hay yields of 6000-7000 kg/ha are possible on relatively poor land, with higher yields on more favorable soils (Hoveland et al. 1990). Sericea yields much less than alfalfa (*Medicago sativa* L.) when both are grown with high fertility and good management. The lower yield of sericea is due mainly to its shorter growing season but also may be related to a lower apparent photosynthesis rate that is about half that of alfalfa (Brown and Radcliffe 1986).

The last hay cutting of sericea should be made no later than mid-August in the southern US so that carbohydrate reserves can accumulate in roots prior to frost. An 8-10 cm stubble should be left after harvest because new growth starts from buds on the stubble (Hoveland and Anthony 1974). If autumn moisture conditions result in good forage growth, it is possible to graze or to make another cutting in mid-October without weakening the stand.

Sericea hay cures rapidly. If atmospheric moisture is low, leaves may shatter if baling is delayed more than 6 h after mowing. It can be cut late in the afternoon, windrowed the next morning before the dew dries, and baled. Ideally, a hay conditioner should be used to crush the stems so they will dry more quickly.

Sericea does not cause bloat problems in cattle and can be a good pasture legume. Grazing should begin in the spring when the plants are 15-20 cm tall. Sericea pastures should not be grazed lower than 10-15 cm since close defoliation will reduce yields, weaken stands, and allow weed encroachment (Hoveland and Anthony 1974).

Where a sericea-grass mixture is grown for soil conservation and is not being grazed, sericea can eliminate stands of grass. On surface-mined land in Kentucky, Interstate sericea seeded with switchgrass (*Panicum virgatum* L.) dominated stands within 3 yr (Henry et al. 1981).

OVERSEEDING INTO SERICEA

Winter annual species can be overseeded into sericea in autumn to provide forage during winter and spring when sericea is dormant (Hoveland and Carden 1971). Rye (*Secale cereale* L.), annual ryegrass (*Lolium multiflorum* Lam.), or crimson clover (*Trifolium incarnatum* L.) overseeded in sericea can furnish an additional 1500 to 2000 kg/ha dry matter and extend the productive season in Georgia and Alabama by at least 2 mo (Hoveland et al. 1990). Planting tall fescue (*Festuca arundinacea* Schreb.) or orchardgrass (*Dactylis golmerata* L.) in sericea can extend the growing season and provide a higher yield than either grass or sericea alone (Hoveland et al. 1975). Because sericea furnishes little N to associated grasses, the grass should be fertilized in autumn with the same level of N the grass would receive if grown alone. Conversely, crimson clover can be used to extend the season and provide fixed N.

FORAGE QUALITY AND ANIMAL PERFORMANCE

Common sericea has large coarse stems and high leaf tannin content that reduce forage quality and intake. The problem of coarse stems has been overcome with the release of fine-stemmed Serala. Serala has a high tannin content but provides satisfactory low-input pasture and hay for beef cows and calves. In Alabama, conception rates, calf-weaning weights, and cow maintenance levels on Serala were similar to those of 'Coastal' bermudagrass (*Cynodon dactylon* [L.] Pers.) (Hoveland et al. 1977).

Tannin reduces digestibility of both dry matter and crude protein (Donnelly et al. 1971; Donnelly and Anthony 1973; Fales 1984). In a feeding trial, Terrill et al. (1989) found tannins reduced intake and decreased forage quality by decreasing fiber and protein digestibility. Insoluble complexes are formed between free tannins and proteins in the digestive tract. Growing animals do not perform as well on high-tannin sericea as they do on low-tannin legume species. In a North Carolina hay-feeding trial, dairy heifers had average daily gains of 0.44, 0.30, and 0.22 kg on crownvetch (*Coronilla varia* L.), alfalfa, and high-tannin sericea, respectively (Burns et al. 1972).

Variability exists among sericea genotypes for leaf tannin content (Donnelly and Anthony 1970). Steers on pasture from April to September for 3 yr in northwest Alabama that ate alfalfa, low-tannin sericea, and high-tannin sericea had average daily gains of 0.98, 0.78, and 0.63 kg and gains of 532, 309, and 278 kg/ha, respectively (Schmidt et al. 1987). The higher gains per hectare on alfalfa were a result of higher-quality forage and a longer productive season. Low-tannin sericea was superior to the high-tannin sericea for growing animals.

Negative correlations between tannin and crude protein and between tannin and digestibility, and the existence of high-yielding, low-tannin breeding lines, indicate that sericea forage quality can be improved while maintaining high yield (Petersen et al. 1991). The problem of high leaf tannin content was greatly reduced in new sericea cultivars such as AU Lotan and AU Donnelly.

Tannin levels in sericea forage are affected by environmental factors; they increase as temperature rises and rainfall decreases (Donnelly 1959). The effect of temperature on tannin content has been demonstrated in controlled environmental chambers (Fales 1984) and in the field where tannin content of both high- and low-tannin cultivars increased from June to August and then declined by October (Windham et al. 1988).

Field-drying of sericea hay reduced the tannin content to 17% of that in uncut forage and increased intake when fed to sheep (Terrill et al. 1989). High-tannin sericea is not readily grazed, but when fed as hay, it is highly palatable, indicating a decline in tannin content. In vitro digestibility of low- and high-tannin sericea forage can be improved by protein or polyvinylpyrrolidone supplements that overcome inhibition by condensed tannins of the enzymes associated with cell wall digestion (Petersen and Hill 1991).

SEED PRODUCTION

Sericea seed yields are highest if no forage is harvested during the year of seed harvest (Adamson and Donnelly 1973). A single early hay cutting reduces seed yields less than later or more frequent cuttings (Hoveland and Anthony 1974). Irrigation in Alabama during the critical seed-setting months of August and September increased seed yields (Donnelly and Patterson 1969; Adamson and Donnelly 1973). Dodder (*Cuscuta campestris* Yunck.) is a serious weed problem in sericea seed production, so it is important to locate seed production on areas free of this pest. Herbicides are available for dodder control (Smith and Calvert 1987).

Proper combine adjustment is important, as seed loss can be high during harvest. Proper timing of seed harvest also is important. Harvesting is done by direct combining in autumn when most, but not all, of the seed is brown. If many green seeds and leaves are present at harvest time, the seed should be spread out to dry or dried artificially to prevent overheating.

DISEASES AND PESTS

Sericea is relatively free from insect damage. This does not appear to be caused by inhibiting effects of tannins in high-tannin cultivars as low-tannin breeding lines are similarly unattractive to fall armyworm (*Pseudaletia frugiperda* [J. E. Smith]) and other insects (Buntin and Wiseman 1990). Grasshoppers can reduce seed yields by defoliating plants in late summer but can be controlled with insecticides.

Sericea normally is not affected by diseases. However, southern blight can cause stand loss (Wells et al. 1969). *Rhizoctonia* spp. may destroy leaf tissue of low-tannin sericea lines in late summer and autumn, but genetic sources of disease resistance have been identified and incorporated into new cultivars (Mosjidis and Donnelly 1989).

Root-knot nematodes (*Meloidogyne* spp.) can be serious pests on light-textured soils, reducing yields and often stands of sericea. The discovery of genetic lines resistant to root-knot nematodes has made it possible to incorporate a high level of resistance in newer cultivars (Minton and Donnelly 1971).

Other Lespedeza Species

Several shrubby species, such as *L. bicolor* Turcz., *L. thunbergii* (DC.) Nakai, *L. cyrtobotrya* Miq., and *L. japonica* L. Bailey, have value as ornamentals, for soil conservation, and as food and shelter for wildlife. *Lespedeza bicolor* has been planted widely as food for wildlife, as have the annual lespedezas (Ball et al. 1991).

QUESTIONS

1. What factors limit the range of adaptation of annual lespedeza in the US?
2. Why are annual lespedezas valuable legumes for low-input pasture systems?
3. Why are annual lespedezas so desirable for improving summer grazing?
4. How can you distinguish striate lespedeza from korean lespedeza?
5. What factors are responsible for lower animal performance on common sericea lespedeza as compared with that on improved cultivars.
6. What factors may be responsible for slow establishment of sericea lespedeza?
7. How should sericea lespedeza be managed in late summer or fall to ensure good regrowth the next spring?
8. What forage species can be overseeded into sericea lespedeza pastures to extend the production season?

REFERENCES

Adamson, HC, and ED Donnelly. 1973. Effect of Cutting and Irrigation on Seed Yield of Interstate Sericea. Ala. Agric. Exp. Stn. Leafl. 87.

Baldridge, JD. 1957. The lespedezas: II. Culture and utilization. Adv. Agron. 9:122-41.

Ball, DM, CS Hoveland, and GD Lacefield. 1991. Southern Forages. Atlanta, Ga: Potash and Phosphate Institute.

Beuselinck, PR, and RL McGraw. 1990. Registration of 'Marion' annual lespedeza. Crop Sci. 30:423-24.

———. 1991. Seed production of annual lespedeza harvested for herbage. J. Appl. Seed Prod. 9:64.

Brown, RH, and DE Radcliffe. 1986. A comparison of apparent photosynthesis in sericea lespedeza and alfalfa. Crop Sci. 26:1208-11.

Buntin, GD, and BR Wiseman. 1990. Growth and development of two polyphagous lepidopterans fed high- and low-tannin sericea lespedeza. Entomol. Exp. Appl. 55:69-78.

Burns, JC, RD Mochrie, and WA Cope. 1972. Responses of dairy heifers to crownvetch, sericea lespedeza, and alfalfa forages. Agron. J. 64:193-95.

Cope, WA. 1957. The lespedezas: III. Lespedeza breeding and improvement. Adv. Agron. 9:142-57.

Davis, DK, RL McGraw, and PR Beuselinck. 1991. Nonstructural carbohydrate accumulation in annual lespedeza. In Am. Soc. Agron. Abstr. Madison, Wis., 141.

Donnelly, ED. 1959. The effect of season, plant maturity, and height on the tannin content of sericea lespedeza, *L. cuneata*. Agron. J. 51:71-73.

———. 1965. Registration of Serala sericea lespedeza. Crop Sci. 5:605.

———. 1971. Registration of Interstate sericea lespedeza. Crop Sci. 11:601-2.

———. 1979. Selection for chasmogamy in sericea lespedeza. Crop Sci. 19:528-31.

———. 1981. Registration of AU Lotan sericea. Crop Sci. 21:474.

Donnelly, ED, and WB Anthony. 1970. Effect of genotype and tannin on dry matter digestibility in sericea lespedeza. Crop Sci. 10:200-202.

———. 1973. Relationship of sericea lespedeza leaf and stem tannin to forage quality. Agron. J. 65:993-94.

Donnelly, ED, and NA Minton. 1979. Registration of Serala 76 and Interstate 76 sericea lespedeza. Crop Sci. 19:929.

Donnelly, ED, and RM Patterson. 1969. Effect of irrigation and clipping on seed production and chasmogamy of sericea genotypes. Agron. J. 61:501-2.

Donnelly, ED, WB Anthony, and JW Langford. 1971. Nutritive relationships in high- and low-tannin sericea lespedeza under grazing. Agron. J. 63:749-51.

Fales, SL. 1984. Influence of temperature on chemical composition and in vitro dry matter disappearance of normal- and low-tannin sericea lespedeza. Can. J. Plant Sci. 64:637-42.

Gray, E, and JH Reynolds. 1970. Effect of stage of maturity on yield and leaf content of annual lespedeza. Tenn. Farm Home Sci. Prog. Rep. 74:25-28.

Hanson, CH, WA Cope, and JL Allison. 1956. Tar spot of korean lespedeza caused by *Phyllachora* sp.: Losses in yield and deferential susceptibility of strains. Agron. J. 48:369-70.

Henry, DS, and NL Taylor. 1981. Registration of Appalow sericea lespedeza. Crop Sci. 21:144.

Henry, DS, WF Kuenstler, and SA Sanders. 1981. Establishment of forage species on surface mined land in Kentucky. J. Soil and Water Conserv. 36:111-13.

Henson, PR. 1957. The lespedezas: I. Origin, history, and development of lespedeza in the United States. Adv. Agron. 9:113-22.

Hoveland, CS, and WB Anthony. 1974. Cutting management of sericea lespedeza for forage and seed. Agron. J. 66:189-91.

Hoveland, CS, and EC Carden. 1971. Overseeding winter annual grasses in sericea lespedeza. Agron. J. 63:333-34.

Hoveland, CS, GA Buchanan, and ED Donnelly. 1971. Establishment of sericea lespedeza. Weed Sci. 19:21-24.

Hoveland, CS, WB Anthony, EL Carden, JK Boseck, and WB Webster. 1975. Sericea-Grass Mixtures. Ala. Agric. Exp. Stn. Circ. 221.

Hoveland, CS, WB Anthony, JA McGuire, WA Griffey, and HE Burgess. 1977. Forage Systems for Beef Cows and Calves in the Piedmont. Ala. Agric. Exp. Stn. Bull. 497.

Hoveland, CS, WR Windham, DL Boggs, RG Durham, GV Calvert, JF Newsome, JW Dobson, Jr., and M Owsley. 1990. Sericea Lespedeza Production in Georgia. Ga. Agric. Exp. Stn. Res. Bull. 393.

Joost, RE, and CS Hoveland. 1985. Mechanism of aluminum tolerance in sericea lespedeza. In Proc. 15th Int. Grassl. Congr., Jap. Soc. Grassl. Sci., Tochigi-ken, Japan, 394-96.

———. 1986. Root development of sericea lespedeza and alfalfa in acid soils. Agron. J. 78:711-14.

Joost, RE, BE Mathews, and CS Hoveland. 1989. Phosphorus response of sericea lespedeza on a

Georgia ultisol. Commun. Soil Sci. Plant Anal. 20:837-49.
Joost, RE, CS Hoveland, ED Donnelly, and SL Fales. 1986. Screening sericea lespedeza for aluminum tolerance. Crop Sci. 26:1250-51.
Kenney, R. 1932. Korean Lespedeza. Ky. Agric. Exp. Stn. Circ. 258.
Logan, RH, CS Hoveland, and ED Donnelly. 1969. A germination inhibitor in the seedcoat of sericea (*Lespedeza cuneata* [Dumont] G. Don). Agron. J. 62:265-66.
McGraw, RL, PR Beuselinck, and CA Roberts. 1989. Growth and forage quality of annual lespedeza. In Am. Soc. Agron. Abstr. Madison, Wis., 171.
McGraw, RL, PR Beuselinck, RD Hammer, and CA Roberts. 1990. Growth and quality during the seeding year of alfalfa, red clover, and annual lespedeza. In Am. Soc. Agron. Abstr. Madison, Wis., 189-90.
McNair, AD, and WB Mercier. 1911. Lespedeza, or Japan Clover. Farmers' Bull. 441. Washington, D.C.: US Gov. Print. Off.
Minton, NA, and ED Donnelly. 1971. Reaction of field-grown sericea lespedeza to selected *Meloidogyne* spp. J. Nematol. 3:369-73.
Mosjidis, JA. 1990. Daylength and temperature effects on emergence and early growth of sericea lespedeza. Agron. J. 82:923-26.
Mosjidis, JA, and ED Donnelly. 1989. Registration of 'AU Donnelly' sericea lespedeza. Crop Sci. 29:237-38.
Nakata, S. 1952. Photoperiod response of lespedeza. Plant Physiol. 27:644-47.
Offutt, MS. 1963. Registration of Summit lespedeza. Crop Sci. 3:368.
———. 1968. Some effects of photoperiod on the performance of korean lespedeza. Crop Sci. 8:309-13.
Offutt, MS, and J Baldridge. 1956. Inoculation Studies Related to Breeding for Resistance to Bacterial Wilt in Lespedeza. Mo. Agric. Exp. Stn. Res. Bull. 603.
Ohashi, H, RM Polhil, and BG Schubet. 1981. Desmodieae. In RM Polhill and PH Raven (eds.), Advances in Legume Systematics, vol. 2, pt. 1, Proc. Int. Legume Conf., Royal Botanic Gardens, Kew, England, 292-300.
Peterson, JC, and NS Hill. 1991. Enzyme inhibition by sericea lespedeza tannins and the use of supplements to restore activity. Crop Sci. 31:827-32.
Peterson, JC, NS Hill, JA Mosjidis, and WR Windham. 1991. Screening sericea lespedeza germplasm for herbage quality. Agron. J. 83:581-88.
Pieters, AJ. 1934. The Little Book of Lespedeza. Strasburg, Va.: Shenandoah Publishing House.
———. 1939. Lespedeza Sericea and Other Perennial Lespedezas for Forage and Soil Conservation. USDA Circ. 534.
Polhill, RM, and PH Raven (eds.). 1981. Advances in Legume Systematics. Vol. 2, pt. 1, Proc. Int. Legme Conf., Royal Botanic Gardens, Kew, England.
Powell, JL, RI Barnhisel, ML Ellis, and JR Armstrong. 1983. Suitability of various legume species for revegetation of acid surface-mined coal spoils in western Kentucky. In Symp. Surf. Min., Hydrol., Sedimentol., and Reclam., Univ. of Kentucky, Lexington, 121-29.
Schmidt, SP, CS Hoveland, ED Donnelly, JA McGuire, and RA Moore. 1987. Beef Steer Performance on Cimarron Alfalfa and Serala and AU Lotan Sericea Lespedeza Pastures. Ala. Agric. Exp. Stn. Circ. 288.
Smith, AE, and GV Calvert. 1987. Weed Control in Sericea Lespedeza. Ga. Agric. Exp. Stn. Res. Bull. 357.
Smith, GE. 1941. The effect of photo-period on the growth of lespedeza. J. Am. Soc. Agron. 33:231-36.
Terrill, TH, WR Windham, CS Hoveland, and HE Amos. 1989. Forage preservation method influences on tannin concentration, intake, and digestibility of sericea lespedeza by sheep. Agron. J. 81:435-39.
Webster, JM. 1972. Economic Nematology. London: Academic Press.
Wells, HD, NA Minton, and ED Donnelly. 1969. *Lespedeza cuneata,* a new host for *Sclerotium rolfsii.* Plant Dis. Rep. 53:528-29.
Wiersema, JH, JH Kirkbride, Jr., and CR Gunn. 1990. Legume (Fabaceae) nomenclature in the USDA germplasm system. In USDA Tech. Bull. 1757.
Wilson, CM. 1955. The effect of soil treatments on the tannin content of lespedeza sericea. Agron. J. 47:83-86.
Windham, WR, SL Fales, and CS Hoveland. 1988. Analysis for tannin concentration in sericea lespedeza by near infrared reflectance spectroscopy. Crop Sci. 28:705-8.
Wright, DL, RE Blaser, and JM Woodruff. 1978. Seedling emergence as related to temperature and moisture tension. Agron. J. 70:709-12.

22 Other Temperate Legumes

DARRELL A. MILLER
University of Illinois

CARL S. HOVELAND
University of Georgia

MANY temperate tropical and subtropical legume species have been introduced and cultivated in the US to meet a diversity of climates and soils. Although certain species have a narrow range of adaptation, they are quite valuable in that area. Tropical and subtropical legume species are presented in Chapter 23. Legumes are utilized not only for forage but also for erosion control and green manure in rotation with other crops.

PERENNIAL LEGUMES

Crownvetch. Crownvetch, *Coronilla varia* L., a native of central Europe, is now widely distributed throughout most of Europe and extends eastward to the Caucasus and Asia Minor (Duke 1981). It was used mainly as an ornamental in Europe and was available commercially for similar purposes in the US by 1890. The species appears to be well adapted to a wide area ranging from northern Georgia to eastern Canada and west to eastern Nebraska. Although established plants are tolerant of moderately acid and infertile soils, it is best adapted to well-drained, fertile soils with a pH 6.0 or above.

Crownvetch, a long-lived perennial, spreads by creeping underground rootstocks. It has a deeply penetrating taproot and numerous lateral roots. Stems are hollow, angular, decumbent to ascending, and up to 1. 2 m in length (Fig. 22. 1). Flowers, variegated white to purple, are borne on umbels or tightly contracted racemes that arise from leaf axils. Seeds are borne in indehiscent pods that are transversely divided into 3-7 single-seeded segments that break into sections when dry. Yellow ochre to mahogany in color, the seeds are rod-shaped and about 3.5 mm long and 1 mm in diameter. There are about 245 seeds/g. Seeds have a hard seed coat and must be scarified before planting.

Crownvetch is widely used in the eastern US for beautification, erosion control, highway embankments, mine spoil areas, and other disturbed areas. It has a number of desirable pasture attributes such as spreading by rhizomes, a relatively long productive season, freedom from insects and diseases, and absence of bloat in grazing animals. However, seedling vigor is poor, and establishment is slow, so weeds can be a problem since crownvetch is less tolerant of herbicides than alfalfa. Crownvetch can be successfully seeded in prepared land and into Kentucky bluegrass (*Poa pratensis* L.) sod, but seeding into tall fescue (*Festuca arundinacea* Schreb.) swards has not been successful.

Crownvetch yields of 7-9 mt ha^{-1} can be obtained when the legume is cut twice a year for hay (Mays and Evans 1972; Hart et al. 1977). Cutting more frequently decreases forage yield and stand persistence of crownvetch. Close cutting to a 5-cm stubble decreases stands, with the cultivar 'Penngift' surviving better than 'Chemung' under this treatment (Hart et al. 1977). Seasonal trends of carbohydrate reserves in the roots are similar to those for alfalfa. The importance of regrowth sites such as axillary and crown buds depends

DARRELL A. MILLER. *See Chapter 4.*

CARL S. HOVELAND. *See Chapter 20.*

Fig. 22.1. Crownvetch is a useful and beautiful perennial legume for erosion control on flood prevention dams or highway slopes. *SCS photo.*

on the stage of growth when the forage is harvested because the proportion of axillary buds producing regrowth increases with a delay in the first harvest until full bloom.

In North Carolina crownvetch pasture was found to provide high daily gains for beef cows and calves, but under continuous stocking, stand persistence declined more rapidly than under frequent clipping (Burns et al. 1977). In another study lamb gains were excellent on Penngift crownvetch pasture in the Transvaal of South Africa and not affected by rotational stocking (Barnes and Dempsey 1992). However, continuous and frequent rotational stocking resulted in severe invasion by bermudagrass (*Cynodon dactylon* [L.] Pers.), which was not a problem when a rest interval of 30 d between grazings was used. Animals that have grazed crownvetch previously accept the species more readily than do animals that have not been acclimated (Bryant et al. 1977).

Gustine (1979) reveiwed the antiquality components of crownvetch and found that the glucose esters of 3-nitropropanoic acid (NPA) are the primary compounds. Crownvetch may be safely fed to ruminants, but if fed to nonruminants, its concentration should not exceed 5% of the ration.

Seed yields of crownvetch are low because of uneven maturity due to an indeterminate flowering habit, shattering of seed, and excessive vegetative growth (Al-Tikrity et al. 1974). A yield of 530 kg ha^{-1} has been obtained from hand-harvested experimental plots, but commercial seed yields of about 100 kg ha^{-1} are common. A cross-pollinated species, crownvetch is pollinated by honeybees and bumblebees. It appears to be relatively free of insect and disease pests.

Penngift, 'Emerald', and Chemung developed as ecotypes in Pennsylvania, Iowa, and New York, respectively, are the most widely used cultivars of crownvetch. There are about 20 species of *Coronilla*.

Cicer Milkvetch. Cicer milkvetch, *Astragalus cicer* L., is distributed from the Caucasus Mountains across southern Europe to Spain. It was introduced into the US in the 1920s and is used as a pasture, hay, or conservation species in the Great Plains and western US and in adjacent areas of Canada. Cicer milkvetch, a very winter-hardy species, is adapted to a wide range of conditions, ranging from irrigated land to dry lands that receive more than 400 mm of annual precipitation. It also is adapted to a wide range of soil types and tolerates slight acidity to moderate alkalinity.

Cicer milkvetch, a long-lived perennial, spreads by rhizomes (Fig. 22.2). It has a branched taproot and ascending or prostrate hollow stems that may reach a length of 1.3 m under favorable conditions. Leaves are pinnately compound with 8-17 pairs of leaflets plus one terminal leaflet. Up to 50 pale yellow to white flowers are borne on racemes that arise from leaf axils. Bumblebees are the principal pollinator of this cross-pollinated species. At maturity the black bladderlike seedpods contain up to 12 seeds that are generally bright yellow. There are about 260 seeds/g. The seeds have a shiny, hard seed coat and must be scarified before planting.

Forage quality of cicer milkvetch appears to be equal to that of alfalfa (McGraw and Marten 1986). It does not cause bloat and tolerates grazing well. However, in Minnesota poor palatability and low intake by grazing heifers was found to result in daily gains about one-half that on birdsfoot trefoil (*Lotus corniculatus* L.) or common sainfoin (*Onobrychis viciifolia* Scop.) (Marten et al. 1987). Photosensitization of ruminants grazing cicer milkvetch suggests a possible phototoxin in the plant. (See Chap. 9, Vol. 2.)

When grown with adequate moisture and harvested for hay, cicer milkvetch generally yields 75%-80% of that of alfalfa. Seasonal trends of carbohydrate reserves in the roots and rhizomes are similar to those for birdsfoot trefoil. When harvested frequently under simulated grazing conditions, forage yields compare favorably with those of other pasture legumes (Townsend et al. 1978). Seed yields in excess of 500 kg ha^{-1} are common. No major disease or insect pests have been noted.

Four cultivars have been developed. 'Lutana' was released in Montana and Wyoming

Fig. 22.2. Cicer milkvetch is a rhizomatous perennial forage legume adapted to drier regions of the US. *ARS, USDA-Colorado State Univ. photo.*

in 1970; 'Oxley' was licensed for use in Canada in 1971; and in Colorado 'Monarch' was released in 1980 and 'Windsor' in 1992.

Sainfoin. Common sainfoin, *Onobrychis viciifolia* Scop, is native to and has been grown for centuries in Europe and western Asia (Ditterline and Cooper 1975). Although sainfoin was tested widely in the US during the early 1900s, it has not attained the status of a major forage crop. It is adapted to the dry calcareous soils of the western US and to adjacent areas of Canada that receive at least 330 mm of annual precipitation, as well as to sites with limited irrigation. In comparison with other legumes, sainfoin grows well on soils low in phosphorus (P). Nodulation with *Rhizobium* is not always effective, and nitrogen (N) fixation may not be adequate.

Sainfoin has a deep, branched taproot. Erect, hollow stems about 1 m in length arise from the crown. Leaves are pinnately compound, with 11-29 leaflets. Up to 80 pink flowers are borne on an erect raceme. A single seed is produced in a bilaterally compressed pod. Seeds are kidney-shaped and in colors ranging from olive to brown or black. The weight of milled (pods removed) seed is about 1.5 g/100 seeds. Seed scarification is not necessary.

Sainfoin has excellent seedling vigor, but it should not be seeded with a companion crop (Ditterline and Cooper 1975). Hay yields of sainfoin are generally less than that of alfalfa, with production concentrated in spring. Water use efficiency of sainfoin is lower than that of alfalfa (Bolger and Matches 1990). Thus, in dryland areas, sainfoin's greatest utility is for early-season irrigated pasture or hay. Forage crude protein content is lower than, and digestibility similar to, alfalfa (McGraw and Marten 1986). Cattle performance on sainfoin hay has compared favorably with that obtained on alfalfa (Parker and Moss 1981). Sainfoin, a nonbloating legume, has furnished excellent daily gains when grazed with cattle or sheep (Marten et al. 1987; Karnezos and Matches 1991). Since close continuous grazing will result in stand losses, sainfoin stand persistence is improved by light to medium grazing intensities at the bud or flower stages (Mowrey and Matches 1991).

Seed yields are excellent and range up to 1450 kg ha^{-1}. Seed is threshed by means of swathing and drying the forage in windrows. The seed has potential for use as a protein supplement for monogastric animals. Honeybees are the principal pollinator of this cross-pollinated species. Crown and root rots associated with *Fusarium* spp. cause loss of stand under both dryland and irrigated conditions. However, several bacterial pathogens have been found in this disease complex under irrigation. Although insect pests have not been a problem in the US, the three most important pest species are the *Rhizobium*-nodule-eating weevil (*Sitona scissifons* Say), the lygus bugs (*Lygus elisus* van Duzee, *L. hesperus* Knight), and the sainfoin bruchid (*Bruchudius unicolor* Oliver). Sainfoin is resistant to the alfalfa weevil, *Hypera postica* Gyll.

Five cultivars of sainfoin have been developed in the US and Canada. 'Eski' and 'Remont' were released in Montana in 1964 and 1971, respectively. 'Melrose' and 'Nova' were released in Canada in 1969 and 1980, respectively. 'Renumex' was released in New Mexico in 1977. There are many species of the genus *Onobrychis*. *Onobrychis arenaria* (Kit.) DC. and *O. transcaucasica* Grossh. are cultivated in the former USSR, and *O. sativa* Lam. is cultivated in Great Britain.

Kudzu. Kudzu, *Pueraria lobata* (Willd.) Ohwi, is native to Japan, China, and Korea (Duke 1981). It was introduced to the US in 1876 and used as an ornamental until it became widely planted for soil conservation during the 1920s and 1930s. Kudzu is best adapted south of Virginia and Kentucky and west to eastern Oklahoma and Texas. It is a warm weather plant, growing from early spring until late fall. The aboveground parts are killed by frost, and deep freezing of soil will kill the entire plant. It grows on a wide range of soil types and is tolerant of soil acidity and low fertility but not poor drainage. Kudzu is a deep-rooted, drought tolerant, rapidly growing, coarse, hirsute, stoloniferous vine. Stems may attain a length of 20 m in a season, are 1 cm or less in diameter, and climb high in trees or other objects. The green trifoliolate leaves resemble grape leaves and have leaflets hirsute on the upper surface. The leaves are very white beneath, 10-15 cm in length and width, and often three-lobed. Racemes are 10-20 cm long with many reddish purple flowers that rarely set seed before frost except in the extreme southeastern US. Propagation is usually by transplanted seedlings or crowns.

Kudzu has been used primarily to prevent soil erosion (Lynd and Ansman 1990). It can be a pest as it spreads into pine plantations and other areas but is easily controlled by close continuous grazing or herbicides. The forage is of high nutritive quality and is palatable to livestock. Hay yields are relatively low, but the plant retains leaves well during harvest. Animal performance on kudzu pasture is good, but it will not tolerate close continuous grazing, so light or rotational stocking is desirable (Duke 1981).

BIENNIAL LEGUMES

Sweetclover. Sweetclover, *Melilotus,* is Eurasian in origin and ranges from central Europe to as far east as Tibet (Duke 1981). Some species have become naturalized in North America and other continents. Sweetclover was first reported growing in the US in 1739; the two principal types are yellow, *M. officinalis* Lam., and white, *M. alba* Medik. By 1900 sweetclover was well recognized for its soil improvement properties. Later it was widely used as a field crop in the Corn Belt and Great Plains. Sweetclover is adapted to a wide range of soil and climatic conditions. It does not, however, tolerate acid soils. It is drought tolerant and winter-hardy. In recent years the hectarage of sweetclover has declined. An excellent review of all phases of sweetclover improvement, culture, and utilization was published by Smith and Gorz (1965).

Generally, the cultivated forms of sweetclover are biennial, but some annual forms are also cultivated. In the seedling year the biennial form produces a plant with a single well-branched stem. Sweetclover has a deeply penetrating taproot, and several buds form at the crown near the end of the first growing season. The root then enlarges and serves as a storage site for food reserves. Coarse, vigorous stems that may reach a height of 2.8 m for *M. officinalis* arise from the crown buds in the spring of the second year. Leaves are trifoliolate, and the leaflets tend to be toothed around the margin. Numerous flowers are borne on racemes. Seeds have a hard seed coat and should be scarified before planting.

In addition to its remarkable value as a soil-improving crop, sweetclover is used also for pasture, hay, or silage. Its use for pasture, however, far exceeds its use for hay or silage. Coumarin, an aromatic compound, affects the palatability of sweetclover forage until the animals become accustomed to the bitter taste (Duke 1981). Dicoumarol, a toxic substance derived from coumarin during the heating and spoilage of sweetclover hay or silage, reduces the blood-clotting ability of the animals consuming the forage; death may result (see Chap. 9, Vol. 2). This problem has been overcome by the development and release of low-coumarin cultivars.

Sweetclover is an excellent seed producer, but losses are high because of the indeterminate growth habit and the loose attachment of seedpods. Insect pollinators are needed for *M. officinalis* and *M. alba* but not for *M. indica*. Seed yields average about 225 kg ha^{-1}.

In addition to the 21 cultivars developed prior to 1965, 'Polara', a biennial white-flowered cultivar, and 'Norgold', a biennial yellow-flowered cultivar, have been released in Canada. 'Denta' is a biennial yellow-flowered cultivar developed in Nebraska. All three cultivars are low in coumarin. There are a total of 20 *Melilotus* spp.

Diseases of sweetclover include root rot caused by *Phytophthora cactorum* (Led. & Cohn) Schroet., black stem caused by *Ascochyta meliloti* Trus. and *Cercospora davisii* El. & Ev., and stem canker caused by *A. caulicola* Laub. The major insect pest of sweetclover is the adult sweetclover weevil, *Sitona cylindricollis* Fåhr. Other insect pests are the sweetclover root borer, *Walshia miscecolorella* (Chambers), the sweetclover aphid, *Therioaphis riehmi* (Boerner), and blister beetles, *Epicauta* spp.

ANNUAL LEGUMES

Vetches. The genus *Vicia* contains about 150 species, including 15 that are native to the US. However, most vetch species are native to the Mediterranean region of southern Europe (Duke 1981). Vetches are widely grown in temperate areas of the world for green manure and forage (Fig. 22.3).

Cold-hardy vetch species such as hairy vetch, *V. villosa* Roth, are adapted over a wide

Fig. 22.3. Plowing down winter vetch can supply the entire N needs for a crop of sorghum or cotton. *Auburn Univ. photo.*

area of the US. Common vetch, *V. sativa* L., is less cold-hardy and can be grown as a winter legume only in areas with milder winter temperatures. Hairy vetch and *V. angustifolia* L. are more tolerant of poorly drained soils than are common vetch or hybrids (Hoveland and Donnelly 1966). Vetches in general are more tolerant to acid soil conditions than are most legume crops but have a relatively high requirement for P.

Cultivated vetches are viny annuals with stems attaining a length of 60-180 cm. The stems bear leaves with pinnate leaflets and terminate in tendrils that attach themselves to stems of other plants such as small grains, cotton, or sorghum stalks. White or purple flowers, depending on the species, are borne in a cluster or raceme. Seed and pod characteristics vary with species, but in general the seed is round or oval, and the pods are elongated and compressed. In the southern US, vetch flowers during April or May, and seed matures in late May or June.

Vetches, seeded on prepared land in autumn, can produce 2000 kg ha^{-1} of forage over the winter-early spring season to provide over 100 kg N ha^{-1} for subsequent crops. In addition to use as green manure in a crop rotation, vetches can be overseeded on warm-season grass sods to extend the grazing season and provide good beef steer gains (Harris et al. 1972). Grazing of vetch should not begin until the plants are at least 15 cm tall. Close grazing below the lowest leaf axil will remove axillary buds, resulting in slow regrowth.

Vetch seed should be harvested when most of the seedpods have turned brown. Hairy vetch will shatter quickly after pods are mature, and delay in harvesting may cause heavy shattering losses. Vetch seed yields range from 400 kg ha^{-1} to over 1500 kg ha^{-1}.

Insects are the main problem with vetches, particularly in seed production. Pea aphids, *Acyrthosiphon pisum* (Harris); corn earworm, *Heliothis zea* Boddie; fall armyworm, *Spodoptera frugiperda* (J. E. Smith); and spider mites, *Tetranychus* spp., can damage vetch. Hairy vetch is susceptible to the vetch bruchid or weevil, *Bruchus brachialis* Fahr, which destroys the interior of the seed and which may not emerge for several weeks after harvest, leaving only empty seed coats. This insect pest is largely responsible for the usual poor natural reseeding of hairy vetch in pastures.

Most of the hairy vetch hectarage is of the common type, but there is one high-yielding hardy cultivar, 'Madison', developed in Nebraska. 'Lana' is a nonhardy cultivar of winter vetch, *V. villosa*. Common vetch, *V. sativa*, cultivars include 'Willamette' and 'Warrior'. 'Woodford' is a cultivar of bigflower vetch, *V. grandiflora*, released in Kentucky for use on less-developed pastures. Interspecific hybrids, *V. sativa* × *V. cordata*, developed in Alabama include 'Cahaba White', 'Vantage', and 'Nova II'. 'Vangard' is a hybrid of *V. sativa* × *V. serratifolia*. All of the hybrid vetches are resistant to the bruchid insect.

Lupines. Many species of *Lupinus* are native to America, but the three cultivated large-seeded species originated in the Mediterranean region (Duke 1981). Blue lupine, *L. angustifolius* L., has been the most commonly grown species in the US; white lupine, *L. albus* L., is the most winter-hardy and has less seed shattering; and yellow lupine, *L. luteus* L., is grown as a summer annual in northern Europe. Lupines were an important winter green manure and grazing crop in northern Florida, southern Georgia, and Alabama during the 1940s and 1950s, but disease problems and cheap N fertilizer virtually eliminated the crop in this area. Lupines are grown as a grain legume crop in Australia, New Zealand, and Europe. In recent years, there has been increasing interest in summer production of lupines for grain in the upper Midwest.

Lupines are erect-growing winter annuals, 1 m or more in height, with coarse stems, fingerlike leaves, and a mass of colorful white, yellow, or blue flowers. Lupines require a growing season of 5 mo or more free from serious moisture stress and with monthly maximum temperatures between 15° and 25°C. Most cultivated lupines are intolerant of winter temperatures below –5° to –10°C.

Lupines thrive on moderately acid sandy as well as loam soils with good drainage. The seeds are large, so seeding rates of 50-180 kg ha^{-1} are needed. Most lupines are high in alkaloids, which repel insects and make the plants unattractive to grazing animals (Lopez-Bellido and Fuentes 1986). Lupines high in alkaloids are considered bitter and are avoided by grazing animals. "Sweet," or low-alkaloid, lupine seed is widely grown for grain production in western Australia.

Major disease problems of lupines are viruses, gray leaf spot (*Stemphylium solani*), anthracnose (*Glomerella cingulata* Ston.), brown spot (*Pleiochaeta setosa*), powdery mildew (*Erysiphe polygoni* DC.), *Ascochyta* stem canker, and root rot (*Pythium de-*

baryanum). Root-knot nematodes (*Meloidogyne* spp.), white-fringed beetle (*Graphognathus leucoloma*), and lupine maggot (*Hylemya lupini*) are additional pests of lupines. Because of the many pests of lupines, it is not desirable for lupines to be planted on the same land two seasons in a row.

Blue lupine cultivars developed in south Georgia include the low-alkaloid, more winter-hardy, disease-resistant 'Frost' and 'Tifblue-78', which is similar to Frost but also is nonshattering. 'Hope', developed in Arkansas, is a high-alkaloid, cold tolerant cultivar.

Roughpea. Roughpea, *Lathyrus hirsutus* L., is also known as *caleypea* or *singletary pea*. This winter annual, a native of the Mediterranean area, is a weak-stemmed, decumbent plant, with a high percentage of hard seed that favors volunteering (Duke 1981). Roughpea grows well on wet clay soils of the lower Mississippi Delta area where many other legumes do not thrive. It also grows well on calcareous clay soils of the Alabama and Mississippi Black Belt.

It is a palatable, nutritious forage, and livestock gain well on it. Seeds of roughpea are toxic, and cattle grazing at plant maturity may exhibit stiffness or lameness from eating the seed. However, animals recover rapidly when removed from the pasture. Roughpea sod-seeded in dallisgrass (*Paspalum dilatatum* Poir.) can sharply increase forage yield and extend the grazing season. This, together with its reseeding ability and tolerance to wet soil conditions, makes roughpea a useful legume where other legumes do not thrive.

Field Pea. The field pea, *Pisum sativum* L. subsp. *sativum* var. *arvense* (L.) Poir., is similar to the garden pea except the latter has a sweeter and more delicate flavor (Duke 1981). In the past, field peas were grown primarily for soil improvement, and there has been a renewed interest in this practice. In general, the field pea is used as a winter annual in the South and as a summer annual in the North. However, the field pea is used as a winter annual in the Palouse region of the Pacific Northwest, where it is grown primarily as a seed crop for export. When used for forage, the field pea is usually grown with a small grain.

'Austrian Winter', a common cultivar, was widely grown for years, but it has been replaced by improved cultivars such as 'Melrose' and 'Glacier', which were released by the Idaho Agricultural Experiment Station.

Cowpea. The cowpea, *Vigna unguiculata* (L.) Walp. ssp. *unguiculata*, was widely grown for hay and pasture in the southeastern US for many years but currently is little used (Duke 1981). It is viny with weak stems, large leaves, and curved pods. The kidney-shaped seed varies in color. This summer annual is well adapted to acid soils and low fertility. Delayed leaf senescence traits allow cowpea to survive and recover from midseason drought (Gwathmey and Hall 1992). Hay yields of 5000 kg ha^{-1} can be obtained, and forage quality is good. The enormous genetic diversity of this species suggests that breeding for forage production would result in improved cultivars.

Pigeonpea. Pigeonpea, *Cajanus cajan* (L.) Millsp., is a perennial but is often grown as an annual. It grows to a height of 4 m in Africa. Its main use is for human food (Whiteman et al. 1985). Pigeonpeas are also grown as a green manure crop and for grazing. Dry matter yield and quality of the foliage are important. Dry matter yields have been reported as high as 51 mt ha^{-1} in Australia and up to 57 mt ha^{-1} in Colombia, based on two harvests per year (Whiteman et al. 1985). A cutting schedule of 8- to 12-wk intervals will give adequte yield and quality forage. Live weight gains as high as 1 kg head^{-1} d^{-1} has been recorded for beef grazing at the fruiting stage (Nene et al. 1990). Seed husks and pod walls are commonly fed to cattle, and the green leaves are used as cattle feed in the trashing process. In the tropics, after the pods are harvested, plants are often left in the field, and cattle graze the new green leaves of the harvested plants (Nene et al. 1990).

Velvetbean. Velvetbean, *Mucuna pruriens* (L.) DC. (formerly *Stizolobium deeringianum* Bort.), is a native of India and was introduced into Florida in the mid-1800s and used as an ornamental until it became widely grown for green manure and pasture (Piper and Morse 1938). During the late 1930s over 800,000 ha were grown annually in the southeastern US, mainly on well-drained sandy soils.

Velvetbean is a vigorous summer viny legume attaining a length of 8 m or more, with trifoliolate leaves having large ovate leaflets. The white to dark purple flowers are borne in long clusters. The hairy seedpods are 5-15 cm long with 3-6 seeds per pod. Velvetbean is relatively free of pests except for the velvetbean caterpillar (*Anticarsia gemmatilis*).

Velvetbean can be grown for green manure as far north as Virginia and Kentucky but generally will not mature seed. Planting should be delayed until danger of frost is past. It tolerates soil acidity, but it responds well to P fertilization. Forage digestibility and crude protein content are high, making it valuable for grazing in late summer and autumn when other pasture is of low quality. As a green manure crop, it forms a dense cover and smothers or shades out bermudagrass and yellow nutsedge (*Cyperus esculentus* L.).

Sesbania (Hemp Sesbania). Hemp sesbania, *Sesbania exaltata* (Raf.) Rydb. ex A.W. Hill, is a warm-season legume native to the southern coastal US and west to southern California (Duke 1981). This semiwoody plant grows up to 4 m tall with narrow, oblong leaflets arranged pinnately. Flowers are yellow with purple spots. Sesbania produces high yields and is used as a summer green manure crop with winter truck crops. It is disliked by livestock and has no value as forage. (Also see Chap. 23.)

Guar. Guar, *Cyamopsis tetragonoloba* (L.) Taub., is probably native to India and Pakistan (Duke 1981). It is a bushy warm-season annual 3 m tall, trifoliolate, with leaflets 5-10 cm long. The clustered pods contain 5-12 hard seeds that produce guar gum for a variety of industrial uses. It is a productive green manure crop and furnishes high-quality grazing for livestock. Production in the US is located in the southwestern states.

Florida Beggarweed. Florida beggarweed, *Desmodium tortuosum* (Sw.) DC., is native to the West Indies but found in Mexico and Central and South America. It is grown from North Carolina south to Florida and along the Gulf Coast into Texas. A few hectares are grown on the southeastern coastal plains for pasture, hay, green manure, and quail feed. It has been considered a weed in the past (Cardina and Brecke 1989).

QUESTIONS

1. Why is crownvetch so desirable for erosion control on highway embankments?
2. List the advantages of cicer milkvetch over that of other legumes. Where is it best adapted?
3. Why would sainfoin be a more desirable legume crop than alfalfa in certain areas?
4. Sweetclover was once an important legume in the US. What reasons might account for the reduced use of this legume in recent years?
5. If kudzu prevents soil erosion, why isn't it being grown more extensively?
6. Why are vetches so useful as green manure for summer row crops?
7. What problems have limited the usefulness of lupines? How have plant breeders improved lupine species?
8. Why would one not graze a mature sward of roughpea?
9. Describe the roles of field peas, cowpeas, pigeonpeas, and velvetbeans as they relate to a forage program.

REFERENCES

AL-Tikrity, W, GW McKee, WW Clarke, RA Peiffer, and ML Risius. 1974. Seed yield of *Coronilla varia* L. Agron. J. 66:467-68.

Barnes, DL, and CP Dempsey. 1992. Towards optimum grazing management for sheep production on crownvetch (*Coronilla varia* L.). J. Grassl. Soc. S.Afr. 9:83-89.

Bolger, TP, and AG Matches. 1990. Water-use efficiency and yield of sainfoin and alfalfa. Crop Sci. 30:143-48.

Bryant, HT, RC Hanmer, RE Blaser, and JP Fontenot. 1977. Evaluation of acceptability by beef cattle of crownvetch grazed at several stages of maturity. J. Anim. Sci. 45:939-44.

Burns, JC, WA Cope, and ER Barrick. 1977. Cow and calf performance, per hectare productivity, and persistence of crownvetch under grazing. Agron. J. 69:77-81.

Cardina, J, and BJ Brecke. 1989. Growth and development of Florida beggarweed (*Desmodium tortuosum*) selections. Weed Sci. 37:207-10.

Ditterline, RL, and CS Cooper. 1975. Fifteen Years with Sainfoin. Mont. Agric. Exp. Stn. Bull. 681.

Duke, JA. 1981. Handbook of Legumes of World Economic Importance. New York: Plenum.

Gustine, DL. 1979. Aliphatic nitro compounds in crownvetch: A review. Crop Sci. 19:197-203.

Gwathmey, CO, and AE Hall. 1992. Adaptation to midseason drought of cowpea genotypes with contrasting senescence traits. Agron. J. 32:773-78.

Harris, RR, EM Evans, JK Boseck, and WB Webster. 1972. Fescue, Orchardgrass, and Coastal Bermudagrass Grazing for Yearling Beef Steers. Ala. Agric. Exp. Stn. Bull. 432.

Hart, RH, AJ Thompson III, and WE Hungerford. 1977. Crownvetch-grass mixtures under frequent cutting: Yields and nitrogen equivalent values of crownvetch cultivars. Agron. J. 69:287-90.

Hoveland, CS, and ED Donnelly. 1966. Response of *Vicia* genotypes to flooding. Agron. J. 58:342-45.

Karnezos, TP, and AG Matches. 1991. Lamb production on wheatgrasses and wheatgrass-sainfoin mixtures. Agron. J. 83:278-86.

Killinger, GB. 1968. Pigeon peas (*Cajanus cajan* [L.] Druce), a useful crop for Florida. In Proc. Soil and Crop Sci. Soc. Fla. 28:162-67.

Lopez-Bellido, L, and M Fuentes. 1986. Lupine crop as an alternative source of protein. Adv. Agron. 40:239-95.

Lynd, JQ, and TR Ansmon. 1990. Exceptional forage regrowth, nodulation and nitrogenase activity of kudzu (*Pueraria lobata* [Willd.] Ohwi) grown

on eroded Dougherty loam subsoil. J. Plant Nutr. 13:861-85.
McGraw, RL, and GC Marten. 1986. Analysis of primary spring growth of four pasture legume species. Agron. J. 78:704-10.
Marten, GC, FR Ehle, and EA Ristan. 1987. Performance and photosensitization of cattle related to forage quality of four legumes. Crop Sci. 27:138-45.
Mays, DA, and EM Evans. 1972. Effects of variety, seeding rate, companion species, and cutting schedule on crownvetch yield. Agron. J. 64:283-85.
Mowrey, DP, and AG Matches. 1991. Persistence of sainfoin under different grazing regimes. Agron. J. 83:714-16.
Nene, YL, SD Hall, and VK Sheila. 1990. The Pigeonpea. Wallingford, U.K.: CAB International, 47-301.
Parker, RJ, and BR Moss. 1981. Nutritional value of sainfoin hay compared with alfalfa hay. J. Dairy Sci. 64:206-10.
Piper, CV, and WJ Morse. 1938. The Velvetbean. USDA Farmers Bull. 1276.
Smith, WK, and HJ Gorz. 1965. Sweetclover. Adv. Agron. 17:163-231.
Townsend, CE, DK Christensen, and AD Dotzenko. 1978. Yield, quality, and persistence of cicer milkvetch as influenced by cutting frequency. Agron. J. 70:109-13.
Whiteman, PC, DC Byth, and ES Wallis. 1985. Pigeonpea (*Cajanus cajan* [L.] Millsp.). In RJ Summerfield and EH Roberts, Grain Legume Crops. London: Collins, 658-98.

23 Tropical and Subtropical Forages

ALBERT E. KRETSCHMER, JR.
University of Florida

WILLIAM D. PITMAN
University of Florida

THE tropics and subtropics represent a large portion of the earth's land area, with many different climates, soils, and vegetation. The tropics may be defined in geographical terms as the region between the Tropics of Cancer and Capricorn (23.5°N and 23.5°S latitudes, respectively), but this does not reflect biological boundaries. Within the Tropics and subtropics, highlands exist in which temperate and/or frigid regions prevail and the vegetation matches the prevailing climatic environments.

The tropics also can be delineated by the mean annual temperatures (20°, 21.5°, and 23°C have been proposed) and/or the mean temperature of the coldest months (8°, 20°, and 21°C have been proposed) (Blumenstock 1957). These temperatures are somewhat lower for the subtropics, which have been delineated as being between 23.5° and 30°N and 23.5° and 30°S latitudes. Land areas under US jurisdiction that may be classified as tropical or subtropical include Florida, the coastal areas of the Gulf of Mexico, Hawaii, Puerto Rico, the US Virgin Islands, and the Trust Territories of the Pacific, Guam, and American Samoa.

ALBERT E. KRETSCHMER, JR., is Professor of Agronomy, University of Florida, Fort Pierce Research and Education Center, Fort Pierce, Florida. He received the PhD degree from Rutgers University. He specializes in tropical forage plant collection and evaluation, pasture management, and germplasm conservation.

WILLIAM D. PITMAN is Associate Professor of Agronomy, University of Florida, Ona Research and Education Center, Ona, Florida. He received the MS degree from Texas Tech University and the PhD degree from Texas A&M University. He specializes in tropical forage management and utilization.

The authors thank Dr. Peter P. Rotar, University of Hawaii (retired), for his contributions to this chapter.

TROPICAL AND SUBTROPICAL GRAZING LANDS

Substantial portions of the agricultural land in the tropics and subtropics are best suited for grazing only. Vast regions in Africa, Asia, Latin America, and Oceania are used as grazing lands (Table 23.1) but vary greatly in topography, elevation, soil type, natural fertility, type of vegetation, available water or rainfall, and climate. A highly developed technology has increased productivity of grazing lands in developed countries in temperate latitudes. Similar technology is being applied to problems of grasslands in the tropics and subtropics.

Compared with temperate species, much less is known about tropical grasses and legumes, especially their potential as forage crops, and the number of tropical genera and species is much greater. Research is needed on their productivity, on their chemical and nutritional composition (Minson 1980), and on management (agronomic and cultural) practices under various tropical climatic conditions.

Pasture improvement in the tropics and subtropics varies from nil, inclusion of a legume into existing grasses with or without fertilizer, to a complete replacement of the existing vegetation with improved grasses and legumes that require annual inputs of fertilizer to remain productive. On a world scale, much less than 5% of grazing lands of the

TABLE 23.1. Land areas, grazing lands, and numbers of cattle in the tropics and subtropics

Area	Total area (ha × 10^6)	Grazing lands		Cattle	
		ha × 10^6	%	× 10^6	Number/1000 ha grazing land
Africa	1750	493	28	113	230
America					
Southern US	62	11	18	11	980
Central America	272	81	30	48	590
South America	1430	290	20	151	520
Asia	762	31	4	236	7600
Oceania					
Northern Australia	175	142	80	8	54
Papua New Guinea	46	0.1	0.2	0.1	300
Pacific Islands	9	0.5	6	0.6	1180
Total or Mean	4506	1048	23	567	540

Source: 't Mannetje (1978).

tropics have been improved ('t Mannetje 1978), even though pasture improvement could increase beef production per unit of land area sixfold (Stobbs 1976).

Although more than half the world's cattle population is in the tropics and subtropics, only about 20%-30% of the meat is produced in the tropics. This can be attributed to many social, political, and cultural problems as well as unproductive soils, low-quality forages, inadequate animal disease control, and lack of production incentives.

EFFECTS OF CLIMATE ON GROWTH AND DEVELOPMENT

Annual rainfall in the tropics varies from 50 to 100 mm in some areas to well over 3000 mm in others. In equatorial regions, rainfall is typically plentiful and fluctuates little during the year; the major limiting factor to plant growth is low light intensity due to the nearly continuous cloud cover. In most areas outside the equatorial regions (from 5°-10° to 30°N and 5°-10° to 30°S latitudes) seasonal changes in rainfall occur. The following seasonally wet regions may be defined (Webster and Wilson 1966) as (1) areas with 1000-2000 mm annual rainfall with two rainy seasons and one short and another hardly noticeable dry season, (2) areas with similar total annual rainfall to that of the first area but with a bimodal distribution (two well-pronounced dry seasons), (3) areas with about 750-1250 mm annual rainfall with one fairly long dry season, and (4) areas with less than 750 mm rainfall with a very short wet season and one very long dry season. In the seasonally dry areas, herbage dry matter (DM) production is dependent upon the length of the rainy season; animal production is relatively good during the wet season, and there is usually a large weight loss during the dry season because of low feed value of the existing forage. Lack of energy is the greatest problem, closely followed by the low crude protein (CP) concentration that can be less than 3% in some species.

Environmental conditions, high growth rates, and high DM production affect the quality of tropical grasses. They are usually higher in fiber and lignin, mature at much faster rates, and hence are lower in quality than are temperate grasses at comparable stages of growth and development (Minson 1980).

In the equatorial region (between 5°-10°N and S of the equator) photoperiods or daylengths are nearly constant year-round. Farther away from the equator these effects are felt: at 23.5°N and S latitudes, daylength varies from 13 h 40 min in summer to 10 h 20 min in winter. A majority of the tropical and subtropical pasture species are either day-neutral or short-day plants. In the subtropics these effects become more pronounced and may limit tropical forage production and reproduction, especially by short-day plants dependent upon seed production during the winter when frosts may severely limit flowering and seed set.

Tropical grasses have a higher temperature optimum for DM production than temperate grasses. Optimum temperatures are in the order of 30°-40°C as compared with 20°-30°C for many Festucoid or temperate grasses. Tropical legumes have lower optimum temperatures (25°-30°C) than tropical grasses, while for many temperate legumes optimum growth temperatures are in the range of 20°-25°C. Most tropical grass and legume growth is substantially reduced below 15°C, and tops are usually not tolerant to frost (McWilliam 1978). Although foliage of tropical species is killed by frost, many can survive temperatures of about –5°C and below. Subtropical areas with a cold season and frosts, which experience annual alternation of tropical and temperate conditions on the same site, are among the most difficult environments in which to develop improved pastures.

Photosynthesis in leaves of temperate grasses responds to increased solar radiation up to about 60% of full sunlight whereas tropical grasses typically respond up to full sunlight. Temperate and tropical legumes respond to increased light similarly to temperate grasses. Consequently, due to radiation and temperature responses, tropical grasses have photosynthetic and growth rates that are nearly double those of temperate grasses (Cooper 1970).

CHARACTERISTICS OF TROPICAL AND SUBTROPICAL FORAGES

Legumes. The family Leguminosae consists of three subfamilies, Caesalpiniodeae, Mimosoideae, and Papilionoideae, and 112 tribes, which include about 651 genera and 17,250 species. The main centers of origin and diversity for the tropical legumes occur in Brazil, Mexico, eastern Africa, and the Sino-Himalayan region. A majority of all commercially available tropical legumes are native to tropical America. Most of the commercial cultivars are in the legume tribes Indigofereae, Aeschynomeneae, Desmodieae, and Phaseoleae.

The diversity of tropical legumes, from herbs to trees, provides the germplasm potential for most edaphic and climatic environments in the tropics and subtropics with annual rainfall above about 500 mm. Research in the use of tropical legumes in association with permanent grass pastures began after World War II, but intensive work on a wide scale began in the early 1960s. Presently, in most tropical countries, researchers appreciate the potential value of tropical legumes in providing nitrogen (N), improving forage quality, and increasing carrying capacity compared with grasses alone without N fertilization. Native or naturalized tropical forage legumes in many instances have contributed to the grazing system without notice by the cattle producer.

Cutting experiments with tropical legume-grass associations have shown that as much as 300-400 kg N ha^{-1} yr^{-1} can be fixed by some tropical legumes. Realistically, 100-200 kg N ha^{-1} could be expected from an appropriately managed tropical pasture consisting of 30% legume and 70% grass. From 100 to 250 kg ha^{-1} yr^{-1} of fertilizer N has been required on pure grass swards to obtain DM yields similar to those from the same grass mixed with tropical legumes, and 100-500 kg N ha^{-1} yr^{-1} has been required to obtain similar CP yields (Kretschmer 1980). Absolute responses depend on yields of both the grass and legume species, their N concentrations, and the percentage of legume in the association. Some grasses are more productive than others at a given level of soil N. Tropical legume N accumulation, and thus potential N contribution to the pasture system, is positively correlated with the legume herbage yield, since as little as 6% to 10% of the total N in legumes is located in the roots (Bushby and Lawn 1992).

In spite of positive research results with tropical legumes, the limited success of a large number of commercial plantings illustrates that the lack of legume survival is the single most important constraint to sustainable legume productivity. This is true in spite of the agronomic and morphologic diversity of officially released tropical legumes for commercial use.

Although many tropical legume species have been evaluated in almost all tropical countries, major success with tropical legume use has been in Australia, Florida, and Hawaii and more recently Colombia, Venezuela, and tropical Asia. Brazil, one of the major centers of diversity of the tropical legumes, has not had the success previously expected when commercial cultivars have been tried. In general, legumes in small plots subjected to clipping have shown greater persistence than when they are subjected to grazing. A number of factors influencing persistence of legume stands includes growth cycle, annuals vs perennials; growth habit, twining or climbing legumes vs prostrate or creeping types; grazing pressure and length of rest period; distribution, consistency, and effectiveness of annual rainfall; nutrient limitations of most tropical soils, particularly low levels of phosphorus (P), low pH, high levels of aluminum (Al) and manganese (Mn), and lack of calcium (Ca); lack of or unavailability of adapted *Rhizobium* strains; insects and diseases; competition from companion grasses; and a lack of understanding of the interrelationships between legumes/grasses and livestock. One or more of these may contribute to lack of, or decline of, established legumes in tropical pastures.

Grasses. Of the estimated 7,000-10,000 species of Gramineae, less than 40 are used to any appreciable extent in the establishment of sown pastures in the tropics and subtropics. Species are mainly from the warm-season tribes Andropogoneae, Paniceae, Chlorideae, and Eragrosteae and, at higher elevations, from the temperate tribes Festuceae and Agrosteae.

The majority of sown tropical pasture species originated in Africa, with a smaller number from tropical and subtropical Central and South America, a few from India and southern Asia, but none from Australia.

The high photosynthetic rates in C_4 grasses lead to high crop growth rates and high quantities of DM production. With high N rates and a long interval between harvests, *Pennisetum purpureum* Schumach. (elephantgrass), produced over 85,000 kg DM ha^{-1} yr^{-1} (Vincente-Chandler et al. 1959). Many others produce 20,000 to 30,000 kg ha^{-1}. In contrast, *Lolium multiflorum* Lam. (annual ryegrass), a temperate grass, produced a maximum of 21,400 kg DM ha^{-1} yr^{-1} with high N rates in its adapted environment (Hughes 1970). The potential for high DM production gives tropical grasses a marked competitive advantage in utilization of solar energy over their associated legumes, which utilize C_3 photosynthesis.

Maintenance of a legume in a tropical pasture must be achieved by N limitation to the potential grass productivity and/or grazing and defoliation management to favor the legume. *Pueraria phaseoloides* (Roxb.) Benth. (tropical kudzu) is an example of a very persistent viny legume, even under intensive grazing, because of its low acceptability by cattle. However, most viny legumes require lenient grazing pressure and defoliation to grow and compete with their associated tropical grasses. Under such management regimes, viny legumes are capable of climbing and displaying their leaf canopies above the grasses; however, under heavy grazing pressure essentially all growing points of the legumes may be removed, resulting in loss of their competitive advantage and eventual loss of the legume stand. Prostrate legumes and some legumes that branch readily may benefit from more intense grazing pressures that prevent shading by the associated grass.

Dry matter digestibility of tropical grasses is lower than that of temperate grasses. This is due in part to the growth characteristics of the tropical grasses under high-temperature regimes as well as to their anatomy (Chap. 3). Digestibility and N concentrations of most tropical grasses are greatly reduced after about 8 wk of regrowth and during the last part of the growing season. Hence pasture management considerations affecting nutritional value are especially important in livestock production (Minson 1990).

Addition of as little as 10% legume to a grass diet containing less than 4% CP has overcome CP deficiency and led to large increases in total herbage intake by sheep (Minson and Milford 1966). Well-managed tropical grass/legume pastures can be highly productive, carrying three to four animal units per hectare during the growing season. Legumes provide a valuable source of N that cannot be economically supplied in most tropical areas.

TROPICAL AND SUBTROPICAL FORAGE SPECIES

Information on the species that follow was derived from many sources, including O'Reilly (1975), Bogdan (1977), Skerman (1977), Humphreys (1980), Summerfield and Bunting (1980), Polhill and Raven (1981), Stinton (1987), and Skerman and Riveros (1990). Whiteman (1980) discusses agronomic characteristics of the tropical species, and Kretschmer (1989) provides a tabular summary of growth habits and environmental adaptation of the more important tropical forage legumes. A summary of legume adaptability to various Australian climatic zones is presented by Gramshaw et al. (1989). The latest work on tropical forage species descriptions and adaptations is presented by 't Mannetje and Jones (1992). It provides a broad botanical and agronomic review of many of the species listed.

In the following presentation of characteristics of individual tropical legumes and grasses, the following information is provided in parentheses after the botanical name: common names, cultivars or varieties and locations where available, and number of species in the genus.

Legumes

***Aeschynomene americana* L.** (American jointvetch and aeschynomene; commercial [annual] in Florida and cv. 'Glenn' [annual] and 'Lee' [perennial] in Australia; about 160 spp. in the genus [Rudd 1955; Kretschmer and Bullock 1980; Kretschmer and Snyder 1983; Kretschmer et al. 1986]). Native to all tropical areas in the Americas between about 30°N and 30°S latitudes (up to about 2500 m), this annual or short-lived perennial herb has glandular to subglabrous stems, with prostrate to erect growth habit to 1-2 m. Pinnately compound leaves are from 2 to 7 cm long with 20-60 leaflets. Flower color is mauve to yellow normally with red or purple stripes formed on axillary racemes; pods have four to eight articulations (joints). American jointvetch is adapted to areas with warm, humid climates and will withstand waterlogging and temporary flooding. It has been successfully used for grazing in mixtures with grasses in Florida (Hodges et al. 1982) and Australia (Bishop et al. 1988), and it has also been used in the US Gulf Coast states as wildlife feed.

Aeschynomene falcata **(Poir.) DC.** (jointvetch; cv. 'Bargoo' in Australia). Native to altitudes of 1800 m in northwestern and east central South America. A prostrate perennial herbaceous legume with pubescent stems up to 1 m; normally leaves have five to seven leaflets. *Aeschynomene falcata* has small, 10-mm-long, yellow flowers; it is drought tolerant but can survive temporary flooding. It is used for grazing in Australia (Wilson et al. 1982).

Other *Aeschynomene* species. Other *Aeschynomene* species with potential forage value include *A. villosa, A. evenia, A. sensitiva,* and *A. indica*. These have been evaluated for adaptation to specific sites in Florida, Australia, and other locations, with further progress in their development as forage cultivars anticipated.

Alysicarpus vaginalis **(L.) DC.** (alyceclover, buffalo clover, and one-leaf clover; commercial in Florida and Australia [annuals]; 30 spp. in the genus). This annual/perennial legume herb is native to tropical Asia but has been cultivated and is naturalized in most tropical Latin America countries. It can reach a height of about 150 cm, with stems glabrous to pubescent. Adventitious rooting can occur from prostrate stems. Leaves are unifoliolate, lanceolate to ovate, and glabrous to puberulous. Flower color is pink to orangish to purple. Alyceclover is adapted to a wide range of soil (pH from 4.5 to above neutral) and rainfall conditions, although best production occurs with adequate drainage and annual rainfall, predominately during summer, of 1000 mm or more. It reestablishes adequately from seed. The commercial ecotypes are very susceptible to root-knot nematodes (*Meloidogyne* spp.) and grow best when soil pH is above 5.0. There are two distinct phenological types: those with an erect growth habit and those prostrate types that can reach a height of 10 to 30 cm and root from the stolons. The erect type is much more productive, while the prostrate type can survive heavy grazing pressures and tends to perennate. Alyceclover is readily eaten by cattle (Gramshaw et al. 1987).

Arachis glabrata **Benth.** (rhizoma peanut; cv. 'Florigraze', 'Arb', 'Arblick', and 'Arbrook' in Florida). Native to Brazil, Argentina, and Paraguay between about 13°S and 28°S, this long-lived herb produces a thick mat of rhizomes. Glabrous, tetrafoliolate leaves with linear-lanceolate to obovate or cuneate leaflets are up to 4 cm by 2 cm. Stems are about 35 cm long and 2-3 mm thick. Flower color varies from yellow to orange. Very little seed is produced; thus planting is done using rhizomes. Commercial plantings of this highly nutritious species primarily have been of Florigraze, although limited plantings of Arbrook have been made on drier sites. Depending on the density of planting material and climatic variations, satisfactory stands usually occur in the second or third year after planting. It has been used for hay and grazing in Florida and southern Georgia (Prine et al. 1981) and probably is commercially adaptable to other southern states and tropical countries.

Arachis pintoi **Krap. & Greg. nom. nud.** (pinto peanut; cv. 'Amarillo' in Australia). It is native to valleys of central Brazil, first collected in 1954, and distributed to most Central and South American countries, Australia, the US, and elsewhere in the tropics and subtropics. Pinto is a stoloniferous perennial herb with prostrate to erect stems, up to 20 cm long in dense swards, and tetrafoliolate leaves with leaflets up to 4.5 cm long and 3.5 cm wide. Flower color is yellow, producing large nut yields (seeds are 8-10 mm by 4-6 mm). Seedlings of Amarillo develop rapidly, and initial grazing can occur in the planting year. Inoculation with a specific *Rhizobium* strain usually is necessary for effective N fixation, but even then, plant color is often a pale green. Pinto is tolerant of heavy grazing pressure but competes well in association with tall grasses that are not grazed heavily (Grof 1985; Cook and Franklin 1988; Oram 1990a).

Calopogonium mucunoides **Desv.** (calopo; eight spp. in the genus). This annual to short-lived perennial is indigenous to tropical America and the West Indies but is widespread in the tropics of Asia and Africa through introduction in the early 1900s. This vigorous vine, twining up to 1 m or more, has densely pilose stems and trifoliolate leaves, with leaflets elliptic or ovate or rhomboid-ovate. Flowers, initiated by short days, are blue to purple and produce brownish pods containing three to eight compressed squarish seeds. Calopo grows rapidly in the true tropics, but in higher latitudes where frost occurs, its growth is reduced and seed production limited. Calopo is most suited for areas with above 1000 mm of annual rainfall and up to about 1500 m elevation. Calopo is not readily grazed by cattle; thus it is rather stable as a component in grass mixtures in its ideal environment. Similar to tropical kudzu, it is used as a cover crop in plantation crops. Calapo is susceptible to viruses.

***Centrosema* species** (32 named spp.). There have been recent compilations of the vast amount of research work done with the species in this genus (Schultze-Kraft et al. 1989; Schultze-Kraft and Clements 1990). The world's collection of germplasm is about 3700 accessions, which includes 32 named species, 2 unnamed species, and several hybrids. As of 1990, the largest collections were of *C. pubescens* (about 1325), *C. virginianum* (426), *C. plumieri* (445), *C. pascuorum* (110), *C. brasilianum* (371), *C. acutifolium* (77), and *C. macrocarpum* (445). Most *Centrosema* species are twining perennials with trifoliolate leaves, with or without short stolons. Large showy flowers are purple to white and even reddish. A short description of the most important forage species follows.

Centrosema acutifolium Benth. (cv. 'Vichada' in Colombia) is native to 4°-6°N latitude in Colombia, Venezuela, and west central and southeast Brazil. This perennial twining legume has recently been studied as a grazed or harvested forage. It flowers very late in the subtropics and probably is restricted to the hot, subhumid tropics with annual rainfall of 1000 to 2500 mm and with an up to 5-mo dry season. It is similar to *C. pubescens* in morphology, general appearance, and nutritive value, but seed is cylindrical, 5-7 mm by 3 mm, and greenish yellow with fine blackish mottled marks and is formed from light purple flowers. The inflorescence is an axillary raceme with up to about 24 flowers in pairs along the rachis. Leaflets are 4.0-8.5 cm long and 2.5-4.0 cm wide. Unlike *C. pubescens,* it tolerates highly acid soil pH down to about 4.3 and high levels of Al and Mn. *Centrosema acutifolium* has a low fertilizer requirement; however, its productivity is less than that of *C. macrocarpum*.

Centrosema macrocarpum Benth. occurs naturally between 20°S and N latitudes from northern Brazil to Venezuela, Colombia, and all of Central America to Mexico (Schultze-Kraft 1986). This robust perennial vine with taproots, rooting sporadically from stem nodes, has similar tolerances to soil restraints as does *C. acutifolium*. Leaflets are about 8-13 cm long and 3-8 cm wide and almost glabrous to pubescent. Up to about 30 flowers in pairs are borne on axillary racemes. Petals are cream colored with purple centers. The seed is yellowish brown, either unicolored or mottled or marbled. It is very short-day photoperiod sensitive, and its area of adaptation probably is restricted to the humid and subhumid tropics with more than about 1000 mm of annual rainfall. *Centrosema macrocarpum* is the most tolerant species to the economic diseases typically affecting other *Centrosema* species. As with most *Centrosema* species, attack by leaf-eating insects can be temporarily severe, but damage can be reduced by grazing or cutting.

Centrosema pascuorum Mart. ex Benth. (centurion; 'Cavalcade' [a hybrid] and 'Bundey' in Australia) is a herbaceous, twining annual native to tropical South and Central America, mainly in semiarid regions or those with long dry seasons. It is utilized in the Northern Territory of Australia as a grazing legume where there is a short but reliable wet season (Oram 1990b). It can tolerate prolonged flooding as well as drought but requires a reliable 4- to 6-mo, 700- to 1500-mm annual rainfall. Adventitious roots can be produced on prostrate stems. Leaflets are 2-15 cm long and 0.3-1.7 cm wide, with one or two flowers formed at the end of 0.5- to 2.0-cm peduncles. Flowers are wine red, and they are 15-25 mm long and wide. The highly palatable forage is well utilized in the wet and dry seasons, but sufficient seed must be produced for self-regeneration.

Centrosema pubescens Benth. (centro and butterfly pea; commercial in Australia) is a vigorous climbing perennial legume that has been used or tried for many years for grazing or as cover in plantation crops. It competes well with grasses, but its persistence under intense grazing is unreliable. Although it can persist under about 750-mm annual rainfall, it is more productive in higher-rainfall areas in the humid and subhumid tropics. It tolerates temporary waterlogging and flooding as well as a 3- to 5-mo dry season. Unlike *C. acutifolium* and *C. macrocarpum*, it will not tolerate low soil pH or high Al and Mn soils. Leaflets are elliptical, ovate-oblong to ovate-lanceolate (1-7 cm long by 0.5-4.5 cm wide). Pods are linear, 4-17 cm long by 6-7 mm wide and flattened, with dark brown seeds. Flower color varies from almost white to deep purple. Cultivar 'Belalto' (*C. schiedeanum* L.), once considered to be *C. pubescens,* has been most successfully used in the humid tropics of Australia (Teitzel and Burt 1976). In south Florida, summertime production of commercial *C. pubescens* has been excellent when associated with grasses, but seed production is not reliable because of its late flowering.

Centrosema virginianum Benth. has many of the same characteristics as *C. pubescens,* but leaflets and overall plant appearance of *C. virginianum* are generally smaller, and plants are less vigorous than *C. pubescens. Centrosema virginianum* is native from Maryland west to Oklahoma and south into Argentina. Although there have been extensive testing and breeding programs with this species, there is no commercial use.

***Chamaecrista rotundifolia* (Pers.) Greene** (formerly *Cassia rotundifolia* Pers.; roundleaf cassia; cv. 'Wynn' in Australia). This annual or short-lived perennial semierect to prostrate herb is native from Mexico and the Caribbean region extending into Brazil and Uruguay. It is commercialized in Australia, where it is successful in seasonally dry areas with 700-1400 mm of rainfall. Leaves are bifoliolate with subrotund to obovate leaflets 0.5-1.0 cm long. Yellow flowers are produced in the late summer and early fall at higher latitudes (Strickland et al. 1985; Cook 1988).

***Clitoria ternatea* L.** (butterfly pea; cv. 'Milgarra' in Australia). This perennial twining herb with a woody crown area has white to purple flowers that occur throughout the year in the tropics. Productivity is best with about 1500 mm of annual well-distributed rainfall; however, plants can survive in the seasonally dry tropics with about 500 mm. It does not tolerate flooding. It is adapted to clay but not infertile soils. It has been tried sparingly in Mexico and elsewhere in Latin America, Australia, and Southeast Asia (Reid and Sinclair 1980; Hall 1985).

***Desmanthus virgatus* (L.) Willd.** (desmanthus; cv. 'Marc' [early-flowering], 'Bayamo' [midseason], and 'Yuman' [late] in Australia). Probably originating in Mexico and now widespread in the Americas from Arizona, Texas, Florida, and the Caribbean Islands, throughout Central and South America to Argentina, this legume is a perennial and sometimes annual shrub or herb. It is erect to prostrate, 0.5-3.0 m high, taprooted, with stems unbranched or moderately branching. Bipinnate leaves, 2-8 cm long, have one to seven pairs of primary branches up to 7 cm long, each with 10-25 pairs of linear to linear-oblong leaflets, 4-9 mm long and 1-2 mm wide. The whitish inflorescence is similar to that of leucaena but smaller, being a globose 6- to 10-flowered head. There are four recognized botanical varieties, with most accessions belonging to var. *virgatus*. This variety has been collected from annual rainfall areas of 250-2000 mm, and up to about 2000-m elevation, but most accessions are best adapted to near-neutral to alkaline soils. Plant crowns of some accessions can survive frost, but foliage is susceptible. Selected accessions of *D. virgatus* var. *virgatus* have grown well in some experimental areas in Florida but have failed to persist in others. Three cultivars of *D. virgatus* var. *virgatus* are in the early stages of use in Australia, and other genotypes have been used as cut forage in India. Desmanthus is adaptable to grazing although the thicker stems are not eaten.

***Desmodium* species.** There are about 300 species in this genus. Several have been used successfully as forages, if only in limited climatic zones, while many others have been studied experimentally. As is typical of the tribe, Desmodeae, which contains about 27 mostly tropical genera, most *Desmodium* species are native to the Sino-India region (Ohashi 1973).

Desmodium heterocarpon (L.) DC. (carpon desmodium; cv. 'Florida' in Florida) has several forms and is native to, or distributed throughout, tropical, subtropical, and warm regions of Southeast and East Asia, India, Pacific islands, and Australia (Ohashi 1973). This long-lived perennial herb has ascending or creeping stems to 1 m and more. Flower color ranges from pink, purple, mauve, and violet to white, and flowers are 4-7 mm long. The herbaceous cv. Florida (var. *heterocarpon*) was released in 1979 (Kretschmer et al. 1979); it has pinkish flowers on dense inflorescences that form from axillary or terminal racemes. A concentrated seed set occurs in late October that has produced seed yields of over 300 kg ha^{-1} in south Florida. Unifoliolate leaves are found on seedlings and lower stems, while typically three broadly elliptic, ovate, or obovate leaflets are found on mature plants, with terminal leaflets sometimes being twice as large as lateral leaflets.

Carpon desmodium has been successfully used in south Florida where there are an estimated 5000 ha mixed with most of the commonly used grasses, particularly bahiagrass (*Paspalum notatum* Flugge). Although the quality is lower than that of some of the other tropical legumes, carpon desmodium is readily grazed along with all associated grasses. Initial seedling growth is slow, but long-term persistence of individual plants and regeneration from seed provide distinct advantages over the annual tropical legumes such as aeschynomene, which are currently available for peninsular Florida (Pitman and Kretschmer 1993). In bahiagrass mixtures, the carpon desmodium component has maintained stable plant populations for more than 14 yr, thus improving forage quality and productivity. Nutrient requirements of this legume are much less than those of the temperate legumes, but P and lime are important. Soil pH should be between about 5.0 to 6.0 for maximum growth. Plants survive temperatures as low as –5°C, and also several frosts annually that kill the foliage. Carpon desmodium can survive continuous, heavy grazing pressure or lenient, rotational grazing, but grazing after July reduces seed yields in Florida.

Desmodium heterocarpon (L.) DC. ssp. *ovalifolium* (Prain) Ohashi (cv. 'Itabela' in Brazil) is native in tropical Southeast Asia from northern Thailand to southern Sumatra and probably also in the Burma-Laos-Vietnam-Cambodia area. Recently the subspecies ranking was assigned rather than the species classification, *D. ovalifolium* (Ohashi 1991). It has spread to most parts of the tropics because of its value as a cover crop in tree plantations and for grazing (Schultze-Kraft and Benavides 1988). Flower color varies from purple to dark pink, turning to bluish when flowers are wilting. Seedlings and young plants are unifoliolate, whereas older plants are typically trifoliolate. This low-growing, stoloniferous perennial herb can produce ascending growth in undisturbed swards. Compared with carpon desmodium, it is less palatable to cattle, has lower-quality forage because of its higher tannin concentration, is more of a tropical species, and flowers up to 2 mo later than carpon desmodium in the subtropics.

Desmodium heterophyllum (Willd.) DC. (hetero desmodium; cv. 'Johnstone' in Australia) is a perennial prostrate or slightly ascending herb with a woody rootstock; it is native to Southeast Asia and surrounding islands and has spread to many other humid tropical areas where annual rainfall exceeds about 1500 mm. Stems are freely branching, can form adventitious roots, and grow to about 1 m long. Leaves are trifoliolate. Pink flowers are 3-5 mm long, and pods have three to six joints. Because of its indeterminant flowering pattern and fruit that is formed below the upper foliage canopy, mechanized seed harvesting is difficult; thus seed is not commercially available. Hetero desmodium competes well with associated grasses and withstands very heavy grazing pressure (Partridge 1986).

Desmodium incanum DC. (syn. *D. canum* Schinz & Thell.; creeping beggarweed and kaimi clover) is native to southern North America and tropical America; this species is now widely distributed to many tropical areas of the world, and as with many *Desmodium* species, it has seedpods that easily attach to humans and animals. This perennial herb with woody stems and rootstock is an invader in many pastures, growing to about 20 cm or more in height. Leaves are trifoliolate with terminal leaflets 2-9 cm long and about 1.5-4.5 cm wide. Flowers are blue, red, or purple and about 6 mm long. Its palatability is low because of high tannin concentration, but plants withstand heavy grazing pressure. It is adapted to annual rainfall areas of above 1000 mm.

Desmodium intortum (Mill.) Urb. (greenleaf desmodium; cv. 'Greenleaf' in Australia and 'Tengeru' in Tanzania) is indigenous to uplands of 800-2500 m elevation from Mexico to southern Brazil. This large perennial trailing, but not twining, herb with erect or ascending reddish brown branched stems requires 1000-1500 mm of annual rainfall. It can grow at lower elevations in the subtropics. Trifoliolate leaves with usually ovate leaflets, 3-12 cm by 1.5 cm, often have sparse reddish brown marks on the upper surface. Pink to purple flowers, 8 mm long, are formed on dense terminal or axillary panicles up to 30 cm long. It is tolerant of acid soils with pH of 5.0 or above but does not tolerate sustained waterlogging. It combines well with numerous grasses. This productive legume has a restricted growing area of commercial adaptation mostly in Australia, probably because of its lack of persistence under heavy grazing pressure and its late flowering and limited seed production in areas subject to frost.

Desmodium uncinatum (Jacq.) DC. (silverleaf desmodium; cv. 'Silverleaf' in Australia) is native to Central and South America, from Mexico to northern Argentina and Uruguay in the 500- to 2000-m elevation range. This trailing perennial legume has many of the general characteristics of greenleaf desmodium. However, leaflets have silvery striped watermarks, and in the subtropics it flowers earlier than does greenleaf desmodium. It is adapted to annual rainfall areas above about 1000 mm. Its area of actual commercial use is less than that of greenleaf desmodium.

***Gliricidia sepium* (Jacq.) Kunth ex Wap.** (gliricidia). This small (up to 12-m) hardwood deciduous tree is native to the seasonally dry Pacific coast of Central America from sea level to about 1200-m elevation. It has become naturalized from Mexico to northern South America and has been transported into the Caribbean, West Africa, and then to Asian countries. Leaves are alternate, pinnate, 15-40 cm long with 7-17 opposite leaflets per leaf. Flowers are whitish pink or purple. The narrow, flat pods, about 12 cm by 1.3 cm, contain 4-10 seeds, which are produced in the dry season. Used primarily in a cut-and-carry forage system of wild trees or live fence posts, gliricidia has recently been cultivated for use in a protein bank system. Also, it can be used for firewood or as a dual-purpose crop. It is well adapted to humid, tropical climates and acid, infertile soils, has some of the same attributes as leucaena, but is not as well adapted to grazing (Withington et al. 1987).

***Indigofera hirsuta* L.** (hairy indigo; commercial in Florida). Hairy indigo is an annual erect-growing herbaceous species that develops woody stems with maturity. Leaves are pinnately compound with five to seven leaflets. Both stems and leaves are covered with short bristly hairs. Flowers are red and arranged in dense elongate racemes. Rectangular pods contain small cube-shaped seeds. Although native to tropical Africa and Asia, it has been planted as a forage and cover crop in Florida for the past 50 yr. Two distinct types, early- and late-flowering, have been used in Florida. Hairy indigo is particularly adapted to well-drained, sandy soils. Due to extreme hardseededness, seed may lie dormant for 10 yr or more and germinate when grass sod is disturbed. Although commonly used for grazing, hairy indigo is sometimes rejected by cattle, or they require a period of acclimation to the forage.

***Lablab purpureus* (L.) Sweet** (lablab bean and hyacinth bean; cv. 'Highworth' and 'Rongai' in Australia). This annual/biennial native to the Old World tropics is an erect or climbing legume up to 1 m high, with longer stems in climbing types. Leaves are trifoliolate, with leaflets 3-15 cm long and 1.5-14 cm wide. Inflorescences are axillary, 4-20 cm long, with white to purple flowers. Large pods are of different sizes and shapes depending on the subspecies. Lablab bean is adapted to warm, humid regions, is drought tolerant, and will grow in areas with less than 500 mm annual rainfall. Germplasm of 'Tift 1', an annual, was registered in Tennessee by Fribourg et al. (1984). Lablab bean tolerates light frosts and grows on a wide variety of soils, from slightly acid to alkaline, but is not tolerant of waterlogging. Although it is normally grown alone as forage, in the Bahama Islands and elsewhere small landholders utilize the edible beans that must be cooked prior to eating. Unpasturized milk from cows grazing lablab bean may have an undesirable flavor. Highworth is earlier flowering than Rongai, which sometimes does not produce seed in south Florida because of early frost.

***Leucaena leucocephala* (Lamb.) de Wit.** (leucaena, ipil-ipil, and koa haole; cv. 'Peru', 'El Salvador', and 'Cunningham' in Australia and 'K8' among other K numbers in Hawaii; about 40 spp. in the genus). This species is native to Guatemala and surrounding Central American countries. The common growth form, a shrubby tree up to 8 m high, is probably indigenous to the Yucatan Peninsula, while the more erect form, ie., El Salvador, grows to about 16 m and probably originated in El Salvador, Guatemala, and Honduras. Both forms were widely distributed throughout Mexico, Central America, and northern South America more than 400 yr ago (Brewbaker 1987). Presently, leucaena is commonly found throughout Southeast Asia and may be the dominant vegetation on coral islands. Leaves are bipinnate with four to nine pairs of primary branches, each containing 11-22 pairs of leaflets. Numerous flowers form in white globose heads that are 2-5 cm in diameter. This moderately hardwood species is used as firewood and can be used as flooring although it does not resist termites. Leucaena is planted in rows spaced 1-5 m apart in association with the common grass used in the area of adaptation. Because it can survive temperatures of about −7°C or less, leucaena has a wide range of adaptation; however, soil pH should be >5.5, and nonwaterlogging areas should be used for maximum production. Leucaena is being evaluated in Florida and is being used commercially in many tropical countries.

Normally because of slow seedling growth and weed competition, grazing of leucaena is delayed for up to a year after planting seed or transplanting seedlings. Once established, individual plants are long-term persistent. Excess stem growth can be controlled by cutting or grazing with mature cattle. There has been a problem with a sucking insect, the leucaena psyllid (*Heteropsylla cubana*), which can kill leaves and large stems, markedly reducing productivity. Populations of this insect have been controlled by grazing leucaena stands. Leucaena foliage contains a toxic amino acid, mimosine, which limits its use for nonruminant animals to 5% or less of the diet DM (Jones and Megarrity 1986). In ruminants, mimosine is degraded to 3-hydroxy-4(IH)-puridine (DHP), which can cause goiter and other maladies causing poor animal performance. In many areas, however, rumen microbes are present in cattle that can detoxify DHP, thus preventing toxicity. These rumen bacteria have been isolated and successfully transferred to Australia and other countries, thereby solving the toxicity problem (Jones 1979, 1985).

***Macroptilium atropurpureum* (DC.) Urb.** (siratro; cv. 'Siratro' in Australia; about 20 spp. in the genus). Probably originating in Mexico, this perennial herbaceous, twining legume extends from southern Texas to Argentina and northern Brazil. Siratro has been evaluated for potential use in most of the tropical areas of the world. Leaves are trifoliolate, with ovate leaflets that are 3-8 cm long and 2-5 cm wide (mostly lobed lateral leaflets). Typically, 3-13 purple flowers, 2 cm long, are formed on axillary peduncles. Pods are 4-8 cm long and 4-6 mm wide. They dehisce vigorously when the 3- to 4-mm seeds are ripe. Siratro, bred from wild Mexican ecotypes, has been used successfully in Australia for many years (Jones and Jones 1978). It persists well in southern Florida when cut every 45-60 d but does not survive under the commonly used grazing practices. It must be rotationally grazed for maximum persistence. Siratro grows best with annual rainfall from 700 to 1500 mm, but it is not well suited to the continuously wet or to the very hot, dry tropics. Plants can survive single occurrence freezes to about –5°C and several annual frosts that kill the foliage. Rust (*Uromyces appendiculatus*) and foliar leaf blight (*Rhizoctonia solani*) can cause severe leaf damage and weaken plants severely (Sonoda et al. 1991). Isolation studies in Florida resulted in identification of several rust-resistant accessions from Mexico, one of which was registered as 'IRFL 4655' (Jones and Jones 1978; Kretschmer et al. 1985, 1992a, 1992b).

***Macroptilium lathyroides* (L.) Urb.** (phasey bean; cv. 'Murray' in Australia and commercial in Florida). Naturalized in tropical America (including southern Florida), this annual/biennial and sometimes perennial is widely distributed throughout the tropics and subtropics. Herbaceous plants are erect, 60-80 cm high, and branching, but they can develop a twining habit as they mature or are shaded. Trifoliolate leaves have ovate-lanceolate to occasionally elliptical leaflets, 4-8 cm long and 1.0-3.5 cm wide. Red-purple flowers, about 2.5 cm long, are formed on racemes. Flowering is day-neutral; thus seed normally is produced before the first frost in the more northern extension of its area of adaptation. It is adapted to a wide range in rainfall, 500-3000 mm, and can grow in slightly acid to alkaline soils. Some ecotypes have potential for use as annuals in the humid southern US. *Macroptilium lathyroides* is a pioneer-type legume that produces a large growth during the initial year, with plant populations diminishing over the years. Foliage is palatable and readily consumed by cattle, but older stems are rejected (Pitman et al. 1986).

***Macroptilium longepedunculatum* (Benth.) Urb.** (llanos macro; cv. 'Maldonado' in Australia). Naturally occurring from Mexico to southern Brazil, this herbaceous perennial, closely resembling but generally smaller than siratro, is adapted to monsoonal lowland regions with sporadic flooding or waterlogging. It has only recently been cultivated in the Northern Territory of Australia. It survives in southern Florida but appears to be an annual or biennial that is neither as vigorous nor as persistent as siratro (Cameron 1990).

***Macrotyloma axillare* (E. Mey.) Verdc.** (axillaris and perennial horse gram; cv. 'Archer' in Australia; about 24 spp. in the genus). Naturally occurring in the sub-Saharan region from Ethiopia to Senegal and south to the Transvaal and Natal, this perennial twining herb has been successfully used for grazing and stockpiling in small areas of Australia. Because of its bitter taste, livestock do not accept it readily at first grazing. It is not well adapted for southern Florida probably because of seasonal waterlogging (Cameron 1986).

***Neonotonia wightii* (Wight & Arn.) Lackey**

(glycine and perennial soybean; cv. 'Tinaroo', 'Cooper', 'Clarence', and 'Malawi' in Australia; one sp. in the genus). Originally from tropical and subtropical Africa, India, and Asia, perennial soybean is a climbing or trailing perennial herb that has a general resemblance to siratro. Trifoliolate leaflets are 1.5-15 cm long and 1-12 cm wide. Small white flowers are produced on racemes and are initiated by short days in the subtropics. Because it is tolerant of drought, yet survives temporary waterlogging, perennial soybean has persisted under grazing in some areas of Latin America and is used in isolated regions in Australia. It requires fertile soils with near-alkaline pH (above 5.5). It is moderately tolerant to salinity and cold temperature and highly tolerant of soil Mn. Optimum annual rainfall is 1000-2000 mm. Some accessions produce seed in south Florida prior to frost, and plants overwinter if frosts are not too severe (Cameron 1984).

***Pueraria phaseoloides* (Roxb.) Benth.** (tropical kudzu and puero; 20 spp. in the genus). This robust, trailing, perennial legume is native to East and Southeast Asia, where it has been used for years as a cover crop (sometimes grazed) in rubber and oil palm plantations. It has been widely distributed in the high-rainfall areas of tropical America, but because of late-flowering, it does not produce seed in south Florida and is not productive where frequent frosts occur because high night temperatures are required for maximum growth. It can withstand waterlogging and light frosts and will survive long dry periods, although optimum growth occurs with above 1500-mm annual rainfall. It is tolerant of slightly acid soils. Stems up to 9 m long and trifoliolate leaves with leaflets 5-12 cm long and to about 11 cm wide can form a dense, smothering mat of foliage that dominates most grasses and weeds if not grazed heavily or periodically cut. Small blue flowers form on long axillary racemes producing linear pods up to 13 cm long. Tuberous roots are edible. Seed is difficult to obtain in Latin America (it is more plentiful in Asia), and vegetative planting is common in the Americas. Although tropical kudzu is reported to be very acceptable to cattle in some areas, in many areas of Latin America it is not well accepted. This has resulted in overgrazing of the grass component, producing almost pure stands of the legume.

***Sesbania sesban* (L.) Merrill** (sesbania; about 40 spp. in the genus). Sesbania is a heterogeneous warm-season annual/biennial or short-lived perennial that is widely cultivated throughout tropical Africa, Asia, and America. This shrub or short-lived tree, which is up to 8 m high with stems up to 12 cm in diameter, has pinnately compound leaves, with 6-27 linear oblong leaflet pairs up to 26 mm by 5 mm. Racemes are up to 20 cm long with flowers, 2 to 20, mostly yellow with purple spots. Straight to slightly curved almost cylindrical pods, with 10-50 seeds each, are up to 30 cm long by 5 mm wide. Adapted to annual rainfall of 500-2000 mm and adapted to up to 2300-m elevation, sesbania cannot withstand heavy frost. It tolerates waterlogging and temporary flooding. It can be grazed but normally is used in a cut-and-carry system. Also, it is used as firewood to provide shade for coffee (*Coffea arabica* L.) and other crops and as a green manure crop (Evans and Rotar 1987; Evans and Macklin 1990). (Also see Chap. 22.)

***Stylosanthes* species** (stylo; about 30 spp.). More species from this genus have been successfully used as forages than from any other tropical legume genus. The range of this genus is primarily tropical America. There are three species from Africa and *S. sundaica* from Asia (Stace and Edye 1984). *Stylosanthes biflora* is restricted from about 27° to 41°N latitude in eastern North America, while *S. calcicola* and *S. hamata* are found in south Florida. Leaves are trifoliolate and flowers are mostly small and yellowish. Plants are herbaceous to subshrubs, and most are vulnerable to sustained waterlogging. The primary impediment to more widespread use of the commercial species, particularly *S. guianensis,* is susceptibility to anthracnose (*Colletotrichum gloeosporioides*) (Irwin et al. 1984).

Stylosanthes capitata Vog. (cv. 'Capica' in Colombia) is a perennial herb or subshrub originating in subhumid and arid areas of northeastern Venezuela and Brazil. It is mostly many branched, is semierect to erect, has a distinct taproot, and is a heavy seed producer. Leaflets are elliptical to sometimes obovate. Inflorescences are in dense clusters, flowers are about 4-6 mm long, and fruits have two sections. It is adapted to acid soils, tolerates normally toxic levels of Al and Mn, and can grow in infertile soils. Best growth occurs in savanna areas with annual rainfall above 1000 mm, but productivity is impaired when soil pH is near or above neutral. Its growth in Florida has been poor because of an apparent soil nutrient imbalance. It has had very limited commercial success in Latin America.

Stylosanthes guianensis (Aubl.) Sw. (stylo; cv. 'Schofield', 'Cook', 'Endeavor', 'Graham', and 'Oxley' fine stem in Australia, 'Bandeirante' in Brazil, and 'Savanna' in Florida) has seven botanical varieties, separated by morphologic and ecologic characteristics with var. *guianensis* used or studied worldwide more than the others. This perennial herb or subshrub, up to 1.5 m high, originated in Brazil and is found naturally in South and Central America through Mexico. Plants have trifoliolate leaves with elliptical to lanceolate leaflets that are 0.5-4.5 cm long and 2-20 mm wide. Inflorescences typically have more than four yellow to orange 4- to 8-mm flowers each. Stylo grows best in slightly acid (pH >5.0) soils but does not grow well under continuously waterlogged conditions, being moderately drought tolerant. In the subtropics, because of frost and occasional freezes, *S. guianensis* may function as an annual or short-lived perennial.

This semierect to erect, and sometimes prostrate, legume is used mostly for grazing, although it is used as a cover crop in plantations and as a green manure crop. It can be continuously or rotationally grazed. Severe anthracnose damage, black lesions on leaves and stems that cause plant death, has reduced commercial use in Australia and elsewhere (Irwin et al. 1984). There are several races of the disease, so selection and breeding are not very effective tools in developing resistant cultivars. In south Florida the pathogen is present but does not develop disease symptoms until flowering commences in the cool autumn weather. A cool temperature is believed to prevent rapid disease development, which limits plant damage. This usually short-day species grows rapidly when rainfall and summer temperatures exist, thus producing large quantities of forage by fall if not grazed. It competes well with grasses and weeds. Quality is reduced during the flowering period because of the bristly texture of the canopy and plant maturation.

Stylsanthes hamata (L.) Taub. (caribbean stylo; cv. 'Verano' and 'Amiga', both tetraploids in Australia) is native to the drier or better-drained areas from southern Florida and the Caribbean Islands, in some Central American countries, and into Colombia, Venezuela, and Brazil. It is a short-lived perennial herb, prostrate to erect, with trifoliolate leaves about the size of *S. guianensis*. Seedpods are double articulated, and flowers are yellowish with reddish marks on the petals. This legume has been very successful in northern Australia as a pasture legume. It has also been used in Thailand, India, and several West African countries. In southern Florida, where summer growth is excellent, Verano is mostly an annual and fails to persist well, probably because of frequent high water tables. A main attribute is its resistance to anthracnose, which contributes to its widespread use in the more tropical areas of Queensland, Australia. *Stylsanthes hamata* has replaced *S. humilis* in many of the anthracnose susceptible growing areas.

Stylsanthes humilis Kunth (townsville stylo; cv. 'Patterson' [early], 'Lawson' [midseason], and 'Gordon' [late-flowering] in Australia and 'Khon Kaen' in Thailand) is native or naturalized in tropical America from Mexico into Brazil but not in Florida or the Caribbean Islands. It is now widely distributed in tropical Australia, where it was introduced unknowingly in the early 1900s, and elsewhere in Africa and Asia. It is an erect or prostrate, short-day annual herb that resembles *S. hamata,* but only the upper articulation of the fruit is fertile. In Florida, commercial plantings were made in the 1970s, but lack of long-term persistence prevented its continued use. Evaluation of available germplasm, which includes types that flower as early as August in Florida, may prove valuable for warmer areas in the lower-southern US. Except for Khon Kaen, cultivars are not resistant to anthracnose. In Florida, anthracnose was found to be a serious problem only with types that flower as early as August.

Stylosanthes scabra (Vog.) (shrubby stylo; cv. 'Seca', 'Fitzroy', and 'Siran' in Australia) is a shrubby legume that grows in Bolivia, Colombia, and Venezuela and is widespread in Brazil. An erect to suberect perennial to about 2 m high, shrubby stylo has a strong taproot system that can penetrate soils to several meters. Trifoliolate leaves have terminal leaflets that are 2.0-3.3 cm long and 4-12 mm wide. The yellow several-flowered inflorescence is up to about 30 mm long with loment fruit that usually produces two fertile seeds. This is one of the most drought tolerant tropical legumes and is adapted to annual rainfall areas of 500-2000 mm. It tolerates light frosts, but crowns can be killed by temperatures that fall below about –3°C. It is adapted to moderately acid sandy soils of low fertility, but it is not adapted to clay soils. Seca and Siran are resistant to anthracnose, which has been a major problem with Fitzroy in northern Queensland, Australia, where most of the commercial use has occurred.

***Vigna adenantha* (G. Mey.) Marechal, Marcherpa, & Stainer** (about 150 spp. in the genus). This species is native from Mexico through Argentina, occurs in tropical East and West Africa, and has spread as far as Taiwan and other islands. This nearly glabrous legume is a moderately lived perennial, is vigorously viny with prostrate to climbing stems up to 4 m or more long, and can root along prostrate stems. Trifoliolate leaves have leaflets about 70 cm long and 45 cm wide. Inflorescences are erect and mostly 10-20 cm long, with 5-15 whitish purple flowers. Pods are mostly linear, 7-12 cm long and 8-10 mm wide. When grazed in Florida, it has persisted better than siratro, which has a similar appearance. *Vigna adenantha* is much more tolerant of waterlogging than siratro and is not attacked by rust. It mixes well with bahiagrass but does not tolerate continuous high stocking rates (Muir and Pitman 1991). Protection from grazing during autumn is believed to help prevent loss of stand. Because it flowers late and has a growth habit where leaves cover maturing pods, it is difficult to mechanize seed harvesting. Thus, no seed is commercially available.

Vigna parkeri (creeping vigna; cv. 'Shaw' in Australia). This twining, blue-flowered short perennial is a high-quality legume originating from East and central Africa and has become naturalized in the highlands of Papua, New Guinea. Three subspecies have been identified. Trifoliolate leaves are made up of round, ovate, to ovate-lanceolate leaflets that are 1-9 cm long and 1-5 cm wide. Frequently, a crescent-shaped whitish watermark appears on the leaflets. The axillary inflorescence normally bears two to five flowers that produce pods that are 1-3 cm long and 4.5-5.5 mm wide. Adapted to moist sites and tolerant to infertile acid soils with pH down to about 5.0, Shaw has an ecological niche similar to that for greenleaf desmodium. Its use in Australia and elsewhere is limited, probably because of its indeterminant flowering that prevents concentrated seed set and results in low harvestable seed yields. It has persisted well in south Florida under varied management (Pitman et al. 1988), but seed is very difficult to obtain.

Zornia latifolia (zornia; 80 spp. in the genus). Representatives of the genus are found in tropical and warm temperate areas of the world, primarily in the Americas and Africa. The taxonomy of the genus is still in flux. *Zornia latifolia* is a heterogeneous species with wide adaptation to soils and climate. An accession collected from Brazil and being evaluated in Florida is erect to prostrate (grazed or mowed) and has many branches from stems and from the crown. Leaves are bifoliolate with leaflets 2-3 cm long and 0.5-1.5 cm wide. Alternate yellow flowers are produced on 10- to 30-cm spikes and are sequentially placed from base to terminal. There are one to seven articulated lomentum, which also dehisce sequentially. Although many ecotypes require short days for flowering, the Florida accession is day neutral, thus producing flowers in the summer. This may be of value in moving the species farther north. Florida results indicate that *Z. latifolia* is adaptable to moderately drained soils that are waterlogged only infrequently. The prostrate growth resulting from grazing or mowing is indicative of *Z. latifolia*'s ability to withstand high grazing pressure. This legume is compatible with bahiagrass and produces seed that is easily harvested.

Grasses

***Andropogon gayanus* Kunth** (gambagrass; cv. 'Kent' in Australia and several cultivars in Latin America). Gambagrass occurs naturally throughout tropical Africa from 15°N to 25°S, with annual rainfall of about 400 to 1500 mm, except in very dry or very humid regions. It is a tufted perennial with short rhizomes and stems usually 1-3 m high. Leaves are either glabrous or hairy, with blades up to 100 cm long and 4-30 mm wide with a white midrib. The inflorescence is a panicle with as many as six groups of branches, each terminating in a pair of racemes. The botanical variety *bisquamulatus* (one of four recognized) is the most important for use as forage, and all Latin American cultivars were developed from one Centro Internacional de Agriculture Tropical (CIAT) introduction of this variety. Gambagrass has a short-day flowering response. It may be vegetatively propagated from rooted tillers but is usually planted from seed. It is best adapted to warm areas with 400- to 1500-mm annual rainfall. It is capable of withstanding long dry seasons, tolerates flooding, is tolerant of grass fires, and grows on most soils except heavy clays. It associates well with twining tropical legumes and *Stylosanthes* species. (Toledo et al. 1990).

TABLE 23.2. Some agronomic characteristics of selected tropical legume species

Scientific name	Plant characters					Tolerance to						
	Life Cycle[a]	Growth Habit[b]	Grazing Habit[c]	Rainfall[d]	Acceptance[e]	Drought[f]	Frost[f]	Water[f]	Soil pH[g]	Grazing[h]	Disease[h]	Insect[i]
Aeschynomene americana	A	E	E	1000	H	F	P	E	M	H	H	M
A. falcata	P	D	D	700	H	G	F	F	LM	H	H	M
Alysicarpus vaginalis	AP	D	ED	700	H	G	P	F	MH	H	H*	H
Arachis glabrata	P	ED	D	750	H	E	P	P	MH	H	H	H
A. pintoi	P	EDS	D	1000	H	F	P	G	LM	H	H	H
Calopogonium mucunoides	ABP	TD	D	1000	VL	P	P	F	MH	H	M	H
Centrosema acutifolium	P	TS	E	1000	M	E	P	G	LM	M	M	M
C. macrocarpon	P	TS	TS	1000	M	E	P	G	LM	M	H	M
C. pascuorum	A	TD	TD	700	M	G	P	G	H	M	H	H
C. pubescens	P	TS	D	1250	M	G	P	F	MH	M	M	M
C. virginianum	AP	TS	D	1000	H	F	P	G	MH	M	H	H
Chamaecrista rotundifolia	AB	D	D	700	L	E	P	P	M	H	H	H
Clitoria ternatea	P	ET	E	400	M	G	G	F	MH	M	H	H
Desmanthus virgatus	P	E	E	250	H	E	P	F	H	H	H	H
Desmodium heterocarpon	P	ED	DE	1000	M	F	P	F	M	H	H*	M
D. heterocarpon ssp. *ovalifolium*	P	D	D	1250	L	G	P	G	LM	H	L*	H
D. heterophyllum	P	DS	D	1500	M	F	P	H	LM	H	H	H
D. incanum	P	ERS	D	1000	L	E	G	G	MH	H	H	H
D. intortum	P	ET	ES	900	M	F	F	F	MH	L	H	M
D. uncinatum	P	ET	ES	900	M	F	F	F	M	L	H	M
Indigofera hirsuta	A	E	E	750	L	E	F	F	MH	H	H	H
Lablab purpureus	AB	ET	E	500	M	G	F	P	MH	L	M*	M
Leucaena leucocephala	P	E	E	500	H	E	F	P	H	H	H	L
Lotononis bainesii	AP	DS	DS	1000	H	E	E	F	LM	H	M	H
Macroptilium atropurpureum	P	TS	D	750	H	E	F	F	MH	L	L*	M
M. lathyroides	AB	ET	E	600	H	G	P	G	MH	M	H	H
M. longepedunculatum	AB	T	D	900	M	G	P	E	MH	M	M	H
Macroptyloma axillare	P	T	T	700	L	G	P	P	H	M	M*	M
Neonotonia wightii	P	T	D	700	M	G	F	P	H	M	M	H
Pueraria phaseoloides	P	T	TD	1000	VL	P	P	G	M	H	H	H
Sesbania sesban	P	E	E	1000	M	P	P	G	M	L	H	H
Stylosanthes capitata	P	E	E	1000	M	E	P	P	LM	H	L	M
S. guianensis	P	E	ED	700	M	G	F	F	MH	M	L	H
S. hamata	AB	E	ED	750	M	E	P	P	M	H	M	H
S. humilis	A	ED	D	500	M	E	P	F	M	H	L	H
S. scabra	P	E	E	500	L	E	F	P	LM	H	M	H
Vigna adenantha	P	T	T	1000	M	P	P	G	MH	L	M	M
V. parkeri	P	TS	TSD	1000	H	P	P	G	M	M	M	H
Zornia latifolia	P	E	D	1000	H	G	P	F	M	H	H	H

[a]Life cycle: A = annual; B = biennial or short-lived perennial; P = perennial.
[b]Growth habit: when undisturbed or undefoliated; D = decumbent/prostrate; E = erect; T = twining/climbing; S = stoloniferous/adventitious rooting R = rhizomatous.
[c]Grazing growth habit: after intense grazing; same symbols as undisturbed growth habit.
[d]Minimum annual rainfall (mm).
[e]Acceptance = relative acceptability or palatability by cattle: VL = very low; L = low; M = moderate; H = high.
[f]Relative tolerance to: Drought; Frost damage to foliage; Waterlogging or temporary flooding; P = poor; F = fair; G = good; E = excellent.
[g]Soil pH adaptability range: L = 4.0-5.0; M = 5.0-6.5; H = 6.0-8.3.
[h]Grazing = grazing pressure constraints on mature plants assuming annual plants are permitted to produce seed: L = survival only under intermittent grazing with 5 to 6 wk rest periods; M = survival with continuous (0.75 AU (animal units)/ha) or intermittent intense grazing pressure for 4-8 wk alternated

***Axonopus affinis* A. Chase.** (See Chap. 35.)

***Brachiaria brizantha* (A. Rich.) Stapf** (palisadegrass; cv. 'Marandu' in Brazil and 'La Libertad' in Colombia). A more palatable grass than *B. decumbens*, this tufted, prostrate or semierect to erect perennial has short rhizomes and stems 30-200 cm tall and linear leaf blades 10-100 cm long and 3-20 mm wide, from glabrous to hairy. Sometimes *B. brizantha* is difficult to distinguish from *B. decumbens*. The areas of adaptation of these two species are similar, but *B. decumbens* is less competitive, and Marandu is somewhat resistant to spittlebugs. At present this is the preferred grass by cattlemen in Brazil, followed by *B. decumbens*.

***Brachiaria decumbens* Stapf** (signalgrass; cv. 'Basilisk' in Australia). Native to East Africa but becoming naturalized in Latin America and Asia, signalgrass is a trailing perennial with 30-60 flowering stems that extend up to 150 cm tall. Leaf blades are hairy, lanceolate, and 5-25 cm long and 7-20 mm wide. Signalgrass has stolons and rhizomes that can spread rapidly under favorable conditions to produce a dense turf. In Florida, it tends to be tufted with short stolons/rhizomes. It flowers profusely and is propagated from seed. Under field conditions it is difficult to differentiate from *B. brizantha*. *Brachiaria decumbens* is best adapted to the humid and subhumid tropics with annual rainfall above 1250 mm and with no more than 5 mo of dry season. It grows best on moist, fertile soils and withstands seasonal flooding. However, it also grows well on acid, highly Al-saturated soils. Signalgrass tolerates heavy grazing pressures. It is very competitive with weeds and associated legumes. Signalgrass is damaged badly by spittlebugs (*Anaeolamia, Deois,* and *Zulia* genera), which have damaged and killed signalgrass in areas of Brazil and elsewhere in Latin America. Because of this, growers have sought other species such as *B. humidicola, B. brizantha,* and *B. dictyoneura*. Signalgrass is acceptable to ruminants but not to horses and can cause photosensitization in sheep, goats, and young cattle.

***Brachiaria dictyoneura* (Fig. & De Not.) Stapf** (cv. 'Llanero' in Colombia). Native to eastern and southern Africa, this perennial creeping to tufted grass is very similar to *B. humidicola,* having a similar adaptive range and use. It has a moderate nutritive value, slightly higher than *B. humidicola,* withstands droughts of 4-5 mo as well as waterlogging, and is tolerant to acid, Al-saturated soils. It is less affected by spittlebugs or at least recovers more rapidly from damage than *B. humidicola*. Thus, its potential in its adaptive areas is good.

***Brachiaria humidicola* (Rendle) Schweick.** (koroniviagrass and creeping signalgrass; cv. 'Tully' in Australia and 'INIAP-701' in Ecuador). *Brachiaria humidicola* originated in eastern and southern Africa and has since spread widely in Southeast Asia, the Pacific area, and tropical Latin America. This highly stoloniferous, prostrate grass has linear to narrowly lanceolate leaf blades, 4-30 cm long and 3-10 mm wide. It is sometimes confused with *B. dictyoneura,* which appears more tufted and is less stoloniferous. *Brachiaria humidicola* is very competitive with associated legumes, but it also retards weed invasion. It can withstand very heavy grazing pressure, grows on soils of low fertility, is more resistant or at least recovers more rapidly from spittlebug attack than *B. decumbens,* and tolerates waterlogging and temporary flooding. Its acceptability to cattle is lower than signalgrass's, and intake may be limited because of its low CP concentration.

***Brachiaria mutica* (Forssk.) Stapf** (paragrass). Probably of tropical African origin, this grass has been naturalized worldwide in most tropical and near-tropical regions where annual rainfall is 1500 mm or more or in near-waterlogged or flooded sites. It is very palatable, is stoloniferous with short rhizomes, has ascending to decumbent stems up to about 2 m long, and has leaf blades 6-30 cm long and 5-20 mm wide. The inflorescence is a panicle, 6-30 cm long with 5-20 densely flowered racemes that are 2-15 cm long and separated by several millimeters. In the Americas it is planted vegetatively, while seed is available in Australia. This is one of several grasses used in the "ponded pasture system" in Australia (Anning and Kernot 1989).

***Chloris gayana* Kunth** (rhodesgrass; cv. 'Bell' [diploid] in Texas and 'Callide', 'Samford' [tetraploids], 'Pioneer', and 'Katambora' [diploids] in Australia and 'Pokot', 'Nzoia', 'Masaba', and 'Karpedo' in Africa). Rhodesgrass is native to most tropical and subtropical areas of Africa and has been distributed worldwide to similar climatic zones. Normally stems extend 0.5-2.0 m high, with leaf blades glabrous, 15-20 cm long and 2-20 mm wide, and with two to four leaves on stems originating from stolon nodes. The digitate, ascending panicle produces 3-20 spikes, each 4-15 cm long. This very heterogeneous, stoloniferous, sometimes tufted, high-quality perennial grass has more cold tolerance than most other tropical species, is able to survive dry seasons of up to 6 mo, can withstand periodic flooding, is tolerant of fire, tolerates soils from heavy clays to sands with pH above 4.5, and tolerates high soil salinity. Temperature range for growth is about 0° to 50°C, with optimum annual rainfall of 600-1500 mm. Also, it responds well to high levels of fertilizer N. It has been used for many years (Bell) in Texas, although susceptibility to rhodesgrass scale, lack of winterhardiness, and the introduction of other grasses such as buffelgrass (*Cenchrus ciliaris* L.) have limited its use there. Rhodesgrass has recently been planted commercially in southern Florida (Callide), where it has been well accepted by growers for use on virgin soils. It does not compete well with common bermudagrass (*Cynodon dactylon* [L.] Pers.) or torpedograss (*Panicum repens* L.). It associates well with legumes. In Australia and Africa it has been one of the more important cultivated species.

***Cynodon aethiopicus* Clayton & Harlan, *C. nlemfuensis* Vanderyst var. *nlemfuensis* and var. *robustus* Clayton & Harlan, and *C. plectostachyus* (K. Schum.) Pilger** (stargrass; cv. 'McCaleb', 'Ona', 'Florico', and 'Florona' in Florida). Stargrasses are of East African origin, are drought tolerant and widely adapted to tropical to near-temperate regions with low temperatures to about −4°C and with more than about 650-mm annual rainfall. These stoloniferous grasses have been used successfully for many years in south Florida (Hodges et al. 1984) and in Latin America, even though they are planted vegetatively. On infertile sites, stargrasses require rather intensive management, especially regular fertilization, for high levels of production and stand persistence. Animal performance on stargrass pastures is highly dependent upon grazing pressure. Stocking to utilize only the upper leafy canopy can produce excellent animal performance, while grazing pressure that produces utilization of the lower stemmy stubble results in much lower animal responses (Pitman 1991). Much of the literature dealing with *C. plectostachyus* prior to about 1970 refers to this species as *C. nlemfuensis*. (Also see Chap. 33.)

***Dichanthium annulatum* (Forssk.) Stapf** (hindigrass, shedagrass, and Kleberg bluestem). This perennial tufted and stoloniferous species occurs through Africa and Asia and has been distributed widely to other tropical areas with annual rainfall of 400-1000 mm. Stems are up to about 1 m long with leaf blades up to about 30 cm long and 2-8 mm wide. *Dichanthium annulatum* is tolerant of high grazing pressure and is drought tolerant. Kleberg bluestem and other *Dichanthium* species (Pretoria-90 bluestem, Angleton grass, and Medio bluestem) have shown promise in south Texas, but they are restricted in range due to lack of winterhardiness. Kleberg bluestem provides poor quality forage, while Pretoria-90 produces good quality forage but very little seed (Conrad and Holt 1983). Other species in use are *D. caricosum* (L.) A. Camus (Nadi bluegrass and Angleton grass) and *D. mucronulatum* Jansen. These are naturalized in North Africa, India, Burma, and the Malaysian region.

***Digitaria eriantha* Steud.** (See Chap. 35.)

***Echinochloa polystachya* (H.B.K.) Hitchc.** (alemangrass; cv. 'Amity' in Australia). This aquatic or subaquatic tropical perennial grass is found from Louisiana and Brownsville, Texas, through the West Indies to Buenos Aires, Argentina. It has rather coarse erect or stoloniferous culms 1-2 m long, generating from a spreading crown with rooting at glabrous or slightly pubescent nodes. This highly nutritious grass has leaf blades 30-40 cm long and 1.5-2.5 cm wide. Panicles of this almost completely sterile grass are 15-25 cm long. Alemangrass forms extensive colonies in seasonal swamps and on less wet areas. It is used primarily in Latin America for grazing, silage, and hay. It has been used for grazing sparingly on organic and sand soils of south Florida, but because of its vegetative method of propagation, it has not been readily accepted by growers. It could become a weed problem if planted near drainage canals. It has been reported successful in accelerating drying of colloidal phosphate settling ponds in central Florida. In Australia, Amity has been used to replace paragrass in ponded pasture management systems because it can thrive with deeper water levels and is a more productive grass. Diseases and insects generally are of little economic importance; however, spittlebug damage has been noted on a planting in Florida where little grazing had occurred. Other species used are *E. colona* (L.) Link (jungle rice) and *E. crus-galli* (L.) Beauv. (barnyard millet); however, both can become serious weeds in rice fields.

Eriochloa puctata **(L.) Desv. ex Hamillt.** = ***Eriochloa polystachya*** **Kunth.** (caribgrass). Native to tropical and subtropical America, including the West Indies, this perennial has stems up to 1.5 m high with glabrous leaves, 10-30 cm long and 4-12 mm wide. The erect panicles are about 15 cm long, but viable seed production is low. Its growth habit and area of adaptation and use are very similar to those of paragrass.

Hemarthria altissima (Poir.) **Stapf & C. E. Hubb.** (See Chap. 35.)

Heteropogon contortus **(L.) Beauv. ex Roem. & Schult** (speargrass). This perennial native or naturalized grass in the South Pacific islands, Australia, India, and Asia is now widespread in the tropics and subtropics. This tufted, glabrous perennial has slender, often branched stems 0.5-1.5 m high. Leaf blades are mostly 7-20 cm long and 3-7 mm wide. The inflorescence is a single raceme or a scant spathate panicle with slender rays 4-15 cm long and with a single peduncle supporting dense cylindrical racemes 3-8 cm long. The base of the spikelets has a pronounced callus that forms a very sharp point that is barblike. *Heteropogon contortus* is polymorphic, has been classified into several geographical types, and is adapted to annual rainfall areas of 600-1100 mm. It is seldom cultivated because of difficulties in seed production. However, because of its widespread use (ie., over 30 million ha in Australia), it is of economic importance. Its quality is poor during the dry season. It is a dominant species and highly tolerant of fire. It has been oversown with Townsville stylo and other legumes.

Hymenachne amplexicaulis **(Rudge) Nees** (hymenachne; cv. 'Olive' in Australia). This grass occurs widely in tropical South and Central America and also is native or naturalized in southwest Florida. It is found in swampy and seasonally flooded soils. Hymenachne is a robust, rhizomatous perennial with glabrous stems up to about 2 m. Stems can be erect or ascending. Leaf blades are mostly lanceolate, cordate at the base, and markedly narrower in the upper half, 10-45 cm long and up to 3 cm wide. The panicle is narrow and spikelike, cylindrical, and 20-40 cm long. Spikelets, 3-5 mm long, are lanceolate and upright. In Queensland, Australia, hymenachne, cv. Olive, is being used in the ponded pasture system, and even if plants die from extreme drought, regeneration by seed is rapid when rainfall begins. Olive grows better, but it is less palatable than Amity alemangrass. Olive does not survive as well under continuous flooding and is, reputedly, less drought tolerant than paragrass.

Hyparrhenia rufa **(Nees) Stapf** (jaraguagrass). This native perennial, sometimes annual, from Africa has been planted extensively in Latin America, but it has been replaced by newer grass species during the past 10-20 yr. Tufts of erect stout to slender stems up to 2.5 m high produce leaf blades that are 30-60 cm long and 2-8 mm wide. The spathate panicle has paired racemes 2-2.5 cm long. In Latin America, jaraguagrass is usually planted from seed collected by hand. It is adapted over a wide range of climates and soil types. In Africa it is grown below 1800 m in subhumid areas with up to 1500-mm annual rainfall in areas with a distinct dry season of up to 5 mo. In south Florida, it grows wild to a limited extent but is nonproductive. As the season progresses into the dry season, CP concentration of foliage decreases to 5% or less.

Imperata cylindrica **(L.) Raeusch** (cogongrass and bladygrass). This low-quality native of the Old World tropics and subtropics of Southeast Asia, Africa, India, and Australia also occurs in tropical and subtropical America. Cogongrass is a robustly rhizomatous perennial that forms tufts of leaves. Culms are erect to 2-3 m tall, unbranched, solid, and pubescent at nodes. Leaf blades are linear-lanceolate, 10-180 cm long and 5-25 mm wide, flat, erect, and spreading or drooping and when old have serrated, cutting edges. The cylindrical speciform panicle is about 6-30 cm long and 2 cm wide. Cogongrass grows in areas of annual rainfall of about 500-5000 mm and can survive waterlogging but not continuous flooding. Its upper elevation limit is about 2000 m. Although cogongrass is palatable when young, the serrate leaf edges limit intake of mature leaves. Seed is produced in abundance and effectively dispersed, contributing to weed problems from this grass. Once established, it is almost impossible to eliminate.

Melinis minutiflora **Beauv.** (molassesgrass). This usually tufted perennial is indigenous in nearly all parts of tropical Africa at elevations of 800-2500 m. Stems up to 1.5 m high may be erect or ascending and often are branched. Leaf blades are 5-20 cm long and covered with viscid hairs that impart a distinct molasses-like odor. Purple panicles are 10-30 cm long. It is adapted to cool, moderately moist climates with annual rainfall above 1000 mm. Tolerant to infertile soils, but not to flooding, molassesgrass is a strong competitor with associated legumes. Molassesgrass has been used extensively in Latin America but, like jaraguagrass, presently is not usually planted because better-adapted species are now available.

***Panicum coloratum* L. var. *makarikariense* Goossens** (makarikarigrass; cv. 'Bambatsi', 'Pollock', and 'Burnett' in Australia). A native of tropical Africa, this is a robust, tufted perennial, erect, ascending, or spreading by long creeping stems. Stems reach 40-150 cm in height. Leaf blades, 5-40 cm long and 4-14 mm wide, have a green to bluish color. Makarikarigrass is best adapted to subtropical regions with annual rainfall of 500-1000 mm. It thrives on heavy clay soils, tolerates seasonal flooding, and is frost tolerant. See Chapter 31 for *Panicum coloratum* var. *coloratum* (kleingrass).

***Panicum maximum* Jacq.** (guineagrass and green panic; there are many cv., including 'Hamil', 'Riversdale', 'Gatton', 'Petrie', and 'Makueni' in Australia and 'Colonao' and 'Vencedor' in Brazil). This variable species, with about a dozen named botanical varieties, is native to fertile soils of Africa, but it is now one of the most widely distributed species in the tropics and subtropics of the world. It was introduced into the West Indies in the middle 1700s and probably later to other European-controlled areas in Latin America. It reached Singapore in the late 1800s and the Philippines in about 1900. It can be found from the southern US into the subtropics of South America. Guineagrass is a very nutritive, erect perennial bunchgrass, rhizomatous at the base or rooting at the nodes, with slender to stout mostly erect stems that are 0.5-4.5 m high. Leaf blades are 15-100 cm long and up to 35 mm wide. Panicles are loose, 15-60 cm long, up to 25 cm wide, and much branched, with the lowermost branches in a distinct whorl. It can be established from seed or crown pieces. Seed shatters freely on ripening.

Guineagrass is best adapted to warm regions with over 900 mm of annual rainfall but will survive long droughts on deep soils and is shade tolerant. It responds to higher-fertility conditions, but it is not well adapted to waterlogged soils or to soils low in pH (below about 5.0) or high in Al and Mn. It is considered to be an excellent "fattening" grass for beef cattle. In Florida and Texas a vigorous tall guineagrass is a severe weed pest in citrus groves. However, guineagrass does not persist when sown on lowland pasture areas in Florida, presumably because of lower soil fertility and prolonged waterlogging.

Colonao is an example of the larger types used for grazing or harvesting, while Petrie, 'Sabi', Makueni, and 'Ember' are included in the smaller, low-growing types used mainly for grazing. One of the most distinctive commercial types of *P. maximum* is green panic, cv. Petrie (*P. maximum* var. *trichoglume* Eyles), which is a bunchgrass with ascending habit, the crown expanding by short horizontal stems. Leaves are narrow and soft and borne on slender stems, 1.5-2.0 m high. It can grow with annual rainfall as low as 560 mm but is better adapted to about 1000 mm. It survives light frosts and grows well in all soils except deep sands and heavy clays, and it tolerates heavy grazing pressure.

***Panicum repens* L.** (torpedograss). This rhizomatous perennial is widespread in Indonesia, Malaysia, Thailand, and other Southeast Asian countries, and it grows in wetter tropical areas in the world. It is naturalized in the Gulf Coast states from Florida to Texas, and years ago in Florida it was widely planted, vegetatively (Hodges and Jones 1950), because seed production is very poor. Presently, it is used as a native pasture and is infrequently fertilized. Torpedograss has very sharp, pointed rhizomes and sometimes stolons, has linear-acuminate leaves 7-25 cm long and 2-8 mm wide, and has a narrowly oblong panicle inflorescence that is 5-20 cm long. It is very competitive and very tolerant of acid, waterlogged, or flooded soils. It is no longer planted in the US, although it is cultivated in Asia.

***Paspalum plicatulum* Michx.** (See Chap. 34.)

***Pennisetum clandestinum* Hochst. ex Chiov.** (kikuyugrass; cv. 'Breakwell', 'Crofts', 'Noonan', and 'Whittet' in Australia). Naturally occurring across the elevated plateau of East and central Africa, it has been distributed to subtropical and highland tropi-

TABLE 23.3. Some agronomic characteristics of selected tropical forage grass species

	Plant characters			Tolerance to					
Scientific name	Growth Habit[b]	Rainfall[d]	Accept-ance[e]	Drought[f]	Frost[f]	Water[f]	Soil pH[g]	Disease[i]	Insect[i]
Andropogon gayanus	E	400	M	E	P	G	LM	M	MH
Axonopus affinis	D	1000	M	P	P	G	MH	H	H
Brachiaria brizantha	EDR	750	M	G	P	P	M	H	M
B. decumbens	ESR	1250	M	G	P	F	LM	M	L
B. dictyoneura	DSR	750	M	G	P	G	LM	H	M
B. humidicola	DSR	1000	L	G	P	G	LM	M	M
B. mutica	ESR	1000	H	P	P	E	MH	H	M
Cenchrus ciliaris (see Chap. 31)	ER	300	LM	E	F	P	H	H	M
Chloris gayana	ES	650	H	G	F	G	MH	H	M
Cynodon dactylon (see Chap. 33)	DSR	750	LM	F	P	F	MH	M	M
C. nlemfuensis	ES	750	M	G	F	F	MH	H	M
Dichanthium annulatum	ES	400	M	G	P	P	M	H	H
Digitaria eriantha (see Chap. 35)	DSE	1000	H	F	P	G	MH	L	L
Echinochloa polystachya	ES	1000	H	F	P	E	M	H	H
Eriochloa puctata	ES	1000	H	F	P	E	MH	H	H
Hemarthria altissima (see Chap. 35)	ES	1000	HML	P	G	G	M	H	M
Heteropogon contortus	E	600	L	G	F	P	L	H	H
Hyparrhenia rufa	E	600	L	G	P	F	M	H	H
Imperata cylindrica	ER	500	VL	G	P	G		H	H
Melinis minutiflora	E	900	M	F	P	P	M	H	H
Panicum coloratum var. *makarikariense*	E	600	H	E	G	G	M	H	H
P. maximum var. *maximum* (see Chap. 35)	E	900	H	G	P	P	MH	H	H
P. maximum var. *trichoglume*	E	600	M	G	G	P	H	H	H
P. repens	ESR	1000	LM	F	F	E	LMH	H	H
Paspalum notatum (see Chap. 34)	DR	1000	L	E	G	G	MH	H	H
P. plicatulum (see Chap. 34)	E	750	H	F	F	G	M	H	H
Pennisetum clandestinum	DR	900	H	G	E	F	LM	M	M
P. purpureum (see Chap. 35)	E	1000	H	G	F	F	MH	H	H
Setaria sphacelata	ER	900	H	F	G	G	M	H	H
Stenotaphrum secundatum	DS	1000	M	G	G	F	MH	L	L

Note: See footnotes of Table 23.2 (acceptability of grasses is based on mature herbage).

cal areas of the world that have humid climates. Although used for lawns and recreational areas, its primary use is for grazing. It has become a severe weed in highland, 1500- to 3000-m, vegetable-producing areas of Latin America. Kikuyugrass is a vigorous, stoloniferous, rhizomatous, deep-rooted, long-lived perennial with short culms 8-15 cm high, arising from long prostrate stolons, multibranching and rooting from the nodes. Leaf blades are up to 20 cm long and 1-5 mm wide, and flowering stems are usually hidden beneath the sterile shoots at the base of the sward. The sterile shoots inhibit flowering. Kikuyugrass can be propagated by seed or vegetatively. Optimum growth occurs with 25°C day and 20°C night temperatures. Thus, it is not adapted to lowland tropical areas. Foliage can tolerate –2°C frosts. It survives heavy grazing pressure and tolerates soil pH of below 5.0, but it is not adapted to moderate and high-fertility soils with annual rainfall above 1000 mm. It is not adapted to the warm climate of south Florida.

***Pennisetum purpureum* Schumach.** (See Chap. 35.)

***Setaria sphacelata* (Schum.) Stapf & C. E. Hubb. ex M. B. Moss** (setaria, golden timothy; cv. 'Nandi' [diploid], 'Narok', 'Kazungula', 'Solander', and 'Splenda' [tetraploids] in Australia). Originating and widely distributed in tropical and subtropical Africa, it was first brought into cultivation as a pasture plant in Kenya and later cultivated in most of the wet, lowland, tropical world. Two recognized botanical varieties of this highly heterogeneous species are *sericea* and *spendida*. It can be propagated vegetatively or by seed. This tufted perennial is rarely rhizomatous and normally has erect stems to 3 m high, flattened young tillers, and leaf blades up to 10-70 cm long and 8-20 mm wide. Spikelike panicles are dense, cylindrical, and 10-50 cm long with two to three spikelets on short branches. Each spikelet is subtended by 5-15 stiff bristles. Seed production of this cross-pollinated species is poor on some cultivars. It tolerates seasonal waterlogging, is somewhat frost tolerant, is compatible with numerous legumes, persists with annual rainfall above about 1000 mm, and is adapted to most soil types except highly acid (<5.0) or alkaline soils. It may contain high oxalate concentrations that reduce its normally high quality. Kazungula has been very popular in the more moist areas of Brazil. *Setaria sphacelata* tolerates temporary heavy grazing pressure, but rotational grazing is preferred (Jones and Jones 1989). The species can persist in Florida, but plants do not form a dense sward. Thus, *S. sphacelata* does not perform as well as many other grasses.

***Stenotaphrum secundatum* (Walter) O. Kuntze.** (See Chap. 35.)

QUESTIONS

1. Why are tropical grasses usually lower in quality than temperate grasses?
2. In terms of plant growth, how would you define the tropics? The subtropics?
3. The general quality of tropical forage legumes is less than that of temperate legumes. Which commonly measured factor is primarily responsible for this difference?
4. Which world regions contain the most tropical legume diversity? The most tropical grasses?
5. What are two primary benefits expected from legumes in tropical grass pastures?
6. Many of the commercial tropical grass cultivars are much more productive than temperate cultivars. Is this the reason for the tropical having poorer quality? Explain.
7. Name three general types of growth forms within tropical legumes presently being used commercially.
8. What is the most important characteristic needed in tropical legumes used for grazing?

REFERENCES

Anning, P, and J Kernot (eds.). 1989. North Queensland Beef Production Series 1. Ponded Pastures. Townsville, Queensland, Australia: Queensland Department of Primary Industries.

Bishop, HG, BC Pengelly, and DH Ludke. 1988. Classification and description of a collection of the legume genus *Aeschynomene*. Trop. Grassl. 22:160-75.

Blumenstock, DI. 1957. Distribution and characteristics of tropical climates. In Proc. 9th Pac. Sci. Congr. 20:3.

Bogdan, AV. 1977. Tropical Pasture and Fodder Plants. London: Longmans.

Brewbaker, JL. 1987. Species in the genus *Leucaena*. Leucaena Res. Rep. 7:6-20.

Bushby, HVA, and RJ Lawn. 1992. Accumulation and partitioning of nitrogen and dry matter by contrasting genotypes of mungbean (*Vigna radiata* [L.] Wilczek). Aust. J. Agric. Res. 43:1609-28.

Cameron, AG. 1984. Tropical and subtropical pasture legumes. 4. Glycine (*Neonotonia wightii*): An outstanding but soil specific legume. Queensl. Agric. J. 10:311-16.

———. 1986. Tropical and subtropical pasture legumes. 10. Axillaris (*Macrotyloma axillare*): A legume with limited roles. Queensl. Agric. J. 112:59-63.

———. 1990. Maldonado—A New Legume for the Top End. Agnote 429. Darwin, Australia: Department of Primary Industries of Fish.

Conrad, BE, and EC Holt. 1983. Year-round Grazing of Warm-Season Perennial Pastures. Tex. Agric. Exp. Stn. MP-1540.

Cook, BG. 1988. Persistent new legumes for heavy grazing. 2. Wynn round-leafed cassia. Queensl. Agric. J. 114:119-21.

Cook, BG, and TG Franklin. 1988. Crop management and seed harvesting of *Arachis pintoi* Krap. et Greg. nom. nud. J. Appl. Seed Prod. 6:26-30.

Cooper, JP. 1970. Potential production and energy conversion in temperate and tropical grasses. Herb Abstr. 40:1-15.

Evans, DO, and B Macklin (eds.). 1990. Perennial Sesbania Production and Use. A Manual of Practical Information for Extension Agents and Development Workers. Waimanalo, Hawaii: Nitrogen Fixing Tree Association.

Evans, DO, and PP Rotar. 1987. Sesbania in Agriculture. London: Westview.

Fribourg, HL, JR Overton, WW McNeill, EW Culvahouse, MJ Montgomery, M Smith, RJ Carlisle, and NW Robinson. 1984. Evaluations of the potential of hyacinth bean as an annual warm-season forage in the mid-South. Agron. J. 76:905-10.

Gramshaw, D, JW Reed, WJ Collins, and ED Carter. 1989. Sown pastures and legume persistence: An Australian overview. In GC Marten, AG Matches, RF Barnes, RW Brougham, RJ Clements, and GW Sheath (eds.), Persistence of Forage Legumes, Proc. Trilateral Workshop. Madison, Wis.: American Society of Agronomy, 1-22.

Gramshaw, D, BC Pengelly, FW Muller, WAT Harding, and RJ Williams. 1987. Classification of a collection of the legume *Alysicarpus* using morphological and preliminary agronomic attributes. Aust. J. Agric. Res. 38:355-72.

Grof, B. 1985. Forage attributes of the perennial groundnut *Arachis pintoi* in a tropical savanna environment in Colombia. In Proc. 15th Int. Grassl. Congr. Kyoto, Japan, 168-70.

Hall, TJ. 1985. Adaptation and agronomy of *Clitoria ternatea* L. in northern Australia. Trop. Grassl. 19:156-63.

Hodges, EM, and DW Jones. 1950. Torpedo Grass. Fla. Agric. Exp. Stn. Circ. S-14.

Hodges, EM, P Mislevy, LS Dunavin, OC Ruelke, and RL Stanley, Jr. 1984. 'Ona' Stargrass. Fla. Agric. Exp. Stn. Circ. S-268A.

Hodges, EM, AE Kretschmer, Jr., P Mislevy, RD Roush, OC Ruelke, and GH Snyder. 1982. Production and Utilization of the Tropical Legume *Aeschynomene*. Fla. Agric. Exp. Stn. Circ. S-290.

Hughes, R. 1970. Factors involved in animal production from temperate pastures. In Proc. 11th Int. Grassl. Congr., Surfers Paradise, Queensland, Australia, A31-38.

Humphreys, LR. 1980. A Guide to Better Pastures for the Tropics and Subtropics. 4th ed. Ermington, New South Wales, Australia: Wright Stephenson.

Irwin, JAG, DF Cameron, and JM Lenne. 1984. Responses of stylosanthes to anthracnose. In HM Stace and LA Edye (eds.), The Biology and Agronomy of *Stylosanthes*. New York: Academic Press, 295-310.

Jones, RJ. 1979. The value of *Leucaena leucocephala* as a feed for ruminants in the tropics. World Anim. Rev. 31:13-22.

———. 1985. Leucaena toxicity and the ruminant degradation of mimosine. In AA Seawright, MP Hegarty, LF James, and RF Keeler (eds.), Plant Toxicology, Proc. Aust.-USA Poisonous Plants Symp., Brisbane, Australia, 111-19.

Jones, RJ, and RM Jones. 1978. The ecology of Siratro-based pastures. In JR Wilson (ed.), Plant Relations in Pastures. Melbourne, Australia: CSIRO, 353-67.

———. 1989. Liveweight gain from rotationally and continuously grazed pastures of Narok setaria and Samford rhodesgrass fertilized with nitrogen in southeast Queensland. Trop. Grassl. 23:135-42.

Jones, RJ, and RG Megarrity. 1986. Successful transfer of DHP-degrading bacteria from Hawaiian goats to Australian ruminants to overcome the toxicity of leucaena. Aust. Vet. J. 63:69-78.

Kretschmer, AE, Jr. 1980. Use of Legumes in Pastures as Sources of Nitrogen. Ft. Pierce AREC Res. Rep. RL-1980-1.

———. 1989. Tropical forage legume development, diversity, and methodology for determining persistence. In GC Marten, AG Matches, RF Barnes, RW Brougham, RJ Clements, and GW Sheath (eds.), Persistence of Forage Legumes, Proc. Trilateral Workshop. Madison, Wis.: American Society of Agronomy, 117-38.

Kretschmer, AE, Jr., and RC Bullock. 1980. *Aeschynomene* spp.: Distribution and potential use. In Proc. Soil and Crop Sci. Soc. Fla. 39:145-51.

Kretschmer, AE, Jr., and GH Snyder. 1983. Potential of *Aeschynomene* species for pastures in the tropics. In JA Smith and VW Hays (eds.), Proc. 14th Int. Grassl. Congr., Lexington, Ky. Boulder, Colo.: Westview, 783-86.

Kretschmer, AE, Jr., JB Brolmann, GH Snyder, and SW Coleman. 1979. 'Florida' Carpon Desmodium (*Desmodium heterocarpon* [L.] DC.). A Perennial Tropical Forage Legume for Use in South Florida. Fla. Agric. Exp. Stn. Circ. S-260.

Kretschmer, AE, Jr., RM Sonoda, RC Bullock, GH Snyder, TC Wilson, R Reid, and JB Brolmann. 1985. Diversity in *Macroptilium atropurpureum* (DC.) Urb. In Proc. 15th Int. Grassl. Congr., Kyoto, Japan, 155-57.

Kretschmer, AE, Jr., GH Snyder, and TC Wilson. 1986. Productivity and persistence of selected *Aeschynomene* spp. In Proc. Soil and Crop Sci. Soc. Fla. 45:174-78.

Kretschmer, AE, Jr., JB Brolmann, RM Sonoda, GH Snyder, RC Bullock, and BJ Boman. 1992a. A Tropical Forage Crops Bibliography. Ft. Pierce AREC Res. Rep. FTP-92-2.

Kretschmer, AE, Jr., RM Sonoda, RC Bullock, and TC Wilson. 1992b. Registration of IRFL 4655 *Macroptilium atropurpureum* germplasm. Crop Sci. 32:836.

McWilliam, JR. 1978. Response of pasture plants to temperature. In JR Wilson (ed.), Plant Relations in Pastures. Melbourne, Australia: CSIRO, 17-34.

't Mannetje, L. 1978. The role of improved pastures

for beef production in the tropics. Trop. Grassl. 12:1-9.

't Mannetje, L, and RM Jones. 1992. Plant Resources of South-east Asia. Pudoc Sci. Publ. Wageningen, Netherlands.

Minson, DJ. 1980. Nutritional differences between tropical and temperate pastures. In FHW Morley (ed.), Grazing Animals, Amsterdam: Elsevier Science Publishing, 143-57.

———. 1990. Forage in Ruminant Nutrition. London: Academic Press.

Minson, DJ, and R Milford. 1966. The energy values and nutritive value indices of *Digitaria decumbens, Sorghum almum,* and *Phaseolus atropurpureum.* Aust. J. Agric. Res. 17:411-23.

Muir, JP, and WD Pitman. 1991. Grazing tolerance of warm-season legumes in peninsular Florida. Agron. J. 83:297-302.

Ohashi, H. 1973. Ginkoana—Contributions to the Flora of Asia and the Pacific Region. No. 1. Tokyo: Academia Scientific Books.

———. 1991. Taxonomic studies in *Desmodium heterocarpon* (L.) DC. (Leguminosae). J. Jap. Bot. 66:14-25.

Oram, RJ. 1990a. B. Legumes 21. *Arachis* (a) *Arachis pintoi* Krap et Greg. nom-nud. (Pinto peanut) cv. Amarillo. Aust. J. Exp. Agric. 30:445-46.

———. 1990b. Register of Australian Herbage Plant Cultivars. Queensland, Australia: CSIRO.

O'Reilly, MV. 1975. Better Pastures for the Tropics. Rockhampton, Queensland: Arthur Yates.

Partridge, IJ. 1986. Effect of stocking rate and super-phosphate level on oversown fire climax grassland of mission grass (*Pennisetum polystachion*) in Fiji. Trop. Grassl. 20:166-73.

Pitman, WD. 1991. Management of Stargrass Pastures for Growing Cattle Using Visual Pasture Characteristics. Fla. Agric. Exp. Stn. Bull. 884.

Pitman, WD, and AE Kretschmer, Jr. 1993. Carpon Desmodium for Peninsular Florida Pastures: Considerations for Establishment and Use. Fla. Agric. Exp. Stn. Circ. S-385.

Pitman, WD, CG Chambliss, and AE Kretschmer, Jr. 1988. Persistence of tropical legumes on peninsular Florida flatwoods (Spodosols) at two stocking rates. Trop. Grassl. 22:27-33.

Pitman, WD, AE Kretschmer, Jr., and CG Chambliss. 1986. Phasey Bean, a Summer Legume with Forage Potential for Florida Flatwoods. Fla. Agric. Exp. Stn. Circ. S-330.

Polhill, RM, and PH Raven (eds.). 1981. Advances in Legume Systematics. Vol. 2, Pts. 1 and 2. Proc. Int. Legume Conf., Royal Botanic Gardens, Kew, Surrey, England.

Prine, GM, LS Dunavin, JE Moore, and RD Roush. 1981. 'Florigraze' Rhizoma Peanut. A Perennial Forage Legume. Fla. Agric. Exp. Stn. Circ. S-275.

Reid, R, and DF Sinclair. 1980. An Evaluation of a Collection of *Clitoria ternatea* for Forage and Grain Production. Genet. Res. Commun. 1. Brisbane, Australia: CSIRO.

Rudd, VE. 1955. American species of *Aeschynomene.* US Natl. Herb. 32:1-172. Smithsonian Institute, Washington, D.C.

Schultze-Kraft, R. 1986. Natural distribution and germplasm collection of the tropical pasture legume *Centrosema macrocarpum* Benth. Angew Botanik 60:407-19.

Schultze-Kraft, R, and G Benavides. 1988. Germplasm Collection and Preliminary Evaluation of *Desmodium ovalifolium* Wall. Genet. Res. Commun. 12. Brisbane, Australia: CSIRO.

Schultze-Kraft, R, and RJ Clements. 1990. *Centrosema:* Biology, Agronomy, and Utilization. Cali, Colombia: CIAT

Schultze-Kraft, R, RJ Williams, L Coradin, JR Lazier, and AE Kretschmer, Jr. 1989. 1989 World Catalog of *Centrosema* Germplasm. Cali, Colombia: CIAT; Rome: IBPGR.

Skerman, PJ. 1977. Subtropical Forage Legumes. hu Plant Prod. Prot. Ser. 2. Rome: FAO.

Skerman, PJ, and F Riveros. 1990. Tropical Grasses. Rome: FAO.

Sonoda, RM, AE Kretschmer, Jr., and TC Wilson. 1991. Evaluation of *Macroptilium atropurpureum* (DC) Urb. germplasm for reaction to foliar diseases. In Proc. Soil and Crop Sci. Soc. Fla. 51:25-27.

Stace, HM, and LA Edye (eds.). 1984. The Biology and Agronomy of *Stylosanthes.* New York: Academic Press.

Stinton, CH. 1987. Advances in Legume Systematics. Pt. 3. Proc. Int. Legume Conf., Royal Botanic Gardens, Kew, Surrey, England.

Stobbs, TH. 1976. Beef production from sown and planted pastures in the tropics. In AJ Smith (ed.), Beef Cattle Production in Developing Countries. Edinburgh, Scotland: Univ. of Edinburgh Press, 164-83.

Strickland, RW, RG Greenfield, GPM Wilson, and GL Harvey. 1985. Morphological and agronomic attributes of *Cassia rotundifolia* Pers., *C. pilosa* L. and *C. trichopoda* Benth., potential forage legumes for northern Australia. Aust. J. Exp. Agric. 25:100-108.

Summerfield, RJ, and AH Bunting (eds.). 1980. Advances in Legume Science. Proc. Int. Legume Conf. Vol 1. Royal Botanic Gardens, Kew, Surrey, England.

Teitzel, JK, and RL Burt. 1976. *Centrosema pubescens* in Australia. Trop. Grassl. 10:5-14.

Toledo, JM, R Vera, C Lascano, and JM Lenne. 1990. *Andropogon gayanus* Kunth: A Grass for Tropical Acid Soils. Cali, Colombia: CIAT.

Vincente-Chandler, J, S Silva, and J Figarella. 1959. Effect of nitrogen fertilization and frequency of cutting on the yield and composition of napiergrass in Puerto Rico. J. Agric. Univ. Puerto Rico 43:215-27.

Webster, CC, and PN Wilson (eds.). 1966. Agriculture in the Tropics. London: Longmans.

Whiteman, PC. 1980. Tropical Pasture Science. New York: Oxford Univ. Press.

Wilson, GPM, RM Jones, and BG Cook. 1982. Persistence of jointvetch (*Aeschynomene falcata*) in experimental sowings in the Australian subtropics. Trop. Grassl. 16:155-56.

Withington, D, N Glover, and J Brewbaker. 1987. *Gliricidia sepium* (Jacq.) Walp.: Management and Improvement. Spec. Publ. 87-01. Waimanolo, Hawaii: Nitrogen Fixing Tree Association.

24 Timothy

ALBERT R. McELROY
Agriculture Canada

H. T. KUNELIUS
Agriculture Canada

TIMOTHY (*Phleum pratense* L.) is a cool-season forage grass, widely grown in the cool, moist regions of the US, Canada, and Europe. It is classified in the tribe Agrostideae and is the only *Phleum* of major economic importance. A native of the Old World, timothy was brought under cultivation in the early 1700s in North America, where it was called *Herd's grass* (Nath 1967). Cultivated strains were later reintroduced into Britain from North America. However, at that time it was being cultivated in Sweden under the name *Ängkampe*.

DISTRIBUTION AND ADAPTATION

Timothy is widely adapted in temperate, moist environments. It does not persist in droughty conditions. In the US, timothy is grown east of the Great Plains south through Missouri, Kentucky, Virginia, and adjacent states and in the Pacific Northwest (Fig. 24.1). In Canada, it is a major forage grass in the Atlantic provinces, Quebec, and Ontario. Timothy is well adapted to the conditions of northern Europe and is also produced in the temperate areas of South America, Australia, and northern Japan. As an example of the widespread use of timothy, there are 70 cultivars from 19 countries included in the *List of Cultivars Eligible for Certification* (1992).

Timothy is one of the most winter-hardy cool-season forage grasses. It is particularly popular in areas subject to severe winter conditions due to its reliability as a forage producer. Consequently, its area of utilization extends further north than orchardgrass (*Dactylis glomerata* L.), for example, which is less winter-hardy. Naturalized North American populations of timothy are found as far north as James Bay, Ontario, at 52°N (Dore and McNeill 1980).

Seed can be produced in most areas where timothy is grown as a forage. However, in North America, most certified seed is grown in the higher rainfall areas north of the Great Plains and the Peace River region of Alberta.

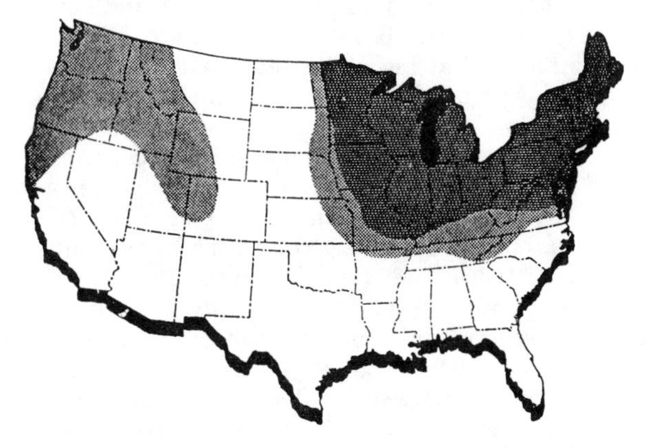

Fig. 24.1. Timothy-producing areas in the contiguous US.

ALBERT R. McELROY is Project Leader of the Forage Genetics Program, Plant Research Centre, Agriculture Canada, Ottawa. He received the MSc from Université Laval and the PhD from the Univeristy of Guelph. His research has centered on breeding methodology for autopolyploid forages and genetics of forage quality.

H. T. KUNELIUS is a Research Scientist at the Agriculture Canada Research Station, Charlottetown, Prince Edward Island, where he conducts research on forage and pasture management. He has the MSc from the University of Helsinki and the PhD from the University of Manitoba.

Seed of European cultivars is routinely produced in North America, with approximately one-third of the Canadian crop going to European markets. Seed multiplication outside the area of development does not appear to alter population structure or forage yield (Simon and Kastenbauer 1979).

PLANT DESCRIPTION

Timothy is a perennial bunchgrass that reaches approximately 1 m in height. It has no rhizomes, forms an open sod, and is not an aggressive grass.

The leaf blades are flat, often twisted, and hairless except for sparse cilia on the blade collar. The ligule is blunt, up to 6 mm long, and distinctly notched. Auricles are absent. The compact and swollen lowest internodes, referred to as the *haplocorm* or *corm,* act as a repository for carbohydrate reserves and play an important role in regrowth and persistence following defoliation. Individual tillers are annuals. New shoots, which develop from nodes near the haplocorm, perpetuate the plant as a perennial.

The panicles are dense, cylindrical, and spikelike, 5-10 cm long and 6-10 mm thick. The single flower per spikelet is attached to the main axis by a short (1-mm) branch. The root system in timothy is fibrous but very shallow. Up to 80% of root mass has been found in the top 5 cm of the soil profile (Garwood 1967). Regrowth is typically poor under conditions of moisture stress, partly a result of the shallow root system and the fact that new tillers must develop their own root system.

Timothy is an obligatory long-day plant with no cold period requirement for floral induction. The critical photoperiod for flower induction can vary with cultivar (Hay and Pedersen 1986; Heide 1982), and strains developed at higher latitudes tend to have a longer critical photoperiod. The development of timothy is related to growing degree-days (GDD). In Atlantic Canada, timothy cultivars require 350-450 GDD (>5°C) to reach early heading stage (Bootsma 1984). Heading occurs in regrowth after the initial harvest, but many tillers may be vegetative.

Timothy is very winter-hardy. It exhibits substantial tolerance to both cold temperatures (Fig. 24.2) and ice encasement, a major factor affecting survival under winter stress. Cultivar differences have been reported for both these traits, which are not closely associated in this species (Andrews and Gudleifsson 1983). However, winterhardiness is not an important distinguishing feature among cultivars, and differences in field trials are not usually evident except under extreme conditions. For example, cultivars developed at higher latitudes tend to be more winter-hardy in Alaska than those selected in southern areas (Klebesadel and Helm 1986).

IMPORTANCE AND USE

Timothy is a very popular forage grass in areas where it is adapted. It is easy to establish, associates well with other species in mixtures, and produces good yields of high-quality forage when harvested at the early heading stage. Seed is easy to produce and is usually in abundant supply, with the price being competitive with other forage grasses. In the northeastern US, between 815 and 990 mt of timothy seed were sold annually between 1987 and 1991 (Pardee 1992). In eastern Canada, the volume of certified timothy seed sold is more than double that of all other forage grasses combined.

Timothy's ability to withstand low temperatures and ice encasement make it suitable for forage production where legumes are subject to winter damage. In northern areas it is often included in forage mixtures as an insurance against crop failure due to winter stress. In cases where alfalfa (*Medicago sativa* L.) is eliminated from a sward by heaving, low temperatures, or ice encasement, the surviving timothy will usually provide a reasonable forage crop, provided that a spring application of nitrogen (N) fertilizer is made. In areas of ex-

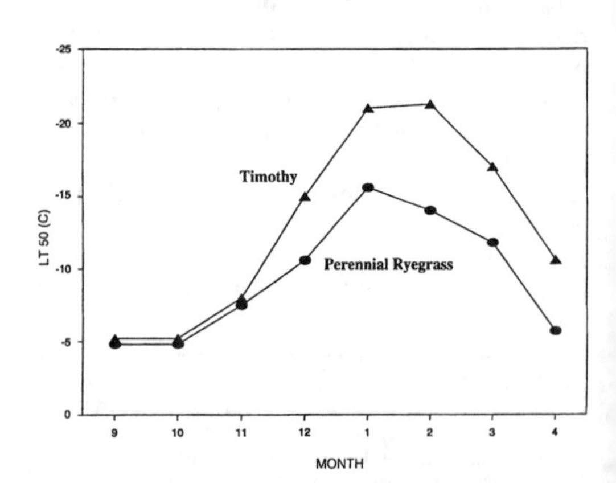

Fig. 24.2. LT50 of timothy and perennial ryegrass sampled throughout the fall and winter. The LT50 is the temperature at which 50% of a population regrows following cold stress. (Adapted from Suzuki 1993.)

treme winter stress, timothy is frequently grown in monoculture. Timothy is grown primarily for use as conserved feed, either for hay or silage. It is commonly grown in association with a legume, usually alfalfa, red clover (*Trifolium pratense* L.), or birdsfoot trefoil (*Lotus corniculatus* L.). The legume component reduces the need for N fertilization of established stands and improves forage quality. Although the quality of legume forages is generally acknowledged to be higher than that of grasses, a legume-grass forage mixture is frequently more desirable than pure legume in dairy and beef rations.

The dry matter yield of timothy is comparable with that of other forage grasses when the first cut is taken around the heading stage. It also compares favorably in feeding value, although direct comparisons with other species must be interpreted with some caution due to maturity differences. Both stage of maturity and age of plants are known to affect forage quality, but the effects are often confounded. Timothy tends to mature later than other forage grasses. It was found to be higher in in vitro dry matter digestibility (IVDMD) than orchardgrass, tall fescue (*Festuca arundinacea* Schreb.), smooth bromegrass (*Bromus inermis* Leyss.), and reed canarygrass (*Phalaris arundinacea* L.) when all were harvested on the same dates (Collins and Casler 1990b). The timothy cultivars headed later than the other species and therefore were less advanced phenologically or in maturity at each date. Conversely, Seoane and Gervais (1982) found that the digestibility and dry matter intake by sheep of timothy were lower than those of smooth bromegrass when cultivars were harvested at early heading. The timothy headed approximately 2 wk later than smooth bromegrass. In both studies, herbage N content was similar among species.

Timothy is generally considered to produce high-quality silage (Table 24.1) when harvested at early heading (Burgess and Grant 1974). However, the levels of water-soluble carbohydrates have been found to be low for good silage fermentation in some cases (Kunelius and Halliday 1989).

Timothy is used to some extent for grazing. It may be included in pasture mixtures, particularly for land with variable drainage where less winter-hardy species tend to be eliminated from the sward. However, timothy is sensitive to frequent defoliation and does not persist well in grazed swards. The apex is removed in grazing due to the upright growth habit, and the shallow root system makes it susceptible to moisture stress. The quality of timothy pasture is very high. In the United Kingdom, the digestibility of timothy under pasture management has been found to be equal to that of orchardgrass but better than that of tall fescue, although the liveweight gain by lambs was greater with timothy than with orchardgrass, perennial ryegrass (*Lolium perenne* L.), and tall fescue (Davies and Morgan 1982a, 1982b).

CULTURE AND MANAGEMENT

Timothy is usually grown in association with a legume such as alfalfa or red clover. It is often grown in pure stands in areas where legume persistence is limited due to winter stresses or imperfect drainage. Nitrogen with phosphorus (P) and potassium (K) must be applied to maintain productivity of timothy stands (Bélanger et al. 1989).

Establishment. Timothy is generally easy to establish when sown in the spring or early fall under conditions of adequate moisture and moderate temperatures. Drought and high temperatures during seedling establishment can be very deleterious to the young plants. Shallow (1.25-cm) seeding in a firm seedbed is ideal. Seeding rates vary from approximately 2 to 10 kg ha^{-1}.

Timothy dry matter yields are typically low in the establishment year. When seeded with alfalfa, the timothy component of the sward is low in the establishment year, so it is usually planted with a cereal nurse crop that can be harvested as forage or grain.

TABLE 24.1. Feed intake, milk yield, and milk composition of cows fed silages made from first-cut timothy or corn

	Timothy	Corn
Silage DM intake (kg d^{-1} cow^{-1})	10.9	11.5
Grain DM intake (kg d^{-1} cow^{-1})	6.5	7.1
Milk yield (kg d^{-1} cow^{-1})	22.3	22.9
Milk fat (%)	3.1	3.4
Milk protein (%)	2.6	2.9

Source: Adapted from Burgess and Grant (1974).

Production. In established stands, growth begins early in the season, and the rate of dry matter accumulation is similar to that of orchardgrass and smooth bromegrass (Darisse et al. 1980). Timothy is most productive in spring and early summer under conditions of cool temperatures and abundant moisture. Smith (1972) determined the optimum day/night temperatures to be 21°/15°C.

Dry matter yields increase to anthesis, and maximum seasonal yields can be obtained when the initial harvest is made at this stage (Darisse et al. 1980). However, the nutritional quality declines rapidly with advancement in maturity. Therefore, the harvest schedule is a compromise between yield, forage quality, and persistence.

Harvesting at early to midheading stage of development results in quality forage and good persistence of timothy (Brown et al. 1968). Food reserves in corms are adequate, and new shoots start developing (Sheard 1968). In practice, ideal timing of harvest is not always possible, and cutting at stem elongation, a critical period for persistence, may be necessary. Frequent cutting during stem elongation of primary growth can result in stand deterioration (Brown et al. 1968), but under infrequent harvesting it is possible to maintain good timothy stands longer than 4 yr (Kunelius and McRae 1986). Casler and Walgenbach (1990) found that early-maturing timothy cultivars persisted better with alfalfa than did late-maturing cultivars. Late-maturing cultivars are often grown with red clover or birdsfoot trefoil, which are typically harvested later than alfalfa.

The forage quality of timothy in spring growth declines markedly as the plants mature. The crude protein (CP) content drops from over 20% in early vegetative stages to approximately 10% at heading. Similar declines occur in IVDMD (Darisse et al. 1980; Mason and Lachance 1983). The declines in CP and IVDMD continue after heading, although less rapidly, but the in vivo digestibility of both CP and DM have been shown to continue to decline linearly (Waldie et al. 1983).

The effect of declining quality with maturation can be partially offset by growing cultivars of differing maturity. When harvested at a particular date, late-maturing cultivars tend to have higher quality than early ones, presumably because they are less advanced phenologically. Therefore, the use of both early- and late-maturing cultivars and harvesting each at heading will help maintain forage quality.

The regrowth of timothy is typically variable and is strongly influenced by temperature and moisture availability. Under conditions of high temperatures (>30°C) and severe moisture stress, regrowth may be negligible. Some cultivar differences exist, but regrowth is typically slower than that of most other forage species. As a consequence, the regrowth of predominantly timothy stands is often grazed rather then conserved as hay or silage.

DISEASES AND PESTS

Disease and insect pests do not normally pose any serious limitations to timothy forage or seed production. Stem rust, *Puccinia graminis* var. *phlei-pratensis,* was found to reduce the winter survival of some susceptible populations (Myers and Chilton 1941), but the subsequent development of resistant cultivars has effectively controlled this disease. Timothy generally has good resistance to organisms causing crown rust (*P. coronata*), leaf spot (*Helminthosporium, Drechsleria, Pyrenophora,* and *Bipolaris*), snow mold (*Typhula, Sclerotinia,* and *Fusarium*), and eyespot (*Heterosporium phlei*), although some infections occur (Braverman 1986).

Silver top, characterized by whitening and premature death of the upper internode and head, is commonly found. In Quebec, where timothy is the most popular forage grass, the proportion of infected heads was found to be 9% in a survey of 12 fields from four different regions (Gagné et al. 1985). *Fusarium poae* was thought to be the causal organism because it was isolated from all infected heads, but artificial inoculation did not produce symptoms.

Choke disease, caused by *Epichloë typhina,* has been reported in Hokkaido, Japan (Shimanuki and Sato 1983). The incidence of the disease, which begins at the jointing stage of the first growth, was found to fluctuate markedly over a 7-yr period. Endophytic mycelia were in found in haplocorms, leaves, culms, and young tillers, but not in the roots, of diseased and apparently healthy plants.

Several insect pests attack timothy, but the effect on forage yield or quality is generally considered to be minimal. For example, Mackun and Baker (1990) showed some benefit to forage yield by controlling insects in a birdsfoot trefoil-timothy stand, but it was too small to justify economically. The European skipper, *Thymelicus lineola,* is found in many areas, but the degree of infestation has been shown to vary considerably (McNeil et al. 1975). Early harvesting of forage will control populations since eggs are laid in the stems in late

June and early July. Chemical and biological insecticides have been shown to be effective in controlling the pest and improving forage yield (Thompson 1977).

Timothy is a good host plant for root-lesion nematodes, *Pratylenchus penetrans* (Kimpinski and Willis 1982). The root-lesion nematode has only a limited effect on the growth of timothy. However, there can be an increase in nematode population in fields growing timothy, which may have an adverse effect on succeeding crops.

NEW CULTIVAR PRODUCTION

Timothy cultivars, like those of other cool-season forage grasses, are heterogenous populations. Variability among individuals within a given population is typically small for simply inherited traits, such as maturity, but more extensive for complex traits like yield or quality. Genetic variability both within and among populations is exploited in the development of new cultivars. The vast majority of cultivars are hexaploid, although a few diploids such as 'S-50', developed at the Welsh Plant Breeding Station, Aberystwyth, are used for pasture production in Europe.

Ecotypes and landrace strains have been used for timothy production. A trial of 72 strains was seeded at Cornell University in 1920 (Smith and Myers 1934), and it is likely that many of these were local landraces. Under adverse conditions, some ecotypes are more productive than selected cultivars (Cenci 1980). A few cultivars, such as 'Korpa' from Iceland and 'Toro' from Italy, were developed by mass selection of local landraces. However, most timothy cultivars are synthetics, i.e., the advanced generation of two to many parental clones that were interpollinated. Parental clones are selected on the basis of phenotype, the performance of the individual plants, or combining ability, which is determined through progeny testing.

The most distinguishing feature among North American timothy cultivars is maturity. Koch (1976) reports a range in heading date of 20 d among adapted cultivars. However, maturity differences among cultivars vary with latitude. In Ontario, 15 cultivars were recommended for use in 1993. The range in heading date among these cultivars is approximately 14 d in the southern part of the province but only about 5 d for the same cultivars grown approximately 6° farther north.

Dry matter yield remains a major selection criterion, and modest gains continue to be made. Most cultivars now have adequate resistance to foliar diseases and environmental stresses as a result of continued selection for these traits. Regrowth is a particular weakness of timothy. Some cultivar differences are evident, but efforts to improve this trait have given only modest results. Even the best cultivars have poorer regrowth than orchardgrass or tall fescue.

Forage quality improvement is the focus of many breeding programs. Although cultivars developed specifically for improved feeding value have yet to appear on the market, studies indicate that improvement of quality traits should be possible through breeding. Selection for forage quality appears to have begun as early as the 1930s. Evans and Ely (1936) note that some timothy plants tend to maintain the normal green color of their leaves with advance in maturity. The leaves of an early and a late population, selected over three generations, remained green for a longer time than those of ordinary timothy. Since then, genetic variation for quality parameters has been reported (Berg and Hill 1983; Surprenant et al. 1990a, 1990b), including rate of IVDMD decline with advance in maturity (McElroy and Christie 1986). Crude protein content, neutral detergent fiber (NDF), and IVDMD are common selection criteria used to improve the forage quality of timothy. Although digestibility differences of up to 100 g kg^{-1} among genotypes within cultivars or experimental populations are routinely found in breeding programs, developing improved cultivars is complicated by the fact that digestibility is a complex trait, affected by many factors. The heritability of quality traits is much lower than maturity, for example, and consequently the response to selection is lower.

Some variation among cultivars for quality attributes has been reported, although these differences are sometimes confounded with maturity or harvest date. For example, when harvested at early heading, early-maturing cultivars tend to degrade faster in the rumen than late-maturing lines (Seoane et al. 1981). When harvested at particular dates, regardless of growth stage, later-maturing cultivars had higher IVDMD (Collins and Casler 1990a). NDF increased linearly with time for all cultivars, although the increase was most rapid in late cultivars.

QUESTIONS

1. What is the general area of adaptation of timothy?
2. Describe the general appearance and growth habit of timothy.

3. What are the main uses of timothy?
4. Discuss the advantages and disadvantages of timothy relative to other cool-season forage grasses.
5. Why is timothy not particularly well suited to pasture production?
6. What is the major distinguishing feature among timothy cultivars? Of what practical significance are these differences?
7. What specific factors make timothy resistant to winter injury?
8. How important are diseases to timothy production?
9. Identify forage legume species that are compatible with timothy for producing hay, silage, or pasture.

REFERENCES

Andrews, CJ, and BE Gudleifsson. 1983. A comparison of cold hardiness and ice encasement tolerance of timothy grass and winter wheat. Can. J. Plant Sci. 63:429-35.

Bélanger, G, JE Richards, and RB Walton. 1989. Effects of 25 years of N-fertilization, P-fertilization and K-fertilization on yield, persistence and nutritive-value of a timothy sward. Can. J. Plant Sci. 69:501-12.

Berg, CC, and RR Hill, Jr. 1983. Quantitative inheritance and correlations among forage yield and quality components in timothy. Crop Sci. 23:380-84.

Bootsma, A. 1984. Forage crop maturity zonage in the Atlantic region using growing degree-days. Can. J. Plant Sci. 64:329-38.

Braverman, SW. 1986. Disease resistance in cool-season forage range and turf grasses. II. Bot. Rev. 52:1-112.

Brown, CS, GA Jung, KE Varney, RC Wakefield, and JB Washko. 1968. Management and Productivity of Perennial Grasses in the Northeast. IV. Timothy. W.Va. Univ. Agric. Exp. Stn. Bull. 570T.

Burgess, PL, and EA Grant. 1974. A grassland system for dairy cattle based on ensiled timothy (*Phleum pratense*) cultivars. In VG Iglovikov and AP Movsisyants (eds.), Proc. 12th Int. Grassl. Congr., Moscow, 3:78-86.

Casler, MD, and RP Walgenbach. 1990. Ground cover potential of forage grass cultivars mixed with alfalfa at divergent locations. Crop Sci. 30:825-31.

Cenci, CA. 1980. Evaluation of differently adapted types of *Phleum pratense* L. Z Pflanzenzüchtung 85:148-56.

Collins, M, and MD Casler. 1990a. Forage quality of five cool-season grasses. I. Cultivar effects. Anim. Feed Sci. Technol. 27:197-207.

———. 1990b. Forage quality of five cool-season grasses. II. Species effects. Anim. Feed Sci. Technol. 27:209-18.

Darisse, JPF, P Gervais, and JC St-Pierre. 1980. Influence du stade de croissance sur le rendement et la composition chimique de deux cultivars de la fléole des prés, du brome et du dactyle. Naturaliste Can. 107:55-62.

Davies, DA, and TEH Morgan. 1982a. Performance of ewes and lambs on perennial ryegrass, cocksfoot, tall fescue and timothy pastures under upland conditions. J. Agric. Sci., Camb. 99:145-51.

———. 1982b. Herbage characteristics of perennial ryegrass, cocksfoot, tall fescue and timothy pastures and their relationship with animal performance under upland conditions. J. Agric. Sci., Camb. 99:153-61.

Dore, WG, and J McNeill. 1980. Grasses of Ontario. Res. Branch Agric. Can. Monogr. 26. Quebec: Canadian Government Publishing Centre.

Evans, MW, and JE Ely. 1936. Timothy selection for improvement in quality of hay. Agron. J. 28:941-47.

Gagné, S, C Richard, and C Gagnon. 1985. Présence et causes possibles de la coulure des graminées chez la fléole des prés au Québec. Can. Plant Dis. Surv. 65:17-21.

Garwood, EA. 1967. Some effects of soil water conditions and soil temperature on the roots of grasses. 1. The effects of irrigation on the weight of root material under various swards. J. Br. Grassl. Soc. 22:176-81.

Hay, RKM, and K Pedersen. 1986. Influence of long photoperiods on the growth of timothy (*Phleum pratense* L.) varieties from different latitudes in northern Europe. Grass and Forage Sci. 41:311-17.

Heide, OM. 1982. Effects of photoperiod and temperature on growth and flowering in Norwegian and British timothy cultivars (*Phleum pratense* L.). Acta Agric. Scand. 32:241-52.

Kimpinski, J, and CB Willis. 1982. Influence of soil temperature and pH on *Pratylenchus penetrans* and *P. crenatus* in alfalfa and timothy. J. Nematol. 13:333-38.

Klebesadel, LJ, and D Helm. 1986. Food reserve storage, low-temperature injury, winter survival, and forage yields of timothy in subarctic Alaska as related to latitude-of-origin. Crop Sci. 26:325-34.

Koch, DW. 1976. In vitro digestibility in timothy (*Phleum pratense* L.) cultivars of different maturity. Crop Sci. 16:625-26.

Kunelius, HT, and L Halliday. 1989. Nutritive value and production of cool season grasses under two harvest regimes. In D Desroches (ed.), Proc. 16th Int. Grassl. Congr., 4-11 Oct., Nice, France. Montrouge: Dauer, 827-28.

Kunelius, HT, and KB McRae. 1986. Effect of defoliating timothy cultivars during primary growth on yield, quality and persistence. Can. J. Plant Sci. 66:117-23.

List of Cultivars Eligible for Certification. 1992. Paris: Organization for Economic Development.

McElroy, AR, and BR Christie. 1986. Variation in digestibility decline with advance in maturity among timothy (*Phleum pratense* L.) genotypes. Can. J. Plant Sci. 66:323-28.

Mackun, IR, and BS Baker. 1990. Insect populations and feeding damage among birdsfoot trefoil-grass mixtures under different cutting schedules. J. Econ. Entomol. 83:260-67.

McNeil, JN, RM Duchesne, and A Comeau. 1975. Known distribution of the European skipper

Thymelicus lineola (Lepidoptera: Hesperiidae) in Quebec. Can. Entomol. 107:1221-25.

Mason, W, and L Lachance. 1983. Effects of initial harvest date on dry matter yield, in vitro dry matter digestibility and protein in timothy, tall fescue, reed canarygrass and Kentucky bluegrass. Can. J. Plant Sci. 63:675-85.

Myers, WM, and SJP Chilton. 1941. Correlated studies of winterhardiness and rust reaction of parents and inbred progenies of orchardgrass and timothy. Agron. J. 33:215-20.

Nath, J. 1967. Cytogenetical and related studies in the genus *Phleum* L. Euphytica 16:267-82.

Pardee, WD. 1992. 1991 Report: Northeast Seed Use Survey of Small Seeded Grasses and Legumes. N.Y. State Coll. Agric. Life Sci. PB 92-1. Ithaca, N.Y.: Cornell Univ.

Seoane, JR, and P Gervais. 1982. Valeur nutritive des foins de luzerne (Iroquois), de brome (Saratoga) et de fléole (Timfor et Champ) pour les moutons. Naturaliste Can. 109:103-7.

Seoane, JR, MC Moreno, and P Gervais. 1981. Evaluation nutritionnelle de six cultivars de la fléole des prés utilisé dans l alimentation des ovins. Naturaliste Can. 108:263-69.

Sheard, RW. 1968. Influence of defoliation and nitrogen on the development and the fructan composition of the vegetative reproductive system of timothy (*Phleum pratense* L.). Crop Sci. 8:55-66.

Shimanuki, T, and T Sato. 1983. Occurrence of the choke disease on timothy caused by *Epichloë typhina* (Pers. ex Fr.) Tul. in Hokkaido and location of the endophytic mycelia with plant tissue. Hokkaido Nat. Agric. Exp. Stn. Res. Bull. 138:87-97.

Simon, U, and A Kastenbauer. 1979. Growth type and yield comparisons of forage species after seed multiplication in Germany and in the United States. II. Meadow fescue, timothy and perennial ryegrass. Crop Sci. 19:209-13

Smith, D. 1972. Effect of day-night temperature regimes on growth and morphological development of timothy plants derived from winter and summer tillers. J. Br. Grassl. Soc. 27:107-10.

Smith, HF, and CH Myers. 1934. A biometrical analysis of yield trials with timothy varieties using rod rows. Agron. J. 26:117-28.

Surprenant, J, R Michaud, and G Allard. 1990a. Effect of one cycle of divergent phenotypic selection for crude protein, digestibility and digestible yield in timothy. Can. J. Plant Sci. 70:757-65.

Surprenant, J, R Michaud, G Allard, and CA St-Pierre. 1990b. Heritability of physical properties and other quality traits in timothy. Can. J. Plant Sci. 70:683-89.

Suzuki, M. 1993. Fructans in crop production and preservation. In M Suzuki and NJ Chatterton (eds.), Science and Technology of Fructans. Boca Raton, Florida: CRC Press, 227-55.

Thompson, LS. 1977. Field tests with chemical and biological insecticides for control of *Thymelicus lineola* on timothy. J. Econ. Entomol. 70:324-26.

Waldie, G, SBM Wright, and RDH Cohen. 1983. The effects of advancing maturity on crude protein and digestibility of meadow foxtail (*Alopecurus pratensis*) and timothy (*Phleum pratense*). Can. J. Plant Sci. 63:1083-85.

25 Smooth Bromegrass

MICHAEL D. CASLER
University of Wisconsin

IRVING T. CARLSON
Iowa State University

THE genus *Bromus* comprises about 100 species. It is grouped with the small grains in the grass subfamily Pooideae (Gould and Shaw 1983). These grasses are adapted to cool climates or to regions in which cool seasons prevail during parts of the growing season. Such cool-season grasses produce vegetative growth during the early part of the season and mature seed in the long days of early summer.

The bromegrasses vary greatly in adaptation and use and include several important forage and range grasses. Their aggressive reproduction, either through self-seeding or vegetative spread, provides conservation value for establishment of plant cover but also causes them to be classified as weeds in some situations. Hitchcock (1971) lists 42 species of this genus in the US. These include native and introduced perennials and a large group of introduced annuals. Most bromegrasses are readily grazed and are relatively nutritious during periods of rapid early growth.

Smooth bromegrass, *B. inermis* Leyss., is the most widely used of the cultivated bromegrasses. Known as *bromegrass, Austrian brome, Hungarian brome,* and *Russian brome,* it has been cultivated in the US since the early 1880s. It has gained considerable prominence for pasture and hay production, as a component in grass-legume mixtures, and as a means of erosion control.

The first recorded introduction of smooth bromegrass was in California in 1884. The seed may have been obtained from Hungary, where this grass had been grown experimentally for about 30 yr and was beginning to be utilized as a crop. Trial seed packets were distributed widely by the California Agricultural Experiment Station. Smooth bromegrass was grown in the Midwest by the late 1890s. Nebraska recommended it as "Hungarian bromegrass" as early as 1897-98 on the basis of yield tests (Lyon 1899).

The Canadian provinces began growing smooth bromegrass from seed initially imported from northern Germany in 1888. A large seed shipment was received in 1898 from the Penza region of Russia (53°N) by N. E. Hanson of South Dakota. Large-scale distribution of seed imported from northern Europe and successful seed production in Canada and the Dakotas led to early predominance of this northern type in the seed trade.

Following the drought years of the 1930s, attention was directed to smooth bromegrass as the principal survivor among the introduced grasses in the Midwest. Large quantities of smooth bromegrass seed were produced, much of it from the earliest plantings of smooth bromegrass in the Midwest. Demand for seed to replace the overstocked and depleted Kentucky bluegrass (*Poa pratensis*

MICHAEL D. CASLER is Professor of Agronomy, University of Wisconsin at Madison. He received the MS and PhD degrees from the University of Minnesota. His research is focused on the breeding and genetics of traits related to cell wall development and ruminal degradation of cell walls in smooth bromegrass and on the breeding and genetics of traits related to compatibility of grass-legume mixtures.

IRVING T. CARLSON is Emeritus Professor of Agronomy, Iowa State University. He received the MS degree from Washington State University and the PhD from the University of Wisconsin at Madison. His research concerned breeding and evaluation of cool-season forage crops, including smooth bromegrass, orchardgrass, reed canarygrass, tall fescue, alfalfa, and birdsfoot trefoil.

L.) and native pastures also promoted the importation of large quantities of smooth bromegrass seed from Canada.

Field plot tests were established in 1939 in Nebraska to explain differences among new plantings. Seed fields of smooth bromegrass from southern Kansas to Calgary, Canada, were sampled for strain testing at this location. Differences in plant type, establishment, and yield were soon attributed to different seed sources (Newell and Keim 1943). Adapted strains derived from old fields in Nebraska, Kansas, Iowa, and Missouri were superior in field tests in these states. Subsequent tests indicated these strains were adapted throughout the southern part of the Corn Belt. Seed of this southern type soon constituted a major portion of seed sold annually in the US.

DISTRIBUTION AND ADAPTATION

Smooth bromegrass, native to Europe and Asia, is adapted to most temperate climates. The region of major adaptation in North America is centered in the Corn Belt and adjacent areas northward and westward into Canada. Its range of distribution and use extends throughout these latitudes on favorable sites or sites with irrigation (Fig. 25.1).

Smooth bromegrass survives periods of drought and extremes in temperature. In dry summer periods it becomes dormant until the return of cool short days and fall precipitation. It grows successfully as far north as Fairbanks, Alaska (Irwin 1945). Laughlin et al. (1973) states that smooth bromegrass is the dominant and most dependable perennial forage crop in Alaska.

Smooth bromegrass can be grown on a variety of soil types, including sandy loams. Growth is best on deep fertile soils of well-drained silt loam or clay loam. It is deep rooted and fills the surface soil with many roots and rhizomes.

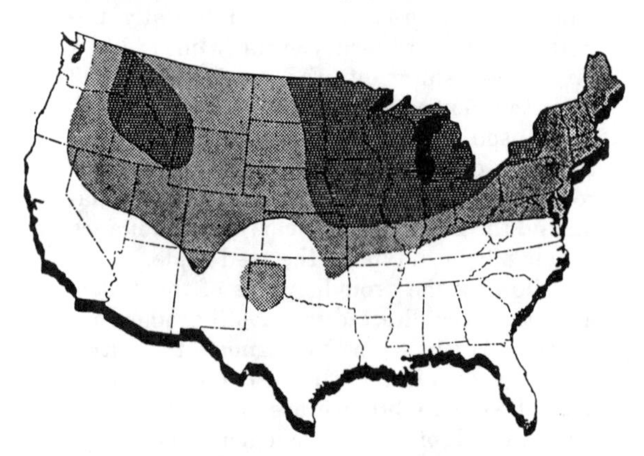

Fig. 25.1. Areas of major use and distribution of smooth bromegrass.

PLANT DESCRIPTION

The smooth bromegrass plant is a leafy, tall-growing, sod-forming perennial. It has a closed leaf sheath and a characteristic W-shaped constriction at approximately the midpoint of the leaf blade. It spreads vegetatively by rhizomes and is readily propagated by seed. For panicle production, plants require a period of growth under cool short days before being subjected to long days (Newell 1951). Large open panicles emerge from the upper leaf sheath following stem elongation in late spring or early summer. Numerous spikelets of the inflorescence are each made up of 5-10 florets. Plants are cross-pollinated by wind. At anthesis, visible clouds of pollen are disseminated at intervals over a period of several days.

Smooth bromegrass is a productive grass. When compared with orchardgrass (*Dactylis glomerata* L.), reed canarygrass (*Phalaris arundinacea* L.), and tall fescue (*Festuca arundinacea* Schreb.) in Iowa trials, it was the most winter-hardy species and usually the highest-yielding species in the spring (Wedin et al. 1966; Carlson and Wedin 1974). A major disadvantage of smooth bromegrass is its slow recovery after cutting, which contributes to low aftermath yields and poor seasonal distribution of yield. Regrowth is slow because most or all tiller apices are removed by cutting, and as a consequence, regrowth must develop from buds at belowground nodes.

The high quality of smooth bromegrass forage was demonstrated by superior average daily gains of cattle or sheep on smooth bromegrass compared with (1) orchardgrass, tall fescue, and reed canarygrass in Iowa (Wedin et al. 1970); (2) orchardgrass and reed canarygrass in Minnesota (Marten and Jordan 1974); (3) orchardgrass, tall fescue, and a turf-type perennial ryegrass (*Lolium perenne* L.) in West Virginia (Reid et al. 1978); and (4) alfalfa (*Medicago sativa* L.) and timothy (*Phleum pratense* L.) in Quebec (Seoane et al. 1981). In the Iowa and Minnesota trials superior average daily gains compensated for the lower carrying capacity of smooth bromegrass pastures to give high liveweight gains per hectare.

IMPORTANCE AND USE

Smooth bromegrass in its area of adaptation is one of the best cool-season grasses for hay, pasture, silage, and green chop. It is usually grown with a legume, preferably alfalfa or red clover (*Trifolium pratense* L.), for forage uses; however, it can be grown alone and yields well with annual applications of N fertilizer.

Smooth bromegrass has been a principal component of grass-legume mixtures for both irrigated and nonirrigated pastures. It is frequently planted with a legume for hay, to be followed by use as pasture. The N supplied by the legume keeps the grass productive for several years. In combination with a legume, smooth bromegrass makes an excellent sod crop for use in crop rotations. Grass roots improve both the organic matter content and soil structure. Decay of legume roots and sloughing of nodules from legume roots maintain a balance of available N that aids in the nutrition of living grass roots and decomposition of dead grass roots and release of their N to the subsequent crop.

Smooth bromegrass is also valuable for erosion control. It is commonly used for protection of roadsides and even steep road cuts. In many areas it is the principal grass to be used alone or in mixture with other grasses for constructed waterways and terraces. Fertilizer N is a necessary requirement for the establishment of any grass cover on exposed subsoils or eroded slopes. Seedlings must be established when soil erosion by water can be avoided.

CULTIVARS

Smooth bromegrass cultivars differ in seedling vigor, stand establishment, spring vigor, aftermath and total forage yield, seasonal distribution of yield, disease resistance, persistence, forage quality, and yield and quality of seed. Cultivars are classified into northern, intermediate, and southern types based on area of adaptation (Table 25.1). The northern type is adapted to the northern Great Plains, western Canada, and Alaska; the southern type is adapted to the central Great Plains, Corn Belt, northeastern US, and eastern Canada (Hanson 1972).

Northern and southern types differ in morphological and physiological traits. Leaves of the southern type are borne at a lower level on culms and are wider, coarser, and more glaucous than leaves of the northern type (Knowles and White 1949). Panicles of the southern type are more contracted. The southern type is more frost tolerant, it starts growth earlier in spring, and it is more resistant to leaf diseases. In addition, southern strains form a denser sod than northern strains.

Southern cultivars produce more forage in the US (below 49°N latitude, excluding Alaska) than do northern cultivars (Fig. 25.2). The yield advantage of southern cultivars increases from north to south by approximately 0.1 Mg/ha for each 1° drop in latitude. Above approximately 51°N, about 200 km north of the US-Canada border, northern cultivars generally are superior in forage yield to southern cultivars. Southern cultivars are generally superior in seedling vigor, establishment capacity, drought tolerance, and soil conservation than northern cultivars in the US. In western Canada, at latitudes ranging from 49° to 55°N, northern cultivars produced about twice as much seed as southern cultivars (Knowles and White 1949).

TABLE 25.1. Smooth bromegrass cultivars

Cultivar	Originator	Year released	Type
Lincoln	USDA, Nebraska AES	1942	Southern
Manchar	SCS, Washington and Idaho AES	1943	Intermediate
Saratoga	New York AES	1955	Southern
Carlton	Agriculture Canada, Saskatoon	1961	Northern
Baylor	Rudy-Patrick Co.	1964	Southern
Blair	Rudy-Patrick Co.	1964	Southern
Polar	USDA, Alaska AES	1965	Northern
Fox	Minnesota AES	1968	Southern
Magna	Agriculture Canada, Saskatoon	1968	Intermediate
Barton	Land O'Lakes, Inc.	1973	Southern
Beacon	Land O'Lakes, Inc.	1973	Southern
Tempo	Agriculture Canada, Ottawa	1975	Southern
Rebound	South Dakota AES	1978	Southern
Bromex	Northrup King Co.	1979	Southern
Jubilee	Maple Leaf Mills, Ltd., Ontario	1979	Northern
Signal	Agriculture Canada, Saskatoon	1983	Southern
York	New York AES	1989	Southern
Radisson	Agriculture Canada, Saskatoon	1990	Southern
Badger	Wisconsin AES	1992	Southern
Alpha	Wisconsin AES	1993	Southern

Sources: Hanson (1972); cultivar registration and extension papers; personal communications.

Breeding programs have emphasized forage and seed production, aftermath production, leafiness, disease resistance, seed quality, seedling vigor, forage quality, persistence under intensive management with alfalfa, and adaptation to local conditions. Suitable cultivars are available for most parts of the smooth bromegrass region. Seed is produced under certification programs of several states and Canadian provinces.

Most cultivars have been developed in the north central states. The old cultivar 'Lincoln', developed in Nebraska, was derived from fields originating from the earliest introductions of seed from central Europe. 'Lyon' was developed in Nebraska from plants selected from Lincoln for seed quality and forage type. 'Baylor', 'Blair', 'Barton', and 'Beacon' are high-yielding proprietary cultivars developed in Iowa. They have improved brown leaf spot resistance compared with Lincoln. 'Fox' and 'Bromex', developed in Minnesota, also have improved brown leaf spot resistance.

'Saratoga', developed in New York, is a leading cultivar in the northeastern states. It has shown superiority in aftermath production (Wright et al. 1967; Carlson and Wedin 1974). 'Rebound' was developed in South Dakota by selection in Saratoga for increased aftermath production. It is similar to Saratoga in leaf disease resistance, and it has a restricted rhizomatous spreading habit.

'Manchar', developed from a Manchurian introduction in Washington, is grown in the Pacific Northwest. Although it ranks high in second-harvest yields in the north central states, it is susceptible to leaf diseases, is relatively poor in stand establishment, and ranks low in total forage yield (Thomas et al. 1958; Carlson and Wedin 1974). 'Tempo' and 'Jubilee' are southern- and northern-type cultivars, respectively, developed in Ontario, Canada.

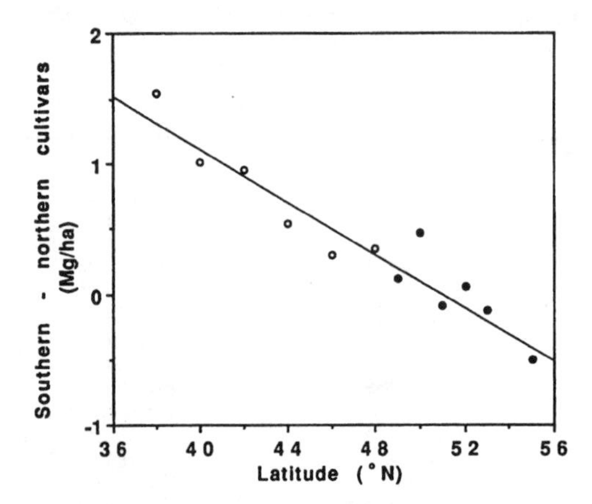

Fig. 25.2. Mean differences in forage yield between southern and northern smooth bromegrass cultivars tested at various latitudes in North America. (Open symbols are from Thomas et al. 1958, and closed symbols are from Knowles and White 1949.)

'Carlton' and 'Magna' were both selected for high forage and seed production in Saskatchewan, Canada. Carlton is a typical northern type that had low forage yield in Iowa trials (Carlson and Wedin 1974). Magna was derived mostly from the old southern-type cultivar 'Fischer'. It yields more seed of higher quality than southern-type cultivars in western Canada.

'Polar' was developed in Alaska from 16 clones selected primarily for yield and winterhardiness (Hodgson et al. 1971); 11 of the 16 parents trace back to hybrids between *B. pumpellianus* Scribn. and *B. inermis*. Polar is consistently superior to other cultivars in yield and winterhardiness in Alaska.

'Signal', 'York', and 'Radisson' were selected for improved seed production, seed fertility, or seed quality. Signal and Radisson were developed largely from germplasm selected from Magna, while York derives from Saratoga. Radisson has been the highest-yielding cultivar in forage yield trials in eastern Canada and Saskatchewan, but it is low to intermediate in seed production.

'Alpha' and 'Badger' were developed by selection for improved in vitro dry matter digestibility (IVDMD). They have shown higher IVDMD than all cultivars and forage yield equal to Rebound and Saratoga. Alpha was also selected for improved persistence in mixtures with alfalfa.

CULTURE AND MANAGEMENT

Time and method of establishment, fertilization, and management of pure stands and grass-legume mixtures for pasture and hay are important considerations in the use of smooth bromegrass.

Establishment. A moist, fertile, firm seedbed is required for grass and grass-legume mixtures. In the eastern and northern parts of its adaptive region, smooth bromegrass is planted in the spring, often with a legume and companion crop. Conditions for establishment of grass-legume mixtures are usually more favorable in the spring than in the fall. In these areas fall-sown mixtures may suffer winter injury, especially the legume. In the midcontinent and farther south, late summer and fall seedings are greatly preferred. Seedlings that develop in the cool days of autumn grow rapidly, the winters are less severe in these regions, and seedlings become fully established before another hot season. There is usually less weed competition with late summer or fall establishment.

In the drier parts of the smooth bromegrass belt the companion crop usually is omitted or its seeding rate is reduced by up to 50%. Experiments in Pennsylvania show competition from the companion crop may be detrimental even in areas of ample precipitation (Lueck et al. 1949).

Smooth bromegrass seed may be either drilled or broadcast. Drilling is much preferred because it permits more accurate control of seeding rate and a uniform shallow depth of planting. Smooth bromegrass and some other grass seed can be planted with grain drills that have attached seed boxes for legume seed or with special grassland drills with seed boxes for planting both chaffy and free-flowing seed (Moore and Haiwick 1969). Most hopper-type fertilizer spreaders can be calibrated to broadcast seed if drills are not available. The optimum seeding depth is 0.5 cm, and seed should be placed no deeper than 1.0 to 1.3 cm (Leuck et al. 1949; Moore and Haiwick 1969). It is essential to cover the seed.

Rate of planting will vary depending on the planting objective, components of the mixture, climate, and soil type. Most recommended seeding rates range from 11 to 22 kg/ha for smooth bromegrass alone, 5 to 13 kg/ha smooth bromegrass with 3 to 11 kg/ha alfalfa, 5 to 9 kg/ha smooth bromegrass with 4 to 7 kg/ha alfalfa and 3 to 7 kg/ha red clover, and 7 to 9 kg/ha smooth bromegrass with 3 to 4 kg/ha orchardgrass and 4 to 9 kg/ha alfalfa (Fuelleman et al. 1943; Rather and Harrison 1944; Schaller and Carlson 1977).

Management. Weeds or companion crops may retard establishment from spring sowing. Competition for moisture, light, and nutrients should be reduced by early and repeated mowing at a height that does not severely defoliate or cover the seedlings. Stands established in the fall may be harvested for seed or cut for hay the following year but will not attain maximum productivity until the second or third year.

Smooth bromegrass requires careful management to maximize yield and maintain thick stands. The growth stage at the time of grazing or mowing is an important consideration. A critical time in the spring is during stem elongation (jointing) and early-heading stages. Thinning of stands and reduction of subsequent regrowth have occurred when the first harvest was taken at these stages, particularly when all or most shoot apices were removed by cutting (Reynolds and Smith 1962; Eastin et al. 1964; Smith et al. 1973). Results from the northeastern US show the severe yield reductions that accompany such early spring harvests (Table 25.2). These growth stages are characterized by a low carbohydrate reserve level and an absence of new tillers (Reynolds and Smith 1962; Eastin et al. 1964). Second-harvest smooth bromegrass yields increase as the first harvest is delayed from jointing to postanthesis. Reserve carbohydrate levels are highest and new tillers are emerging through the soil surface at the postanthesis stage of spring growth.

The nutritive value of smooth bromegrass is high in the spring and declines as the crop matures (Fig. 25.3; Wright et al. 1967). Total digestible dry matter (DM) increases to about midbloom, while digestible protein percentage declines steadily past this stage. Because smooth bromegrass does not usually produce heads after first harvest, remaining vegetative through summer and fall, digestibility and protein estimates for summer and fall growth are often similar to those for the spring growth boot stage (Wright et al. 1967).

Regrowth following second and subsequent harvests is also favored by high reserve carbohydrate status, presence of new tillers, and

TABLE 25.2. Dry matter production of smooth bromegrass for first-cutting harvests at various stages of maturity

Maturity of 1st harvest each year	Forage Production			3-yr average
	1st yr	2d yr	3d yr	
	mt/ha (t/A)			
Prejoint	7.62b (3.40)	6.95bc (3.09)	2.05bc (0.91)	5.54c (2.47)
Early head	9.06ab(4.04)	6.73c (3.00)	0.85c (0.37)	5.55c (2.48)
Early bloom	9.69a (4.32)	8.79ab (3.92)	3.27b (1.45)	7.25b (3.23)
Past bloom	9.84a (4.38)	10.22a (4.55)	5.36a (2.39)	8.47a (3.77)

Source: Wright et al. (1967).

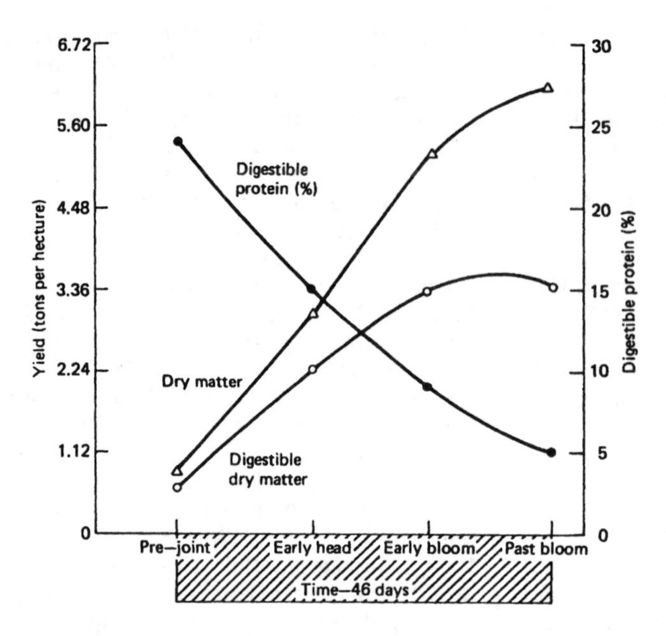

Fig. 25.3. Yield, digestible dry matter, and digestible protein of smooth bromegrass harvested at various stages of maturity (Miller 1984).

cutting above some of the shoot apices. In contrast to some other cool-season grasses, e.g., orchardgrass and tall fescue, stem elongation occurs in regrowth of smooth bromegrass, and elevated growing points are subject to removal by cutting or grazing. If all or most of the growing points are removed, regrowth must again come from buds at belowground nodes.

Grazing usually is not as severe as mowing in terms of removing shoot apices; however, consideration of growth stage is still important. Early spring grazing is detrimental, as is pasturing late in the season. Overgrazing too early in the season frequently results if smooth bromegrass is the only pasture available. Smooth bromegrass pastures to be stocked continuously should be understocked during the early part of the season to allow growth to accumulate for later use. Forage quality, however, decreases with advance in maturity. In either situation, additional pasture is required for the midsummer period, especially in the midlatitudes where temperatures are higher.

Rotational stocking of smooth bromegrass pastures is superior to continuous stocking. It provides better timing for utilization of forage throughout the grazing season. Smooth bromegrass pastures may be advantageously stocked heavily in the spring after growth has been sufficient to restore reserves and before seed formation. Rest periods between defoliations should be longer in the summer when growth is slower. Fall growth of smooth bromegrass may be grazed judiciously or allowed to accumulate for use after growth ceases. Protection from close defoliation during the early fall allows for accumulation of reserves and ensures vigorous growth the following spring. At least 20 to 30 cm of growth should be present before grazing and at least 8 to 12 cm of residue should be left to aid recovery of the pasture. Smooth bromegrass can be stockpiled in the fall for winter grazing.

Smooth bromegrass in a mixture with alfalfa appears to be a good combination for hay and pasture. The mixture is productive and is ideally suited for hay harvests in favorable years with an excess of early pasture. The grass in the mixture is important for decreasing bloat danger when used as pasture, and it also extends the season for grazing. Smooth bromegrass, however, tends to be most productive in May and June. The legume improves acceptability and feeding value, with better animal performance resulting.

Smooth bromegrass may be a poor choice for use in mixtures with alfalfa managed for high-quality hay. It has not persisted well when such mixtures are cut three times each year (early June, mid-July, and late August) in the northern Great Lakes area (Smith et al. 1973). Stand losses, which were much more severe with a 4-cm than with a 10-cm stubble height, were attributed to taking the first harvest each year during the critical period between stem elongation and heading growth stages. The slow and weak regrowth of smooth bromegrass following such harvests is suppressed by the more rapid regrowth of alfalfa (Paulsen and Smith 1968). Smith et al. (1973) also observed severe stand losses of smooth bromegrass mixed with alfalfa cut four times each year at the 4-cm stubble height. In contrast, smooth bromegrass persisted well in western Canada in a mixture with alfalfa cut four times each year regardless of time of first cutting from vegetative to early-heading stages (McElgunn et al. 1972).

Some progress has been made in breeding smooth bromegrass cultivars to be more persistent in mixtures with alfalfa. In a mixture with alfalfa, cultivars selected for improved persistence had 40% greater ground cover and 42% faster recovery than unselected cultivars after cutting (Casler 1988). The cultivar Alpha had the greatest ground cover averaged over six experiments, 10% greater than the second-ranked cultivar (Casler 1988; Casler and Walgenbach 1990).

Fertilization. Smooth bromegrass requires regular nitrogen (N) fertilization when grown in pure stands. Old stands may become sodbound, resulting in reduced productivity, but this can be easily corrected by N fertilization. Minimum recommended rates for adequate forage yield and protein concentration vary from 45 to 112 kg N/ha. Seasonal distribution of yield can be improved by timing N applications prior to the time when forage is needed (Krueger and Scholl 1970; Narasimhalu et al. 1981). At least 50 to 80 kg N/ha should be applied in early summer to smooth bromegrass pastures to reduce the yield decline due to summer slump. Beef production on smooth bromegrass pasture was increased by N fertilization, 0.5 kg beef/kg N in Indiana (Lechtenberg et al. 1974) and 1.3 kg beef/kg N in Nebraska (Rehm et al. 1971).

Smooth bromegrass forage yields generally respond linearly to increasing rates of N fertilizer up to 100 to 504 kg N/ha (Table 25.3). The rate at which forage yield begins to level off varies considerably among locations, even among some that are close to each other geographically and in soil type. Below these values, forage yield increases by 10 to 35 kg DM/kg N at most sites. These responses also vary, but most values are between 15 and 25 kg DM/kg N. In a 22-yr study at Fargo, North Dakota, forage yield increased by 37 kg DM/kg N between 0 and 75 kg N/ha (Meyer et al. 1977). Forage yield in the North Dakota study did not peak but continued to increase at a rate of 9 kg DM/kg N between 75 and 300 kg N/ha.

Economically effective rates of N fertilization vary considerably with the prices of N fertilizer and the hay crop. Economic models have been developed to estimate optimal N fertilization rates (Malhi et al. 1987; McCaughey et al. 1990; Zentner et al. 1989). These studies also show optimal rates vary considerably from one site to another, with water availability as the major factor influencing optimal N rates. Reductions in April-June precipitation by as little as 5.4 cm can reduce the economic optimum N rate by as much as 116 kg N/ha (McCaughey et al. 1990).

Rates of N fertilization on smooth bromegrass as low as 112 kg N/ha can lead to significant increases in nitrate N in the soil (Harapiak et al. 1992; Malhi et al. 1991; Nut-

TABLE 25.3. Effect of N fertilization on smooth bromegrass forage yields

Location of study	N rate at which forage yield was maximized (*kg/ha*)	Forage yield response to N (*kg DM/kg N*)	Authors
Wisconsin			
Arlington	314	25.4	Smith 1972
Madison	314	16.6	Smith 1972
Texas			
Bushland	504	15.2	Eck et al. 1981
Nebraska			
Nebraska Agricultural Experiment Station	180	16.2	Rehm et al. 1971
Hankins	180	5.6	Rehm et al. 1971
Spence	135	11.6	Rehm et al. 1971
Alberta			
Lacombe	200	34.7	Malhi et al. 1986
Joffre	150	31.2	Malhi et al. 1986
Botha	100	25.2	Malhi et al. 1986
Rocky Mountain House	150	27.4	Malhi et al. 1986
Ellerslie	>300	18.0	Penney et al. 1990
Vimy	>300	19.3	Penney et al. 1990

tall et al. 1980). One study shows smooth bromegrass forage yield and fertilizer N recovery were greater for ammonium nitrate than for urea (Ukrainetz and Campbell 1988). However, fertilization with ammonium nitrate can result in greater buildup of nitrate N in the upper 60 cm of the soil profile (Harapiak et al. 1992) and greater acidification of the upper 5 cm of the soil profile (Malhi et al. 1991) than fertilization with urea. Fall applications of ammonium nitrate led to greater nitrate N accumulation than spring applications after an 11-yr period (Harapiak et al. 1992). Rates above about 200 kg N/ha can increase the risk of grass tetany and nitrate N toxicity to animals grazing lush pasture.

Adequate phosphorus (P) is also important to maintain productive stands of smooth bromegrass. In one study three annual applications of 20 kg P/ha increased forage yield of smooth bromegrass pastures by about 0.7 Mg/ha (Nuttall et al. 1980). This yield increase was maintained for an additional 3 yr without additional P fertilizer. Over a 5-yr period, a single 60-kg P/ha application and annual 10-kg P/ha applications gave the same annual yield increase over that from not applying P: 1.4 Mg/ha (Malhi et al. 1992). Fertilization with potassium (K) from three different sources had little effect on smooth bromegrass forage yield when applied at a rate of 125 kg K/ha annually for 3 yr (George et al. 1979).

SEED PRODUCTION

In the US, most smooth bromegrass seed is produced in the central Great Plains and the Pacific Northwest. In Canada, most seed is produced in the southern portions of the Prairie Provinces. Seeds are generally planted in rows 60 to 90 cm apart with 80 to 100 seeds/m of row. Seed yields of 350 to 550 kg/ha in the central Great Plains, 700 to 800 kg/ha in the Pacific Northwest, and 400 to 500 kg/ha in the Prairie Provinces are typical.

Responses of tillering and panicle formation to photoperiod, temperature, and N availability have been shown in greenhouse studies (Newell 1951). Tillers on which panicles are formed are produced during the short days of fall and very early spring. For panicles to be formed, soil N must be readily available during short cool days prior to elongation of the culms. Fertilizer N should be applied in either the fall or very early spring. Later applications increase vegetative growth, and their effect on seed production decreases as the season advances.

Applications of fertilizer N are required for maximum seed production (Anderson et al. 1946). Seed yields increase with increasing rates of N, but lodging can be severe with rates greater than 100 kg N/ha. Heads of lodged stems fail to develop caryopses. Rates required for seed production vary with the degree of depletion of available soil N. Old stands usually require more N than do new stands (Canode and Law 1978). Rates of 45 to 90 kg N/ha frequently are recommended (Knowles et al. 1951); however, fall applications up to 135 kg/ha have increased seed yields from older stands (Canode and Law 1978).

Harvesting. Seed of smooth bromegrass should be allowed to ripen fully before harvesting with the combine-harvester. To reduce the amount of green material in the seed, the crop is harvested when the culms have dried below most of the heads but before shattering occurs. Direct harvesting is seldom satisfactory until this stage is reached. The combine may be set to cut ripened heads and to leave green stems and leaves. To avoid loss from early shattering, the crop may be mowed and picked up by the combine after the seed has dried in the swath (Knowles et al. 1951). Burning or removal of postharvest residue has been found to increase smooth bromegrass seed yields (Canode and Law 1978; Fulkerson 1980), with burning being superior to mechanical removal in eastern Washington. Late August burning of residue has been found to increase the vigor of autumn growth.

Weeds. Smooth bromegrass for seed production should be planted in fields free of primary noxious weed seeds. The chief problem weed in northern states is quackgrass, *Elytrigia repens* (L.) Nevski. Other wheatgrasses may be troublesome to separate from mixtures at harvest. Weeds on roadsides and field boundaries should be mowed to prevent seed formation. Winter annual bromegrasses are also principal invaders. These cool-season grasses mature seed that is harvested with the seed crop and is difficult to remove. Burning postharvest residue eliminated a downy bromegrass (*B. tectorum* L.) weed problem in eastern Washington (Canode and Law 1978).

DISEASES AND PESTS

In subhumid areas grasshoppers and seedling blight are factors affecting grass establishment. Foliar diseases of smooth bromegrass are more prevalent and serious in humid areas and seasons. Gross et al. (1975) found that leaf diseases reduced IVDMD of smooth bromegrass. Some progress in selection of cultivars with leaf disease resistance has been made. In some years seed production of smooth bromegrass has been seriously affected by the bromegrass seed midge, *Stenodiplosis bromicola* Marikovsky & Agafonova (Nielsen and Burks 1958), and by thrips.

OTHER SPECIES

The genus *Bromus* includes grasses varying from early-maturing winter annuals to the slower-maturing, more productive perennials. The annuals for the most part are considered weeds where more valuable perennials are grown. In some areas they provide valuable ground cover or good forage for short periods.

Meadow bromegrass (*B. riparius* Rehm., but sometimes incorrectly called *B. erectus* Huds.) is a long-lived perennial that offers promise for nonirrigated or irrigated pasture. 'Regar' is a cultivar selected at the Idaho Agricultural Experiment Station from the Turkish introduction 'PI 172390'. It is a relatively early-maturing bunchgrass with moderate spread and good regrowth (Foster et al. 1966). In Montana the contribution of Regar to forage yields of grass-legume mixtures was significantly greater than that of Manchar smooth bromegrass at three locations (Cooper et al. 1978). Its superiority was most evident in aftermath harvests. 'Paddock' and 'Fleet' are two recent cultivars that are similar to Regar in forage characteristics but have improved seed production (Knowles et al. 1993).

Native perennial bromegrasses are important forage plants in the mountain regions of the western US. Among these are nodding brome (*B. anomalus* Rupr.), fringed brome (*B. ciliatus* L.), and pumpelly brome (*B. pumpellianus* Scribn.). They are abundant up to 3350 m and furnish high-quality forage. California brome, *B. carinatus* Hook. & Arn., and the closely related mountain brome, *B. marginatus* Nees ex Steud., are short-lived perennials. Both produce deep, well-branched root systems and make a good leafy growth. They are utilized by grazing animals and are good seed producers. 'Cucumonga' is a cultivar of California bromegrass. 'Bromar' is a mountain bromegrass cultivar released in Washington for the intermountain area and the Pacific Northwest. It has been grown with sweetclover in short rotations. 'Blizzard' and 'Bosir', two cultivars of *B. sitchensis,* a species native to mountain meadows of the Pacific Coast, from Washington to Alaska, are used for pastures in northern Europe.

The winter annuals are widespread in their distribution because of prolific seed production. The best known of these is chess, *B. secalinus* L., occurring in waste places and grainfields. Downy bromegrass, Japanese chess (*B. japonicus* Thunb. ex Murr.), and hairy chess (*B. commutatus* Schrad.) are widely distributed in pastures and rangelands. These annuals become prominent when they replace perennial grasses depleted by grazing. Downy bromegrass is widespread over the ranges of the West and in some areas constitutes the major source of feed for livestock in the spring (Klemmedson and Smith 1964; Murray et al. 1978). Such annuals as soft chess, *B. mollis* L., and ripgutgrass, *B. rigidus* Roth, are acceptable in their early stages and produce lush pasturage for a short time in the spring. The presence of awns on the matured seed of many of these annuals constitutes a serious health hazard to grazing.

Rescuegrass, *B. unioloides* (Willd.) H.B.K., is a native of Argentina, introduced into the South before the Civil War. It is also commonly cited as *B. catharticus* Vahl and *B. willdenowii* Kunth (Gould and Shaw 1983). In areas with mild winters this grass makes excellent winter pasture. Several cultivars have been developed.

Rescuegrass germplasm was reintroduced into the northern US in 1986 in the form of cultivars, and experimental populations developed primarily in New Zealand and France. 'Matua' prairiegrass is the only cultivar of this group that has been extensively tested and utilized in the US. It is persistent and provides high-yield and quality pasture in the Pacific Northwest, the southern Great Plains (under irrigation), and parts of the northeast US where winters are mild. It has not survived well in the north central US, where winters can be severe and snow cover is often unreliable. Improper fall management, insect and disease infestations, and summer drought can also result in severe stand depletion of rescuegrass (Hume et al. 1990). In Pennsylvania, multiple fall harvests and/or a cutting height above 8 cm led to greatly improved persistence of rescuegrass compared with a single fall harvest at a cutting height of 8 cm (Jung and Shaffer 1990).

Field bromegrass, *B. arvensis* L., a winter annual, is adapted to the Corn Belt and eastward. It was introduced into the US in the late 1920s. Seeded in late summer, this annual grows late in the fall and for a long period the following spring. Its greatest value is as a winter cover and green manure crop. Its superiority to other annual cover crops is in its rapidly developed and very extensive fibrous root growth with great soil-holding capacity. Field bromegrass has not proved troublesome as a weed.

QUESTIONS

1. Where in North America is smooth bromegrass an important forage crop?
2. Explain the origin of northern and southern types of smooth bromegrass in relation to the original introduction of this grass into North America.
3. What are the characteristics of smooth bromegrass that contribute to its usefulness as a forage crop? In soil conservation?
4. Why does the yield advantage of southern-type over northern-type cultivars increase from north to south in the north central region of the US?
5. Discuss objectives of smooth bromegrass breeding programs and how their achievement contributes to the value of this grass.
6. Indicate the importance of some species of *Bromus* other than smooth bromegrass, and tell how they are utilized.
7. What are the factors that contribute to successful stand establishment of smooth bromegrass?
8. Why do older stands of smooth bromegrass require more N fertilizer for maximum forage and seed production than new stands?
9. How and why does time of applying N fertilizer to smooth bromegrass affect yields of forage and seed?
10. Why is growth stage at the time of grazing or mowing an important consideration in proper management of smooth bromegrass?

REFERENCES

Anderson, KL, RE Krenzin, and JC Hide. 1946. The effect of nitrogen fertilizer on bromegrass in Kansas. J. Am. Soc. Agron. 38:1058-67.

Canode, CL, and AG Law. 1978. Influence of fertilizer and residue management on grass seed production. Agron. J. 70:543-46.

Carlson, IT, and WF Wedin. 1974. Iowa Smooth Bromegrass Performance Tests, 1961-1972. Iowa State Univ. Coop. Ext. Serv. AG90.

Casler, MD. 1988. Performance of orchardgrass, smooth bromegrass, and ryegrass in binary mixtures with alfalfa. Agron. J. 80:509-14.

Casler, MD, and RP Walgenbach. 1990. Ground cover potential of forage grass cultivars mixed with alfalfa at divergent locations. Crop Sci. 30:825-31.

Cooper, CS, LE Welty, SB Laudert, and LE Weisner. 1978. Evaluation of Regar Meadow Bromegrass in Montana. Mont. Agric. Exp. Stn. Bull. 702.

Eastin, JD, MR Teel, and R Langston. 1964. Growth and development of six varieties of smooth bromegrass (*Bromus inermis* Leyss.) with observations on seasonal variation of fructosan and growth regulators. Crop Sci. 4:555-59.

Eck, HV, T Martinez, and GC Wilson. 1981. Tall fescue and smooth bromegrass. I. Nitrogen and water requirements. Agron. J. 73:446-52.

Foster, RB, HC McKay, and EW Owens. 1966. Regar Bromegrass. Idaho Agric. Exp. Stn. Bull. 470.

Fuelleman, RF, WL Burlison, and WG Kammlade. 1943. Bromegrass and Bromegrass Mixtures. Culture and Utilization. Ill. Agric. Exp. Stn. Bull. 496.

Fulkerson, RS. 1980. Seed yield responses of three grasses to post-harvest stubble removal. Can. J. Plant Sci. 60:841-46.

George, JR, ME Pinheiro, and TB Bailey, Jr. 1979. Long-term potassium requirements of nitrogen-fertilized smooth bromegrass. Agron. J. 71:586-91.

Gould, FW, and RB Shaw. 1983. Grass Systematics. 2d ed. College Station: Texas A&M Univ. Press.

Gross, DF, CT Mankin, and JG Ross. 1975. Effect of diseases on in vitro digestibility of smooth bromegrass. Crop Sci. 15:273-75.

Hanson, AA. 1972. Grass Varieties in the United States. USDA Agric. Handb. 170, rev.

Harapiak, JT, SS Malhi, M Nyborg, and NA Flore. 1992. Soil chemical properties after long-term nitrogen fertilization of bromegrass: Source and time of nitrogen application. Commun. Soil Sci. Plant Anal. 23:85-100.

Hitchcock, AS. 1971. Manual of the Grasses of the United States. 2d ed. New York: Dover.

Hodgson, HJ, AC Wilton, RL Taylor, and LJ Klebesadel. 1971. Registration of Polar bromegrass. Crop Sci. 11:939.

Hume, DE, RE Fallon, and RE Hickson. 1990. Productivity and persistence of prairie grass (*Bromus willdenowii* Kunth.). 2. Effects of natural reseeding. N.Z. J. Agric. Res. 33:395-403.

Irwin, DL. 1945. Forty-seven Years of Experimental Work with Grasses and Legumes in Alaska. Alaska Agric. Exp. Stn. Bull. 12.

Jung, GA, and JA Shaffer. 1990. Fall management studies with Matua prairie grass. In Am. Soc. Agron. Abstr. Madison, Wis., 147.

Klemmedson, JO, and JG Smith. 1964. Cheatgrass (*Bromus tectorum* L.). Bot. Rev. 30:226-62.

Knowles, RP, and WJ White. 1949. The performance of southern strains of bromegrass in western Canada. Sci. Agric. 29:437-50.

Knowles, RP, HA Friesen, and DA Cooke. 1951. Brome Grass Seed Production in Western Canada. Can. Dept. Agric. Publ. 866.

Knowles, RP, VS Baron, and DH McCartney. 1993. Meadow Bromegrass. Agric. Can. Publ. 1889/E. Swift Current, Saskatchewan: Agriculture Canada Research Station.

Krueger, CR, and JM Scholl. 1970. Performance of Bromegrass, Orchardgrass, and Reed Canarygrass Grown at Five Nitrogen Levels and with Alfalfa. Wis. Agric. Exp. Stn. Res. Rep. 69.

Laughlin, WM, PF Martin, and GR Smith. 1973.

Potassium rate and source influences on yield and composition of bromegrass forage. Agron. J. 65:85-87.

Lechtenberg, VL, CL Rhykerd, GO Mott, and DA Huber. 1974. Beef production on bromegrass (*Bromus inermis* Leyss.) pastures fertilized with anhydrous ammonia. Agron. J. 66:47-50.

Lueck, AG, VG Sprague, and RJ Garber. 1949. The effects of a companion crop and depth of planting on the establishment of smooth bromegrass, *Bromus inermis* Leyss. Agron. J. 41:137-40.

Lyon, TL. 1899. Hungarian Brome Grass (*Bromus inermis*). Nebr. Agric. Exp. Stn. Bull. 61.

McCaughey, WP, EG Smith, and ATH Gross. 1990. Economics of N-fertilization of dryland grass for hay production in southwestern Manitoba. Can. J. Plant Sci. 70:559-63.

McElgunn, JD, DH Heinrichs, and R Ashford. 1972. Effects of initial harvest date on productivity and persistence of alfalfa and bromegrass. Can. J. Plant Sci. 52:801-4.

Malhi, SS, DK McBeath, and VS Baron. 1986. Effects of nitrogen application on yield and quality of bromegrass hay in central Alberta. Can. J. Plant Sci. 66:609-16.

Malhi, SS, VS Baron, and DK McBeath. 1987. Economics of N fertilization of bromegrass for hay in central Alberta. Can. J. Plant Sci. 67:1105-9.

Malhi, SS, M Nyborg, JT Harapiak, and NA Flore. 1991. Acidification of soil in Alberta by nitrogen fertilizers applied to bromegrass. In RJ Wright et al. (eds.), Plant-Soil Interactions at Low pH. Amsterdam: Kluwer, 547-53.

Malhi, SS, DK McBeath, and M Nyborg. 1992. Effect of phosphorus fertilization on bromegrass hay yield. Commun. Soil Sci. Plant Anal. 23:113-22.

Marten, GC, and RM Jordan. 1974. Significance of palatability differences among *Phalaris arundinacea* L., *Bromus inermis* Leyss., and *Dactylis glomerata* L. grazed by sheep. In VG Iglovikov and AP Movsisyants (eds.), Proc. 12th Int. Grassl. Congr., Moscow, 3:305-12.

Meyer, DW, JF Carter, and FR Vigil. 1977. Bromegrass fertilization at six nitrogen rates: Long and short term effects. N.Dak. Farm Res. 34:13-17.

Miller, DA. 1984. Smooth bromegrass. In Forage Crops. New York: McGraw-Hill, 391.

Moore, RA, and GB Haiwick. 1969. Establishing pasture and forage crops. S.Dak. FarmHome Res. 20:6-9.

Murray, RB, HF Mayland, and PJ Van Soest. 1978. Growth and Nutritional Value to Cattle of Grasses on Cheatgrass Range in Southern Idaho. US For. Serv. Res. Pap. INT-199.

Narasimhalu, P, WN Black, KB McRae, and KA Winter. 1981. Effects of annual rate and timing of N fertilization on production of timothy, bromegrass, and reed canarygrass. Can. J. Plant Sci. 61:619-23.

Newell, LC. 1951. Controlled life cycles of bromegrass, *Bromus inermis* Leyss., used in improvement. Agron. J. 43:417-24.

Newell, LC, and FD Keim. 1943. Field performance of bromegrass strains from different regional seed sources. J. Am. Soc. Agron. 35:420-34.

Nielsen, EL, and BD Burks. 1958. Insect infestation as a factor influencing seed set in smooth bromegrass. Agron. J. 50:403-5.

Nuttall, WF, DA Cooke, J Waddington, and JA Robertson. 1980. Effect of nitrogen and phosphorus fertilizers on a bromegrass and alfalfa mixture grown under two systems of pasture management. I. Yield, percentage legume in sward, and soil tests. Agron. J. 72:289-94.

Paulsen, GM, and D Smith. 1968. Influences of several management practices on growth characteristics and available carbohydrate content of smooth bromegrass. Agron. J. 60:375-79.

Penney, DC, SS Malhi, and L Kryanowski. 1990. Effect of rate and source of N fertilizer on yield, quality and N recovery of bromegrass grown for hay. Fert. Res. 25:159-66.

Rather, HC, and CM Harrison. 1944. Alfalfa and Smooth Bromegrass for Pasture and Hay. Mich. Agric. Exp. Stn. Circ. Bull. 189.

Rehm, GW, WJ Moline, EJ Schwartz, and RS Moomaw. 1971. The effect of fertilization and management on the production of bromegrass in northeast Nebraska. Univ. Nebr. Res. Bull. 247:1-27.

Reid, RL, K Powell, JA Balasko, and CC McCormick. 1978. Performance of lambs on perennial ryegrass, smooth bromegrass, orchardgrass and tall fescue pasture. J. Anim. Sci. 46:1493-1502.

Reynolds, JH, and D Smith. 1962. Trend of carbohydrate reserves in alfalfa, smooth bromegrass, and timothy grown under various cutting schedules. Crop Sci. 2:333-36.

Seoane, JR, M Cote, P Gervais, and JP LaForest. 1981. Prediction of the nutritive value of alfalfa (Saranac), bromegrass (Saratoga), and Timothy (Champ, Climax, Bounty) fed as hay to growing sheep. Can. J. Anim. Sci. 61:403-13.

Schaller, FW, and IT Carlson. 1977. Forage Crop Varieties and Seeding Mixtures. Iowa State Univ. Coop. Ext. Serv. Pam. 564, rev.

Smith, D. 1972. Influence of Nitrogen Fertilization on the Performance of an Alfalfa-Bromegrass Mixture and Bromegrass Grown Alone. Wis. Agric. Exp. Stn. Res. Rep. R2384.

Smith, D, AVA Jacques, and JA Balasko. 1973. Persistence of several temperate grasses grown with alfalfa and harvested two, three, or four times annually at two stubble heights. Crop Sci. 13:553-56.

Thomas, HL, EW Hanson, and JA Jackobs. 1958. Varietal Trials of Smooth Bromegrass in the North Central Region. North Cent. Reg. Publ. 93, Minn. Agric. Exp. Stn. Misc. Rep. 32.

Ukrainetz, H, and CA Campbell. 1988. N and P fertilization of bromegrass in the dark brown soil zone of Saskatchewan. Can. J. Plant Sci. 68:457-70.

Wedin, WF, IT Carlson, and RL Vetter. 1966. Studies on nutritive value of fall-saved forage, using rumen fermentation and chemical analyses. In Proc. 10th Int. Grassl. Congr., Helsinki, Finland, 424-28.

Wedin, WF, RL Vetter, and IT Carlson. 1970. The potential of tall grasses as autumn saved forages under heavy nitrogen fertilization and intensive grazing management. In Proc. N.Z. Grassl. Assoc. 32:160-67.

Wright, MJ, GA Jung, CS Brown, AM Decker, KE Varney, and RC Wakefield. 1967. Management and Productivity of Perennial Grasses in the Northeast: II. Smooth Bromegrass. Northeast Reg. Res. Publ., W.Va. Agric. Exp. Stn. Bull. 554 T.

Zentner, RP, H Ukrainetz, and CA Campbell. 1989. The economics of fertilizing bromegrass in Saskatchewan. Can. J. Plant Sci. 69:841-59.

26 Orchardgrass

BERTRAM R. CHRISTIE
Agriculture Canada

ARTHUR R. McELROY
Agriculture Canada

ORCHARDGRASS (*Dactylis glomerata* L.) is an important forage species in many parts of the world. It is native to Europe, North Africa, and parts of Asia. The species was introduced into the US, probably in the 1750s, and in 1763 seed was shipped to Great Britain from Virginia (Leafe 1988). Although orchardgrass is native to Great Britain, it was not utilized as a forage there until seed of improved strains was imported from the US.

Orchardgrass is often found growing in shady areas, and it is likely that in colonial times it was found growing in orchards; hence the name *orchardgrass*. In Great Britain, the common name is *cocksfoot,* which is probably derived from the characteristic shape of the inflorescence (Fig. 26.1).

DISTRIBUTION AND ADAPTATION

Throughout the world, orchardgrass is found in areas of moderate to high rainfall, moderate winters, and warm summers. It is grown in the eastern US, from Maine to the Gulf States, and from the Atlantic coast to the eastern Great Plains (Fig. 26.2).

In the Northeast and the north central states, it is one of the major grass species for pasture (Hoveland 1992). Orchardgrass is also grown in the Pacific Northwest in areas of high rainfall. In Canada, it is utilized in the eastern provinces and in high rainfall areas west of the Rocky Mountains. It is also grown in northwestern Europe, New Zealand, eastern Australia, and northern Japan. In many of these areas it also can be found growing along roadsides, in orchards, and in various waste places.

Environment

TEMPERATURE. The optimum temperature for growth depends upon a number of other factors, such as variation between day and night temperatures, photoperiod, and moisture. With adequate moisture, and with a 16-h photoperiod, the optimum temperature is 20°-22°C (Eagles 1967a; Mitchell and Lucanus 1962). Lower night temperatures will increase the rate of tillering, while lower day temperatures will reduce the weight of individual tillers (Mitchell and Lucanus 1960). Among a wide range of temperature combinations, the most topgrowth occurred at 22°/12°C, day/night (Baker and Jung 1968). With a reduction in photoperiod, the rate of tiller initiation increases, but the optimum temperature for growth is reduced only slightly.

The increase in growth with an increase in temperature has been attributed to an increase in photosynthetic activity of the leaves, rather than an increase in leaf area (Eagles 1967a). Orchardgrass plants grown at 25°/15°C had larger leaves and a higher rate of leaf appearance than those grown at 15°/5°C (Van Esbroeck et al. 1989).

Temperatures above 28°C greatly reduce growth and tillering (Baker and Jung 1968; Knievel and Smith 1973). Such temperatures

BERTRAM R. CHRISTIE is Research Scientist, Agriculture Canada, Charlottetown, Prince Edward Island, and Adjunct Professor, Nova Scotia Agricultural College. He received the PhD degree from Iowa State University. His research interests are forage crop breeding and quantitative inheritance.

ARTHUR R. McELROY. *See Chapter 24.*

Fig. 26.1. Orchardgrass is a fast-growing, cool-season perennial that has numerous basal leaves and a "cocksfoot"-shaped inflorescence. *C. C. Berg photo.*

also greatly decrease the nucleic acid content of leaves, which appears to occur at temperatures lower than those that suppress growth (Baker and Jung 1970).

Orchardgrass is more heat tolerant than timothy (*Phleum pratense* L.) or Kentucky bluegrass (*Poa pratensis* L.) but is less so than smooth bromegrass (*Bromus inermis* Leyss.) or tall fescue (*Festuca arundinacea* Schreb.) (Baker and Jung 1968). Orchardgrass grows rapidly at cool temperatures and is especially productive in early spring. It is reasonably productive in late fall, although less so than tall fescue (Archer and Decker 1977).

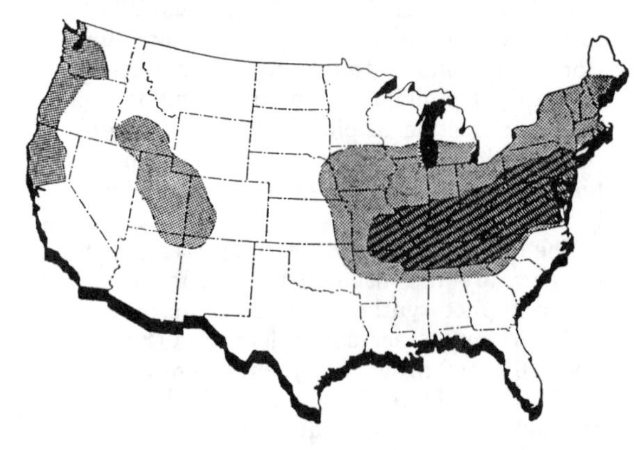

Fig. 26.2. Adaptation and production area of orchardgrass in the US.

Orchardgrass is only moderately winterhardy, and it will not survive northern climatic conditions if snow cover is lacking. For this reason timothy and smooth bromegrass are grown farther north and at higher elevations than orchardgrass.

LIGHT. Orchardgrass is productive over a wide range of light intensities and does well in full sunlight. However, it is very shade tolerant and is one of the few forage species that will grow and produce in reduced light. Orchardgrass has been found to produce as well when shaded to 36% of the incident light as when in full sunlight (Blake et al. 1966). Under the same conditions, the production of ladino clover (*Trifolium repens* L.) was reduced in the first year, and by the end of the second year, there was no clover left in the stand.

A leaf-area index (LAI) between 3 and 8 is considered optimal for orchardgrass (Pearce et al. 1965). Each fully expanded leaf maintains a high rate of photosynthesis for 15-20 d; then the rate declines rapidly (Treharne et al. 1968). This pattern of change with leaf age is not affected by temperature. The rate appears to be related to the balance between synthesis and degradation of chlorophyll. Genotypes have been found to differ in rates of photosynthesis (Treharne et al. 1968).

After harvesting, orchardgrass regrows rapidly and can reach an LAI of 5, intercepting 95% of the light, within 15 d (Pearce et al. 1965).

SOIL MOISTURE. Moisture.Orchardgrass is more drought tolerant than either timothy or Kentucky bluegrass. Smooth bromegrass and tall fescue, however, are more drought tolerant than orchardgrass and are better adapted to areas that have a combination of low rainfall and high temperature. Orchardgrass makes more efficient use of soil water than does perennial ryegrass (*Lolium perenne* L.), probably because of a more extensive root system, especially at depths of 40-60 cm from the soil surface (Evans 1978). At depths below 70 cm, tall fescue has more roots than does orchardgrass (Garwood and Sinclair 1979). Orchardgrass does not tolerate flooding or wet soils as well as reed canarygrass (*Phalaris arundinacea* L.). However, it persists and grows well on soils that have moderately poor drainage, especially in areas with mild winters.

PLANT DESCRIPTION

Orchardgrass is a cool-season grass that grows in clumps or tufts and is an example of a "bunch" grass. It starts growth early in the spring, develops rapidly, and flowers during late May or early June in the northern US and southern Canada. The actual date of flowering depends on daylength, temperature, and cultivar. Orchardgrass seedlings grow more rapidly than those of smooth bromegrass or tall fescue but not as vigorously as those of reed canarygrass (Newman and Moser 1988). It is not a competitive species in the seedling stage but is very competitive later in its life cycle (Spedding and Diekmahns 1972).

Leaves of orchardgrass are folded in the bud and in cross section appear V-shaped. The sheath is distinctly flattened and keeled. These leaf features are the most distinguishing characteristic of orchardgrass and make the species easy to identify. The leaf collar is divided and glabrous, the ligule is membranous and 2-8 mm long, and the auricles are absent (Clarke et al. 1944). The leaf blade is normally 2-10 mm wide and 5-25 cm long, although it may reach a length of 1 m. Leaves vary in color from light green to dark blue-green.

The inflorescence of orchardgrass is developed by differentiation of the terminal meristem at the base of the plant, and it is moved upward by elongation of internodes. Flowering culms are usually 1.0-1.3 m high and have 2-4 nodes. The inflorescence is 8-15 cm long, and the spikelets contain 2-5 florets each. The lowermost branches of the inflorescence are longer and have more branching than ones near the top. Orchardgrass reproduces sexually by seed formation and asexually by tiller formation.

Tillering occurs almost continuously throughout the growing season. Therefore, on any one date, tillers will be at many different stages of development. A large number of tillers within a clump remain vegetative, retaining the growing point close to the ground, and produce only leaves. This gives orchardgrass, even when headed, many basal leaves and a leafy appearance. The continuous production of vegetative tillers results in rapid recovery after cutting—another distinguishing feature of this species.

Like other grasses, orchardgrass produces a fibrous root system. The root system is extensive and deeper than those of Kentucky bluegrass and timothy but not as well distributed as that of smooth bromegrass (Table 26.1).

Carbohydrates are stored in the lower part of the leaf blade, tiller bases, and roots, but the greatest concentration is in the plant base near the soil surface (Table 26.2). Glucose, fructose, sucrose, and some starch are stored, but by far the largest stored carbohydrate

TABLE 26.1. Mass and distribution of roots for cool-season forage grasses in a Rayne silt loam

Species	Weight to 46 cm (kg/ha)	Root mass below 7.6 cm (%)
Orchardgrass	5234	20
Smooth bromegrass	4689	49
Kentucky bluegrass	4292	12
Timothy	2572	11

Source: Adapted from Gist and Smith (1948).

fraction is fructan (Okajima and Smith 1964). When grown at low temperatures, fructans accumulate in the leaves, and especially in the leaf sheaths (Eagles 1967b). These stored carbohydrates are considered to be important for winter survival.

Forage Nutritive Value. In the vegetative stage, orchardgrass is similar in digestibility to timothy and smooth bromegrass (Table 26.3). As the plant inflorescence emerges and reaches anthesis, orchardgrass and smooth bromegrass are of similar digestibility, but timothy is lower. Following anthesis, the digestibility of orchardgrass drops rapidly, and at early seed set is about 51% (Table 26.3). When compared on the same date, orchardgrass, because of its earlier maturity, is lower in digestibility and protein than timothy and smooth bromegrass. Later-maturing cultivars of orchardgrass are higher in digestibility and protein than early cultivars on the same date, but they are very similar when compared at the same stage of maturity (Gervais 1991).

As orchardgrass grows in the spring and matures, the yield of dry matter increases, while quality, as measured by digestibility and protein content, decreases (Table 26.4). The general recommendation is to cut for hay or silage at the heading stage or when the forage is about 65% digestible.

However, most orchardgrass used for hay or silage is grown with a legume, so it is important to select a cultivar that is similar to the legume in maturity. It is difficult to predict quality for a specific orchardgrass cultivar, as the proportion of vegetative versus reproductive tillers varies with environmental conditions (Buxton and Marten 1989). The digestibility of orchardgrass could be improved by breeding (Stratton et al. 1979; Shenk and Westerhaus 1982).

Under field conditions, rates of leaf formation and senescence are similar after 35 d of regrowth. Dead leaves accumulate in the lower canopy when cutting is delayed, and there is little advantage, in terms of forage yield, in separating harvests by more than 5-6 wk. High rates of N fertilization will delay senescence.

Mineral composition of forages is of concern to the livestock nutritionist. There appears to be sufficient heritability of mineral concentration to improve this aspect of orchardgrass quality through breeding (Stratton and Sleper 1979). In the first crop, the content of many minerals (potassium [K], phosphorus [P], calcium [Ca], magnesium [Mg], sulfur [S], zinc [Zn], copper [Cu], and iron [Fe]) declines as the plant matures (Gervais 1991). The second cutting of forage is higher in P, Ca, Mg, manganese (Mn), and Fe than is the first crop. On the other hand, K, Zn, and Cu are about the same as the season progresses (Gervais 1991). High rates of N fertilization may result in excessive uptake of N and K and reduced Mg uptake. This can be of considerable importance in areas where soil and forage Mg levels are too low to meet livestock needs. Magnesium deficiency in cattle causes a condition commonly known as *grass tetany,* a nu-

TABLE 26.2. Distribution of carbohydrates in orchardgrass tissues

Part of plant	Weight of entire plant (%)	Carbohydrates (% dry wt)		
		Reducing sugars	Sucrose	Fructan
Upper 2/3 of leaf blades	14.0	1.4	8.4	7.6
Lower 1/3 of leaf blades	12.1	1.2	5.8	22.0
Upper 1/2 of tiller bases	9.4	1.9	3.6	23.7
Lower 1/2 of tiller bases	23.6	0.7	2.6	36.2
Roots	40.9	1.2	8.9	8.2

Source: Sprague and Sullivan (1950).

TABLE 26.3. Yield, in vitro digestibility (IVD), and crude protein (CP) of three grasses at five stages of maturity

Stage	Orchardgrass			Timothy			Smooth Bromegrass		
	Yield (kg/ha)	IVD (%)	CP (%)	Yield (kg/ha)	IVD (%)	CP (%)	Yield (kg/ha)	IVD (%)	CP (%)
Early Vegetative	2016	76.4	23.2	2177	79.7	23.0	2960	77.9	22.4
Boot	2874	74.7	13.3	5181	70.9	10.6	4225	75.1	13.4
Heading	3911	71.2	11.0	6593	64.1	8.9	5861	69.1	10.0
Anthesis	5115	61.3	8.2	8212	56.9	6.9	8194	59.4	6.7
Early Seed	5958	51.8	6.6	8602	53.1	5.7	8307	59.7	5.8

Source: Fulkerson (1983).
Note: Species were grown at Guelph, Canada.

TABLE 26.4. Yield, proportion of leaves, in vitro digestibility (IVD), and crude protein (CP) of spring forage of orchardgrass, grown at Guelph, Canada

Date	Yield (*kg/ha*)	Leaf (%)	IVD (%)	CP (%)
May 7	535	100	76.4	28.0
May 14	1099	100	75.0	23.2
May 22	2016	90	76.4	16.0
May 28	2849	77	76.0	13.0
June 4[a]	4008	59	72.0	10.8
June 11	4627	52	65.0	9.2
June 18	4997	53	61.9	8.4
June 25	5473	47	56.9	7.5
July 3	5570	47	53.5	6.9
July 9	5760	50	52.4	6.6
July 11	5707	49	49.7	6.1
July 23	5185	53	46.3	6.2

Source: Fulkerson (1983).
[a]50% headed.

tritional disease of common occurrence in many areas. Orchardgrass is a relatively poor accumulator of Mg, especially at cool temperatures. Rate and type of fertilizer can influence the amount of forage consumed and contribute significantly to animal performance (Reid et al. 1966)

Flowering. Temperature and photoperiod play important roles in the flowering of orchardgrass. Natural induction, i.e., vernalization, occurs in the fall when days are cool and photoperiods are short. Maximum heading occurs after orchardgrass plants have been grown for 6 wk at a 10-h photoperiod and a temperature of 10°C (Kozumplik and Christie 1972). Short photoperiods seem to be more important than low temperatures, although some reports indicate that both conditions are required for induction (Gardner and Loomis 1953). Genetic differences have been found for induction requirements (Kozumplik and Christie 1972). Under field conditions floral initiation occurs in the spring, when days become warmer and photoperiod increases. Application of N in the fall increases the number of heads per unit area in the next year.

IMPORTANCE AND USE

In the US, orchardgrass was tested at many experiment stations but was not widely accepted until about 1940 (Smith et al. 1986). In recent years, the use of orchardgrass has declined, and fewer cultivars have been developed. This decline may be due in part to a decline in use of pastures for intensive grazing in those areas where orchardgrass is adapted. However, in addition to pasture, orchardgrass is utilized as green chop, silage, or hay, when grown either alone or in a mixture with legumes.

Orchardgrass is easy to establish, and the seedlings are able to compete with other crop plants and weeds (Borman et al. 1990). Once established, it will survive for many years if properly managed. Orchardgrass recovers well after cutting or grazing and produces good second- and third-harvest growth. Consequently, it is more productive during the summer than either timothy or smooth bromegrass. Although competitive, it is fairly compatible with such legume species as alfalfa (*Medicago sativa* L.) and ladino clover. Alfalfa tends to be less persistent in hay stands with orchardgrass than with timothy or smooth bromegrass.

Orchardgrass is less winter-hardy than timothy or smooth bromegrass, and this limits its utilization in the colder regions of North America and Europe. Unless properly managed, it matures quickly in the early summer, becoming coarse and unpalatable. In mixtures, it must be properly managed to maintain the desired grass:legume balance.

Pasture. Orchardgrass is an important pasture grass species in the northeast and north central states (Hoveland 1992). It is usually grown in a mixture with one or more legumes. Because of its rapid growth and good tillering ability, orchardgrass is well suited for early spring pastures and better suited to rotational stocking than to continuous stocking. Frequent mowing to simulate grazing has been shown to result in a thinner stand of orchardgrass (Davies 1988; Parish et al. 1990). When stocked continuously, animals tend to regraze the same areas until the plants are weakened by frequent removal of leaf tissue. In some cases, especially with sheep, tillers may be eaten so close to the ground that the stem base tissue used for food storage is removed. Under frequent grazing or clipping, orchardgrass has sustained growth better than ryegrasses, timothy, or tall fescue (Davies 1988), but it could be gradually replaced by more prostrate-growing species such as Kentucky bluegrass.

Although orchardgrass is more competitive than some of the other forage grasses when grown in mixture with legumes (Parish et al. 1990; Shaeffer et al. 1990), it is often grown with white clover for pasture. White clover, especially the ladino type, can be maintained in the mixture with proper management. Clover provides N for the grass, and the mixture provides a palatable and productive forage, which can be maintained for several years. If grazing is delayed, the grass may provide excessive competition and result in the elimination of the clover. Wagner (1952) found that clipping to stimulate grazing when the orchardgrass was 20-30 cm tall provided a good balance between yield and persistence of white clover. If the white clover is reduced or eliminated from a mixture with orchardgrass, it can be reestablished without conventional tillage. The orchardgrass should be weakened in the late summer or fall by heavy stocking or by herbicides. The white clover should be overseeded in the fall or winter. It is important to weaken the orchardgrass, as the white clover will not establish if the orchardgrass is too competitive. Clover seedlings are especially sensitive to shading; thus the orchardgrass should be grazed or clipped early to let light get to those seedlings.

Hay. Orchardgrass may be grown for hay either in pure stands or with legumes. When pure stands are utilized, N must be applied in combination with other nutrients for maximum production and a higher protein content. However, if orchardgrass is grown with a legume, fertilization with N is rarely required. In the US, alfalfa is commonly grown with orchardgrass for hay. Orchardgrass is more competitive with alfalfa than many other grasses are (Casler 1988; Shaeffer et al. 1990). If the desired mixture is to have a large grass component, then orchardgrass would be recommended (Shaeffer et al. 1990). Since orchardgrass matures early in the season, it combines well in a mixture with an early-flowering alfalfa cultivar. Mixtures of alfalfa with different cultivars of orchardgrass, or with other grasses, allow the optimum time of harvest to be spread over several weeks.

CULTIVARS

Cultivars available in orchardgrass exhibit a wide range in maturity, i.e., date of anthesis, probably more so than those of any of the other common forage species. Early-maturing cultivars tend to have higher yields at each cut and to recover faster after cutting than later-maturing cultivars (Casler 1988; Kochanowska-Bukowska 1992). On a given calendar date, later-maturing cultivars are usually more leafy, higher in digestibility, and considered to be more palatable. However, at the same stage of growth, the early-maturing cultivars are often higher in quality (Gervais 1991).

In selecting cultivars for a hay mixture, it is important to combine an early-maturing cultivar of orchardgrass with an early one of alfalfa; in other words the cultivars of the mixture components must be of similar maturity. If the cultivars are quite different in maturity, it will be difficult to select an optimum harvest date. As a result, the quality of the harvested forage may be reduced, and because one of the components may be cut at a critical stage, maintenance of that component in the mixture may be difficult. Since the quality of grasses declines rapidly upon anthesis, it would be advantageous to harvest an alfalfa-orchardgrass sward when the orchardgrass heads out.

The same orchardgrass cultivar may be recommended for production of both hay and pasture. However, a cultivar that performs well when managed for hay or silage may not be the best cultivar for a pasture (Price et al. 1991).

New cultivars of orchardgrass continue to be produced by forage breeders employed by governments, universities, and commercial companies. Modest gains in yield have been achieved, as well as improvements in disease resistance and other traits (Stratton et al. 1985; McClain 1986). In northern areas a major goal is increased resistance to winter stress.

CULTURE AND MANAGEMENT

Orchardgrass is usually established with little or no difficulty. The usual recommendation is to seed at 8-12 kg/ha in the early spring or late summer. Good stands have been obtained with seeding rates as low as 4.5 kg/ha (Sprague et al. 1963). In more northerly areas, it is recommended that seeding be done in the early spring. Delayed seeding in the spring results in a decrease in forage yield in the year after seeding (Fraser and Kunelius 1991).

When orchardgrass is seeded with a legume, the seeding rate is usually decreased by 50% or more. Orchardgrass, either alone or in a mixture with a legume, is often seeded with a cereal companion crop, which is harvested as silage, hay, or grain. The orchardgrass is harvested the next year. In the year of seeding, orchardgrass is not sufficiently productive to warrant harvesting.

The seedbed should be loose on top and firm underneath. The seed should be planted no deeper than 0.6 cm. Orchardgrass emergence decreases very rapidly as the depth of seeding increases. The use of press wheels directly over the seed or a culti-packer ensures that the seed will be in close contact with the soil.

Freshly harvested seed has a low germination for 3-4 wk due to a growth inhibitor found in the lemma, palea, and caryopsis of the seed. Germination can be enhanced by removal of the lemma and palea or by storage at temperatures near freezing to decrease the amount of inhibitor (Fendall and Canode 1971). The germination of freshly harvested seed will also be enhanced by exposing the seed to alternating temperatures from 15° to 39°C or from 10° to 30°C in combination with an 8-h photoperiod (Sprague 1940). Emergence of orchardgrass seedlings is promoted by short days and moderate temperatures. Temperatures of 30°C or higher reduce emergence (Sprague 1944).

Plants grown at the same radiant energy develop leaves and roots faster when exposed to short photoperiods than when exposed to long ones (Templeton et al. 1969), but individual leaves may be smaller (Robson et al. 1988). However, this response may depend upon the origin of the plant material, as cultivars from more northerly regions may show an increase in dry matter with longer photoperiods (Solhaug 1991). With equal photoperiods, leaves appear more rapidly with higher light intensity, but they become senescent faster (Taylor et al. 1968). The total leaf area formed is greater when plants are subjected to low light intensity, especially with cool temperatures. Short photoperiods, low light intensity, and cooler temperatures generally enhance leaf development, depending on many other factors, such as plant age, cultivar, and associated climate.

Fertilization. Orchardgrass yield is very responsive to N applications, responding to applications as high as 672 kg/ha (Reynolds et al. 1971; Donohue et al. 1981). However, applying large amounts of N fertilizer may reduce winter survival of plants and hence reduce yield the following year (Meister and Lehman 1990). A total annual application of 450 kg N/ha was found to adversely affect the persistence of orchardgrass.

For hay production, split applications of N reduce lodging and give a better distribution of forage over the growing season. Nitrogen rates of 56-84 kg/ha in the early spring and after each cutting are generally recommended. The number of applications depends on the number of cuts, which varies with the length of the growing season and the need for additional forage. It is of course imperative that N be used in combination with other nutrients for high yields. For mixtures with legumes, N is not usually required. If the management favors the legume, it may be necessary to add some N in order to maintain the orchardgrass (Kanyama-Phiri et al. 1990).

Fertilization with animal manure will promote the growth of orchardgrass and reduce the legume component of a mixture. Rates of poultry manure as low as 2 mt/ha may reduce the white clover content in mixtures with orchardgrass (Bomke and Lowe 1991). The total yield of an orchardgrass plus white clover mixture was found to increase with increasing rates of poultry manure up to 10 mt/ha. Dry matter yields began to decrease when the rate of manure application reached 40 mt/ha. Analyses for various minerals indicated that none accumulated to toxic levels at any of the rates of manure application.

Harvest Schedule

GROWTH STAGE. Total seasonal DM yields generally increase as harvest date of the first crop is delayed from the vegetative stage to anthesis. If cut at head emergence, yields are 15%-25% lower than when cut at early anthesis (Washko et al. 1967; Fulkerson 1983). Delaying first harvest until early anthesis was found to reduce aftermath yields by 15%-25%. Delaying the first cut beyond anthesis decreased aftermath yields still more. Generally, the later the first cut is taken after head emergence, the lower the total seasonal yield (Fulkerson 1983; Kochanowska-Bukowska 1992).

The proper time to harvest first-growth orchardgrass depends on the use to be made of the forage. Harvesting can be delayed until anthesis or later and still provide adequate protein and energy levels for maintenance of some animals. When high quality is important, orchardgrass should be cut at head emergence. Harvesting may be spread over several weeks by planting early-, intermediate-, and late-maturing cultivars; however, early-maturing cultivars are generally more productive.

Harvesting the first crop at different growth stages has little influence on stand longevity. If the first harvest is taken late, and especially if N is readily available, the stands become thin and exceedingly bunchy. Cutting the plants in the month of September in Ontario, Canada, has been shown to reduce yields in the following year (Fulkerson 1983).

HARVEST FREQUENCY. Both DM yield and root weight vary inversely with frequency of harvest. Orchardgrass can tolerate frequent cutting and, under frequent cutting, has been found to persist better than forage species such as perennial ryegrass, tall fescue, and timothy (Davies 1988). Highest DM yields are usually obtained in the northern US and Canada when orchardgrass is cut two or three times per year. The height at which the plants should be cut has been found to vary with frequency of cutting. The more frequent the cutting, the higher the cutting height necessary to ensure stand persistence. Orchardgrass has been found to persist well when cut two to three times per season at a height of 8 cm (Kading 1990). If cut five times per season, it persists better if cut at 15 cm. During the summer, especially if moisture is limiting, the height of cutting should be increased (Seo et al. 1988).

Regrowth of orchardgrass is a function of energy from photosynthetic activity, carbohydrate reserves, or both. If carbohydrate reserve levels are high, defoliation does not result in much stress because new leaves are produced at the expense of reserves (Smith et al. 1986). Frequent cutting at ground level or continuous close grazing almost always results in a reduction of reserves and a loss of stand, especially with high rates of N fertilization. Other factors such as low soil moisture, high temperature, and presence of disease also make orchardgrass more sensitive to cutting frequency and height.

SEED PRODUCTION

Orchardgrass seed has been produced commercially in the US for nearly 150 yr. Today most of the seed is produced in Oregon. Areas favored for grass seed production usually have low rainfall and high levels of sunshine after head emergence. For seed production, orchardgrass should be grown in rows spaced 22-40 cm apart. Since few heads are produced in the year of seeding, the first seed harvest would be taken in the year after seeding. Fertilization with N is essential for good seed yields. Nitrogen applied in the late fall increases the number of heads per unit area, while an application in the early spring increases the size of the heads (Falkowski et al. 1992; Fulkerson 1970). For maximum yields, N should be applied in the late fall and again in the early spring. Early-maturing cultivars tend to be better seed producers than later ones.

DISEASES AND PESTS

Many pathogens can attack orchardgrass, but there is little evidence regarding the economic losses due to disease. Most of the common diseases are found on the leaves, where part or all of the leaf blades are destroyed. For example, purple leaf spot, a fungal disease caused by *Stagonospora arenaria* (Sacc.), is frequently found in the northeastern US. Severe infections have been shown to reduce yield (Zeiders et al. 1984) and forage quality (Mainer and Leath 1978). Breeding for disease resistance is a practical means of controlling the disease (Berg et al. 1992). Control of leaf diseases by dusting or spraying is generally not economical although in Oregon spraying seed production fields with a fungicide has given higher seed yields in some years (Welty 1989). Control by crop rotation and management has been only partly successful. The best way to control diseases is with resistant cultivars.

More than 30 insect species have been identified in orchardgrass stands (Hardee et al. 1963). Insect damage can result in losses in yield, quality. and stand longevity, but there is little information on the economic losses incurred. In production, potential losses from insects are generally ignored.

QUESTIONS

1. What characteristics would you use to identify orchardgrass?
2. What factors affect the nutritive value of orchardgrass? How can these factors be influenced by the farmer?
3. How does orchardgrass compare with other cool-season grasses in terms of adaptation, yield, seasonal distribution of forage, feeding value, and persistence?
4. What are the flowering requirements of orchardgrass?
5. What characteristics of orchardgrass make it a suitable species for pasture?
6. How would you manage an orchardgrass-legume pasture? A hayfield? Why?
7. What management factors are important in the production of orchardgrass seed?
8. What factors determine the distribution of orchardgrass? Why?

9. How would you manage orchardgrass for hay in order to produce the maximum dry matter over the season? The maximum amount of aftermath?
10. How are diseases controlled under orchardgrass field conditions?

REFERENCES

Archer, KA, and AM Decker. 1977. Autumn-accumulated tall fescue and orchardgrass. I. Growth and quality as influenced by nitrogen and soil temperature. Agron. J. 69:601-5.

Baker, BS, and GA Jung. 1968. Effect of environmental conditions on the growth of four perennial grasses. I. Response to controlled temperature. Agron. J. 60:155-58.

———. 1970. III. Nucleic acid concentration as influenced by day-night temperature combinations. Crop Sci. 10:376-78.

Berg, CC, RT Sherwood, and RR Hill, Jr. 1992. Inheritance of resistance to Stagonospora leaf spot in a diallel cross of orchardgrass. Crop Sci. 32:1123-26.

Blake, CT, DS Chamblee and WW Woodhouse, Jr. 1966. Influence of some environmental and management factors on the persistence of ladino clover in association with orchardgrass. Agron. J. 58:487-89.

Bomke, AA, and LE Lowe. 1991. Trace element uptake by two British Columbia forages as affected by poultry manure application. Can. J. Soil Sci. 71:305-12.

Borman, MM, WC Krueger, and DE Johnson. 1990. Growth patterns of perennial grasses in annual grassland type of southwest Oregon. Agron. J. 82:1093-98.

Buxton, DR, and GC Marten. 1989. Forage quality of plant parts of perennial grasses and relationship to phenology. Crop Sci. 29:429-35.

Casler, MD. 1988. Performance of orchardgrass, smooth bromegrass and ryegrass in binary mixtures with alfalfa. Agron. J. 80:509-14.

Clarke, SE, JA Campbell, and W Shevkenek. 1944. The Identification of Certain Native and Naturalized Grasses by Their Vegetative Characters. Tech. Bull. 50, publ. 762. Ottawa, Canada: Agriculture Canada.

Davies, A. 1988. The regrowth of grass swards. In MB Jones and A Lazenby (eds.), The Grass Crop. London: Chapman and Hall, 85-127.

Donohue, SJ, RJ Bula, DA Holt, and CL Rhykerd. 1981. Morphological development, yield, and chemical composition of orchardgrass at several soil nitrogen levels. Agron. J. 73:5-9.

Eagles, CF. 1967a. The effect of temperature on vegetative growth in climatic races of *Dactylis glomerata* in controlled environments. Ann. Bot. 31:31-39.

———. 1967b. Variation in soluble carbohydrate content of climatic races of *Dactylis glomerata* (Cocksfoot) at different temperatures. Ann. Bot. 31:645-51.

Evans, PS. 1978. Plant root distribution and water use patterns of some pastures and crop species. N.Z. J. Agric. Res. 21:261-65.

Falkowski, M, S Kozlowski, and I Kukulka. 1992. Growth and development characteristics of *Dactylis glomerata* in seed plantations. In Proc. 14th Gen. Meet. Eur. Grassl. Fed., 8-11 June, Lahti, Finland, 443-44.

Fendall, RK, and CL Canode. 1971. Dormancy-related growth inhibitors in seeds of orchardgrass (*Dactylis glomerata* L.). Crop Sci. 11:727-30.

Fraser, J, and HT Kunelius. 1991. Influence of seeding time on the performance of white clover/orchardgrass mixtures in Atlantic Canada. In Proc. 9th Eastern Forage Improv. Conf., Charlottetown, PEI, Canada, 117-18.

Fulkerson, RS. 1970. Producing Grass Seed in Ontario. Inf. Leafl. Toronto, Canada: Ontario Ministry of Agriculture and Food.

———. 1983. Research Review of Forage Production. Guelph, Ontario, Canada: Crop Science Department, University of Guelph.

Gardner, FP, and WE Loomis. 1953. Floral induction and development in orchardgrass. Plant Physiol. 28:201-17.

Garwood, EA, and J Sinclair. 1979. Use of water by six grass species. 2. Root distribution and use of soil water. J. Agric. Sci., Camb. 93:25-35.

Gervais, P. 1991. Composition morphologique et chimique, a trois stades de croissance, de certains cultivars de quatre graminees fourrageres perennes cultives au Quebec. Bull. tech. 16. Comite des Plantes Fourrageres, Ministere de l'Agriculture, Des Pecheries et de l'Alimentation, Quebec, Canada.

Gist, GR, and RM Smith. 1948. Root development of several common forage grasses to a depth of eighteen inches. J. Am. Soc. Agron. 40:1036-42.

Hardee, DD, HY Forsythe, Jr., and GG Gyrisco. 1963. A survey of the Hemiptera and Homoptera infesting grasses (Gramineae) in New York. J. Econ. Entomol. 56:555-59.

Hoveland, CS. 1992. Grazing systems for humid regions. J. Prod. Agric. 5:23-27.

Kading, H. 1990. Einfluss der Schnitthohe auf Ertrag, Leistungsdauer und Inhaltsstoffe verschiedener Grasarten. Wirtschaftseigene Futter 36:31-40 (Herb. Abstr. 62[1992]:3303).

Kanyama-Phiri, GY, CA Raguse, and KL Taggard. 1990. Responses of a perennial grass-legume mixture to applied nitrogen and differing soil textures. Agron. J. 82:488-95.

Knievel, DP, and D Smith. 1973. Influence of cool and warm temperatures and temperature reversal at inflorescence emergence on growth of timothy, orchardgrass, and tall fescue. Agron. J. 65:378-83.

Kochanowska-Bukowska, Z. 1992. Persistence and yield of several varieties of *Dactylis glomerata* L. as affected by different time of harvest. In Proc. 14th Gen. Meet. Eur. Grassl. Fed., 8-11 June, Lahti, Finland, 427-28.

Kozumplik, V, and BR Christie. 1972. Heading response of orchardgrass seedlings to photoperiod and temperature. Can. J. Plant Sci. 52:369-73.

Leafe, EL. 1988. The history of improved grasslands. In MB Jones and A Lazenby (eds.), The Grass Crop. London: Chapman and Hall, 1-23.

McClain, EF. 1986. Registration of 'Piedmont' orchardgrass. Crop Sci. 26: 835-36.

Mainer, A, and KT Leath. 1978. Foliar diseases alter carbohydrate and protein levels in leaves of alfalfa and orchardgrass. Phytopathol. 68:1252-55.

Meister, E, and J Lehman. 1990. Leistungs-und Qualitatsmerkmale verschiedener Graser bei steigender Stickstoffdungung. Landwirtschaft Schweiz 3:125-30 (Herb. Abstr. 62[1992]:3369).
Mitchell, KJ, and R Lucanus. 1960. Growth of pasture species in controlled environment. II. Growth at low temperatures. N.Z. J. Agric. Res. 3:647-55.
———. 1962. Growth of pasture species under controlled environment. III. Growth at various levels of constant temperature with 8 and 16 hours of uniform light per day. N.Z. J. Agric. Res. 5:135-44.
Newman, PR, and LE Moser. 1988. Seedling root development and morphology of cool-season and warm-season forage grasses. Crop Sci. 28:148-51.
Okajima, H, and D Smith. 1964. Available carbohydrate fractions in the stem bases and seed of timothy, smooth bromegrass, and several other northern grasses. Crop Sci. 4:317-20.
Parish, R, R Turkington, and E Klein. 1990. The influence of mowing, fertilization, and plant removal on the botanical composition of an artificial sward. Can. J. Bot. 68:1080-85.
Pearce, RB, RH Brown, and RE Blaser. 1965. Relationships between leaf area index, light interception and net photosynthesis in orchardgrass. Crop Sci. 5:553-56.
Price, MA, YA Papadopoulos, LF Laflamme, GM Hunter, KB McRae, and NR Fulton. 1991. Protocol for assessing forage varieties for pasture use. In Proc. 9th Eastern Forage Improv. Conf., Charlottetown, PEI, Canada (abstr.), 121-22.
Reid, RL, GA Jung, and SJ Murray. 1966. Nitrogen fertilization in relation to the palatability and nutritive value of orchardgrass. J. Anim. Sci. 25:636-45.
Reynolds, JH, CR Lewis, and KF Laaker. 1971. Chemical Composition and Yield of Orchardgrass Forage Grown under High Rates of Nitrogen Fertilization and Several Cutting Managements. Tenn. Agric. Exp. Stn. Bull. 479.
Robson, MJ, GJA Ryle, and J Woledge. 1988. The grass plant—its form and function. In MB Jones and A Lazenby (eds.), The Grass Crop. London: Chapman and Hall, 25-83.
Seo, S, MS Park, YC Han, and JK Lee. 1988. [Studies on cutting management of pasture during the mid-summer season V.] Korean J. Anim. Sci. 30:212-17 (Herb. Abstr. 60[1990]:1124).
Shaeffer, CC, DW Miller, and GC Marten. 1990. Grass dominance and mixture yield and quality in perennial grass-alfalfa mixtures. J. Prod. Agric. 3:480-85.
Shenk, JS, and MO Westerhaus. 1982. Selection for yield and quality in orchardgrass. Crop Sci. 22:422-25.
Smith, D, RJ Bula and RP Walgenbach. 1986. Forage Management. Dubuque, Iowa: Kendall/Hunt.
Solhaug, KA. 1991. Influence of photoperiod and temperature on dry matter production and chlorophyll content in temperate grasses. Norw. J. Agric. Sci. 5:365-83 (Herb. Abstr. 62[1992]: 3502).
Spedding, CRW, and EC Diekmahns (eds.). 1972. Grasses and Legumes in British Agriculture. Bull. 49, Commonwealth Bureau of Pastures and Field Crops. Farnham Royal, Bucks, England: Commonwealth Agricultural Bureaux.
Sprague, MA, MM Hoover, Jr., MJ Wright, HA MacDonald, BA Brown, AM Decker, JB Washko, VG Sprague, and KE Varney. 1963. Seedling Management of Grass-Legume Associations. N.J. Agric. Exp. Stn. Bull. 804.
Sprague, VG. 1940. Germination of freshly harvested seeds of several *Poa* species and of *Dactylis glomerata*. J. Am. Soc. Agron. 32:715-21.
———. 1944. The effects of temperature and daylength on seedling emergence and early growth of several pasture species. Soil Sci. Soc. Am. Proc. 8:287-94.
Sprague, VG, and JT Sullivan. 1950. Reserve carbohydrates in orchardgrass clipped periodically. Plant Physiol. 25:95-102.
Stratton, SD, and DA Sleper. 1979. Genetic variation and interrelationships of several minerals in orchardgrass herbage. Crop Sci. 19:477-81.
Stratton, SD, DA Sleper, and AG Matches. 1979. Genetic variation and interrelationships of in vitro dry matter disappearance and fiber content in orchardgrass herbage. Crop Sci. 19:329-33.
Stratton, SD, CW Edminster, and RR Ronnenkamp. 1985. Registration of 'Rancho' orchardgrass. Crop Sci. 25:366.
Taylor, TH, JP Cooper, and KJ Treharne. 1968. Growth response of orchardgrass (*Dactylis glomerata* L.) to different light and temperature environments. I. Leaf development and senescence. Crop Sci. 8:437-40.
Templeton, WC, Jr., JL Menees, and TH Taylor. 1969. Growth of young orchardgrass (*Dactylis glomerata* L.) plants in different environments. Agron. J. 61:780-82.
Treharne, KJ, JP Cooper, and TH Taylor. 1968. Growth response of orchardgrass (*Dactylis glomerata* L.) to different light and temperature environments. II. Leaf age and photosynthetic activity. Crop Sci. 8:441-45.
Van Esbroeck, GA, JR King, and VS Baron. 1989. Effects of temperature and photoperiod on the extension growth of six temperate grasses. In D Desroches (ed.), Proc. 16th Int. Grassl. Congr., 4-11 Oct., Nice, France, 459-60.
Wagner, RE. 1952. Yield and Botanical Composition of Four Grass-Legume Mixtures under Differential Cutting. USDA Tech. Bull. 1063.
Washko, JB, GA Jung, AM Decker, RC Wakefield, DD Wolf, and MJ Wright. 1967. Management and Productivity of Perennial Grasses in the Northeast. 3. Orchardgrass. W.Va. Agric. Exp. Stn. Bull. 557T.
Welty, RE. 1989. The effect of fungicide application on seed yield of *Dactylis glomerata*. In D Desroches (ed.), Proc. 16th Int. Grassl. Congr., 4-11 Oct., Nice, France, 671-72.
Zeiders, KE, CC Berg, and RT Sherwood. 1984. Effect of recurrent phenotypic selection on resistance to purple leaf spot in orchardgrass. Crop Sci. 24:182-85.

27 Reed Canarygrass

CRAIG C. SHEAFFER
University of Minnesota

GORDON C. MARTEN
Agricultural Research Service, USDA, and University of Minnesota

REED canarygrass, *Phalaris arundinacea* L., was cultivated in Sweden by 1749 and in other parts of northern Europe by 1850 (Martin 1985). Its cultivation in the US probably occurred shortly afterward along the north Atlantic Coast; an Oregon planting was recorded in 1885. Some credit the name *canarygrass* to the fact that seed of the annual species *P. canariensis* L. is used as food for canary birds. Others believed it was named after the Canary Islands, where *P. canariensis* L. is native.

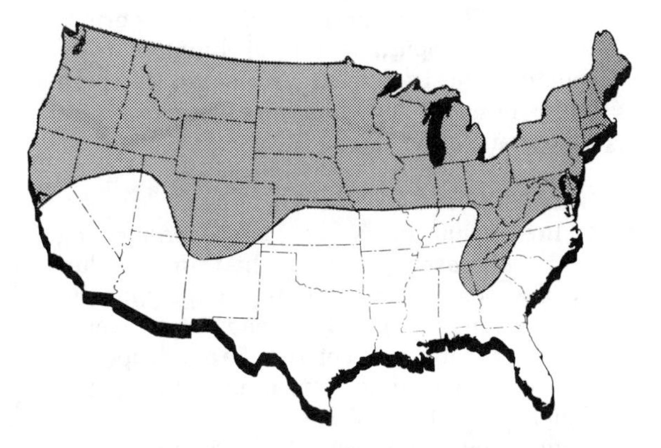

Fig. 27.1. Areas of the US in which reed canarygrass is adapted. In the dry western areas it must be irrigated.

DISTRIBUTION AND ADAPTATION

Members of the genus *Phalaris* have been collected on every major land mass except Antarctica and Greenland (Anderson 1961). Reed canarygrass is indigenous to the temperate portions of all five continents and is well adapted to the northern half of the 48 states and southern Canada. The largest hectarages are found in the Pacific Northwest and the humid north central states (Fig. 27.1).

While its natural habitat is poorly drained and wet areas, reed canarygrass is as drought tolerant as many cool-season grasses grown in humid and subhumid regions. When grown under moisture deficits in Minnesota, its yield and persistence were equal to smooth bromegrass, *Bromus inermis* Leyss, and orchardgrass, *Dactylis glomerata* L., and superior to timothy, *Phleum pratense* L. (Sheaffer et al. 1992). Dry matter production per unit of water used for reed canarygrass in North Dakota was 30 kg ha^{-1} mm^{-1}, which was less efficient than that of crested wheatgrass and similar to that of western wheatgrass (Frank et al. 1985).

Reed canarygrass is very tolerant of flooding, especially if water is moving and temperatures are low. With spring flooding in Canada, mature plants tolerated inundation for more than 49 d and seedings for 35 to 49 d (McKenzie 1951). No damaging effects were

CRAIG C. SHEAFFER. *See Chapter 16.*

GORDON C. MARTEN is Associate Director, Beltsville Agricultural Research Center, ARS, USDA, and Adjunct Professor of Agronomy, University of Minnesota. He received the MS and PhD degrees from the University of Minnesota and had postdoctoral study at Purdue University. His research emphasized pasture management, forage quality as influenced by environmental and management factors, and the development of techniques for forage evaluation.

found when this species was grown in pots with 25 mm of water over the soil surface for 3 mo. Reed canarygrass is only moderately adapted to saline soils but tolerates a pH range of 4.9 to 8.2.

Leaves of reed canarygrass were found to be killed by air temperatures not low enough to kill those of timothy or Kentucky bluegrass, *Poa pratensis* L. (Evans and Ely 1941). Reed canarygrass suffered more winter injury than smooth bromegrass when irrigated in Minnesota (Sheaffer et al. 1992). However, it had less winter injury than orchardgrass and timothy. In Alaska, reed canarygrass cultivars such as 'Venture' and 'Palaton', which were developed in the north central US, lacked winterhardiness (Klebesadel and Dofing 1991). However, Norwegian strains from 69° to 70°N latitude survived through several winters. Reed canarygrass does not perform well in subtropical or tropical climates. Its net photosynthesis is maximum at air temperatures of about 20°C and is reduced to 80% of maximum at 38°C.

PLANT DESCRIPTION

Reed canarygrass is a coarse, sod-forming, tall cool-season perennial. In thin or volunteer stands it often grows in clumps up to 1 m or more in diameter. It spreads underground by short scaly rhizomes that form a heavy sod in well-managed solid seedings. It ranges in height from 60 to 240 cm, and reproductive stem lodging may occur. Flowers are borne in semidense, spikelike panicles of 5 to 20 cm. The light gray to gray-black waxy seeds (about 3 mm long) shatter quickly at ripening; they mature from the top of the panicle downward. Reed canarygrass requires vernalization for flowering; therefore, only the initial spring regrowth produces seed. If harvested, regrowth is vegetative and short culmed. The leafy culms of vegetative regrowth are usually sufficiently stout to prevent lodging. Individual plants vary greatly in agronomic characteristics.

Forage Quality

NUTRITIVE VALUE. Forage nutritive value of reed canarygrass declines with maturity because of decreased leafiness and increased structural components. In Iowa, herbage digestibility declined linearly at rates of 3.7 to 6.3 g kg^{-1} d^{-1} after May 1 as reed canarygrass matured from elongating stem to seed-ripening stages (Buxton and Marten 1989). At jointing, stem and leaf digestibility averaged 581 and 646 g kg^{-1}, respectively, and declined to 435 and 578 g kg^{-1}, respectively, at anthesis. Decreases in digestibility were associated with increases in cell wall components (Buxton 1990). Environmental factors such as drought or the use of growth regulators such as mefluidide [N-[[2,4-dimethyl-5-{[(trifluoromethyl)sulfonyl]amino}phenyl]] acetamide] can enhance herbage nutritive value by delaying maturity and enhancing leafiness (Sheaffer and Marten 1986; Sheaffer et al. 1992).

Reed canarygrass has equivalent or greater digestibility compared with other perennial cool-season grasses. Its concentrations of acid detergent fiber, neutral detergent fiber, and lignin are similar to those of smooth bromegrass, timothy, tall fescue (*Festuca arundinacea* Schreb.), orchardgrass, and quackgrass (*Elytrigia repens* [L.] Nevski) (Buxton and Marten 1989; Buxton 1990; Sheaffer et al. 1990b). Like other cool-season grasses, reed canarygrass is more digestible than perennial warm-season grasses. Actively growing fall-saved tall fescue pasture was more digestible than dormant reed canarygrass in Iowa (Wedin et al. 1966). The digestibility of vegetative reed canarygrass is often equal to that of first-bloom alfalfa (Marten 1985).

Like digestibility, the crude protein (CP) concentration of reed canarygrass is higher in leaves than in stems and decreases as plants mature. Nitrogen (N) fertilization can have a major effect on forage CP concentration (Malzer and Schoper 1984); the total forage can range from 7% CP with low N to 30% CP with high N fertilization. Reed canarygrass is a superior extractor of N from soils, and it often contains more CP than other cool-season grasses managed and harvested the same way (Marten 1985).

PALATABILITY AND ALKALOIDS AND THEIR SIGNIFICANCE. Lack of palatability, i.e., lack of selection when a choice is offered, is the most frequently cited reason why this agronomically superior species has not become a leading forage grass in its area of adaptation. Lack of relative palatability of 'Rise' reed canarygrass was associated with the lower average daily gains by lambs that grazed it without choice compared with gains by lambs that grazed more palatable smooth bromegrass or orchardgrass (Marten and Jordan 1974). Only the lambs grazing reed canarygrass had recurrent diarrhea. On the other hand, the preference of dairy heifers for smooth bromegrass over reed canarygrass was of no practical significance when the two grasses were grazed without choice, as heifers gained the same on both species (Marten and Donker 1968).

Total alkaloid concentration in reed canarygrass genotypes was highly correlated with palatability to sheep when a wide range of alkaloid concentrations existed (Marten et al. 1973). In later studies, palatability scores for both sheep and cattle were correlated with the indole alkaloid concentration of eight clones of reed canarygrass, but the correlation was higher for sheep than it was for cattle. These results indicate that sheep are often more sensitive to variations in forage palatability caused by indole alkaloids than are cattle.

The threshold indole alkaloid level beyond which sheep will consistently reject reed canarygrass genotypes in cafeteria offerings is approximately 0.6% dry weight, although sometimes rejection will occur at 0.4% dry weight. Indole alkaloid concentrations of 0.4% and greater rarely occur in total herbage of reed canarygrass cultivars although numerous individual genotypes within the population can exceed this concentration. A negative association was also found to exist between voluntary intake (when no choice was offered) by sheep and relative palatability of genotypes of reed canarygrass (Marten et al. 1976). Reduced palatability may also be caused by buildup of structural components during maturation.

Marten et al. (1976) also discovered that total indole alkaloid concentration of eight diverse genotypes of reed canarygrass was highly negatively associated with average daily gains by both lambs and steers confined to pastures of individual genotypes. Both classes of animals had more diarrhea when they grazed plants high in total alkaloids and when they grazed tryptamine-carboline-containing compared with gramine-containing plants.

In another Minnesota experiment (Marten et al. 1981), sheep grazed exclusively either Rise, which contained a mixture of gramine and tryptamines-carbolines (means of 0.32% to 0.28% dry weight), 'Vantage', which contained gramine as its only indole alkaloid (0.33% to 0.20% dry weight), or the experimental strain 'MN-76', which also contained only gramine (0.12% to 0.09% dry weight).

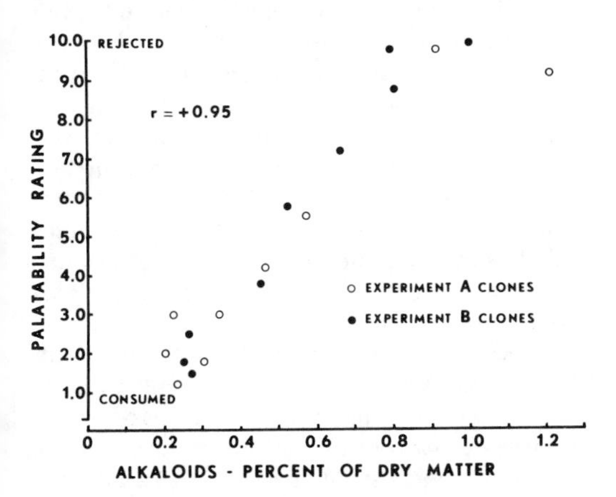

Fig. 27.2. Relationship between palatability rating (high rating = low palatability) and alkaloid concentration of 18 reed canarygrass clones from two experiments (Marten 1985).

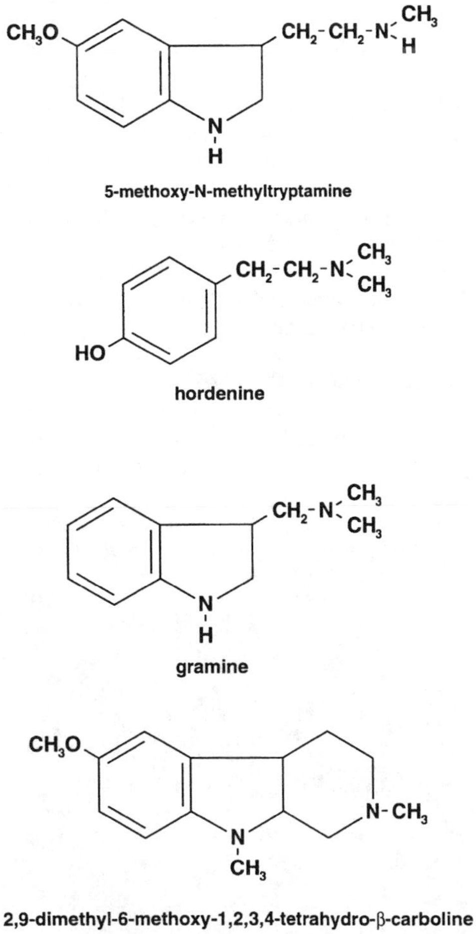

CH_3O, $CH_2-CH_2-N<^{CH_3}_{H}$, N, H

5-methoxy-N-methyltryptamine

$CH_2-CH_2-N<^{CH_3}_{CH_3}$, HO

hordenine

$CH_2-N<^{CH_3}_{CH_3}$, N, H

gramine

CH_3O, N-CH_3, N, CH_3

2,9-dimethyl-6-methoxy-1,2,3,4-tetrahydro-β-carboline

Fig. 27.3. Structures of some alkaloids found in reed canarygrass. Alkaloids usually have the following properties: (1) chemically basic; (2) nitrogen containing; (3) of plant origin; (4) complex molecular structure (nitrogen atom usually in a heterocyclic ring); and (5) pharmacological activity (Marten 1973).

Lambs grazing MN-76 gained from 51% to 95% more per day than did those grazing Rise, and from 14% to 87% more than those grazing Vantage. The correlation between total indole alkaloid concentration of the grasses and lamb gain was –0.97 in year one and –0.66 in year two. The researchers concluded that lambs have reduced gain if the indole alkaloid concentration in reed canarygrass is at or above 0.2% dry weight and that diarrhea in lambs grazing reed canarygrass is due to tryptamine-carboline alkaloids.

INTAKE. Actual intake, measured by units of grass dry matter (DM) consumed/animal unit/day, of reed canarygrass has been compared with that of other forages (Marten 1985). In one trial, reed canarygrass hay was inferior to both birdsfoot trefoil, *Lotus corniculatus* L., and alfalfa, *Medicago sativa* L., in both total and digestible dry matter intake by sheep, but its intake was usually not different from that of smooth bromegrass or timothy. Grazing steers usually consumed more reed canarygrass than tall fescue in Iowa. Delaying harvest of reed canarygrass hay from the first week in June in New Hampshire until 12-15 d later caused a 20% to 24% lower DM intake by cattle and sheep.

IMPORTANCE AND USE

Reed canarygrass is successfully used for pasture, hay, and silage. It is a vital part of many soil conservation programs and has become the most popular species to be used as a pollution control measure in irrigation of sewage effluent from municipal and industrial sources.

Fig. 27.4. Extreme differences in herbage removal of reed canarygrass clones by grazing sheep indicate differences in relative palatability and the possibility of breeding for its improvement. *Minn. Agric. Exp. Stn. photo.*

Pasture. For best quality pasture, grazing should be initiated when reed canarygrass is from 15 to 30 cm tall. Mature grass will be poorly utilized and will give poor animal intake and performance. Reed canarygrass begins growth in the spring similar to how smooth bromegrass and timothy do and, if moisture and fertility are not limiting, provides consistent growth until the killing frost.

Rotational stocking, with heavy grazing pressure for short periods, provides the best utilization of reed canarygrass. With continuous light grazing pressure, young growth from unfolding shoots is prominent and selectively grazed, leaving mature growth. With season-long grazing, a stiff, short stubble of reproductive stems and vegetative regrowths may build up, even under heavy rotational stocking, and cattle will not graze below this stubble. A single close (5- to 7-cm) clipping in midsummer will solve the problem (Marten and Donker 1968).

Cattle gains of 0.6 to 0.9 kg d^{-1} have been obtained on unsupplemented reed canarygrass pasture (Marten and Donker 1968). Also, lambs that grazed low-alkaloid MN-76 reed canarygrass in Minnesota gained as well as those that grazed smooth bromegrass (0.1 kg d^{-1}).

Reed canarygrass is one of the highest-yielding perennial grasses used for grazing in its area of adaptation, yielding more than smooth bromegrass in upland pasture trials in Iowa, Minnesota, and Nebraska (Wedin et al. 1966; Marten and Donker 1968). Reed canarygrass is similar to orchardgrass in that both regrow well throughout the pasture season. It is not as useful as tall fescue for fall-saved winter pasture because reed canarygrass leaves are not frost resistant, whereas tall fescue continues to grow after frost and retains its quality better into the winter (Wedin et al. 1966).

Reed canarygrass develops a sod sufficiently firm to support grazing animals even on wetland. Farm machinery is supported on sod in areas that were impassable before reed canarygrass was established. It is well adapted to seepy hillsides and soils with poor internal drainage as well as droughty areas.

Hay. The high yield capacity of reed canarygrass in diverse environments is documented by Marten (1985). In four northeastern states and in Minnesota DM yields over 13 mt ha^{-1} were obtained with adequate fertilizer and favorable cutting management. Reed canarygrass has yielded DM up to 17 mt ha^{-1} in West Virginia.

Best quality hay will result if the first crop is harvested before heading. With this cutting management, three excellent quality crops can be obtained (with adequate N and moisture) in all but the northernmost areas of adaptation, where one or two crops can be harvested annually. Reed canarygrass will yield more when cut two or three times per year than when cut four times (Marten 1985; Sheaffer et al. 1990a, 1990b). In low wet areas, which cannot be harvested by machinery before seed formation, reed canarygrass provides a high yield of bedding straw.

Silage. Reed canarygrass makes a palatable and nutritious silage when harvested before heading, to ensure adequate sugar concentrations, and when finely chopped (Pitt 1990). Preservation of reed canarygrass as silage is similar to that of other perennial grasses. Incomplete or poor quality fermentation often occurs at DM concentrations less than 20%.

Sewage Effluent Renovation. Reed canarygrass is unsurpassed among cool-season grasses for land utilization of N and other nutrients that occur in municipal and industrial waste effluents (Marten 1985; Quin 1979). In New Zealand, the drainage water under effluent-treated land planted to reed canarygrass maintained much lower nitrate-N (NO_3) levels than that planted to ryegrass-clover. Further, the rhizome system of reed canarygrass continued to extract N even when temperatures were too cold to promote herbage growth (Quin 1979).

Reed canarygrass removes more N from soil than does corn, *Zea mays* L., or other cool-season grasses when treated with large amounts of wastewater effluent (Marten 1985). Corn can be successfully grown in reed canarygrass sods that are temporarily suppressed by herbicides in effluent renovation systems.

SOIL CONSERVATION. Reed canarygrass is frequently used for gully control and for the maintenance of grassed waterways, riparian strips, stream channel banks, and edges of farm ponds because its vigorous, spreading growth prevents soil erosion. For control and healing of gullies, small pieces of sod can be embedded at intervals of 30 to 60 cm in gullies when the soil is moist, either in early spring or late summer. Great volumes of water will pass over properly embedded sod pieces without removing them. The shoots will penetrate up to 20 cm of sediment.

Where large gullies are filled and shaped into waterways, sod pieces can be planted with a manure spreader in the center of the waterway at a rate of 35 to 45 mt ha^{-1}. They should be disked in, followed by firming with a roller. Grass seed should be distributed over the entire waterway at the same time that sod pieces are planted.

Reed canarygrass will also produce roots and shoots from the nodes of freshly cut, well-jointed reproductive culms that have been covered with 3 to 5 cm of moist soil. Mature stems cut at post-anthesis to seed shattering produce more vigorous plants than less mature stems (Casler and Hovin 1980). In very wet, muddy conditions this method may be superior to sod planting. Culms of reproductive growth should be used from reed canarygrass stands that are at least 3 yr old.

Reed canarygrass should not be used in drainage ditches or along slow-running shallow streams because silting may occur. It is a problem weed along irrigation banks and ditches.

CULTIVARS

Few cultivars of reed canarygrass have been developed by plant breeders and even fewer are available in the US and Canada as commercial seed. Among the earlier cultivars were 'Superior', developed in Oregon for use on upland sites; 'Ioreed', released in Iowa in 1946; 'Auburn', released in Alabama in 1952; and 'Frontier', released by the Canadian Department of Agriculture in 1959. Rise, selected for improved disease resistance and high seed retention, was developed by the Rudy-Patrick Seed Company in Iowa. It was also licensed in Canada in 1971. 'Vantage' was developed at Iowa State University and released in 1972. It has better seed retention and heads 2 to 3 d earlier than Rise in Iowa.

The Minnesota Agricultural Experiment Station and the USDA released the low-alkaloid (gramine-type) germplasm population, MN-76, in 1983 (Hovin and Marten 1983). Venture and Palaton were released and marketed in 1985. They are the first commercial cultivars with enhanced palatability, low gramine, and no tryptamine and carboline alkaloids. Another low-alkaloid cultivar, 'Rival', was released in 1985 by the University of Manitoba in Canada. However, Rival has higher concentrations of hordenine and gramine alkaloids than Venture or Palaton (Wittenberg et al. 1991). Variability in agronomic and forage quality characteristics among reed canarygrass genotypes suggests great opportunity for the development of improved cultivars (Marten 1985; Jones et al. 1989).

CULTURE AND MANAGEMENT

Seeding. Establishing a good stand can be a problem in growing reed canarygrass, as the seed may germinate slowly and irregularly (Marten 1985). Low seed germination may be caused by dormancy; immaturity; overheating during drying; and susceptibility of broken, aged, and dehulled seeds to molds after planting. A seeding rate of 6 to 10 kg ha^{-1} (pure live seed) is usually adequate on a firm seedbed.

Either spring or late summer seedings are satisfactory, if sufficient moisture is available. Plants must have at least five leaves before onset of winter because reed canarygrass seedlings are more susceptible to killing by cold temperatures than are those of most other cool-season grasses. Very late fall seeding may be required on poorly drained areas, and it can be successful if seed does not germinate until spring. Seed and seedlings can withstand 35 d or more of flooding (McKenzie 1951).

Mixtures with Other Species. Reed canarygrass is well adapted to mixtures with legumes when harvested at vegetative to early-flowering stages. However, if allowed to provide excessive shade through infrequent cutting, reed canarygrass can eliminate legumes from mixtures. In contrast, if reed canarygrass is subject to excessive shade by vigorous weeds or legumes, it may fail to establish in mixtures.

Reed canarygrass usually is not a dominant component of mixtures with alfalfa but once established persists well. Reed canarygrass generally comprises less than 50% of a mixture; however, long-term mixture composition is maintained more consistently under diverse cutting managements and locations with reed canarygrass than with orchardgrass and smooth bromegrass (Sheaffer et al. 1990a; Casler and Walgenbach 1990). Persistence and yield of reed canarygrass in mixtures with alfalfa have been enhanced by using alternate row-seeding patterns instead of broadcast and mixed within-row seedings (Sheaffer and Marten 1992).

For poorly drained soils, either alsike, *Trifolium hybridum* L., or ladino clover, *T. repens* L., or birdsfoot trefoil mixes well with reed canarygrass. On upland soils, reed canarygrass was less competitive with birdsfoot trefoil than were many other cool-season grasses (Sheaffer et al. 1984). When reed canarygrass was mixed with either orchardgrass or tall fescue, yields of the mixtures never exceeded those of reed canarygrass in monoculture; reed canarygrass became the dominant species in these mixtures (Sheaffer et al. 1981).

Fertilization. On upland soil, reed canarygrass becomes "sod-bound" and unproductive within several years unless it is well fertilized, especially with N. Increased yields may occur with added phosphorus (P) and potassium (K) when soils are deficient in these elements or when high N rates are applied (Kroth et al. 1976; Allison et al. 1992). More than 200 kg ha^{-1} yr^{-1} of N may be required to obtain a maximum reed canarygrass yield (Malzer and Schoper 1984; Allison et al. 1992). Split applications of N during the growing season result in more uniform production and lengthen the productive period of reed canarygrass. In West Virginia, fertilization of reed canarygrass with N at rates up to 383 kg ha^{-1} was economically profitable at N and forage prices of \$0.44 and \$0.60 kg^{-1}, respectively (Colyer et al. 1977).

Cutting. Reed canarygrass in Minnesota is less susceptible to stand depletion under cutting frequencies designed to produce high forage yield or high-quality forage (two to four cuts per year, respectively) than are smooth bromegrass, orchardgrass, tall fescue, or timothy (Marten 1985; Sheaffer et al. 1990a, 1990b). Under good growing conditions, a stubble height of 4 cm is sufficient to achieve good stand persistence and yield (Smith et al. 1973). Cutting at greater heights (8 cm or greater) may be advantageous in stressful environments or with frequent cutting (Horrocks and Washko 1971). Because all regrowths of reed canarygrass have culms, frequent close cutting will eventually reduce stands and yields. This is probably because the terminal meristem is removed, leaf area for photosynthesis is reduced, and levels of carbohydrate reserves in rhizomes and stem bases are insufficient to support rapid regrowth.

SEED PRODUCTION

Reed canarygrass seed matures from the top of the panicle downward, and early-maturing seed shatters before the bulk of the crop is ready for harvest. Shattering is a two-stage process. First, the rachilla becomes disjointed about 12 d after flowering, and then the seeds are released from the glumes. In low-shattering genotypes, the glumes retain the seed within the spikelet (Bonin and Goplen 1963). Although most newer cultivars have enhanced shattering resistance, shattering remains a problem and is a major reason for the continued high price of reed canarygrass seed. Yields of seed vary from less than 60 up to 400 kg ha^{-1}.

Commercial seed is harvested from solid stands. However, maximum seed production occurs in rows about 90 cm apart. Wide spacing promotes tillering and allows cultivation for weed control. Most reed canarygrass is combined following swathing. Swathing greatly reduces shattering potential and allows drying of the crop. About 50% of the seed should be dark gray or brown before swathing. Removing the stubble and residue after seed harvest by burning or short clipping is critical to maintain high-yielding seed production.

For Bonin and Goplen (1966), heritability estimates for seed yield in reed canarygrass were 91% for seed shattering, 74% for seed weight per panicle, and 70% for number of panicles per plant. The heritability for theoretical seed yield was 36%. They concluded that seed yield could be improved by selection for both low seed shattering and high seed weight per panicle. Seed yield was not correlated with number of panicles per plant.

Reed canarygrass may be induced to flower and set seed in greenhouses by first subjecting plants to 12 wk of cold treatment (less than 6°C) in a growth chamber having an 8-h photoperiod (Heichel et al. 1980).

DISEASES AND INSECTS

Several disease and insect pests have been identified as limiting reed canarygrass yield (see Marten 1985 for references on this topic). A fungal pathogen, *Helminthosporium giganteum* Heald & Wolf, sometimes attacks leaves of reed canarygrass. Another fungal disease, "tawny blotch" (*Stagonospora foliicola* [Bres.] Buba), occurs on reed canarygrass, especially when sewage effluent is applied on mature grass. Disease severity could be reduced by cutting three times instead of twice per year. Genetic diversity in susceptibility of reed canarygrass plants to tawny blotch and to *H. cateranium* Drechs. indicates good prospects for developing cultivars resistant to both pathogens. A *Septoria* disease of reed canarygrass in Pennsylvania has been identified, and a reed canarygrass germplasm source that contained resistance to *Septoria* leaf disease has been released.

In Connecticut reed canarygrass was infested with frit fly, *Oscinella frit* (L.), which gives the characteristic symptom of the central leaf dying ("flagging") in young shoots, while other leaves remain green. Yields of reed canarygrass were reduced 0.52 mt ha^{-1} in the second harvest due to frit fly infestation, and fly control using insecticides increased regrowth yields by as much as 32%. However, in a nationwide survey in 1965, frit fly was seldom found on reed canarygrass.

Reed canarygrass irrigated with municipal sewage effluent in Pennsylvania had fewer culms killed by frit fly and fewer larvae of cereal leaf beetle, *Oulema melanopus* L., than that which was not irrigated. Armyworms, *Pseudaletia unipunctata* (Haworth), sometimes preferentially attack reed canarygrass. Within a 72-h period armyworms nearly completely devoured a 10-ha field of regrowth reed canarygrass in Minnesota; they avoided adjacent fields of corn and soybeans. Armyworms can be controlled by spraying with an insecticide as soon as the first leaf-notching (symptom of armyworms) or larvae are observed.

OTHER SPECIES

According to Anderson (1961), 15 species have been identified within the genus *Phalaris*. He gives keys, synonymies, descriptions, illustrations, and distributions of these species.

An interesting taxonomic problem has developed regarding the incorrect naming of the species commonly referred to as *P. tuberosa* var. *stenoptera* (Hack.) Hitchc. in the US and simply *P. tuberosa* L. in Australia. Its common name is *hardinggrass* in the US and *Toowoomba canarygrass* or *phalaris* in Australia. Other incorrect scientific names for the species have been *P. bulbosa* L. and *P. nodosa* Murr. These names were antedated by *P. aquatica* L., published in 1755, which is the oldest available legitimate name. This name has priority, according to the International Code of Nomenclature, and should therefore be used in all future references to this species. Hardinggrass may be a hybrid of *P. aquatica* × *P. arundinacea;* if this is true, the correct name is *P. stenoptera* Hack. (E. E. Terrell, personal communication, 1974). If hardinggrass were to be considered a variety, its correct name would be *P. aquatica* var. *stenoptera* (Hack.) Burk. It is a cool-season perennial, grown in the southwest and the southeast US for winter pasture and hay. An improved cultivar of hardinggrass, 'Oasis', was developed in Alabama (Hoveland et al. 1982).

Koleagrass is an introduction from Morocco classified as *P. tuberosa* var. *hirtiglumis* Batt. & Trab., but it was indistinguishable from introductions labeled *P. aquatica* in an Alabama study (Hoveland and Anthony 1971). The Soil Conservation Service in California increased another plant introduction from Morocco, received as *P. tuberosa* var. *stenoptera* and later reidentified as *P. tuberosa* var. *hirtiglumis,* which the SCS called 'Perla'. It resembles other cultivars of *P. aquatica* but has better seedling vigor and more winter forage production in Alabama.

Ronphagrass, *P. aquatica* L. (*P. tuberosa* L.) × *P. arundinacea* L., is a sterile hybrid between hardinggrass and reed canarygrass that was developed in South Africa. Hybrids of hardinggrass and reed canarygrass often have the high winter production of hardinggrass and the winterhardiness characteristics of reed canarygrass (Allison and Starling 1963). Hexaploid plants of the first generation backcross of the hybrid to hardinggrass were also more winter-hardy than hardinggrass and did not shatter mature seed as readily as reed canarygrass. Embryo abortion following interspecific pollination of these species was attributed to failure of the endosperm to differentiate normally. Ronphagrass may produce "staggers" in sheep similar to those caused by *P. aquatica* in Australia. Supplementation of the diet with cobalt apparently prevents this condition.

Canarygrass, *P. canariensis* L., is an annual grown for bird food. The species and name should not be confused with reed canarygrass, a perennial. Two cultivars, 'Keet' and 'Alden', have been released by the Minnesota Agricultural Experiment Station.

Littleseed canarygrass, *P. minor* Retz, is a winter annual used for pasture in Brazil, Uruguay, and Argentina. It has become a problem weed in wheat in the Punjab region of India.

A distinct variant within reed canarygrass is ribbongrass or gardener's garters, *P. arundinacea* var. *picta* (L.) Asch. & Graebn. (Anderson 1961). The blades have white longitudinal stripes, making it an attractive ornamental.

QUESTIONS

1. Describe how reed canarygrass is adapted to diverse soil and climatic conditions.
2. Compare the feeding quality of reed canarygrass with that of other forage species.
3. What is the significance of the occurrence of alkaloids in reed canarygrass? How was the problem solved?
4. Why does reed canarygrass have great potential as a forage species?
5. List and discuss the problems associated with the establishment, production, and utilization of reed canarygrass as forage.
6. Why is reed canarygrass widely used in sewage effluent renovation and in soil conservation?
7. What are the recommended practices in growing and harvesting reed canarygrass seed?
8. Why is there relatively little concern about reed canarygrass diseases and insects?

REFERENCES

Allison, DC, and JL Starling. 1963. Cytogenetic studies of the BC_1 and BC_2 generations from interspecific hybrids between *Phalaris arundinacea* and *P. tuberosa*. Crop Sci. 3:154-57.

Allison, DW, K Guillard, MM Rafey, JH Grabber, and WM Dest. 1992. Response of reed canarygrass to nitrogen and potassium fertilization. J. Prod. Agric. 5:595-601.

Anderson, DE. 1961. Taxonomy and distriubtion of the genus *Phalaris*. Iowa State J. Sci. 36:1-96.

Bonin, SG, and BP Goplen. 1963. A histological study of seed shattering in reed canarygrass. Can. J. Plant Sci. 43:200-205.

———. 1966. Heritability of seed yield components and some visually evaluated characteristics in reed canarygrass. Can. J. Plant Sci. 46:51-58.

Buxton, DR. 1990. Cell-wall components in divergent germplasms of four perennial forage grass species. Crop Sci. 30:402-8.

Buxton, DR, and GC Marten. 1989. Forage quality of plant parts of perennial grasses and relationship to phenology. Crop Sci. 29:429-35.

Casler, MD, and AW Hovin. 1980. Genetics of vegetative stand establishment characteristics in reed canarygrass, clones. Crop Sci. 20:511-15.

Casler, MD, and RP Walgenbach. 1990. Ground cover potential of forage grass cultivars mixed with alfalfa at divergent locations. Crop Sci. 30:825-31.

Colyer, D, FL Alt, JA Balasko, PR Henderlong, GA Jung, and V Thang. 1977. Economic optima and price sensitivity of N fertilization for six perennial grasses. Agron. J. 69:514-17.

Evans, MW, and JE Ely. 1941. Growth habits of reed canarygrass. J. Am. Soc. Agron. 33:1018-27.

Frank, AB, JD Berdahl, and RE Barker. 1985. Morphological development and water use in clonal lines of four forage grasses. Crop Sci. 25:339-44.

Heichel, GH, AW Hovin, and KI Henjum. 1980. Seedling age and cold treatment effects on induction of panicle production in reed canarygrass. Crop Sci. 20:683-87.

Horrocks, RD, and JB Washko. 1971. Studies of tiller formation in reed canarygrass (*Phalaris arundinacea* L.) and 'Climax' timothy (*Phleum pratense* L.). Crop Sci. 11:41-45.

Hoveland, CS, and WB Anthony. 1971. Winter forage production and in vitro digestibility of some *Phalaris aquatica* introductions. Crop Sci. 11:461-63.

Hoveland, CS, RL Haaland, CD Berry, and JF Pedersen. 1982. Oasis Phalaris—A New Cool Season Perennial Grass. Ala. Agric. Exp. Stn. Circ. 259.

Hovin, AW, and GC Marten. 1983. MN-76 low-alkaloid reed canarygrass germplasm. Crop Sci. 23:1017-18.

Jones, TA, IT Carlson, and DR Buxton. 1989. Legume compatibility of reed canarygrass clones related to agronomic and other morphological traits. Crop Sci. 29:1-7.

Klebesadel, LJ, and SM Dofing. 1991. Reed Canarygrass in Alaska: Influence of Latitude-of-Adaption on Winter Survival and Forage Productivity and Observations on Seed Production. Alaska Agric. and For. Exp. Stn. Bull. 84.

Kroth, EM, L Meinke, and RF Dudley. 1976. Establishing reed canarygrass on low fertility soil. Agron. J. 68:791-94.

McKenzie, RE. 1951. The ability of forage plants to survive early spring flooding. Sci. Agric. 31:358-67.

Malzer, GL, and RP Schoper. 1984. Influence of time and rate of N application on yield and crude protein of three cool season grasses grown on organic soils. Can. J. Plant Sci. 64:319-28.

Marten, GC. 1973. Alkaloids in reed canarygrass. In AG Matches (ed.), Antiquality Components of Forages. Madison, Wis.: Crop Science Society of America, 5-31.

———. 1985. Reed canarygrass. In ME Heath, RF Barnes, and DS Metcalfe (eds.), Forages: The Science of Grassland Agriculture, 4th ed. Ames: Iowa State Univ. Press, 207-16.

Marten, GC, and JD Donker. 1968. Determinants of pasture value of *Phalaris arundinacea* L. vs. *Bromus inermis* Leyss. Agron. J. 60:703-5.

Marten, GC, and RM Jordan. 1974. Significance of palatability differences among *Phalaris arundinacea* L., *Bromis inermis* Leyss., and *Dactylis glomerata* L. grazed by sheep. In VG Iglovikov and AP Movsisyants (eds.), Proc. 12th Int. Grassl. Congr., Moscow, 3:305-12.

Marten, GC, RF Barnes, AB Simons, and FJ Wooding. 1973. Alkaloids and palatability of *Phalaris arundinacea* L. grown in diverse environments. Agron. J. 65:199-201.

Marten, GC, RM Jordan, and AW Hovin. 1976. Biological significance of reed canarygrass alkaloids and associated palatability variation to grazing sheep and cattle. Agron. J. 68:909-14.

———. 1981. Improved lamb performance associated with breeding for alkaloid reduction in reed canarygrass. Crop Sci. 21:295-98.

Pitt, RE. 1990. Silage and Hay Preservation. NREAS-5. Ithaca, N.Y.: Northeast Regional Agricultural Engineering Service, Cooperative Extension.

Quin, BF. 1979. A comparison of nutrient removal by harvested reed canarygrass and ryegrass *Lolium*-clover in plots irrigated with treated sewage effluent. N.Z. J. Agric. Res. 22:291-302.

Sheaffer, CC, and GC Marten. 1986. Effect of mefluidide on cool-season grass forage yield and quality. Agron. J. 78:75-79.

———. 1992. Seeding patterns affect grass and alfalfa yield in mixtures. J. Prod. Agric. 5:328-32.

Sheaffer, CC, AW Hovin, and DL Rabas. 1981. Yield and composition of orchardgrass, tall fescue, and reed canarygrass mixtures forage production. Agron. J. 73:101-6.

Sheaffer, CC, GC Marten, and DL Rabas. 1984. Influence of grass species on composition, yield, and quality of birdsfoot trefoil mixtures. Agron. J. 76:627-32.

Sheaffer, CC, DW Miller, and GC Marten. 1990a. Grass dominance and mixture yield and quality in perennial grass-alfalfa mixtures. J. Prod. Agric. 3:480-85.

Sheaffer, CC, DL Wyse, GC Marten, and PH Westra. 1990b. The potential of quackgrass for forage production. J. Prod. Agric. 3:256-59.

Sheaffer, CC, PR Peterson, MH Hall, and JB Stordahl. 1992. Drought effect on yield and quality of perennial grasses in the north central United States. J. Prod. Agric. 5:556-61.

Smith, D, AVA Jacques, and JA Balasko. 1973. Persistence of several temperature grasses grown with alfalfa and harvested two, three, or four times annually at two stubble heights. Crop Sci. 13:553-56.

Wedin, WF, IT Carlson, and RL Vetter. 1966. Studies on nutritive value of fall-saved forage, using rumen fermentation and chemical analyses. In Proc. 10th Int. Grassl. Congr., Helsinki, Finland, 424-28

Wittenberg, KM, GW Duynisveld, and H Tosi. 1991. Cultivar evaluation of reed canarygrass. In Proc. Manit. Agri-Forum, 9-10 Jan., Winnipeg, Manitoba. Winnipeg: Manitoba Agriculture Forum, 29-36.

28 The Fescues

DAVID A. SLEPER
University of Missouri

ROBERT C. BUCKNER
Retired from Agricultural Research Service, USDA, and University of Kentucky

THE genus *Festuca* includes about 80 species that are adapted to temperate or cool zones (Willis 1973). The species vary greatly in morphology, cytology, and growth habit. The genus can be classified taxonomically into broad-leafed types (sections Bovinae and Scariosae) and fine-leafed types (section *Festuca* [Ovinae]). Many of the species are important and are widely used for forage, turf, and conservation purposes. Valuable fine-leafed species are sheep fescue, *F. ovina* L.; red fescue, *F. rubra* L.; Idaho fescue, *F. idahoensis* Elmer.; and greenleaf fescue, *F. viridula* Vasey. The important and widely used broad-leafed species are meadow fescue, *F. pratensis* Huds., and tall fescue, *F. arundinacea* Schreb.

Tall fescue was introduced from Europe into North and South America. Western Europe is the main center of variation of the tribe Festuceae (Terrell 1979). Various *Festuca* species are also found in North Africa (Borrill 1972). Meadow fescue and tall fescue were recognized as separate and distinct species by Schreber in 1771, who describes tall fescue as being more robust than meadow fescue, and therefore he designated tall fescue as *F. arundinacea*. However, confusion continued as to the difference between the two species, and tall fescue was regarded by Hitchcock (1935) as *F. elatior* var. *arundinacea* (Schreb.) Winn. During the late 1940s two tall fescue cultivars, 'Alta' and 'Kentucky-31', were becoming widely used and described once again by Hitchcock (1951) as *F. elatior* L. (meadow fescue) and *F. arundinacea* Schreb. (tall fescue).

The exact time of introduction of tall fescue into the US is unknown. Terrell (1979) reports the earliest recorded specimen is a collection in the US National Herbarium that was found in ballast of ships in Camden, New Jersey, in 1879.

Early performance tests were conducted during the late 1800s at the Utah and Kentucky Agricultural Experiment Stations and at the Bureau of Plant Industry, Washington, D.C. Results showed tall fescue to be taller, to be more drought and cold tolerant, to form denser stands, to be more competitive with weeds, and to thrive on a wider range of soils than other *Festuca* species. However, tall fescue did not attain prominence until the release of the Kentucky-31 and Alta cultivars during the early 1940s by the Kentucky and Oregon Agricultural Experiment Stations, respectively (Buckner et al. 1979a).

DISTRIBUTION AND ADAPTATION

Climatic (rainfall and temperature), edaphic (soil texture and moisture), and geographic (latitude and elevation) factors are primarily instrumental in determining the distribution of tall fescue. Generally tall fescue is well adapted to the humid, temperate areas of the

DAVID A. SLEPER is Professor of Agronomy at the University of Missouri, Columbia. He an MS degree from Iowa State University and a PhD in plant breeding and genetics from the University of Wisconsin, Madison. His research emphasizes tall fescue breeding and genetics.

ROBERT C. BUCKNER is retired Research Agronomist, ARS, USDA, and Emeritus Adjunct Professor of Agronomy, University of Kentucky. He holds an MS degree from the University of Kentucky and a PhD in plant breeding from the University of Minnesota. His research emphasized tall fescue breeding and genetics.

US and the world (Burns and Chamblee 1979).

Distribution of tall fescue into the more northern latitudes is restricted by its ability to survive cold winter temperatures; therefore, as it moves into either higher latitudes (Scandinavia and Canada) or higher altitudes (the Alpine zone of Switzerland and eastern Turkey), its occurrence is greatly restricted (Borrill et al. 1976). The response of tall fescue to warm temperatures appears to be associated mainly with soil moisture availability. When soil moisture is adequate, tall fescue remains green and continues some growth, whereas stand thinning occurs if water deficit stress becomes severe. When growing at elevations of less than 1500 m (Hafenrichter et al. 1949), its adaptability and forage production on nonirrigated land is dependent on an annual rainfall of at least 450 mm.

Tall fescue is grown from northern Florida to southern Canada and is extensively used as the grass component of mixtures in irrigated pastures in the western intermountain region of the US (Fig. 28.1). It is the major cool-season grass species in a region of the humid eastern US that is defined approximately as 32°N to 40°N latitude, and from the meridian of eastern Kansas (95°W longitude) to the eastern edge of the Piedmont area (from south of New England across the mid-Carolinas, Alabama, and Georgia). This is a transition zone between cool, temperate climates and subtropical climates, defined on the western boundary by limited rainfall. Tall fescue occupies approximately 12-14 million ha in pure and mixed stands in the US and is considered to be the predominant cool-season grass species.

Although tall fescue grows best on deep, moist soils that are heavy to medium in texture and are high in organic matter, it can grow on soils that vary from strongly acid (pH 4.7) to alkaline (pH 9.5) (Cowan 1956). It persists and helps conserve thin topsoils on droughty slopes, yet it forms dense sods and produces excellent growth on poorly drained soils where few other cool-season grasses survive. Tall fescue growth and persistence on the varied soils, except on deep, coarse, droughty sands, perhaps explain its rapid acceptance as a cool-season grass in the transition zone. The massive rooting of tall fescue is frequently cited as the factor allowing its wide adaptation and growth on the many different soil types. The ability of roots to grow into these soils decreases bulk density, improves structure, and reduces soil erosion.

Tall fescue is adapted throughout Europe and parts of northern Africa. It occurs on the Baltic coasts, throughout the Caucasus, and in western Siberia, extending into China. The species has been introduced into North and South America, Australia, Japan, New Zealand, and South and East Africa (Burns and Chamblee 1979).

PLANT DESCRIPTION

Tall fescue is a perennial tufted bunchgrass. Individual plants may or may not have short rhizomes. Genetic variation exists for rhizome morphology (Jernstedt and Bouton 1985). Tall fescues with long rhizomes have been developed that are likely to be more persistent under heat and drought stress (De Battista and Bouton 1990).

Although it is essentially a bunchgrass, if kept mowed or grazed, thick stands produce an even sod. Tall fescue has numerous shiny, ribbed, dark green leaves. It has ciliated auri-

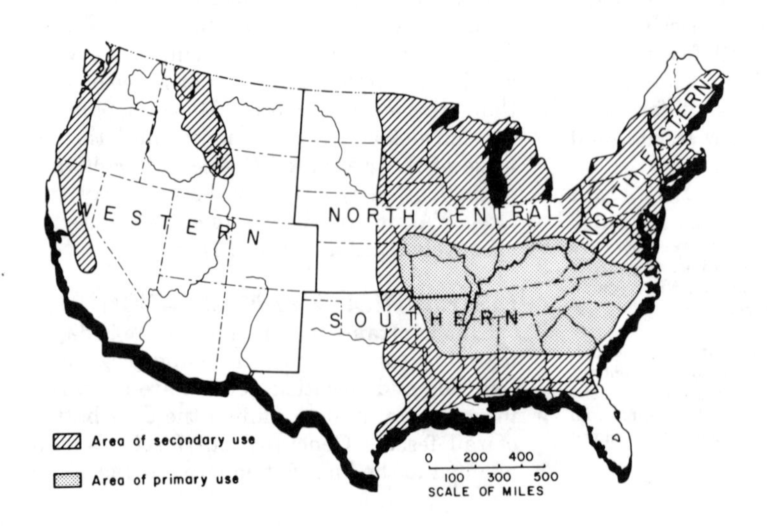

Fig. 28.1. Adaptation and use of tall fescue in the US. (Reproduced from *Tall Fescue,* ASA Monograph No. 20, 1979, pp. 9-30, by permission of ASA, CSSA, and SSSA.)

cles and collars and numerous stomata on the lower epidermis of leaves, with two to three rows of stomata adjoining the sclerenchymatous tissue (Crowder 1956).

Reproductive culms usually are erect, stout, and smooth, attaining a maximum height of 2 m (Terrell 1979). The branched, panicle-type heads are 10-35 cm long, varying from broad and loosely branched to rather narrow with short branches bearing several spikelets. Each floret has three stamens and one pistil. Five to seven seeds are produced per spikelet. Vernalization and long days are necessary for floral induction and initiation. Throughout most of the US tall fescue reaches anthesis during May to early June except at high elevations in northern areas, where it may reach anthesis in early July. It generally flowers later than orchardgrass (*Dactylis glomerata* L.) and earlier than smooth bromegrass (*Bromus inermis* Leyss.) when planted at the same location.

Tall fescue is an allohexaploid (2n = 6x = 42 chromosomes) synthesized from two progenitor species, *F. pratensis* (2n = 2x = 14) and *F. arundinacea* var. *glaucescens* Boiss. (2n = 4x = 28) (Sleper 1985) (Table 28.1). Genetic relatedness among tall fescue and its relatives has been demonstrated using a *Pst*I-genomic DNA library established from hexaploid tall fescue (Xu et al. 1991). Through restriction fragment length polymorphism, DNA probes have verified that *F. pratensis* and *F. arundinacea* var. *glaucescens* are the progenitors of tall fescue. The probes from the same tall fescue DNA library also hybridized with DNA from perennial ryegrass (*Lolium perenne* L.), which indicates a high degree of genetic relatedness between tall fescue and perennial ryegrass (Xu et al. 1992).

Numerous F_1 hybrids have been produced artificially or naturally among (1) species within *Festuca* section Bovinae; (2) section Bovinae × section *Festuca* (Ovinae), e.g., tall × red fescue; (3) section Bovinae × section Scariosae; (4) *Festuca* × *Vulpia*, and (5) *Festuca* × *Lolium* (Terrell 1979). The closest relationships are among the four cultivated species, *F. pratensis, F. arundinacea, L. perenne,* and *L. multiflorum* (Terrell 1979). Gene exchange among these taxa occurs freely, and little structural chromosomal differentiation exists.

Productivity. Tall fescue produces approximately two-thirds of its annual growth in the spring during culm elongation (reproductive growth). The vegetative regrowth often ceases during midsummer under heat and drought stress, but it resumes again in the fall. Tall fescue is renowned among cool-season grasses for active fall growth that remains green after frosts to give good quality forage for winter grazing. Highest forage yields of tall fescue are obtained by infrequent (two or three) defoliations to short stubble heights of approximately 5 cm. Infrequent defoliations may result in sparse, clumpy stands, whereas frequent defoliation stimulates tillering and the production of dense, turflike sods. Dry matter production of tall fescue varies from approximately 3 to 10 mt ha^{-1}, depending upon location and management practices (Matches 1979).

Nutritive Value. Tall fescue is a species with high nutritive value. Factors that govern nutritive value of tall fescue appear to be age of leaves, fertility of soil in which the plant is grown, season of the year, and genetic variation. Generally, nutritive value is improved if the plants are kept grazed rather closely, clipped to prevent accumulation of old leaves and stems, and properly fertilized for good growth, or the total forage is improved if grown with legumes. Nutritive value is lowest during summer and highest during fall (Bughrara et al. 1991) (Fig. 28.2). Quality during the spring is intermediate between summer and fall and is influenced by maturity, with early-maturing cultivars having lower quality on a given date than later-maturing cultivars. Crude protein for total accumulated growth of tall fescue in Missouri ranged from approximately 100 to slightly more than 200 g kg^{-1} in the spring, but due to accumulated

TABLE 28.1. Tall fescue and its progenitors

Species	Common name	Genomic designation	Chromosome number
Festuca pratensis Huds.	Meadow fescue	PP	2n = 2x = 14
Festuca arundinacea var. *glaucescens* Boiss.	None given	$G_1G_1G_2G_2$	2n = 4x = 28
Festuca arundinacea var. *genuina* Schreb.	Tall fescue	$PPG_1G_1G_2G_2$	2n = 6x = 42

material, it was reduced to approximately 80 g kg^{-1} in the late fall (Matches 1979). High forage quality of young tissue during the fall is attributed to an increase in soluble carbohydrates, while crude fiber and lignin remain at about the same level as they did at earlier stages (Brown et al. 1963).

The ability of tall fescue to support animal growth, reproduction, and health is its best measure of forage quality. Early on, tall fescue was considered to have lower quality than other cool-season grasses despite having comparable measurements of nutritive value. Many investigators compared animal performance on tall fescue pastures in pure seedings and in tall fescue-legume mixtures with that of other grasses. Average daily gains of cattle (*Bos taurus*) grazing tall fescue were superior to those of other cool-season grasses in some tests and inferior in others. Based on most recent studies, this inconsistent animal performance has been attributed to the presence or absence of the endophytic fungus *Acremonium coenophialum* Morgan-Jones & Gams (previously *Epichloë typhina* [Pers.] Tul.) (Bacon et al. 1977).

THE ENDOPHYTE AND TALL FESCUE

The biology of the *Acremonium*-tall fescue association is fairly well understood, and for further information the reader is referred to Bacon 1993; Ball et al. 1993; Clay 1993; Hoveland 1993; Siegel et al. 1987; and White and Morrow 1993. The life cycle of the endophyte follows that of tall fescue. The endophyte is not transmitted from plant to plant in the field but rather is seed transmitted. As the seed germinates, the fungal mycelia grow within the intercellular spaces into the developing shoots. During vegetative growth, the endophyte grows into the leaf sheath only and does not go into the blade. However, toxins produced in the sheath are translocated to the blade, which is consumed by ruminants. During the reproductive phase of tall fescue, the endophyte rises up the culm with the developing inflorescence and later colonizes the aleurone layer of the developing seed. When the infected seed is harvested and eventually germinates, the life cycle of the endophyte is complete.

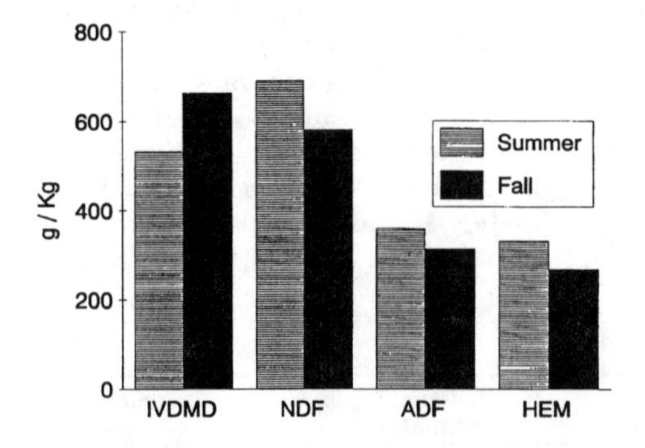

Fig. 28.2. In vitro dry matter digestibility (IVDMD), neutral detergent fiber (NDF), acid detergent fiber (ADF), and hemicellulose (HEM) of tall fescue in the summer and fall (Bughrara et al. 1991).

Cattle consuming tall fescue infected with the endophyte may succumb to three major disorders: (1) fescue foot, (2) bovine fat necrosis, (3) and fescue toxicosis, which is also called *summer syndrome* (Schmidt and Osborn 1993).

Fescue foot usually occurs in cattle grazing tall fescue in the winter. Symptoms first appear as a tenderness of the hind legs while walking, then proceed to gangrene and necrosis at the tail, ears, and feet. In extreme cases, the tail or hoof may be lost. Blood vessels in animals' extremities normally constrict in response to cold ambient temperatures to conserve body heat. Ergopeptine alkaloids present in endophyte-infected tall fescue are vasoconstrictors and thus may accentuate the response to cold and critically limit blood flow (Garner et al. 1993). Direct proof that endophyte toxins cause fescue foot has not been established (Schmidt and Osborn 1993).

Symptoms of fescue foot resemble but are not identical to those of ergot or selenium poisoning. Sporadic outbreaks of fescue foot have been reported in most southeastern states. Compared with the total number of cattle pastured on tall fescue, the incidence of acute cases of fescue foot is extremely low; however, severe economic losses to a producer may result when it occurs within a herd. Fescue foot has not been reported for livestock other than cattle. Most cases of fescue foot occur among cattle that graze pure stands of tall fescue during late fall and winter and that have received large amounts of nitrogen (N) fertilizer.

Bovine fat necrosis is associated with digestive disturbance, bloating, reduced passage of digesta, kidney dysfunction, or other physiological disorders resulting from the presence of necrotic or hard fat lesions in the abdominal cavity (Schmidt and Osborn 1993). Fat necrosis generally is found in cattle continuously stocked on tall fescue pastures in areas of the southeastern US that are fertilized annually with 13-25 mt ha^{-1} of poultry litter or with high rates of ammonium nitrate (Wilkinson et al. 1983). To reduce the potential for fat necrosis and other problems associated with high-nutrient inputs from poultry litter, tall fescue pastures should not receive more than 9 mt ha^{-1} annually. Again, bovine fat necrosis is hypothesized to be caused by the endophyte, but no conclusive evidence is available.

Fescue toxicosis of cattle grazing tall fescue is associated with the symptoms of rough hair coat, reduced rate of gain and/or milk production, rapid breathing, excess salivation, increased body temperature, depressed serum-prolactin levels, standing in shade and water, poor conception rates, and a general unthrifty condition during the warmest grazing season (Stuedemann and Hoveland 1988) (Fig. 28.3). Although this disorder is promoted by high ambient temperature and humidity, it can occur at any time of year. Cattle weight gain is reduced by 45 g d^{-1} for each 10 percentage point increase in endophyte infection (Schmidt and Osborn 1993).

Fescue toxicosis has been directly linked to an accumulation of ergopeptine alkaloids, with ergovaline being the most predominant type (Garner et al. 1993; Lyons et al. 1986). Other alkaloids such as pyrrolizidines and lysergic acid amide are also present in endophyte-infected tall fescue; however, their role in fescue toxicosis is poorly understood. Pyrrolizidines are considered to be more important as insect deterrents for the plants rather than as toxins affecting cattle (Bush et al. 1993).

Fescue toxicosis is not usually fatal; however, poor performance of the large number of cattle grazing tall fescue has a much greater cost to animal production than do the fescue foot and bovine fat necrosis problems. Indeed, around 90% of tall fescue stands in the US are highly infected with the endophyte (Shelby and Dalrymple 1987). Hoveland (1993) estimates that fescue toxicosis causes a $609 million loss annually to US livestock producers.

Fig. 28.3. An animal grazing tall fescue infected with the endophyte. Note the rough hair coat and the general unthrifty condition of the grazing ruminant. *Photo courtesy of R. E. Munson, Mo. Agric. Exp. Stn.*

IMPORTANCE AND USE

In 1940, tall fescue grew on approximately 16,000 ha in the US; by 1973 it had become the predominant cool-season perennial grass, occupying an estimated 12-14 million ha (Bucker et al. 1979a). The rapid acceptance and increase of tall fescue in the US is attributed to excellent seed production, good yields of forage, excellent persistence, adaptation to a wide range of soil conditions, compatibility with low-input management, a comparatively long grazing season, and the absence or near absence of susceptibility to serious disease and insect damage (Hanson 1979).

The principal use for tall fescue in the US is in cow-calf enterprises because of its ease of use and length of grazing season (Van Keuren and Stuedemann 1979). Calves are typically born in the spring, when tall fescue growth is abundant, and weaned in the late fall. Calf-weaning weights have often been reported to be poor because of the effects of the endophyte, but they can be improved by including legumes in mixtures with tall fescue, feeding energy or hay supplements, or stocking on warm-season grasses during the summer (Schmidt and Osborn 1993; Morrow et al. 1988; Tucker et al. 1989). Dry cows are overwintered on tall fescue pastures and supplemented with other types of hay depending upon the quality of forage available. The dense root system of tall fescue produces a sod that resists treading damage by livestock during extended periods of wet weather.

Tall fescue is used little by dairy farmers because the presence of the endophyte reduces milk production (Hemken et al. 1979). However, studies have shown that milk production of dairy animals consuming endophyte-free tall fescue is equivalent to that from consumption of other cool-season grasses (Strahan et al. 1987).

Pure stands of tall fescue fertilized with N support a longer grazing season, higher stocking rate, and greater gain per hectare than pure stands of orchardgrass. Fertilization with N may depress average daily gain slightly, but it nearly doubles carrying capacity of tall fescue pastures and greatly increases liveweight gain per hectare (Gross et al. 1966; Mott et al. 1971).

Relative to other cool-season grasses tall fescue tolerates continuous stocking, which is the predominant type of grazing management in the transition zone. In pastures, it is always advisable to seed a legume with tall fescue for N fixation and to improve animal performance (Fig. 28.4). Annual lespedeza (*Kummerowia* spp.), white (*Trifolium repens* L.) and red clover (*T. pratense* L.), birdsfoot trefoil (*Lotus corniculatus* L.), and alfalfa (*Medicago sativa* L.) have been used successfully with tall fescue. Alsike clover (*T. hybridum* L.) has been used successfully with tall fescue on some of the wetter soils. The choice of legume depends upon adaptation of the legume and the method for utilizing the forage.

A management system using tall fescue as pasture or as hay and pasture during spring and early summer maximizes the use of tall fescue when it is at its highest quality (Fig. 28.5). Cattle can be moved to other forages during midsummer, when production and quality of tall fescue are low. Pastures should be fertilized in late summer so that fall grazing can begin again in 3-4 wk until growth ceases in winter. Alternatively, the growth can be permitted to accumulate without grazing from late summer to mid-to late fall. Cattle can then graze the stockpiled forage into winter. This is sometimes referred to as *deferred grazing*. Nitrogen fertilizer increases forage yield of the stockpiled tall fescue, with growth response increasing as the length of the accumulation period increases (Collins and Balasko 1981). In zones having mild to moderate winters, such as in the southern Corn Belt and the upper South, stockpiled tall fescue provides a low-cost feed during late fall and winter when pasture growth has ceased. Tall fescue is probably the best cool-season grass available for stockpiling for winter pasture because it maintains its nutritive value longer in the transition zone than any other cool-season grass.

Like many other cool-season grasses, tall fescue produces most abundantly under conditions of high irrigation and fertility. Where managed well and grazed closely, and with a proper legume balance or high rate of fertilizer N, it provides productive pasture (Matches 1979).

Starting in the mid-1940s the USDA Soil Conservation Service encouraged the use of tall fescue because its wide adaptation and long-season growth make the crop particularly valuable for conservation. Its high persistence and density stabilize eroded and disturbed soil and help restore organic matter

Fig. 28.4. Growth during spring of a well-managed mixture of tall fescue and ladino clover. It is always advisable to seed a legume with tall fescue when used for pasture. *Ky. Agric. Exp. Stn. photo.*

Fig. 28.5. Round bales in a field of stockpiled tall fescue for use as late fall and winter feed. A management system using tall fescue as pasture, or as hay and pasture, during spring and early summer maximizes the use of tall fescue at its highest quality. *Ky. Agric. Exp. Stn. photo.*

and productivity to degraded land. The ecological value of tall fescue in controlling erosion adds substantially to its economic value in sustaining beef production in the transition zone.

Tall fescue has also become valuable throughout the transition zone for general purpose turf and low-maintenance grass cover. Its use has increased greatly as grass cover on play areas, athletic fields, airfields, parks, highway rights-of-way, and industrial sites and public institutions. Acceptance and use of tall fescue on commercial and residential lawns has been rapid during the past 10 yr (Watson et al. 1992), especially with the development of low-growing, high-tiller density types. Improved management practices and the future development of cultivars with improved texture, density, greenness under environmental stress, rhizomatous growth habit, and mowability suggest that the use of tall fescue for turf will continue to increase.

CULTIVARS

Most tall fescue pastures are currently seeded to Kentucky-31 in the transition zone because it was the major seed source available during its rapid adoption in the 1950s and 1960s. Some seed of Alta was also available, but soon seed sources of the two cultivars were indistinguishable. With the markedly increased popularity of tall fescue came demand for new cultivars with improved forage quality and adaptation to more specific environmental and management regimes (Asay et al. 1979). Currently, most new cultivars are being released from the private sector. Cultivar development in the private sector increased greatly with the passage of the Plant Variety Protection Act in 1970. Seed of improved cultivars is being produced in the Pacific Northwest, where yields are high and where seed production can be better controlled.

Emphasis is currently placed on developing endophyte-free tall fescue cultivars with over 20 cultivars available (Pedersen and Sleper 1988). In the future, breeders are likely to introduce novel endophytes into tall fescue, those that continue to improve stand persistence and resistance to pests but do not decrease animal performance. This is based on the premise that deleting livestock toxins from the tall fescue-endophyte association will not reduce plant persistence or pest resistance.

Breeding procedures that increase the frequency of desirable alleles are used to improve desirable traits for tall fescue, which is naturally cross-pollinated. Common breeding approaches are modifications of mass selection and recurrent selection. Considerable genetic variation exists for many important traits such as digestibility, neutral detergent fiber, acid detergent fiber, carbon exchange rate, leaf growth parameters, mineral contents, seed yield, forage yield, and succulence (Buckner et al. 1979b; Sleper 1985). Most of these traits have been developed by the public sector, whereas private sector breeders have concentrated mainly on forage yield and turf characteristics.

MANAGEMENT PRACTICES

Tall fescue is normally established during fall from mid-August to mid-October and during spring from March through May. Seeding rates vary from 10 to 18 kg ha^{-1} for pure seedings. Rates should be reduced to about half when seeding with a legume. Seed should be planted at depths of 6-20 mm by either broadcast or drill equipment into a clean, weed-free, firm seedbed. Seed of tall fescue germinates slower and seedling growth is slower than for several other cool-season grasses. Endophyte-infected tall fescue can be established under more adverse soil conditions than endophyte-free cultivars; however, all cultivars do best on high-fertility soils. Therefore, adequate levels of N, phosphorus (P), and potassium (K) should be available to establish new stands. Although tall fescue may survive on soils varying in pH from 4.5 to 9.5, pH values of 6.0 to 7.0 should be maintained for best soil fertility conditions for companion legumes.

Forage yield of tall fescue responds readily to high rates of N (Templeton and Taylor 1966). Matches (1979) summarized several trials and found an average response of about 30 kg forage dry matter per kg N applied. Two-thirds of the annually applied N should occur in the spring and the remaining one-third in the fall to correspond to the seasonal growth pattern of tall fescue. Lack of N fertilization may cause an unproductive, sod-bound condition. Legumes persist in properly grazed mixtures with tall fescue, but liming to pH 6.0 to 7.0 and fertilizing with P and K are necessary to maintain good vigorous growth of the companion legume.

Maintaining legumes in mixtures with tall fescue is a major problem because the grass can form a sod and is relatively competitive. When legumes disappear from the stand, they may be reestablished through renovation. Renovation involves correcting soil fertility deficiencies, making a seedbed by destroying or suppressing a portion of the grass sod, and distributing the inoculated legume seed evenly over the soil, as in an original planting. Fields may be renovated in early fall or spring. Generally spring seedings of legumes on a renovated sod are more satisfactory because droughts are encountered more frequently in early fall, which reduces or makes germination of the legume more uncertain. The legume must become well established to survive the winter. The grass should be grazed moderately after seeding during the spring to prevent undue light competition to the legume (Taylor et al. 1979). Studies have shown that seedlings of red clover are most tolerant of shade, followed by alfalfa, with birdsfoot trefoil being least tolerant (Gist and Mott 1957).

Because a high proportion of the current tall fescue fields are infected with the endophyte, there is interest in converting infected pastures to cultivars without the endophyte (DeFelice and Henning 1990). Fribourg et al. (1988) recommends that pastures should be replaced if they are infected 30% or more. Munson and Bailey (1990) have shown that infected stands can be successfully replaced with endophyte-free cultivars by treating infected pastures with a herbicide such as Gramoxone Extra® (1,1'-Dimethyl-4,4'-bipyridinium) followed by seeding a summer annual such as grain sorghum (*Sorghum bicolor* [L.] Moench), sorghum-sudan (*S. bicolor* [L.] Moench), or dwarf pearlmillet (*Pennisetum americanum* [L.] Leeke). Winter annuals such as rye (*Secale cereale* L.), wheat (*Triticum aestivum* [L.] emend. Thell.), barley (*Hordeum vulgare* L.), or annual ryegrass (*Lolium multiflorum* Lam.) may be seeded in the fall after application of a herbicide to infected tall fescue pastures. Annuals can smother infected tall fescue plants not killed by the herbicide while providing interim forage crops. More than one herbicide treatment coupled with sequential annual croppings may be required to completely eliminate infected stands before seeding with a noninfected cultivar.

Endophyte-free tall fescue is generally less

persistent than infected tall fescue in the drought- and heat-stressed environments of the Ozarks (West et al. 1988), coastal plains (Joost and Coombs 1988), and the southern Piedmont (Hill et al. 1991). In these environments, livestock producers may choose to use legumes with tall fescue to offset the endophyte effect, to graze infected tall fescue predominantly when temperatures are cool and the endophyte produces less toxins, or to use other forage species better adapted to these environments such as bermudagrass (*Cynodon dactylon* [L.] Pers.) or annual ryegrass.

Grazing Management. Because tall fescue germinates and establishes slower than some other cool-season grasses, new stands can be seriously damaged by overgrazing or grazing too soon. Animal traffic should be minimized on new stands, especially while the ground is wet. Grazing well-established stands to heights of 5-10 cm, however, promotes high production, quality, and longevity (Matches 1979). Tillering of tall fescue is stimulated by frequent defoliation (Sleper and Nelson 1989). Once established, the plants can tolerate heavy animal traffic better than other cool-season species.

Care in grazing endophyte-infected tall fescue is necessary to minimize animal health hazards, as discussed earlier. Animals should not be allowed to graze infected pastures during periods of high temperature and humidity as these conditions exacerbate the toxic effects of the endophyte. In Missouri, cows grazing infected tall fescue during summer lost 32.7 kg more weight, produced 1.7 kg less milk per day, and weaned a calf that was 24 kg lighter than those grazing endophyte-free 'Mozark' tall fescue (Peters et al. 1992). Cattle should not be allowed to graze seed heads as seeds are a potent source of toxins in infected pastures. In addition, having a good complement of legumes or other grasses in association with infected tall fescue tends to dilute the detrimental effects of the endophyte to grazing ruminants.

SEED PRODUCTION

The major portion of tall fescue seed is produced on fields managed for both forage and seed, with seed being a co-product of the overall forage program. Approximately 90% of the total US seed crop is produced in the central transition zone, mostly in Missouri, and is composed largely of uncertified seed. The remainder is produced in the Pacific Northwest, particularly the Willamette River valley in Oregon, and consists largely of certified seed of proprietary forage and turf cultivars.

Seed production in cultivated rows spaced 45-60 cm apart is common in dry regions of the Pacific Northwest. Seeding rates are 3.5-6.0 kg ha^{-1}. Herbicides are used for weed control in establishing stands. Although there are many instances of seed yields exceeding 1100 kg ha^{-1}, the average in Missouri is approximately 335 kg ha^{-1} when tall fescue is used for seed and pasture. Seed yields in Oregon are approximately 2.5 times the national average (Youngberg and Wheaton 1979). Generally, seed yields progressively decline as stands age, with a much smaller magnitude of decline in row plantings than in solid stands. Decline of seed yields as the stand ages may be reduced by planting in rows, maintenance of high soil fertility, annual herbicide applications, and fall grazing.

Use of commercial fertilizers, especially N, is essential for seed production. Phosphorus and K should be maintained in the medium range. Nitrogen is applied to solid stands of the transition zone in late fall or early winter at 90-135 kg ha^{-1}. Fertilizer rates of 56-67 kg ha^{-1} in the fall and 100-112 kg ha^{-1} in late February or March are recommended for seed production in Oregon (Youngberg and Wheaton 1979).

Tall fescue seed may be harvested either by direct combining of standing grass or by windrowing and then combining at a later date. Combining should begin when 85%-95% of the seed is mature (Youngberg and Wheaton 1979). Seed do not mature evenly, making timing of harvest difficult, but seed moisture content relates well with maximum yield of pure live seed (PLS). Maximum yield of PLS per hectare is produced by cutting the crop when seed moisture in the standing crop is 43% and allowing the seed to mature in the windrow.

Management during summer and fall is critical for development of vigorous tillers that will bear seed heads the following spring. The stubble should be clipped after seed harvest and the refuse removed to permit light penetration to stimulate development of new tillers that will produce seed in the following year (Kroth et al. 1977). Grazing pressure should be light to moderate during early fall, but it may be increased during November to remove all growth by January 15. Tillers need to be vernalized by cold temperatures, and this is generally complete by January 1.

DISEASES AND PESTS

The vast majority of recognized diseases of tall fescue are caused by fungi. Crown rust (*Puccinia coronata* Corda. var. *coronata*), cercospora leaf spot (*Cercospora apii* Fres.), and brown patch in turf (*Rhizoctonia solani* Kuehn) are the most conspicuous fungal diseases of tall fescue (Chapman 1979). Stem rust caused by *P. graminis* is widespread in the Willamette River valley in Oregon and reduces seed yields (Welty and Barker 1992). Tall fescue hosts several virus diseases, but damage is minimal.

Nematodes have been shown to reduce stands and persistence of tall fescue pastures, particularly in the southern US. The degree of damage from nematodes is inversely related to the presence of the endophyte. Kimmons et al. (1990) show that the endophyte curtailed the numbers of *Pratylenchus scribneri* Steiner and *Meloidogyne marylandi* Jepson & Golden in tall fescue. In another study (West et al. 1990), the endophyte in tall fescue reduced soil numbers of *P. scribneri, P. projectus* Jenkins, and *Tylenchorhynchus acutus* Allen.

Foliage-feeding insects have generally not been a problem with tall fescue, probably because most pastures in the US are infected with the endophyte. Numerous studies have shown that insect damage to tall fescue is considerably reduced when the endophyte is present (Funk et al. 1993).

OTHER SPECIES

Meadow Fescue. Morphological similarities of tall and meadow fescue have caused difficulty among taxonomists trying to separate the two grasses. Meadow fescue is diploid (2n = 2x = 14 chromosomes), and tall fescue is hexaploid (2n = 6x = 42). Meadow fescue generally has shorter stature and narrower leaves than tall fescue. It is a perennial species that flourishes in deep rich soils and can be grown in about the same region as timothy (*Phleum pratense* L.). It performs well when grazed, has been used to some extent for hay, and is recognized as an excellent pasture grass in western Europe. Meadow fescue has not been used extensively in the US, probably because of its high susceptibility to disease, and it offers little advantage over tall fescue. In areas where adapted, meadow fescue has been used in pasture mixtures on wetlands.

Meadow fescue does contain an endophyte identified as *Acremonium uncinatum* (Gams et al. 1990). The endophyte in meadow fescue has been shown by DNA sequence analysis to be different from that found in tall fescue (Schardl and Siegel 1993). At present, it is not known if the endophyte in meadow fescue confers similar attributes as the endophyte found in tall fescue.

Fine-leafed Fescues. Fine-leafed fescues are species within section *Festuca* (Ovinae). The two most important species are sheep fescue (*F. ovina* L.) and red fescue (*F. rubra* L.). Both species are used primarily for lawns and turf, especially in shaded sites. Leaf size, plant height, and genetic growth characters are somewhat more comparable to Kentucky bluegrass (*Poa pratensis* L.) than to tall fescue.

There is considerable interest in incorporating endophytes into fine-leafed fescues used for turf to give resistance to several insect pests (Funk et al. 1993). The *Acremonium* endophyte has been incorporated into hard fescue (*F. longifolia* Thuill), chewings fescue (*F. rubra* L. subsp. *commutata* Gaud), strong creeping red fescue (*F. rubra* L. subsp. *rubra*), and blue fescue (*F. glauca* Lam.).

Other species of importance as pasture species under range conditions in the western US are Idaho fescue (*F. idahoensis* Elmer), Arizona fescue (*F. arizonica* Vasey), and greenleaf fescue (*F. viridula* Vasey). These species generally occur as components in a range ecosystem and are rarely planted in pure stands.

QUESTIONS

1. What is considered the center of origin of tall fescue?
2. What factors determine the distribution of tall fescue?
3. Why is tall fescue so widely adapted?
4. Describe the management practices that produce the highest yields.
5. Name and describe three major physiological disorders of cattle grazing tall fescue.
6. What is the primary use of tall fescue?
7. Describe renovation procedures to convert infected tall fescue in pastures to noninfected cultivars.
8. Name three other species of *Festuca*.

REFERENCES

Asay, KH, RV Frakes, and RC Buckner. 1979. Breeding and cultivars. In RC Buckner and LP Bush (eds.), Tall Fescue, Am. Soc. Agron. Monogr. 20. Madison, Wis., 111-39.

Bacon, CW. 1993. Abiotic stress tolerances (moisture, nutrients) and photosynthesis in endophyte-infected tall fescue. Agric. Ecosyst. Environ. 44:123-41.

Bacon, CW, JK Porter, JD Robbins, and ES Luttrell. 1977. *Epichloe typhina* from tall fescue grasses. Appl. Environ. Microbiol. 35:576-81.

Ball, DM, JF Pedersen, and GD Lacefield. 1993. The tall fescue endophyte. Am. Sci. 81:370-79.

Borrill, M. 1972. Studies in *Festuca*. III. The contribution of *F. scariosa* to the evolution of polyploids in sections Bovinae and Scariosae. New Phytol. 71:523-32.

Borrill, M, BF Tyler, and WG Morgan. 1976. Studies in *Festuca*. 7. Chromosome atlas (pt. 2). An appraisal of chromosome race distribution ecology, including *F pratensis* var *apennina* (De Not) Hack,-tetraploid. Cytol. 41:219-36.

Brown, RH, RE Blaser, and JP Fontenot. 1963. Digestibility of fall grown Kentucky 31 fescue. Agron. J. 55:321-24.

Buckner, RC, JB Powell, and RV Frakes. 1979a. Historical development. In RC Buckner and LP Bush (eds.), Tall Fescue, Am. Soc. Agron. Monogr. 20. Madison, Wis., 1-8.

Buckner, RC, LP Bush, and PB Burrus II. 1979b. Succulence as a selection criterion for improved forage quality in *Lolium-Festuca* hybrids. Crop Sci. 19:93-96.

Bughrara, SS, DA Sleper, and GF Krause. 1991. Genetic variation in tall fescue digestibility estimated using a prepared cellulase solution. Crop Sci. 31:883-89.

Burns, JC, and DS Chamblee. 1979. Adaptation. In RC Buckner and LP Bush (eds.), Tall Fescue, Am. Soc. Agron. Monogr. 20. Madison, Wis., 9-30.

Bush, LP, FF Fannin, MR Siegel, DL Dahlman, and HR Burton. 1993. Chemistry, occurrence and biological effects of saturated pyrrolizidine alkaloids associated with endophyte-grass grasses. Agric. Ecosyst. Environ. 44:81-102.

Chapman, RA. 1979. Diseases and nematodes. In RC Buckner and LP Bush (eds.), Tall Fescue, Am. Soc. Agron. Monogr. 20. Madison, Wis., 307-18.

Clay, K. 1993. The ecology and evolution of endophytes. Agric. Ecosyst. Environ. 44:39-64.

Collins, M, and JA Balasko. 1981. Effects of N fertilization and cutting schedules on stockpiled tall fescue. I. Forage yield. Agron. J. 73:803-7.

Cowan, JR. 1956. Tall fescue. Adv. Agron. 8:283-320.

Crowder, LV. 1956. Morphological and cytological studies in tall fescue (*Festuca arundinacea*) and meadow fescue (*F. elatior* L). Bot. Gaz. 117:214-23.

De Battista, JP, and JH Bouton. 1990. Greenhouse evaluation of tall fescue genotypes for rhizome production. Crop Sci. 30:536-41.

DeFelice, MS, and JC Henning. 1990. Renovation of endophyte (*Acremonium coenophialum*) -infected tall fescue (*Festuca arundinacea*) pastures with herbicides. Weed Sci. 38:628-33.

Fribourg, HA, SR Wilkinson, and GN Rhodes, Jr. 1988. Switching from fungus-infected to fungus-free tall fescue. J. Prod. Agric. 1:122-27.

Funk, CR, RH White, and JP Breen. 1993. Importance of *Acremonium* endophytes in turf-grass breeding and management. Agric. Ecosyst. Environ. 44:215-32.

Gams, W, O Petrini, and D Schmidt. 1990. *Acremonium uncinatum,* a new endophyte in *Festuca pratensis*. Mycotaxon. 37:67-71.

Garner, GB, GE Rottinghaus, CN Cornell, and H Testereci. 1993. Chemistry of compounds associated with endophyte/grass interaction: Ergovaline- and ergopetine-related alkaloids. Agric. Ecosyst. Environ. 44:65-80.

Gist, GR, and GO Mott. 1957. Some effects of light intensity, temperature, and soil moisture on the growth of alfalfa, red clover and birdsfoot trefoil seedlings. Agron. J. 49:33-46.

Gross, HD, L Goode, WB Gilbert, and GL Ellis. 1966. Beef grazing systems in Piedmont, North Carolina. Agron. J. 58:307-10.

Hafenrichter, AL, LA Mullen, and RL Brown. 1949. Grasses and Legumes for Soil Conservation in the Pacific Northwest. USDA Misc. Publ. 678.

Hanson, AA. 1979. The future of tall fescue. In RC Buckner and LP Bush (eds.), Tall Fescue, Am. Soc. Agron. Monogr. 20. Madison, Wis., 341-44.

Hemken, RW, LS Bull, JA Boling, E Kane, LP Bush, and RC Buckner. 1979. Summer fescue toxicosis in lactating dairy cows and sheep fed experimental strains of ryegrass-tall fescue hybrids. J. Anim. Sci. 49:641-46.

Hill, NS, DP Belesky, and WC Stringer. 1991. Competitiveness of tall fescue as influenced by *Acremonium coenophialum*. Crop Sci. 31:185-90.

Hitchcock, AS. 1935. Manual of the Grasses of the United States. USDA Misc. Publ. 200.

Hitchcock, AS, rev. by A Chase. 1951. Manual of the Grasses of the United States. 2d ed., USDA Misc. Publ. 200. New York: Dover.

Hoveland, CS. 1993. Importance and economic significance of the *Acremonium* endophytes to performance of animals and grass plant. Agric. Ecosyst. Environ. 44:3-12.

Jernstedt, JA, and JH Bouton. 1985. Anatomy, morphology, and growth of tall fescue rhizomes. Crop Sci. 25:539-42.

Joost, RE, and DF Coombs. 1988. Importance of *Acremonium* presence and summer management to persistence of tall fescue. Am. Soc. Agron. Abstr. Madison, Wis., 130.

Kimmons, CA, KD Gwinn, and EC Bernard. 1990. Nematode reproduction on endophyte-infected and endophyte-free tall fescue. Plant Dis. 74:757-61.

Kroth, E, R Mattas, L Meinke, and A Matches. 1977. Maximizing production potential of tall fescue. Agron. J. 69:319-22.

Lyons, PC, RD Plattner, and CW Bacon. 1986. Occurrence of peptide and clavine ergot alkaloids in tall fescue grass. Sci. 232:487-89.

McMurphy, WE, KS Lusby, SC Smith, SH Muntz, and CA Strasia. 1990. Steer performance on tall fescue pastures. J. Prod. Agric. 3:100-102.

Matches, AG. 1979. Management. In RC Buckner and LP Bush (eds.), Tall Fescue, Am. Soc. Agron. Monogr. 20. Madison, Wis., 171-99.

Morrow, RE, JA Stricker, GB Garner, VE Jacobs, and WG Hires. 1988. Cow-calf production on tall fescue-ladino clover pastures with and without nitrogen fertilization or creep feeding: Fall calves. J. Prod. Agric. 1:145-48.

Mott, GO, CJ Kaiser, RC Peterson, R Peterson, Jr., and CL Rhykerd. 1971. Supplemental feeding of steers on *Festuca arundinacea* Schreb pastures fertilized at three levels of nitrogen. Agron. J. 63:751-54.

Munson, RE, and WC Bailey. 1990. Renovation of *Acremonium coenophialum* infected tall fescue pastures using annual forage rotations. In SS Quisenberry and RE Joost (eds.), Proc. Int. Symp. *Acremonium*-Grass Interactions. Baton Rouge: Louisiana Agricultural Experiment Station, 39-40.

Pedersen, JF, and DA Sleper. 1988. Considerations in breeding endophyte-free tall fescue forage cultivars. J. Prod. Agric. 1:127-32.

Peters, CW, KN Grigsby, CG Aldrich, JA Paterson, RJ Lipsey, MS Kerley, and GB Garner. 1992. Performance, forage utilization, and ergovaline consumption by beef cows grazing endophyte fungus-infected or endophyte fungus-free tall fescue or orchardgrass pastures. J. Anim. Sci. 70:1550-61.

Schardl, CL, and MR Siegel. 1993. Molecular genetics of *Epichloë typhina* and *Acremonium coenophialum*. Agric. Ecosyst. Environ. 44:169-85.

Schmidt, SP, and TG Osborn. 1993. Effects of endophyte-infected tall fescue on animal performance. Agric. Ecosyst. Environ. 44:233-62.

Shelby, RA, and LW Dalrymple. 1987. Incidence and distribution of the tall fescue endophyte in the United States. Plant Dis. 71:783-86.

Siegel, MR, GCM Latch, and MC Johnson. 1987. Fungal endophytes of grasses. Annu. Rev. Phytopathol. 25:293-315.

Sleper, DA. 1985. Breeding tall fescue. In J Janick (ed.), Plant Breeding Reviews, vol. 3. Westport, Conn.: AVI Publishing, 313-42.

Sleper, DA, and CJ Nelson. 1989. Productivity of selected high and low leaf area expansion *Festuca arundinacea* strains. Proc. 16th Int. Grassl. Congr., Nice, France, 379-80.

Strahan, SR, RW Hemken, JA Jackson, Jr., RC Buckner, LP Bush, and MR Siegel. 1987. Performance of lactating dairy cows fed tall fescue forage. J. Dairy Sci. 70:1228-34.

Stuedemann, JA, and CS Hoveland. 1988. Fescue endophyte: History and impact on animal agriculture. J. Prod. Agric. 1:39-44.

Taylor, TH, WF Wedin, and WC Templeton, Jr. 1979. Stand establishment and renovation of old sods for forage. In RC Buckner and LP Bush (eds.), Tall Fescue, Am. Soc. Agron. Monogr. 20. Madison, Wis., 155-70.

Templeton, WC, and TH Taylor. 1966. Yield response of a tall fescue-white clover sward to fertilization with nitrogen, phosphorus, and potassium. Agron. J. 58:319-22.

Terrell, EE. 1979. Taxonomy, morphology, and phylogeny. In RC Buckner and LP Bush (eds.), Tall Fescue, Am. Soc. Agron. Monogr. 20. Madison, Wis., 31-39.

Tucker, CA, RE Morrow, RR Gerrish, CJ Nelson, GB Garner, VE Jacobs, WG Hires, JJ Shinkle, and JR Forwood. 1989. Forage systems for beef cattle: Effect of winter supplementation and forage system on reproductive performance of cows. J. Prod. Agric. 2:217-21.

Van Keuren, RW, and JA Stuedemann. 1979. Tall fescue in forage-animal production systems for breeding and lactating animals. In RC Buckner and LP Bush (eds.), Tall Fescue, Am. Soc. Agron. Monogr. 20. Madison, Wis., 201-32.

Watson, JR, HE Daerwer, and DP Martin. 1992. The turfgrass industry. In DV Waddington, RN Carrow, and RC Shearman (eds.), Turfgrass. Madison, Wis.: American Society of Agronomy, 30-88.

Welty, RE, and RE Barker. 1992. Latent-period responses of stem rust in tall fescue incubated at four temperatures. Crop Sci. 32:589-92.

West, CP, E Izekor, DM Oosterhuis, and RT Robbins. 1988. The effect of *Acremonium coenophialum* on the growth and nematode infestation of tall fescue. Plant Soil 112:3-6.

West, CP, E Izekor, RT Robbins, R Gergerich, and T Mahmood. 1990. *Acremonium coenophialum* effects on infestations of barley yellow dwarf virus and soil-borne nematodes and insects in tall fescue. In SS Quisenberry and RE Joost (eds.), Proc. Int. Symp. *Acremonium*-Grass Interactions. Baton Rouge: Louisiana Agricultural Experiment Station, 196-98.

White, JF, Jr., and AC Morrow. 1993. Taxonomy, life cycle, reproduction, and detection of *Acremonium* endophytes. Agric. Ecosyst. Environ. 44:13-37.

Wilkinson, SR, JA Stuedemann, and DJ Williams. 1983. Animal performance on tall fescue: Fat necrosis. In Proc. Tall Fescue Toxicosis Workshop, Univ. of Georgia, Athens, 15-18.

Willis, JC, rev. by HKA Shaw. 1973. A Dictionary of Flowering Plants and Ferns. 8th ed. London: Cambridge Univ. Press.

Youngberg, H, and HN Wheaton. 1979. Seed production. In RC Buckner and LP Bush (eds.), Tall Fescue, Am. Soc. Agron. Monogr. 20. Madison, Wis., 141-53.

Xu, WW, DA Sleper, and DA Hoisington. 1991. A survey of restriction fragment length polymorphisms in tall fescue and its relatives. Genome 34:686-92.

Xu, WW, DA Sleper, and S Chao. 1992. Detection of RFLPs in perennial ryegrass, using heterologous probes from tall fescue. Crop Sci. 32:1366-70.

29 Bluegrasses, Ryegrasses, and Bentgrasses

JOHN A. BALASKO
West Virginia University

GERALD W. EVERS
Texas A&M University

ROBERT W. DUELL
Rutgers University

SEVERAL short- to medium-height cool-season grasses occur in heterogeneous mixtures in US grasslands and produce significant amounts of high-quality forage. Common among them are the bluegrasses, ryegrasses, and bentgrasses, species of the genera *Poa, Lolium,* and *Agrostis,* respectively. Bluegrasses and bentgrasses usually volunteer in pastures and meadows of their regions of adaptation, whereas ryegrasses are less persistent and are seeded more frequently.

Bluegrasses

The genus *Poa* includes more than 200 species that occur in cool, humid regions throughout the world. Apomictic reproduction within the genus has given rise to numerous closely related taxa and has made difficult the tracing of species movement around the world. *Bluegrass* seems to be a word of American origin because in England Kentucky bluegrass, *P. pratensis* L., is known as *smooth-stalked meadow grass*. The introduction and dissemination of Kentucky bluegrass in the US is speculative partly because of the multiplicity of common names given to it by colonists, farmers, and explorers (Carrier and Bort 1916). According to Carrier and Bort, it was probably introduced to the East Coast of the US by early colonists and preceded English settlers to the Ohio River valley either by natural dissemination or by introduction of seed by French settlers and missionaries from Canada.

JOHN A. BALASKO is Professor of Agronomy at West Virginia University. He received the MS degree from West Virginia University and the PhD from the University of Wisconsin. He teaches and conducts research in the areas of forage management and forage quality.

GERALD W. EVERS. *See Chapter 20.*

ROBERT W. DUELL, Professor Emeritus, Rutgers University, conducted research and taught subjects involving pastures, turf, and conservation. He received the MS degree from the University of Connecticut and the PhD from Rutgers University. He has been involved in the development of cultivars of several grass species and has served as a consultant in the US, Brazil, and Africa.

DISTRIBUTION AND ADAPTATION

Hitchcock (1951) describes the distribution and adaptation of 69 species of bluegrass in the US. A more recent list of *Poa*s occurring in the US and Canada includes 84 species and 20 infraspecies/subspecies (Soil Conservation Service 1982). Among the *Poa*s, Kentucky bluegrass is the most common and important forage and turf species. It is now ubiquitous in permanent pastures in humid, temperate regions of the US (Fig. 29.1), occurring in every state and extending into Canada. It volunteers readily from seed and dormant rhizomes in fertile soils in the temperate North and in subhumid portions of eastern prairie states. It persists poorly in humid southern latitudes and in semiarid regions of the West and Southwest.

Other bluegrasses provide significant amounts of forage in other regions of the US, but they lack the widespread distribution, abundance, and importance of Kentucky bluegrass. For example, Canada bluegrass, *P. compressa* L., occurs as a forage species in moderately acidic, shallow, and infertile soils of the US Northeast and Canada. Rough bluegrass, *P. trivialis* L., occurs over the same area as Kentucky bluegrass and usually is found in wet soils and shady areas; inland bluegrass, *P. interior* Rybd., is most common in Rocky Mountain meadows, and Canby bluegrass, *P. canbyi* (Scribn.) Piper, and big bluegrass, *P. ampla* Merr., are valuable species in western rangelands (Young et al. 1981). Texas bluegrass, *P. arachnifera* Torr., is native to southern Kansas, Oklahoma, western Arkansas, and Texas. Mutton bluegrass, *P. fendleriana* (Steud.) Vasey, is one of the few bluegrasses of forage importance in the Southwest. Annual bluegrass, *P. annua* L., is a common weed in pastures, meadows, and lawns in the humid, temperate North. It begins growth and matures earlier in spring than other bluegrasses.

PLANT DESCRIPTION

Kentucky bluegrass is a persistent, low-growing, sod-forming, winter-hardy cool-season species. Aerial shoots originate from nodes in tiller bases and from apices of rhizomes when favorable growing conditions occur. In established sods, axillary rhizome buds are usually suppressed, but branched rhizomes may result when terminal buds are injured (Etter 1951). There are usually three to four live leaves on a culm. The leaves are folded in the bud shoot, are parallel sided, and terminate in a boat-shaped tip that is a characteristic of the *Poa*. When held to the sun, two translucent lines appear on each side of the leaf midrib. The ligule of Kentucky bluegrass is short and truncate. Periods of short days and cool temperatures are required for induction of floral primordia. The induced primordia are initiated to flower by the long days and warm temperatures of early spring (Peterson and Loomis 1949). Reproductive tillers terminate in a panicle inflorescence in spring. The open panicles are pyramidal or oblong-pyramidal in shape and are usually 10-20 cm long (Fig. 29.2). Glumes at the base of spikelets are shorter than the lowermost lemma, and spikelets contain three to five florets. Lemmas have cottony pubescence at their bases. Rooting is shallow, and roots and rhizomes may be matted in mature stands. Rhizomes form mostly in late spring after inflorescences have formed; they also form in summer and early fall.

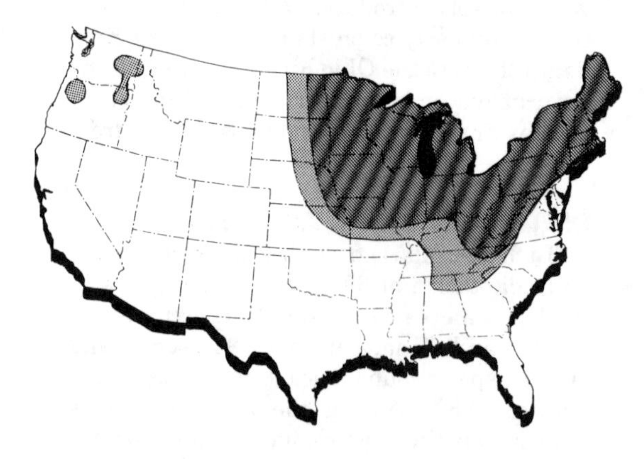

Fig. 29.1. Kentucky bluegrass is present in all states of the US, including Alaska, but is most important as a forage in the north central and northeastern regions. It is important as a seed crop in parts of Idaho, Washington, and Oregon.

Canada bluegrass is shorter than Kentucky bluegrass, has flattened culms that are decumbent at the base, and spreads by rhizomes. It has shorter dull bluish green leaves that taper to the tip, a somewhat longer ligule, and a shorter more compact panicle, 3-10 cm long.

Rough bluegrass is shade tolerant and usually has paler green, finer-textured leaves than Kentucky bluegrass. It lacks rhizomes but may spread when decumbent culms root at nodes near the base. Leaf sheaths of rough bluegrass are scabrous, which makes them rough to the touch.

Annual bluegrass lacks rhizomes and forms a bunch type of growth. Its leaves are contracted and wrinkled and terminate in rounded boat-shaped tips. Its ligule is longer than those of the perennials discussed earlier. It can flower throughout the growing season; however, it is very sensitive to hot, dry weather and usually dies during summer. Perennial types may persist for many years in intensively managed turfs and in cool, moist environments.

IMPORTANCE, USE, AND PRODUCTIVITY

Bluegrasses growing in association with

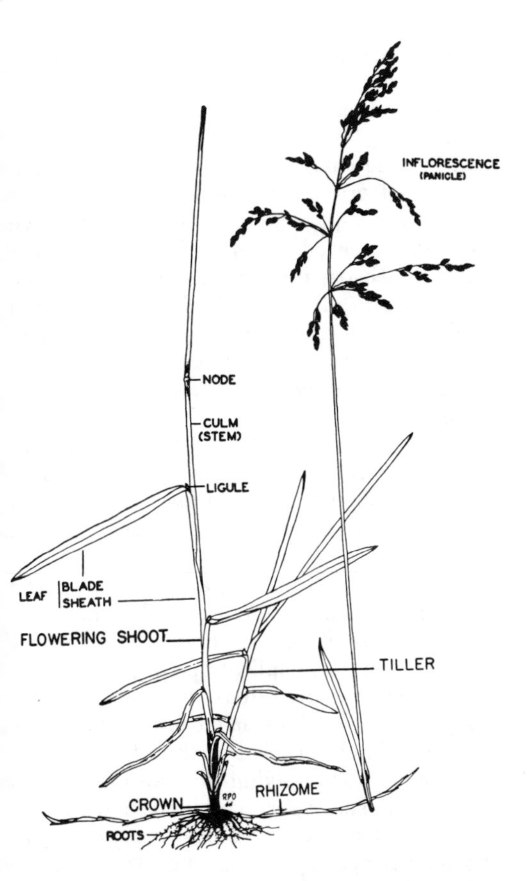

Fig. 29.2. A mature Kentucky bluegrass plant showing the three types of shoots—reproductive or flowering, tiller or vegetative, and rhizome—that originate from the crown (Etter 1951).

other cool- and warm-season grasses and legumes are most important in pastures of the northeast and north central regions of the US. Common cool-season species in these pastures include bluegrasses, orchardgrass (*Dactylis glomerata* L.), timothy (*Phleum pratense* L.), meadow fescue (*Festuca pratensis* Huds.), tall fescue (*F. arundinacea* Schreb.), redtop (*Agrostis gigantea* Roth), sweet vernalgrass (*Anthoxanthum odoratum* L.), common velvetgrass (*Holcus lanatus* L.), and white (*Trifolium repens* L.), red (*T. pratense* L.), alsike (*T. hybridum* L.), and hop (*T. agrarium* L.) clovers. Typically these pastures are bimodal in their production with the highest peak occurring in spring and the lowest in fall (Rayburn 1991). Common warm-season species in these pastures include crabgrass (*Digitaria* spp.), broomsedge (*Andropogon virginicus* L.), povertygrass (*Danthonia spicata* [L.] Beauv.), nimblewill (*Muhlenbergia schreberi* J. F. Gmel.), and purpletop (*Tridens flavus* [L.] Hitchc.) and in southern regions common (striate) lespedeza (*Kummerowia striata* [Thunb.] Schindler) and korean lespedeza (*K. stipulacea* [Maxim.] Makino). In a weedlike manner, bluegrasses also invade tall-growing warm-season prairie grasses in overgrazed rangeland of the east central Great Plains (Samson and Moser 1982; Waller and Schmidt 1983; Masters et al. 1992).

The botanical composition of pastures including bluegrasses varies over and within growing seasons and is affected primarily by environmental factors and grazing pressure. Usually, if high temperatures, low soil fertility, lack of precipitation, or prolonged overgrazing limit growth, the percentages of bluegrasses, white clover, and other desirable cool-season species decrease, and less productive species such as annual bromegrasses (*Bromus* spp.), povertygrass, and broomsedge become dominant. On fertile soils, the amount of white clover and the forage quality of bluegrass pastures is positively correlated with precipitation amounts. Incidence of bloat increases when the white clover component contributes more than one-third of the herbage mass.

Tall-growing grass and legume combinations are usually preferred as pasture because under managed systems they have higher average carrying capacity. However, Kentucky bluegrass pastures are useful as holding areas for livestock during periods when tall-growing species are under environmental stress; the former have higher percentages of their leaf area close to the soil surface and are better able to absorb the abuses of traffic and close grazing. Also, when tall-growing grasses and legumes decline in fertile pastures, as a result of abusive management and/or lack of persistence, bluegrass and associated species often volunteer, provide high-quality forage, and prevent soil erosion. Because of better tolerance to close and frequent defoliation, Kentucky bluegrass is frequently found where horses or sheep are kept under more-or-less continuous grazing.

Livestock production data from some typical bluegrass-white clover pasture experiments are shown in Table 29.1. Data and conclusions from these experiments indicate that bluegrass-white clover associations are relatively unproductive in midsummer, particularly during dry growing seasons, and that grazing days per year and animal gains per hectare are less for bluegrass than for tall-growing grass and legume mixtures. Possible exceptions occur at high elevations where temperature, rainfall, and soil fertility do not limit the competitiveness of bluegrass (Vicini et al. 1982; Burns and Harvey 1986). In the north central US, where annual precipitation is less than in the Northeast, Kentucky bluegrass is less desirable because it competes strongly with more productive grasses and legumes during cool parts of the growing season and contributes little to productivity in summer. Kentucky bluegrass has the advantage of not containing *Acremonium* endophytes (Sun and Breen 1993) that have been associated with animal toxicosis, especially from grazing tall fescue and perennial ryegrass.

CULTURE AND MANAGEMENT

Kentucky bluegrass usually volunteers in pastures; however, when required, seed is applied at the rate of 6-12 kg ha^{-1} in early fall, except in the coldest areas of adaptation, where spring seedings are recommended. Seed of compatible legumes such as white, red, and alsike clovers and birdsfoot trefoil (*Lotus corniculatus* L.) are often included in mixtures to improve summer productivity, to improve forage quality, and to provide nitrogen (N) for the grass. Grasses such as orchardgrass, timothy, and tall fescue may be included in pasture mixtures where one or more hay harvests are made each year in late spring before grazing begins. Tall-growing species typically decline and require reseeding after years of grazing (Burns and Harvey 1986; Bryan and Prigge 1990). (Renovation of unimproved pastures is discussed in Chap. 3, Vol. 2.)

In Connecticut, Kentucky bluegrass volunteered and prospered when growth of taller species was kept short by grazing, when pastures were mowed occasionally to control unpalatable weeds, and when lime and phosphorus (P) were added (Table 29.2). Potassium (K) did not affect the botanical composition of grazed pastures significantly, so data from treatments including P and P plus K were averaged.

In the study, complete fertilizer applications were required to maintain desirable botanical composition if forage was cut and removed as hay. Addition of N fertilizer in spring stimulated early growth, allowed grazing to begin 10 d earlier, and increased the pasture area occupied by Kentucky bluegrass at the expense of clovers and weeds. It is best to apply N in split applications in cool seasons when bluegrass growth is most rapid. Kentucky bluegrass-white clover pastures may be maintained indefinitely and their forage quality improved by applying lime and fertilizers according to soil test recommendations. In Indiana, lime was used to adjust soil pH to 6.5, and four annual applications of 67 kg ha^{-1} of P_2O_5 and 34 kg ha^{-1} of K_2O were applied to Kentucky bluegrass pastures. Compared with control pastures, yield of total digestible nutrients, carrying capacity, and beef production

TABLE 29.1. Comparative livestock production data from representative experiments that included blue grass-white clover pastures

West Virginia (Vicini et al. 1982)			*Adjusted 205-d calf weights*
Native bluegrass pasture			211 kg
Creep-grazing into native pasture from tall fescue			193 kg
Tall fescue pasture			182 kg
North Carolina (Burns and Harvey 1986)			*Steer average daily gain, Apr. to Oct.*
Hill land pasture			0.77 kg d^{-1}
Kenhy tall fescue (endophyte free)			0.49 kg d^{-1}
Virginia (Blaser et al. 1969)	*Steer days per hectare ($d\ ha^{-1}$)*	*Daily gain per steer ($kg\ d^{-1}$)*	*Gain per hectare ($kg\ ha^{-1}$)*
Bluegrass-white clover	638	0.55	349
Orchardgrass-ladino clover	635	0.58	368
Orchardgrass-224 kg N/ha	768	0.49	373
Ky-31 tall fescue-ladino clover	749	0.46	346
Ky-31 tall fescue-224 kg N/ha	996	0.41	411
Virginia (Allen et al. 1992)	*Calf-weaning weight*		*Cow body weight*
Fescue (endophyte free)-ladino clover	250 kg		484 kg
Bluegrass-white clover	244 kg		460 kg
Iowa (Wedin et al. 1967)	*Cow gains*	*Calf gains (% of unimproved pasture)*	*Steer gains*
Bluegrass-white clover—unimproved	100	100	100
Bluegrass-white clover + N and P	143	133	158
Bluegrass-white clover + birdsfoot trefoil	164	121	267

TABLE 29.2. Effects of fertilizer and lime on pasture botanical composition

Treatment	Clovers	Kentucky bluegrass	Bent-grasses	Weeds
		(% area occupied)		
No P	3	5	22	34
P and PK	11	17	37	19
PL and PLK	16	31	24	15
PKN and PLKN	8	52	23	7

Source: Brown et al. (1960).
Note: P = superphosphate; K = potash; L = limestone; N = soluble nitrogen.

were increased by 50%. Additional annual applications of 134 kg ha^{-1} of N increased beef production by another 39% during a 4-yr period (Mott et al. 1953).

Bluegrass pastures are often undergrazed in spring, which gives rise to an accumulation of mature, low-quality forage that is often available for grazing later when growth is slower. It is best to stock heavily early in the growing season when bluegrass is most productive and to reduce the stocking rate by moving some animals to other pastures before bluegrass is weakened and weed species invade. The ratio of bluegrass to clover in permanent pastures can be influenced by grazing management. As the clover component decreases, the pasture can be grazed more closely so the grass competes less with the clover. If the bluegrass component is too low, the pasture can be allowed to grow to a height of 20-30 cm so that the grass competes more heavily with the legume. Nitrogen fertilizer favors the grass component and may also be used to change the ratio of grass to legume (Table 29.2).

In hilly areas, grazing should begin on south-facing slopes, which warm first in spring (Bennett et al. 1972). Maintaining a stubble height of 5-10 cm in spring promotes tillering, which helps keep a dense sod. Excessive defoliation often results in shallow rooting, an open sod, invasion by weeds, and damage by insects that feed on roots and rhizomes. These effects are particularly damaging in summer when Kentucky bluegrass is less able to recover. In spite of severe mismanagement, Kentucky bluegrass is usually able to recover or regenerate and be productive with the return of good cultural practices.

CULTIVARS

Poas have been difficult to classify taxonomically because of the gradations and overlapping combinations of morphological and anatomical characters that occur within the genus. The widespread distribution and variability in bluegrasses have been attributed to the hybridization that occurs within the genus (Hartung 1946; Clausen 1961).

Kentucky bluegrass is a pseudogamous, facultative apomict and occurs as a series of aneuploids (x = 7), with frequently published dysploid chromosome numbers in the sporophytic generation ranging from 28 to ±124. Higher numbers have been reported for *Poa*, but they are uncommon. Dysploid chromosome numbers in races of Kentucky bluegrass had a distinct mode around 68 and others at 49-50 and 56. In the sporophytic generation, almost all chromosome numbers from 42 to 87 were represented (Hartung 1946). Apomictic reproduction tends to perpetuate genotypes with irregular chromosome numbers that do not conform to a regular polyploid series. Apomictic bluegrass cultivars are true breeding because they are capable of producing seed asexually without the union of male and female gametes, but aberrant plants, which are capable of sexual reproduction, occur. The latter are usually reduced in size and lack the vigor of plants originating from maternally produced apomictic seed (Jacklin et al. 1990).

Apomictic reproduction is beneficial because it provides the agronomic advantage of establishment by seed and the genetic advantage of propagating desirable genotypes by asexual reproduction. Aberrant off-type plants in certified seed production fields are rogued in order to prevent genetic shift within cultivars.

Intraspecific (Pepin and Funk 1971) and interspecific hybridizations of *Poa* are possible and have been proposed as a means of producing true-breeding hybrids that possess the desirable characteristics of both parents. For example, Nitzsche (1983) describes a procedure for hybridizing Kentucky bluegrass and fowl bluegrass, *P. palustris* L., an obligate apomict induced to reproduce sexually by X-radiation. A resulting hybrid had the rhizomatous character lacking in fowl bluegrass and was void of the cottony pubescence that occurs on Kentucky bluegrass seed and interferes with seed processing. Other interspecific hybrids between bluegrasses occur.

Most cultivars of bluegrass have been developed for use as turf. 'Merion', the first Kentucky bluegrass cultivar registered by the American Society of Agronomy, originated as a single plant selection. This cultivar has good resistance to leaf spot, a low growth habit, and improved tolerance of close mowing (Myers 1952). Of the nearly 50 bluegrass cultivars registered by the Crop Science Society of America between 1952 and 1993, 4 ('Sherman' big bluegrass, *P. ampla* Merr., and 'Park', 'Troy', and 'Ginger' Kentucky bluegrass) were described as forage or dual forage-turf-type cultivars. Two other cultivars, 'Tundra' glaucous bluegrass (*P. glacua* Vahl.) and 'Gruening' alpine bluegrass (*P. alpina* L.), were specifically developed in Alaska for erosion control and reclamation in subarctic and arctic regions. Turf cultivars described as having low maintenance requirements such as Troy, 'Arboretum', Park, 'Kenblue', and 'South Dakota Certified' may be most useful for pasture in that they are similar to common Kentucky bluegrass. Many cultivars developed for turf use require high rates of N, irrigation, close mowing, dethatching, and pest control.

SEED PRODUCTION

More than 90% of the bluegrass seed produced in the US is grown in contiguous counties of Idaho and Washington and in regions of west central and eastern Oregon (1987 Census of Agriculture 1989). About 25% of the hectarage is irrigated. An important, but smaller, amount of seed is produced in northwest Minnesota. In Oregon, most bluegrass seed is produced outside of the Willamette River valley in contrast to ryegrass seed production. Most of the seed produced in the US is certified. The US both exports and imports Kentucky bluegrass seed.

In the major seed production areas, bluegrass is established in rows, and specific management procedures have been developed for producing high yields of quality seed. Timings of N fertilizer applications, irrigation, pest control measures, and harvesting are critical to the production of maximum economic yields (Wirth and Burt 1976). Seed yields vary widely from year to year and from field to field because of weather, diseases, insects, weeds, differences in fertility and residue management, and cultivar. Seed of improved turf-type cultivars, in comparison with seed of common Kentucky bluegrass, sells for a premium because of strong buyer demand. With good management, seed yields ranging from 336 to 784 kg ha^{-1} are common. Under irrigation, yields in excess of 1340 kg ha^{-1} have been achieved.

After swathing, seed fields are harvested with self-propelled combines modified to increase rubbing of the seed. Residue left on fields following harvest is removed mechanically or burned. Residue removal reduces disease, insect, weed, and thatch problems in succeeding seed crops and increases the ratio of reproductive to vegetative culms. Burning is more effective than mechanical means of straw removal and promotes large-diameter healthy tillers that lead to good seed yields the following year (Hickey and Ensign 1983). Recent restrictions on burning have promoted research into alternative methods of residue removal. Per hectare yields of bluegrass seed are expected to decrease as burning is phased out.

DISEASES AND PESTS

Bluegrasses are susceptible to many of the same diseases as other cool-season forage grasses. Persistence in pastures is usually not affected, but reductions in forage yield and quality occur. Some common diseases of bluegrass include leaf spot and crown, root, and rhizome rots, *Drechsleria poae* (Bandys) Shoem.; leaf, stem, and stripe rusts, *Puccinia* spp.; stripe smut, *Ustilago striiformis* (Westend.) Niessl.; and powdery mildew, *Erysiphe graminis* DC. (Smiley et al. 1992). Fungicides are commonly used to control diseases in seed fields but are too costly to apply to pastures.

Grubs cause the most serious damage to bluegrass pastures. Adults of the Japanese beetle (*Popillia japonica* Neuman), May beetles (*Phyllophaga* spp.), green June beetle (*Contis nitida* L.), northern masked chafer (*Cyclocephala borealis* Arrow), and European chafer (*Rhizotrogus* [*Amphimallon*] *majalis* Razoumowsky) oviposit in thin overgrazed pastures, and their larvae feed on roots and rhizomes. Damage is most severe and recovery is poor during dry years. Other insects that may feed on top growth include a number of species of billbug, *Sphenophorus* spp.; sod webworms, which include species of several genera of the subfamily Crambinae; chinch bugs, *Blissus* spp.; and several species of cutworms and armyworms (Tashiro 1987). Control of insect damage in bluegrass pastures is best achieved through good grazing and cultural management. Renovation with legumes has also reduced grub damage.

Annual Ryegrass

Annual ryegrass, *Lolium multiflorum* Lam. (2n = 14), also called *Italian* or *English ryegrass,* is a cool-season grass that is probably indigenous to southern Europe. It is self-incompatible but will readily cross-pollinate with other *Lolium* species and with the genus *Festuca*. There are many different plant types because of natural hybridizations within the *Lolium* genus. It is generally considered an annual in the US, but some types may persist for several growing seasons in the British Isles and Europe, where it is considered a short-rotation ryegrass (Spedding and Diekmahns 1972). It is estimated that there are over 1.2 million ha of annual ryegrass in the US, with about 90% used for winter pasture in the southeastern US. The wide use of annual ryegrass is due to its attributes: ease of establishment, high-quality forage, higher yields and later maturity than small grains, good reseeding ability, and adaptability to a wide range of soil types.

DISTRIBUTION AND ADAPTATION

The largest area of annual ryegrass production is from eastern Texas and Oklahoma to the Atlantic Coast and from the Gulf Coast north to the transition zone of warm- and cool-season species (Fig. 29.3). Minor production areas are along the Pacific Coast where annual rainfall exceeds 50 cm and in the southwestern US where it is grown under irrigation. Besides the British Isles and Europe, annual ryegrass is grown in north central Mexico, Australia, New Zealand, and South America. In Canada it is grown as a summer annual (Kunelius and Narasimhalu 1983).

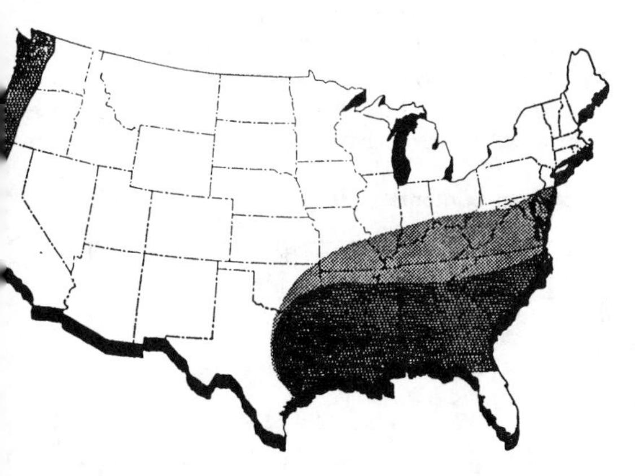

Fig. 29.3. Annual ryegrass-producing areas in the US.

Annual ryegrass is adapted to all soils from sand to clay with soil pH 5.5-8.0. Optimum growth occurs at pH 5.7 and above because of poor nutrient availability and possible aluminum (Al) toxicity at lower pHs. Annual ryegrass is better adapted to poorly drained soils than other cool-season annual grasses because of its ability to produce adventitious roots on, or near, the soil surface. However, biomass production is greatest on fertile, well-drained soils.

PLANT DESCRIPTION

Annual ryegrass normally is considered an annual, but in temperate climates some types may behave as biennials or short-lived perennials. The root system is highly branched and dense with many fibrous adventitious roots. Vegetative culms are erect at first but may become decumbent as the primary axis grows. Culms may reach a height of 100-120 cm. Leaf sheaths are glabrous and red to purple at the base of the stem. Leaf blades are rolled in the sheath and unfold flat with the adaxial side strongly ribbed and the abaxial side glossy green with a prominent midrib. Ligules are up to 4 mm long, and auricles usually are present. Culms have 4-5 nodes below a spike-type inflorescence that is about 30 cm long. An inflorescence contains about 40 sessile spikelets arranged alternately, with 10-20 fertile florets per spikelet. Slender awns up to 15 mm long are usually attached to the lemma. There are approximately 500,000 seeds/kg. Minimum daylength necessary for inflorescence development in this long-day plant varies with ryegrass type and geographical origin (Cooper 1952). Vernalization is not essential for annual ryegrass, but exposure to low temperatures will make the plant more responsive to photoperiod (McLean and Watson 1992).

IMPORTANCE AND USE

Over 90% of the annual ryegrass hectares are used for winter pasture. About 80% of this ryegrass pasture is established by overseeding into warm-season perennial grasses to extend the grazing season and to produce a high-quality forage. Annual ryegrass establishes easily, but grass sods are usually cut or grazed closely prior to seeding in order to improve seed-to-soil contact and to reduce plant competition. Annual ryegrass is also used on poorly drained soils where small grains are not adapted. A low seeding rate, 10 kg ha^{-1}, of annual ryegrass is used with small-seeded annual clovers as a carrier for planting in a small grain seedbox and to reduce the incidence of bloat from the clover. Late-maturing cultivars of annual ryegrass can be mixed with small grains to extend the grazing season. Minor uses are for hay, lawns, golf courses, roadside stabilization, and a cover crop. Annual ryegrass is frequently planted in southern states to provide forage for beef cow herds during winters when hay supplies are low (Evers 1981).

The endophytic fungus *Acremonium* has been found in some annual ryegrass cultivars (Nelson and Ward 1990). The hyphae are thinner than *A. lolii* Latch, Christensen, & Samuels, which infects perennial ryegrass (*L. perenne* L.). Although present in the seed, the endophyte was not found in leaf sheaths until spring, which indicates warm temperatures may be required for growth. No significant animal health problems have been associated with the the use of annual ryegrass as forage in the US.

Annual ryegrass produces some of the highest-quality forage that can be grown in the southeastern US (Ellis and Lippke 1976). Dry matter (DM) digestibility may be near 80% early in the season, particularly in more temperate areas. As the season advances, digestibility decreases but may still be greater than 65% for much of the grazing season. The reason for the apparently high digestibility lies in the cellular structure of the plant. Concentration of neutral detergent fiber (NDF), mostly cell walls, may range from nearly 30% of the total DM early in the season to 60% or somewhat greater late in the season (Table 29.3). Thus, concentration of neutral detergent solubles, considered to be mostly cell contents and essentially totally digestible, is high for much of the grazing season. If annual ryegrass is fertilized with N or grown with clover, its protein concentration usually exceeds that required for most classes of livestock. Applications of N increase DM production and protein concentration but do not significantly affect digestibility. Tetraploid cultivars had a slightly lower mean NDF percentage than diploid cultivars when the tetraploid cultivars were grown as autumn-seeded annuals in northeast Texas (Table 29.3). Ryegrass pastures are used for stocker cattle, replacement heifers, and lactating dairy cows and are limit grazed, i.e., grazed 2-4 h d^{-1} with fall-calving beef cows.

CULTIVARS

Most annual ryegrass cultivars released in the US are listed in Table 29.4. 'Gulf', the first cultivar released, had crown rust resistance and higher yields than common annual ryegrass (Weihing 1963b). Since then, selection has been on improved cold tolerance and higher yields. Tetraploid cultivars have wider leaves but are not necessarily higher yielding than diploids. Hybrids between annual ryegrass and perennial ryegrass are primarily grown in Europe and the state of Washington. 'Kenhy' and 'Johnstone' tall fescue resulted from hybridization of tall fescue and annual ryegrass.

CULTURE AND MANAGEMENT

Annual ryegrass will germinate well when day temperatures range from 10° to 30°C, although the rate of germination decreases as temperature decreases (Young et al. 1975; Nelson et al. 1992). Seedling density and early biomass production increase as seeding rates increase (Fig. 29.4). When overseeding warm-season perennial grasses, annual ryegrass is seeded at 23-34 kg ha^{-1} in pure stands, 22-28 kg ha^{-1} when mixed with small grains, and 11-22 kg ha^{-1} when mixed with annual clovers (Evers et al. 1992a). Higher

TABLE 29.3. Mean neutral detergent fiber (NDF) and protein contents of forage from *L. multiflorum* Lam.

Component	December	March	April	Early May	Late May
		(Mean of diploid cultivars)			
NDF (%)	33.5	37.9	46.8	59.8	58.0
Protein (%)	25.5	22.2	17.0	12.7	13.4
		(Mean of tetraploid cultivars[a])			
NDF (%)	30.4	36.2	42.2	51.7	55.7
Protein (%)	24.2	22.5	17.5	12.3	14.5

Source: Nelson and Rouquette (1981).
[a]Tetraploid cultivars studied tended to be later maturing.

TABLE 29.4. Annual ryegrass cultivars released in the US

Cultivar	Originator	Release year	Maturity	Cold tolerance	Crown rust resistance
Gulf	USDA, Texas AES	1958	intermediate	average	good
Florida Rust Resistant	Florida AES	1965	early	average	good
Magnolia	USDA, Mississippi AES	1965	intermediate	good	average
Marshall	Mississippi AES	1980	late	excellent	poor
Florida 80	Florida AES	1982	early	good	good
Rustmaster	DLF Trifolium Seed Co.	1985	early	average	good
King's Alamo[a]	Douglass King Seed Co.	1987	late	average	good
Beefbuilder[a]	Forbes Seed Co.	1988	late	average	good
Jackson	Mississippi AES	1989	intermediate	good	good
Surrey	Florida AES	1989	intermediate	good	good
TAM-90	Texas AES	1991	intermediate	good	good
Rio	Olsen-Fennell Seeds, Inc.	1991	intermediate	average	good

Sources: Personal communication L.R. Nelson, G.M. Prine, and C.E. Watson.
[a]tetraploid.

seeding rates of 56-112 kg ha^{-1} will improve early forage production but are usually economical only when grazed by lactating dairy cattle. A planting depth of 0.6-1.3 cm is best when drilling annual ryegrass into the soil. Good seed-to-soil contact is essential for good stands when broadcasting the seed on a grass sod. Pulling a drag or spike-tooth harrow over the pasture after seeding will help move seed to the soil surface. Annual ryegrass should be planted in prepared seedbeds 8 wk before the first killing frost or overseeded 4 wk before the first killing frost. Disking the perennial grass sod lightly before overseeding will permit earlier planting and improve annual ryegrass stands, which enhances early forage production (Fig. 29.4). Late plantings are susceptible to winter-killing if freezing temperatures are preceded by 2-3 wk of mild weather.

Growth rate of Gulf annual ryegrass is greatest when the mean daily temperature is 15°-20°C (Weihing 1963a). Studies on the influence of temperature and soil-water potential on leaf expansion rate in Ireland support these data (Keatinge et al. 1980). Annual ryegrass yields are highest in the Gulf Coast region, where winters are mild. Growth during the winter months is suppressed farther north; therefore, yields decrease. Rye, *Secale cereale* L., and wheat, *Triticum aestivum* L. emend. Thell., which have more cold tolerance than annual ryegrass, can be included in seeding mixtures to improve forage production during the winter months.

Annual ryegrass yields increased in response to N amounts up to 560 kg ha^{-1} without irrigation in Louisiana (Robinson et al. 1987) and to 1120 kg ha^{-1} under irrigation in California (Ehlig and Hagemann 1982). High nitrate concentrations were found in forage grown at high N rates in the latter study. Re-

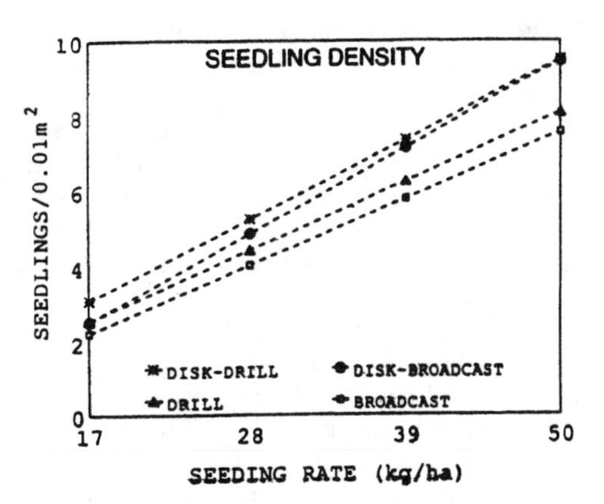

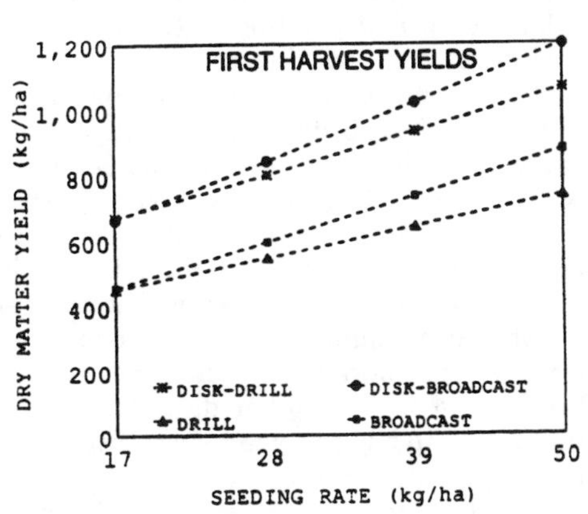

Fig. 29.4. Influence of planting method and seeding rate on overseeded annual ryegrass seedling density and first-harvest yields. (Two-year averages from Evers et al. 1992a.)

sponse to N is dependent on available soil moisture, light, and temperature (Weihing 1963a). High N assimilation rates per unit root weight have been documented by Hull and Mooney (1990). They also report a significant positive correlation between leaf N content and photosynthetic rate. Hillard et al. (1992) report the yield response of annual ryegrass to lime on an acid soil was due more to the increase in soil calcium (Ca) than to elevated soil pH. Lime increased soil P availability and increased plant tissue P, Ca, and magnesium (Mg) concentrations. Aluminum toxicity can be a problem on acid soils because it inhibits the uptake of Ca, Mg, K, and sodium (Na) (Rengel and Robinson 1989). Tolerance to soluble Al varies among annual ryegrass cultivars. Fertilizer rates for pure stands or mixtures with small grains should be based on soil test results, and N fertilizer should be split in two to three equal applications during the growing season.

The initial N fertilizer application on annual clover-ryegrass mixtures overseeded on warm-season perennial grasses should be delayed until after ryegrass and clover emergence (Evers et al. 1992b). This reduces the risk of N loss through leaching and uptake by warm-season perennial grasses, which would also increase plant competition to the emerging seedlings. The early spring N application recommended for pure stands of annual grass pastures can be eliminated on clover-ryegrass mixtures if 40%-60% of the mixture is clover.

Annual ryegrass can be grazed when it is approximately 15-20 cm tall. This plant height may be reached from late November to as late as April, depending on the interaction of planting date, seeding rate, weather, and location. Pastures should be stocked continuously at 700 kg body weight/ha in autumn and winter. Stocking rate can be increased two- to threefold during the spring flush of growth. Annual ryegrass should be defoliated no lower than 5-7 cm so that sufficient residual leaf area is left for regrowth. The most efficient use of annual ryegrass pastures by beef cows is achieved by limiting grazing to 2-3 h d^{-1} or 4 h every other day. Cattle must have access to hay or some other forage when not on ryegrass pasture.

Annual ryegrass will reseed itself if managed properly. Grazing should be terminated when inflorescences begin to appear, which is usually in April in the lower South. After ryegrass seed matures and shatters, livestock can be returned to graze the ryegrass stubble, new growth of warm-season perennial grasses, and other volunteer grasses and weeds. An after-ripening requirement and high-temperature seed dormancy limit summer germination following spring seed production (Prine et al. 1982). A light disking immediately before the time annual ryegrass would normally germinate will improve stand establishment. Grazing can continue after annual ryegrass germinates in autumn as long as excessive trampling or grazing does not damage the young seedlings. All viable ryegrass seed germinates in the autumn, so a new seed crop must be produced each year.

SEED PRODUCTION

About 95% of the commercial seed production of annual ryegrass occurs in the Willamette River valley in Oregon. In 1991, an estimated 44,000 ha of annual ryegrass were harvested for seed, with an average yield of 2000 kg ha^{-1} (Miles 1992). Nearly all seed is certified.

The proper method and time of harvest are essential for the production of the maximum yield of high-quality seed. The moisture content of the seed in the standing crop should be 35%-40%. Ryegrass is cut, windrowed, and allowed to dry to 12% moisture before threshing. The proper time to cut and windrow ryegrass is when one or two florets on the majority of the inflorescences dislodge when the inflorescences are lightly pulled between two fingers. If the ryegrass is allowed to stand until seed can be harvested from standing inflorescences, a single rain can cause shattering and loss of an entire seed crop.

DISEASES AND PESTS

Most recent cultivars of annual ryegrass are resistant to crown rust, *P. coronata* (Pers.) Cda., which was once the major disease problem (Prine 1991). Minor problems are caused by leaf spot, *Drechslería* spp., barley yellow dwarf virus, and blast, *Piriculvaria grisea* (Cke.) Sacc. Diseases in seed production are blind seed disease, *Philea temulenia* Prill. & Delacr., and ergot, *Claviceps purpurea* (Fr.) Tul., which have been controlled by postharvest burning of straw and stubble.

There are no insect problems unique to annual ryegrass. Mole crickets, *Scapteriscus* spp., and fall armyworms, *Spodoptera frugiperda* J. E. Smith, attack young seedlings of annual ryegrass as well as other cool-season annual forages in the southeastern US.

Perennial Ryegrass

Perennial ryegrass, *Lolium perenne* L. (2n = 14), is a temperate perennial grass that is indigenous to Europe, temperate Asia, and North Africa. High forage quality is the major attribute of perennial ryegrass, which makes it the choice for pasture where adapted. It is grown in North and South America, Europe, New Zealand, and Australia.

DISTRIBUTION AND ADAPTATION

Perennial ryegrass is more sensitive to temperature extremes and drought than annual ryegrass. It is the predominant grass in pastures of wet, mild-temperate climates such as those of Great Britain and New Zealand. Optimum growth occurs between 20° and 25°C (Spedding and Diekmahns 1972). Perennial ryegrass grows best on fertile, well-drained soils. The two areas of perennial ryegrass production for forage in the US are the Northeast, with about 50,000 ha, and the Pacific Coast in Oregon and Washington, with about 50,000 ha (Fig. 29.3). Even where generally well adapted in the US, it is likely to be less persistent than other perennial temperate grass species such as orchardgrass, tall fescue, timothy, smooth bromegrass (*B. inermis* Leyss.), and Kentucky bluegrass.

PLANT DESCRIPTION

Perennial ryegrass is a cool-season bunchgrass that can behave as an annual, short-lived perennial, or perennial depending on environmental conditions. The root system is highly branched and produces adventitious roots from the basal nodes of tillers. Vegetative culms can reach a length of 20 cm, and under intensive grazing the rooting of these vegetative tillers produces a stoloniferous-type growth habit (Spedding and Diekmahns 1972). Leaves are folded in the bud in contrast to annual ryegrass, where the leaves are rolled. The ligule is membranous, from 0.5 to 2.0 mm long, and may be toothed near the apex. Auricles are small and clawlike. Leaf sheaths are usually compressed but sometimes are cylindrical, glabrous, and red to purple at the base. Leaf blades have a prominently ridged upper or adaxial surface and a shiny glabrous lower or abaxial surface. Leaf blades are approximately 4 mm wide and 180 mm long, increasing in size from the first to the seventh leaf on a tiller. However, a tiller rarely has more than three live leaves at any one time. The spike-type inflorescence, up to 30 cm long, has 5-40 alternate, sessile, awnless spikelets positioned edgewise to the rachis. Spikelets contain 3-10 florets, with two glumes present in the terminal spikelet and with the inner glume missing in the other spikelets.

A vernalization period of cold temperatures is necessary prior to photoperiodic (>13 h) floral initiation for most cultivars. Mediterranean types do not require cold temperatures and initiate inflorescences at daylengths of 9-10 h. Perennial ryegrass produces seed once a year in late spring in the northeastern US. It is self-incompatible but will readily cross-pollinate. Natural hybrids with other *Lolium* species and intergeneric hybrids with *Festuca* have been formed, mostly to improve the forage quality of tall fescue or to improve the persistence of the perennial ryegrass. Perennial ryegrass seed is ovate to oblong with a shallow groove along its ventral surface (Spedding and Diekmahns 1972). Seed length is 5-8 mm, and width at the midpoint is 1.0-1.5 mm. Seed of tetraploid cultivars is larger than that of diploid cultivars. There are approximately 550,000 seeds/kg.

IMPORTANCE AND USE

Perennial ryegrass is considered the premier quality forage grass species throughout the world. Pysher and Fales (1992) found that perennial ryegrass had a higher in vitro dry matter digestibility (IVDMD) than other temperate perennial grass species. Tetraploid cultivars were slightly higher in IVDMD than diploid cultivars. Total nonstructural carbohydrate (TNC) concentrations were higher in 'Pennfine' perennial ryegrass, a turf-type ryegrass, than in seven other cool-season perennial grass species grown in Pennsylvania (Jung et al. 1976). Because of its high quality, the primary use of perennial forage-type ryegrass in the US is for lactating dairy cows on pasture. It is the principal feed source for dairy cows in New Zealand and is important for dairy production in Great Britain and Europe. Perennial ryegrass is suitable for all classes of livestock, especially those with high nutrient requirements such as young growing animals.

Perennial ryegrass can be harvested for hay

or silage. There are more hectares of perennial ryegrass planted for turf in the US than for pasture because ryegrass has good seedling vigor. Presently, lack of persistence is the main limitation to expanded use of perennial ryegrass for pasture in the US.

CULTIVARS

Because of self-incompatibility and crosses with other *Lolium* species and *Festuca*, there are many ryegrass types. Persistence of perennial × annual hybrids is generally intermediate between annual and perennial cultivars, and these crosses have been referred to as short-rotation ryegrasses. There are diploid and tetraploid forage-type cultivars. Tetraploids have fewer, but larger, tillers with wider leaves, which results in more open-type sods, and tetraploids are generally less persistent than diploids. Both the seed and seedlings of tetraploids may be larger, but growth rate following emergence is often greater for diploids. The tetraploid cultivars 'Barvestra', 'Reveille', 'Bastion', 'Citadel', and 'Taptoe' and the annual × perennial hybrid 'Bison' have performed well in the northeastern US. Diploids are used for turf because of their dense sods.

CULTURE AND MANAGEMENT

In Pennsylvania, perennial ryegrass can be seeded in April through May or in early August at a depth of 0.6-1.3 cm (Hall 1992). Recommended seeding rates on a clean, well-prepared seedbed are 16-20 kg ha^{-1} in pure stands and 5-9 kg ha^{-1} in mixtures with legumes. When seeding with a no-till drill or surface seeding, the existing sod should be mowed or grazed short to reduce competition. Annual digestible dry matter production was maximized when alfalfa (*Medicago sativa* L.) and perennial ryegrass were seeded in a mixture at 15 and 11 kg ha^{-1}, respectively (Jung et al. 1991).

Perennial ryegrass is more compatible with alfalfa than are other cool-season perennial grass species (Jung et al. 1982). Perennial ryegrass-alfalfa mixtures were found to produce more protein and digestible dry matter per ha than orchardgrass-alfalfa mixtures. Animal preference was perennial ryegrass > alfalfa > orchardgrass. After 2 yr of grazing, ryegrass pastures contained 53% more alfalfa plants than did orchardgrass pastures.

New stands of perennial ryegrass should be well established and 25-30 cm tall before grazing or harvesting for hay (Hall 1992). Established stands can be grazed when 5-7 cm tall but should not be grazed shorter than 4 cm. Under mechanical defoliation, perennial ryegrass yields were higher when cutting height was increased from 4 to 7 cm (Motazedian and Sharrow 1986). Yield and persistence of perennial ryegrass are better under rotational than continuous stocking. Ryegrass should be 20-25 cm tall before grazing to a 5-cm stubble. Following grazing, pastures should be rested for a minimum of 2 wk. When cut for hay, perennial ryegrass requires a longer drying time than other grasses because it has a higher moisture percentage.

Perennial ryegrass will grow on soils with pHs of 5-8, but forage production is best at pHs of 6-7. Fertilization should be based on a soil test. Perennial ryegrass is very responsive to N, but economical levels probably do not exceed annual applications of 170 kg ha^{-1} yr^{-1} (Hall 1992). Applications of N should be split so that no single application exceeds 60 kg ha^{-1}. If perennial ryegrass is planted in a mixture with a legume, N fertilizer should be reduced or even eliminated.

SEED PRODUCTION

In 1991 an estimated 45,120 ha of perennial ryegrass were harvested for seed in the Willamette River valley of Oregon (Miles 1992). Average seed yield was 1300 kg ha^{-1}. Most seed produced is of turf-type cultivars and is certified.

DISEASES AND PESTS

There are no disease or insect problems specific to perennial ryegrass, but it is susceptible to pest problems common to all cool-season perennial grasses in the US Northeast. In New Zealand the Argentine stem weevil, *Listronotus bonariensis* Kuschel, is a major problem in perennial ryegrass. Perennial ryegrass is resistant to this weevil if infected by the fungal endophyte *A. lolii* (Popay et al. 1990). However, presence of the endophyte is linked to the occurrence of a neurological disorder in livestock known commonly as *ryegrass staggers*.

Bentgrasses

Agrostis species of some agricultural importance include redtop, *A. alba* L. = *A. gigantea* Roth. (Soil Conservation Service 1982); colonial bentgrass, *A. tenuis* Sibth.; creeping bentgrass, *A. stolonifera* L. var. *palustris* (Huds.) Farw., syn. *A. palustris* Huds.; and velvet bentgrass, *A. canina* L. The latter three species are sometimes found in closely grazed pastures in the same regions and soils as redtop, but they are more widely used as turfgrasses. Redtop is a perennial grass with leaves that are flat and pointed. Ligules are 0.6 cm long and pointed. Its highly branched panicles are reddish, hence its common name. Spikelets of redtop, like those of other bentgrasses, contain a single floret. Distribution of redtop in the US is similar to that of Kentucky bluegrass, but redtop is better adapted to acidic soils, poor clayey soils of low fertility, and poorly drained land.

Before 1940 redtop was considered second ranked to Kentucky bluegrass as a pasture grass in the US. Its use declined, being replaced largely by orchardgrass and tall fescue, and it is now considered of minor importance. It may be sown at 2.5-5.5 kg ha^{-1} on poor or wet land with annual lespedezas or alsike clover for hay or pasture. It is also used to vegetate reclaimed surface-mined land. The cultivar 'Streaker' was released recently for use in pasture, low-maintenance turf, and reclamation areas (Jacklin et al. 1989).

White et al. (1992) report a recent colonization of *A. alba* L. by the endophytic fungus *Acremonium starrii* near Pluckmin, New Jersey; however, only 1 of 32 specimens from across North America was infected. They also tested 10 specimens of *A. palustris* and 14 of *A. tenuis,* but none were infected.

QUESTIONS

1. What factors contribute to the persistence of bluegrasses in pastures?
2. Discuss the significance and use of bluegrasses, ryegrasses, and bentgrasses in pastures.
3. Why is the genus *Poa* so diverse?
4. Why is seed of the grasses of this chapter produced primarily in the Pacific Northwest?
5. How would you distinguish between annual and perennial ryegrass?
6. Discuss the grazing management of bluegrasses and ryegrasses.
7. What are the beneficial and adverse effects of removing straw and stubble from seed fields?
8. Describe the inflorescences of Kentucky bluegrass, annual and perennial ryegrass, and redtop.

REFERENCES

Allen, VG, JP Fontenot, DR Notter, and RC Hammes, Jr. 1992. Forage systems for beef production from conception to slaughter. I. Cow-calf production. J. Anim. Sci. 70:576-87.

Bennett, OL, EL Mathias, and PR Henderlong. 1972. Effects of north- and south- facing slopes on yield of Kentucky bluegrass (*Poa pratensis* L.) with variable rate and time of nitrogen application. Agron. J. 64:630-35.

Blaser, RE, HT Bryant, RC Hammes, Jr., RL Boman, JP Fontenot, and CE Polan. 1969. Managing Forages for Animal Production. Va. Polytech. Inst. Res. Div. Bull. 45.

Brown, BA, RI Munsell, and AV King. 1960. Fertilization and Renovation of Grazed, Permanent Pastures. Storrs Agric. Exp. Stn. Bull. 350.

Bryan, WB, and EC Prigge. 1990. Effects of stocking rate and overseeding with red clover on productivity of native pasture continuously grazed by yearling steers. J. Agron. and Crop Sci. 165:273-80.

Burns, JC, and RW Harvey. 1986. Steer performance and productivity of 'Kenhy' fescue and persistence of fescues in the central Appalachians. Agron. J. 78:283-87.

Carrier, L, and KS Bort. 1916. The history of Kentucky bluegrass and white clover in the United States. J. Am. Soc. Agron. 8:256-66.

Clausen, J. 1961. Introgression facilitated by apomixis in polyploid *Poa* species. Euphytica 10:87-94.

Cooper, JP. 1952. Studies on growth and development in *Lolium*. III. Influence of season and latitude on ear emergence. J. Ecol. 40:352-79.

Ehlig, CF, and RW Hagemann. 1982. Nitrogen management for irrigated annual ryegrass in southwestern United States. Agron. J. 74:820-23.

Ellis, WC, and H Lippke. 1976. Nutritional values of forages. In EC Holt (ed.), Grasses and Legumes in Texas—Development, Production, and Utilization, Tex. Agric. Exp. Stn. RM-6C, 26-66.

Etter, AG. 1951. How Kentucky bluegrass grows. Ann. Mo. Bot. Gard. 38:293-375.

Evers, GW. 1981. Substituting Ryegrass for Hay. Tex. Agric. Exp. Stn. PR-3909.

Evers, GW, LR Nelson, JL Gabrysch, and JM Moran. 1992a. Ryegrass establishment in east Texas. In Forage Research in Texas—1992, Tex. Agric. Exp. Stn. PR-5029, 41-42.

Evers, GW, VA Haby, JL Gabrysch, JM Moran, JV Davis, and AT Leonard. 1992b. Nitrogen fertilization of arrowleaf-ryegrass mixtures. In Field Day Report—1992 Overton, Tex. Agric. Exp. Stn. Tech. Rep. 92-1, 59-60.

Hall, MH. 1992. Ryegrass. Penn. State Univ. Agron. Facts 19.

Hartung, ME. 1946. Chromosome numbers in *Poa, Agropyron,* and *Elymus*. Am. J. Bot. 33:516-31.
Hickey, VG, and RD Ensign. 1983. Kentucky bluegrass seed production characteristics as affected by residue management. Agron. J. 75:107-10.
Hillard, JB, VA Haby, and FM Hons. 1992. Annual ryegrass response to limestone and phosphorus on an ultisol. J. Plant Nutr. 15:1253-68.
Hitchcock, AS, rev. by A Chase. 1951. Manual of the Grasses of the United States. 2d ed., USDA Misc. Publ. 200. New York: Dover.
Hull, JC, and HA Mooney. 1990. Effects of nitrogen on photosynthesis and growth rates of four California annual grasses. Acta Ecol. 11:453-68.
Jacklin, AW, AD Brede, and RH Hurley. 1989. Registration of 'Streaker' redtop. Crop Sci. 29:1089.
Jacklin, AW, AD Brede, LA Brilman, RW Duell, and CR Funk. 1990. Registration of 'Freedom' Kentucky bluegrass. Crop Sci. 30:1356-57.
Jung, GA, RE Kocher, CF Gross, CC Berg, and OL Bennett. 1976. Nonstructural carbohydrate in the spring herbage of temperate grasses. Crop Sci. 16:353-59.
Jung, GA, LL Wilson, PJ LeVan, RE Kocher, and RF Todd. 1982. Herbage and beef production from ryegrass-alfalfa and orchardgrass-alfalfa pastures. Agron. J. 74:937-42.
Jung, GA, JA Shaffer, and JL Rosenberger. 1991. Sward dynamics and herbage nutritional value of alfalfa-ryegrass mixtures. Agron. J. 83:786-94.
Keatinge, JDH, MK Garrett, and RH Stewart. 1980. Response of perennial and Italian ryegrass cultivars to temperature and soil water potential. J. Agric. Sci. 94:171-76.
Kunelius, HT, and P Narasimhalu. 1983. Yields and quality of Italian and Westerwolds ryegrasses, red clover, alfalfa, birdsfoot trefoil, and persian clover grown in monocultures and ryegrass-legume mixtures. Can. J. Plant Sci. 63:437-42.
McLean, SD, and CE Watson, Jr. 1992. Divergent selection for anthesis date in annual ryegrass. Crop Sci. 32:847-51.
Masters, RA, KP Vogel, and RB Mitchell. 1992. Response of central plains tallgrass prairies to fire, fertilizer and atrazine. J. Range Manage. 45:291-95.
Miles, SD. 1992. Oregon County and State Agricultural Estimates. Oreg. State Univ. Spec. Rep. 790.
Motazedian, I, and SH Sharrow. 1986. Defoliation effects on forage dry matter production of a perennial ryegrass-subclover pasture. Agron. J. 78:581-84.
Mott, GO, RE Smith, WM McVey, and WM Beeson. 1953. Grazing Trials with Beef Cattle on Permanent Pastures at Miller-Purdue Memorial Farm. Purdue Agric. Exp. Stn. Bull. 581.
Myers, WM. 1952. Registration of varieties and strains of bluegrass (*Poa* spp.). Agron. J. 44:155.
Nelson, LR, and FM Rouquette, Jr. 1983. Neutral detergent fiber and protein levels in diploid and tetraploid ryegrass forage. In JA Smith and VW Hays (eds.), Proc. 14th Int. Grassl. Congr., Lexington, Ky. Boulder, Colo.: Westview, 234-37.
Nelson, LR, and SL Ward. 1990. Presence of fungal endophyte in annual ryegrass. In SS Quisenberry and RE Joost (eds.), Proc. Int. Symp. *Acremonium*-Grass Interactions. Baton Rouge: Louisiana Agricultural Experiment Station, 41-43.
Nelson, LR, SL Ward, and GW Evers. 1992. Germination response of annual ryegrass lines to alternating temperatures. In Am. Soc. Agron. Abstr. Madison, Wis., 163.
1987 Census of Agriculture. 1989. Part 51: United States Summary and State Data. US Department of Commerce, Bureau of the Census. Washington, D.C.
Nitzsche, W. 1983. Interspecific hybrids between apomictic forms of *Poa palustris* L. × *Poa pratensis* L. In JA Smith and VW Hays (eds.), Proc. 14th Int. Grassl. Congr., Lexington, Ky. Boulder, Colo.: Westview Press, 155-57.
Pepin, GW, and CR Funk. 1971. Intraspecific hybridization as a method of breeding Kentucky bluegrass (*Poa pratensis* L.) for turf. Crop Sci. 11:445-48.
Peterson, ML, and WE Loomis. 1949. Effects of photoperiod and temperature on growth and flowering of Kentucky bluegrass. Plant Physiol. 24:31-43.
Popay, AJ, RA Prestidge, DD Rowan, and JJ Dymock. 1990. The role of *Acremonium lolii* mycotoxins in insect resistance of perennial ryegrass (*Lolium perenne*). In SS Quisenberry and RE Joost (eds.), Proc. Int. Symp. *Acremonium*-Grass Interactions. Baton Rouge: Louisiana Agricultural Experiment Station, 44-48.
Prine, GM. 1991. Evaluation of crown rust susceptibility and breeding of annual ryegrass at the University of Florida. In Proc. Soil and Crop Sci. Soc. Fla. 50:30-36.
Prine, GM, LS Dunavin, P Mislevy, KJ McVeigh, and RL Stanley, Jr. 1982. Florida 80 Ryegrass. Fla. Agric. Exp. Stn. Circ. S-291.
Pysher, D, and S Fales. 1992. Production and quality of selected cool-season grasses under intensive rotational grazing by dairy cattle. In Proc. Am. Forage and Grassl. Conf., Grand Rapids, Mich., 32-36.
Rayburn, EB. 1991. Forage Quality of Intensive Rotationally Grazed Pastures, 1988-1990. Anim. Sci. Memo 151. Ithaca, N.Y.: Cornell Univ.
Rengel, Z, and DL Robinson. 1989. Aluminum effects on growth and macronutrient uptake by annual ryegrass. Agron. J. 81:208-15.
Robinson, DL, KG Wheat, and NL Hubbert. 1987. Nitrogen Fertilization Influences on Gulf Ryegrass Yields, Quality, and Nitrogen Recovery from Olivier Silt Loam Soil. La. Agric. Exp. Stn. Bull. 784.
Samson, JF, and LE Moser. 1982. Sod-seeding perennial grasses into eastern Nebraska pastures. Agron. J. 74:1055-60.
Smiley, RS, PH Dernoeden, and BB Clarke. 1992. Compendium of Turfgrass Diseases. 2d ed. St. Paul, Minn.: American Phytopatholoy Society Press.
Soil Conservation Service. 1982. National List of Scientific Plant Names. USDA-SCS TP-159, Ecological Science Staff.
Spedding, CRW, and EC Diekmahns (eds.). 1972. Grasses and Legumes in British Agriculture. Bull. 49, Commonwealth Bureau of Pastures and Field Crops. Farnham Royal, Bucks, England:

Commonwealth Agriculture Bureaux.

Sun, S, and JP Breen. 1993. Inhibition of *Acremonium* endophyte by Kentucky bluegrass. In DE Hume, GCM Latch, and HS Easton (eds.), Proc. 2d Int. Symp. on *Acremonium*/Grass Interactions. Palmerston North, New Zealand: AgResearch, Grassland Research Centre, 19-22.

Tashiro, H. 1987. Turfgrass Insects of the United States and Canada. Ithaca, N.Y.: Cornell Univ. Press.

Vicini, JL, EC Prigge, WB Bryan, and GA Varga. 1982. Influence of forage species and creep grazing on a cow-calf system. II. Calf production. J. Anim. Sci. 55:759-64.

Waller, SS, and DK Schmidt. 1983. Improvement of eastern Nebraska tall-grass range using atrazine or glyphosate. J. Range Manage. 36:87-90.

Wedin, WF, RL Vetter, JM Scholl, and WR Woods. 1967. An evaluation of birdsfoot trefoil (*Lotus corniculatus*) in pasture improvement. Agron. J. 59:525-28.

Weihing, RM. 1963a. Growth of ryegrass as influenced by temperature and solar radiation. Agron. J. 55:519-21.

———. 1963b. Registration of Gulf annual ryegrass. Crop Sci. 3:366.

White, JF, Jr., PM Halisky, S Sun, G Morgan-Jones, and CR Funk, Jr. 1992. Endophyte-host associations in grasses. XVI. Patterns of endophyte distribution in species of the tribe Agrostideae. Am. J. Bot. 79:472-77.

Wirth, ME, and LA Burt. 1976. Estimated Costs of Establishing and Producing Kentucky Bluegrass Seed in the Inland Pacific Northwest. Wash. State Univ., Coll. Agric. and Res. Cent. Circ. 597.

Young, JA, RA Evans, and BL Kay. 1975. Germination of Italian ryegrass seeds. Agron. J. 67:386-89.

Young, JA, RA Evans, RE Eckert, Jr., and RD Ensign. 1981. Germination- temperature profiles for Idaho and sheep fescue and canby bluegrass. Agron. J. 73:716-20.

30 Wheatgrasses and Wildryes: The Perennial Triticeae

KAY H. ASAY
Agricultural Research Service, USDA

WHEATGRASSES and wildryes are some of the most important grasses in the temperate regions of western US and Canada. These drought-resistant perennial species are excellent sources of forage and habitat for livestock and wildlife and are also valued for soil stabilization and watershed management. As members of the Triticeae tribe, they are related to and have been hybridized with cultivated cereal crops including wheat (*Triticum* spp.), barley (*Hordeum* spp.), and rye (*Secale cereale* L.). The taxonomy of the wheatgrasses and wildryes has been somewhat unsettled for several years. In North America, the wheatgrasses traditionally have been included in the genus *Agropyron,* and the wildryes have largely been treated as species in the genus *Elymus* (Hitchcock 1951; Bowden 1965). More recently, however, taxonomic realignments have been proposed that are based on genomic or biological relationships as well as plant morphology (Barkworth et al. 1983; Dewey 1983, 1984; Tzvelev 1976).

Wheatgrasses

The new taxonomic treatments (Table 30.1) limit *Agropyron* to the crested wheatgrass complex. Self-fertile, caespitose species such as slender wheatgrass and bearded wheatgrass, previously *A. trachycaulum* and *A. subsecundum,* respectively, now are included in *Elymus,* a genus based on combinations of the 'S', 'H', 'Y', and 'P' genomes. Dewey (1984) also includes thickspike wheatgrass in *Elymus* as *E. lanceolatus*. This long-anthered, cross-pollinating species was formally treated as *A. dasystachyum*. Intermediate wheatgrass and tall wheatgrass have been included in the genus *Thinopyrum* as *T. intermedium* and *T. ponticum,* respectively. These species were previously treated as *A. intermedium* and *A. elongatum* ssp. *ruthenicum*. The genus *Thinopyrum* is made up of species with the 'J-E' genome or, as some have designated, the 'J' and 'E' genomes. Bluebunch wheatgrass, traditionally treated as *A. spicatum,* now is included with other species containing the 'S' genome in the genus *Pseudoroegneria* as *P. spicata*. Western wheatgrass, formally treated as *A. smithii* now is *Pascopyrum smithii*. This octoploid species, which comprises the 'S', 'H', 'N', and an unknown genome, is the only member of the genus *Pascopyrum*. Because Dewey's taxonomic treatment is more consistent with biological relationships and is in general agreement with that of Eurasian botanists working in the native habitat of

KAY H. ASAY is Research Geneticist with the ARS, USDA, at Utah State University. He received the MS degree from the University of Wyoming and the PhD from Iowa State University. His major research interests include breeding improved grasses and legumes for semiarid range, and his research includes genetics and sterility trends in interspecific hybrids.

many of these grasses, this system will be applied in this chapter. The term *wheatgrass* will refer to all species in the traditional *Agropyron*, even though most of them now are in *Elymus, Elytrigia*, or *Thinopyrum*.

DISTRIBUTION AND ADAPTATION

Depending on the taxonomic authority, from 100 to 150 wheatgrass species have been described. More than two-thirds of these are native to Eurasia. From 22 to 30 species are considered to be native to North America (Rogler 1973; Cronquist et al. 1977). A few species are found in South America, New Zealand, and Africa. Most wheatgrasses are adapted to steppe or desert areas with subhumid to arid climatic conditions. Both native and introduced forms are used in North American range improvement programs. The most important of these are listed in Table 30.1.

Native wheatgrasses are prevalent in the

TABLE 30.1. Common names, scientific names, and characteristics of important wheatgrasses (WG) and wildryes (WR)

Common name	Proposed scientific name[a]	Old scientific name[b]	Native (N) or introduced (I)	Bunch (B) or sod (S)	Cross- (C) or self- (S) pollinated	Chromosome Number (2n)
Wheatgrasses						
Western WG	*Pascopyrum smithii* (Rydb.) Löve	*Agropyron smithii* Rydb.	N	S	C	56
Thickspike WG	*Elymus lanceolatus* (Scribn. & Smith) Gould	*A. dasystachyum* (Hook.) Scribn.	N	S	C	28(42)[c]
Streambank WG	*E. lanceolatus* (Scribn. & Smith Gould	*A. riparium* Scribn. & Smith ex Piper	N	S	C	28(42)
Bluebunch WG	*Pseudoroegneria spicata* (Pursh) Löve	*A. spicatum* (Pursh) Scribn. & Smith	N	B	C	14,28
Beardless WG	*Ps. spicata* (Pursh) Löve	*A. inerme* (Scribn. & Smith) Rydb.	N	B	C	14,28
Snake River WG	*E. lanceolatus* ssp. *wawawai*	*A. spicatum* (Pursh) Scribn. & Smith	N	B	C	28
Slender WG	*E. trachycaulus* (Link) Gould ex Shinners	*A. trachycaulum* (Link) Malte ex H. F. Lewis	N	B	S	28
Crested WG (Fairway)	*A. cristatum* (L.) Gaertner	*A. cristatum* (L.) Gaertner	I	B	C	14,28 (42)
Crested WG (Siberian)	*A. fragile* (Roth) Candargy	*A. sibiricum* (Willd.) Beauv.	I	B	C	28
Crested WG (Standard)	*A. desertorum* (Fisch. ex Link) Schultes	*A. desertorum* (Fisch. ex Link) Schultes	I	B	C	28
Intermediate WG	*Thinopyrum intermedium* (Host) Barkworth & D. R. Dewey	*A. intermedium* (Host) Beauv.	I	S	C	42
Pubescent WG	*T. intermedium* (Host) Barkworth & D. R. Dewey ssp. *barbulatum*	*A. trichophorum* (Link) Richt.	I	S	C	42
Tall WG	*T. ponticum* (Podp.) Barkworth & D. R. Dewey	*A. elongatum* (Host) Beauv.	I	B	C	70
Quackgrass	*Elytrigia repens* (L.) Nevski	*A. repens* (L.) Beauv.	I	S	C	42
Wildryes						
Great Basin WR	*Leymus cinereus* (Scrib. & Merr.) Löve	*Elymus cinereus* Scrib. & Merr.	N	B	C	28,56
Beardless WR	*L. triticoides* (Buckl.) Pilger	*E. triticoides* Buckley	N	S	C	28
Canada WR	*E. canadensis* L.	*E. canadensis* L.	N	B	S	28
Blue WR	*E. glaucus* Buckley	*E. glaucus* Buckley	N	B	S	28
Russian WR	*Psathyrostachys juncea* (Fisch.) Nevski	*E. junceus* Fisch.	I	B	C	14(28)
Altai WR	*L. angustus* (Trin.) Pilger	*E. angustus* Trin.	I	B[d]	C	84(28, 42,56)
Dahurian WR	*E. dahuricus* Turcz. ex Grieseb.	*E. dahuricus* Turcz. ex Grieseb.	I	B	S	42

[a] Tzvelev (1976), Dewey (1984), Barkworth et al. (1983), Carlson (1986).
[b] Hitchcock (1951), Bowden (1964).
[c] Bracketed chromosome number indicates less common.
[d] Occasional short rhizomes.

vast prairies of the northern Great Plains of the US and Canada. They also occur in the higher altitudes of the Rocky Mountains and in the more xeric areas in the intermountain West. The general area of distribution for the wheatgrasses is shown in Figure 30.1. In their native habitat, wheatgrasses occasionally occur as pure stands, but they are more often found in association with other perennial grasses, sedges, forbs, and shrubs. The composition of these plant communities reflects a wide range of soil and climatic conditions. For example, western wheatgrass and thickspike wheatgrass occur in varying combinations with 21 other grasses and sedges in the mixed prairie communities of southern Saskatchewan and Alberta (Coupland 1950). In general, the introduced wheatgrasses are seeded and used in the same areas of adaptation as the native species.

Environmental concerns with biological diversity along with restoration and preservation of native ecosystems have prompted some to propose that only native species be used in seeding programs on western rangelands, particularly federally owned lands. Proponents of this point of view argue that these areas should be restored to their native or pristine state. It should be noted, however, that man has significantly altered the environment. It follows that the optimum vegetative climax for a particular site will be adjusted as well. Different combinations of grass, forb, and shrub species will be required to meet the multiple demands imposed on rangelands in terms of conservation and aesthetics as well as for habitat and forage for livestock and wildlife.

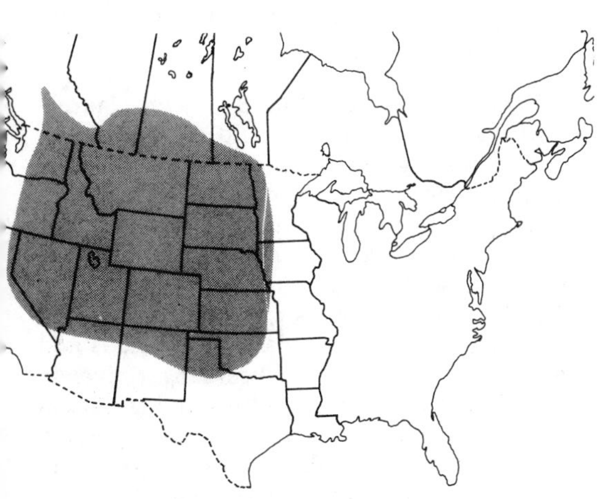

Fig. 30.1. Primary distribution of wheatgrasses in the US and Canada.

Grasses indigenous to Asia and Europe have evolved under more intense management than their North American counterparts, and they have demonstrated these adaptations on American rangelands as well. In general, the introduced species are easier to establish and are more productive and persistent when subjected to close defoliation and the other uses now imposed on western rangelands. On the other hand, it would be a mistake not to use native plants in rangeland improvement programs. Improved cultivars are being developed, and hybrids between native and introduced grasses have been made to develop new species or cultivars with selected attributes of their parental species (Asay 1992). It is evident that additional demands will be made on rangelands in the future. Improved ecosystems consisting of a balance of native and introduced species must be developed to meet these needs.

PLANT DESCRIPTION

Most wheatgrasses are cool-season perennial species, and both caespitose (bunch) and rhizomatous (sod-forming) types occur. With few exceptions, such as in some rhizomatous species, they produce abundant high-quality seed. Although a few species such as slender wheatgrass are self-pollinated, most wheatgrasses are cross-pollinating polyploids. Considerable intraspecific variation is present for vegetative and floral characteristics, and genetic introgression between species provides additional sources of variation. The inflorescence is a terminal spike with multiple florets per spikelet. The spikelets are sessile and solitary (rarely in pairs) and are flatwise on the rachis. Complete botanical descriptions of the wheatgrasses are provided by Hitchcock (1951) and Cronquist et al. (1977).

SPECIES AND CULTIVARS

Native Wheatgrasses

WESTERN WHEATGRASS (*Pascopyrum smithii* [Rydb.] Löve). This rhizomatous, long-lived perennial produces an open but uniform sod. It is characterized by glaucus stems and leaves that are typically bluish green. The leaves also are distinctly ribbed. Its spikes are from 7 to 15 cm long, and the glumes are rigid and gradually taper from near the base into a short awn. Spikes with both single and double spikelets per node occur. It is morphologically similar to and often confused with thickspike wheatgrass; however, it can be distinguished on the basis of its asymetrical glume shape (Cronquist et al. 1977). It is the only octoploid (2n = 56) wheatgrass that is native to North America, and the species is known to be of hybrid origin. Based on cytological evidence, as well as morphological, ecological, and reproductive characteristics of closely related species, Dewey (1975b) concludes that thickspike wheatgrass and beardless wildrye (*Leymus triticoides* [Buckl.] Pilger), or closely related taxa, were progenitors of western wheatgrass. According to Dewey (1975b, 1984), its genome constitution is 'SSHHJJNN', but recent evidence suggests that the 'J' genome may not be present (Zhang and Dvorak 1991; Wang and Jensen 1994).

Western wheatgrass occurs across the western two-thirds of the US and Canada, from Ontario to Alberta in the North (Hitchcock 1951) and into New Mexico and the Texas panhandle in the South (Hanson 1972). It is a prominent grass in the northern and central Great Plains, where it is often found in association with blue grama, *Bouteloua gracilis* (H.B.K.) Lag. ex Steud., and the needlegrasses, *Stipa* spp. In the intermountain region, it typically grows in communities with bluebunch wheatgrass, thickspike wheatgrass, and various shrubs. Considerable ecotypic variation occurs in western wheatgrass, and the species is adapted to a wide range of soil types. It is often found on heavy alkaline soils characteristic of the bottoms of intermittent swales, shallow lake beds, or areas subjected to periodic flooding (Rogler 1973). The species may be the most alkali tolerant of the North American wheatgrasses (Beetle 1955). Although established stands of western wheatgrass can withstand extended periods of drought, it apparently does not utilize available water as efficiently as some other prevalent grasses in the Great Plains and intermountain region. Its water use efficiency (WUE) was found to be substantially less than that of blue grama, which may explain why it is more prevalent on lowland sites of the shortgrass steppe (Monson et al. 1986). Similar findings are reported by Frank and Karn (1988). They found the WUE of western wheatgrass to be substantially lower than that of crested or intermediate wheatgrass.

Because of its sod-forming characteristics, western wheatgrass is well adapted for stabilizing soils on sites subject to excessive erosion. Vegetative propagation has been successfully used to establish stands on special purpose sites such as terrace outlets (Rogler 1973). The species has been recommended for revegetation of rangelands disturbed by surface-mining operations, construction, overgrazing, brush control, and fires. It was one of the most promising of 174 grass, forb, and shrub species evaluated by the USDA-SCS in Wyoming and Montana for reclaiming saline seeps and other problem sites (Scheetz et al. 1981). In Nebraska trials, western wheatgrass effectively controlled wind erosion in sand blowouts by forming a comparatively rapid vegetative cover (Malakouti et al. 1978).

Western wheatgrass is a cool-season grass and is considered to be an excellent source of forage during the spring and early summer. However, the quality of its forage declines rapidly as plants mature later in the season, and animals will tend to prefer other species grown in association with it. This may explain why previously mixed stands are often dominated by western wheatgrass after an extended period of continuous grazing (Hafenrichter et al. 1968). Because its leaves cure well on the stem, it provides moderately good forage during the fall and early winter. As with other rhizomatous wheatgrasses, it is a relatively poor seed producer, and stands are slow to establish from seed. Extensive rhizome development of established plants along with delayed germination of dormant seed in the soil often can transform a poor stand into a relatively good one after 2 or 3 yr.

In the past, most seed of western wheatgrass was harvested from native stands, but improved cultivars have been released from breeding programs. The cultivars 'Barton', 'Rosana', 'Arriba', and 'Flintlock' were released during the 1970s, and two cultivars, 'Walsh' and 'Rodan', have been released since then (Table 30.2).

THICKSPIKE WHEATGRASS (*Elymus lanceolatus* [Scribn. & Smith] Gould). This perennial sod-forming grass, called *northern wheatgrass* in Canada, is morphologically similar to western wheatgrass, but it is more drought resistant and less productive. It is an allotetraploid with a genome constitution of 'SSHH'. This species is cross-pollinating and has long anthers. It tends to be glaucous and has spikes that are 6 to 12 cm long. Spikelets have from three to eight florets and are loosely to closely imbricate. It is widely distributed on the plains and dry sandy deserts and shores from Michigan to British Columbia, south to Illinois, and west to Nebraska, Colorado, Nevada, and Oregon (Hitchcock 1951).

Streambank, Montana, and Wyoming wheatgrasses are so similar to thickspike wheatgrass that the four grasses are listed as

Table 30.2. Wheatgrass cultivars

Cultivar	Year released	Release agencies	Adaptation characteristics
Western wheatgrass			
Barton	1970	ARS, SCS, Kansas AES	Pasture, range, hay, and conservation in western Kansas, central and western Nebraska, and adjoining areas in Oklahoma and Colorado.
Rosana	1972	SCS, Montana AES	Irrigated hay or pasture in overflow sites and for range seedings in northern Rocky Mountains and adjacent Great Plains.
Arriba	1973	SCS, New Mexico and Colorado AES	Revegetation of range and problem sites, especially in southwestern US.
Flintlock	1975	ARS, Nebraska AES	Dryland hay or pasture and soil conservation, particularly in lower elevations of Central Great Plains.
Walsh	1982	Agriculture Canada, Alberta	Conservation and forage in areas typical of Canadian prairies; tolerant of drought, salinity, and spring flooding.
Rodan	1983	ARS, SCS, North Dakota AES	Forage and conservation in northern Great Plains of western Dakotas and eastern Montana and Wyoming; upland type.
Thickspike wheatgrass			
Sodar	1954	SCS, Idaho and Washington AES	Glabrous, low-growing type originally from Grant County, Oregon; special purpose grass for sod cover on arid disturbed sites and arid turf.
Critana	1971	SCS, Montana AES	Collected from roadside cuts in north central Montana; ground cover type for reclamation of dry disturbed sites; possible introgression with slender wheatgrass.
Elbee	1980	Agriculture Canada, Alberta	Originally from plains of Alberta and Saskatchewan; improved forage and seed yield; for revegetation of disturbed sites.
Snake River wheatgrass			
Secar	1981	SCS, Washington, Oregon, Montana, and Wyoming AES	Bearded tetraploid (2n = 28) originally from near Lewiston, Idaho; early maturing; drought resistant; adapted to low elevations of Pacific Northwest.
Bluebunch wheatgrass			
Whitmar	1946	SCS, Idaho and Washington AES	Beardless, diploid (2n = 14) form; indigenous to Palouse prairies of Washington; relatively poor seedling vigor; good forage and seed yields; drought resistant.
Goldar	1989	SCS, ARS, Idaho and Utah AES	Diploid (2n = 14) collected from Asotin County, Washington; for range and reclamation on bluegrass wheatgrass sites in intermountain area of Idaho, Utah, and Nevada.
Slender wheatgrass			
Primar	1946	SCS, Washington and Idaho AES	From native stands near Beebe, Montana; selected for early maturity and tolerance to alkali; for use in legume-grass conservation mixtures for pasture, hay, and green manure.
Revenue	1970	Agriculture Canada, Saskatchewan	Original collections from near Revenue, Saskatchewan; easier to establish and more productive and salt tolerant than Primar in Canadian trials.
San Luis	1984	SCS and several public agencies	Original collection from San Luis Valley, Colorado, at 2280-m elevation; improved persistence and good seedling vigor; for reclamation, hay, and pasture, primarily at higher elevations.
Pryor	1988	SCS, Montana and Wyoming AES	From collection made between the Pryor and Beartooth Mountains of south central Montana; good salinity tolerance and seedling vigor; recommendated for hay or pasture in short-term rotation.
Crested wheatgrass			
Fairway	1932	Univ. of Saskatchewan, Saskatoon	Diploid (2n = 14) selected from Siberian introduction (PI 19536); leafier but less drought resistant than Standard; dominant in Canadian seed trade.
Nordan	1953	ARS, North Dakota AES	Standard tetraploid (2n = 28) type originally from cold, dry plains of the former USSR; erect with relatively large seed and good seedling vigor.
P-27	1953	SCS, Idaho AES	Siberian type native to former USSR; narrow spikes and fine, leafy stems; adapted to light, droughty soils.
Summit	1953	Agriculture Canada, Saskatchewan	Standard type; general acceptance has been impeded by problems with seed processing.

Table 30.2. (*Continued*)

Cultivar	Year released	Release agencies	Adaptation characteristics
Parkway	1969	Agriculture Canada, Saskatchewan	Diploid selected from Fairway on basis of improved vigor, height, and leafiness.
Ephraim	1983	USFS, Utah Division, Wildlife Research, SCS, and several state AES	Derived from collection made near Ankara, Turkey; rhizomatous with adequate soil moisture.
Ruff	1974	ARS, Nebraska AES	Diploid derived from Fairway; recommended for low-precipitation areas of Great Plains and semiarid regions.
Hycrest	1984	ARS, SCS, Utah AES	Derived from hybrid between induced tetraploid Fairway type and Standard; produces abundant forage and seed on semiarid range sites; excellent stand establishment vigor under environmental stress.
Kirk	1987	Agriculture Canada, Saskatoon	Tetraploid form of *A. cristatum;* selected for improved vigor, fertility, seed retention, and reduced awn development; spike characteristics of *A. cristatum,* but more upright.
Intermediate wheatgrass			
Greenar	1945	SCS, Washington, Idaho, and Oregon AES	Original collection (PI 98568) from former USSR; moderately rhizomatous with dark green leaves; good seedling vigor.
Amur	1952	SCS, New Mexico AES	Derived from Manchurian collection; pubescent plants are prevalent.
Chief	1961	Agriculture Canada, Saskatchewan	Selected from collection from former USSR (PI 98568); excellent seed and forage yield in Canadian trials.
Oahe	1961	South Dakota AES	Originally from former USSR (PI 98568); selected for improved seed and forage yield and resistance to rust.
Luna	1963	SCS, New Mexico AES	Pubescent form; originally from Caspian region of former USSR; good drought resistance and seedling vigor.
Topar	1964	SCS, Washington, Idaho, Oregon, and California AES	Originally introduced from central Asia; recommended for shallow soils with low fertility.
Greenleaf	1966	Agriculture Canada, Alberta	Pubescent form of intermediate wheatgrass; selected for increased seedling and vegetative vigor, earliness of spring growth, winterhardiness, dark green foliage, and pubescent spikelets.
Tegmar	1968	SCS, Washington and Idaho AES	Derived from Turkish introduction; late-maturing dwarf type; for soil stabilization.
Slate	1969	ARS, Nebraska AES	Selected from PI 98568 (USSR) and Amur strain; strongly rhizomatous; slate green in color; adapted to central Great Plains.
Clarke	1980	Agriculture Canada, Saskatchewan	A 20-clone synthetic developed from four introductions; adapted to northern Great Plains; good yields of seed and forage; resistant to drought and cold.
Reliant	1991	ARS, SCS, North Dakota AES	A 6-clone synthetic derived from 24 different cultivars and experimental lines; selected for improved persistence, forage quality, and yield of forage and seed; particularly adapted to northern Great Plains.
Manska	1992	ARS, North Dakota and Nebraska AES	Pubescent form of intermediate wheatgrass; noted for high nutritive value, based on IVDMD and animal performance.
Tall wheatgrass			
Largo	1937	SCS, New Mexico AES	Parent germplasm introduced from Turkey; large coarse culms with blue-green leaves.
Alkar	1951	SCS, Idaho and Washington AES	Originally introduced from former USSR (PI 98526); productive, alkali tolerant.
Jose	1965	SCS, New Mexico AES	Indigenous to Eurasia; earlier in maturity and not as coarse as other strains of tall wheatgrass.
Orbit	1966	Agriculture Canada, Saskatchewan	Traces to germplasm from USSR (Neb. 98526); excellent winterhardiness; tolerant of salinity.
Tyrell	1981	Victorian Department of Agriculture, Australia	Developed from Largo; adapted to saline soils typical of northwest Victoria, Australia.
Platte	1972	ARS, Nebraska AES	Derived from Neb. 98526 and another breeding line; blue-green leaves; adapted to alkaline sites, particularly to lower valleys of Platte River drainage.

Sources: Asay et al. (1985b), Berdahl et al. (1992), Douglas and Ensign (1954), Elliott and Bolton (1970), Hanson (1972), Knowles (1990), Lawrence (1967, 1981), Morrison and Kelly (1981), Morrison and Wolfe (1957), Newell (1974), N.Mex. Coop. Ext. Serv. (1977), Oram (1981), Smoliak and Johnston (1980, 1984), Stroh et al. (1972), USDA Ext. Serv. (1978), Wilson and Smoliak (1977), Wolfe and Morrison (1957a, 1957b).

a polymorphous group of species by Hitchcock (1951), and they are not recognized as separate species by Dewey (1983) (Table 30.1). Streambank wheatgrass is similar to thickspike wheatgrass except that it has more imbricate spikelets and glabrous lemmas (Hitchcock 1951). Dewey (1983) considers it to be a glabrous form of thickspike wheatgrass. Based on data from cytological studies, Dewey (1970) confirms earlier conclusions by Bowden (1965) that Montana wheatgrass and Wyoming wheatgrass were the products of genetic introgression between thickspike wheatgrass and bluebunch wheatgrass. Montana wheatgrass, which has the rhizomatous characteristic of thickspike wheatgrass and the divergent awns of bluebunch wheatgrass, is considered by Dewey as a subspecies (*E. lanceolatus* ssp. *albicans*). Wyoming wheatgrass is apparently the glabrous form of Montana wheatgrass and probably originated from hybrids between bluebunch wheatgrass and streambank wheatgrass. It is treated as variety *griffithsii* of subspecies *albicans*. Snake River wheatgrass is a new taxon, morphologically similar to bluebunch wheatgrass but genomically aligned with thickspike wheatgrass ('SSHH' genomes). It has been tentatively treated as a subspecies of thickspike wheatgrass (*E. lanceolatus* ssp. *wawawaiensis*) (Carlson 1986). Snake River wheatgrass is most prevalent along the lower Snake River drainage.

Two basic growth forms of thickspike wheatgrass occur across its adaptive range: forage and ground cover types. The ground cover types are more procumbent and are less productive, but they are better sod formers than the forage types (Dubbs et al. 1974). The ground cover types of thickspike wheatgrass are valued primarily for special purpose applications. They are particularly well suited for soil stabilization on disturbed range sites and other dry areas subject to erosion such as surface mines, roadsides, airports, recreational areas, and construction sites. The forage types are most productive during the early summer when earlier grasses such as crested wheatgrass are more mature and lower in nutritional value.

The development of improved cultivars of thickspike wheatgrass has been done primarily by Agriculture Canada, USDA-ARS, and USDA-SCS. Two cultivars of thickspike wheatgrass, 'Critana' and 'Elbee', and one of the streambank wheatgrass form, 2n = 'Sodar', have been released from these programs (Table 30.2). An advanced generation amphiploid hybrid between bluebunch wheatgrass and thickspike wheatgrass was released as the SL-1 germplasm pool by the USDA-ARS, Logan, Utah, in 1990 (Asay et al. 1991b). Research also has been initiated by the USDA-ARS at Logan, Utah, to develop improved cultivars of Snake River wheatgrass. In this program, Snake River wheatgrass has been hybridized with a promising forage type of thickspike wheatgrass (USDA-SCS T-21076), with the goal of combining the rust resistance and leafiness of Snake River wheatgrass with the vegetative vigor and awnless inflorescence of thickspike wheatgrass (Jones et al. 1991).

BLUEBUNCH WHEATGRASS (*Pseudoroegneria spicata* [Pursh] Löve). This cool-season perennial bunchgrass is one of the most valuable native grasses in the intermountain region and the Pacific Northwest. Both diploid (2n = 14) and autotetraploid (2n = 28) forms occur, although the tetraploids apparently are limited primarily to eastern Washington and adjacent northwest Idaho (Hartung 1946) and Canada. Bluebunch wheatgrass contains the 'S' genome and apparently has been involved in the phylogeny of several other associated grasses on the western range (Dewey 1966). It is closely related to, and interfertile with, beardless wheatgrass, previously designated as *A. inerme,* but not given a separate species status by Dewey (1983, 1984). Beardless wheatgrass lacks the prominent divergent lemma awns (a simply inherited recessive trait) that are characteristic of bluebunch wheatgrass.

Bluebunch wheatgrass is generally caespitose under arid and semiarid conditions; however, mildly rhizomatous types occur under more optimum moisture conditions. The presence of rhizomes may be the result of genetic introgression with thickspike wheatgrass (Dewey 1970). Typical plants are glaucus, from 60 to 100 cm tall with flat to loosely involute leaf blades. The spikes are narrow from 8 to 15 cm long with distant spikelets. Glumes are infrequently awned and are about half the length of the spikelet. Lemmas are about 1 cm long and, with the exception of the beardless wheatgrass form, have a strongly divergent awn. Bluebunch wheatgrass is predominantly cross-pollinated, although varying degrees of self-fertility have been observed in the tetraploids (Jensen et al. 1990).

In general, the species is adapted to the same soil and climatic conditions as crested wheatgrass. It is often found on dry plains, on rocky slopes and hills, and in open or partial shade but seldom on wet soils. Bluebunch wheatgrass is widely distributed across the western US, north into Canada and south to the northern edge of the Sonoran Desert. Where it is well adapted, it often occurs in nearly pure stands. In the intermountain region, it is often the codominant species with sagebrush (*Artemisia* spp.) on sites receiving from 20 to 40 cm of annual precipitation. It often grows in association with Idaho fescue (*Festuca idahoensis* Elmer) and Sandberg bluegrass (*Poa secunda* J. S. Presl) in the Palouse prairie of eastern Washington (Miller et al. 1987; Cronquist et al. 1977).

Bluebunch wheatgrass is often preferentially grazed by livestock and wildlife in mixed stands, and stands are often depleted under heavy grazing pressure. Careful management is essential to maintain productive stands (Hafenrichter et al. 1968; Mueggler 1975). Close defoliation can be particularly damaging during the early spring when energy reserves are low. It is recommended that grazing be delayed until the late boot stage, when energy reserves are at more optimum levels (Daer and Willard 1981). Grazing pressure should be adjusted to remove only 50% of the current forage production (Hafenrichter et al. 1968). Bluebunch wheatgrass responds to nitrogen (N) fertilization, but in mixed stands, aggressive species such as cheatgrass (*Bromus tectorum* L.) also become more competitive (Mason and Miltimore 1959).

Two cultivars, 'Whitmar' and 'Goldar', have been released by the USDA-SCS. Whitmar, a beardless type derived from a population collected in the Palouse of eastern Washington, was released in the 1940s (Hanson 1972). Goldar was made available in 1989, although it traces to a collection made much earlier in 1934 in Asotin County, Washington (Gibbs et al. 1991). The cultivar 'Secar', originally identified as bluebunch wheatgrass, has since been described as thickspike wheatgrass.

SLENDER WHEATGRASS (*Elymus trachycaulus* [Link] Gould ex Shinners). This self-fertile bunchgrass was the first native grass to be generally used in revegetation programs in the western US and Canada (Rogler 1973), and it is one of the few self-fertile wheatgrasses to be of economic importance. Considerable morphological variation exists in the species. Four taxa described by Hitchcock (1951) are now treated as varieties of or in synonymy with *E. trachycaulus* (Cronquist et al. 1977; Gleason and Cronquist 1991). One of these variants, bearded wheatgrass (syn. *A. subsecundum* var. *andinum*), has distinctive awns and is considered to be more robust and longer lived than typical slender wheatgrass (Hafenrichter et al. 1968). Although slender wheatgrass is predominantly a bunchgrass, rhizomatous plants occasionally occur in some ecotypes. It is a short-lived perennial often found on dry to moderately wet roadsides, streambanks, meadows, and woodlands from the valley bottoms to subalpine and alpine elevations. It is widely distributed in North America from Newfoundland to Alaska; south to the central US; and to New Mexico, Arizona, and California in the Southwest (Rogler 1973). It is an excellent seed producer, and the seed is usually of good quality (Hafenrichter et al. 1968). It is less drought tolerant than most of the wheatgrasses, including crested and bluebunch wheatgrass, and therefore, it is generally seeded on dryland sites receiving at least 35 cm of annual precipitation. The species has relatively good seedling vigor; productive stands are often obtained 1 or 2 yr after seeding. Because of this attribute and its tendency to be short-lived, it has been widely used as a cover or nurse crop during the establishment of more permanent forage crops. Slender wheatgrass has moderate salinity tolerance and has demonstrated potential for reclaiming problem sites such as saline seeps and mine spoils and for vegetating watersheds and roadsides (Rauser and Crowle 1962; Scheetz et al. 1981). The lack of longevity in slender wheatgrass may be associated with its susceptibility to insects such as the grass billbug (*Sphenophorus* spp.) (Asay et al. 1983).

Four cultivars, 'Primar', 'Revenue', 'Pryor', and 'San Luis', are presently available commercially (Table 30.2). Primar was released in 1946 by the USDA-SCS, Revenue was licensed by Agriculture Canada at Saskatoon, Saskatchewan, in 1970. Pryor, which was released by the USDA-SCS in 1988, has demonstrated improved salinity tolerance and seedling vigor (Majerus et al. 1991a). San Luis, also released by the USDA-SCS, was collected from the San Luis valley of Colorado at an altitude of 2280 m. It is considered to be more persistent than other slender wheatgrass cultivars and is recommended for soil stabilization at higher elevations.

Introduced Wheatgrasses

CRESTED WHEATGRASS (*Agropyron* spp.). This complex of species is native to the steppe region of European Russia and southwestern

Siberia, and it was first successfully introduced to North America in 1906. Early field testing initiated by the USDA at Mandan, North Dakota, in 1915 led to the acceptance of crested wheatgrass in the northern Great Plains (Dillman 1946). Figure 30.2 shows a nursery for breeding crested wheatgrass. In the present alignment, species of the crested wheatgrass complex are the sole members of the genus *Agropyron*. Although subspecies have been described, Dewey (1983, 1984) limits this complex to 'Fairway' (*A. cristatum*), 'Standard' (*A. desertorum*), and 'Siberian' (*A. fragile*). The type species is *A. cristatum* (L.) Gaertn. The crested wheatgrass complex is based on the 'P' genome and consists of an autoploid series of diploid (2n = 14), tetraploid (2n = 28), and hexaploid (2n = 42) forms. The tetraploids are the most common and are widely distributed from central Europe and the Middle East through central Asia to Siberia, China, and Mongolia (Dewey, 1986; Tzvelev 1976). The diploids have a similar range of adaptation; however, they occur much more sporadically. The hexaploid forms are the least common; however, they are known to occur in Turkey, Iran, and Kazakhstan (Dewey and Asay 1975; Asay et al. 1990). In North America, the Standard cultivars are most common in the US, and the Fairway form is the most prevalent in Canadian seed channels.

In North America, crested wheatgrass is particularly well adapted to the northern and central Great Plains and the more arid intermountain region. The crested wheatgrasses are long-lived perennials adapted to temperate rangelands receiving from 20 to 45 cm of annual precipitation. Crested wheatgrass was named in reference to its linear pectinate (comblike) spike that has an oblongate or ovate shape. Although ecotypes with rhizomes are occasionally found, plants are more often caespitose with erect culms from 15 to more than 100 cm tall.

Floral characteristics are used extensively to differentiate among species in the complex. The diploid Fairway form has broad spikes with spikelets arranged in a comblike manner, whereas the tetraploid Standard spikes are longer and more slender with spikelets at an acute angle to the rachis. Fairway is more apt to have awn tips on the seed than Standard. Also, the upper leaf surface is usually

Fig. 30.2. Crested wheatgrass (*Agropyron* spp.) breeding nursery.

pubescent in Fairway in contrast to the smooth (glaucous) condition in Standard. Spikes of Siberian crested wheatgrass are similar to those of Standard but are longer and more narrow. Colchicine-induced tetraploids of the Fairway type have been produced at Logan, Utah, and Saskatoon, Saskatchewan, and have been used in breeding programs. Broad-spiked tetraploids also occur in nature and are interfertile with the Standard and Siberian tetraploids. These intermediate hybrids are the source of some taxonomic confusion. There is some evidence that the Standard form arose from hybrids between Siberian and tetraploid Fairway (Asay et al. 1992).

The first notable impact of crested wheatgrass in North America was during the early settlement of the northern Great Plains of the US and the Prairie Provinces of Canada. During the dust bowl and depression period of the 1930s, seedings of crested wheatgrass stabilized and saved the soil on vast areas of deteriorated rangeland and abandoned cropland (Lorenz 1986). From these early seedings, many of which remain today, this durable and productive grass has expanded from the Great Plains to other semiarid and arid regions. During the 1940s, the annual poisonous forb halogeton (*Halogeton glomeratus* [Bieb.] C. A. Mey.) infested intermountain rangelands and was a serious threat to grazing animals. Crested wheatgrass was effectively used to suppress this weed on many sites (Mathews 1986; Young and Evans 1986). Gomm (1981) estimates that 5.1 million ha had been seeded to crested wheatgrass in western North America, and this area has undoubtedly increased since then.

In general, crested wheatgrass has excellent drought resistance and has been used successfully on sites receiving from 20 to 40 cm annual precipitation. Siberian and Standard cultivars are more drought tolerant than Fairway, although the latter type appears to be more persistent in more humid environments. It has excellent seedling vigor, and stands are relatively easy to obtain, often under harsh environmental conditions. Depending on the latitude, the upper elevation limit is approximately 2500 m. The grass will not endure more than 7 to 10 d of spring flooding without severe loss of stands. It is less tolerant of soil salinity than tall wheatgrass, western wheatgrass, and quackgrass (*E. repens* [L.] Nevski); however, breeding for improved salinity tolerance appears to be feasible (Dewey 1962). It is extremely long-lived in its area of adaptation. Productive stands more than 30 yr old exist in the intermountain region and Great Plains (Dillman 1946; Cook 1966; Looman and Heinrichs 1973).

Crested wheatgrass has excellent seed characteristics: yields are good, little processing of the seed is required, and the seed flows easily through the drills. Although seed often is produced commercially in 30- to 60-cm row spacings, it is recommended that row spacings from 90 to 110 cm be used for sustained production. Fall and spring applications of N fertilizer are recommended to maximize seed yields.

Crested wheatgrass produces abundant forage and is more widely used for grazing than for hay. The grass is particularly noted for its production of nutritious forage during the early spring; however, leaf senescence later in the season leads to a rapid decline in forage quality. Although the quality and seasonal distribution of its forage is improved significantly by application of N fertilizer and by the associated growth of legumes and shrubs (Rumbaugh et al. 1982), it is generally recommended that it be grazed most heavily during the spring and early summer. Other grasses including intermediate wheatgrass, smooth bromegrass (*Bromus* spp.), and several native species should be used to complement crested wheatgrass later in the season.

Since its introduction, crested wheatgrass has become a major component of the temperate rangelands of North America. Other native wheatgrasses such as bluebunch wheatgrass and thickspike wheatgrass are often considered to be more ecologically acceptable. But the tolerance of crested wheatgrass to intense grazing management can not be disputed. It consistently increases the grazing capacity of the native stands, and the improvements are realized over an extended period. For example, Smoliak and Dormaar (1985) found that this durable grass produced 113% more forage than native rangeland during a 25-yr period in southern Alberta. Most of the environmental concerns regarding crested wheatgrass were provoked by the early practice of seeding it in monocultures. In its native environment, crested wheatgrass is an integral part of genetically diverse ecosystems, and recently, it has been included in mixtures with other introduced and native species in North American seedings as well.

Early crested wheatgrass seedings often consisted of unimproved strains, and some were mixtures of both Fairway and Standard types. The first product of North American breeding programs was the diploid *A. cristatum* cultivar Fairway, which was distributed in 1927 by the University of Saskatchewan at Saskatoon. This cultivar is still widely used in range seeding programs, particularly in Canada. Crested wheatgrass breeding in the US was initiated by the USDA-ARS at Mandan, North Dakota. The tetraploid Standard cultivar, 'Nordan', was released from this program in 1953, and it has been a principal cultivar in US seed trade. Several cultivars, including 'Hycrest', 'Ephraim', 'Ruff', and 'Kirk', since released from other public breeding programs are now making a major impact (Table 30.2).

INTERMEDIATE WHEATGRASS (*Thinopyrum intermedium* [Host] Barkworth & D. R. Dewey). The first introduction of intermediate wheatgrass (PI-20639) was from Trans Ural Siberia in 1907 (Dewey 1978b). In its native habitat, it is distributed from the steppes to the lower mountain areas of southern Europe, through the middle East and central Asia, and to western Pakistan (Tzvelev 1976). Few if any collections of intermediate wheatgrass have been made north of 30°N. It is productive on sites up to 3000-m elevation, sites that receive at least 35 cm of annual precipitation, which is between the adaptive range of smooth bromegrass and that of crested wheatgrass. In North America, it is widely grown for pasture and hay from Nebraska to Manitoba, Canada, and west to Washington and California. As presently treated, intermediate wheatgrass includes grasses previously designated as *A. intermedium* (Host) Beauv., *A. trichophorum* (Link) K. Richt. (pubescent wheatgrass), and *A. pulcherrimum* Grossh. Intermediate wheatgrass is considered to be a segmental autoallohexaploid (2n = 42) with the 'E_1', 'E_2', and 'S' genomes (R. R-C. Wang, personal communication).

This cross-pollinating grass has moderately creeping rhizomes, and it has relatively large seeds that are borne on erect stalks. It has good seedling vigor and is fairly easy to establish, even when seeded with a cereal companion crop. Because of its seed size, the grass also has been proposed as a potential perennial grain crop, although this objective has yet to materialize (Wagoner 1990). It is most persistent on soils that are well drained, and it has moderate resistance to saline and alkaline soils. It is well suited to be grown with alfalfa (*Medicago sativa* L.) under dryland conditions and limited irrigation. Its erect stems help prevent alfalfa from lodging, and it usually is not at an advanced stage of maturity when alfalfa is ready to harvest (Lawrence 1977). Because it is from 1 to 2 wk later than smooth bromegrass and crested wheatgrass, it provides a later period of summer grazing. It also retains its forage quality relatively well after frosts in the fall.

Despite its wide adaptation and productivity, intermediate wheatgrass has not been used as extensively as either smooth bromegrass or crested wheatgrass. Failure of early cultivars to persist more than 4 or 5 yr, particularly under intense management, marred its reputation. Also, initial cultivars were not good seed producers; however, this deficiency has since been overcome by breeding. Intermediate wheatgrass has some morphological similarities to quackgrass, which may have discouraged some from using it. The grass, in fact, is easy to eradicate. Careful management is essential to the longevity of intermediate wheatgrass stands. It is sensitive to continuous grazing or close defoliation, particularly at the earlier stages of phenological development (Currie and Smith 1970).

The pubescent form of intermediate wheatgrass, often identified as *pubescent wheatgrass,* is characterized by pubescence on the spikes, seed, and occasionally leaves. Pubescent wheatgrass coexists with the typical intermediate form in Old World and North American habitats (Dewey 1978b), and the two grasses are interfertile. Variation among plants in the field indicates that considerable hybridization and genetic introgression has occurred between them. The pubescent form is considered to be better adapted to the more southern limits of the species adaptive range in Asia (Sinskaja 1961) and the US (Cornelius 1965). It appears to be better suited to droughty, infertile soils and saline sites than typical intermediate wheatgrass (Hafenrichter et al. 1968).

The first significant interest in the development of improved cultivars of intermediate wheatgrass was prompted by the introduction of a particularly promising accession (PI-98568) from Maikop, USSR, in 1932. This accession was released by the South Dakota Agricultural Experiment Station as the cultivar 'Ree', and selections from this population later were used in the development of the cultivars 'Greenar', 'Chief', 'Oahe', and 'Slate'. Subsequent plant exploration led to the release of several other cultivars. The pubescent wheatgrass cultivars 'Topar', 'Luna', and 'Greenleaf' were released prior to 1966. More recently, the pubescent cultivar 'Manska' was cooperatively released by the USDA and state experiment stations in North Dakota and Nebraska (Table 30.2).

TALL WHEATGRASS (*Thinopyrum ponticum* [Podp.] Barkworth & D. R. Dewey). This hardy, erect perennial bunchgrass is native to southern Europe and Asia Minor and was introduced into North America from Turkey in 1909 (Weintraub 1953). It is a decaploid (2n = 70) genomically related to intermediate wheatgrass (Dewey 1984). In its native habitat, it is commonly found on saline meadows and seashores (Beetle 1955). Tall wheatgrass now is grown throughout the intermountain West and the northern Great Plains and as far south as northern New Mexico and Arizona. It has long erect leaves that tend to be coarse, distinctly ribbed, and blue-green. It is the latest-maturing wheatgrass adapted to the temperate areas of western North America and is one of the most productive. The species is adapted to range sites receiving at least 35 to 40 cm of annual precipitation or on subirrigated or irrigated land. Tall wheatgrass is particularly noted for its capacity to produce forage and persist in areas that are too alkaline or saline for other productive crops. Although the grass becomes quite coarse as it approaches maturity, it retains its green color longer than other wheatgrasses. Thus, it is a good source of pasture and hay during the late summer, when forage often is in short supply. It also has been used successfully as a silage crop. Tall wheatgrass has large seed that is easy to harvest and plant. It has good seedling vigor, and established plants have an exceptionally deep root system, which contributes to its resistance to drought (Weintraub 1953; Hafenrichter et al. 1968).

Because of its late maturity, it is usually recommended that tall wheatgrass be seeded alone. A 20-cm stubble should be left at season's end to prevent animals from grazing too close the following year. Grazing should not be initiated until at least 25 cm of new growth have accumulated above the stubble.

Although the gene base of tall wheatgrass in most North American breeding programs is relatively narrow, several cultivars of tall wheatgrass have been released (Table 30.2). In addition, the species has been widely used in wheat breeding programs as a source of resistance to salinity, drought, and disease (Dewey 1984; Sharma and Gill 1983).

QUACKGRASS (*Elytrigia repens* [L.] Nevski). This cool-season, persistent perennial is noted for its aggressive rhizomes and was probably introduced into North America from Eurasia (Hitchcock 1951). It is a widely adapted hexaploid (2n = 42) distributed in the temperate areas of every continent in the world (Rogler 1973). Quackgrass is commonly found in waste areas and along ditch banks and roadsides; because of its aggressive growth habit, it is often a problem weed in cultivated croplands, gardens, lawns, and orchards. The grass is tolerant of saline and alkaline soils as well as flooding. In spite of its weedy characteristics, quackgrass is an excellent soil binder and often is a valuable source of early-season forage in humid, temperate regions and under irrigation in the West. It is used to a limited extent in the northeastern and midwestern US as a rough turfgrass in lawns and golf courses.

Although quackgrass has been used successfully as a parent in interspecific hybridization programs, very little breeding work has been done within the species. Consequently, no improved cultivars are presently available. However, it is a highly variable species, and ample genetic variability appears to be available to improve its agronomic potential through breeding. Selection for types that are less rhizomatous appears to be a realistic breeding objective (Asay 1992).

INTERSPECIFIC HYBRIDS. Hybridization among perennial Triticeae species occurs frequently in nature. Genetic exchange among several species has been documented, with thickspike wheatgrass, bluebunch wheatgrass, and slender wheatgrass as notable examples (Dewey 1984). Natural hybrids between thickspike wheatgrass and slender wheatgrass are reported to be widespread in the US and Canada (Bowden 1965; Dewey 1975a). Genetic introgression involving the thickspike wheatgrass complex and bluebunch wheatgrass is discussed earlier. Most wheatgrasses are allopolyploids of hybrid origin, and some have been artificially synthesized through hybridization and chromosome doubling. Dewey (1972) produced an amphiploid from a hybrid between *P. libanotica* and *E. caninum* that was morphologically and cytologically identical to *E. leptourus*. Stebbins (1955) obtained artificial microspecies of glaucus wildrye (*E. glaucus*) from its interspecific hybrids with *Hordeum jubatum*. The hybrid origin of western wheatgrass (Dewey 1975b) is discussed earlier under "western wheatgrass."

Interspecific hybridization is beginning to have an impact on breeding programs aimed at developing improved cultivars of wheatgrass and wildrye; however, it is evident that several obstacles must be overcome before significant genetic advances can be achieved. Interspecific hybrids are usually meiotically irregular and are partially to completely sterile. Artificially doubling the chromosome number through colchicine treatment has been effective in achieving fertility in some hybrids, but sterility recurs in subsequent generations, especially at higher ploidy levels (Asay and Dewey 1976). Moreover, most interspecific hybrids are inferior to naturally occurring species, and large numbers of hybrids must be screened to identify potential breeding populations.

Perhaps the wheatgrass hybrid with the greatest potential for western rangelands was obtained by crossing Fairway crested wheatgrass with the Standard type. Even though these two grasses are in the same genus, Fairway is a diploid (2n = 14) and Standard is a tetraploid (2n = 28), and the F_1 hybrid between them is usually a sterile triploid (2n = 21). Fertile hybrids were obtained by doubling the chromosome number of the Fairway parent prior to the cross (Dewey and Pendse 1968). A breeding program was initiated with this hybrid population in 1974 by the USDA-ARS at Logan, Utah, and the cultivar Hycrest was released in 1984. This cultivar is more robust and productive than the parental species, especially during seedling establishment (Asay et al. 1985b).

The hybrid between quackgrass and bluebunch wheatgrass is another notable example. The F_1 progenies from this cross were pentaploid (2n = 35) and morphologically variable (Dewey 1976). However, because the population was partially fertile, generations were advanced without chromosome doubling. Six cycles of selection yielded a meiotically regular population with 42 chromosomes and characteristics of both parental species. Genetic variability for degree of rhizome development was evident, and the character was effectively altered through selection. The cultivar 'NewHy' was released from this hybrid breeding population in 1989 (Asay et al. 1991a). This cultivar has been especially productive on alkaline and saline range sites.

Interspecific hybridization with perennial Triticeae species is proving to be a valuable procedure in wheat breeding programs. Several wheatgrass species have been crossed with diploid, tetraploid, and hexaploid wheat to (1) transfer genes such as those conditioning resistance to disease and environmental stress from the perennial grasses to wheat and (2) facilitate studies of phylogeny and genome structure. Tall and intermediate wheatgrasses have been frequently included in crossing schemes; however, quackgrass, thickspike wheatgrass, crested wheatgrass, slender wheatgrass, and others also have been involved. Application of special techniques such as embryo culture has provided a means of obtaining viable hybrids that were previously not possible (Sharma and Gill 1983).

Wildryes

In his taxonomic realignment of the perennial Triticeae grasses, Dewey, (1983, 1984) included the wildrye grasses in the genera *Elymus, Leymus,* and *Psathyrostachys* (Table 30.1). Many of the wildryes are genomically similar to the wheatgrasses. This is particularly true for self-pollinating species with the 'SH' genomes such as Canada wildrye (*E. canadensis*) and slender wheatgrass (*E. trachycaulus*). The most noteworthy taxonomic changes involved the transfer of Russian wildrye from *Elymus,* where it was *E. junceus,* to *Psathyrostachys* as *P. juncea* and the transfer of Great Basin wildrye, Altai wildrye, and related species from *Elymus* to *Leymus*. Great Basin wildrye is now treated as *L. cinereus* and Altai wildrye as *L. angustus*. The genus *Psathyrostachys* is based on the 'N' genome, and *Leymus* consists of species with the 'N' genome combined with an unidentified genome.

DISTRIBUTION AND ADAPTATION

The wildryes, a diverse group of perennial forage grasses, are widely distributed in the temperate areas of the world, primarily in Asia, Europe, and North America. Their general area of adaptation is similar to the wheatgrasses (Fig. 30.1). As with the wheatgrasses, several wildryes indigenous to Asia and Europe have been introduced into North America and have been instrumental in rangeland improvement programs. The most noteworthy of these introduced species are Russian wildrye, Altai wildrye, and Dahurian wildrye. The most predominant native wildryes are Great Basin wildrye (Fig. 30.3), blue wildrye, Canada wildrye, and beardless wildrye.

SPECIES AND CULTIVARS

Native Wildryes

GREAT BASIN WILDRYE (*Leymus cinereus* [Scrib. & Merr.] Löve). This long-lived perennial is widely distributed in the north from British Columbia to Saskatchewan, south to California, and east to Arizona and Colorado (Fig. 30.3). It is often found in low-lying areas such as along streams, valley bottoms, and roadsides. It is moderately tolerant of alkaline and saline soils and is generally adapted to areas receiving from 25 to more than 40 cm of precipitation annually. The species was earlier treated as *Elymus cinereus,* but since has been moved to *Leymus* (Löve 1980). Two ploidy levels, tetraploid (2n = 28) and octoploid (2n = 56), have been documented. The tetraploid form is most often observed in the western US, whereas the octoploid appears to be more common in Canada. It is closely related to giant wildrye, *L. condensatus* (Presl.) A. Löve, a grass that appears to be indigenous to the coastal regions of California. Great Basin wildrye is cross-pollinated and is known to hybridize with other *Leymus* species in nature; hybrids with beardless wildrye are particularly common. Although substantial variation occurs among ecotypes, it is characterized as a tall bunchgrass ranging in height from 0.7 to 2.0 m and is occasionally taller under more optimum conditions. Its glabrous stems and largely basal leaves tend to become relatively coarse with advancing maturity. As with other *Leymus* species, spikes have multiple spikelets and tend to be erect, from 11 to 20 cm long (Cronquist et al. 1977).

Great Basin wildrye is valued particularly as a forage for grazing animals during the fall and winter. It also has been used to complement tall wheatgrass on alkaline and saline sites and as a component of seed mixtures on mine spoils and other disturbed areas. Although it is robust and vigorous, this native grass is sensitive to heavy grazing pressure, and stands have been severely damaged by

Fig. 30.3. Great Basin wildrye (*Leymus cinereus*).

overgrazing. To maintain productive stands, a 30-cm stubble should be left when grazed and about 25 cm when harvested for hay. Because of its tall upright growth habit, Great Basin wildrye is a good habitat for pheasants, waterfowl, and other wildlife species. It also provides a windbreak and protection for young calves and lambs in a ranching operation (Howard 1979). Because of a relatively high degree of seed dormancy and poor seedling vigor, Great Basin wildrye is difficult to establish (Evans and Young 1983). Two growing seasons may be required to obtain satisfactory stands. Ergot (*Claviceps purpurea*) can be a problem under certain environmental conditions. When black sclerotia are abundant in the spikes, these fungi are poisonous to livestock and have been known to cause abortions (Cronquist et al. 1977).

Two cultivars of Great Basin wildrye have been released, both by the USDA-SCS (Table 30.3). The octoploid (2n = 56) cultivar 'Magnar' was released in 1979 (Howard 1979). This was followed in 1991 by the release of 'Trailhead', a tetraploid (2n = 28). Trailhead is reported to be more productive and persistent under drought conditions than Magnar (Majerus et al. 1991b). The USDA-ARS at Logan, Utah, has produced several artificial hybrids between Great Basin wildrye and several other *Leymus* species. These include beardless wildrye; *L. simplex* (Scrib. & Willam) D. Dewey; Altai wildrye (*L. angustus* [Trin.] Pilger); and *L. multicaulus,* Kar. & Kir., Tzvelev (Dewey 1984).

BLUE WILDRYE (*Elymus glaucus* Buckley). This caespitose, self-fertile grass is indigenous to the higher elevations of North America. It is most often found in timbered areas from the Pacific Coast to the Rocky Mountains. It is regarded as an allotetraploid (2n = 28) with the 'S' and 'H' genomes (Dewey 1982). It is characterized by erect to slightly nodding spikes with awn-pointed glumes and awned lemmas. Seed usually must be processed to remove the awns, and seed yields are often reduced by shattering. Blue wildrye's leaf blades are flat and somewhat lax. As with several other 'SH', self-fertile species, it tends to be relatively short lived; however, it has excellent seedling vigor, and stands are maintained through natural reseeding. It is noted for its tolerance of shade and compatibility with woody plants. It is often used as a companion crop to minimize competition from weeds during the establishment of seedling trees. It also is used to provide a grass cover in woodlots and other tree and shrub plantings. It is often found on burned over forestlands and in disturbed areas at higher elevations (Hafenrichter et al. 1968).

Considerable variation exists among ecotypes of blue wildrye in nature, but because it is self-fertile, genetic uniformity within strains and ecotypes is common. Hence, it is not difficult to maintain genetic purity in populations collected from different sites.

Although several strains have been collected and evaluated, very little breeding has been done with blue wildrye. The USDA-SCS at Pleasanton, California, developed an improved strain from a collection made from a winter annual rangeland in western California. Although this strain was tentatively given the name 'Lomas', it was never officially released (Hanson 1965). A second strain was developed by the USDA-SCS at Pullman, Washington, from a collection made from the Wenatchee National Forest in Washington. This strain was designated as P-2662, and seed was increased for testing, but it was not released as a cultivar (Hanson 1972).

CANADA WILDRYE (*Elymus canadensis* L.). This self-fertile bunchgrass is widely distributed across the US and Canada (Hitchcock 1951; Bowden 1964). It is characterized by tufted culms that are from 1.0 to 1.5 m tall and dense and awned spikes that are often nodding. The spikes have multiple (two to four) spikelets per node, and the prominent awns are divergently curved at maturity (Hitchcock 1951). It is an allotetraploid (2n = 28) with the 'S' and 'H' genomes (Dewey 1967).

Canada wildrye has a wide range of ecotypic diversity associated with its adaptation to multiple habitats. It tends to have a relatively coarse texture; however, it remains green later in the summer than many temperate grasses. It has demonstrated the capacity to produce abundant forage, particularly under subirrigated conditions. It also is a good seed producer, although threshing and seed cleaning are complicated by its long awns (Hafenrichter et al. 1949). It has not been used extensively in range- or pasture-seeding programs.

Although accessions of Canada wildrye have been collected and evaluated, very little breeding has been done with the species. The only known cultivar is 'Mandan 419', which was released in 1946 by the USDA-ARS at Mandan, North Dakota (Table 30.3). Canada wildrye has been hybridized with several other related species in cytotaxonomic studies

Table 30.3. Wildrye cultivars

Cultivar	Year released	Release agencies	Adaptation characteristics
Great Basin Wildrye			
Magnar	1979	SCS, Idaho AES	Octoploid (2n = 56) derived from accession (P-5979) obtained from Saskatoon, Saskatchewan in 1939; selected for general vigor, increased germination, and uniformity; blue-green leaves; mildly rhizomatous.
Trailhead	1991	SCS, Montana and Wyoming AES	Tetraploid (2n = 28); originally collected from silty subirrigated drainage in central Montana; more productive and persists better on arid sites than Magnar.
Canada wildrye			
Mandan 419	1946	ARS, North Dakota AES	Derived from a collection made near Mandan, North Dakota; shorter in stature, softer-textured leaves, and more persistent than common Canada wildrye.
Beardless wildrye			
Shoshone	1980	SCS, Montana and Wyoming AES	Originally collected in Wyoming in 1958; has potential for stabilizing saline seeps; difficult to establish; actually may be *Leymus multicaulus*, an Asian relative of beardless wildrye.
Russian wildrye			
Vinall	1960	ARS, North Dakota and several other AES	Five-clone synthetic; first cultivar of Russian wildrye to be released in North America; selected for improved seed yield and resistance to lodging.
Sawki	1963	Agriculture Canada, Saskatchewan	Ten-clone synthetic; selected for seed and forage yield and erect growth habit.
Mayak	1971	Agriculture Canada, Saskatchewan	Broad-based synthetic; improved forage and seed yield and resistance to leaf spot diseases.
Cabree	1976	Agriculture Canada, Alberta	Six-clone synthetic; noted for resistance to seed shattering and productivity of forage and seed; also selected for culm strength and seedling vigor.
Swift	1978	Agriculture Canada, Saskatchewan	Broad-based synthetic; derived from Sawki and North Dakota accession 1546 based on selection for seedling vigor, disease resistance, and yield of forage and seed.
Bozoisky-Select	1984	ARS, SCS, Utah AES	Derived from PI-440627 (Bozoisky) from Kazakhstan; improved stand establishment characteristics; exceptional productivity on intermountain range sites.
Tetracan	1988	Agriculture Canada, Saskatchewan	Derived from colchicine-induced tetraploids (2n = 28) from several diploid sources; larger seeds and spikes and wider leaves than diploids; improved seedling vigor.
Mankota	1991	ARS, SCS, North Dakota AES	Parental germplasm selected from 29 cultivars, strains, and accessions; substantially improved forage yield; recommended for wide range of environments in northern Great Plains.
Altai wildrye			
Prairieland	1976	Agriculture Canada, Saskatchewan	Broad-based synthetic from USSR introductions (SC 15011 and 15012); selected for improved seed and forage yield, seed quality, and resistance to leaf spot; 50% glaucous plants.
Eejay	1991	Agriculture Canada, Saskatchewan	Selected from same germplasm base as Prairieland for high seed and forage yield, seedling vigor, and resistance to leaf spot; 75% of plants are nonglaucous (green).
Pearl	1991	Agriculture Canada, Saskatchewan	Derived from same germplasm base as Prairieland and Eejay; 80% glaucous plants; better establishment, vigor, and seed yield than Prairieland, but slightly lower in dry matter yield.
Dahurian wildrye			
Arthur	1989	Agriculture Canada, Saskatchewan	Hexaploid, short-lived; excellent seedling vigor and forage yields during establishment year, and good seed yields; erect green spikes; recommended as companion crop in seedings with longer-lived grass species.
James	1989	Agriculture Canada, Saskatchewan	Heads 2 d later than Arthur; otherwise similar.

Sources: Asay, et al. (1985a), Berdahl et al. (1992), Hanson (1972), Howard (1979), Lawrence (1967, 1972, 1976, 1979), Lawrence et al. (1990a, 1990b, 1991a, 1991b), Majerus et al. (1991b), Smoliak (1976).

(Dewey 1984). It has been hybridized with slender wheatgrass in a breeding program at the University of Alberta at Edmonton (Aung and Walton 1990). The objectives are to combine the cold tolerance and persistency of Canada wildrye with agronomic characteristics of slender wheatgrass.

BEARDLESS WILDRYE (*Leymus triticoides* [Buckl.] Pilger). Beardless wildrye is a rhizomatous, moderately long-lived perennial often found on poorly drained alkaline sites. It is distributed in western US from Texas and California north to Washington and Montana. Although it is morphologically distinct from Great Basin wildrye, the two grasses have the same basic genomes, and they hybridize readily (Dewey 1984). Seed dormancy and poor seedling vigor have impeded the general acceptance of beardless wildrye in seeding mixtures. Delayed seed germination apparently is associated with impermeability of the outer seed coat. Germination has been significantly improved by oxygen application and long-term storage of up to 6 yr (Hafenrichter et al. 1949; Knapp and Wiesner 1978). It is cross-pollinated, and considerable variation has been observed among ecotypes for vegetative characteristics, seed quality, and resistance to disease. Two types are known to occur in nature. The most promising from an agronomic point of view is tall, coarse, and leafy with vigorous rhizomes, and it is a fair seed producer. The other type has shorter culms, fine stems, narrow leaves, and small spikes, and it is a poor seed producer (Hafenrichter et al. 1949).

Several accessions of beardless wildrye have been evaluated, and the cultivar 'Shoshone' was released by the USDA-SCS from a collection made in Wyoming in 1958. This cultivar has demonstrated potential for use in irrigated pastures and stabilizing saline seeps. Shoshone in fact may be *L. multicaulus,* Kar. & Kir., Tzvelev, an Asian relative of beardless wildrye that was accidently introduced to the collection site in a pasture seed mixture (J. R. Carlson 1990, personal communication).

Introduced Wildryes

RUSSIAN WILDRYE (*Psathyrostachys juncea* [Fisch.] Nevski). Russian wildrye is the only species of *Psathyrostachys* that has proven to be an important forage grass. In its native habitat, it is widely distributed on open slopes and steppes from the Middle East and Russia across Central Asia to northern China. Although it was introduced into North America in 1926, its potential in range improvement was not fully recognized until the 1950s. It since has become an important forage grass on semiarid rangelands of the intermountain West and northern Great Plains. It is typically a diploid ($2n = 14$); however, autotetraploid ($2n = 28$) forms have been artificially induced, and such types may occur in nature. The species contains the 'N' genome, which also is found in combination with other genomes in *Leymus* and *Pascopyrum* (Dewey 1984).

Russian wildrye is a cross-pollinated, caespitose, long-lived perennial with erect, naked stems. It has abundant basal leaves that have good nutritional value and are palatable to grazing animals. It has a dense spike with multiple spikelets per node and a fragile rachis. It is noted for its extensive root system that may extend to 2.5 m or more. However, the preponderance of its roots occur within 30 cm of the soil surface (Rogler and Schaaf 1963). It is often found on loam and clay soils, and it is moderately tolerant of saline and alkaline conditions. The species does not appear to be well adapted to extremely sandy soils. Observations under arid range conditions in northwest Utah indicate that drought resistance of established plants compares favorably with that of crested wheatgrass.

Long-term studies by Smoliak and Dormaar (1985) have shown that Russian wildrye is considerably more productive than native range. Over a 25-yr period, it produced 47% more forage than the native *Stipa-Bouteloua* plant association in the mixed prairie region of Alberta. Due mainly to its basal leaves, it is usually recommended for grazing instead of hay. Although it is a cool-season grass, it retains its nutritive value during the late summer and fall better than many other grasses, including crested and intermediate wheatgrass. It may, however, be less palatable to grazing animals than crested wheatgrass in the early spring. Its forage cures well and has been used for grazing during the fall and winter (Knipfel and Heinrichs 1978).

As with crested wheatgrass, early seedings of Russian wildrye often were made as a monoculture or with a legume. Environmental concerns now dictate that a more-diversified seed mix be used that includes other grasses, shrubs, and legumes. In the northern Great Plains, optimum row spacings were found to be 90 cm for seed production and 60 cm for production of forage (Lawrence and Heinrichs 1968; Leyshon et al. 1981). Results from these long-term (20-yr) studies indicate that, although forage yields during the first harvest year may be higher in narrower (30-cm) row spacings, wider spacings (60 to 90 cm) provided a more optimum environment during subsequent seasons. In these studies, Russian wildrye was more productive in association with alfalfa than when seeded alone.

Russian wildrye has two major deficiencies. First, because of poor seedling vigor, it is difficult to establish, particularly on harsh range sites. Second, commercial seed production is impeded by the tendency of its seed to shatter soon after maturity. Problems with stand establishment often are magnified by weed competition, soil crusting, and excessive depth of planting. Preparation of a firm weed-free seedbed and accurate placement of seed are critical management practices. To avoid loss of seed due to shattering, swathing in lieu of direct combining is usually recommended. Russian wildrye also has a tendency to become strongly caespitose and more widely spaced in older stands. Because these plants are strong competitors, seedlings are unable to become established in adjacent areas, and excessive erosion can result (Rogler and Schaaf 1963).

Several cultivars of Russian wildrye have been developed and released (Table 30.3). Objectives of most breeding programs are aimed at correcting the obvious limitations, particularly poor seedling vigor. Genetic progress for improved seedling vigor has been effectively attained by first screening in the laboratory and greenhouse on the basis of individual seed weight, coleoptile length, and seedling emergence from deep seedings (5 cm or more). Final selections then are based on stand establishment under actual range conditions (Lawrence 1963; Asay and Johnson 1980; Berdahl and Barker 1984). Induced tetraploids have been shown to have larger seed and better seedling vigor than their diploid counterparts (Berdahl and Barker 1991). The tetraploid cultivar 'Tetracan' was recently released by the Agriculture Canada breeding program at Swift Current (Lawrence et al. 1990b), and tetraploid germplasm has been included in USDA-ARS breeding programs at Mandan, North Dakota, and Logan, Utah, as well.

ALTAI WILDRYE (*Leymus angustus* [Trin.] Pilger). Altai wildrye is a moderately rhizomatous perennial that is naturally distributed in the Altai mountain region of western Siberia and in western Mongolia and Kazakhstan. In its native habitat, it is found on steppes, semideserts, alkaline meadows, and river and lake bottoms (Lawrence 1983). The species typically has 2n = 84 chromosomes, but types with 28, 42, and 56 also are known to occur (Dewey 1978a). Altai wildrye is noted for its excellent winterhardiness and extensive root system, which contributes to its resistance to drought. Like Russian wildrye, its leaves are mostly basal; however, it is larger and more robust than Russian wildrye. Plants are from 60 to 120 cm tall with linear, erect spikes from 10 to 20 cm long. Two spikelets occur per node and the seeds are two to three times larger than that of Russian wildrye (Nevski 1934; Lawrence 1983).

Altai wildrye has been recommended for grazing during the late fall and winter (Lawrence 1976). Its leaves remain on the plant and cure relatively well, and its erect culms and leaves protrude above the snow better than those of most range grasses. Although it has a relatively coarse growth habit, its palatability in sheep grazing trials compared favorably with crested and intermediate wheatgrass (Lawrence 1983). The tolerance of Altai wildrye to saline and alkaline conditions approaches that of tall wheatgrass (McElgunn and Lawrence 1973).

Seed dormancy and poor seedling vigor are the most serious weaknesses of Altai wildrye. Although selection for improved seedling vigor and more rapid germination has been effective (Lawrence 1977; Lawrence et al. 1991a, 1991b), it is still more difficult to establish than most wheatgrasses. Although its seed does not shatter as easily as Russian wildrye's, it is usually recommended that the seed crop be swathed in the firm dough stage and combined a week or more later. The optimum row spacing for seed production in the northern Great Plains was found to be approximately 120 cm (Lawrence 1980).

Altai wildrye was not recognized as an important forage grass until relatively recently. The first commercial cultivar, 'Prairieland', was released by Agriculture Canada, Swift Current, Saskatchewan, in 1976 (Table 30.3) (Lawrence 1976). In 1991, the cultivars 'Eejay' and 'Pearl' were released from the breeding program at Swift Current (Lawrence 1991a, 1991b). The USDA-ARS has initiated a breeding program with Altai wildrye at Logan, Utah. In this program, the species has been hybridized with Great Basin wildrye and mammoth wildrye, *L. racemosus* (Lam.) Tzvelev (Asay 1992).

DAHURIAN WILDRYE. (*Elymus dahuricus* Turcz. ex Grieseb.). This short-lived perennial, which is native to Siberia, Mongolia, and China, has yet to become an important forage grass in North America. It is a self-pollinating hexaploid (2n = 42) with the 'S', 'H', and 'Y' genomes (Lu and Bothmer 1992). Dahurian wildrye is a bunchgrass with lax wide leaves and spikes with two to four spikelets per node. It has awns from 10 to 20 mm long. Although it has more erect spikes, Dahurian wildrye is otherwise similar to Siberian wildrye (*E. sibiricus* L.).

Dahurian wildrye has good seedling vigor and is relatively productive during the usually short life of the stand. It also produces good aftermath yields after clipping or grazing. As with slender wheatgrass, it is recommended as a short-rotation forage grass or as a component of mixed seedings to provide forage while slower-growing, longer-lived perennials are becoming established (Lawrence and Ratzlaff 1985, 1989).

Only two cultivars, 'James' and 'Arthur', are commercially available. Both were licensed by Agriculture Canada, Swift Current, Saskatchewan, in 1989 (Lawrence et al. 1990a).

QUESTIONS

1. What species have arisen through genetic introgression between thickspike wheatgrass and bluebunch wheatgrass?
2 What are the three major types of crested wheatgrass?
3 Which wheatgrasses would you recommend for alkaline or saline sites?
4. What are the two basic growth forms of thickspike wheatgrass?
5. How would you use slender wheatgrass in a management system?
6. What are the major morphological and adaptation differences between intermediate wheatgrass and crested wheatgrass?
7. What cultivar was developed from the hybrid between quackgrass and bluebunch wheatgrass?
8. List the two most serious limitations of Russian wildrye?
9. What is the major difference between the Great Basin wildrye cultivars Magnar and Trailhead?
10. Describe the native habitat of Altai wildrye.

REFERENCES

Asay, KH. 1992. Breeding potentials in perennial Triticeae grasses. Hereditas 116:167-73.

Asay, KH, and DR Dewey. 1976. Fertility of 17 perennial colchine-induced Triticeae amphiploids through four generations. Crop Sci. 16:508-13.

Asay, KH, and DA Johnson. 1980. Screening for improved stand establishment in Russian wild ryegrass. Can. J. Plant Sci. 60:1171-77.

Asay, KH, JD Hansen, BA Haws, and PO Currie. 1983. Genetic differences in resistance of range grasses to the bluegrass billbug (*Sphenophorus parvulus* Coleoptera: Curculionidae). J. Range Manage. 36:771-72.

Asay, KH, DR Dewey, FB Gomm, DA Johnson, and JR Carlson. 1985a. Registration of 'Bozoisky-Select' Russian wildrye. Crop Sci. 25:575.

———. 1985b. Registration of 'Hycrest' crested wheatgrass (Reg. No. 16). Crop Sci. 25: 368-69.

Asay, KH, KB Jensen, DR Dewey, and CH Hsiao. 1990. Genetic introgression among 6x, 4x, and 2x ploidy levels in crested wheatgrass. In Am. Soc. Agron. Abstr. Madison, Wis., 79.

Asay, KH, DR Dewey, WH Horton, KB Jensen, PO Currie, NJ Chatterton, WT Hansen II, and JR Carlson. 1991a. Registration of 'NewHy' RS hybrid wheatgrass. Crop Sci. 31:1384-85.

Asay, KH, DR Dewey, KB Jensen, WH Horton KW Maughan, NJ Chatterton, and JR Carlson. 1991b. Registration of *Pseudoroegneria spicata* × *Elymus lanceolatus* hybrid germplasm SL-1. Crop Sci. 31:1391.

Asay, KH, KB Jensen, CH Hsiao, and DR Dewey. 1992. Probable origin of standard crested wheatgrass, *Agropyron desertorum* Fisch. ex Link, Schultes. Can. J. Plant Sci. 72:763-72.

Aung, T, and PD Walton. 1990. Morphology and cytology of the reciprocal hybrids between *Elymus trachycaulus* and *Elymus canadensis*. Genome 33:123-30.

Barkworth, ME, DR Dewey, and RJ Atkins. 1983. New generic concepts in the Triticeae of the intermountain region: Key and comments. Great Basin Nat. 43:561-72.

Beetle, AA. 1955. Wheatgrasses of Wyoming. Wyo. Agric. Exp. Stn. Bull. 336.

Berdahl, JD, and RE Barker. 1984. Selection for improved seedling vigor in Russian wild ryegrass. Can. J. Plant Sci. 64:131-38.

———. 1991. Characterization of autotetraploid Russian wildrye produced with nitrous oxide. Crop Sci. 31:1153-55.

Berdahl, JD, RE Barker, JF Karn, JM Krupinsky, RJ Naas, DA Tober, and IM Ray. 1992. Registration of 'Mankota' Russian wildrye. Crop Sci. 32:1073.

Bowden, WM. 1964. Cytotaxonomy of the species and interspecific hybrids of the genus *Elymus* in Canada and neighboring areas. Can. J. Bot. 42:547-601.

———. 1965. Cytotaxonomy of the species and interspecific hybrids of the genus *Agropyron* in Canada and neighboring areas. Can. J. Bot. 43:1421-48.

Carlson, JR. 1986. A study of morphological variation within *Pseudoroegneria spicata* (Pursh) A. Löve (Poaceae: Triticeae). MS thesis, Oregon State Univ., Corvallis.

Cook, CW. 1966. Development and Use of Foothill Ranges in Utah. Utah Agric. Exp. Stn. Bull. 461.

Cornelius, DR. 1965. Latitude as a factor in wheatgrass variety response on California rangeland. In Proc. 9th Int. Grassl. Congr. 1:471-73.

Coupland, RT. 1950. Ecology of mixed prairie in Canada. Ecol. Monogr. 20:271-315.

Cronquist, A, AH Holmgren, NH Holmgren, JL Reveal, and PK Holmgren. 1977. The Monocotyledons. Vol. 6, Intermountain Flora Vascular Plants of the Intermountain West, U.S.A. New York: Columbia Univ. Press.

Currie, PO, and DR Smith. 1970. Response of Seeded Ranges to Different Grazing Intensities. USDA For. Serv. Prod. Res. Rep. 112.

Daer, T, and EE Willard. 1981. Total nonstructural carbohydrate trends in bluebunch wheatgrass related to growth and phenology. J. Range Manage. 34:377-79.

Dewey, DR. 1962. Breeding crested wheatgrass for salt tolerance. Crop Sci. 2:403-7.

———. 1966. Synthetic *Agropyron-Elymus* hybrids. I. *Elymus canadensis* × *Agropyron subsecundum*. Am. J. Bot. 53:87-94.

———. 1967. Synthetic hybrids of *Elymus canadensis* × *Sitanion hystrix*. Bot. Gaz. 128:11-16.

———. 1970. The origin of *Agropyron albicans*. Am. J. Bot. 57:12-18.

———. 1972. The origin of *Agropyron leptourum*. Am. J. Bot. 59:836-42.

———. 1975a. Introgression between *Agropyron dasystachyum* and *A. trachycaulum*. Bot. Gaz. 136:122-28.

———. 1975b. The origin of *Agropyron smithii*. Am. J. Bot. 62:524-30.

———. 1976. Derivation of a new forage grass from *Agropyron repens* × *Agropyron spicatum* hybrids. Crop Sci. 16:175-80.

———. 1978a. Advanced generation hybrids between *Elymus giganteus* and *E. angustus*. Bot. Gaz. 139:369-76.

———. 1978b. Intermediate wheatgrasses of Iran. Crop Sci. 18:43-48.

———. 1982. Genomic and phylogenetic relationships among North American perennial Triticeae. In JR Estes, RJ Tyrl, and JN Brunken (eds.), Grasses and Grasslands: Systematics and Ecology. Norman: Univ. of Oklahoma Press, 51-88.

———. 1983. New nomenclatural combinations in the North American perennial Triticeae (Gramineae). Brittonia 35:30-33.

———. 1984. The genomic system of classification as a guide to intergeneric hybridization with the perennial Triticeae. In JP Gustafson (ed.), Gene Manipulation in Plant Improvement, Proc. 16th Stadler Genet. Symp. New York: Plenum, 209-79.

———. 1986. Taxonomy of the crested wheatgrasses. In KL Johnson (ed.), Crested Wheatgrass: Its Values, Problems and Myths, Symp. Proc. Utah State Univ., 3-7 Oct. 1983, Logan, 31-44.

Dewey, DR, and KH Asay. 1975. The crested wheatgrasses of Iran. Crop Sci. 15:844-49.

Dewey, DR, and PC Pendse. 1968. Hybrids between *Agropyron desertorum* and induced-tetraploid *Agropyron cristatum*. Crop Sci. 8:607-11.

Dillman, AC. 1946. The beginnings of crested wheatgrass in North America. J. Am. Soc. Agron. 38:237-50.

Douglas, DS, and RD Ensign. 1954. Sodar Wheatgrass. Idaho Agric. Exp. Stn. Bull. 234.

Dubbs, AL, RT Harada, and JR Stroh. 1974. Evaluation of Thickspike Wheatgrass for Dryland Pasture and Range. Mont. Agric. Exp. Stn. Bull. 677.

Elliott, CR, and JL Bolton. 1970. Licensed Varieties of Cultivated Grasses and Legumes. Can. Dept. Agric. Publ. 1405.

Evans, RA, and JA Young. 1983. 'Magnar' basin wildrye—germination in relation to temperature. J. Range Manage. 36:395-98.

Frank, AB, and JF Karn. 1988. Growth, water-use efficiency, and digestibility of crested, intermediate, and western wheatgrass. Agron. J. 80:677-80.

Gibbs, JL, G Young, and JR Carlson. 1991. Registration of 'Goldar' bluebunch wheatgrass. Crop Sci. 31:1708.

Gleason, HA, and A Cronquist. 1991. Manual of Vascular Plants of Northeastern United States and Adjacent Canada. 2d ed. Bronx: New York Botanical Garden.

Gomm, FB. 1981. Letter on the value of crested wheatgrass. Encl. 3-7. In G Ferry, D Luman, R Ross, W Sandau, and V Schulze (eds.), Review of the Bureau's Oregon and Washington Range Seeding Program. Portland, Oreg.: USDI Bureau of Land Management.

Hafenrichter, AL, LA Muellen, and RL Brown. 1949. Grasses and Legumes for Soil Conservation in the Pacific Northwest. USDA Misc. Publ. 678.

Hafenrichter, AL, JL Schwendiman, HL Harris, RS McLauchlan, and HW Miller. 1968. Grasses and Legumes for Soil Conservation in the Pacific Northwest and Great Basin States. USDA Agric. Handb. 339. Washington, D.C.: US Gov. Print. Off.

Hanson, AA. 1965. Grass Varieties in the United States. USDA Agric. Handb. 170. Washington, D.C.: US Gov. Print. Off.

———. 1972. Grass Varieties in the United States. USDA Agric. Handb. 170. Washington, D.C.: US Gov. Print. Off.

Hartung, ME. 1946. Chromosome numbers in *Poa, Agropyron,* and *Elymus*. Am. J. Bot. 33:516-31.

Hitchcock, AS, rev. by A Chase. 1951. Tribe 3. Hordeae. In Manual of Grasses of the United States, 2d ed., USDA Misc. Publ. 200. New York: Dover, 230-80.

Howard, CG. 1979. 'Magnar' basin wildrye (*Elymus cinereus* Scribn. and Merr.) description, adaptation, use, culture, management, and seed production. In Proc. 19th Annu. Meet. Nev. Comm. on Conserv. of Plant Matter, Nevada Agricultural Experiment Station, Univ. of Nevada, Reno, 28-31.

Jensen, KB, YF Zhang, and DR Dewey. 1990. Mode of pollination of perennial species of the Triticeae in relation to genomically defined genera. Can. J. Plant Sci. 70:215-25.

Jones, TA, DC Nielson, and JR Carlson. 1991. Developing a grazing-tolerant native grass for bluebunch wheatgrass sites. Rangelands 13:147-50.

Knapp, AD, and LE Wiesner. 1978. Seed dormancy of beardless wildrye (*Elymus triticoides* Buckl.). J. Seed Technol. 3:1-9.

Knipfel, JE, and DH Heinrichs. 1978. Nutritional quality of crested wheatgrass, Russian wild ryegrass, and Altai wild ryegrass throughout the grazing season in southwestern Saskatchewan. Can. J. Plant Sci. 58:581-82.

Knowles, RP. 1990. Registration of Kirk crested wheatgrass. Crop Sci. 30:749.

Lawrence, T. 1963. A comparison of methods of evaluating Russian wild ryegrass for seedling vigor. Can. J. Plant Sci. 43:307-12.

———. 1967. Sawki, Russian wild ryegrass. Can. J. Plant Sci. 47:612-13.

———. 1972. Mayak, Russian wild ryegrass. Can. J. Plant Sci. 52:121-22.

———. 1976. Prairieland, Altai wild ryegrass. Can. J. Plant Sci. 56:991-92.

———. 1977. Effects of selection for speed of germination on establishment, vigor, and yield in Altai wild ryegrass. Can. J. Plant Sci. 57:1085-90.
———. 1979. Swift, Russian wild ryegrass. Can. J. Plant Sci. 59:515-18.
———. 1980. Seed yield of Altai wild ryegrass as influenced by row spacing and fertilizer. Can. J. Plant Sci. 60:249-53.
———. 1981. Clarke intermediate wheatgrass. Can. J. Plant Sci. 61:467-69.
———. 1983. Altai wildryegrass. Agric. Can. Publ. 1602. Swift Current, Saskatchewan: Agriculture Canada Research Station.
Lawrence, T, and DH Heinrichs. 1968. Long-term effects of row spacing and fertilizer on the productivity of Russian wild ryegrass. Can. J. Plant Sci. 48:75-84.
Lawrence, T, and DC Ratzlaff. 1985. Evaluation of fourteen grass populations as forage crops for southwestern Saskatchewan. Can. J. Plant Sci. 65:951-57.
———. 1989. Emergence of Dahurian wild ryegrass influenced by depth of seeding. Forage Notes 34:46-49.
Lawrence, T, PG Jefferson, and CD Ratzlaff. 1990a. James and Arthur, two cultivars of Dahurian wild ryegrass. Can. J. Plant Sci. 70:1187-90.
———. 1991a. Eejay, Altai wild ryegrass. Can. J. Plant Sci. 71:551-53.
———. 1991b. Pearl, Altai wild ryegrass. Can. J. Plant Sci. 71:547-49.
Lawrence, T, AE Slinkard, CD Ratzlaff, NW Holt, and PG Jefferson. 1990b. Tetracan, Russian wild ryegrass. Can. J. Plant Sci. 70:311-13.
Leyshon, AJ, MR Kilcher, and JD McElgunn. 1981. Seeding rates and row spacings for three forage crops grown alone or in alternate grass-alfalfa rows in southwestern Saskatchewan. Can. J. Plant Sci. 61:711-17.
Looman, J, and DH Heinrichs. 1973. Stability of crested wheatgrass pastures under long-term pasture use. Can. J. Plant Sci. 53:501-6.
Lorenz, RJ. 1986. Introduction and early use of crested wheatgrass in the northern Great Plains. In KL Johnson (ed.), Crested Wheatgrass: Its Values, Problems and Myths, Symp. Proc. Utah State Univ., 3-7 Oct. 1983, Logan, 9-20.
Löve, AA. 1980. IOPB chromosome number reports. LXVI. Poaceae-Triticeae-Americanae. Taxon. 29:163-69.
Lu, Bao-Rong, and R von Bothmer. 1992. Interspecific hybridization between *Elymus himalayanus* and *E. schrenkianus,* and other *Elymus* species (Triticeae: Poaceae). Genome 35:230-37.
McElgunn, JD, and T Lawrence. 1973. Salinity tolerance of Altai wild ryegrass and other forage grasses. Can. J. Plant Sci. 53:303-7.
Majerus, ME, JG Scheetz, and LK Holzworth. 1991a. Pryor Lender Wheatgrass, a Conservation Plant for Montana and Wyoming. Leafl. Bridger, Mont.: USDA, Soil Conservation Service, Plant Materials Center.
———. 1991b. Trailhead Basin Wildrye, a Conservation Plant for Montana and Wyoming. Leafl. Bridger, Mont.: USDA, Soil Conservation Service, Plant Materials Center.
Malakouti, MJ, DT Lewis, and J Stubbendieck. 1978. Effect of grasses and soil properties on wind erosion in sand blowouts. J. Range Manage. 31:417-20.
Mason, JL, and JE Miltimore. 1959. Increase in yield and protein content of native bluebunch wheatgrass from nitrogen fertilization. Can. J. Plant Sci. 39:501-4.
Mathews, WL. 1986. Early use of crested wheatgrass seedings in halogeton control. In KL Johnson (ed.), Crested Wheatgrass: Its Values, Problems and Myths, Symp. Proc. Utah State Univ., 3-7 Oct. 1983, Logan, 27-28.
Miller, RF, JM Seufert, and MR Haferkamp. 1987. The Ecology and Management of Bluebunch Wheatgrass (*Agropyron spicatum*): A Review. Oreg. State Univ. Agric. Exp. Stn. Bull. 669. Corvallis.
Monson, RK, MR Sackschewsky, and GJ Williams III. 1986. Field measurements of photosynthesis, water-use efficiency, and growth in *Agropyron smithii* (C_3) and *Bouteloua gracilis* (C_4) in the Colorado shortgrass steppe. Oecol. 68:400-409.
Morrison, KJ, and CA Kelley. 1981. Secar Bluebunch Wheatgrass. Wash. State Univ. Ext. Serv. EB-0991.
Morrison, KJ, and HH Wolfe. 1957. Greenar Intermediate Wheatgrass. Wash. State Univ. Ext. Circ. 272.
Mueggler, WF. 1975. Rate and pattern of vigor recovery in Idaho fescue and bluebunch wheatgrass. J. Range Manage. 28:198-204.
Nevski, SA. 1934. Tribe XIV. Hordeae Benth. In VL Komarov (ed.), Flora of the USSR, vol. 2. Jerusalem: Israel Program for Scientific Translation, 469-579.
Newell, LC. 1974. Registration of Slate intermediate wheatgrass (Reg. No. 10). Crop Sci. 14:340-41.
New Mexico Cooperative Extension Service. 1977. Arriba western wheatgrass. Circ. 475.
Oram, RN. 1981. Register of Australian herbage plant cultivars. J. Aust. Inst. Agric. Sci. 47:179-80.
Rauser, WE, and WL Crowle. 1962. Salt tolerance of Russian wild ryegrass in relation to tall wheatgrass and slender wheatgrass. Can. J. Plant Sci. 43:397-407.
Rogler, GA. 1973. The wheatgrasses. In ME Heath, DS Metcalfe, and RF Barnes (eds.), Forages: The Science of Grassland Agriculture. 3d ed. Ames: Iowa State Univ. Press, 221-30.
Rogler, GA, and HM Schaaf. 1963. Growing Russian Wildrye in the Western States. USDA Leafl. 313.
Rumbaugh, MD, DA Johnson, and GA Van Epps. 1982. Forage yield and quality in a Great Basin shrub, grass, and legume pasture experiment. J. Range Manage. 35:604-9.
Scheetz, JG, ME Majerus, and JR Carlson. 1981. Improved plant materials and their establishment to reclaim saline seeps in Montana. In Am. Soc. Agron. Abstr. Madison, Wis., 96.
Sharma, HC, and BS Gill. 1983. Current status of wide hybridization in wheat. Euphytica 32:17-31.
Sinskaja, EN. 1961. The levels of group adaptation in plant populations. Plant Breed. Abstr. 31:763-64.
Smoliak, S. 1976. Cabree Russian wild ryegrass. Can. J. Plant Sci. 56:993-96.
Smoliak, S, and JF Dormaar. 1985. Productivity of Russian wildrye and crested wheatgrass and

their effect on prairie soils. J. Range Manage. 38:403-5.
Smoliak, S, and A Johnston. 1980. Elbee northern wheatgrass. Can. J. Plant Sci. 60:1473-75.
———. 1984. Registration of Walsh western wheatgrass. Crop Sci. 24:1216.
Stebbins, GL. 1955. Experimental origin of a reproductively isolated population in the grass genus *Elymus*. Sci. 121:625.
Stroh, JR., AA Thornburg, and DE Ryerson. 1972. Registration of Critana thickspike wheatgrass. Crop Sci. 12:394.
Tzvelev, NN. 1976. Tribe 3. Triticeae Dum. In Poaceae USSR. Leningrad: Nauka Publishing House, 105-206.
USDA Extension Service. 1978. New crop cultivars. ESC 584 13:209-11.
Wagoner, P. 1990. Perennial grain new use for intermediate wheatgrass. J. Soil and Water Conserv. 45:81-82.
Wang, R R-C, and KB Jensen. 1994. Absence of the J genome in *Leymus* species (Poaceae: Triticeae): Evidence from DNA hybridization and meiotic pairing. Genome 37:231-35.
Weintraub, FC. 1953. Grasses Introduced into the United States. USDA Agric. Handb. 58. Washington, D.C.: US Gov. Print. Off.
Wilson, DB, and S Smoliak. 1977. Greenleaf pubescent wheatgrass. Can. J. Plant Sci. 57:289-91.
Wolfe, HH, and KJ Morrison. 1957a. Whitmar Beardless Wheatgrass. Wash. State Univ. Ext. Circ. 273.
———. 1957b. Nordan Crested Wheatgrass. Wash. Ext. Circ. 274.
Young, JA, and RA Evans. 1986. History of crested wheatgrass in the intermountain area. In KL Johnson (ed.), Crested Wheatgrass: Its Values, Problems and Myths. Symp. Proc. Utah State Univ., 3-7 Oct. 1983, Logan, 21-25.
Zhang, H-B, and J Dvorak. 1991. The genome origin of tetraploid species of *Leymus* (Poaceae: Triticeae) inferred from variation in repeated nucleotide sequences. Am. J. Bot. 78:871-84.

31 Grasses of the Plains and Southwest

PAUL W. VOIGT
Agricultural Research Service, USDA

W. CURTIS SHARP
Soil Conservation Service, USDA

GRASSLANDS OF THE US

THE prairies of the Great Plains form the largest area of true grass range in the world. These natural grasslands, which evolved over time in response to the earth's variable climate, extend from Mexico into Canada and provide a major forage base for animal agriculture in the US.

These grasslands have changed over geologic time, ranging in character from woodland during moist, cool eras to grasslands in more arid periods (Axelrod 1985). Between 4000 and 8000 yr ago a drying trend extended an arm of the prairies between the Ohio River and the Great Lakes all the way to the Appalachian Mountains (Transeau 1935). As wetter times returned, the prairie again retreated westward as the forest advanced, leaving behind soils of grassland origin and relict prairie communities. Change continues today in response to the environment. For example, black grama, *Bouteloua eriopoda* (Torr.) Torr., in parts of southern New Mexico is being replaced by honey mesquite, *Prosopis glandulosa* Torr. var. *glandulosa* (Gibbens et al. 1992). The effects of humans, through increasing levels of atmospheric carbon dioxide (CO_2) that can favor C_3 woody and herbaceous plants over C_4 grasses (Chap. 3), are a new factor that can interact with precipitation and temperature changes and affect grassland composition and function (Mayeux et al. 1992).

The major grasslands of the US, where grasses discussed in this chapter have evolved or have been introduced, are the tall- and short-grass regions of the Great Plains and the desert-grassland region of the Southwest (Fig. 31.1). The tall grasses, e.g., big bluestem (*Andropogon gerardii* [L.] Vitman), switchgrass (*Panicum virgatum* L.), and indiangrass (*Sorghastrum nutans* [L.] Nash) (Chap. 32), occur in a 250- to 800-km-wide

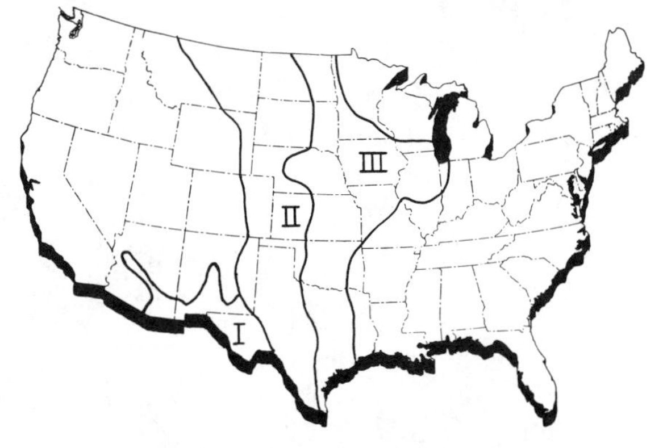

Fig. 31.1. Three major grassland regions of the US are (*I*) desert-grassland (and associated shrubland) of the Southwest, (*II*) short-grass region from the Rocky Mountains to about the 100th meridian, and (*III*) tall-grass region continuing to the forest edge. Cropping, overgrazing, and woody plant encroachment have greatly altered large parts of these grasslands.

PAUL W. VOIGT, ARS, USDA, received the MS and PhD degrees from the University of Wisconsin at Madison. He has spent most of his career in the southern Great Plains. His primary research interests are the study of apomixis and the improvement of lovegrasses, but he has also worked with bluestems, gramas, and other grasses.

W. CURTIS SHARP is SCS National Plant Materials Specialist, SCS, USDA, Washington, D.C. He received his MS degree from Pennsylvania State University. He has worked at local, state, regional, and national levels in developing plants for conservation since 1957.

strip along the eastern edge (zone III), which has been protected from advancing woody (C_3) vegetation by fire. In the drier region to the west lies the short-grass prairie (zone II). Its most important warm-season grasses include blue grama, *B. gracilis* (H.B.K.) Lag. ex Steud., and buffalograss, *Buchloe dactyloides* (Nutt.) Engelm.

The east to west transition from tall to short grasses is not abrupt, although relatively sharp transitions can occur across soil boundaries because of differences in plant-soil-water relations. Sandy soils from western Nebraska to eastern New Mexico provide habitat favorable for tall grasses. Finer-textured soils in this region cannot support tall grasses under grazing (see Voigt and MacLauchlan 1985). Southwest of the short-grass region lies the desert grassland (Cable 1975; Martin 1975). This discontinuous grassland, interspersed with shrublands, extends from southwest Texas to southern Arizona and far into northern Mexico. Major grasses include black grama, tobosagrass (*Hilaria mutica* [Buchl.] Benth.), and Arizona cottontop (*Digitaria californica* [Benth.] Henr.).

GRASS ECOTYPES

Traveling south to north the length of the Great Plains, little bluestem (*Schizachyrium scoparium* [Michx.] Nash), sideoats grama (*B. curtipendula* [Michx.] Torr.), blue grama, and other grasses can be followed from Mexico to Canada. These warm-season grasses are adapted to a wide climatic range, but the genetic makeup of regional strains differs. When grown in Kansas, seed of blue grama from North Dakota and other northern states produced plants that flowered earlier, were shorter, and produced fewer tillers than seed from ecotypes from Texas. Southern ecotypes grew continuously throughout the growing season; some did not mature seed prior to frost. Ecotypes from the central section ceased growth about October 1; those from the North ceased growth at the end of August (Riegel 1940). These differences resulted from the response of the different ecotypes to photoperiod (Fig. 31.2). Thus, a North Dakota sideoats grama consisted largely of long-day plants with a critical photoperiod of about 14 h. Populations from south Texas consisted almost entirely of short-day or intermediate plants (Olmstead 1944). Numerous studies

Fig. 31.2. Adaptation of warm-season grass ecotypes found throughout the Great Plains is controlled by response to photoperiod. Sideoats grama strains grown at Manhattan, Kansas, were from (*A*) North Dakota, (*B*) Nebraska, (*C*) Kansas, (*D*) Oklahoma, and (*E*) Texas. Measuring stick shows height in feet. *SCS photo.*

have shown similar responses in growth and flowering of the most important warm-season but not the cool-season grasses of the Great Plains (McMillan 1959). Thus, warm-season grasses such as the gramas that have their centers of diversity in the Southwest (see Voigt and MacLauchlan 1985) have evolved a response to photoperiod that allowed them to extend their range to the north and develop the grasslands we know today.

GRASS BREEDING

Because of ecotypic gradients in photoperiod response, different cultivars of warm-season grasses are needed for use in different latitudes of the Great Plains. Cultivars cannot be moved very far south of their areas of origin because of reduced productivity from early-flowering and early fall dormancy. They cannot be moved too far north because seed necessary for long-term stand maintenance may not mature because cultivars have insufficient winterhardiness (see Voigt and MacLauchlan 1985). Moving germplasm too far east can result in increased disease severity because of higher humidity (Zeiders 1982). Drought can adversely impact eastern ecotypes when moved west.

Impetus for much of the genetic work with grasses native to the Great Plains came from the dustbowl that developed as a result of plowing rangeland for crop production and the drought of the 1930s (Fig. 31.3). Initially, seed production depended on harvests from remaining grass stands. However, such harvests, although still important for some species, are not reliable because of their dependence on natural rainfall and fertility. The SCS plant materials program and programs of ARS and state agricultural experiment stations have produced many cultivars (Table 31.1) (Alderson and Sharp 1993). These cultivars, when planted in rows, fertilized, and irrigated, can provide a more reliable, uniform, and higher-quality source of seed than is frequently collected from rangeland harvests.

GRASS SEED PRODUCTION

Although grass seed production, harvesting, and processing can present special problems, techniques for many grasses are well developed (see Voigt and MacLauchlan 1985). Problems caused by uneven maturity, seed shattering, awns or other seed appendages, and seed dormancy are common. Removal of awns by use of a hammermill or debearder is a common practice to improve purity and seed flow characteristics. Estimating purity of chaffy-seeded grasses, notably the bluestems, gramas, and indiangrass is important in determining the value of seed lots and their seeding rates. Purity of chaffy-seeded species can vary greatly because of stem pieces and other trash, seed appendages, and the structure of the seed unit (Fig. 31.4). In sideoats grama the seed unit itself can vary from a

Fig. 31.3. Plowing of rangeland coupled with drought resulted in wind erosion that destroyed soil, crops, and people's lives. Gradual climatic shifts have sometimes limited natural redevelopment of rangelands with original species. *ARS photo.*

TABLE 31.1. Perennial grass cultivars

Species	Cultivars[a]
	Warm-season grasses
Bluestem, little	Aldous, Blaze, Cimarron, Pastura
Bluestem, Old World	
Bothriochloa	Caucasian, Ganada, King Ranch, Plains, WW Iron Master, WW Spar
Dichanthium	Gordo, Kleburg, Medio, PMT-587
Buffalograss (seeded)	Bison, Comanche, Plains, Texoka, Topgun, Sharp's Improved
Buffelgrass	T-4464 (common), Llano, Nueces
Crabgrass	Red River
Galleta	Viva
Gamagrass, eastern	Iuka, Pete, Shepherd Select
Grama, black	Nogal
Grama, blue	Alma, Hachita, Lovington
Grama, sideoats	Butte, El Reno, Haskell, Kildeer, Niner, Pierre, Trailway, Vaughn
Kleingrass	Selection 75, Verde
Lovegrass, atherstone	Cochise
Lovegrass, boer	A-84 (common), Catalina
Lovegrass, lehmann	A-68 (common)
Lovegrass, sand	Bend, Mason, Neb. 27
Lovegrass, weeping	A-67 (common), Ermelo, Morpa, Renner
Lovegrass, wilman	Palar
Muhley, spike	El Vado
Panic, blue	A-130 (common)
Rhodesgrass	Bell
Sacaton, alkali	Saledo, Saltalk
Sandreed, prairie	Goshen, Pronghorn
Sprangletop, green	Van Horn
	Cool-season grasses
Foxtail, creeping	Garrison, Retain
Foxtail, meadow	Dan, Mountain
Needlegrass, green	Lodorm
Ricegrass, indian	Nezpar, Paloma

[a]Cultivars followed by "common" are usually available as common seed, i.e., cultivar unstated, although all common seed may not be of that cultivar. Most listed cultivars are commercially available.

naked seed to an entire spike containing at least one seed. Seeding rates must be based on the number of pure live seed (PLS) per area. The PLS content is the percentage pure seed times the percentage germination divided by 100. For example, if a lot of bulk seed of big bluestem contains 65% by weight of seed, and 35% is trashy material, and the germination percentage of the pure seed is 60%, the weight of PLS is calculated as

$$PLS = (65 \times 60)/100 = 3900/100 = 39\%$$

of the total bulk.

PERENNIAL WARM-SEASON GRASSES

Bluestems

BIG BLUESTEM AND SAND BLUESTEM. See Chapter 32 for big bluestem (*Andropogon gerardii* Vitman) and sand bluestem (*A. gerardii* var. *paucipilus* [Nash] Fern.) (syn. *A. hallii* Hack.).

LITTLE BLUESTEM. *Schizachyrium scoparium* (Michx.) Nash, a primary grass of the tall-grass region, is a bunchgrass with or without rhizomes that grows to a height of about 1 m. It is a morphologically diverse and taxonomically complex species (see Voigt and MacLauchlan 1985) and is found throughout the contiguous US except for Nevada and the Pacific Coast states. Little bluestem is readily grazed before heading of the reproductive tillers, but it is not well accepted by livestock at later stages of growth. However, repeated close defoliations during the growing season can result in loss of stand (Mullahey et al. 1990). Little bluestem is more drought resistant and can persist under grazing on heavier soils in drier climates than the other North American tall grasses. Creeping bluestem, *S. stoloniferum* Nash, is a related species important in parts of the southeastern US (Kalmbacher et al. 1986).

OLD WORLD BLUESTEMS. These bluestems, part of a complex of species in the genera *Bothriochloa, Capillipedium,* and *Dichanthium,* reproduce predominantly by apomixis (deWet and Harlan 1970) and are best adapted to fine-textured soils. They are native to Africa, the Middle East, and southern Asia. Yellow bluestem, *B. ischaemum* (L.) Keng., because of its ease of establishment, drought resistance, and winterhardiness (Sims and Dewald 1982), is the most important species of this complex in the US. Caucasian

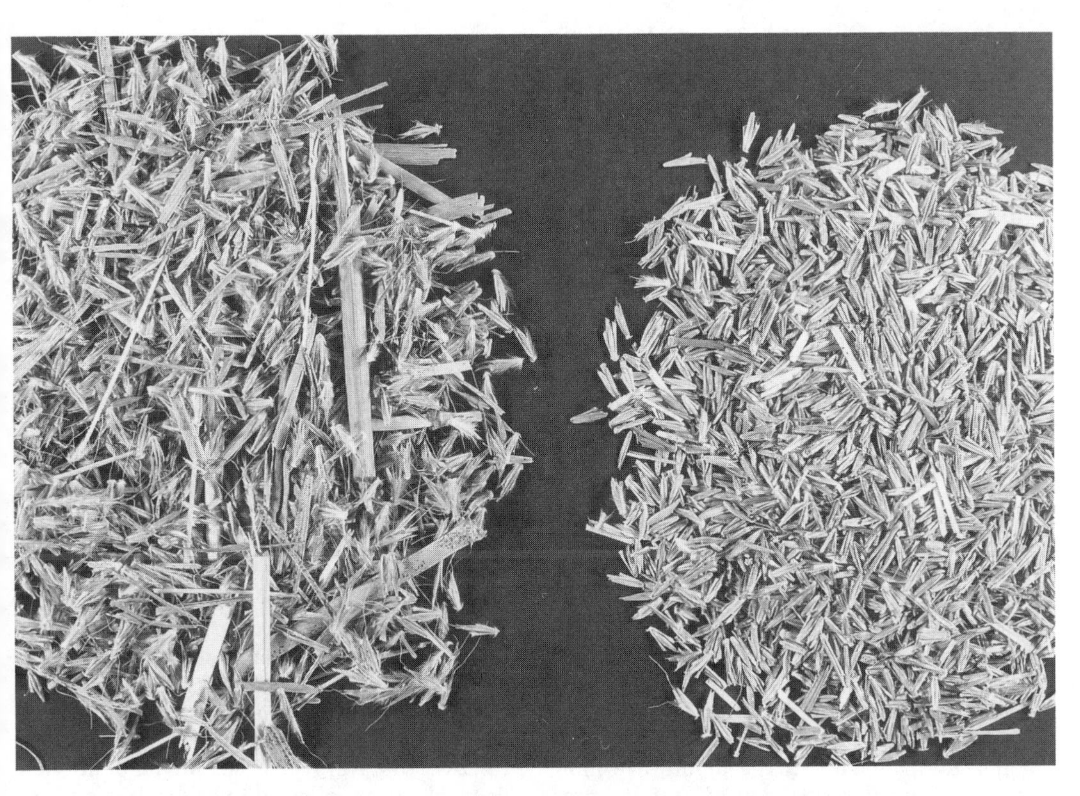

Fig. 31.4. *Left,* commercial big bluestem; *right,* debearded and cleaned big bluestem. Debearding removes appendages from the seed, and cleaning increases the percentage of pure seed. Clean seed flows easier through handling and seeding equipment. *SCS photo.*

bluestem, *B. caucasia* (Trin.) C.E. Hubb., is slightly more winter-hardy than most of the yellow bluestems but may be less drought resistant than some of them (Coyne and Bradford 1986). Caucasian bluestem is adapted to the southern Corn Belt, where it provides high yields of medium quality forage for midsummer grazing. Recently released cultivars of yellow bluestem (Table 31.1) have supplanted 'King Ranch' bluestem because of their greater winterhardiness and resistance to leaf rust. There are no specific cultivars of caucasian bluestem. The weedy nature of the Old World bluestems attests to their aggressive establishment ability.

Because of crown structure, the Old World bluestems are tolerant of continuous stocking at high stocking rates (Christiansen and Svejcar 1987, 1988; Forwood et al. 1988). Their crown consists of densely packed stem bases, containing numerous sites for tiller initiation, and leaf material that, because of its protected location, escapes defoliation. Stocking at high rates can result in more tillers per meter than stocking at low rates. Because of an indeterminate growth habit (Anderson and Matches 1983), the Old World bluestems, when grown in the southern Great Plains, are frequently more responsive to added nitrogen (N) fertilizer and late-season precipitation than big bluestem or switchgrass (Taliaferro et al. 1975).

The forage quality of Old World bluestems declines rapidly as they mature (Dabo et al. 1987, 1988). To achieve optimum animal performance for growing livestock, forage must be kept in a vegetative and actively growing condition. In general, the Old World bluestems are less palatable to livestock than many of the grasses native to North America, and it is recommended they be planted as pure stands so they can be fenced and managed as separate pastures (Dewald et al. 1985). They are frequently used in grazing systems to provide summer pasture in areas where perennial cool-season grasses are low in summer production.

Although *Dichanthium* species are of considerable importance worldwide (see Chap. 23), they are of limited interest in the US because of a lack of adequate winterhardiness and other problems. 'Kleberg' bluestem, *D.*

annulatum (Forssk.) Stapf., is well adapted to parts of South Texas but provides poor quality forage (Mutz and Drawe 1983). 'Pretoria 90' has good forage quality (Conrad and Holt 1983) but produces very little seed.

Seed production of the Old World bluestems is difficult because of indeterminate flowering, unreliable seed set, and seed appendages. However, seed production methods are known (see Voigt and MacLauchlan 1985), and new seed-harvesting technology (Dewald et al. 1985) has increased availability of recent cultivars. Seed production problems have prevented commercialization of cane bluestem, *B. barbinodis* (Lag.) Herter, a North American species.

Buffalograss. *Buchloe dactyloides* (Nutt.) Engelm. is a major species of the short-grass region from Texas to South Dakota. It is frequently found on soils of higher clay content and does not do well on sandy soils. The species is low growing, drought resistant, and spreads by numerous stolons to form a dense sod. Buffalograss is predominantly dioecious, with male and female plants usually in equal frequency. About 10% of the plants are monoecious (Wenger 1943). Sex of buffalograss plants appears to be under polygenic control (Huff 1991). Male flowers extend 3-6 cm above the foliage while female flowers occur in burlike clusters of three to five spikelets within the leafy canopy. Buffalograss withstands heavy grazing and has an excellent reputation as cured winter feed. However, it can be less palatable than other warm-season grasses and is lower in crude protein than blue grama (see Voigt and MacLauchlan 1985).

Despite extensive research, seed production of buffalograss used to be erratic because seed was harvested from rangelands that produced only 60-110 kg/ha of burs (Wenger 1943). The cultivar 'Texoka', selected to produce a higher than normal percentage of seed-bearing plants in its offspring, can produce 1100 kg/ha of burs (see Voigt and MacLauchlan 1985). Seed dormancy can be a serious problem affecting buffalograss stand establishment. However, soaking seed in a 0.5% solution of KNO_3 for 24 h, followed by chilling at 5°C for 6 wk, will effectively break dormancy (Wenger 1943). Caryopses germinate readily when extracted from the bur.

Buffalograss persists well with minimum levels of management and has excellent drought resistance, which has led to increased interest for using it as turf. Some buffalograss turf cultivars are propagated vegetatively.

Buffelgrass. *Cenchrus ciliaris* L. is native from southern Africa to India and was introduced to south Texas in 1946. Common buffelgrass, 'T-4464', is a drought tolerant bunchgrass that can grow to a height of 1 m or more. Although seed harvesting can be difficult, because of uneven maturity and fluffy seed, and dormancy can hinder establishment, buffelgrass has become the most important seeded grass in south Texas and in northern Mexico (Hanselka 1985). Buffelgrass is best adapted to sandy loam soils and is not adapted to deep sands or soils with poor surface drainage (Williamson and Pinkerton 1985). Buffelgrass is productive, high in forage quality, and is readily grazed by livestock (Hanselka 1985). While it is more tolerant of grazing than more upright grasses, it cannot tolerate continuous heavy stocking. It establishes relatively easily and has the potential to spread into unseeded areas. In south Texas it is subject to injury from sudden winter freezes. Discovery of a sexual plant in this apomictic grass allowed hybridization and selection of new apomictic cultivars (Taliaferro and Bashaw 1966). Cultivars developed in this way such as 'Nueces' have gained acceptance because of the increased protection from cold temperatures resulting from rhizomes and because of the increased forage production.

Eastern Gamagrass. *Tripsacum dactyloides* (L.) L. has long had a reputation as a valuable forage grass. However, because of difficulties in seed production, seed quality (caused by in part by sterility), and establishment (Ahring and Frank 1968), it has received little research attention and commercial use until recently. Gamagrass is a highly variable species. Diploids reproduce sexually, while polyploids are highly apomictic (Burson et al. 1990). In the US, gamagrass is found from Texas north to Kansas and east to Massachusetts. This tall grass of wet habitats has high photosynthetic rates and water use efficiency when moisture is adequate (Coyne and Bradford 1985).

Gamagrass seeds are formed within a hard cupulate fruit case that increases seed dormancy. A fall planting that relies on natural stratification during the winter to reduce seed dormancy, resulting in spring emergence, is a good seeding practice (Ahring and Frank 1968).

Interest in this grass was stimulated by discovery of a gynomonoecious mutant that changes the upper part of the inflorescence from male to female (Dewald et al. 1987). This recessive characteristic allows plants to produce 10-25 times more seed than the normal inflorescence (Jackson et al. 1992). Unfortunately, that seed is only 30% to 60% as large as those produced on normal plants. This reduction in seed size results in greatly reduced establishment of the slow-developing seedlings.

To achieve maximum sustainable yields, gamagrass must be fertilized with N (Kalmbacher et al. 1990) and receive adequate moisture. Whole plant digestibility of gamagrass varies from low, 400-480 g/kg (Kalmbacher et al. 1990), to moderate, 520-570 g/kg (Horner et al. 1985). However, diets selected by cattle grazing 'Pete' gamagrass were very high in digestibility, 773 g/kg (Burns et al. 1992). Average daily gains for steers grazing gamagrass exceeded those for steers grazing flaccidgrass, *Pennisetum flaccidum* Griseb., or bermudagrass, *Cynodon dactylon* (L.) Pers. The leafiness of gamagrass was an important factor in its excellent performance. These results suggest that eastern gamagrass could become an important forage grass for summer grazing in parts of the Corn Belt and areas to the south. In drier areas, irrigation would be required for successful use.

Gramagrasses

BLUE GRAMAGRASS. *Bouteloua gracilis* (H.B.K.) Lag. ex Steud. is found in the short-grass region from Texas to Canada and is of importance also in parts of the desert grassland. In the northern part of its range, blue grama may be less important than the cool-season grasses with which it is associated, but in the southern part of the short-grass region it can occur in almost pure stands. Blue grama is a bunchgrass with narrow, primarily basal leaves and a mature culm height of 25-60 cm. It produces a large number of vegetative shoots, only a small portion of which become reproductive. Thus, it is well adapted to continuous stocking because regrowth from active shoot apices located low in the canopy can occur quickly if growing conditions are favorable. Quality of blue grama forage is excellent (see Voigt and MacLauchlan 1985).

Blue grama stand establishment has been particularly difficult because of its pattern of seedling development (Hyder et al. 1971; also see Chap. 2). The soil surface must remain moist for 2-4 d during initial germination, emergence, and seedling growth. A second period of 2-4 d with a moist soil surface about 2-8 wk later is required for growth of the adventitious roots that are essential for establishment (Wilson and Briske 1979). The fact that blue grama has not become reestablished in Central Plains fields that were abandoned in the 1930s suggests the climate has changed since those grasslands evolved.

Genetic variation among blue grama collections appears more extensive in the Southwest (Harlan 1958) than in the northern US (Barker et al. 1983). This discovery is in accord with the suggestion of a southern origin for gramagrasses. Seedling growth parameters from New Mexican germplasm were superior to those from Kansas at warm temperatures and were not inferior at cooler temperatures (Wilson 1981). This variation could be useful in improving establishment of this difficult to establish grass.

SIDEOATS GRAMAGRASS. *Bouteloua curtipendula* (Michx.) Torr. is more widely distributed than blue grama, being found throughout all three grassland regions. This grass grows to a height of 0.7-1.0 m and usually is found in mixed stands. Sideoats is divided into two taxonomic groups, *B. curtipendula* var. *curtipendula* and var. *caespitosa*. Plants of var. *curtipendula* are found throughout the tall- and short-grass regions (Gould 1959) and spread readily from rhizomes. They are tetraploid, 40 chromosomes, and reproduce sexually. Plants of var. *caespitosa* are bunch types without rhizomes, have 85-101 chromosomes, and reproduce apomictically (Harlan 1949). The rhizomatous type contains the ecotypic variation in response to photoperiod seen in many warm-season grasses native to the Great Plains (Hopkins 1941), but the bunch-type sideoats grama found in the Southwest has not evolved to be photoperiod sensitive (McMillan 1961).

Cultivars of both bunch and spreading types have been released, but the bunch-type cultivars, e.g., 'Premier', are no longer available. Part of the reason for the lack of popularity of the bunch cultivars is their poor palatability compared with spreading types. The abundance of stems, responsible for the high seed production of the bunch types, may be the cause. Large clones of spreading cultivars produce few stems. Where winterhardiness is not a problem, bunch types can be more productive than spreading ecotypes.

BLACK GRAMAGRASS. *Bouteloua eriopoda* (Torr.) Torr. is one of the most important species of the desert-grassland region. It grows in open stands on coarse-textured soils (see Voigt and MacLauchlan 1985). Despite the presence of stolons that are necessary for long-term perennation, black grama is considered a bunchgrass because few stolons are produced. Black grama flowering culms are 20-60 cm long and leaves are short, thin, and inconspicuous. Individual tufts spread slowly by tillering. Seedlings rarely become established in stands (Valentine 1970).

Black grama stands are susceptible to damage by extreme drought and heavy stocking. Thinned stands fill in only slowly through establishment of new propagules from stolons. Extensive damage cannot be easily repaired because stolon length is usually limited to a few centimeters (Valentine 1970). Further, two successive favorable growing seasons are required for vegetative propagation, one for formation of the stolon and the new plant, the second for rooting and establishment (see Voigt and MacLauchlan 1985). Average life span of black grama plants in south central New Mexico was only 2.2 yr (Wright and Van-Dyne 1976). Climate change appears to be a major factor in the replacement of some black grama stands with honey mesquite (Hennessy et al. 1983).

Hilaria

TOBOSAGRASS. *Hilaria mutica* (Buckl.) Benth. is a productive grass that spreads by rhizomes. It is best adapted to fine-textured soils and predominates in depressions and swales. It occurs in the desert grassland and parts of the southern short-grass regions and will grow any time during the frost-free season that soil moisture is available. Tobosa culms grow to a height of 30 to 75 cm but produce few fertile seeds. Controlled burning is an important tool for removing old unpalatable growth and improving productivity (Neuenschwander et al. 1975). Rotation stocking produces more liveweight gain than continuous stocking (Anderson 1988).

OTHER *HILARIAS*. Galleta, *H. jamesii* (Torr.) Benth., is similar to tobosa in general characteristics. However, it is more widely distributed as it occurs from Texas to Arizona and as far north as Wyoming. Its palatability to livestock is low unless the grass is succulent. It is adapted to fine-textured soils receiving runoff from adjacent areas (see Voigt and MacLauchlan 1985).

Curly mesquite, *H. belangeri* (Steud.) Nash, an important component of south and west Texas rangelands, is only 10-25 cm tall and is adapted to the same general region as tobosa. It cures well as standing forage and is readily accepted by all classes of livestock. It spreads by stolons, is quite drought resistant, and withstands close defoliation (see Voigt and MacLauchlan 1985).

Lovegrasses

WEEPING LOVEGRASS. *Eragrostis curvula* (Schrad.) Nees was introduced into the US from Africa (Crider 1945), and it is part of a large apomictic complex. Weeping lovegrass is a bunchgrass with long, narrow, primarily basal leaves and a mature plant height of 1-1.5 m. It is winter-hardy to northern Oklahoma and best adapted to sandy soils. It responds well to low rates of N fertilizer and has excellent regrowth potential from its many vegetative tillers. Its long growing season, for a warm-season grass, is important in parts of Oklahoma and Texas where there are few cool-season perennial grasses. Weeping lovegrass's relatively poor forage quality is overcome with intensive management (McIlvain and Shoop 1970; Dahl and Cotter 1984). 'Morpa', a cultivar with a more favorable lignin:cellulose ratio produces higher animal gains than common weeping lovegrass (Voigt et al. 1970). Weeping lovegrass is most effectively used when it is grown in monoculture to serve as one component of a multiforage grazing system (McIlvain 1976). Seed of weeping lovegrass is easy to produce, and stand establishment is not difficult.

BOER LOVEGRASS. *Eragrostis curvula* (Schrad.) Nees var. *conferta* Nees is more drought resistant and less winter-hardy than weeping lovegrass and grows to a height of 0.7-1.0 m. Boer lovegrass is most useful for revegetation in the desert-grassland region (Cox et al. 1982). Rare sexual strains of boer lovegrass will hybridize readily with weeping lovegrass (Voigt and Bashaw 1972), demonstrating the close relationship between the two grasses. Boer lovegrass is the most palatable member of the *E. curvula* apomictic complex (Voigt et al. 1986) and has potential as a grass for rangelands.

LEHMANN LOVEGRASS. *Eragrostis lehmanniana* Nees differs from boer and weeping lovegrass in that its procumbent stems easily root from the nodes and it frequently does not form the sturdy crown typical of the *E. curvula* complex. This weakly stoloniferous growth habit produces an open sod. Lehmann lovegrass is less winter-hardy but is easier to establish than boer lovegrass. Lehmann lovegrass is adapted to certain climatic zones of the desert-grassland region where it has become naturalized (Cox and Ruyle 1986). Within its zone of adaptation, stands persist and can expand in size through establishment of new seedlings. It can invade stands of mesquite or grasses (Cable 1971). Management methods to effectively use this forage resource are still being developed. Lehmann lovegrass reproduces by apomixis, although sexual types were identified (Voigt et al. 1992).

SAND LOVEGRASS. *Eragrostis trichodes* (Nutt.) Wood is the most important lovegrass of those native to the US. It is found on sandy soils in the southern and central part of the tall-grass region. It is more palatable than weeping lovegrass but cannot tolerate close defoliation (Moser and Perry 1983). It is frequently included in mixtures of grasses planted on sandy soils within its area of adaptation, but it is usually short-lived.

OTHER LOVEGRASSES. The atherstone lovegrass cultivar 'Cochise', although classified as *E. trichophora* × *E. lehmanniana* (Holzworth 1980), is part of the *E. curvula* complex. It has good drought resistance and is more winter-hardy than boer lovegrass, but it is very stemmy. Its tendency to naturally reseed (Jordan 1981) suggests it may have the ability to spread.

Plains lovegrass, *E. intermedia* Hitch., a range grass native to the southern plains and Southwest, has not been well studied. Wilman lovegrass, *E. superba* Peyr., another lovegrass native to Africa, shows promise in central and south Texas because of its establishment ability and relatively good forage quality.

Panicgrasses

SWITCHGRASS. (*Panicum virgatum* L.) See Chapter 32.

KLEINGRASS. *Panicum coloratum* L. is a bunchgrass native to South and East Africa. The type adapted from south Texas to the Oklahoma border is fine stemmed, leafy, 0.8 to 1.4 m tall, and variable in color and pubescence, and it ranges in growth habit from semiprostrate to erect. The species contains much genetic diversity (Lloyd and Thompson 1978), but the type adapted to the US (sometimes referred to as *P. coloratum* var. *coloratum*) represents only a small fraction of that variability. Kleingrass has relatively good forage quality and produces better animal gains than 'Coastal' bermudagrass (Conrad and Holt 1983) and rhodesgrass, *Chloris gayana* Kunth (McCawley and Dahl 1980).

Kleingrass is indeterminate in growth habit and produces tillers throughout the growing season. Thus, it tolerates heavy utilization (Evers and Holt 1972) and has excellent regrowth capacity. Kleingrass is best adapted on sandy loam to clay soils, grows earlier in the spring and later in the fall than many warm-season grasses, and is drought resistant (see Voigt and MacLauchlan 1985). Seed shattering is a serious seed production problem (Hearn and Holt 1969), but resistance was recently reported (Young 1991). Kleingrass establishment can be difficult because of its poor seedling growth (Tischler and Voigt 1983). Germplasm with improved seedling vigor has been developed (Hussey and Holt 1986; Young et al. in press). Photosensitization of sheep and hepatoxicosis of horses grazing kleingrass has occurred in central Texas in occasional years (Bridges et al. 1987) and is associated with the presence of saponins.

OTHER PANICGRASSES. Blue panicgrass, *P. antidotale* Retz., is native to southern Asia. It is a robust bunchgrass that grows from stout rhizomes to a height of 1.5-2.5 m. It is more winter-hardy than kleingrass and can survive as far north as northern Oklahoma. Blue panic is usually not used as a pasture grass because of its high N requirement (Holt 1967). It is most frequently used as a component of grass mixtures for rangeland revegetation. Vine mesquite, *P. obtusum* H.B.K., an apomictic (Anderson and Wright 1974), sod-forming grass that spreads from long tough stolons is found in the southern short- and desert-grassland regions.

OTHER PERENNIAL WARM-SEASON GRASSES

Dropseeds and Sacatons

Sand dropseed, *Sporobolus cryptandrus* (Torr.) A. Gray, is widely distributed throughout the US but is particularly important in the Great Plains and Southwest because it is readily accepted by livestock and has produced excellent animal performance on unseeded, abandoned croplands. The period of active growth of strains of sand dropseed is related to the length of growing season at their place of origin (Quinn and Ward 1969). In contrast to most warm-season grasses of the Great Plains, sand dropseed is primarily self-pollinated. Seeds are frequently produced while the inflorescences are still enclosed in the leaf sheaths. Mesa dropseed, *S. flexuosus* (Thurb.) Rydb., is an important dropseed of parts of the desert-grassland region (Gibbens 1991).

Alkali sacaton, *S. airoides* (Torr.) Torr., is more widely distributed than big sacaton, *S. wrightii* Munro ex Scribn., but both species can be highly productive and have the ability to efficiently exploit extra water for forage production. Spring grazing is the most effective way to use big sacaton (Haferkamp 1982; Cox et al. 1989). The sacatons can be difficult to establish.

Muhlys. Although the muhlys are not well known as forage plants, several make important contributions to the productivity of the rangelands where they occur in mixed stands. Bush muhly, *Muhlenbergia porteri* Scribn., can be an important resource in the desert-grassland region, though it is not a preferred forage grass and is sensitive to grazing (Miller and Donart 1981). Spike muhly, *M. wrightii* Vasey, and mountain muhly, *M. montana* (Nutt.) Hitch., are important in higher-elevation ranges.

Sandreeds. Prairie sandreed, *Calamovilfa longifolia* (Hook.) Scribn., is the most important grass of this North American genus of four species. Its forage importance is limited to sandy soils of the central to northern Great Plains. Although productive, it is sensitive to repeated defoliations (Mullahey et al. 1991). Forage quality of prairie sandreed is lower than that of sand bluestem, *A. gerardii* var. *paucipilus* (Nash) Fern., and similar to little bluestem (Burzlaff 1971). Establishment can be difficult (Masters et al. 1990), but addition of vesicular-arbuscular mycorrhizae can increase the probability of success (Brejda et al. 1993). Big sandreed, *C. gigantea* (Nutt.) Scribn. and Merr., a southern analogue to prairie sandreed, is important for soil conservation.

Arizona cottontop, *Digitaria californica* (Benth.) Henr., is a very important grass of the desert grasslands that is tolerant of relatively heavy stocking (Cable 1982). Its indeterminate growth habit and light, fluffy seeds have made commercialization difficult. In contrast, green sprangletop, *Leptochloa dubia* (H.B.K.) Nees, is available for seeding because seed is easy to produce, it establishes easily, and it provides good quality grazing. However, this native of the southern plains and Southwest is a short-lived perennial and is used primarily to provide forage while more slowly developing species become established. Plains bristlegrass, *Setaria macrostachya* H.B.K., native to the same region, is also short-lived. Its propagation is made difficult by seed dormancy.

WARM-SEASON ANNUAL GRASSES

Crabgrass. Hairy, *Digitaria sanguinalis* (L.) Scop., and southern, *D. ciliaris* (Retz.) Koel, crabgrass are introduced, decumbent annual grasses whose stems readily branch and root at the nodes. Both have been frequently grazed when opportunity or necessity dictated. Management practices necessary for use of crabgrass as a naturally reseeding annual include seeding, if needed to achieve initial stands; winter tillage to a depth of 8 to 10 cm; and use of 120 kg/ha of N fertilizer. With appropriate stocking and grazing management average daily gains of stockers in southern Oklahoma were about 240 kg/ha during 122 d of grazing (Dalrymple et al. 1991).

COOL-SEASON GRASSES

Foxtails. Meadow foxtail, *Alopecurus pratensis* L., is a weakly rhizomatous species native to temperate parts of Eurasia. Creeping foxtail, *A. arundinaceus* Poir., is strongly rhizomatous. Each contains types that vary widely in rate of spread, coarseness, and inflorescence color. Foxtails contain the same genomes, i.e., sets of chromosomes, and are morphologically very similar (Sieber and Murray 1981). The foxtails are adapted to wet meadows from the central US to the Pacific Northwest. They start growth early in spring, produce a spring seed crop, and recover quickly after harvesting because only about 20% of all shoots are reproductive (Rumberg and Siemer 1976). Their best use is as pasture where they can support a high stocking rate and generally produce good animal performance. However, depressed animal performance, suggesting the presence of an unidentified antiquality component, has been reported (Rode and Pringle 1986). Seed shattering is a seed production problem, but genetic resistance has been identified (Boe and Ross 1991) and the cultivar 'Retain' released. Establishment can be difficult because of weak, slow-growing seedlings.

Indian Ricegrass. *Oryzopsis hymenoides* (Roem. & Schult.) Ricker is a self-pollinated (Jones and Nielson 1989), highly variable, and drought-resistant bunchgrass that is widely distributed across the west from Canada to Mexico. It is best adapted to coarse soils but tends to be short-lived. Thus, stand maintenance requires frequent recruitment of new seedlings (Robertson 1976). Indian ricegrass is closely related to the genus *Stipa*. Sterile hybrids between indian ricegrass and 11 different *Stipa* species have been reported (Johnson 1972). Indian ricegrass is considered high in forage quality and is used primarily for winter grazing. Seed quality and dormancy are important factors adversely impacting stand establishment. Genetic resistance to seed shattering has been identified (Jones and Nielson 1992) and should result in improved seed yields and quality. Dormancy treatments such as acid or mechanical scarification and stratification are used to reduce seed dormancy and improve prospects for successful stand establishment. Variation for characteristics such as drought and grazing tolerance (Orodho and Trlica 1990) suggests that, in time, indian ricegrass could become an even more important forage plant.

Needlegrasses. Several species of the genus *Stipa* are important range grasses, but only green needlegrass, *S. viridula* Trin., has been successfully commercialized for seed production. However, green needlegrass has not been as productive or persistent in areas to which it is adapted as other cool-season grasses such as Altai wildryegrass, *Leymus angustus* (Trin.) Pilger, and Russian wildryegrass, *Psathyrostachys juncea* (Fisch.) Nevski (Lawrence and Lodge 1975). Combinations of scarification and stratification can reduce seed dormancy and result in increased germination. Selection for reduced seed dormancy resulted in lower postharvest dormancy and faster germination (see Voigt and MacLauchlan 1985).

Texas wintergrass, *S. leucotricha* (L.) Trin & Rupr., is a very important range grass in parts of central and north Texas where it is the only cool-season perennial grass of major importance. Seed of this grass is produced either by self- or cross-pollination. The balance between pollination mechanisms depends on environmental conditions (Brown 1952). Basal axillary florets that are not exerted from leaf sheaths may be essential to stand longevity, especially under heavy stocking (Call and Spoonts 1989). Needle-and-thread, *S. comata* Trin. & Rupr., is another important needlegrass whose seed is not commercialized.

Tall Oatgrass. *Arrhenatherum elatius* (L.) Presl. is a short-lived perennial bunchgrass introduced from Europe. It is widely adapted to well-drained soils across the northern US. Tall oatgrass establishes easily and in some areas can maintain its stand by volunteer seeding (Hull and Holmgren 1964). Its forage is high in quality, but seed shattering is a serious problem.

QUESTIONS

1. Define the terms *hybrid, cultivar, ecotype, photoperiod, genome, diploid, polyploid, dioecious, monoecious, apomixis, adventitious root, rhizome, stolon, indeterminate growth,* and *photosensitization.*
2. How did warm-season grasses such as sideoats grama become adapted to the northern plains? Why do ecotypes of these grasses have a limited range of latitudes to which they are adapted?
3. What are chaffy-seeded grasses? Define PLS. Why is PLS important?
4. Select an area within the three grasslands discussed in this chapter. Which warm-season grasses are adapted there? Suggest how these grasses might be planted and managed.
5. Why are some grasses better for controlling erosion than others? List several species that could be most useful for controlling water erosion. Wind erosion. What plant characteristics are most important for erosion control?
6. Select several grasses including those native to North America and elsewhere. List their advantages and disadvantages. Can you suggest any physiological mechanisms, morphological characteristics, or other basic reasons for any of the differences you have listed?

REFERENCES

Ahring, RM, and H Frank. 1968. Establishment of eastern gamagrass from seed and vegetative propagation. J. Range Manage. 21:27-30.

Alderson, JS, and WC Sharp. 1993. Grass Varieties of the United States. USDA Agric. Handb. 170. Washington, D.C.: US Gov. Print. Off.

Anderson, B, and AG Matches. 1983. Forage yield, quality, and persistence of switchgrass and caucasian bluestem. Agron. J. 75:119-24.

Anderson, CA, and LN Wright. 1974. Cytology and cytogenetics of vine mesquite (*Panicum obtusum* H.B.K.). II. Asexual mode of reproduction. J. Ariz. Acad. Sci. 9:91-96.

Anderson, DM. 1988. Seasonal stocking of tobosa managed under continuous and rotation grazing. J. Range Manage. 41:78-83.

Axelrod, DI. 1985. Rise of the grassland biome, central North America. Bot. Rev. 51:163-201.

Barker, RE, JD Berdahl, JM Krupinsky, and ET Jacobson. 1983. Collections of western wheatgrass and blue grama and associated nematode genera in the western Dakotas. In JA Smith and VW Hays (eds.), Proc. 14th Int. Grassl. Congr., Lexington, Ky. Boulder, Colo.: Westview, 237-40.

Boe, A, and JG Ross. 1991. Combining abilities of creeping foxtail parents selected for seed retention. Crop Sci. 31:624-25.

Brejda, JJ, DH Yocom, LE Moser, and SS Waller. 1993. Dependence of 3 Nebraska sandhills warm-season grasses on vesicular-arbuscular mycorrhizae. J. Range Manage. 46:14-20.

Bridges, CH, BJ Camp, CW Livingston, and EM Bailey. 1987. Kleingrass (*Panicum coloratum* L.) poisoning in sheep. Vet. Pathol. 24:525-31.

Brown, RR, J Henry, and W Crowder. 1983. Improved processing for high quality seed of big bluestem, *Andropogon gerardii,* and yellow indiangrass, *Sorghastrum nutans*. In JA Smith and VW Hays (eds.), Proc. 14th Int. Grassl. Congr., Lexington, Ky. Boulder, Colo.: Westview, 272-74.

Brown, WV. 1952. The relationship of soil moisture to cleistogamy in *Stipa leucotricha*. Bot. Gaz. 113:438-44.

Burns, JC, DS Fisher, KR Pond, and DH Timothy. 1992. Diet characteristics, digestion kinetics, and dry matter intake of steers grazing eastern gamagrass. J. Anim. Sci. 70:1251-61.

Burson, BL, PW Voigt, RA Sherman, and CL Dewald. 1990. Apomixis and sexuality in eastern gamagrass. Crop Sci. 30:86-89.

Burzlaff, DF. 1971. Seasonal variations of the in vitro dry-matter digestibility of three sandhill grasses. J. Range Manage. 24:60-63.

Cable, DR. 1971. Lehmann lovegrass on the Santa Rita Experimental Range, 1937-1968. J. Range Manage. 24:17-21.

——— 1975. Range Management in the Chaparral Type and Its Ecological Basis: The Status of Our Knowledge. USDA For. Serv. Res. Pap. RM-155.

———. 1982. Partial defoliation stimulates growth of Arizona cottontop, *Trichachne californica*. J. Range Manage. 35:591-93.

Call, CA, and BO Spoonts. 1989. Characterization and germination of chasmogamous and basal axillary cleistogamous florets in Texas wintergrass. J. Range Manage. 42:51-55.

Christiansen, S, and T Svejcar. 1987. Grazing effects on the total nonstructural carbohydrate pools in caucasian bluestem. Agron. J. 79:761-64.

———. 1988. Grazing effects on shoot and root dynamics and above- and below-ground nonstructural carbohydrate in caucasian bluestem. Grass and Forage Sci. 43:111-19.

Conrad, BE, and EC Holt. 1983. Year-round Grazing of Warm-Season Perennial Pastures. Tex. Agric. Exp. Stn. MP-1540.

Cox, JR, and GB Ruyle. 1986. Influence of climatic and edaphic factors in the distribution of *Eragrostis lehmanniana* Nees in Arizona. J. Grassl. Soc. S. Afr. 3:25-29.

Cox, JR, HL Morton, TN Johnsen, Jr., GL Jordan, SC Martin, and LC Fierro. 1982. Vegetation Restoration in the Chihuahuan and Sonoran Deserts of North America. USDA-ARS Rev. Man. ARM-W-28.

Cox, JR, RL Gillen, and GB Ruyle. 1989. Big sacaton riparian grassland management: Seasonal grazing effects on plant and animal production. Appl. Agric. Res. 4:127-34.

Coyne, PI, and JA Bradford. 1985. Comparison of leaf gas exchange and water-use efficiency in two eastern gamagrass accessions. Crop Sci. 25:65-75.

———. 1986. Biomass partioning in 'Caucasian' and 'WW-Spar' Old World bluestems. J. Range Manage. 39:303-10.

Crider, FJ. 1945. Three Introduced Lovegrasses for Soil Conservation. USDA Circ. 730.

Dabo, SM, CM Taliaferro, SW Coleman, FP Horn, and PL Claypool. 1987. Yield and digestibility of Old World bluestem grasses as affected by cultivar, plant part, and maturity. J. Range Manage. 40:10-15.

———. 1988. Chemical composition of Old World bluestem grasses as affected by cultivar and maturity. J. Range Manage. 41:40-48.

Dahl, BE, and PF Cotter. 1984. Management of Weeping Lovegrass in West Texas. Tex. Tech. Range and Wildl. Manage., n. 5.

Dalrymple, RL, J Baker, and S Swigert. 1991. Crabgrass Seminar and Field Day Report. Publ. CG-91. Ardmore, Okla.: Noble Foundation.

Dewald, CL, WA Berg, and PL Sims. 1985. New seed technology for old farmland. J. Soil and Water Conserv. 40:277-79.

Dewald, CL, BL Burson, JMJ deWet, and JR Harlan. 1987. Morphology, inheritance, and evolutionary significance of sex reversal in *Tripsacum dactyloides*. Am. J. Bot. 74:1055-59.

deWet, JMJ, and JR Harlan. 1970. Apomixis, polyploidy, and speciation in *Dichanthium*. Evol. 24:270-77.

Evers, GW, and EC Holt. 1972. Effects of defoliation treatments on morphological characteristics and carbohydrate reserves in kleingrass. Agron. J. 64:17-20.

Forwood, JR, AG Matches, and CJ Nelson. 1988. Forage yield, nonstructural carbohydrate levels, and quality trends of caucasian bluestem. Agron. J. 80:135-39.

Gibbens, RP. 1991. Some effects of precipitation patterns on mesa dropseed phenology. J. Range Manage. 44:86-90.

Gibbens, RP, RF Beck, RP McNeely, and CH Herbel. 1992. Recent rates of mesquite establishment in the northern Chihuahuan desert. J. Range Manage. 45:585-58.

Gould, FW. 1959. Notes on apomixis in sideoats grama. J. Range Manage. 12:25-28.

Haferkamp, MR. 1982. Defoliation impacts on quality and quantity of forage harvested from big sacaton (*Sporobolus wrightii* Munro). J. Range Manage. 35:26-31.

Hanselka, CW. 1985. Grazing management strategies for buffelgrass (*Cenchrus ciliaris*). In Buffelgrass: Adaptation, Management, and Forage Quality, Tex. Agric. Exp. Stn. MP-1575.

Harlan, JR. 1949. Apomixis in side-oats grama. Am. J. Bot. 36:495-99.

———. 1958. Blue grama types from west Texas and New Mexico. J. Range Manage. 11:84-87.

Hearn, CJ, and EC Holt. 1969. Variability in components of seed production in *Panicum coloratum* L. Crop Sci. 9:38-40.

Hennessy, JT, RP Gibbens, JM Tromble, and M Cardenas. 1983. Vegetation changes from 1935 to 1980 in mesquite dune lands and former grasslands of southern New Mexico. J. Range Manage. 36:370-74.

Holt, EC. 1967. Sustained production of blue panicgrass, *Panicum antidotale* Retz., as influenced by management practices. Agron. J. 59:309-11.

Holzworth, LK. 1980. Registration of 'Cochise' atherstone lovegrass. Crop Sci. 20:823-24.

Hopkins, H. 1941. Variations in the growth of side-oats grama grass at Hays, Kansas, from seed produced in various parts of the Great Plains region. Trans. Kans. Acad. Sci. 44:86-95.

Horner, JL, LJ Bush, GD Adams, and CM Taliaferro. 1985. Comparative nutritional value of eastern gamagrass and alfalfa hay for dairy cows. J. Dairy Sci. 68:2615-20.

Huff, DR. 1991. Sex ratios and inheritance of anther and stigma color in diploid buffalograss. Crop Sci. 31:328-32.

Hull, AC, Jr., and RC Holmgren. 1964. Seeding Southern Idaho Rangelands. USDA For. Serv. Res. Pap. INT-10.

Hussey, MA, and EC Holt. 1986. Selection for increased seed weight in kleingrass. Crop Sci. 26:1162-63.

Hyder, DN, AC Everson, and RE Bement. 1971. Seedling morphology and seeding failures with blue grama. J. Range Manage. 24:287-92.

Jackson, LL, CL Dewald, and CC Bohleu. 1992. A macromutation in *Tripsacum dactyloides* (Poaceae): Consequences for seed size, germination, and seedling establishment. Am. J. Bot. 79:1031-38.

Johnson, BL. 1972. Polyploidy as a factor in the evolution and distribution of grasses. In The Biology and Utilization of Grasses. New York: Academic Press, 19-35.

Jones, TA, and DC Nielson. 1989. Self-compatability in 'Paloma' indian ricegrass. J. Range Manage. 42:187-90.

———. 1992. High seed retention of indian ricegrass, PI 478 833. J. Range Manage. 45:72-74.

Jordan, GL. 1981. Range Seeding and Brush Management on Arizona Rangelands. Ariz. Agric. Exp. Stn. T81121.

Kalmbacher, RS, FG Martin, and WD Pitman. 1986. Effect of grazing stubble height and season on establishment, persistence, and quality of creeping bluestem. J. Range Manage. 39:223-27.

Kalmbacher, RS, LS Dunavin, and FG Martin. 1990. Fertilization and harvest season of eastern gamagrass at Ona and Jay, Florida. In Proc. Soil and Crop Sci. Soc. Fla. 49:166-73.

Lawrence, T, and RW Lodge. 1975. Grazing seed field aftermath of Russian wild ryegrass, Altai wild ryegrass and green needlegrass. Can. J. Plant Sci. 55:397-406.

Lloyd, DL, and JP Thompson. 1978. Numerical analysis of taxonomic and parent-progeny relationships among Australian selections of *Panicum coloratum*. Queensl. J. Agric. Anim. Sci. 35:35-46.

McCawley, PF, and BE Dahl. 1980. Nutritional characteristics of high yielding exotic grasses for seeding cleared south Texas brushland. J. Range Manage. 33:442-45.

McIlvain, EH. 1976. Interrelationships in management of native and introduced grasslands. In The Grasses and Grasslands of Oklahoma, Ann. Okla. Acad. Sci. 6:61-74.

McIlvain, EH, and MC Shoop. 1970. Grazing weeping lovegrass for profit—8 keys. In Proc. 1st Weeping Lovegrass Symp. Ardmore, Okla.: Nobel Foundation.

McMillan, C. 1959. The role of ecotypic variation in the distribution of the central grassland of North America. Ecol. Monogr. 29:285-308.

———. 1961. Nature of the plant community. VI. Texas grassland communities under transplanted conditions. Am. J. Bot. 48:778-85.

Martin, SC. 1975. Ecology and Management of Southwestern Semidesert Grass-Shrub Ranges: The Status of Our Knowledge. USDA For. Serv. Res. Pap. RM-156.

Masters, RA, KP Vogel, PE Reece, and D Bauer. 1990. Sand bluestem and prairie sandreed establishment. J. Range Manage. 43:540-44.

Mayeux, HS, HB Johnson, and HW Polley. 1992. Global change and vegetation dynamics. In Proc. Natl. Noxious Range Weed Conf. Boulder, Colo.: Westview, 62-74.

Miller, RF, and GB Donart. 1981. Response of *Muhlenbergia porteri* Scribn. to season of defoliation. J. Range Manage. 34:91-94.

Moser, LE, and LJ Perry. 1983. Yield, vigor, and persistence of sand lovegrass (*Eragrostis trichodes* [Nutt.] Wood) following clipping treatments on a range site in Nebraska. J. Range Manage. 36:236-38.

Mullahey, JJ, SS Waller, and LE Moser. 1990. Defoliation effects on production and morphological development of little bluestem. J. Range Manage. 43:497-500.

———. 1991. Defoliation effects on yield and bud and tiller numbers of two sandhill grasses. J. Range Manage. 44:241-45.

Mutz, JL, and DL Drawe. 1983. Clipping frequency and fertilization influence herbage yields and crude protein of four grasses in south Texas. J. Range Manage. 36:582-85.

Neuenschwander, LF, SH Sharrow, and HA Wright. 1975. Review of tobosa grass (*Hilaria mutica*). Southwest Nat. 20:255-63.

Olmstead, CE. 1944. Growth and development in range grasses. IV. Photoperiodic responses in twelve geographic strains of side-oats grama. Bot. Gaz. 106:46-74.

Orodho, AB, and MJ Trlica. 1990. Clipping and long-term grazing effects on biomass and carbohydrate reserves of indian ricegrass. J. Range Manage. 43:52-57.

Quinn, JA, and RT Ward. 1969. Ecological differentiation in sand dropseed (*Sporobolus cryptandrus*). Ecol. Monogr. 39:61-78.

Riegel, A. 1940. A study in the variations in the growth of blue grama grass from seed produced in various sections of the Great Plains region. Trans. Kans. Acad. Sci. 43:155-71.

Robertson, JH. 1976. The autecology of *Oryzopsis hymnenoides*. Mentzelia 2:18-21, 25-7.

Rode, LM, and WL Pringle. 1986. Growth, digestibility and voluntary intake by yearling steers grazing timothy (*Phleum pratense*) or meadow foxtail (*Alopecurus pratensis*) pastures. Can. J. Anim. Sci. 66:463-72.

Rumberg, CB, and EG Siemer. 1976. Growth of ver-

nalized and nonvernalized creeping foxtail. Crop Sci. 16:172-74.
Sieber, WK, and BG Murray, 1981. Hybridization between tetraploid species of *Alopecurus* (Poaceae): Chromosome behavior of natural and artificial hybrids. Bot. J. Linn. Soc. 83:293-310.
Sims, PL, and CL Dewald. 1982. Old World Bluestems and Their Forage Potential for the Southern Great Plains. A Review of Earlier Studies. USDA-ARS Agric. Rev. Man. ARM-S-28.
Taliaferro, CM, and EC Bashaw. 1966. Inheritance and control of obligate apomixis in breeding buffelgrass, *Pennisetum ciliare*. Crop Sci. 6:473-76.
Taliaferro, CM, FP Horn, BB Tucker, R Totusek, and RD Morrison. 1975. Performance of three warm-season perennial grasses and a native range mixture as influenced by N and P fertilization. Agron. J. 67:289-92.
Tischler, CR, and PW Voigt. 1983. Seedling characteristics and rates of seed reserve utilization of wilman lovegrass and kleingrass. Crop Sci. 23:953-55.
Transeau, EN. 1935. The prairie peninsula. Ecol. 16:423-37.
Valentine, KA. 1970. Influence of Grazing Intensity on Improvement of Deteriorated Black Grama Range. N.Mex. Agric. Exp. Stn. Bull. 553.
Voigt, PW, and EC Bashaw. 1972. Apomixis and sexuality in *Eragrostis curvula*. Crop Sci. 12:843-47.
Voigt, PW, and RS MacLauchlan. 1985. Native and other western grasses. In ME Heath, RF Barnes, and DS Metcalfe (eds.), Forages: The Science of Grassland Agriculture, 4th ed. Ames: Iowa State Univ. Press, 177-87.
Voigt, PW, WR Kneebone, EH McIlvain, MC Shoop, and JE Webster. 1970. Palatability, chemical composition, and animal gains from selections of weeping lovegrass, *Eragrostis curvula* (Schrad.) Nees. Agron. J. 62:673-76.
Voigt, PW, LI Croy, and FP Horn. 1986. Forage quality of winterhardy lovegrasses. J. Range Manage. 39:276-80.
Voigt, PW, BL Burson, and RA Sherman. 1992. Mode of reproduction in cytotypes of lehmann lovegrass. Crop Sci. 32:118-21.
Wenger, LE. 1943. Buffalo Grass. Kans. Agric. Exp. Stn. Bull. 321.
Williamson, J, and B Pinkerton. 1985. Buffelgrass establishment. In Buffelgrass: Adaptation, Management, and Forage Quality, Tex. Agric. Exp. Stn. MP-1575, 25-29.
Wilson, AM. 1981. Air and soil temperature effects on elongation of adventitious roots of blue grama seedlings. Agron. J. 73:693-97.
Wilson, AM, and DD Briske. 1979. Seminal and adventitious root growth of blue grama seedlings on the central plains. J. Range Manage. 32:209-13.
Wright, GR, and GM VanDyne. 1976. Environmental factors influencing semidesert grassland perennial grass demography. Southwest Nat. 21:259-74.
Young, BA. 1991. Heritability of resistance to seed shattering in kleingrass. Crop Sci. 31:1156-58.
Young, BA, CR Tischler, and PW Voigt. 1993. Improving seedling establishment in *Panicum coloratum*. In Proc. 17th Int. Grassl. Congr., Palmerston North, New Zealand, and Australia, 426-27.
Zeiders, KE. 1982. Leaf spots of big bluestem, little bluestem, and Indiangrass caused by *Ascochyta brachypodii*. Plant Dis. 66:502-5.

32 Switchgrass, Big Bluestem, and Indiangrass

LOWELL E. MOSER
University of Nebraska

KENNETH P. VOGEL
Agricultural Research Service, USDA and University of Nebraska

SWITCHGRASS (*Panicum virgatum* L.), big bluestem (*Andropogon gerardii* Vitman), and indiangrass (*Sorghastrum nutans* [L.] Nash) are tall warm-season (C_4) grasses that predominated the North American tall-grass prairie (Weaver 1968). Although they are generally associated with the natural vegetation of Great Plains and the western Corn Belt, they occur widely in grasslands and nonforested areas throughout North America east of the Rocky Mountains and south of 55°N latitude (Stubbendieck et al. 1991) (Fig. 32.1). They have been seeded in mixtures in the Great Plains for over 50 yr as pasture and range grasses. In the past 20 yr they have become increasingly important as pasture grasses in the central and eastern US because of their ability to be productive during the hot months of summer when cool-season grasses are relatively unproductive (Fig. 32.2). Although there are differences among cultivars, in a specific adaption zone switchgrass is generally the earliest of these grasses to be available for grazing and the earliest in flowering, and indiangrass is the latest (Gerrish et al. 1987) (Fig. 32.2). Although the grasses belong to different genera, they have similar areas of adaptation, uses, and management requirements. In this chapter specific differences among the species are explained when they affect management practices.

LOWELL E. MOSER. *See Chapter 2.*

KENNETH P. VOGEL is Plant Geneticist and Research Leader with the Agricultural Research Service, USDA, and Adjunct Professor of Agronomy at the University of Nebraska at Lincoln. He received his MS degree from Colorado State University and the PhD from the University of Nebraska at Lincoln. He researches breeding and genetics of switchgrass, big bluestem, and other grasses adapted to the central Great Plains.

DISTRIBUTION AND ADAPTATION

Although these grasses were native to most of North America, they are most abundant in the Great Plains states. Plants of these species are all photoperiod sensitive. Their photoperiod requirement is based on the latitude where they evolved. Switchgrass, for example, requires short days to initiate flowering (Benedict 1941). Apparently in nature, flowering is induced by decreasing daylength during early summer. When grown at a common location in the central Great Plains,

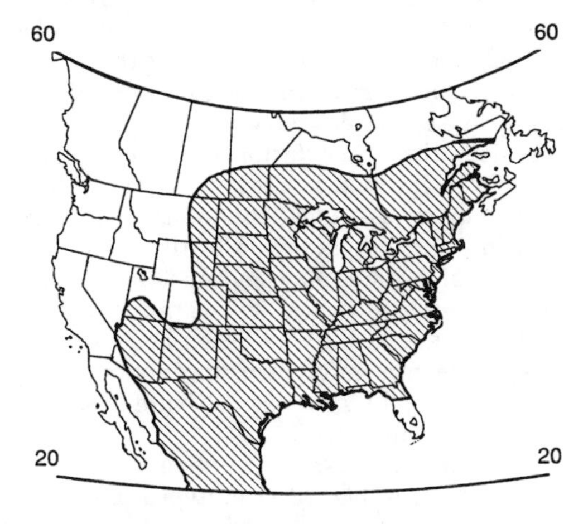

Fig. 32.1. General area of adaptation for switchgrass, big bluestem, and indiangrass. (Adapted from Stubbendieck et al. 1991.)

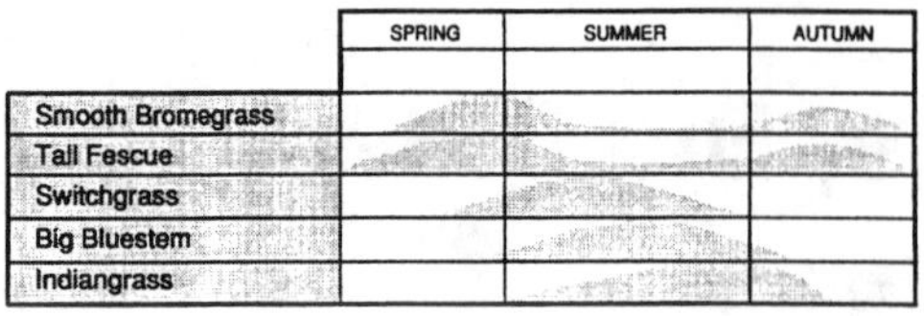

	SPRING	SUMMER	AUTUMN
Smooth Bromegrass			
Tall Fescue			
Switchgrass			
Big Bluestem			
Indiangrass			

Fig. 32.2. Seasonal yield distribution of switchgrass, big bluestem, and indiangrass compared with that of the cool-season grasses, smooth bromegrass, and tall fescue. (Modified from Waller et al. 1986.)

switchgrasses from the Dakotas (northern ecotypes) flower and mature early and are short in stature, while those from Texas and Oklahoma (southern ecotypes) flower late and are tall (Cornelius and Johnston 1941; McMillian 1959). Moving northern ecotypes south gives them a shorter than normal photoperiod, and they flower early. The opposite occurs when southern ecotypes are moved north, and as a result they stay vegetative longer and produce more forage than northern strains moved south (Newell 1968).

Flowering of the tall bluestems, big and sand bluestem (*A. gerardii* var. *paucipilus* [Nash] Fern.), syn. (*A. hallii* Hack.), and indiangrass responds to photoperiod in a similar manner. The response to photoperiod can be modified by growing degree days to some extent since the flowering date of cultivars will vary from year to year.

The photoperiod response also appears to be associated with winter survival. Southern types moved too far north will not survive winters because they continue growth too late in the fall and do not winter harden properly. As a general rule, these grasses should not be moved more than 500 km north of their area of origin because of the possibility of stand losses due to winter injury. In addition to photoperiod, the other factor that determines specific adaptation is response to precipitation and the associated humidity. Cultivars that are naturally adapted to the more arid Great Plains states may develop foliar disease problems when grown in the more humid eastern states. Cultivars based on naturally adapted eastern germplasm may not be as tolerant to drought stress as cultivars based on western germplasm.

Since these grasses are native to most of North America (Fig. 32.1), ecotypes or strains evolved that were adapted to specific geographic regions, and the total range of each species is represented by a combination range of adapted, native ecotypes. Breeding programs of the USDA and some state experiment stations have resulted in the development and release of cultivars that are adapted to specific regions of the central and eastern US. Cultivar development programs for these grasses have received increased emphases in recent years.

The two primary factors determining area of adaptation of specific cultivars are response to photoperiod and precipitation and the associated humidity, both of which are indicated by the origin of the germplasm used to develop specific cultivars. Although much of the prairie and grasslands that were once occupied by these grasses has been plowed and converted into cropland, remanent prairie sites still exist in most areas and are an invaluable germplasm resource. Some extensive native tall-grass prairies are, most notably, the flint hills in Kansas, the Osage prairie in Oklahoma, and the sandhills of Nebraska.

PLANT DESCRIPTION

Switchgrass. Switchgrass is an erect warm-season (C_4) perennial grass. It grows from 0.5 to 2.0 m tall, and most tillers produce a seedhead when moisture is adequate. Although the plant resembles a loose bunchgrass, it has short rhizomes, and a stand has the potential to thicken and form a sod. The depth of switchgrass roots can be up to 3 m (Weaver 1968). The inflorescence is a diffuse panicle, 15-55 cm long, with spikelets at the end of long branches. Spikelets are two flowered, with the second floret being fertile and the first one sterile or staminate. The seed unit is a fertile floret. It is smooth and slick with an indurate lemma and palea that adhere tightly to the caryopsis. The seed threshes clean and is easy to process and plant. On average, there are 860,000 seeds kg^{-1}, but large seed weight differences exist among cultivars. Johnson and Boe (1982) found 100 seed weights from 103 to 201 mg. Switchgrass is a cross-pollinated plant that is largely self-incompatible (Talbert et al. 1983). Switchgrass has a basic chromosome number of nine, and several levels of ploidy exist (Nielsen 1944). Most switchgrass cultivars are either tetraploids or hexaploids (Riley and Vogel 1982).

Switchgrasses have been divided into lowland and upland types. Lowland types are taller, more coarse, and generally more rust (*Puccinia graminis*) resistant, they have a more bunch-type growth, and they may be more rapid growing than upland types. As indicated by the type description, lowland types are found on floodplains and other similar areas while upland types are found in upland areas that are not subject to flooding. Switchgrass can tolerate a wide range of soil conditions. It grows on sand to clay loam soils. Switchgrass tolerates soils with pH values ranging from 4.9 to 7.6 (Duke 1978).

Big Bluestem. Big bluestem was the dominant species of the tall-grass prairie; it made up as much as 80% of the vegetation on some sites (Weaver 1968). Today it can be found in remanent prairies, railroad right-of-ways, or old cemeteries throughout its former range of occurrence. In the tall-grass prairie in the eastern Great Plains it probably is the most abundant and highest-quality species present in good to excellent condition range. Sand bluestem is an ecotype or subspecies adapted to sandy soils such as the Nebraska sandhills. The Old World bluestems (*Bothriochloa* spp.), which were introduced, and little bluestem (*Schizachrium scoparium* [Michx.] Nash), a native, were formerly classified as *Andropogon* spp. and are discussed in Chapter 31. There are 15 other native *Andropogon* species in North America, but big bluestem is by far the most common and widely distributed member.

Big bluestem culms can grow to be 1-2 m tall. Plants often grow in clumps although most plants have short rhizomes (Stubbendieck et al. 1991). Sand bluestem has extensive rhizomes. Big bluestem has numerous basal leaves, and the leafy portions of the canopy seldom are over 50-60 cm high. The root system is very extensive and can be as deep as 2.0-2.5 m. Root mass in the top 10 cm of soil ranges from 7-10 mt ha^{-1} (Weaver 1968). In old plants, the crowns become very dense and tough, making plowing an old stand difficult.

The inflorescence is a purplish colored panicle with generally three digitate racemes, although it may vary from two to nine racemes. A common name for big bluestem is *turkey foot* since the inflorescence resembles a turkey's foot. Spikelets are single flowered and paired; the sessile spikelet is perfect, and under most conditions, the pedicellate one is staminate. Fertile pedicellate spikelets are found on some plants (Boe et al. 1983). The seed unit is the entire fertile, sessile spikelet that includes a rachis joint and the pedicel that supported the pedicellate spikelet. The seed has varying degrees of pubescence and has a twisted awn. These characteristics make unprocessed seed very fluffy and difficult to handle mechanically with conventional seeding equipment. Seed weights typically average 550 seed units/g for unprocessed big bluestem and 230 seed units/g for unprocessed sand bluestem (Wheeler and Hill 1957).

Big bluestem is cross-pollinated and largely self-incompatible (Law and Anderson 1940). The base chromosome number is 10 (Gould 1968). Most cultivars are 2n = 60 (Riley and Vogel 1982). However, 2n chromosome numbers of 60 and 80 occur naturally in the same location (Keeler et al. 1987). Big and sand bluestem plants (2n = 60) are completely interfertile with each other (Newell and Peters 1961). Big bluestem tolerates a wide range of soils except for sands. Sand bluestem grows well on sandy soils. Under native conditions in areas like the Nebraska sandhills, big bluestem grows in the subirrigated meadows where soil texture is finer and organic matter is higher, whereas sand bluestem grows on the sandy hills. On the benches above the subirrigated meadows (the transition between the meadows and the hills) natural hybrids of big bluestem and sand bluestem are found (Barnes 1986).

Indiangrass. Indiangrass is also a tall warm-season grass with short rhizomes. It has a loose, bunch-type growth habit since the rhizomes are generally shorter than 30 mm (McKendrick et al. 1975). Indiangrass generally ranges from 0.5 to 2.0 m tall and has a yellowish brown to black panicle that ranges from 10 to 30 cm in length. It belongs to the Andropogoneae tribe, as does big bluestem, and the spikelet and floret structures of the two species are similar. Spikelets are in pairs on the rachis, with the sessile one being fertile and the pedicellate one being rudimentary or absent (Stubbendieck et al. 1991). As with big bluestem, the spikelets disarticulate below the glumes, the glumes and florets are covered with pubescence, and the seed unit has a twisted awn. The seed units are extremely fluffy, and indiangrass is very difficult to seed unless it is processed. Seed weight of caryopses ranges from 120 to 150 mg/100 seeds. Some indiangrass seed lots have a considerable amount of dormancy (Emal and Conard 1973), depending on cultivar and the season of seed production.

A major portion of indiangrass tillers has been described as biennial (McKendrick et al. 1975). The first year the tillers grow vegetatively, and then after overwintering the same tillers will become reproductive. This characteristic can be easily seen by observation of the earliest growth in the spring. However, the tillers do not appear to be obligate biennials since indiangrass will flower the seeding year. Indiangrass initiates spring growth about the same time as big bluestem and switchgrass but does not develop as quickly. It normally may flower about 4-6 wk later than switchgrass collected in the same area (McKendrick et al. 1975). Although indiangrass has a later heading date than these other warm-season grasses, shoot apices begin to elongate earlier than those of big bluestem (Gerrish et al. 1987), and plants can become rather stemmy.

As the genus name *Sorghastrum* implies, indiangrass appears to be closely related to the sorghums (*Sorghum* spp.). Indiangrass is the only forage grass outside of the sorghum genus that is known to contain cyanogenic glucosides. Indiangrasses occurring in North America and improved cultivars all have 2n = 40 chromosomes (Riley and Vogel 1982). Indiangrass is cross-pollinated, but some plants will produce seed if selfed. Rooting can occur down to about 1.6 m, and it will grow on soils with a pH range from 5.6 to 7.1 (Duke 1978).

IMPORTANCE AND USE

In the past, these grasses have contributed immensely to the native ranges of the central Great Plains. They have additional importance as pasture and reseeded range grasses because of government agricultural programs. Beginning in the 1930s, they have been used to reseed former cropland during programs such as the Soil Bank and the Conservation Reserve Program. In these programs, several million hectares of land were seeded to either monocultures or mixtures containing these grasses. Improved cultivars and improved seeding and management practices have increased their use as pasture grasses independent of farm programs. The reason for their increased use is that they are the best-adapted warm-season grasses that can be used north of the areas where bermudagrass (*Cynodon dactylon* [L.] Pers.) and other subtropical grasses are adapted. In integrated grazing systems, based on cool-season grasses and legumes, they fill a needed niche for productive summer pastures.

In addition to being valuable as forage crops, these grasses also are used for conservation purposes including roadside plantings, waterways, railroad and other right-of-ways, and wildlife cover. Since switchgrass is a high-yielding perennial crop, today there is interest in using switchgrass for a biomass crop for conversion into alcohol fuels (Parrish et al. 1990). Yields up to 16-17 mt ha^{-1} have been reported (Cherney et al. 1990; Parrish et al. 1990). Assuming a 75% extraction efficiency (Dobbins et al. 1990), ethanol yield would be 330 L of ethanol per mt of biomass, which suggests ethanol yields up to 5300-5600 L ha^{-1} could be produced from switchgrass.

CULTIVARS

Breeding work first began on these grasses in the Great Plains states during the 1930s. Initially, numerous accessions (ecotypes or strains) were collected and evaluated (Vogel and Gabrielsen 1986). One or more of the better accessions were selected and increased for evaluation in additional environments, often with the assistance of the Soil Conservation Service. Based on these tests, the accessions often were released directly as cultivars without any additional breeding work. The switchgrass cultivars 'Blackwell' and 'Nebraska 28' were released based on this ecotype evaluation procedure. This same system continues to be used to develop the initial cultivars of these species for geographic or climatic areas where adapted cultivars are not available.

More recently, other cultivars have been developed using more-sophisticated breeding procedures. Currently, grass breeders working on these species are using population improvement breeding procedures including restricted recurrent phenotypic selection (RRPS) and modifications of between and within family selection (Vogel and Pedersen 1993). The switchgrass cultivar 'Trailblazer' was developed using RRPS. Breeding work on these grasses is currently being conducted at several state and USDA research stations. The breeding work emphasizes improving establishment, forage yield and quality, and disease resistance. Genetic studies in each of the species indicates that there is genetic variation for all the traits studied to date and that it should be possible to develop improved cultivars of these grasses (Vogel and Moore 1993).

The principal cultivars that are available for use as forage grasses are listed in Table 32.1. Since latitude of origin and photoperiod response are the primary determinates of area of adaption with growing degrees days having modifying effects, the adaptation zone for cultivars can be based on the USDA plant hardiness zone map (Fig. 32.3). Released cultivars of these grasses are best adapted and most productive in areas where annual precipitation exceeds 450 mm.

CULTURE AND MANAGEMENT

Stand Establishment. These warm-season

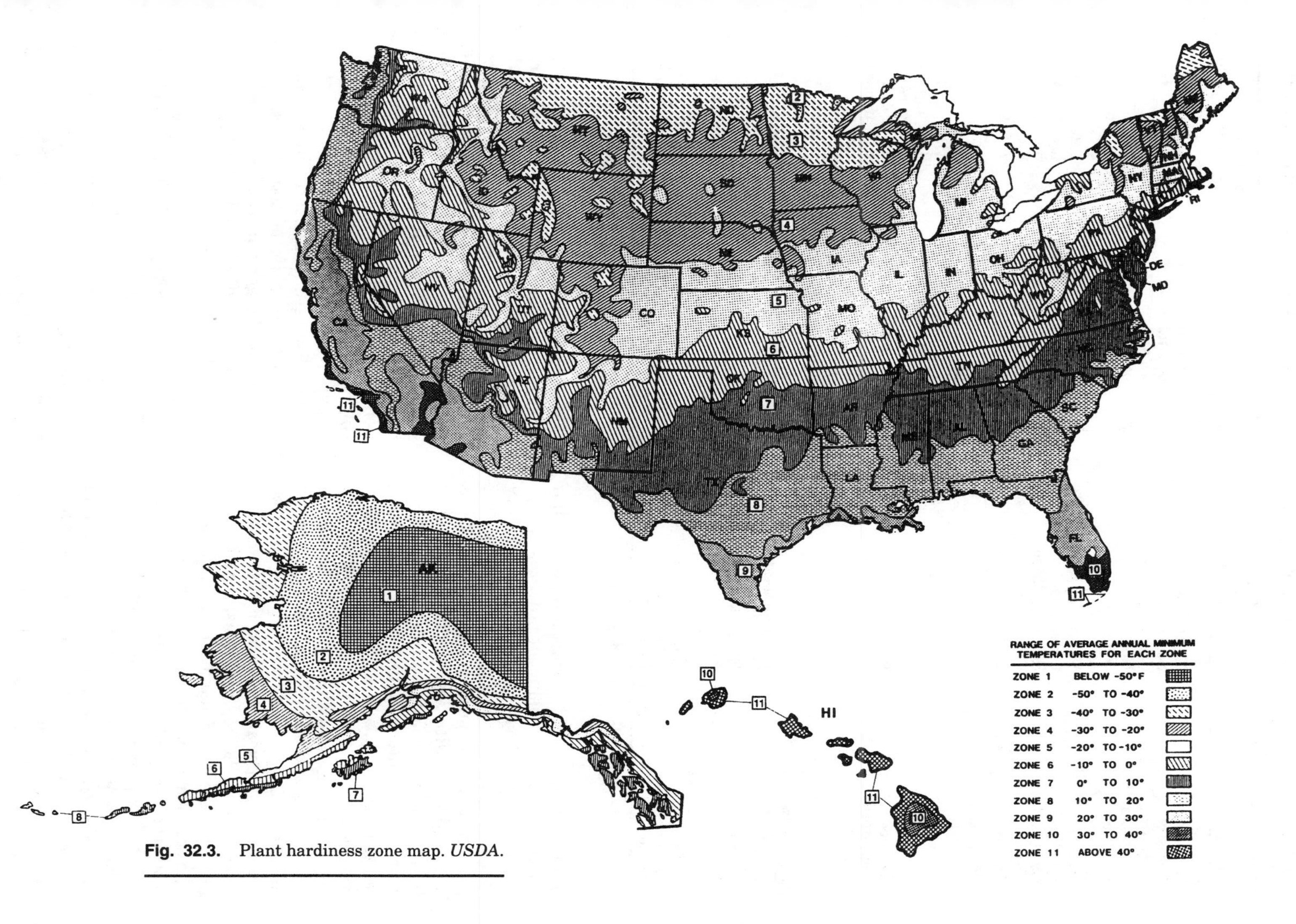

Fig. 32.3. Plant hardiness zone map. *USDA*.

TABLE 32.1. Principal cultivars of switchgrass, big bluestem, and indiangrass and their area or areas of adaptation

Cultivar	Origin of collection	Type	Adaptation zones[a]
Switchgrass			
Dacotah	North Dakota	upland	2, 3, upper 4
Forestburg	South Dakota	upland	3, 4
Sunburst	South Dakota	upland	3, 4
Nebraska 28	Nebraska	upland	3, 4
Summer	Nebraska	upland	3, 4, 5
Pathfinder	Nebraska, Kansas	upland	4, 5
Trailblazer	Nebraska, Kansas	upland	4, 5
Blackwell	Oklahoma	upland	lower 5, 6, 7
Cave-in-Rock	Southern Illinois	upland/lowland	5, 6, 7
Kanlow	Oklahoma	lowland	6, 7
Alamo	Texas	lowland	7, 8, 9
Tall bluestems			
Bison	North Dakota	big bluestem	2, 3
Bonilla	South Dakota	big bluestem	3, upper 4
Niagara	New York	big bluestem	3, 4, upper 5
Champ	Nebraska, Iowa	intermediate	4
Pawnee	Nebraska	big bluestem	lower 4, 5
Roundtree	Iowa	big bluestem	4, 5, upper 6
Kaw	Kansas	big bluestem	lower 5, 6, 7
Goldstrike	Nebraska	sand bluestem	4, 5
Garden	Nebraska	sand bluestem	4, 5
Indiangrass			
Tomahawk	North and South Dakota		3, upper 4
Holt	Nebraska		4
Nebraska 54	Nebraska		lower 4, 5
Oto	Nebraska, Kansas		5, upper 6
Rumsey	Illinois		5, 6
Osage	Kansas, Oklahoma		6, 7
Lometa	Texas		lower 6, 7, 8
Llano	New Mexico		7, 8

[a]Zones are indicated in Figure 32.3.

prairie grasses may be difficult to establish if recommended establishment practices are ignored. The problem in obtaining satisfactory stands may be due to competition with weeds, planting too deep, problems in metering seed, and problems with low-quality or dormant seed. Seed of both big bluestem and indiangrass are considered to be "chaffy" because of pubescence, awns, and other appendages that remain with the seed (Chap. 31). Seed, as harvested, may contain a considerable amount of stem pieces and other inert matter, so cleaning the seed to a high level of purity may be impossible. Special grassland drills with aggressive seed-feeding mechanisms and agitators are needed to handle this chaffy seed. The seed can be aggressively processed to remove the hair and awns (Brown et al. 1983), which makes it possible to plant this chaffy grass seed with conventional seeding equipment. Switchgrass seed is slick and dense and can be cleaned to a very high purity, so seeding with many types of seeders is possible.

These grasses can be seeded as monocultures or as mixtures. Switchgrass is best managed as a monoculture since it tends to be earlier than other warm-season grasses and is very competitive. If switchgrass is used in a mixture, not more than 20% of the mixture by seed count should be switchgrass. Growth of big bluestem and indiangrass is quite compatible, and a mixture of these two diversifies the pasture species base. Seed lots of these grasses, particularly if they have not been processed, can have low purity, and germination may be low due to dormant seed. As a result, seed should be priced and seeded on a pure live seed (PLS) basis.

Pure live seed is calculated by multiplying the purity (the ratio of actual seed to total weight) by the germination. For example, if the seed tag indicated 85% purity and 90% germination,

$$\text{PLS} = 0.85\ (\text{purity}) \times 0.90\ (\text{germination}) = 0.765,$$

or 77% PLS. This means that 1 kg of bulk seed would contain 0.77 kg of PLS. If the desired seeding rate for a grass, like big bluestem, is 7 kg ha^{-1}, then 7 kg $ha^{-1} \div 0.77 =$ 9.1 kg ha^{-1} bulk seed is required. About 9 kg of bulk seed from that particular seed lot would need to be planted per hectare. The actual amount of bulk seed needed to supply the recommended amount of PLS will vary due to seed lot quality. Seeding rates for pure stands of these grasses generally range from 200 to 400 PLS m^{-2}, which is about 2.4-4.8 kg PLS ha^{-1} for switchgrass, 5.7-11.4 kg PLS ha^{-1} for big bluestem, and 5.2-10.4 kg PLS ha^{-1} for indiangrass if an average value is used of 860,000, 350,000, and 385,000 seeds kg^{-1} for switchgrass, big bluestem, and indiangrass, respectively (Anderson 1989). Due to variation in average seed weight (number of seeds per kilogram), and difficulty in determining seed units and germinable seed, Wolf and Parrish (1992) suggest that germinants per unit of bulk seed could be more accurate and be of more use to the producer.

Seed dormancy may be a problem with certain cultivars and seed lots of these grasses. Seeding failures may result if the amount of seed dormancy is not taken into consideration. Although alive, dormant seed will not germinate under normal field conditions. Simple dormancy will be broken if the seed is aged long enough or if it is cold stratified. The normal germination test carried out according to Association of Official Seed Analysts procedures (Justice 1988) includes a period of cold stratification where seed is allowed to imbibe water and then is chilled at 4°C for 2-4 wk before the germination test is conducted at higher temperatures. Therefore, the germination percentage on the seed tag includes dormant seed and does not represent the actual amount of seed that will germinate within 2-4 wk upon planting since the grasses are usually seeded in late spring into warm soil. Producers should run a germination test without chilling if they suspect dormant seed and want to see how much of the seed will actually germinate when they plant it. Much of the dormancy can be broken with aging, so using year-old seed will often overcome dormancy. Seed should have high germination (>75%) and should not be older than 3 yr. Old seed can have good laboratory germination but may have poor seedling vigor under field conditions.

These grasses germinate at higher temperatures than cool-season grasses. Minimum germination temperatures for indiangrass, big bluestem, and switchgrass are 8.6°, 8.9°, and 10.3°C, respectively (Hsu et al. 1985a). Optimum germination temperatures for switchgrass may be lower than those for seedling development (Panciera and Jung 1984). Seedling growth of all three of these grasses at 20° is much slower than at 25° or 30°C (Hsu et al. 1985b). Although seedlings develop slowly, planting in early spring is advantageous even though the soil is cold. Early spring planting will help overcome a dormancy problem. Best stands of switchgrass in Iowa were obtained with early to midspring plantings (Vassey et al. 1985). In the northeastern US, a planting window beginning 3 wk before and extending to 3 wk after the recommended corn (*Zea mays* L.) planting date has been suggested (Panciera and Jung 1984). In some areas, particularly the northern Great Plains, "dormant plantings" are made very late in the fall, late enough that the seed will not germinate. The seed will overwinter, and the cool, moist conditions of spring will put the seed through a natural cold stratification so it will germinate as the weather warms. Spring seedings are normally preferred over dormant plantings. Warm-season grasses should not be planted in late summer because they do not have time to develop sufficiently before winter, and they will winterkill.

Planting too deeply often leads to seeding failures with these warm-season grasses. Switchgrass, big bluestem, and indiangrass should be planted about 1 cm deep and no deeper than 2 cm. If a clean seedbed is used, it should be firm (Chap. 7). No-till seeding into crop residues, particularly sorghum or soybean stubble, or chemically killed sods is often very effective. These grasses have the panicoid type of seedling root development (Chap. 2). The subcoleoptile internode elongates to push the coleoptilar node to the soil surface. Adventitious roots develop from the coleoptilar node if moist conditions exist for several days (Moser and Newman 1988). If surface moisture is not available for adventitious root initiation, establishment may not be successful even though seedlings have emerged (Newman and Moser 1988). Corrective applications of phosphorus (P) or potassium (K) should be made before seeding, but nitrogen (N) applications are generally not made until the grass is established because it will stimulate excessive weed growth during the seeding year. Phosphorus levels should exceed 25 mg kg^{-1} P (Rehm et al. 1976; Rehm 1984). Switchgrass and sand bluestem seedlings have been shown to be mycorrhizal dependent when planted in very sandy soils (Brejda et al. 1993).

Weed competition is one of the major reasons for stand failure of these warm-season prairie grasses. Warm-season grass seedlings do not develop rapidly until conditions are warm, which is the same time that annual weeds develop. Most dicot (broadleaf) weeds can be controlled with 2,4-D (2,4-dichlorophenoxyacetic acid). Generally, 2,4-D should be applied after the seedlings have five leaves and there is a canopy of weeds above the seedlings so most of the spray lands on the weed canopy. Switchgrass and big and sand bluestem have seedling atrazine (6-chloro-N-ethyl-N'-[methylethyl]-1,3,5-triazine-2,4-diamine) tolerance (Martin et al. 1982), and atrazine has been previously approved for use as a preemergence herbicide on these grasses. These grasses have seedling and mature plant tolerance to other herbicides (R. A. Masters and K. P. Vogel, personal communication), but at present, they are not labeled for use on these grasses except for special conditions. Herbicide labels change annually, and a herbicide may not be approved for use on a specific grass even though the grass is tolerant of the herbicide. Producers need to check labels and local sources of herbicide information for recommendations. In many cases the only way weed competition may be reduced is by infrequent clipping or grazing for a short period with a high stocking rate (mob-grazing). A seeding of these grasses generally will not be ready to graze the establishment year, but vegetation may be removed as hay during the establishment year.

Fertilization. These warm-season prairie grasses require adequate soil fertility levels to maintain optimum sustained yields although they can tolerate low-fertility conditions better than most cool-season grasses. On an acid, low-P soil, unfertilized switchgrass and big bluestem were found to yield 50% as much as cool-season grasses that received high levels of lime and fertilizer (Jung et al. 1988). Another study shows switchgrass yields were 12% less on soils with 5 mg kg^{-1} P as compared with soils with 35 mg kg^{-1} P while cool-season grass yields were 35% smaller on the low-P soils (Panciera and Jung 1984). On acidic, low water-holding capacity soils, first-cut switchgrass yields were found to be two to three times greater than those of tall fescue (*Festuca arundinacea* Schreb.) yields on sites with N and four times greater on sites without N, which shows that N use efficiency was greater for switchgrass than for tall fescue (Staley et al. 1991).

These warm-season prairie grasses usually are effective users of organic N in the soil since their greatest N demand comes when the soil is warm and mineralization is proceeding rapidly. The timing of N application is critical in the maintenance of warm-season grass stands. Nitrogen fertilizer should be added in late spring, when the warm-season grass has started active growth and invading cool-season plants have completed most of their spring growth (Rehm et al. 1976). If N is applied too early in the spring or in the previous autumn, cool-season plants will utilize it since the warm-season grasses are not active. The stimulated cool-season invaders will increase rapidly and utilize the soil moisture in spring. Later, during the period of warm-season grass growth, soil moisture will be depleted and the warm-season plants will decline in stand and productivity. Eventually, the reduced vigor allows them to be replaced by cool-season plants. Nitrogen fertilizer should be applied in amounts that will be fully utilized by the warm-season grasses during the growing season, with no carryover N to stimulate cool-season growth in the autumn. On most soils, N and P are the only fertilizers that are required on a routine basis. Nitrogen fertilization rates can be estimated based on annual precipitation. In areas that receive 450 mm of precipitation, 50 kg ha^{-1} of N are often adequate while areas that receive over 750 mm may often require over 100 kg ha^{-1} of N to optimize forage yields.

Grazing and Harvest Management. These grasses will be ready to graze in late spring, about the time the cool-season grasses have completed their spring growth. Since switchgrass is the earliest of these grasses, a common mistake is to begin grazing switchgrass too late. If switchgrass availability exceeds the livestock needs, the plants will mature, producing stemmy reproductive tillers. Forage quality and utilization will be very poor, and animal performance will be unsatisfactory. Big bluestem is later in maturity than switchgrass, and forage quality does not decrease as rapidly with maturity as it does with switchgrass. Indiangrass is generally-less mature at any given calendar date than switchgrass or big bluestem. Grazing of indiangrass pastures should not be delayed, because once indiangrass reaches the heading stage, it becomes rather stemmy and does not retain its quality very well.

Grazing of switchgrass should begin when it is about 30 cm tall. Cattle will graze switchgrass rather uniformly, taking it layer by layer if stocked so consumption nearly equals growth rate. With an appropriate stocking rate, switchgrass can be continuously stocked until the forage is gone. Stocking rate may have to be adjusted to keep the removal rate equal to the rate of growth. This means that switchgrass will provide quality forage for grazing for several months in early summer but will not be available for late summer use. A short period of grazing to partially defoliate switchgrass in late spring can shift a major portion of the yield to later in the summer and improve summer switchgrass quality (George and Obermann 1989; Anderson and Matches 1983).

Switchgrass can be intensively grazed by stocking the pasture sufficiently heavy so the forage is removed in about 3 wk. The switchgrass should be allowed to grow for at least 30 d, depending on moisture, before the regrowth is grazed. This will provide pasture early in the season and again toward the end. If the growing season is sufficiently long, a second regrowth period might provide adequate forage for a third grazing. Switchgrass stands can be damaged by overgrazing. At least 10 cm should be left after grazing during the summer, and 20 cm should be left in the fall after grazing ceases (Mitchell et al. 1994).

Similar management practices should be followed for big bluestem and indiangrass, although big bluestem may be defoliated a little closer. Big bluestem will be grazed in a more patchy manner than switchgrass, so it responds well to rotational stocking. Big bluestem will maintain quality better during the growing season than switchgrass. Switchgrass and the other tall prairie grasses need to have adequate leaf area to produce storage carbohydrates for a period of 4-6 wk before a killing frost. They should not be cut for hay or grazed heavily during this period.

To optimize yield and still obtain adequate forage quality, tall prairie grasses used for hay should be cut when the seed heads begin to emerge. Hay harvested at this stage is usually used for wintering beef cows. If hay is to be used for livestock with higher nutrient requirements, hay should be harvested during the early boot stage. In most areas, the second growth is grazed and is not cut again. The earlier these grasses are grazed or removed as hay, the larger regrowth yields will be (Fig. 32.4).

Indiangrass, like the sorghums, contains a cyanogenic glucoside, dhurrin (Gorz et al. 1979), so pure stands should not be grazed when short or stunted. Upon mastication, prussic acid (HCN) is released from the forage and can be fatal to the grazing ruminant. To our knowledge, there have been no reported cases of prussic acid poisoning of livestock by indiangrass, but this is likely due to indiangrass rarely being used as a monoculture.

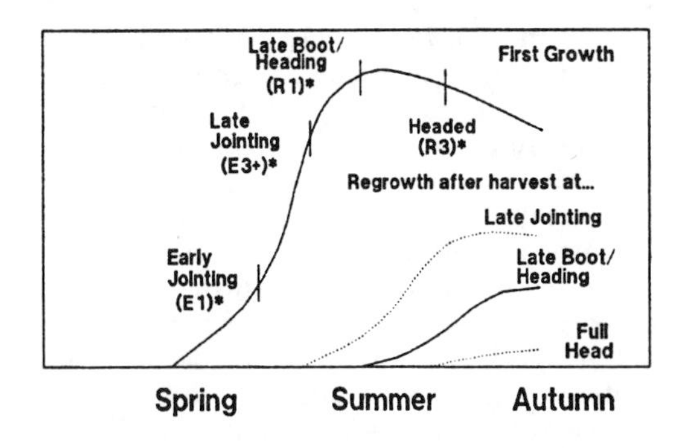

Fig. 32.4. Growth and development of switchgrass (Mitchell et al. 1994; growth stages according to Moore et al. 1991.).

Pure stands of indiangrass should not be grazed until the extended leaf height is 20 cm since the dhurrin concentration falls below the dangerous level at that height (Vogel et al. 1987).

Sheep grazing 'Cave-in-Rock' switchgrass in West Virginia had photosensitization problems that resulted in the death of some lambs (Pouli et al. 1992). Symptoms included marked edema of facial tissue, drooping ears, elevated rectal temperatures, and scab formation on the nose, around the eyes, and on the back of ears. Lambs sought shade and were reluctant to move. Histopathological examinations showed liver damage. The compound causing damage is unknown, but it is expected to be similar to the compound(s) causing a similar problem in kleingrass (*Panicum coloratum* L.). Cattle grazing either switchgrass or kleingrass are apparently not affected by the problem.

Switchgrass, big bluestem, and indiangrass evolved with repeated prairie fires. Prescribed burning is an important management practice that is desirable in many situations. Prescribed or controlled burning, not to be confused with wildfires, may be used in late spring, about the time the growth of the warm-season grass is 2-4 cm, to remove excess plant material. Also, burning is especially important because it keeps cool-season vegetation or brush species from invading (Bragg and Hulbert 1976). Late spring burning may be the only practical way to keep cool-season introduced grasses such as Kentucky bluegrass (*Poa pratensis* L.) and smooth bromegrass (*Bromus inermis* Leyss.) from invading the stand. Forage quality also is increased with spring burning. Periodic prescribed burning will keep these tall warm-season grass stands in excellent condition. Proper burning procedures (Masters et al. 1990) and local burning laws should be followed so prescribed burning is a safe procedure. Burning warm-season grass pastures periodically will aid in maintaining vigorous stands.

Generally, seedings of these grasses have not included legumes since most forage legumes initiate their growth earlier than warm-season grasses. When switchgrass and indiangrass were grown with legumes native to the tall-grass prairie, both yields and crude protein content of the forage were increased. Native legume-grass stands persisted well, but cicer milkvetch (*Astragalus cicer* L.) eliminated the grass stands after 3 yr (Posler et al. 1993). Further research on legume mixtures with switchgrass, big bluestem, and indiangrass is needed.

SEED PRODUCTION

Most seed production practices have resulted from those outlined by Cornelius (1950) and from the experiences of seed producers. Most seed of these grasses is produced in the central Great Plains. Seed fields are usually planted in rows about 1 m apart. In established seed fields 50-112 kg ha^{-1} of N is applied in the spring. Phosphorus is applied if a soil test indicates that it is low. Fields are often burned and cultivated in spring to keep the fields clean and keep the grass in rows, which improves seed yields. Some seed fields are irrigated, but many are not. Approved herbicides are applied as needed for weed control.

Seed is usually harvested by direct combining. The optimum time to combine is when over two-thirds of the spikelets have seed in the hard dough stage and some panicles are beginning to shatter seed (Cornelius 1950). Seed yields of switchgrass often range from 220 to 560 kg ha^{-1} but can be over 1000 kg ha^{-1}. Seed yields of big bluestem and indiangrass range from 150 to over 500 kg ha^{-1}. The seed is usually air dried or dried with minimum heat after harvesting. Processing or debearding the seed of big bluestem and indiangrass after harvesting and drying greatly improves the flowability of the seed in drills and makes it easier to clean and test the seed. The debearding machines used are modified commercial barley debearders (Brown et al. 1981).

DISEASES AND PESTS

There are several diseases that affect these grasses including *Panicum* mosaic virus of switchgrass, various leaf rusts (*Puccinia* spp.), and leaf spot or blotch diseases caused by *Colletotrichum candatum* (Zeiders 1987) (indiangrass) and *Phyllosticta andropogoniva* (Krupinska and Tober 1990) (big bluestem). There are no economical or approved controls for these diseases other than resistant cultivars. Using cultivars that are adapted for specific areas is the best method of reducing losses to diseases. Fortunately, most cultivars and ecotypes of these grasses are genetically diverse and have significant levels of resistance to most diseases that affect these species.

The most serious insect problem that occurs on these grasses is the big bluestem seed midge (*Contarinia wattsi* Gagne). This seed midge can reduce seed yield by over 50% in some years (Carter et al. 1988; Vogel and Manglitz 1990). Presently, there is no control. A chalcidoid wasp, *Tetrastiches nebraskensis* (Girault), parasitizes the midge but does not appear to provide economical levels of control.

QUESTIONS

1. Describe how the growth habits of these grasses are changed if southern ecotypes are moved north and northern ecotypes are moved south.
2. What is the relationship of big and sand bluestems, and what are the conditions of adaptation of each?
3. Why do these grasses play an important role in season-long grazing systems?
4. Describe the barriers that may limit successful stand establishment.
5. Describe the PLS system for calculating seeding rates and why these grasses are seeded on a PLS basis.
6. Describe how seed dormancy is broken in the seed-testing process and how a producer can deal with dormant seed when planting these grasses.
7. Describe N fertilizer management for these grasses and reasons for such management.
8. Describe effective grazing systems for the tall warm-season grasses.
9. Describe management that will keep cool-season grasses from invading warm-season grass stands.
10. Compare the soil conditions required by these warm-season grasses compared with those required by the cool-season grasses.

REFERENCES

Anderson, B. 1989. Establishing Dryland Forage Grasses. Nebr. Coop. Ext. Serv. NebGuide G81-543.

Anderson, B, and AG Matches. 1983. Forage yield, quality, and persistence of switchgrass and caucasian bluestem. Agron. J. 75:119-24.

Barnes, PW. 1986. Variation in the big bluestem

(*Andropogon gerardii*) -sand bluestem (*Andropogon hallii*) complex along a local dune/meadow gradient in the Nebraska sandhills. Am. J. Bot. 73:172-84.

Benedict, HM. 1941. Effect of day length and temperature on the flowering and growth of four species of grasses. J. Agric. Res. 61:661-72.

Boe, AA, JG Ross, and R Wynia. 1983. Pedicellate spikelet fertility in big bluestem from eastern South Dakota. J. Range Manage. 36:131-32.

Bragg, TB, and LC Hulbert. 1976. Woody plant invasion of unburned Kansas bluestem prairie. J. Range Manage. 29:19-24.

Brejda, JJ, DH Yocum, LE Moser, and SS Waller. 1993. Dependence of 3 Nebraska sandhills warm-season grasses on vesicular-arbuscular mycorrhizae. J. Range Manage. 46:14-20.

Brown, RR, J Henry, and W Crowder. 1983. Improved processing for high quality seed for big bluestem (*Andropogon gerardi*) and yellow indiangrass (*Sorghastrum nutans*). In JA Smith and VW Hays (eds.), Proc. 14th Int. Grassl. Congr., Lexington, Ky. Boulder, Colo.: Westview, 272-74.

Carter, MR, GR Manglitz, MD Rethwisch, and KP Vogel. 1988. A seed midge pest of big bluestem. J. Range Manage. 41:253-54.

Cherney, JH, KD Johnson, JJ Volenec, EJ Kladivko, and DK Greene. 1990. Evaluation of Potential Herbaceous Energy Crops onMarginal Croplands: (1) Agronomic Potential. Final Rep.: ORNL/Sub/85-27412/5/P1. Oak Ridge, Tenn: Oak Ridge National Lab.

Cornelius, DR. 1950. Seed production of native grasses. Ecol. Monogr. 20:1-27.

Cornelius, DR, and CO Johnston. 1941. Differences in plant type and reaction to rust among several collections of *Panicum virgatum* L. J. Am. Soc. Agron. 33:115-24.

Dobbins, CL, P Preckel, A Mdafri, J Lowenberg-DeBoer, and D Stucky. 1990. Evaluation of potential herbaceous biomass crops on marginal crop lands: (2) Economic potential. Final Rep.: 1985-89. ORNL/Sub/85-27412/5&P2. Oak Ridge, Tenn.: Oak Ridge National Lab.

Duke, JA. 1978. The quest for tolerant germplasm. In GA Jung (ed.), Crop Tolerance to Suboptimal Land Conditions, Spec. Publ. 32. Madison, Wis.: American Society of Agronomy, 1-61.

Emal, JG, and EC Conard. 1973. Seed dormancy and germination in indiangrass as affected by light, chilling, and certain chemical treatments. Agron. J. 65:383-85.

George, JR, and D Obermann. 1989. Spring defoliation to improve summer supply and quality of switchgrass. Agron. J. 81:47-52.

Gerrish, JR, JR Forwood, and CJ Nelson. 1987. Phenological development of eleven warm-season grass cultivars. In Proc. Forage and Grassl. Conf. Lexington, Ky.: American Forage and Grassland Council, 249-51.

Gorz, HJ, FA Haskins, R Dam, and KP Vogel. 1979. Dhurrin in *Sorghastrum nutans*. Phytochem. 18:20-24.

Gould, FW. 1968. Grass Systematics. New York: McGraw-Hill.

Hsu, FH, CJ Nelson, and AG Matches. 1985a. Temperature effects on germination of perennial warm-season forage grasses. Crop Sci. 25:215-20.

———. 1985b. Temperature effects on seedling development of perennial warm-season forage grasses. Crop Sci. 25:249-55.

Johnson, P, and AA Boe. 1982. Seed size variation in three switchgrass (*Panicum virgatum* L.) varieties. In Proc. S.Dak. Acad. Sci. 61:159.

Jung, GA, JA Shaffer, and WL Stout. 1988. Switchgrass and big bluestem responses to amendments on strongly acid soil. Agron. J. 80:669-76.

Justice, OL. 1988. Rules for testing seeds. J. Seed Tech. 12:1-109.

Keeler, KH, B Kwankin, PW Barnes, and DW Galbraith. 1987. Polyploid polymorphism in *Andropogon gerardii*. Genome 29:374-79.

Krupinsky, JM, and DA Tober. 1990. Leafspot disease of little bluestem, and sand bluestem caused by *Phyllosticta andropogonivara*. Plant Dis. 74:442-45.

Law, AG, and KL Anderson. 1940. The effect of selection and inbreeding on the growth of big bluestem (*Andropogon furcatus* Muhl.). J. Am. Soc. Agron. 32:931-43.

McKendrick, JD, CE Owensby, and RM Hyde. 1975. Big bluestem and indiangrass vegetative reproduction and annual reserve carbohydrate and nitrogen cycles. Agric-Ecosyst. 2:75-93.

McMillian, C. 1959. The role of ecotypic variation in the distribution of the central grassland of North America. Ecol. Monogr. 29:285-308.

Martin, AR, RS Moomaw, and KP Vogel. 1982. Warm-season grass establishment with atrazine. Agron. J. 74:916-20.

Masters, RA, R Stritzke, and SS Waller. 1990. Conducting a Prescribed Burn and Prescribed Burning Checklist. Univ. Nebr. Coop. Exp. Serv. Ext. Circ. EC 90-121.

Mitchell, R, BA Anderson, SS Waller, and LE Moser. 1994. Managing Switchgrass and Big Bluestem for Pasture and Hays. Univ. Nebr. NebGuide G94-1198-A.

Moser, LE, and PR Newman. 1988. Grass seedling development. In JR Johnson and MK Beutler (eds.), Northern Plains Grass Seed Symp., Pierre, S.Dak., 1-37 to 1-47.

Newell, LC. 1968. Effects of strain source and management practice on forage yields of two warm-season prairie grasses. Crop Sci. 8:205-10.

Newell, LC, and LV Peters. 1961. Performance of hybrids between divergent types of big bluestem and sand bluestem in relation to improvement. Crop Sci. 1:370-73.

Newman, PR, and LE Moser. 1988. Grass seedling emergence, morphology, and establishment as affected by planting depth. Agron. J. 80:383-87.

Nielsen, EL. 1944. Analysis of variation in *Panicum virgatum*. J. Agric. Res. 69:327-53.

Panciera, MT, and GA Jung. 1984. Switchgrass establishment by conservation tillage: Planting date responses of two varieties. J. Soil and Water Conserv. 39:68-70.

Parrish, DJ, DD Wolf, WL Daniels, DH Vaughan, and JS Cundiff. 1990. Perennial Species for Optimum Production of Herbaceous Biomass in the Piedmont. Final Rep.: ORNL/Sub/85-274132/5. Oak Ridge, Tenn.: Oak Ridge National Lab.

Posler, GL, AW Lenssen, and GL Fine. 1993. Forage

yield, quality, compatibility, and persistence of warm-season grass-legume mixtures. Agron. J. 85:554-60.

Pouli, JR, RL Reid, and DP Belesky. 1992. Photosensitization in lambs grazing switchgrass. Agron. J. 84:1077-80.

Rehm, GW. 1984. Yield and quality of a warm-season grass mixture treated with N, P, and atrazine. Agron. J. 76:731-34.

Rehm, GW, RC Sorensen, and WJ Moline. 1976. Time and rate of fertilizer application for seeded warm-season and bluegrass pastures. I. Yield and botanical composition. Agron. J. 68:759-64.

Riley, RD, and KP Vogel. 1982. Chromosome numbers of released cultivars of switchgrass, indiangrass, big bluestem, and sand bluestem. Crop Sci. 22:1081-83.

Staley, TE, WL Stout, and GA Jung. 1991. Nitrogen use by tall fescue and switchgrass on acidic soils of varying waterholding capacity. Agron. J. 83:732-38.

Stubbendieck, J, SL Hatch, and CH Butterfield. 1991. North American Range Plants. Lincoln: Univ. of Nebraska Press.

Talbert, LE, DH Timothy, JC Burns, JO Rawlings, and RH Moll. 1983. Estimates of genetic parameters in switchgrass. Crop Sci. 23:725-28.

Vassey, TL, JR George, and RE Mullen. 1985. Early-, mid-, and late-spring establishment of switchgrass at several seeding rates. Agron. J. 77:253-57.

Vogel, KP, and BC Gabrielsen. 1986. Breeding to improve native warm-season grasses. In Warm-season Grasses. Balancing Forage Programs in the Northeast and Southern Corn Belt. Ankeny, Iowa: Soil Conservation Society of America.

Vogel, KP, and GR Manglitz. 1990. Evaluation of furadan and orthene against a bluestem seed midge, 1986 and 1987. Insectic. and Acaracide Tests 15:175-76.

Vogel, KP, and KJ Moore. 1993. Native North American grasses. In J Janick and JE Simon (eds.), New Crops, Proc. 2d. Natl. Symp. New Crops., 6-9 Oct. 1991, Indianapolis, Ind. New York: Wiley, 284-93.

Vogel, KP, and JF Pedersen. 1993. Breeding systems for cross-pollinated perennial grasses. Annu. Rev. Plant Breed. 11:251-74.

Vogel, KP, FA Haskins, and HJ Gorz. 1987. Potential for hydrocyanic acid poisoning of livestock by indiangrass. J. Range Manage. 40:506-9.

Waller, SS, LE Moser and B Anderson. 1986. A Guide for Planning and Analyzing a Year-round Forage Program. Nebr. Coop. Ext. Serv. Bull. EC 86-113-C.

Weaver, JE. 1968. Prairie Plants and Their Environment. Lincoln: Univ. of Nebraska Press.

Wheeler, WA, and DD Hill. 1957. Grassland Seeds. Princeton, N.J.: D. Van Nostrand.

Wolf, DD, and DJ Parrish. 1992. Bluestem germination and seed testing: Meeting the growers needs. Am. Soc. Agron. Abstr. Madison, Wis., 164.

Zeiders, KE. 1987. Leaf spot of indiangrass caused by *Colletotrichum candatum*. Plant Dis. 71:348-50.

33 Bermudagrass

GLENN W. BURTON
Agricultural Research Service, USDA, and University of Georgia

WAYNE W. HANNA
Agricultural Research Service, USDA, and University of Georgia

BERMUDAGRASS, a common name assigned to a number of species of the genus *Cynodon,* probably originated in southeast Africa (Harlan et al. 1970). Of these, *C. dactylon* (L.) Pers., or common bermudagrass, is the only species described as a "ubiquitous, cosmopolitan weed" (Harlan and deWet 1969) because of its worldwide distribution. Their studies suggest that it may have been derived from a cross between two diploid varieties, *aridus* and *afghanicus,* in Asia.

One of the earliest records concerning the introduction of bermudagrass into the US was found in the diary of Thomas Spalding, owner of Sapeloe Island, Georgia, and a prominent antebellum agriculturist. In his diary he made the following entry: "Bermudagrass was brought to Savannah in 1751 by Governor Henry Ellis. If ever this becomes a grazing country it must be through the instrumentality of this grass." Later, writing in *A Geological Account of the United States,* Mease (1807) refers to bermudagrass as one of the most important grasses in the South at the time.

Howard (1881) quotes three prominent southerners to lend weight to his opinion that bermudagrass was at that time the most important pasture grass in the South. Many years later Tracy (1917), representing a new generation, writes, "Bermudagrass is the most common and most valuable pasture plant in the Southern states, being of the same relative importance in that region that Kentucky bluegrass is in the more Northern states."

Although pioneer agriculturists in the South were singing the virtues of bermudagrass, most southern farmers, interested in growing cotton and corn, were trying to destroy the grass that was keeping their fields from washing away. Some states even passed laws prohibiting the introduction and planting of bermudagrass.

DISTRIBUTION AND ADAPTATION

In the US, bermudagrass is best adapted to the states south of a line connecting the southern boundaries of Virginia and Kansas (Fig. 33.1). It grows best when mean daily temperatures are above 24°C. Very little growth is made when these temperatures drop to 6° to 9°C. Temperatures of –2° to –3°C usually kill the stems and leaves back to the ground.

Bermudagrass is more drought resistant than dallisgrass (*Paspalum dilatatum* Poir.), carpetgrass (*Axonopus affinis* Chase), or bahiagrass (*Paspalum notatum* Flugge). It will grow on any moderately well-drained soil provided it has an adequate supply of moisture and plant nutrients. Although it will tolerate flooding for several days, it makes little, if any, growth on waterlogged soils. Better growth generally has been observed on heavy soils than on light sandy soils, probably because heavy soils usually are more fertile and

GLENN W. BURTON is Research Geneticist and Leader, ARS, USDA, and Distinguished Professor of Agronomy, University of Georgia, Tifton. He holds MS and PhD degrees from Rutgers and the University of Nebraska. He has authored over 580 papers describing his research and has bred 40 improved grass cultivars.

WAYNE W. HANNA. *See Chapter 9.*

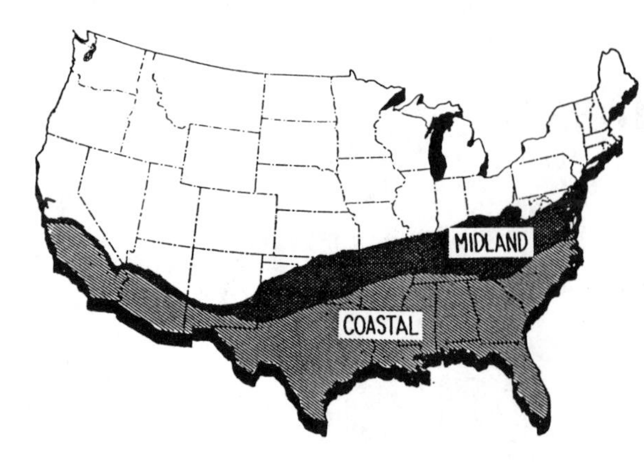

Fig. 33.1. Bermudagrass-producing areas of the US.

retain soil moisture better. When well fertilized and with adequate moisture, bermudagrass has made excellent growth on deep sands. It is affected little by soil pH and has made good growth on both acid and "overlimed" soils. Investigations at the Georgia Coastal Plain Station show that adding lime to soils having a pH below 5.0 favors the growth of this species.

NUTRITIVE VALUE

The nutritive value of any forage plant is greatly influenced by the stage of growth at which it is harvested and the environmental conditions under which it has grown. Unfertilized common bermudagrass cut after it has matured seed will analyze 6%-7% crude protein (CP). Grass from the same plots cut frequently, so that most of the sample is leaves, will contain twice as much CP. At Tifton, Georgia, it has been possible to significantly increase the CP concentration of bermudagrass in almost any stage of growth by heavy applications of nitrogen (N).

Perhaps the best measure of the nutritive value of a grass is the gain that animals make on it. Steers grazing unfertilized common bermudagrass at Tifton, Georgia, for an average of 215 d over a 3-yr period made an average daily gain of 0.39 kg (Stephens 1942). During the same 3 yr, steers on carpetgrass made an average daily gain of 0.2 kg, and on bermudagrass, an average of 112 kg/ha of beef. Carpetgrass produced only 74 kg/ha.

Genotype affects the nutritive value of bermudagrass. Steers grazing 'Coastcross-1' bermudagrass made 12% better average daily gains than those grazing 'Coastal' bermudagrass. Dry matter digestibility (DMD), an excellent index of nutritive value, is a heritable character. The in vitro DMD (IVDMD) of 5-wk-old forage from over 500 genotypes in a bermudagrass world collection grown at Tifton, Georgia, varied from 49% to 69% (Burton and Monson 1972).

CULTIVARS

The following cultivars, which occupy some 10 million ha worldwide, must be planted vegetatively.

Coastal bermudagrass, released in 1943 and named for the Georgia Coastal Plain Station where it was developed, is an F_1 hybrid between 'Tift' bermudagrass (discovered by J. L. Stephens in an old cotton patch near Tifton, Georgia, in 1928) and an introduction from South Africa (Burton 1954). This hybrid, selected as the best of 5000 spaced plants set out in 1938, is characterized by leaves, stems, and rhizomes larger and longer than those of common.

Coastal produces few seed heads, and these rarely contain viable seed. Its much greater resistance to foliage diseases caused by *Helminthosporium cynodontis* Mari. and *H. giganteum* Heald & Wolf enables it to produce more and better quality forage than common bermudagrass. It is immune to rust, *Puccinia cynodontis* Lac ex Desmaz.

Coastal tolerates more frost, makes more growth in the fall, and remains green much later than does common. It is much more drought resistant than common and other grass species generally grown in the South (Burton et al. 1957).

Repeated tests indicate that Coastal will spread faster and maintain a weed-free sod longer than common. Although its stems and leaves are quite coarse, cattle have consistently grazed Coastal in preference to the finer-stemmed common types (Fig. 33.2). When clipped frequently to simulate close grazing, Coastal has produced over twice as much forage as common. In such tests it has been outstanding in its ability to grow and produce more forage in the late summer and fall than other summer-growing grasses with which it was compared.

Over a 5-yr period at Tifton, Georgia, pastures of common and Coastal fertilized with 40 kg N/ha/yr and 0-35-34 (N-P [phosphorus]-K [potassium]) kg/ha every third year carried 2 and 3.2 steers/ha and produced, respectively, 181 and 311 kg/ha of liveweight gain annually (Burton 1954). In these grazing studies, Coastal produced more beef in the late summer and fall than did common bermudagrass, carpetgrass, or bahiagrass.

Coastal grows tall enough to be cut for hay on almost any soil, whereas common usually

Fig. 33.2. Cattle grazing Coastal bermudagrass on a livestock farm near Tifton, Georgia.

is too short to cut. Its fine stems and low moisture content make it easier to cure than any other hay crop adapted to the South. On bright, sunny days, hay cured in the swath or windrow usually is dry enough to bale or stack 24 h after cutting. The four to five cuttings per year afforded by Coastal give the farmer a good chance to save a lot of hay, even though one or more cuttings may be damaged by rain. The uniform annual production of hay obtained from Coastal indicates it is a dependable hay crop. Some 5 million ha of Coastal are grown in the South.

Coastal may become increasingly important in rotations. Crops such as tobacco, tomatoes, and sweet potatoes that suffer from soilborne diseases and nematodes have been remarkably free from such damage when grown in rotation with Coastal bermudagrass. Turning the sod deeply just prior to planting facilitates seedbed preparation, planting the crop, and controlling the bermudagrass that appears later. Such rotations, with the fibrous grass roots reducing soil loss while the planted crop is developing, provide excellent soil and water conservation (Burton et al. 1946; Burton and Johnson 1987).

'Suwanee' bermudagrass is a tall-growing hybrid developed at Tifton (Burton 1962). Its dark green color, red seed heads, and more erect leaves distinguish it from Coastal. It outyields Coastal on light, sandy, infertile soils but is less tolerant of close grazing and is less winter-hardy.

'Midland' bermudagrass is an F_1 hybrid between Coastal and a cold-hardy common from Indiana. Bred in Georgia, Midland was named and released by the Oklahoma station when it greatly surpassed common in tests at Stillwater (Harlan et al. 1954). Although lower in disease resistance and yield than Coastal, Midland is more winter-hardy and can be successfully grown where Coastal winter-kills.

'Coastcross-1' is a sterile F_1 hybrid between Coastal and bermudagrass PI 255445 from Kenya (Burton 1972). It grows taller, has broader, softer leaves, and produces more rapidly spreading aboveground stolons than Coastal, and it develops only an occasional, very short rhizome. It is highly resistant to foliage diseases and the sting nematode, *Belonolaimus longicaudatus* Rau. It makes more growth in the fall but is less winter-hardy than Coastal, which restricts its use to Florida and the lower fourth of the Gulf states. Coastcross-1 is similar in yield to Coastal but produces 30% more live weight gain because it is 12% more digestible (Chapman et al. 1972).

'Alicia', developed by Cecil Greer, Edna, Texas, was reported in Greer's advertising to be a selection made from an African introduction. It is similar to Coastal in growth habit and DM yield but is less winter-hardy, susceptible to rust, and much lower in quality as indicated by many digestibility tests.

'Callie' was selected as an aberrant plant in an old plot of PI 290814 by V. H. Watson, Mississippi State University. It establishes rapidly, gives yields about equal to Coastal, and gives good animal gains when free of rust, but it lacks winterhardiness and is extremely susceptible to rust, which detracts greatly from its quality and reduces its yield.

'Tifton 44' is the best of several thousand F_1 hybrids between Coastal and a common bermudagrass that survived in Berlin, Germany, for 15 yr before it was brought to Tifton in 1966 (Burton and Monson 1978). It is similar to Coastal in DM, yield, and disease resistance, but it is much more winter-hardy, starts growth earlier in the spring, is more digestible, and has given 19% better daily gains when grazed or fed as pellets.

'Ona' (PI 224566) (Hodges et al. 1979), 'Florona' (Mislevy et al. 1980), and 'Florico' (Mislevy et al. 1989) are tall coarse, nonrhizomatous African stargrasses developed at the Ona Agricultural Research Center, Ona, Florida. They lack winterhardiness and are restricted to south Florida.

'Hardie' is a sterile F_1 hybrid that has produced 6% more yield that was 6% more digestible than Midland in several tests in Oklahoma, where it was bred (Taliaferro and Richardson 1980a). Its susceptibility to

Helminthosporium leaf spot restricts it largely to states west of Arkansas.

'Oklan' is a sterile F_1 hybrid that is larger, is 8% more digestible, and has given 12% better average daily gains than Midland (Taliaferro and Richardson 1980b). Bred in Oklahoma, it is resistant to *Helminthosporium* leaf spot but is less winter-hardy than Midland and restricted to areas 100 km south of Stillwater, Oklahoma.

'Brazos' is an F_1 hybrid between materials of African origin produced in Oklahoma (E. C. Holt 1982, personal correspondence). It has been constantly two to four percentage units higher than Coastal in digestibility, and it has produced as well as Coastal on clay and clay loam soils but has produced slightly less on sandy soils.

'Grazer' bermudagrass, an F_1 hybrid between PI 320876 found in the Alps of north Italy and PI 255450 from Kenya, was bred at Tifton and released as 'Tifton 72-84'. Its excellent performance in Louisiana led to its release as Grazer.

'Tifton 68' bermudagrass (*Cynodon nlemfuensis* Vanderyst), a fertile F_1 hybrid (2n = 60) between PI 255450 and PI 293606 from Kenya, is a giant type with large stems, long stolons, and no rhizomes. In years with mild winters, it has yielded 5% more dry matter with a 17% higher digestibility than Coastal bermudagrass. Lack of winterhardiness restricts its use to south Florida, Texas, and the tropics (Burton and Monson 1984).

'Tifton 78' bermudagrass, a sterile F_1 hybrid between Tifton 44 and Callie, is immune to rust, has rhizomes, and is about as winter-hardy as Coastal. Compared with Coastal, it is taller, coarser, spreads faster, establishes more quickly, starts growth earlier in the spring, and in a 3-yr grazing test produced 15% more average daily gains and 36% more live weight gain per hectare (Burton and Monson 1988).

'Tifton 85' bermudagrass, a sterile F_1 hybrid (2n = 50) between PI 290884 from South Africa and Tifton 68, is larger, darker green, faster spreading, and more digestible but a little less winter-hardy than Tifton 78 (Burton et al. 1993). In a 3-yr grazing test, it produced 47% more live weight gain per hectare than Tifton 78 (Hill et al. 1992).

CULTURE

Common bermudagrass may be propagated by planting either seed or vegetative sprigs. Since seed is very small, seeding methods are similar to those for small-seeded grasses and legumes. Bermudagrass seed does not germinate well at low temperatures; hence there is little to be gained from planting before mean daily temperatures of 18°C prevail (Nielsen 1941).

Hulled bermudagrass seed germinates more promptly than unhulled and should be used when early establishment is desired. If the best seeding practices are followed, 5-10 kg/ha of seed should give good stands.

More bermudagrass is propagated by planting sprigs than by seeding; farmers generally have had better success by this method. Poor seed production of the improved bermudas makes it mandatory that they be established from vegetative material. This feature of these improved cultivars has led to the development of labor-saving machinery and methods that make it possible to establish them from sprigs at no greater cost than that of planting common bermudagrass seed.

Many farmers establish bermudagrass nurseries on their farms. Such a nursery can result in a substantial savings in the cost of planting stock and will provide fresh planting material when needed. The nursery should be on land free of common bermudagrass and planted with pure sprigs of the desired cultivar. State crop improvement associations certify vegetatively propagated bermudagrass cultivars to maintain pure sources of planting stock.

A spring-tooth harrow is an excellent tool for digging bermudagrass sprigs (stolons and rhizomes) if a sprig digger is not available. A side-delivery rake will shake the soil from the sprigs and rake them into windrows, from which they can be baled to facilitate handling.

Many different methods have been used in the vegetative establishment of bermudagrass. Nurseries have been planted by using a broomstick (flattened at one end to 3 mm thickness) to push the sprigs into freshly prepared, moist soil so that the green tip of the sprig remains aboveground. Stepping on the sprig after planting firms the soil around it. Commercial planters designed to plant bermudagrass, trees, vegetable plants, or tobacco plants and those developed in farm shops have also been used. Thousands of hectares have been planted to bermudagrass by broadcasting sprigs or freshly cut stems (40+ cm long) by hand from trucks, with manure spreaders, or with other sprig-scattering machines. Sprigs need to be immediately covered with a tandem disk harrow followed by a heavy roller to firm the soil.

Sprigs should be planted at rates to give

one plant every 60-90 cm; 4-6 hL/ha of bulk sprigs (one to two bales) is usually adequate if sprigs are carefully planted in 100-cm rows. Usually 14-18 hL of bulk sprigs are required to give similar stands with broadcast plantings.

Weed control is essential to aid establishment, and the soil should be as weed free as possible. Simazine (2.2-3.3 kg/ha), Diuron (Karmex) (0.89-2.7 kg/ha), or the amine salt of 2,4-D (2.2 kg/ha) should be applied immediately after the sprigs are planted to act as preemergents for both grassy and broadleaf weeds. A second application of 2,4-D 3 to 4 wk after planting may be required if the bermudagrass has not yet covered the ground.

Almost any planting method will succeed if the following rules are observed.

1. Plant only in moist, fertile, weed-free soil.
2. Plant pure live sprigs as soon as possible after harvesting.
3. Plant sprigs at least 6 cm deep to ensure continued soil moisture, but leave tips aboveground.
4. Firm soil around sprigs to keep them moist.
5. Control weeds with herbicides applied immediately after planting.
6. Fertilize to hasten coverage as soon as stolons appear.

Coastal has been successfully established from plantings made every month in the year, but spring and summer plantings are usually best.

MANAGEMENT

In the humid southeastern US, where most soils are acid, bermudagrass sod should receive dolomitic lime (to supply needed magnesium [Mg]) as needed to keep the pH above 5.5 for the grass and above 6.0 if clovers are to be grown with it. Although Coastal is used in most examples, other improved bermudagrasses respond to fertilizer and management practices in a similar manner.

About 2 wk before the last average time for a killing frost in the spring, bermudagrass sods not containing winter legumes should be burned. This practice will control winter weeds, spittlebug (*Philaenus spumarius* L.), and other pests; promote early growth; and frequently increase forage yields (Monson et al. 1974).

Most soils planted to bermudagrass should be limed to pH 5.5 or higher and must be fertilized. Legumes growing in the mixture may satisfy the moderate N needs of the grass. However, bermudagrass, particularly the improved cultivars, can use much more N than the 50-110 kg/ha generally supplied by associated legumes. Annual applications of 448 kg N/ha/yr plus P and K will significantly increase yields over N and double the percentage of crude protein (Table 33.1).

Legumes in Bermudagrass. The clovers (white, *Trifolium repens* L.; crimson, *T. incarnatum* L.; and arrowleaf, *T. vesiculosum* Savi) are grown most frequently with bermudagrass in the South. These legumes extend the grazing season, improve quality of the forage, and supply up to 112 kg N/ha/yr for the bermudagrass. Unfortunately, most soils planted to bermudagrass are too sandy and droughty to grow legumes dependably.

Most any legume will make an excellent growth in association with bermudagrass, provided the soil moisture and plant nutrient requirements of the legume are met and the mixture is managed properly. Lime, P, K, and occasionally some of the secondary and microelements must be applied in generous amounts to many soils to successfully grow legumes with this grass.

Fertilization. In most years, with April 1 to November 1 rainfall between 660 and 760 mm, Coastal bermudagrass yields in south-

TABLE 33.1. Effect of nitrogen on yield and composition of bermudagrass hay

	Total	Average composition (% dry basis)						
Nitrogen (kg/ha)	hay yield (mt/ha)	Crude protein	Ether extract	Crude fiber	N-free extract	Ash	Calcium	Phosphorus
0	1.8	6.9	2.1	29.6	56.3	5.1	0.36	0.31
56	4.7	7.3	2.3	29.3	55.7	5.3	0.46	0.31
112	7.2	8.2	2.3	30.3	53.9	5.2	0.43	0.30
224	10.5	8.8	2.3	30.0	54.2	4.7	0.41	0.28
448	18.8	13.1	2.3	30.9	49.1	4.6	0.46	0.28

Source: Burton (1954).

Note: Based on yields and chemical composition of hay grown at Tifton, Georgia, and taken from plots receiving all N from nitrate of soda applied in March. All plots received 0-49-66 kg/ha at the same time.

east US for any given level of N will be similar. Applications of 112, 224, and 448 kg N/ha may be expected to produce about 11, 16, and 21 mt/ha of hay. In such years irrigation has not significantly increased annual forage yields of Coastal fertilized with 448 kg N/ha/yr.

Drought tolerance of Coastal was well demonstrated in the very dry 1954 season (324 mm rainfall during the growing season), when it produced 13.5 mt/ha of hay with 672 kg N/ha (Fig. 33.3). Its ability to respond to N was shown in 1953: when fertilized with 1008 kg N/ha, it produced 30 mt/ha of hay (Fig. 33.3).

Replicated Coastal pastures fertilized with 56, 112, and 224 kg N/ha/yr (P and K adequate) for a 3-yr period produced annual steer gains of 339, 539, and 766 kg/ha, respectively (Burton 1954).

All N sources are not equally effective when applied as a topdressing to Coastal bermudagrass sod. For example, N from urea or calcium cyanamid was only about 80% as effective as from an ammonium nitrate- or a nitrate of soda-applied broadcast (Burton 1954). In a 5-yr study in which six sources of N, three rates of application, and two frequencies were compared (Burton and Jackson 1962a), ammonium nitrate, ammonium sulfate, ammonium nitrate solution, anhydrous ammonia, urea-ammonium nitrate solution, and urea gave average relative hay yields of 100, 96, 98, 94, 92, and 82, respectively. The relative N recoveries from these different sources were 100, 99, 95, 96, 86, and 74.

Due primarily to placement being in rows instead of broadcast, anhydrous ammonia consistently gave a lag in response that reduced the yield of the first hay cutting by 1.1 mt/ha or more (Jackson and Burton 1962). The poor response from urea topdressing was due to volatilization of urea to ammonia, caused by urease. Since fire destroys urease temporarily, urea N applied soon after the sod is burned in late winter is as effective as other N sources.

Anhydrous ammonia, ammonium nitrate, and ammonium sulfate gave similar responses in percent crude protein of the hay, yields of protein, and N recovery percentages. Urea gave a significantly inferior response in all three of these categories. Percentage recovery of N in the hay decreased significantly as rate of N application increased for all sources except anhydrous ammonia, which gave a similar recovery regardless of rate.

Splitting the N application for cultivated crops grown on sandy soil usually is considered necessary to prevent leaching losses. In this 5-yr study, splitting the application (with half applied in March and half after the second hay cutting, rather than all in March) increased hay yields by 1.3-2.7 mt/ha, except for hay applied with anhydrous ammonia. Splitting the application reduced the protein content of the hay but had no consistent effect on the yield of protein per hectare or the percentage of N recovered. Anhydrous ammonia was equal to the best source of N when the total application was in March but was inferior to most sources when split because of application injury ("burning" or "mechanical" from machinery) and lag in growth response.

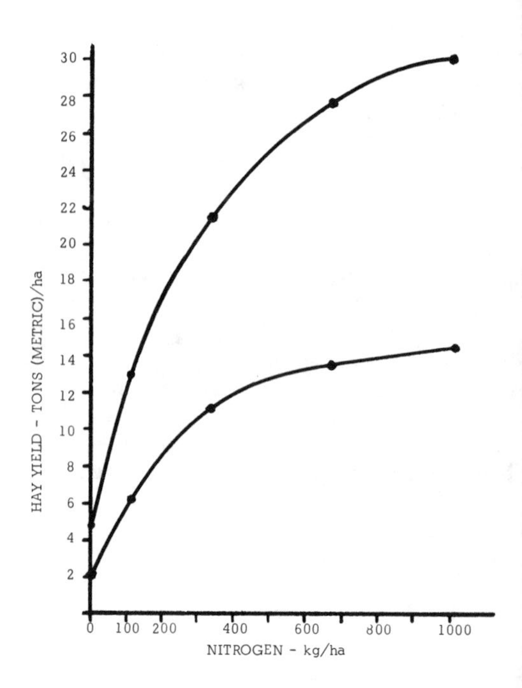

Fig. 33.3. The effect of N rate on the hay production of Coastal bermudagrass cut at 5-wk intervals during a 25-wk period in a wet year (*top*) and a dry year (*bottom*). Rainfall from April 1 to November 1 was approximately 1010 mm in 1953 and 350 mm in 1954.

Anhydrous ammonia probably will always be the cheapest source of N, and it can be applied to sod with little loss. However, the high cost of injecting it into a bermudagrass sod has restricted its widespread use.

Coastal established in 1953 and maintained as a hay crop, with annual N applications of from 0 to 1790 kg/ha through 1954-59, continued to be productive and vigorous through the 6-yr period. Root weights, including rhizomes, which averaged 8500 kg/ha in

1956, were maintained at approximately the same level with all N application rates (Holt and Fisher 1960), but top growth yields increased greatly.

Most sandy soils in the South contain such low reserves of P and K that these elements must be applied annually if much N is used. Maintaining a fertilizer ratio of N:P:K near 9:1:4, respectively, generally has supplied sufficient P and K (Jackson et al. 1959). Although P and K frequently are applied only once a year, splitting the K will reduce luxury consumption and increase the efficiency of this element (Burton and Jackson 1962b). Requirements of calcium (Ca), Mg, and sulfur (S) for bermudagrass are satisfied if 20% superphosphate is used to supply P and dolomitic limestone is applied as needed. Substituting ammonium sulfate for a part of the N in fertilizer blends using triple superphosphate or diammonium phosphate can satisfy S needs at no extra cost. Coastal rarely gives a measurable response to microelements. Generally, annual applications of 35 kg N/ha are required to maintain good stands of Coastal pastures grown without a legume.

Grazing. Management greatly influences the results obtained from any pasture. New growth following close grazing of bermudagrass usually is the most nutritious. Such early grazing improves the nutritive qualities of the forage but sometimes reduces the total quantity produced to the extent that animals are unable to get enough feed for maximum daily gains. This was demonstrated in a stocking rate experiment conducted at Tifton. For example, stocking uniformly fertilized (112-15-56 kg/ha N-P-K) 0.81 ha Coastal pastures at rates of 1.68, 2.24, and 2.8 steers/ha over a 5-yr period gave average daily gains of 0.67, 0.60, and 0.52 kg/steer and annual live weight gains of 392, 469, and 508 kg/ha, respectively (McCormick et al. 1964). Thus, bermudagrass should be grazed close to maximize carrying capacity, yet grazing should be light enough to allow some grass to accumulate for best average daily gain.

Grazing intensity should be adjusted to favor the legume requirements in legume-bermudagrass mixtures. Close grazing benefits low-growing legumes, such as white clover and annual lespedeza (*Kummerowia* spp.). More erect legumes, such as sweetclover (*Melilotus* spp.) and alfalfa (*Medicago sativa* L.), must be given rest periods or be grazed less intensively to keep them in vigorous condition. Close grazing immediately preceding the planting of a legume greatly facilitates its establishment in bermudagrass sod.

Bermudagrass grazed by lactating cows should be well fertilized and mowed frequently to keep an abundance of young succulent grass available at all times. A four-pasture Coastcross-1 bermudagrass rotation where lactating cows grazed new growth for 7 d followed by replacement heifers grazing the stubble for 7 d produced 4550 kg of milk per lactation without any concentrate.

Some farmers remove a cutting of hay from a part of their Coastal pastures during the summer. In addition to producing winter feed, this practice helps to control weeds, to provide new growth, and to scatter animal droppings.

Well-fertilized pastures of improved bermudagrasses usually contain very few weeds if burned in late winter or if overseeded with a winter legume or grass. When broadleaf weeds are present, 2,4-D can be safely used to control them. Winter weeds usually can be controlled by early grazing.

Hay, Silage, and Pellets. Sod of Coastal and similar cultivars should be burned in late winter, sprayed with 2,4-D if additional weed control is needed, and fertilized with about 112 kg/ha N plus P and K in mid-March. Cuttings for hay or silage should be made when the grass is about 40 cm tall and every 4 to 6 wk thereafter. Permitting the grass to grow for more than 6 wk between cuts in summer lowers quality, increases curing time, and does not increase annual hay yields. Allowing the last cutting to grow about 8 wk until the first killing frost in the fall will enable the grass to build up reserves to give more vigorous growth and better stands the next spring. Often this "last cutting" will supply much-needed fall grazing. Hay conditioners will hasten curing so hay can usually be baled or bulk-stacked 24 h after cutting. Fertilizer (56-112 kg N/ha plus P and K) should be applied after each cutting through August to maintain forage yield and protein content.

The frequency of cutting of tall-growing cultivars such as Coastal greatly influences both yield and quality of the hay produced. The effects of these treatments are evident in Table 33.2 (Prine and Burton 1956). A study of these data shows that as the age of the plant increases, leafiness and crude protein decrease, while yield and lignin increase. Digestion trials show that the digestibility of these constituents decreased as the clipping intervals or the age of the grass increased (Knox et al. 1958). Further, dairy heifers fed 4-, 8-, and 13-

wk-old Coastal bermudagrass hays without supplement gained 0.54, 0.41, and 0.0 kg/d, respectively (McCullough and Burton 1962).

Four years of research at Clemson, South Carolina, has shown that silage made from heavily fertilized Coastal (properly ensiled before it is 35 d old) can produce as much milk as corn silage (with 4400 kg/ha of grain) at a savings of $0.47/100 kg of milk (King et al. 1964). For good silage quality, bermudagrass should be cut every 4 to 6 wk, be chopped as short as possible, and be packed well without wilting in an airtight silo. Coastal usually contains 65%-75% moisture when cut at this stage. Adding 40 kg of ground corn or citrus pulp per ton of chopped grass will improve silage quality.

In repeated feeding trials at Tifton, Georgia, steers fed Coastal pellets gained 0.82 kg/d, about 50% more than gains usually obtained from Coastal pastures (McCormick et al. 1965). Pellets were made from dehydrated, chopped forage taken from Coastal fields managed as hay.

Numerous experiments have shown that dehydrated Coastal bermudagrass may be substituted for alfalfa as a source of vitamin A and xanthophyll for poultry feeds. Processors producing these high-quality pellets manage the grass for hay but apply 672 kg N/ha/yr plus P and K and cut the grass every 21-24 d. Yields of 15.7 mt pellets/ha/yr are common.

SEED PRODUCTION

Most commercial seed of common bermudagrass is produced in Arizona and southern California in fields planted for that purpose. Two seed crops are harvested annually with about 450 kg/ha of hulled seed taken in July and about 150 kg/ha taken in November.

CONTROL

Forages that produce hard seed capable of surviving many years in the soil are the most difficult to contain and eradicate. Further, rhizomatous forages are more difficult to kill than stoloniferous or bunch types without rhizomes. Persistence and ability to crowd out other species gives bermudagrass some weed-like characteristics (particularly without care). Rhizomatous seed-producing bermudagrasses, like common, carry a much greater pest potential than the nearly sterile Coastal with its shallow rhizomes. Completely sterile stoloniferous hybrids like Coastcross-1 can be completely eradicated with one shallow cultivation in dry weather.

Growing a high-yielding crop of corn on deeply turned bermudagrass sod followed by disking to kill the few weak survivors after the crop is harvested will usually destroy all but the viable seed left in the soil. Close grazing or mowing followed by disk tilling to a depth of 8-10 cm in dry weather generally kills all plants cut from their roots. One or more additional disk tillings during dry weather to depths great enough to cut the remaining rhizomes from their roots will usually eradicate vegetative bermudagrass. Herbicides such as glyphosate (Roundup) applied at 2.2 kg/ha to actively growing plants with a second application to survivors usually kills all bermudagrass plants.

QUESTIONS

1. What is the range of climatic and soil adaptation of bermudagrass?
2. Enumerate differences between common, Coastal, Coastcross-1, Alicia, Callie, Hardie, Oklan, Brazos, Tifton 44, Tifton 68, Tifton 78, and Tifton 85 bermudagrasses.
3. How are improved bermudagrasses propagated?
4. To what extent can N needed for large yields be obtained from legumes established in bermudagrass?
5. How and to what extent is grazing management changed if legumes are grown in association with bermudagrass?
6. How and to what extent are production and quality of hay influenced by frequency of cutting and N fertilization?

TABLE 33.2. Effect of clipping frequency on yield and quality of Coastal bermudagrass hay

Clipping intervals (wk)	Hay[a] yield (mt/ha)	Leaf[b] (%)	Crude protein[b] (%)	Lignin[b] (%)
1	14.0	...	21.4	...
2	17.4	87.6	20.8	9.4
3	19.3	81.3	18.8	9.6
4	21.7	74.8	17.0	10.3
6	28.1	57.7	13.8	11.2
8	28.0	51.4	12.2	12.0

Sources: Prine and Burton (1956); Knox et al. (1958).

Note: Hay grown at Tifton, Georgia, was fertilized with 672 kg N/ha plus adequate P and K during a 24-wk period.

[a]Hay yield adjusted to 16% moisture.

[b]Percentages are expressed on dry basis.

7. How does bermudagrass rank in importance in comparison with other grasses throughout the South? Why?

REFERENCES

Burton, GW. 1954. Coastal Bermudagrass. Ga. Agric. Exp. Stn. Bull. NS2.

———. 1962. Registration of varieties of bermudagrass. Suwanee (Reg. No. 6). Crop Sci. 2:352-53.

———. 1972. Registration of Coastcross-1 bermudagrass. Crop Sci. 12:125.

Burton, GW, and WH DeVane. 1952. Effect of rate and method of applying different sources of nitrogen upon the yield and chemical composition of bermudagrass, *Cynodon dactylon* (L) Pers., hay. Agron. J. 44:128-32.

Burton, GW, and JE Jackson. 1962a. Effect of rate and frequency of applying six nitrogen sources on Coastal bermudagrass. Agron. J. 54:40-43.

———. 1962b. Single and split potassium applications for Coastal bermudagrass. Agron. J. 54:13-14.

Burton, GW, and AW Johnson. 1987. Coastal bermudagrass rotations for control of root-knot nematodes. J. Nematol. 19:138-40.

Burton, GW, and WG Monson. 1972. Inheritance of dry matter digestibility in bermudagrass, *Cynodon dactylon* (L.) Pers. Crop Sci. 12:375-78.

———. 1978. Registration of Tifton 44 bermudagrass (Reg. No. 10). Crop Sci. 18:911.

———. 1984. Registration of Tifton 68 bermudagrass. Crop Sci. 24:1211.

———. 1988. Registration of Tifton 78 bermudagrass. Crop Sci. 28:187-88.

Burton, GW, CW McBeth, and JL Stephens. 1946. The growth of Kobe lespedeza as influenced by root-knot nematode resistance of the bermudagrass strain with which it is associated. J. Am. Soc. Agron. 38:651-56.

Burton, GW, GM Prine, and JE Jackson. 1957. Studies of drought tolerance and water use of several southern grasses. Agron. J. 49:498-503.

Burton, GW, RN Gates, and GM Hill. 1993. Registration of Tifton 85 bermudagrass. Crop Sci. 33:644-45.

Chapman, HD, WH Marchant, PR Utley, RE Hellwig, and WG Monson. 1972. Performance of steers on Pensacola bahiagrass, Coastal bermudagrass, and Coastcross-1 bermudagrass pastures and pellets. J. Anim. Sci. 34:373-78.

Harlan, JR, and JMJ deWet. 1969. Some variation in *Cynodon dactylon* (L.) Pers. Crop Sci. 9:774-78.

Harlan, JR, GW Burton, and WC Elder. 1954. Midland Bermudagrass, a New Variety for Oklahoma Pastures. Okla. Agric. Exp. Stn. Bull. B-416.

Harlan, JR, JMJ deWet, and KM Rawal. 1970. Geographic distribution of the species of *Cynodon* L. C. Rich. East Afr. Agric. For. J., 230-36.

Hill, GM, RN Gates, and GW Burton. 1992. Grazing performance on the new Tifton 85 bermudagrass hybrid. J. Anim. Sci. 70:suppl., 20.

Hodges, EM, P Mislevy, LS Dunavin, OC Ruelke, and PL Stanley, Jr. 1979. 'Ona', a New Stargrass Variety. Fla. Agric. Exp. Stn. Inst. Food and Agric. Sci. Circ. S-268.

Holt, EC, and FL Fisher. 1960. Root development of Coastal bermudagrass with high nitrogen fertilization. Agron. J. 52:593-96.

Howard, CW. 1881. Manual of Cult Grasses and Forage Plants of the South. Atlanta, Ga.: James Harrison.

Jackson, JE, and GW Burton. 1962. Influence of sod treatment and nitrogen placement on the utilization of urea nitrogen by Coastal bermudagrass. Agron. J. 54:47-49.

Jackson, JE, ME Walker, and RL Carter. 1959. Nitrogen, phosphorus and potassium requirements of Coastal bermudagrass on a Tifton loamy sand. Agron. J. 51:129-31.

King, WA, CC Brannon, and HJ Webb. 1964. Coastal Bermudagrass as a Forage for Dairy Cows. S.C. Agric. Stn. Bull. 516.

Knox, FE, GW Burton, and DM Baird. 1958. Effects of nitrogen rate and clipping frequency upon lignin content and digestibility of Coastal bermudagrass. Agric. Food Chem. 6(2):217-18.

McCormick, WC, WH Marchant, and BL Southwell. 1964. Effects of Stocking Level on Gains of Steers Grazing Coastal. Ga. Agric. Stn. Mimeogr. NS-183.

McCormick, WC, DW Beardsley, and BL Southwell. 1965. Coastal Bermudagrass Pellets for Fattening Beef Steers. Ga. Agric. Stn. Bull. NS-132.

McCullough, ME, and GW Burton. 1962. Quality in Coastal bermudagrass hay. Ga. Agric. Res. 4(1):4-5.

Mease, J. 1807. A Geological Account of the United States, Comprehending a Short Description of Their Animal, Vegetable and Mineral Productions. Philadelphia, Pa.: Birch and Small.

Mislevy, P, WF Brown, LS Dunavin, DW Hall, RS Kalmbacker, AS Overman, OC Ruelke, RM Sonoda, RL Stanley, Jr., and MJ Williams. 1980. Florona Stargrass. Fla. Agric. Exp. Stn. Inst. Food and Agric. Sci. Circ. S-362.

Mislevy, P, WF Brown, R Caro-Costas, J Vicente-Chandler, LS Dunavin, DW Hall, RS Kalmbacker, AJ Overman, OC Ruelke, RM Sonoda, A Sotomayor-Rios, RL Stanley, Jr., and MJ Williams. 1989. Florico Stargrass. Fla. Agric. Exp. Stn. Inst. Food and Agric. Sci. Circ. S-361.

Monson, WG, GW Burton, EJ Williams, and JL Butler. 1974. Effects of burning on soil temperature and yield of Coastal bermudagrass. Agron. J. 66:212-14.

Nielsen, EL. 1941. Establishment of Bermudagrass from Seed in Nurseries. Ark. Agric. Exp. Stn. Bull. 409.

Prine, GM, and GW Burton. 1956. The effect of nitrogen rate and clipping frequency upon yield, protein content and certain morphological characteristics of Coastal bermudagrass. Agron. J. 48:296-301.

Stephens, JL. 1942. Pastures for the Coastal Plain of Georgia. Ga. Coastal Plain Exp. Stn. Bull. 27-57.

Taliaferro, CM, and WL Richardson. 1980a. Registration of Hardie bermudagrass. Crop Sci. 20:413.

———. 1980b. Registration of Oklan bermudagrass. Crop Sci. 20:414.

Tracy, SM. 1917. Bermudagrass. USDA Farmers Bull. 814.

34 Bahiagrass, Dallisgrass, and Other *Paspalum* Species

BYRON L. BURSON
Agricultural Research Service, USDA

VANCE H. WATSON
Mississippi State University

BAHIAGRASS

BAHIAGRASS, *Paspalum notatum* Flugge, a native of South America, is widely distributed in Argentina, Uruguay, Paraguay, Brazil, and the West Indies. It was introduced into the US in 1913 by the Florida Agricultural Experiment Station (Scott 1920).

DISTRIBUTION AND ADAPTATION

Bahiagrass is grown from east Texas to the Carolinas and as far north as northern Arkansas and central Tennessee. It is principally adapted to the coastal area in the southern US (Fig. 34.1) (Burton 1946). It is adapted to a wide range of coastal plain soils but performs best on sandy soils with a pH of 5.5-6.5 (Jones 1971). It grows better on drought-prone soils with relatively low fertility and on sandier soils than do most other pasture grasses.

Fig. 34.1. Bahiagrass-producing areas of the US.

PLANT DESCRIPTION

Bahiagrass is a deep-rooted warm-season perennial. Its short, stout, often exposed rhizomes form dense sod even on sandy soils (Fig. 34.2). The extensive root system limits encroachment from other species. Bahiagrass has many flat or folded basal leaves that are less than 1.25 cm wide, pubescent on the margins, and stiffly spreading. Culms range from 15 to 60 cm in height and occur in dense tufts. The inflorescence is a racemose panicle bearing two or sometimes three racemes. Racemes are curved and ascending, with spikelets occurring in two rows. The single-flowered spikelets are smooth and shiny. Seed is oval, yellowish green, glossy, and about 3 mm in length (Chase 1929).

Bahiagrass exceeds carpetgrass (*Axonopus affinis* Chase), but is usually poorer than bermudagrass (*Cynodon dactylon* [L.] Pers.), in productivity and nutritive value. Forage quality is highest in early spring, but by summer the percentage of cell wall constituents is sufficiently high to suggest animal intake would be limited (Ellzey 1967).

IMPORTANCE AND USE

Bahiagrass is popular in the South because (1) it tolerates a wider range of soil conditions than bermudagrass or dallisgrass (*P. dilata-*

BYRON L. BURSON is Research Geneticist, ARS, USDA, at the Grassland, Soil, and Water Research Laboratory, Temple, Texas. He received the MS and PhD degrees from Texas A&M University. His research involves the cytogenetics and breeding of warm-season grasses, especially the genus *Paspalum*.

VANCE H. WATSON is Head, Research Support Units, and Professor of Agronomy, Mississippi State University. He earned the MS degree from the University of Missouri and the PhD from Mississippi State University. His major research interest is forage production and management.

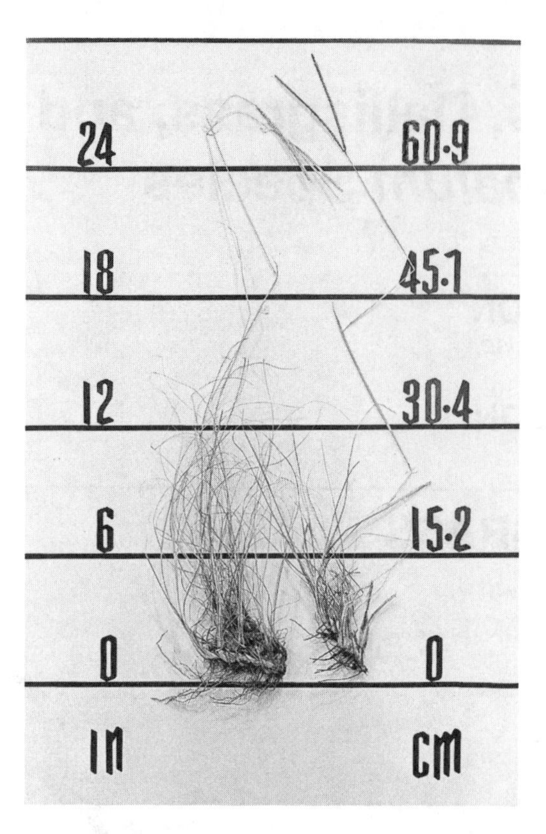

Fig. 34.2. Mature tillers of bahiagrass. Note the elongated culms and panicle with its racemes, and note the numerous new tillers that arise from the short rhizomes that are rooted at the nodes.

tum Poir.), (2) it is resistant to encroachment of weeds, (3) it is established by seed, (4) it is relatively free from damaging insects and diseases, (5) it can persist and produce moderate yields on soils of very low fertility, and (6) it withstands close defoliation.

Bahiagrass is used primarily for permanent pastures, with some use for hay production. Conventional hay-harvesting equipment may leave as much as 60% of the forage uncut because of its morphology and growth habits (Fig. 34.2) (Beaty et al. 1970). Bahiagrass hay is leafy and has few stems and heads (Hoveland 1968).

Beef gains on bahiagrass pasture are intermediate to gains on common and 'Coastal' bermudagrass, and bahiagrass hay is satisfactory for wintering programs (Evans et al. 1961; McCormick et al. 1967). However, neither bahiagrass pasture nor hay is satisfactory as the sole source of forage for high-producing dairy cows (Rollins and Hoveland 1960).

CULTIVARS

Bahiagrass introductions and strains are classified into nine types, of which eight are named cultivars: common, 'Argentine', 'Paraguay', 'Paraguay 22', 'Pensacola', 'Tifhi-1', 'Tifhi-2', and 'Wilmington' (Hanson 1972) and 'Tifton 9' (Burton 1989).

Until the late 1930s the original common bahiagrass (2n = 40) was the most widely grown (Burton 1946). Plants are small with broad leaves and culms that grow 20-45 cm tall. The oval-shaped seeds are covered with a tight, waxy glume that makes scarification necessary for good germination. Common is lower yielding than the other cultivars.

Argentine (2n = 40), introduced into Florida in 1945, has long, broad leaves. The seed is very susceptible to ergot, *Claviceps paspali* Stevens & Hall, which reduces seed production and can produce toxic effects in cattle.

Pensacola (2n = 20), the most popular bahiagrass, was found in Pensacola, Florida, in 1935 by E. H. Finlayson (Finlayson 1941; Burton 1946). It is characterized by long, narrow leaves and long stems and is more winter-hardy than common, Argentine, or Paraguay.

Paraguay (2n = 40) is a short, coarse, narrow-leaved cultivar that is less productive than Pensacola (Burton 1946). It forms a dense sod and provides satisfactory pasture until it becomes tough and unpalatable in midsummer.

Paraguay 22 (2n = 40) is similar to Argentine in growth habit and cold tolerance. It is more productive than Paraguay.

Wilmington (2n = 40) is the most cold-hardy of the bahiagrasses (Burton 1946). It has narrow leaves of medium size. It is less productive than Pensacola and Paraguay.

Tifhi-1 and Tifhi-2 (2n = 20) are selections from Pensacola that are leafier, have more shatter-resistant seed, and are higher yielding than Pensacola. Tifhi-1 has produced more beef per hectare than Pensacola (Hodges et al. 1967).

Tifton 9 is also a diploid (2n = 20). It has longer leaves, more vigor during the seedling stage, more succulence, and equal digestibility. It has yielded 47% more forage than Pensacola over a 3-yr period (Burton 1989).

MANAGEMENT PRACTICES

Bahiagrass generally is sown in early spring after the average date of the last killing frost. Summer plantings usually have severe weed infestations.

Bahiagrass should be planted on a well-prepared seedbed at a depth of 0.6-1.25 cm. Under average farm conditions, 11-17 kg ha^{-1} of seed are required to ensure a good stand. Use of a culti-packer with a seeding attachment or a grain drill is a good method of planting. Soil should be firmed around the seed to allow quick germination and rooting.

Small seedlings of bahiagrass are weak competitors with weeds. For quick establishment of a productive bahiagrass stand, weeds must be controlled. Grazing should be restricted with new plantings because trampling will damage many of the seedlings. Fertilization, preferably according to soil test recommendations, is necessary for satisfactory establishment and production. During establishment, one or two light topdressings of 18-34 kg ha^{-1} nitrogen (N) in mid-June and July will hasten development of young grass plants during their first year (Stephens and Marchant 1960).

After establishment, annual applications of 112-224 kg ha^{-1} N, 29 kg ha^{-1} phosphorus (P), and 56 kg ha^{-1} potassium (K) are required for good production (Hoveland et al. 1971). This will vary with soil type and previous fertilization of the pasture. Fertilize according to soil test and desired forage production. Animals prefer fertilized bahiagrass approximately 10 to 1 over unfertilized bahiagrass (Table 34.1). For better distribution of forage production, N should be applied in three to four applications during the growing season (Blue 1973, 1988; Allen et al. 1977), even though total forage yield is not greatly affected by a split application of N.

The dense, compact sod of bahiagrass generally limits success of associated legumes. However, white (*Trifolium repens* L.), crimson (*T. incarnatum* L.), and arrowleaf (*T. vesiculosum* Savi) clovers can be grown in bahiagrass sod, provided they are well fertilized with P and K and the grass is kept short. The P and K should be applied before planting winter annual legumes. When legume stands are good, the spring topdressing of N should be omitted. However, for high summer grass production, N topdressing should be applied in June and late July (Hoveland et al. 1971).

Summer legumes have been used in bahiagrass pastures in peninsular Florida. On spodosol (flatwoods) sites, the annual legume aeschynomene, *Aeschynomene americana* L., provides grazing of high quality, although natural reseeding of stands is unreliable. Lower-quality forage but greater reliability is provided by the perennial legume carpon desmodium, *Desmodium heterocarpon* (L.) DC. Burning in early spring, grazing moderately to heavily through early summer, and use of a roller chopper to open the bahia sod during the wet summer season are means of enhancing existing legume stands on flatwoods pastures.

SEED PRODUCTION

Except for the cultivar Wilmington, most bahiagrasses are good seed producers. Seeds mature progressively, beginning in early June and continuing during the summer. Seed yields vary depending on the cultivar, time of harvest, weather conditions, and N fertilizer used. Both seed yield and germination increase as the level of N fertilizer is increased up to 200 kg ha^{-1}. Under limited grazing conditions, two or more seed crops can be harvested annually. With good fertilization and combining methods, 112-336 kg ha^{-1} of cleaned seed can be obtained per year (Mancilla 1981).

Since all seeds do not mature simultaneously, close examination of fields before harvesting helps prevent shattering loss. Seed color can be a poor indicator for maturity because green-colored seed may be fully ripe. Proper time for harvest is determined by pulling a gathered handful of seed heads through moderately tightly closed fingers. Mature seeds readily strip off, while immature seeds cling to the seed heads.

Bahiagrass can be easily combined directly because seed heads extend above the leaves and can be cut without including much for-

TABLE 34.1. Application of N and improved grazing time

	Minutes grazed	
Nitrogen (kg ha^{-1})	Common bermudagrass	Pensacola bahiagrass
0	55	30
37	145	75
111	190	125
146	660	275
296	1320	414

Source: Miss. Agric. and For. Exp.Stn. Unpublished data.
Note: Twenty-five cows were placed on pastures for 4 h. All plots received constant rates of P and K.

age. Because combined bahiagrass seed is at various stages of maturity, rapid drying is necessary. Seed should be spread a few centimeters deep on a dry floor and stirred thoroughly once or twice a day. Forced air drying is necessary for large amounts of seed. If heated air is used, the temperature should be kept between 38° and 43°C. Aging of seed up to 1 yr increases percentage germination and tends to diminish dormancy factors that contribute to slow germination (West and Marousky 1989).

DISEASES AND PESTS

Bahiagrass is relatively free from disease and insect pests. However, ergot is a serious problem for the cultivar Argentine. This disease, which attacks the inflorescence and seriously damages the seed, can produce toxic effects in cattle. It is most prevalent in late summer and early fall and can be avoided by mowing the seed heads during that season.

Leaves of most bahiagrass types are susceptible to slight damage by leaf blight, *Helminthosporium micropus* Drechs. Different types of bahiagrass vary in their reaction to the fungus.

Mole crickets are major pests of bahiagrass in Florida (Sailer et al. 1984). Three introduced species of the genus *Scapteriscus* been identified, with tawny mole cricket, *S. nicinus* Scudder, causing the greatest damage in pastures. Damage is more severe on droughty sites in dry seasons. Root damage in these situations can result in essentially a complete loss of bahiagrass stands. Chemical control of mole crickets in pastures is not economically feasible. Progress has been made with development of a biological control. A range of specific and nonspecific natural enemies have been studied, and the most promising means of controlling the mole cricket are from a nematode, *Steinernema scapterisci* Nguyen & Smart, and a fly, *Ormia depleta* Wiedemann.

Dallisgrass

Common dallisgrass, *Paspalum dilatatum* Poir., is native to South America and probably originated in the area of northeastern Argentina, Uruguay, and southern Brazil. The first introduction into the US is unknown, but a herbarium specimen collected from Louisiana was noted in 1842 (Chase 1929). The grass was likely introduced unknowingly into the US in the vicinity of New Orleans. Dallisgrass was named for A. T. Dallis of La-Grange, Georgia, who was an enthusiastic promoter of the grass in about 1900 (Holt 1956).

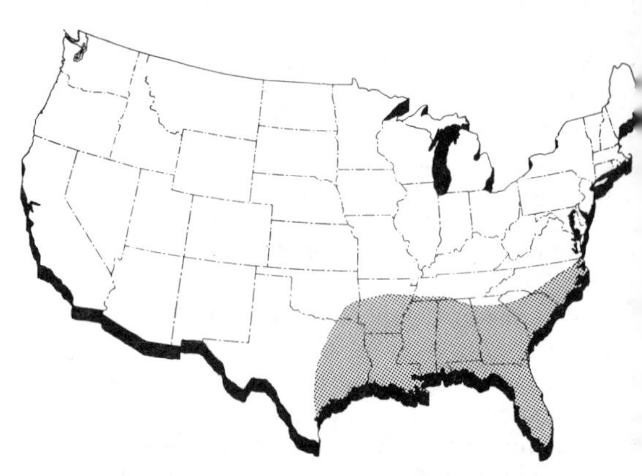

Fig. 34.3. Dallisgrass-producing areas of the US.

DISTRIBUTION AND ADAPTATION

Dallisgrass is distributed from New Jersey to the Gulf of Mexico and westward to Texas and Oklahoma (Fig. 34.3). However, its production is generally confined to areas receiving at least 890 mm annual rainfall, the exception being along streams, ditches, and areas where supplemental moisture is received. It grows on most soils but is best adapted to clay or loam soils that are moist but not wet.

PLANT DESCRIPTION

Dallisgrass is a strongly tufted, leafy, deep-rooted perennial. It usually grows in clumps with few to several culms and numerous leaves emerging from a knotted base composed of very short rhizomes (Fig. 34.4). Culms are ascending to suberect and range in height from 40 to 175 cm. Inflorescences at the top of each culm are racemose panicles commonly composed of three to five ascending to drooping broad racemes, 6-8 cm in length. The single-floret spikelets are arranged in two rows on each side of each raceme; flat and ovate pointed in shape, they range from 3.0 to 3.5 mm in length. Long, white silky hairs occur along the spikelet edge, with sparse pubescence over the surface. The grass has numerous basal leaves. Sheaths commonly are overlapping, compressed, and glabrous except at the very base, where pubescence is prevalent. Leaf blades are flat, ascending to spreading, and normally 10-25 cm long and 3-12 mm wide. They are glabrous except for sparse pubescence at the edges near the base (Chase 1929).

Dallisgrass is palatable, and it produces good quality forage superior to bahiagrass and some bermudagrasses. It grows later into the fall and initiates growth earlier in the spring than most warm-season grasses (Holt 1956), except perhaps limpograss (*Hemarthria altissima* [Poir.] Stapf & C.E. Hubb.) (Chap. 35). Even though the species is adapted to moist-humid areas, it withstands periods of drought. Depending upon plant maturity, soil fertility, and environmental conditions, digestibility ranges from 57% to 63% and crude protein from 4.4% to 23.2% (Committee on Animal Nutrition and Feed Composition 1971).

IMPORTANCE AND USE

Dallisgrass is one of the more important forage grasses in the humid Southeast (Holt 1956). The reasons are (1) it produces good quality forage and often retains its quality late into the summer growing season; (2) it grows well in association with common bermudagrass, white clover, and annual ryegrass (*Lolium multiflorum* Lam.); (3) it is highly palatable; (4) it persists under heavy grazing; and (5) it is well adapted to poorly drained loam and clay soils.

Interest and popularity of dallisgrass has decreased since the 1950s. However, about 8000 ha of improved pastures are still planted with dallisgrass annually. The grass also is quite invasive, spreading by seed, and constitutes a significant part of the botanical composition of both improved and native pastures growing on clay soils, especially those consisting of bermudagrass.

Dallisgrass is primarily used in permanent pastures, where it is grazed by beef cattle and to a lesser extent by dairy animals. In regions where it is the primary forage grass, it is of-

Fig. 34.4. Common dallisgrass produces many leaves and tillers from very short, compact rhizomes, which allow it to withstand heavy grazing. Horizontal lines in background are spaced 34 cm apart.

ten grown in combination with a legume such as white clover. This combination provides grazing over a longer period, reduces the need for N fertilizer, produces higher-quality forage, and reduces spring weeds (Evers 1989).

Dallisgrass is comparable to 'Alicia' bermudagrass in maintaining a beef cow herd under high stocking rates during the summer months in Louisiana. Average daily gain for calves is essentially the same for the two grasses (Hill et al. 1982).

Dallisgrass makes good quality hay before ergot (*Claviceps* spp.) infection is evident, but special management is necessary to avoid baling grass with seed heads heavily infected with ergot. If infected hay is fed to animals, ergot poisoning may result (see Chap. 9, Vol. 2). To avoid this, thresh or combine the seed before baling the vegetative material (Holt 1956).

CULTIVARS

There are at least six different dallisgrass biotypes with chromosome numbers of 2n = 40, 50, and 60 (Bashaw and Forbes 1958; Burson et al. 1991). Common dallisgrass (2n = 50) is the most prevalent and widespread. Improved cultivars of dallisgrass have not been developed through conventional methods (Burton 1962; Bennett et al. 1969) because most biotypes reproduce by apomixis (Bashaw and Holt 1958; Burson et al. 1991). The only reported source of sexuality in dallisgrass is a yellow-anthered biotype (2n = 40) that is a poor forage type (Bashaw and Holt 1958). This sexual biotype was crossed with apomictic common dallisgrass, but their hybrids were inferior to common dallisgrass in forage production (Burton 1962; Bennett et al. 1969). It also was crossed with other *Paspalum* species for phylogenetic studies, but most hybrids were sterile (Burson 1983).

Existing cultivars are selections of naturally occurring ecotypes that differ because of natural mutations. 'B230' and 'B430' were selected from the common biotype in Louisiana because they were more fertile but not sufficiently fertile for commercial seed production. 'Charus', 'Natsugumo', and 'Raki' were released in Uruguay, Japan, and New Zealand, respectively.

Prostrate dallisgrass, *P. dilatatum* var. *pauciciliatum* Parodi (2n = 40), is another apomictic biotype that is adapted to wet, poorly drained soils. It is more prostrate in its growth habit than common dallisgrass. Prostrate produces more forage than common. Its main limitations are poor seed production and susceptibility to ergot. Germplasm of prostrate dallisgrass was released in 1990 (Burton and Wilson 1991).

Uruguayan and Uruguaiana dallisgrasses, *P. dilatatum,* are two hexaploid (2n = 60) apomictic biotypes from Uruguay and Brazil, respectively (Burson et al. 1991). Both have larger inflorescences than common dallisgrass, and the Uruguayan biotype is taller and more erect than common. Some accessions of the Uruguayan biotype produced more forage while maintaining in vitro dry matter digestibility of the forage at a level similar to that of common (Burson et al. 1991). Germplasm of these two hexaploid biotypes is undergoing additional evaluation for forage potential before possible release.

MANAGEMENT PRACTICES

Dallisgrass is difficult to establish because of poor seed quality and slow germination (Holt 1956). It is usually planted in late winter or early spring. However, in the lower South near the Gulf Coast, where winter temperatures are more moderate, seed can be planted successfully during the fall.

For best results, dallisgrass seed is planted into a well-prepared seedbed at a depth of 0.6-1.25 cm. The soil should be in firm contact around the seed; therefore a culti-packer with a seeding attachment or a grain drill should be used. To obtain an adequate stand, 1.7 kg ha^{-1} of pure live seed (PLS) should be planted when the rows are 50 cm apart (Holt and Hutson 1954). However, when the seed is drilled or broadcast, the rate should be 4.5-6.7 kg PLS ha^{-1} (Evers 1988). Most commercial seed is imported from Australia, and its percent germination is often 60% or higher, but seed produced in the US has a much lower germination. Attention should be given to the percent germination of the seed lot, and adjustments should be made in the quantity of seed planted in order to plant the recommended rate in PLS. In the rice-growing areas of Texas and Louisiana, dallisgrass can be broadcast into rice without seedbed preparation after the last draining or in rice stubble after combining in the fall.

Competition from weeds is a problem in establishment because of slow and erratic germination. Planting dallisgrass with a winter annual companion crop such as annual ryegrass in the autumn results in the most successful stands. Annual ryegrass grows during the winter, providing spring weed control. Dallisgrass seed germinates in May through July after the ryegrass dies. Various herbicides can successfully control weed competition and aid in spring establishment of dallisgrass (Evers 1981).

The amount of fertilizer required for establishment or maximum forage production depends upon the soil type and its inherent fertility. The actual fertilizer requirements should be based on soil test recommendations. Nitrogen fertilizer should not be applied when planting dallisgrass. The N is either lost or taken up by competing weeds.

Plant response to fertilization is unclear. Wilkinson and Langdale (1974) indicate that dallisgrass responds to N up to 150 kg ha^{-1} on sandy soils and up to 225 kg ha^{-1} on heavier clay soils. However, animal gain on dallisgrass pastures was not improved at rates above 112 kg ha^{-1} annually (Wilkinson and Langdale 1974). Other studies indicate maximum profit on pasture was obtained from 101 kg ha^{-1} N and 67 kg ha^{-1} P (Holt 1956) or no N when grown with white clover (Evers 1989). Recent findings demonstrate a limited response to P and K (Robinson et al. 1988; Jones and Watson 1991). However, when dallisgrass is grown in association with legumes, fertilizer requirements change. Dallisgrass-white clover mixtures are most profitable when no N is added (Evers 1989). Other nutrients should be applied according to soil test recommendations to meet the clover's requirements.

Since dallisgrass spreads rapidly and volunteers easily, only a limited number of seedlings per square meter are necessary for stand establishment. Plant size increases by tillering, and frequent defoliation greatly favors tillering and competitiveness (Harris et al. 1981). Dallisgrass normally is grown in combination with other forages, and management is extremely important in maintaining a species balance, especially in combination with a legume or a cool-season annual grass (Harris et al. 1981). Shading from a winter annual can delay the growth of dallisgrass in early and midspring.

When grown with bermudagrass, dallisgrass often is selectively grazed, and overgrazing can result. However, dallisgrass persists well under high stocking rates and produces more forage when defoliated to a 7.5-10 cm height rather than to lower or higher levels (Jones 1967; Bryan 1970). Deregibus and Trlica (1990) report that severe but infrequent defoliation increased tillering in dallisgrass.

SEED PRODUCTION

Dallisgrass is noted for low seed yields and poor seed quality. However, with proper management practices, seed yield and quality can be increased. Seed is normally harvested in late spring and late summer. Annual yield is about 160 kg ha^{-1}, with most seed recovered from the late spring harvest (Holt and Bashaw 1963). Besides circumventing disease and insect problems, the early seed crop is produced under environmental conditions for optimum seed set (Pearson and Shah 1981). Application of N at 67 kg ha^{-1} in late winter significantly increases the first seed crop; additional N after the first harvest does not improve the August seed crops. If ungrazed prior to the first harvest, seed yields will be higher (Holt and Bashaw 1963).

Seed should be harvested when inflorescences are brown but before seed shattering becomes severe. Harvesting may be either by direct combining or by mowing, windrowing, and combining (Holt and Bashaw 1963). Because of its indeterminate-flowering habit, a lot of immature seed is harvested. This green material interferes with proper combine threshing and results in material with high moisture. Following harvesting, the seed should be dried immediately with forced air temperatures from 38° up to 60°C (Bennett and Marchbanks 1969) or by spreading seed out in thin layers and stirring frequently until the moisture content is 10%-12%.

DISEASES AND PESTS

The most serious disease of dallisgrass is ergot, *Claviceps paspali* Stevens & Hall. The fungus invades the ovary inside the floret and develops a hard sclerotia body, instead of an embryo, which reduces seed set. The most serious aspect of ergot occurs when livestock consume the sclerotia and poisoning results. Dallisgrass also is attacked by anthracnose, *Colletotrichum graminicola* (Ces.) G. W. Willis, and by leaf blight, *H. micropus* Drechs. Both are foliage diseases that weaken and sometimes kill the plant (Burton 1962). Recently paspalum leaf blight, *Ascochyta paspali* (H. Sydow) Punith, was reported in the north island of New Zealand (Buchanan 1984a) and in the irrigated pastures of northern Victoria and New South Wales in Australia (Williams et al. 1988). The disease has reduced dallisgrass forage yields an estimated 11%-40% in Australia (Williams et al. 1988). The organism is a systemic pathogen that invades the xylem of the vascular tissue, and it can occur in the seed, which may be a vector in its spread (Buchanan 1984b). Because most commercial seed sold in the US is imported from Australia, the introduction and spread of this disease is of concern.

Dallisgrass is sometimes attacked by the sugarcane borer, *Diatraea saccharalis* Fabricius. The insect population normally is not large enough to cause appreciable damage until mid-July (Holt and Bashaw 1963). At that time, the larvae bore into and feed inside the stems. The stem is killed prior to maturity, and seed production is reduced.

Because diseases and insects are not serious problems before midsummer, harvesting the early seed crop tends to circumvent most ergot and sugarcane borer problems. Grazing or cutting the grass for hay during the summer months reduces foliar disease problems.

Other *Paspalum* Species

There are more than 350 species in this genus, and besides bahiagrass and dallisgrass, there are others that contribute to grassland agriculture. Most occur in the native grasslands of tropical and subtropical South America; however, some are found in native pastures in the southern US.

VASEYGRASS

Vaseygrass, *Paspalum urvillei* Steud., is native to the same area of South America as dallisgrass. Taxonomically, both species are very closely related (Chase 1929) and probably have common ancestors (Burson 1983). Vaseygrass was introduced into the southeastern US after dallisgrass but prior to 1880. It has become naturalized in the southern states and is adapted to low wet soils. Its general range is from North Carolina to Florida and west to Texas.

Vaseygrass is a stout erect perennial with many culms, 0.75-2.0 m tall. It is coarser and taller than dallisgrass. The grass is not planted or cultivated but frequently occurs in native pastures in low wet areas. When there is new growth, it is sometimes grazed or cut for hay. However, because of its erect growth habit, it is poorly suited for grazing and is easily eliminated by heavy continuous grazing. Contrary to dallisgrass, vaseygrass has good seed set.

BROWNSEED PASPALUM

Brownseed paspalum, *Paspalum plicatulum* Michx., is a tufted perennial about 50-100 cm tall that grows in small to moderately large clumps. As the name indicates, it has shiny, dark brown seed (florets). It is found in an area from Florida and Georgia west to Texas. The grass grows on a wide range of soils but is best adapted to sandy or sandy loam soils. In pastures on the Texas coastal prairie, cattle readily eat its foliage after frost because it remains green during the winter (Durham and Kothmann 1977). The grass is more popular in Australia, where three cultivars ('Hartley', 'Rodd's Bay', and 'Bryan') were selected and released. Unfortunately, they lack sufficient cold tolerance to survive the winters in the southern US.

KNOTGRASS

Knotgrass, *Paspalum distichum* L., is a rhizomatous and stoloniferous perennial that has become naturalized in the southern states. It grows from Florida to New Jersey west to Oklahoma and Texas and is often found growing along streams and irrigation ditches in the western states extending from California to Washington. Knotgrass is adapted to various soils but prefers wet, fertile soils. Thus, it is commonly found in moist low areas such as wet savannas and along streams and ditches. Since it produces stolons, it often forms a dense sod and is considered a weed in some areas of the world. Knotgrass does not produce large quantities of forage, but it is highly palatable to grazing animals. There is considerable interest in the grass in Japan.

QUESTIONS

1. Where is dallisgrass believed to be native, and when was it first thought to be introduced into the US?
2. What are the characteristics of dallisgrass that make it one of the leading perennial grasses in the South?
3. What is the most serious disease of dallisgrass and why?

4. Why is it easier to grow legumes with dallisgrass than with bahiagrass?
5. What cultivars of bahiagrass are the most widely used?
6. What are the benefits of producing an early seed crop of dallisgrass?
7. Explain why ergot reduces seed fertility in dallisgrass and bahiagrass.
8. Describe two additional *Paspalum* species and compare them with bahiagrass and dallisgrass.

REFERENCES

Allen, M, PE Schilling, EA Epps, CR Montgomery, BD Nelson, and RH Brubacher. 1977. Response of Bahiagrass to Nitrogen Fertilizer. La. Agric. Exp. Stn. Bull. 701.

Bashaw, EC, and I Forbes, Jr. 1958. Chromosome numbers and microsporogenesis in dallisgrass, *Paspalum dilatatum* Poir. Agron. J. 50:441-45.

Bashaw, EC, and EC Holt. 1958. Megasporogenesis, embryo sac development and embryogenesis in dallisgrass, *Paspalum dilatatum* Poir. Agron. J. 50:753-56.

Beaty, ER, RH Brown, and JB Morris. 1970. Response of Pensacola bahiagrass to intense clipping. In Proc. 11th Int. Grassl. Congr., Surfers Paradise, Queensland, Australia, 538-42.

Bennett, HW, and WW Marchbanks. 1969. Seed drying and viability in dallisgrass. Agron. J. 61:175-77.

Bennett, HW, BL Burson, and EC Bashaw. 1969. Intraspecific hybridization in dallisgrass, *Paspalum dilatatum* Poir. Crop Sci. 9:807-9.

Blue, WG. 1973. Role of Pensacola bahiagrass stolon-root systems in fertilizer utilization on fine sand. Agron. J. 65:88-91.

———. 1988. Response of Pensacola bahiagrass on a Florida spodosol to nitrogen sources and times of application. In Proc. Soil and Crop Sci. Soc. Fla. 47:139-42.

Bryan, WW. 1970. Changes in botanical composition in some subtropical sown pastures. In Proc. 11th Int. Grassl. Congr., Surfers Paradise, Queensland, Australia, 636-39.

Buchanan, PK. 1984a. *Ascochyta paspali* a fungal parasite of *Paspalum dilatatum*. N.Z. J. Bot. 22:515-23.

———. 1984b. Systemic growth of *Ascochyta paspali* in paspalum. N.Z. J. Agric. Res. 27:451-57.

Burson, BL. 1983. Phylogenetic investigations of *Paspalum dilatatum* and related species. In JA Smith and VW Hays (eds.), Proc. 14th Int. Grassl. Congr., Lexington, Ky. Boulder, Colo.: Westview, 170-73.

Burson, BL, PW Voigt, and GW Evers. 1991. Cytology, reproductive behavior, and forage potential of hexaploid dallisgrass biotypes. Crop Sci. 31:636-41.

Burton, GW. 1946. Bahiagrass types. J. Am. Soc. Agron. 28:273-81.

———. 1962. Conventional breeding of dallisgrass, *Paspalum dilatatum* Poir. Crop Sci. 2:491-94.

———. 1989. Registration of Tifton 9 bahiagrass. Crop Sci. 29:1326.

Burton, GW, and JP Wilson. 1991. Registration of prostrate dallisgrass germplasm #1. Crop Sci. 31:1392.

Chase, A. 1929. The North American species of *Paspalum*. Contrib. US Natl. Herb. 28(1):310.

Committee on Animal Nutrition and Feed Composition. 1971. Atlas of Nutrition Data on US and Canadian Feeds. Washington, D.C.: National Academy of Science.

Deregibus, VA, and MJ Trlica. 1990. Influence of defoliation upon tiller structure and demography in two warm-season grasses. Acta Oecol. 11:693-99.

Durham, AJ, Jr., and MM Kothmann. 1977. Forage availability and cattle diets on the Texas coastal prairie. J. Range Manage. 30:103-6.

Ellzey, HD. 1967. Effects of Various Levels of Lime, Phosphorus and Potash on the Yield and Quality of Paraguay Bahiagrass. Annu. Prog. Rep. Southeast La. Dairy Pasture Exp. Stn. 30-68.

Evans, EM, LE Ensminger, BD Doss, and OL Bennett. 1961. Nitrogen and Moisture Requirements of Coastal Bermuda and Pensacola Bahia. Ala. Agric. Exp. Stn. Bull. 337.

Evers, GW. 1981. Herbicidal enhancement of dallisgrass establishment. Agron. J. 73:347-49.

———. 1988. Dallisgrass. In Advances. Texas Forage and Grassland Council, 1-4.

———. 1989. Comparison of input levels for upper Gulf Coast pastures. In Forage Research in Texas, Tex. Agric. Exp. Stn. CPR-4731, 1-4.

Finlayson, EH. 1941. Pensacola, a new fine leafed bahia. South. Seedman 4:9-28.

Hanson, AA. 1972. Grass Varieties in the United States. USDA Agric. Handb. 170. Washington, D.C.: US Gov. Print. Off.

Harris, W, BJ Forde, and AK Hardacre. 1981. Temperature and cutting effects on the growth and competitive interaction of ryegrass and paspalum. I. Dry matter production, tiller numbers, and light interception. N.Z. J. Agric. Res. 24:299-307.

Hill, GM, RA Harpel, WB Hallmark, RJ Cormier, and JH Davis. 1982. Alicia, dallisgrass compared for summer grazing. La. Agric. 25(4):10-12.

Hodges, EM, JE McCaleb, WG Kirk, and FM Peacock. 1967. Grazing trails on grass varieties. Annu. Rep. Fla. Agric. Exp. Stn., 50-54.

Holt, EC. 1956. Dallisgrass. Tex. Agric. Exp. Stn. Bull. 829.

Holt, EC, and EC Bashaw. 1963. Factors Affecting Seed Production of Dallisgrass. Tex. Agric. Exp. Stn. MP 662.

Holt, EC, and HC Hutson. 1954. The Establishment of Dallisgrass. Tex. Agric. Exp. Stn. PR 1662.

Hoveland, CS. 1968. Bahiagrass for Forages in Alabama. Ala. Agric. Exp. Stn. Circ. 140.

Hoveland, CS, EL Carden, JR Wilson, and PA Mott. 1971. Summer Grass Residue Affects Growth of Winter Legumes under Sod. Auburn Univ. Highlights Agric. Res. 18(3).

Jones, DW. 1971. Bahiagrass in Florida. Fla. Ext. Serv. Circ. 321A.

Jones, RI. 1967. Comparative effects of differential defoliation of grass plants in pure and mixed stands of two species. S. Afr. J. Agric. Sci. 10:429-44.

Jones, WF, and VH Watson. 1991. Applied phosphorus and potassium effects on yield of dallisgrass-bermudagrass pastures. J. Plant Nutr. 14:585-97.

McCormick, WC, WH Marchant, and BL Southwell. 1967. Coastal Bermudagrass and Pensacola Bahiagrass Hays for Wintering Beef Calves. Ga. Agric. Exp. Stn. Res. Bull. 19.

Mancilla, LE. 1981. Forage and seed production characteristics of bahiagrass under varying management systems. PhD diss., Mississippi State Univ.

Pearson, CJ, and SG Shah. 1981. Effects of temperature on seed production, seed quality and growth of *Paspalum dilatatum*. J. Appl. Ecol. 18:897-905.

Robinson, DL, KG Wheat, NL Hubbert, MS Henderson, and HJ Savoy, Jr. 1988. Dallisgrass yield, quality and nitrogen recovery responses to nitrogen and phosphorus fertilizers. Commun. Soil Sci. Plant Anal. 19:529-42.

Rollins, GH, and CS Hoveland. 1960. Wanted—Good Summer Perennial Grasses for Dairy Cows. Auburn Univ. Highlights Agric. Res. 7(2).

Sailer, RI, JA Reinert, D Boucias, P Busey, RL Kepner, TG Forrest, WG Hudson, and TJ Walker. 1984. Mole Crickets in Florida. Fla. Agric. Exp. Stn. Bull. 846.

Scott, JM. 1920. Bahiagrass. J. Am. Soc. Agron. 13:112-14.

Stephens, JL, and WA Marchant. 1960. Bahiagrass for Pastures. Ga. Agric. Exp. Stn. Bull. NS-67.

West, SH, and F Marousky. 1989. Mechanisms of dormancy in Pensacola bahiagrass. Crop Sci. 29:787-91.

Wilkinson, SR, and GW Langdale. 1974. Fertility needs of the warm-season grasses. In DA Mays (ed.), Forage Fertilization. Madison, Wis.: American Society of Agronomy, Crop Science Society of America, and Soil Science Society of America, 119-45.

Williams, BL, TV Price, and PA Taylor. 1988. Survey of the distribution and perceived losses caused by paspalum leaf blight in Northern Victoria. Crop Prot. 7:28-33.

35 Other Grasses for the Humid South

WILLIAM R. OCUMPAUGH
Texas A&M University

LYNN E. SOLLENBERGER
University of Florida

Limpograss

LIMPOGRASS, *Hemarthria altissima* (Poir.) Stapf & C. E. Hubb., is native to tropical Africa. Its common name was chosen because limpograsses are native to the Limpopo River valley of South Africa.

Limpograss is adapted to Florida and the southern portions of the Gulf Coast states (Fig. 35.1). It is more winter-hardy than digitgrass (*Digitaria eriantha* Steud.) and stargrass (*Cynodon nlemfuensis* Vanderyst) (Quesenberry et al. 1978) and more frost tolerant than bahiagrass (*Paspalum notatum* Flugge) (Quesenberry and Sollenberger 1992). Limpograss will grow well on most soils with a pH of 5.5-6.5. It is best adapted to wet habitats and especially to the seasonally flooded flatwood soils of the southeastern US.

PLANT DESCRIPTION

Limpograss is a robust warm-season perennial forage of the Andropogoneae tribe. Although strongly stoloniferous, erect stems may reach 1.5 m in height but more commonly are 0.3-0.8 m. Leaves are up to 20 cm long and 6 mm wide, with membranous ligules. Spikelike racemes are compressed, 6-10 cm long, with spikelets appearing opposite. Each pair is composed of a bisexual sessile spikelet 5-6 mm long and a smaller pedicelled male spikelet (Ocumpaugh and Rouquette 1985) (Fig. 35.2).

Morphological characteristics like leafiness, internode length, anthocyanin concentration, tiller density, and stem thickness vary among limpograsses and aid in genotype identification. Limpograss produces viable seed, but the percent viability is too low to be useful except in genetic improvement programs (Schank 1972). Limpograss is routinely propagated vegetatively, primarily using mature stems. When conditions and management are favorable, well-established stands are ready for grazing in 3-4 mo.

Limpograss begins growth earlier in the spring and continues to grow later in the autumn than most warm-season grasses. It can

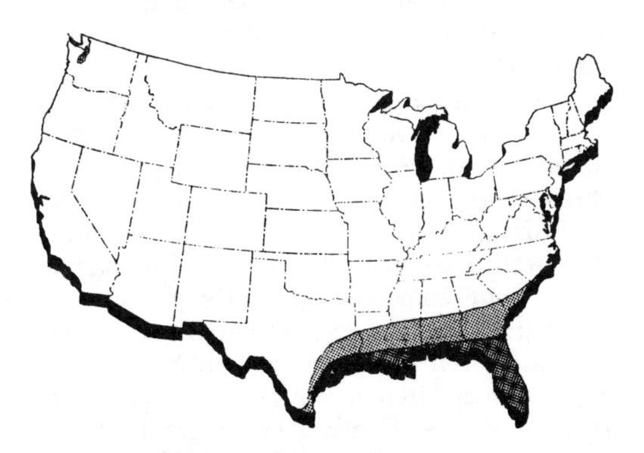

Fig. 35.1. Limpograss- and elephantgrass-producing areas of the US.

WILLIAM R. OCUMPAUGH is Professor of Agronomy, Texas A&M University, and Texas Agricultural Experiment Station, Beeville. He served on the faculty of the University of Florida from 1975 to 1983. He received his MS and PhD degrees from the University of Missouri at Columbia. His specialty is pasture and forage crop management and utilization.

LYNN E. SOLLENBERGER is Associate Professor of Agronomy, University of Florida. He received his MS degree from Pennsylvania State University in 1981 and PhD degree from the University of Florida in 1985. Since 1985 he has taught and conducted research on pasture and forage crop management and utilization.

Fig. 35.2. Vegetative stems and flowering spikelike racemes of limpograss.

provide grazing throughout winter in mild years in southern Florida. Depending upon soil moisture and fertility, dry matter yields up to 29,000 kg/ha have been reported (Christiansen et al. 1988). Annual yields typically range from 12,000 to 20,000 kg/ha (Christiansen et al. 1988; Kalmbacher et al. 1987; Quesenberry et al. 1984). When managed similarly, in vitro organic matter digestibility (IVOMD) is comparable to that of digitgrass and about 6-10 percentage units greater than bahiagrass (Sollenberger et al. 1988a, 1989a). Crude protein (CP) concentration of limpograsses is often lower than that of other tropical grasses. Mature or unfertilized limpograss may be as low as 2%-3% CP (Quesenberry and Ocumpaugh 1980; Sollenberger and Quesenberry 1985), although 12%-14% CP has been reported for immature, well-fertilized limpograss (Kretschmer and Snyder 1979).

CULTIVARS

Four cultivars of limpograss have been released jointly by the Florida Agricultural Experiment Station and the Soil Conservation Service (Quesenberry et al. 1978, 1984). 'Redalta' and 'Greenalta' are diploids (2n = 18), and 'Bigalta' and 'Floralta' are tetraploids (2n = 36). Redalta and Greenalta were so named because of the characteristic pigmentation observed on these forages when subjected to cool temperatures. Bigalta was named for its large stem relative to the diploids. Despite good winterhardiness and persistence (Quesenberry et al. 1984), the diploid cultivars are no longer recommended for new plantings because of low digestibility. Bigalta has higher digestibility than Redalta or Greenalta, but it does not persist well under grazing. Floralta is the most widely planted cultivar and is the one recommended for use in the southern US. It is slightly lower in digestibility than Bigalta but is higher yielding and much more persistent under grazing (Adjei et al. 1988; Quesenberry et al. 1987).

IMPORTANCE AND USE

Limpograss and associations with legumes are used primarily for pasture (Kretschmer and Snyder 1979). The planted area is greatest in Florida, totaling approximately 50,000 ha and expanding rapidly. Because of its early spring and late autumn growth, limpograss may be best utilized as a special purpose forage to extend the grazing season (Quesenberry and Sollenberger 1992). Its digestibility decreases more slowly with maturity than that of other tropical grasses. Based on these characteristics, limpograss is well suited for use as a stockpiled forage for autumn-winter grazing.

MANAGEMENT PRACTICES

Following establishment, limpograss is quite competitive, but chemical weed control may be needed during the establishment phase. Limpograss is very susceptible to 2,4-D (2,4 dichlorophenoxyacetic acid), and complete stand losses have been observed when it was applied. Dicamba (3,6-dichloro-2-methoxybenzoic acid), at rates recommended for controlling most common broadleaf weeds, has been successfully applied at any stage of growth (Quesenberry et al. 1978).

Limpograss will respond to very high rates of nitrogen (N) fertilizer (Christiansen et al. 1988); however, good results have been obtained with 160-180 kg N/ha split in two or three applications between February and August (Sollenberger et al. 1988a). Depending on soil nutrient status and N fertilizer rate, annual applications of phosphorus (P) and potassium (K) may be needed to produce rapid growth.

Because of its relatively poor persistence under grazing, Bigalta is best used as a stockpiled forage for autumn-winter with minimal defoliation during other times of the year. Recommendations for stockpiling limpograsses in north Florida include (1) removal of mature forage by grazing or clipping by early August, (2) application of 60-120 kg N/ha after removing the forage, and (3) no grazing until after mid-October. Animals consuming stockpiled limpograss should be fed protein and mineral supplements (Quesenberry and Ocumpaugh 1980, 1982).

Floralta limpograss is also recommended for stockpiling. Because it is more persistent under grazing, greater use during other times of the year is possible. With continuous stocking over a long period of time, pastures should be grazed to maintain a canopy height of 30 cm or more. Rotational stocking allows closer grazing. Adjei et al. (1988) recommends use of 6 wk or longer rest periods when grazing to a 15-cm postgraze stubble height. However, forage quality of limpograss herbage is generally higher when grazed every 4-5 wk to a 25-cm stubble (Sollenberger et al. 1989b).

Because of the relatively high digestibility, but low CP of limpograss, use of associated legumes or small amounts of CP supplement have been evaluated on limpograss pastures (Sollenberger et al. 1989b). Aeschynomene (*Aeschynomeme americana* L.) associates well with limpograss (Sollenberger and Quesenberry 1985; Sollenberger et al. 1987b). Cattle select aeschynomene leaf and fine stem (Sollenberger et al. 1987a), and gains of yearling steers grazing the mixture were 80% greater (0.70 vs 0.39 kg/d) than on N-fertilized limpograss (Rusland et al. 1988). It was concluded that a portion of the increase in gain was due to elimination of a protein deficiency by the legume. This conclusion was supported by a study in which yearling steers grazing N-fertilized limpograss gained 0.29 kg/d, whereas steers grazing similar pastures and receiving 0.58 kg of ground corn (*Zea mays* L.) plus 0.03 kg urea (21% CP in supplement) per head per day gained 0.53 kg/d (Holderbaum et al. 1991). The low CP concentration of mature limpograss hay and associated poor performance of cattle can be overcome by ammoniation of the hay (Brown et al. 1987). Ammoniation increased CP concentration from 3.2% to 10.3% and digestibility of the hay from 46% to 62%.

DISEASES AND PESTS

All limpograsses are susceptible to sting nematodes, *Belonolaimus longicaudatus* Rau (Quesenberry and Dunn 1978); however, the tetraploid lines seem to be more tolerant than the diploids. This pest can cause foliar chlorosis and will reduce productivity if present in large numbers (Quesenberry and Dunn 1978), but it is generally not a problem on the wet sites to which limpograss is best adapted (Quesenberry et al. 1978). In a test of 17 pasture grass entries, 3 limpograsses were favored hosts for sting nematodes, but yields of other grasses including 'Pangola' digitgrass also were reduced (Boyd and Perry 1970). All limpograsses are somewhat susceptible to the yellow sugarcane aphid, *Sipha flava* Forbes (Oakes 1978). Floralta was the most susceptible cultivar, while most others showed more resistance than digitgrass.

Elephantgrass

Elephantgrass, *Pennisetum purpureum* (L.) Schumach., also known as *napiergrass,* is native to Africa. In 1913 the USDA introduced it in the southern US for experimentation (Ocumpaugh and Rouquette 1985). Elephantgrass has been introduced into nearly all tropical and subtropical regions from sea level to altitudes of 2000 m where rainfall exceeds 1000 mm. Elephantgrass has seen only limited use as a forage plant in the US.

DISTRIBUTION AND ADAPTATION

Elephantgrass grows best when temperatures are hot (30°-35°C), yet it will tolerate cool temperatures down to 10°C before growth ceases (Ocumpaugh and Rouquette 1985). Frost will kill top growth but leave roots unharmed; however, if the soil freezes, root damage will occur and plant loss will result. Periods of drought restrict growth; however, with the onset of rains, plants resume rapid growth. The grass will not tolerate waterlogging or flooded conditions, and it is not adapted to wet soil sites where vaseygrass (*Paspalum urvillei* Steud.) thrives. Elephantgrass can persist in soils that tend to be excessively drained but does not do well in heavy clays. Soil pH should be maintained above 5.5 (Sollenberger et al. 1988b). Its geographic area of adaptation in the US is similar to that of limpograss (Fig. 35.1), but the two plants are best adapted to different soil types.

PLANT DESCRIPTION

Elephantgrass is a robust perennial bunchgrass consisting of many canelike stems up to 3-4 cm in thickness that grow to heights of 2-6 m. Each stem has numerous nodes from which leaf blades originate. The leaves vary in width from 1 to 5 cm and in length from 30 to 120 cm. The leaf blade has serrated margins, smooth upper and lower surfaces, and a midrib that is prominent on the lower side (Ocumpaugh and Rouquette 1985).

'Mott' elephantgrass is a dwarf type that grows to a maximum height of 1.5-2.5 m compared with about 6 m for tall types. Individual bunches produce numerous tillers and produce a basal diameter of 0.5 m or more. The leaf canopy of a well-established bunch can occupy a diameter of 1 m or more. The dwarf trait in elephantgrass is quantitative and is related to reduced internode length, meaning that plant height can be varied through breeding and selection, while number of leaves remains similar.

Elephantgrass is normally propagated with stem cuttings, but Mott (dwarf) elephantgrass is more difficult to propagate than tall elephantgrass. Dwarf elephantgrass is more sensitive to deep planting than tall elephantgrass (Woodard et al. 1985). The optimum planting depth for Mott is 2.5-5.0 cm (with horizontal stem placement), with percent emergence declining as planting depth increases beyond 5 cm. For best establishment use well-fertilized nursery stock with mature stems that are hard with a white and black variegated color. Tall elephantgrass can be propagated with three-node stem pieces, but establishment of dwarf types is best from undefoliated, intact stem cuttings (Sollenberger et al. 1990). Reed (1991) suggests the undefoliated stem cuttings are superior because the leaf blades contain 40%-80% of the N and total nonstructural carbohydrate pools of a tiller. In the subtropical environment, plantings should be made in late summer (August) to allow plants to establish and develop before winter (Woodard et al. 1985; Sollenberger et al. 1991). Alternatively, plantings can be made in late November or December just prior to the onset of freezing temperatures. Under this procedure, the stem cutting is buried and lies dormant until favorable growing conditions in the spring (Ocumpaugh 1989; Woodard et al. 1985).

IMPORTANCE, USE, AND MANAGEMENT

Elephantgrass is very responsive to fertilizer and is one of the fastest-growing, highest-yielding grasses. However, this growth rate also makes management of tall cultivars difficult. Mott dwarf elephantgrass is being adapted, primarily in developing countries, because of its relative ease of management. Additionally, Mott elephantgrass has a high proportion of high-quality leaves (Boddorff and Ocumpaugh 1986) and commonly supports daily gains of nearly 1 kg/d when grazed (Sollenberger and Jones 1989). Milk production was depressed only 3%-7% when Mott silage was substituted for corn silage in the diet of lactating Holstein cows (Ruiz et al. 1992).

For long-term stand maintenance, defoliation intensity and frequency of Mott elephantgrass must be managed. For example, when clipped, defoliation needs to be lenient (a stubble height >20 cm left), or the frequency of defoliation should not be greater than once every 8-9 wk. Since it is easier to regulate partial defoliation under grazing, a stubble height of 45-60 cm and a frequency of defoliation of about 5 wk is recommended for Mott pastures. Mott elephantgrass performs well with rather low rates (100 kg/ha annually) of N fertilizer (Sollenberger and Jones 1989), but the plant will respond to annual rates of 400-500 kg N/ha without any detrimental effects (Woodard and Prine 1990).

CULTIVARS

Many cultivars of tall elephantgrass are used in the Tropics, but the only cultivar being utilized in forage systems in the US is Mott elephantgrass (Sollenberger et al. 1989a). Researchers in both Florida and Georgia are attempting to develop seed-propagated hybrids of pearlmillet, *Pennisetum americanum* (L.) Leeke × elephantgrass. Generally, these hybrids are less winter-hardy than true elephantgrasses but are more productive than pearlmillets.

PESTS

The only insect problem that has been reported on Mott dwarf elephantgrass has been moderate infestations of the two-lined spittlebug (*Prosapia bicincta* Say) (Kalmbacher et al. 1987; Chaparro 1991). In the US, the spittlebug has not been reported to kill entire plants, but it significantly weakens them and may make them more susceptible to defoliation stress. Spittlebugs are difficult to control with pesticides and are generally not a problem in well-managed swards. Burning of frost-killed forage is a good means of spittlebug prevention.

Digitgrass

Digitaria eriantha Steud. is the current classification (Kok 1981, 1984) for all cultivars of digitgrass, or *fingergrass*. The cultivars developed in Florida, and previously classified as *D. decumbens, D. pentzii,* or *D.* x *umfolozi,* are stoloniferous and include the cultivars 'Transvala', 'Taiwan', 'Slenderstem', 'Survenola', and the most widely used Pangola digitgrass (Ocumpaugh and Rouquette 1985). The tufted type (formerly *D. smutsii*) is represented by the Australian cultivars 'Premier' (Oram 1990) and 'Advance' (Hacker 1985). Premier and Advance digitgrass produce viable seed, while the others are planted vegetatively. In Australia the tufted types are being used sparingly. All of the discussion that follows relates to the stoloniferous types.

Pangola was first introduced in the US in 1935. Digitgrass, native to South Africa, is adapted to Florida and extreme southern portions of the Gulf states and southern California. Due to winter damage in more northern areas, the most extensive use of digitgrass is in central and south Florida.

Digitgrass is best adapted to fertile and moist soils, but it will thrive on soils of varying drainage. Digitgrass grows well in full sun to partial shade, from altitudes near sea level to 300 m, on sandy to heavy clay soils ranging in pH from 4.2 to 8.5, and in rainfall of 2500 mm annually. It is less drought tolerant and more sensitive to low N levels than bahiagrass. Digitgrass is high in quality and very palatable and can be lost to weeds if overgrazed (Ocumpaugh and Rouquette 1985). Forage quality of digitgrass is not as high as that of elephantgrass, but it is higher than that of bahiagrass, bermudagrass (*Cynodon dactylon* [L.] Pers.), and stargrass.

PLANT DESCRIPTION

Digitgrass is a stoloniferous perennial that grows to a height of 60-120 cm. It produces many semidecumbent stolons that root at the nodes. The erect seedstalks produce many branches but little viable seed (Ocumpaugh and Rouquette 1985).

IMPORTANCE AND USE

Digitgrass is primarily used in central and south Florida. However, there is less area planted to digitgrass today than in the past because of three factors: (1) digitgrass requires more N fertilizer than bahiagrass to be productive, (2) it is more susceptible to overgrazing and invasion by common bermudagrass than the alternative grasses, and (3) other grass species have become available that are better adapted and more productive than digitgrass.

CULTURE AND MANAGEMENT

Digitgrass is established vegetatively by spreading freshly mowed stems and stolons over a well-prepared seedbed and covering them with a disc harrow. The disc should be adjusted so about half the stems are covered and half are still exposed. Use of a packer following the harrow produces a smoother field, increases moisture retention, and hastens plant establishment. Lime should be applied to soils with a pH of less than 5.0 for more efficient nutrient utilization.

Annual fertilization is required, but N applications should be less than 200 kg/ha. Periodic soil tests should be used as a guide for P and K requirements. Application of K is especially helpful if winter-killing is a possibility. If hay, silage, or plant material is to be harvested, higher rates of fertilizer, especially K, will be required. Digitgrass is sensitive to copper (Cu) deficiency and will benefit from annual applications of 0.05-0.1 kg/ha Cu obtained from cupric oxide (CuO) or copper sulfate ($CuSO_4$).

Digitgrass pastures require an orderly grazing system. Close grazing will weaken the sod unless regrowth is allowed to reach at least 30-45 cm before being grazed again. Digitgrass performs best when rotationally stocked and when 1500 to 2000 kg/ha of residual herbage remain at the end of each grazing period.

DISEASES AND PESTS

The yellow sugarcane aphid is a major pest of digitgrass, especially in south Florida. The nymph and adult forms of the two-lined spittlebug cause damage from July-August until cold weather. Winter burning is considered the best method of preventing spittlebug damage, as chemicals are not only costly but not very effective.

Pangola stunt virus (PSV), a virus-type stunt disease, causes plants to lose vigor and develop dwarfed stems with tufted leaves that frequently exhibit red or yellow coloration. The disease occurs in various parts of the world but has not been reported in the US. Pangola digitgrass is susceptible to PSV, whereas all the other cultivars are resistant (Ocumpaugh and Rouquette 1985).

Other Grasses of Minor Importance

FLACCIDGRASS AND ORIENTAL PENNISETUM

Flaccidgrass, *Pennisetum flaccidum* Griseb., and Oriental pennisetum, *P. orientale* Rich, are tetraploid (2n = 36) apomicts (Chatterji and Timothy 1969a, 1969b) that are adapted to the "transition belt" of the southeastern US, i.e., between the areas where warm- and cool-season perennial grasses form the basis of most pasture systems (Burns et al. 1978). Introductions of flaccidgrass were collected at altitudes of 1600-4300 m in Afghanistan, and those of Oriental pennisetum were collected at altitudes of 600-2700 m in the western Himalayas of India (Burns et al. 1978; Green and Mueller 1992).

'Carostan' flaccidgrass was released by the North Carolina Agriculture Research Service. It is described as an erect-growing warm-season perennial grass with coarse stems that can grow to 1-2 m if unharvested (Green and Mueller 1992). Rhizomes are 3-6 mm thick and 15-30 cm long and contribute to its persistence and spread. Like flaccidgrass, Oriental pennisetums are tall, rhizomatous, deep-rooted perennials (Chatterji and Timothy 1969a, 1969b). The quality of Oriental pennisetums is similar to that of flaccidgrass (Burns et al. 1984), but no cultivars of Oriental pennisetum have been released, primarily because the lines evaluated were not winter-hardy in North Carolina. The species may have potential in warmer areas of the lower Southeast (J. C. Burns, personal communication).

Flaccidgrass is best adapted to well-drained soils with a good moisture supply, although it will grow on droughty soils or moderately wet soils with good surface drainage. Carostan may be propagated by seed, rhizomes, or rooted tillers (Green and Mueller 1992). Care must be taken to completely kill existing plants of common bermudagrass and dallisgrass before planting because they are very competitive with establishing flaccidgrass. Plants grow until frost, which kills them back to ground level. In spring, growth is from tillers and rhizomes and is ready for grazing about 5 wk later than tall fescue, *Festuca arundinacea* Shreb., and about 3 wk earlier than 'Coastal' bermudagrass.

Desirable characteristics of flaccidgrass include high yield but lower neutral detergent fiber (NDF) concentrations and higher digestibility and animal performance with steers than most warm-season perennial grasses (Burns et al. 1984; Green and Mueller 1992). Yields of approximately 12,000 kg/ha have been attained in North Carolina (Burns et al. 1978). Flaccidgrass NDF concentration and in vitro dry matter disappearance (IVDMD) were, respectively, 51.5% and 73.8% in spring and 60.5% and 62.3% in summer (Burns et al. 1984). The summer values compare favorably with NDF and IVDMD of Coastal bermudagrass, which were 65.8% and 55.9%, respectively (Burns et al. 1984). Higher IVDMD and lower NDF of flaccidgrass than Coastal bermudagrass have been associated with greater intake of flaccidgrass than bermudagrass on pasture in June and July (Burns et al. 1991), with greater gains in summer (0.77 vs 0.42 kg/d), respectively (Burns et al. 1984).

For continuous stocking through the grazing season, grazing of flaccidgrass should begin when the grass is 15-20 cm tall, and the stubble height of the pasture should be maintained in the range of 10-18 cm. When rotationally stocked, each grazing period should be initiated when plants are 25-45 cm tall and should end when the stubble is 10-15 cm (Green and Mueller 1992).

ST. AUGUSTINEGRASS

St. Augustinegrass, *Stenotaphrum secundatum* (Walt.) Kuntze, is native to the West Indies, Australia, and southern Mexico. It is also found in South Africa. It has been introduced into southern France and Italy and probably came to the US from Cuba.

St. Augustinegrass is not considered an important pasture plant except in limited areas. Its most widespread use is for lawns in the lower South. It is especially adapted to moist muck soils and is considered the most dependable pasture grass for the organic soils of southern Florida where growth is practically yearlong (Ocumpaugh and Rouquette 1985).

St. Augustinegrass is found along the southern Atlantic and Gulf Coast regions from South Carolina to Florida and Texas. It is not winter-hardy beyond 300 km north of the Gulf. It has good tissue tolerance to frost, is shade tolerant, prefers soil with pH >5.0, and tolerates saline conditions. It is adapted to practically all soil types, especially muck soils, if sufficient moisture is present. The cultivar 'Roselawn' has been commonly used for pasture and turf, but several new cultivars have been released specifically for use in lawns. All cultivars are propagated vegetatively, although a few viable seeds are produced.

JOHNSONGRASS

Johnsongrass, *Sorghum halepense* (L.) Pers., is native to the Mediterranean area of North Africa, South Asia, and southern Europe. It was introduced into the US around 1830. About 1840 Colonel William Johnson carried the grass to the black clay soils of central Alabama, thus the name *johnsongrass* (Ocumpaugh and Rouquette 1985).

Originally introduced as a superior forage crop, johnsongrass is now best known as one of the 10 most noxious weeds in the world (McWhorter 1981). It spreads to distant areas by seed and into adjacent areas by seed and rhizomes. Hard seed in the soil may germinate for years (McWhorter 1981). Cattle fed mature johnsongrass hay, and cattle that graze in fields with ripe johnsongrass seed, will pass viable seed in the feces (Ghersa and Martinez 1985), contributing to the spread over long distances.

Johnsongrass is an erect perennial plant that resembles the annual sorghums. It produces scaly rhizomes that creep extensively, and stems grow to a height of 1-3 m. The inflorescence is an open panicle resembling sudangrass, *S. bicolor* L. Moench. Johnsongrass can be propagated either intentionally or unintentionally by seed, rhizomes, and/or placing mature stem pieces in moist soil. Rhizomes are produced most extensively after johnsongrass forms seed heads. After flowering, a single plant may produce 60-90 m of rhizomes per month (McWhorter 1981).

Since johnsongrass is considered one of the 10 worst weeds in the world (McWhorter 1981), intentional propagation is usually discouraged. However, cropland already heavily infested with johnsongrass can be converted to forage/pastureland with proper management. Sparse stands can be thickened by early spring disking (Ocumpaugh and Rouquette 1985). Johnsongrass should be allowed to become fully established before defoliation. Any defoliation treatment before maturity reduces rhizome production.

Johnsongrass is subject to accumulation of hydrocyanic acid and nitrates when subjected to cold or drought stress. Death of livestock consuming stressed johnsongrass forage can cause significant economic losses.

CARPETGRASS

Carpetgrass, *Axonopus affinis* Chase, a species indigenous to Central America and the West Indies, was introduced into the US during the early colonial period. Another species in the US is tropical carpetgrass, *A. compressus* (Sw.) Beauv. It is less cold-hardy than *A. affinis* and is restricted to lower Florida and Louisiana.

Carpetgrass has spread through the lower coastal plain; it is best adapted to the warmer areas of the Southeast and is most abundant within 150 km of the Gulf of Mexico on sandy and sandy loam soils. It thrives on soils with a high water table. Because of its adaptation to soils of low fertility, it is a ubiquitous invader of infertile upland sites throughout the entire Gulf Coast.

Nutritional value and dry matter yield are low. The forage contains an unusually low concentration of minerals even when well fertilized. Despite its low nutritive value, cattle readily consume carpetgrass. No cultivars have been designated.

RESCUEGRASS

Rescuegrass, *Bromus unioloides* (Willd.)

H.B.K., syn. *B. catharticus* Vahl., is adapted to the South and Southwest where winters are mild and humid. Its agricultural value is limited to areas southward from North Carolina and Tennessee and in areas of the Pacific Coast. It is a short-lived perennial bunchgrass, acts as an annual under cultivation, and will not make a satisfactory growth on low-fertility soils without fertilization. It grows to a height of 60-120 cm and has drooping, open panicles of large, flattened spikelets containing seed almost as large as oat, *Avena sativa* L. Its forage has similar nutritive value but lower palatability than oat when grown on soil of the same fertility. After the grass has reseeded, the area can be used for growing another crop in the summer. It is most commonly found in combination with warm-season perennial grasses such as bermudagrass.

QUESTIONS

1. To what type of land is limpograss best adapted? How would you establish a new stand?
2. What supplements may be needed when livestock are grazing mature limpograss?
3. Why is Floralta the limpograss cultivar recommended for planting in the southern US?
4. Where and why is elephantgrass of considerable importance?
5. How does the quality of dwarf elephantgrass compare with that of most other warm-season perennial grasses?
6. What limits the widespread use of Mott dwarf elephantgrass?
7. Compare digitgrass with other forage grasses grown in its region of adaptation. How would you establish a new stand?
8. Where is the possible origin of flaccidgrass? What are the outstanding characteristics of flaccidgrass?
9. Why is johnsongrass considered a noxious weed in some areas?
10. How should johnsongrass be managed to obtain high yields and maintain stands?
11. What is the principal use of St. Augustinegrass?

REFERENCES

Adjei, MB, P Mislevy, KH Quesenberry, and WR Ocumpaugh. 1988. Grazing-frequency effects on forage production, quality, persistence, and crown total non-structural carbohydrate reserves of limpograsses. Proc. Soil and Crop Sci. Soc. Fla. 47:233-36.

Boddorff, D, and WR Ocumpaugh. 1986. Forage quality of pearl millet x elephantgrass hybrids and dwarf elephantgrass. Proc. Soil and Crop Sci. Soc. Fla. 45:170-73.

Boyd, FT, and VG Perry. 1970. Effects of seasonal temperatures and certain cultural treatments on sting nematodes in forage grasses. Proc. Soil and Crop Sci. Soc. Fla. 30:360-65.

Brown, WF, JD Phillips, and DB Jones. 1987. Ammoniation or cane molasses supplementation of low quality forages. J. Anim. Sci. 64:1205-14.

Burns, JC, DH Timothy, RD Mochrie, DS Chamblee, and LA Nelson. 1978. Animal preference, nutritive attributes, and yield of *Pennisetum flaccidum* and *P. orientale*. Agron. J. 70:451-56.

Burns, JC, RD Mochrie, and DH Timothy. 1984. Steer performance from two perennial *Pennisetum* species, switchgrass, and a fescue-'Coastal' bermudagrass system. Agron. J. 76:795-800.

Burns, JC, KR Pond, and DS Fisher. 1991. Effects of grass species on grazing steers: II. Dry matter intake and digesta kinetics. J. Anim. Sci. 69:1199-1204.

Chaparro, CJ. 1991. Productivity, persistence, nutritive value and photosynthesis response of Mott elephantgrass to defoliation management. PhD diss., Univ. of Florida.

Chatterji, AK, and DH Timothy. 1969a. Apomixis and tetraploidy in *Pennisetum orientale* Rich. Crop Sci. 9:796-98.

———. 1969b. Microsporogenesis and embryogenesis in *Pennisetum flaccidum* Griseb. Crop Sci. 9:219-22.

Christiansen, S, OC Ruelke, WR Ocumpaugh, KH Quesenberry, and JE Moore. 1988. Seasonal yield and quality of 'Bigalta', 'Redalta', and 'Floralta' limpograss. Trop. Agric. (Trinidad) 65:49-55.

Ghersa, CM, and MA Martinez. 1985. The effect of cattle on johnsongrass seed dispersal. Malezas 13:31-51.

Green, JT, Jr., and JP Mueller. 1992. 'Carostan' flaccidgrass. N.C. Forage Crop Prod. Memo 92:1.

Hacker, JB. 1985. Breeding tropical grasses for ease of vegetative propagation and for improved seed production. In Proc. 15th Int. Grassl. Congr., Kyoto, Japan, 251-52.

Holderbaum, JF, LE Sollenberger, JE Moore, DB Bates, WE Kunkle, and AC Hammond. 1991. Protein supplementation of steers grazing limpograss pasture. J. Prod. Agric. 4:437-41.

Kalmbacher, RS, PH Everett, FG Martin, KH Quesenberry, EM Hodges, OC Ruelke, and SC Schank. 1987. Yield and Persistence of Perennial Grasses at Immokalee, Florida: 1981 to 1984. Fla. Agric. Exp. Stn. Bull. 865.

Kok, PDF. 1981. Notes on *Digitaria* in South Africa. Bothalia 13:457.

———. 1984. Studies on *Digitaria* (Poaceae). 1. Enumeration of species and synonymy. S.Afr. J. Bot. 3:184-85.

Kretschmer, AE, Jr., and GH Snyder. 1979. Production and quality of limpograss for use in the subtropics. Agron. J. 71:37-41.

McWhorter, CG. 1981. Johnsongrass . . . as a Weed. USDA Farmers Bull. 1537.

Oakes, AJ. 1978. Resistance in *Hemarthria* species to the yellow sugarcane aphid, *Sipha flava* (Forbes). Trop. Agric. (Trinidad) 55:377-81.

Ocumpaugh, WR. 1989. Establishment of 'Mott' dwarf elephantgrass in a semi-arid climate. In R Desroches (ed.), Proc. 16th Int. Grassl. Congr., 4-11 Oct., Nice, France. Versailles, France: INRA, 539-40.

Ocumpaugh, WR, and FM Rouquette, Jr. 1985. Other grasses for the humid south. In ME Heath, RF Barnes, and DS Metcalfe (eds.), Forages: The Science of Grassland Agriculture, 4th ed. Ames: Iowa State Univ. Press, 263-70.

Oram, RJ. 1990. Register of Australian herbage plant cultivars. Queensland, Australia: CSIRO.

Quesenberry, KH, and RA Dunn. 1978. Differential responses of *Hemarthria* genotypes to sting nematodes in a greenhouse screening trial. Proc. Soil and Crop Sci. Soc. Fla. 37:58-61.

Quesenberry, KH, and WR Ocumpaugh. 1980. Crude protein, IVOMD, and yield of stockpiled limpograss. Agron. J. 72:1021-24.

———. 1982. Mineral composition of autumn-winter stockpiled limpograss. Trop. Agric. (Trinidad) 59:283-86.

Quesenberry, KH, and LE Sollenberger. 1992. Production and utilization of limpograss in the tropics. In Int. Conf. Livest. in the Tropics, IFAS, Univ. of Florida, 8-11.

Quesenberry, KH, LS Dunavin, Jr., EM Hodges, GB Killinger, AE Kretschmer, Jr., WR Ocumpaugh, RO Roush, OC Ruelke, SC Schank, DC Smith, GH Snyder, and RL Stanley. 1978. Redalta, Greenalta and Bigalta Limpograss, *Hemarthria altissima,* Promising Forages for Florida. Fla. Agric. Exp. Stn. Bull. 802.

Quesenberry, KH, WR Ocumpaugh, OC Ruelke, LS Dunavin, and P Mislevy. 1984. Floralta: A Limpograss Selected for Yield and Persistence in Pastures. Fla. Agric. Exp. Stn. Circ. S-312.

———. 1987. Registration of 'Floralta' limpograss. Crop Sci. 27:1087.

Reed, RL. 1991. Factors influencing the establishment of 'Mott' dwarf elephantgrass. MS thesis, Texas A&M Univ.

Ruiz, TM, WK Sanchez, CR Staples, and LE Sollenberger. 1992. Comparison of 'Mott' dwarf elephantgrass silage and corn silage for lactating dairy cows. J. Dairy Sci. 75:533-43.

Rusland, GA, LE Sollenberger, KA Albrecht, CS Jones, Jr., and LV Crowder. 1988. Animal performance on limpograss-aeschynomene and nitrogen-fertilized limpograss pastures. Agron. J. 80:957-62.

Schank, SC. 1972. Chromosome numbers in eleven new *Hemarthria* (limpograss) introductions. Crop Sci. 12:550-51.

Sollenberger, LE, and CS Jones, Jr. 1989. Beef production from nitrogen-fertilized Mott dwarf elephantgrass and Pensacola bahiagrass pastures. Trop. Grassl. 23:129-34.

Sollenberger, LE, and KH Quesenberry. 1985. Factors affecting the establishment of aeschynomene in Floralta limpograss sods. Proc. Soil and Crop Sci. Soc. Fla. 44:141-46.

Sollenberger, LE, JE Moore, KH Quesenberry, and PT Beede. 1987a. Relationships between canopy botanical composition and diet selection in aeschynomene-limpograss pastures. Agron. J. 79:1049-54.

Sollenberger, LE, KH Quesenberry, and JE Moore. 1987b. Effects of grazing management on establishment and productivity of aeschynomene overseeded in limpograss pastures. Agron. J. 79:78-82.

Sollenberger, LE, WR Ocumpaugh, VPB Euclides, JE Moore, KH Quesenberry, and CS Jones, Jr. 1988a. Animal performance on continuously stocked 'Pensacola' bahiagrass and 'Floralta' limpograss pastures. J. Prod. Agric. 1:216-20.

Sollenberger, LE, GM Prine, WR Ocumpaugh, WW Hanna, CS Jones, Jr., SC Schank, and RS Kalmbacher. 1988b. 'Mott' Dwarf Elephantgrass: A High Quality Forage for the Subtropics and Tropics. Univ. Fla. Agric. Exp. Stn. Circ. S-356.

———. 1989a. Registration of 'Mott' dwarf elephantgrass. Crop Sci. 29:827-28.

Sollenberger, LE, GA Rusland, CS Jones, Jr., KA Albrecht, and KL Gieger. 1989b. Animal and forage responses on rotationally grazed 'Floralta' limpograss and 'Pensacola' bahiagrass pastures. Agron. J. 81:760-64.

Sollenberger, LE, CS Jones, Jr., KA Albrecht, and GH Ruitenberg. 1990. Vegetative establishment of dwarf elephantgrass: Effect of defoliation prior to planting stems. Agron. J. 82:274-78.

Sollenberger, LE, MJ Williams, and CS Jones, Jr. 1991. Vegetative establishment of dwarf elephantgrass: Effects of planting date, density and location. Proc. Soil and Crop Sci. Soc. Fla. 50:47-51.

Woodard, KR, and GM Prine. 1990. Propagation quality of elephantgrass stems as affected by the fertilization rate used on nursery plants. Proc. Soil and Crop Sci. Soc. Fla. 49:173-76.

Woodard, KR, GM Prine, and WR Ocumpaugh. 1985. Techniques in the establishment of elephantgrass. Proc. Soil and Crop Sci. Soc. Fla. 44:216-21.

36 Cereals and Brassicas for Forage

SRINIVAS C. RAO
Agricultural Research Service, USDA

FLOYD P. HORN
Agricultural Research Service, USDA

HIGH-QUALITY forages associated with cereal grain production meet or exceed the nutrient requirements of grazing livestock regardless of class or species. With improvements in grain-crop production technology, expanded markets for US grain, and consumer preferences for leaner red meats, the practice of incorporating cereal forages in livestock-growing and -finishing programs has become increasingly important in the US.

Cereal forages fill a unique niche in the US cattle industry; e.g., wherever they are grown, they provide supplemental nutrients for cow-calf herds, support major elements of the stocker cattle industry, and have demonstrated potential to produce acceptable finished beef at several times throughout the year. Cereal crop silages are important for dairy milk production as well. Most cereal grain cultivars have been selected largely for grain yield but also have improved winterhardiness, drought tolerance, pest and disease resistance, morphological characters, and adaptability to a wide array of soil and climatic conditions. The impetus provided by the need to increase cereal grain production has led also to increased forage production that, until tillering is initiated, provides excellent forage at minimal cost.

SRINIVAS C. RAO is a Research Agronomist at the ARS, USDA, Grazinglands Research Laboratory, El Reno, Oklahoma. He received his MS degree from Texas A&M University and PhD from Oklahoma State University. His special interest is the optimum utilization of forage resources to increase red meat production while conserving basic natural resources.

FLOYD P. HORN is a Ruminant Animal Nutritionist and Director of the ARS, USDA, Southern Plains Area, headquartered at College Station, Texas. He received his MS and PhD degrees from West Virginia University. His special interest is the solution of multifaceted, complex agricultural problems while conserving the basic natural resources of soil, water, and air.

Aside from the value afforded by grazing these forages in pure stands, small grains provide farm and ranch operators with versatility in forage-livestock programs. Small grains can often be overseeded or sod-sown in perennial grass or legume pastures, thus providing high-quality forage during winter months when costly energy and protein supplementation would otherwise be necessary. In either pure stands or overseeded pastures, small grains can be used in combinations to increase yield and dependability and to extend the grazing period. Wherever they can be grown, they are an important component of multiforage livestock production systems.

The brassicas have been identified with agriculture since ancient times. Species of this group are valued for their leaves, flowers, roots, and seed and have been used as vegetables, oil crops, fodder plants, and ornamentals. Widespread use of brassicas, particularly root crop species, has diminished somewhat since the turn of the century. In part this could be attributed to the high cost of harvesting and controlling pests in these crops. In the past few years, however, scientists have concentrated some effort on producing root crops that can be harvested directly by grazing animals and that are resistant to both diseases and insect pests.

Small Grains

HISTORY AND ORIGINS

The geographic beginnings of the major crops affecting humans were studied by N. I. Vavilov, the geneticist who directed the All-Union Institute of Plant Industry in Leningrad from 1920 to 1940. He delineated eight "centers of origin" on the basis of crop plant characteristics and observations of wild relatives. Although evidence was later presented to suggest that his concepts were an oversimplification, and that several important crops could have evolved in more than one region, the basic premise that many crops can be traced to their beginnings, if their species and races are clearly distinguishable, has withstood the test of time and provoked massive research efforts to better understand the development of genetic entities.

Wheat (*Triticum* spp.) and barley (*Hordeum* spp.) developed in the Near East. Three kinds of wheat were originally domesticated, none of which are used with any frequency today. Einkorn wheat (*T. monococcum* L.) was a diploid wheat from southeastern Turkey. Emmer wheat (*T. turgidum* L.) was a tetraploid from Palestine and/or southeastern Turkey. *Triticum timopheevii* (Zhuk.) var. *timopheeri,* which was so inconsequential no common name ever emerged, was another tetraploid from Transcaucasian Georgia (Harlan 1971). Wheat cultivars of today developed from none of these. A mutation of emmer is the ancestor of our durum wheat (*T. turgidum* L.). Our bread wheat is a combinant resulting from three primitive domestic wheats plus wild goatgrass, *T. tauschii* (Coss.) Schmal.

Similarly, oat (*Avena* spp.) and winter rye (*Secale* spp.) originated as domesticated crop species in Europe, but development of the cultivars of today has brought about considerable change in the appearance, value, and distribution of these plants (Harlan 1971).

Triticale is a polyploid produced by doubling the chromosome number of the sterile hybrid that resulted from a cross between wheat (*T. aestivum* L. emend Thell [group *aestivum*] or *T. turgidum* L. [group durum]) and rye (*Secale cereale* L.) (Briggle 1969). Since 1960, significant progress has been made by triticale breeders in improving this unique crop. Similar to wheat, triticale is a dual purpose crop that can be grazed during vegetative stages and subsequently still produce grain.

DISTRIBUTION AND ADAPTATION

Plant photothermal requirements and climatic relationships have been documented for wheat, barley, and winter rye by Nuttonson (1955, 1957, 1958).

Production is concentrated between latitudes 30° and 60°N and 27° and 40°S (Nuttonson 1955), but their use as forage is usually limited to south of 40°N latitude. Cereals yield best on medium- to heavy-textured, well-drained soils with well-balanced fertility. They are well suited to soils with a pH range of 5.5 to 7.5; however, good forage growth and production can be obtained with soil pH as low as 5.0 to 5.2 due to their high tolerance of soil acidity (Westerman 1981).

Wheat (*Triticum* spp.). Wheat is grown commercially on a large scale in more than 80 countries throughout the world. Leading countries in wheat production include the republics of the former Soviet Union, the US, China, France, and Canada. The US, Canada, and Australia are major exporters of wheat (USDA 1991). Concentrated areas of US wheat production occur in the Ohio River valley, the central and northern plains, and the Pacific Northwest (Quisenberry and Reitz 1967). Wheat is grown in 42 states, with the leading states in approximate order of area sown being Kansas, North Dakota, Oklahoma, Texas, Montana, Nebraska, and Colorado.

Not all wheat is alike. There are several ways to categorize wheat. For a treatise of kinds of wheat the reader is referred to the 1967 article by Quisenberry and Reitz. Winter wheat is the type most commonly grazed by livestock.

Oat (*Avena sativa* L. and *A. byzantina* K. Koch). Oat is grown commercially on a large scale in the US, the republics of the former Soviet Union, Canada, Germany, France, and Great Britain. These countries produce approximately 75% of the entire world oat crop (Coffman 1961), which in 1991-92 was 42.7 million mt (USDA 1991). Thirty-six states in the US grow oats, but 12 states in the upper Mississippi River valley in the north central US produce 80% of the nation's crop, which is mainly spring seeded. Leading states in order of area sown are South Dakota, Minnesota, Texas, Iowa, and Wisconsin. Nearly all of the world's cultivated oat crop is the species *A. sativa* L., but there are several species of lesser importance produced around the world. Cultivars of *A. byzantina* K. Koch, a winter type, are ideally suited for pasture production in the southern US (Coffman 1961). Winter-type oats are not very cold resistant, limiting adaptation to the Gulf states.

Winter Rye (*Secale* spp.). Winter rye is the most winter-hardy of the small grains and is grown commercially on a large scale in the republics of the former Soviet Union, Poland, Czechoslovakia, Finland, Canada, and the US. The former Soviet Union produces nearly 50% of the world supply. The grain is an important source of concentrate for livestock, and the whole plant provides an invaluable source of green feed. Rye is sown both in autumn and spring, depending upon severity of the winters, but winter rye is much more significant than spring rye as a source of forage for livestock. Winter rye areas of North America extend from northern Canada to the southeastern US. Centers of rye production are found within the spring wheat belt, where hardiness, earliness, and production cost permit rye to compete favorably with wheat. In the northern US winter rye is generally harvested as a cash grain crop, while in the East and South it is grown for pasture, winter cover, or a manure crop.

Triticale (X *Triticosecale* Wittmack). Triticale is currently grown on a total of more than 1 million ha worldwide and is expected to grow in popularity significantly in the next decade as the triticale grain market expands (Varughese et al. 1987). Leading countries in triticale production include the US, the republics of the former Soviet Union, Australia, France, Poland, Argentina, and Canada. Triticale is currently produced primarily in developed countries that are noted for their already high levels of small grain production (Skovmand et al. 1984), but as the knowledge about and adaptability of this crop are enhanced, developing countries should also benefit from its production.

Barley (*Hordeum* spp.). Barley is produced commercially in the republics of the former Soviet Union, Canada, France, Spain, and the US. Thirty-three states produce significant amounts of barley; leading producers include Idaho, North Dakota, California, and Montana. Barley use in pasture programs is uncommon, but it has been especially useful in the intermountain area as a temporary winter pasture or a companion crop to help in weed control during establishment of more permanent crops or grasslands.

PLANT DESCRIPTION

All cereal forages are annuals although they may be planted in fall or spring. They usually tiller from the basal nodes, and thus a single plant is usually composed of numerous tillers. In early growth the tillers are primarily leaf, becoming more stemmy as the plant matures. Approximately 15-25 cm of growth occurs prior to the elongation of internodes; then the apical bud develops into the inflorescence. In cases where a grain crop subsequent to grazing is desired, grazing must stop before jointing or internode elongation begins. If the terminal bud is damaged or removed, no inflorescence develops.

Small grain species in the flowering stage are easily distinguished from one another; however, plants in the vegetative stages are difficult to distinguish. The wheat ligule completely encircles the culm and is sharply curved. Leaves have hairy auricles. Leaf blades from winter rye are coarser and more bluish than those of wheat. Ligules of winter rye are shorter and rounded; auricles may be white, narrow, or absent. The sheath from barley is generally glabrous, and the ligules are short or truncated. These glabrous auricles either partially or completely surround the culm and are larger than those of either wheat or rye. The ligule of oat is well developed and toothed whereas it is blunt in other small grains. Oat has no auricles. In each species there is a wide range of morphological differences among cultivars. To properly distinguish vegetative small grains, one must be familiar with the subtle differences in local cultivars.

CULTIVARS

A wide array of small grain cultivars is available, and many new cultivars are released annually. Attempts have been made to more fully characterize the forage-producing potential of cultivars, and many have been released as forage types.

Small grains are used extensively in the southern US for winter forage in addition to grain production. Relative coldhardiness of cultivars adapted to the southern US is highest in rye and wheat and lowest in triticale and oat. The best source of information on cultivars for forage or forage-grain production is the cultivar tests conducted by the agricultural experiment stations in the cereal-growing states of the US.

MANAGEMENT PRACTICES

Fertilizer. The principal fertilizer need of small grains is normally nitrogen (N). If the cereal is to be used for forage, a fall application of N is particularly critical (Burton and Prine 1958; Dev et al. 1979). However, large fall applications may result in luxury consumption of N, and high forage yields often cannot be sustained without additional N fertilizer later in the year (Colman 1966). While responses to fertilizer are variable depending on the myriad of factors associated with plant nutrient uptake, one can observe that the most common fertilization practice for small grain pasture involves application of N, phosphorus (P), and potassium (K) at planting time and topdressing with N in winter and/or spring (Holt et al. 1969). It is especially important to split N applications when a grain harvest is planned subsequent to grazing (Elder 1967; Altom et al. 1976).

Interseeding of small grains into a dormant or semidormant sod may require a fertilization program similar to that used in a prepared seedbed, except that care must be taken to see that the N fertilizer is not utilized by the existing grasses at the expense of the cereal (Davies 1960; Decker et al. 1969; Denman and Arnold 1970). Control of the existing sod with herbicides reduces such losses (Colman 1966); however, there is a recognized need to develop alternatives to conventional herbicides to reduce costs and ensure protection of the environment. Management practices such as interseeding are especially important where it is desirable to minimize soil disturbance and to reduce soil erosion.

Seeding Dates and Rates. Small grains sown for harvest of both forage and grain should be seeded 3 to 4 wk earlier than for grain production alone (Simmons 1972; Sithamparanathan 1979a, 1979b). Optimum seeding dates differ among production areas and, for grain, range from mid-September in the northern states of the US to early November in the South. Late-seeded crops may not develop adequate winterhardiness and in some cases may not be vernalized. Winter survival is a serious consideration in regions where cold causes high mortality of poorly conditioned plants (Fowler 1982). Plants generally need about 6 wk of growth to become hardy.

Recommended seeding rates range from 30 kg/ha in the dryland plains to 100 kg/ha for areas under irrigation or with more favorable moisture conditions. The basic rate for local conditions is increased by about 50% for irrigated production, 25% for pasture, and 10% per week delay in seeding after the optimum date for that region (Paulsen 1987). Early seeding is frequently affected by the hessian fly (*Mayetiola destructor* [Say]), crown or root rot diseases, or infestations with any of various mosaic viruses. If forage production is an important consideration in planting small grains, it generally is recommended that the seeding rate be increased by 50%-100% or that triticale, which is relatively resistant to insect and disease problems, be selected over wheat (Ehmke 1993).

Planting Small Grain Mixtures. Small grain forages vary in their seasonal production cycles; thus many graziers find it useful to seed mixtures of species. Mixtures extend the normal grazing season in the South while consistently producing high-quality forage. The distribution of forage production is usually just as important as total forage yield. In most studies no single crop or cultivar has shown consistently high forage production throughout the winter growing season (Denman and Arnold 1970).

In selecting species and cultivars to plant, one should consider certain basic needs: (1) Total forage production is of paramount importance because total amount of liveweight produced will be largely dependent upon total amount of forage produced. (2) Distribution of forage should be planned to minimize the need for supplementation. (3) Versatile choices should be included to minimize the hazards associated with low temperatures, drought, insects, bloat, weeds, and disease (Bates 1972). Many of these needs are best met using mixtures, but these are more difficult to manage and do not permit subsequent harvest of a pure grain crop.

Planting in Established Sods. Reduced tillage for crop production is rapidly growing in interest because of the potential it offers for reduced fuel use, increased soil conservation, and increased soil moisture use efficiency. As a result, drilling of small grains into established sods and existing crops has become common (Fig. 36.1). The resulting mixture will usually provide a better distribution of forage quantity and quality throughout an extended grazing season (Vartha 1971; Swain et

al. 1975). Major deterrents to successful sodseeding of small grains include the nonuniformity of stands, inefficient use of fertilizers, and inadequate control of weeds. Use of herbicides during establishment, while increasing the cost of production and requiring extreme caution in selection of chemicals and management of soil residues to protect surface and ground water contamination, provides expanded opportunities for low- or no-till forage production (Murtagh 1971). Even with herbicides, establishment of small grains in sod or standing residue usually requires additional fertilizer to support plant growth and sustain normal decomposition of dead matter.

Stocking Rates and Grazing Pressure. Winter grazing programs for small grain pastures usually are designed to provide supplemental nutrition for cows or complete nutrition for growing animals; 1 ha of properly fertilized and managed pasture of small grain will usually support 2 large cows or 3 small ones on a limited grazing (2 h/d) basis (Simmons 1972). Well-established wheat pasture in Kansas can support 0.8 to 1 grown cow/ha during fall-winter and spring grazing periods, respectively, or 1.4 fall calves and 1 spring yearling per hectare (Swanson 1935). Grazing pressure has little influence on production of grain when fertility levels (primarily N and P) are high and grazing is terminated at the onset of tillering (Fig. 36.2).

Fig. 36.1. Throughout much of the South small grains can be established in an existing warm-season perennial grass sod. The results are reduced need for supplementation and good animal performance. *Samuel Roberts Noble Foundation photo.*

Animal Performance and Use. Excellent levels of performance have been reported for animals grazing small grain pastures. In a comparison of wheat, winter rye, and triticale forages in a beef-finishing program, Patel and Nishimuta (1978) report average daily gains (ADG) of 1.29, 1.41, and 1.01 kg head^{-1}, respectively, but the animals were also fed considerable (and variable) quantities of a high-energy supplement.

For grazed, unsupplemented winter rye pastures in Saskatchewan, Kilcher and Lawrence (1979) report ADG of 0.59 kg in steers and calves with cows. Pastures averaged 48 d of grazing per season and 130 animal unit days of grazing per hectare. Unsupplemented heifers grazing moderately stocked winter wheat pasture in Oklahoma gained an average of 0.55 kg d^{-1} (Horn et al. 1981). In a review of forage-livestock production systems used in the subhumid and semiarid southwestern US, Horn and Taliaferro (1975) report that ADG of stocker calves on small grain pastures frequently exceeded 0.70 kg.

Chemical Composition and Nutritive Value. Small grain forage is lush, high in protein, and low in fiber during most of the winter grazing season. Indeed, crude protein levels normally range from 15% to 34% of dry matter (DM), making this forage an excellent protein supplement for many classes of livestock (Horn 1981). Soluble carbohydrate levels in ungrazed wheat forage vary predictably depending upon season and rate of plant growth. There is no consistent pattern to soluble carbohydrate levels in grazed wheat, although levels in leaves tend to increase as the plant matures. Nitrogen fertilization reduces the soluble carbohydrate content of wheat (Johnson et al. 1974).

The use of small grains for pasture is not without its hazards. Extreme danger from tetany may occur where small grain forage is grazed by cows about to calve or by those that have recently calved. There is convincing evidence (Bohman et al. 1983) that this disorder is not conventional grass tetany and that calcium (Ca) rather than magnesium (Mg) may be the primary mineral deficiency in the forage. Other factors indicative of Ca, P, and Mg metabolic anomalies are also present. For example, wheat pasture is extremely high in K, N, organic acids, and soluble carbohydrates. It is low in Ca compared with P and marginal in Mg. Forage having a ratio of K:(Ca + Mg) greater than 2.2, expressed in milliequiva-

Fig. 36.2. In most years well-established small grain pastures can be grazed all winter with little if any reduction in grain yield.

lents, is usually considered a hazard (Stewart et al. 1981). (See Chap. 9, Vol. 2.)

A high proportion of the N in wheat can be in the form of nonprotein N (NPN). If animals graze rapidly growing, recently fertilized barley, there is a risk of nitrate poisoning (Wright and Davison 1964). Small grain forage is also rapidly digested, and another phenomenon that may be related to either protein or mineral constituents of the forage is bloat (Clay 1973; Horn and Frost 1982). However, small grain pastures can be an extremely valuable resource under careful grazing management.

SEED PRODUCTION

Small grain seed production is relatively free of problems because of the emphasis placed upon grain harvest technology. It is important to note, however, that the landmark advances that have been made in the genetic improvement and widespread adoption of improved cultivars of all small grains are largely the result of the carefully structured system of certified and foundation seed production and distribution.

DISEASES AND PESTS

A number of insects attack small grains on a significant scale, although most serious pest problems relate to grain production when temperatures are warmer. Insects that affect forage production, in descending order of their economic importance, are listed here:

Common Name	*Scientific Name*
Hessian fly	*Mayetiola destructor* (Say)
Grasshopper	*Melanopus* spp.
Russian wheat aphid	*Diuraphis noxia* (Mordvilko)
Armyworm	*Pseudaletia unipuncta* (Haworth)
Greenbug	*Shizaphis graminum* (Rondani)
Chinch bug	*Blissus leucopterus* (Say)
Army cutworm	*Euxoa auxilia* (Grote)

Insecticides and increasingly important biological control agents can provide effective control of nearly all insect pests, but the costs are often high. Genetic resistance to insect attack has proven successful for several of the insects listed, and proper defoliation by grazing animals can markedly reduce forage losses. These practices will reduce the use of chemicals.

Diseases have been a major problem. Most common and destructive diseases are rusts, smut, root rots, powdery mildew, and mosaic and other virus diseases (scab and septoria). Damage from these diseases can be reduced by selection of disease-resistant cultivars, chemical seed treatment, crop rotation, management of infected residues, and avoiding excessively moist soils.

Brassicas

The historical development of the brassicas has not been clearly delineated, but it is suggested that the genus originated in the Mediterranean region and quickly adapted to much of Asia and Europe (Harlan 1971). The systematics, domestication, breeding, and conservation of plants belonging to *Brassica* and allied genera of the Cruciferae have been compiled in an excellent monograph by Tsunoda et al. (1980). Turnip, rape, and kale have been popular livestock fodder for at least 600 yr wherever they can be grown, and for most of that time they were managed as forage.

Important agricultural species of *Brassica* fall into three groups: turnips (*B. rapa* L.), which include common root crops and some rape with turniplike leaves; swedes (*B. napus* L.), which include not only the swedes proper but the common or oil-bearing rapes; and cabbages (*B. oleracea* L.), which include the common vegetable and also the forage crops thousand-headed kale and marrowstem kale.

DISTRIBUTION AND ADAPTATION

Turnip, rape, and kale are distributed over much of Europe, northern Asia, northern North America, and southern Oceania. These crops are grown year-round in cooler and moist climates and as a winter forage in warmer climates. They can be easily grown in fertile soils with little if any cultivation and are frequently used to support beef, dairy, sheep, and even swine production. Most brassicas are cold tolerant, and the leaves can withstand light freezes. Moisture requirements are relatively high for most species, and arid dryland farming seldom produces acceptable yields. Brassicas thrive well on a wide range of soils. Loamy soils are preferred, but even light or peaty soils will produce good yields if rainfall and fertility are adequate. Brassicas are well suited to soils with a pH from 5.5 to 6.5, will not tolerate waterlogged conditions, and may require drainage on heavy soils. In drier areas, soils with moderate organic matter content and moisture-holding capacity are essential for production of these crops.

PLANT DESCRIPTION

Turnip (*Brassica rapa* L.). Turnips generally have large bulbous or tapered roots and are rich in carbohydrates. Selection for use as forage has resulted in changes in root set, such that portions of the roots are exposed and more available to grazing animals. Foliage is erect and succulent during vegetative proliferation, and many turnips can be grazed twice to permit utilization of top growth and roots, respectively. With adequate moisture and fertility, turnips can be expected to provide whole crop yields of 4500 to 7500 kg DM/ha, of which 35% to 50% should be in the storage root (Evans 1979).

Because leaves are higher in nutritional value than roots, new plant species were developed recently to increase the leaf-to-root ratio. These include 'Cyclon' turnip (leaf:root ratio of 4:1), 'Tyfon', a hybrid cross between chinese cabbage (*B. pekinensis* L.) and stubble turnip (*B. rapa* L.) (leaf:root ratio of 9:1). Maximum root dry matter yield is achieved between 80 to 85 d in fall and 90 to 95 d in spring after seeding (Rao and Horn 1986).

Rape (*Brassica napus* L.). Rape has a taproot system that, in contrast to turnips, cannot be harvested by grazing animals. This root system supports an erect, succulent, and highly nutritious foliage that can be grazed and allowed to regrow and grazed again. Rape can be divided into giant and dwarf types (Toosey 1972). Giant types have proportionately larger stems, stand more erect, are taller, and branch less than dwarf types. Dry matter yields and general acceptability of giant types are better than those of dwarf types, although if regrowth is critical, dwarf varieties are often preferable. Under favorable environmental conditions rape can produce forage yields of 5000 to 8000 kg DM/ha. Yield of fall-seeded rape is slightly lower than that of spring-seeded rape. Rape yields generally peak 120 d after seeding.

Kale (*Brassica oleracea* L.). Agricultural kale may be divided into two major groups: "true" (*B. oleracea*) kales, which include marrowstem kales, thousand-headed kales, hybrids, and swedelike kales, which are commonly thought of as rapes but are, in fact, kales (Toosey 1972). Marrowstem kale is a high-yielding crop composed of 60% to 70% stems. The stem contains a pithy center surrounded by an outer skin that is highly digestible when consumed in the immature stage. Generally, increasing the plant density increases the leaf:stem ratio and enhances utilization of the forage. Thousand-headed kale, while producing relatively little stem material, has a woody stem even in premature stages. Forage yield of kale ranges from 5000 to 9000 kg DM/ha, depending on the growing season. Dry matter accumulates slowly during early growth and requires 150 to 180 d to reach maximum yield.

GERMINATION AND EMERGENCE

Rapid germination and seedling establishment of brassicas is of considerable importance in obtaining the high plant population necessary for high yields. Seedling establishment of brassicas can be hindered by low soil temperature and inadequate moisture. Brassica seed can germinate within the temperature range of 5° to 45°C, but a more desirable range is between 10° to 35°C (Wilson et al. 1992). Optimum germination is achieved at 35°C (Tokumasu et al. 1985). The time required for germination increases as the temperature declines.

Soil water content is as important for optimum germination as soil temperature. Rate of germination and percent germination are significantly reduced with decreasing soil moisture. Although the amount of water required for seed germination varies extensively among species, total percent emergence and rate of emergence of all *Brassica* cultivars are highest at soil-water potentials higher than –0.1 megapascal (MPa). Total germination and germination rate decline with decreasing soil-water potential below –0.1 MPa (Rao and Dao 1987).

Cool soil temperature and variable soil-water availability in the early spring result in erratic and reduced seedling emergence. Priming the seed with polyethylene glycol will improve stand establishment in cold soils (Rao et al. 1987).

MANAGEMENT PRACTICES

Seeding Dates and Rates. Brassicas can be established in either fall or spring, but date of sowing, length of growing season, and environmental conditions affect dry matter yield and quality of these high-yielding crops (Harper and Compton 1980; Kalmbacher et al. 1982; Rao and Horn 1986). In North America, maximum production was achieved with a late August planting. If fall planting is delayed, dry matter yield will decrease due to limited time available between planting and harvesting, declining ambient and soil temperatures, and reduced solar radiation levels (Rao and Horn 1986; Harper and Compton 1980).

All forage brassicas are resistant to moderate frost and short periods of moisture stress. However, stands may be killed by prolonged subfreezing temperatures or moisture stress (Sheldrick et al. 1981; Smith 1979). Frozen material can still be consumed by livestock, but once thawed the entire plant spoils quickly (Westover et al. 1933). Seeding date for early spring planting depends on the soil temperature. To achieve reliable stand establishment, soil temperatures should be between 15° to 25°C (Tokumasu et al. 1985).

Rates of seeding vary according to the method of sowing and row spacing. In general, seeding rates range from 2 to 3 kg/ha for turnip and 4 to 5 kg/ha for rape and kale when seeded in rows; the seeding rate should be slightly higher when seed is broadcast. Seed should not be covered with more than 2 cm of soil. Lower seeding rates for turnip as compared with rape and kale is due to smaller seed size. Also, root crops are seeded at high seeding rates to reduce storage root production and increase top production.

Fertilizer and Herbicides. Soil fertility and weed control are crucial to producing high yields of brassicas. Nitrogen and P are the most important elements in forage production. Generally forage brassicas are fertilized with 75 to 120 kg N and 60 kg P/ha for the growing season. If soils are medium to low in K, application of 30 to 60 kg/ha is recommended. Brassicas are poor competitors with other plants, especially during their early growth period. Fertilizer should be applied close to the time of seeding in order to give brassicas a competitive advantage over weeds. Because of their winterhardiness, these crops have a competitive advantage over most weeds after frost. If fertilization is delayed under sod-seeding, the previous crop (sod) will have the advantage over these crops. Fertilizer requirements under sod-seeding are slightly higher compared with those of conventional seeding because the sod tends to utilize nutrients.

Weeds are a major problem during establishment because brassicas are poor competitors. Weeds can compete for light, moisture, and nutrients and can decrease the emergence and establishment of brassicas. Tillage operations before seeding and soil incorporation of preemergence herbicides reduce early-season weed competition for conventional seedings (Cox 1977; Smith 1979), whereas postemergence herbicides are used to suppress sod under sod-seeding (Faix et al. 1979; Jung et al. 1979).

Planting in Established Sods. Reduced tillage has gained widespread acceptance in the last decade as a control for soil erosion, to increase soil-water conservation, and to minimize fossil energy inputs. Seedlings establish easier under no-till management if the existing sod is suppressed before seeding. However, once established, the dense brassica canopy can shade and often suppress competitors (Jung et al. 1984). Direct drilling of brassicas into dormant grass sod has the advantage of rapid changeover from grass to crop and reduces trampling losses when the crop is grazed (Cox 1977; Pascal 1977).

Seedling emergence and establishment of brassicas under conservation tillage methods may be delayed or reduced by cool soil temperatures, especially during spring planting (Jung et al. 1983; Khan et al. 1983). Priming brassica seed improves seedling emergence when soil temperatures are cooler than optimum (Rao et al. 1987). An alternative to priming is to select cultivars that germinate early under unfavorable temperatures. Conservation tillage practices may play a key role in the reintroduction of forage brassicas for livestock production.

Animal Performance and Use. Turnip and rape are readily grazed by cattle and sheep and provide useful supplementary grazing in the midsummer or late fall when warm-season grasses and cereal forages are nonproductive. Due to their high nutritive value, brassicas are especially useful for feeding animals with high nutritive requirements. Such an application might include fattening lambs, flushing ewes, and feeding lactating ewes, dairy cattle, and young beef cattle. Lambs grazing Tyfon turnip gained an average 214 to 249 g/d, whereas lambs fed hay and grain gained 186 to 195 g/d (Koch et al. 1987). Major factors affecting the efficient utilization of these crops are grazing pressure, trampling, and soiling of the crop. Grazing small electrically fenced areas (such as strip-grazing) will improve utilization of brassicas. The degree to which a crop becomes soiled is influenced by soil type and rainfall. Growing crops under sod-seeding reduces soiling and improves utilization. Rape and kale are less susceptible to soiling due to erect growth as compared with root crops.

Animal gain from brassica is minimal during the initial grazing period following transfer of animals from pastures to brassicas. Brassicas are low in dry matter and fiber content. Changing the animal's diet from high fiber to low fiber may result in abnormal ruminal fermentation (Lambert et al. 1987). Lambs grazing brassicas gain slowly due to low fiber content, and antiquality factors inhibit liveweight gain (Marten and Jordan 1982). Supplementation of a dry high-fiber feed along with brassica forage improves the performance of animals as compared with that of animals fed brassica forage alone (Lambert et al. 1987).

Chemical Composition and Nutritive Value. Brassicas are lower in dry matter content but produce greater quantities of total dry matter per unit area than do most cereals and forage grasses. Nutritive value of brassica top growth and roots varies among cultivars and plant parts.

Crude protein levels of turnips normally range from 15% to 20% of DM in leaves, and from 6% to 15% in roots, depending on the size and number of roots per unit area. Crude protein in leafy-stem crops such as rape and kale ranges from 20% to 25% in leaves and averages about 10% in stems (Jung et al. 1979; Kalmbacher et al. 1982; Rao and Horn 1986). Depending on plant parts, brassicas are high in dry matter digestibility, ranging from 75% to 95%, compared with digestibility of good alfalfa (*Medicago sativa* L.) at 70%. Digestibility of the root portion is generally five to seven percentage units higher than that of the leaves of root crops due to the roots' high carbohydrate content.

The nutritional value of brassicas tends to be higher in the fall, and it is retained for a longer period in the fall, as compared with the nutritional value of the same plants in the spring. Retention of nutritive value for fall-seeded brassicas could be attributed (1) to lower dry matter accumulation after brassicas reach their maximum production and (2) to the onset of cooler temperatures (Guillard and Allinson 1984; Rao and Horn 1986).

Although high in nutritional value, brassicas do contain antiquality constituents and elevated mineral concentration that may negatively affect animal performance. Plant concentrations of Ca, Mg, K, copper (Cu), iron (Fe), and manganese (Mn) are greater in summer-grown species than fall-grown (Guillard and Allinson 1989a). Brassica forage exceeds the desired range of Ca:P ratio in ruminant diets, and roots have a similar or slightly lower ratio. Magnesium concentration and availability to ruminants is important due to high K, N, and/or Ca concentration in forages being a factor in the etiology of hypomagnesemia in ruminants (Wilkinson and Stuedemann 1979). Based on the (K + Ca):Mg ratio, hypomagnesemia may be a concern with fall-grown species, particularly when conditions exist for high K and Ca concentrations in herbage (Guillard and Allinson 1989b). Brassicas are also characterized as nitrate-N accumulators. Nitrate-N concentration in fall-grown species is several orders of magnitude higher than in summer-grown species due to cooler soil temperatures, decreased light intensity and reduced photoperiod.

Glucosinolates and s-methyl cysteine sulfoxide in brassica forages are potential toxins that may adversely affect animal production. Glucosinolates release thiocyanate (SCN^-) on hydrolysis, which inhibits thyroid uptake of iodine (I) (Paxman and Hill 1974). Concentration of SCN^- is greater in roots than in foliage, and higher in plants grown in summer than in fall (Guillard and Allinson 1989b). Supplementation of I in the animal's diet will reduce the antithyroid activity of SCN^-.

To avoid potential animal health problems, brassica forage may be fed as only a portion of the total diet. Supplementing with I and P, controlling intake of brassica crops, or adding other forages to the diet should reduce potential problems with mineral imbalances or antigrowth constituents associated with brassica crops (Wikse and Gates 1987).

DISEASES AND PESTS

Brassica crops are subject to insect damage including cabbage flea beetle or striped flea beetle (*Phyllotreta* spp.), armyworm (*Laphygma frugiperda*), cabbage loopers (*Trichoplusia ni*), and aphids (*Brevicoryne brassicae*). Insecticide trials for late summer and early fall seedings suggest that flea beetle control is more important when planting using conventional tillage than when sod-seeding. Need for insecticides is greatly reduced with minimum-till seeding in sods (Jung et al. 1979). When brassica forage is grown for livestock, care should be taken in selecting insecticides that are approved for grazing animals.

The most serious plant diseases are bacterial black rot (*Xanthomonas compestris*) and mildew (*Erysiphe cruciferarum*). These diseases occur primarily on mature crops. Disease damage is more common in the spring than in the fall. Controlling cabbage root maggot (*Hylemya brassicea*), maintaining the crop in a vigorous, active-growing condition, and crop rotation are management methods that reduce the risk of these diseases.

QUESTIONS

1. What are three ways to extend the winter grazing season using small grain pastures?
2. Are there causes for concern when using pure stands of small grains for grazing? What can happen to cows? What about stockers?
3. How does one determine which small grain forage is best for a particular farm system? Where is the best information likely to be found? What should be considered when selecting a species or cultivar?
4. Why is there renewed interest in using root crops as fodder crops?
5. At what time(s) during the year are brassicas likely to be the most important contributors to a grazing program?

REFERENCES

Altom, W, RL Dalrymple, and J Rodgers. 1976. Effect of nitogen rates and dates of application on forage, grain, and straw yields of rye-ryegrass mixture. In RL Dalrymple (ed.), Field Day and Progress Report, Noble Foundation Publ. RR-76, 45-57.

Bates, W. 1972. Forage yields from rye, oat, wheat, barley, triticale varieties and strains. In Samuel Roberts Noble Found. Spec. Publ. R-139.

Bohman, VR, FP Horn, BA Stewart, AC Mathers, and DL Grunes. 1983. Wheat pasture poisoning. I. An evaluation of cereal pastures as related to tetany in beef cows. J. Anim. Sci. 57:1352-61.

Briggle, LW. 1969. Triticale—A review. Crop Sci. 9:197-202.

Burton, GW, and GM Prine. 1958. Forage production of rye, oats, and ryegrass as influenced by fertilization and management. Agron. J. 50:260-62.

Clay, BR. 1973. Stocker cattle losses on small grains pasture. Okla. Vet. 25:15-17.

Coffman, FA. 1961. Oats and Oat Improvement. Am. Soc. Agron. Monogr. 8. Madison, Wis.

Colman, RL. 1966. The effect of herbicides, anhydrous ammonia and seeding rate on winter production of oats sod-sown into grass dominant pasture. Aust. J. Exp. Agric. Anim. Husb. 6:388-93.

Cox, AE. 1977. Husbandry aspects of *Brassica* growing including techniques for sowing and weed control. In JFD Greenhalgh and IH McNaughton (eds.), UK, Brassica Fodder Crops Conf., Scottish Agriculture Development Council, 60-65.

Davies, JG. 1960. Production of winter green feed cereals and renovation of dairy pastures in Canterbury by autumn overdrilling. N.Z. J. Agric. 101:21-29.

Decker, AM, HJ Retzer, FG Swain, and RF Dudley. 1969. Midland Bermudagrass Forage Production Supplemented by Sod-seeded Cool-Season Annu-

al Forages. Md. Agric. Exp. Stn. Bull. 484.
Denman, CE, and J Arnold. 1970. Seasonal Forage Production for Small-Grain Species in Oklahoma. Okla. Agric. Exp. Stn. Bull. B-680.
Dev, G, NS Dhillon, and T Singh. 1979. Effect of rates, sources and timings of nitrogen application on yield and quality of forage oats. Ind. J. Dairy Sci. 32:479-81.
Ehmke, V. 1993. Triticale packs punch for high-plain grazing. Hay and Forage Grow., Apr., 12-13.
Evans, DW. 1979. Crucifers—growth and yield. In Proc. Symp. Sheep Harvested Feeds for the Intermt. West, Colorado State University, Fort Collins, Colo., 59-63.
Faix, JJ, CJ Kaiser, DW Graffis, JM Lewis, ME Mansfield, and GA Jung. 1979. Evaluation of various root crops, kale and rape in sod seeding. Ill. Agric. Exp. Stn. Res. Rep. (Dixon Springs Agric. Ctr.) 7:103-10.
Fowler, DB, 1982. Date of seeding, fall growth, and winter survival of winter wheat and rye. Agron. J. 74:1060-63.
Guillard, K, and DW Allinson. 1984. Evaluation of 'Tyfon' for fall forage production. Proc. Forage and Grassl. Conf., Houston, Tex., 83-87.
———. 1989a. Seasonal variation in chemical composition of forage brassicas. I. Mineral concentrations and uptake. Agron. J. 81:876-81.
———. 1989b. Seasonal variation in chemical composition of brassicas. II. Mineral imbalances and antiquality constituents. Agron. J. 81:881-86.
Harlan, JR. 1971. Agricultural origins: Centers and non-centers. Sci. 174:468-74.
Harper, F, and IJ Compton. 1980. Seeding date, harvest date and yield of forage brassica crops. Grass and Forage Sci. 35:147-57.
Holt, EC, MJ Norris, and JA Lancaster. 1969. Production and Management of Small-Grain for Forage. Tex. Agric. Exp. Stn. Bull. B-1082.
Horn, FP. 1981. Practical alternatives for finishing cattle. In JL Wheeler and RD Mochrie (eds.), Forage Evaluation: Concepts and Techniques. Netley, South Austrialia: American Forage and Grassland Council; CSIRO, 527-33.
Horn, GW, and DF Frost. 1982. Ruminal mortality of stocker cattle grazed on winter wheat pasture. J. Anim. Sci. 55:976-82.
Horn, FP, and CM Taliaferro. 1975. Existing and potential systems of finishing cattle on forages or limited grain rations in the semi-arid southwest. In Forage Fed Beef: Production and Marketing Alternatives in the South, South Coop. Serv. Bull. 220, 401-17.
Horn, GW, TL Mader, SL Armbruster, and RR Frahm. 1981. Effect of monensin on ruminal fermentation, forage intake and weight gains of wheat pasture stocker cattle. J. Anim. Sci. 52:447-54.
Johnson, RR, FP Horn, and AD Tillman. 1974. Influence of Harvest Date, Nitrogen and Potassium Fertility Levels on Soluble Carbohydrate and Nitrogen Fractions in Winter Wheat. Okla. Agric. Exp. Stn. Res. Rep. MP-92.
Jung, GA, WL McClellan, RA Byers, CF Gross, RE Kocher, and HE Reed. 1979. Old forage crops may come back. Crops Soils 31(6):17-19.
Jung, GA, WL McClellan, RA Byers, RE Kocher, LD Hoffman, and HJ Donley. 1983. Conservation tillage for forage brassicas. J. Soil and Water Conserv. 38:227-30.
Jung, GA, RE Kocher, and A Glica. 1984. Minimum tillage forage turnip and rape production on hill land as influenced by sod suppression and fertilizer. Agron. J. 76:404-8.
Kalmbacher, RS, PH Everett, FG Martin, and GA Jung. 1982. The management of brassica for winter forage in the sub-tropics. Grass and Forage Sci. 37:219-25.
Khan, AA, NH Peck, AG Taylor, and C Samimy. 1983. Osmoconditioning of beet seeds to improve emergence and yield in cold soils. Agron. J. 75:788-94.
Kilcher, MR, and T Lawrence. 1979. Spring and summer pastures for southwestern Saskatchewan (comparison of *Elymus junceus, Elymus angustus* and fall rye). Can. J. Plant Sci. 59:339-42.
Koch, DW, FC Ernst, Jr., NR Leonard, RR Hedberg, TJ Blenk, and JR Mitchell. 1987. Lamb performance on extended-season grazing of tyfon. J. Anim. Sci. 64:1275-79.
Lambert, MG, SM Abrams, HW Harpster, and GA Jung. 1987. Effect of hay substitution on intake and digestibility of forage rape fed lambs. J. Anim. Sci. 65:1639-46.
Marten, GC, and RM Jordan. 1982. Double-cropped annual forages for autumn pasture. In Am. Soc. Agron. Abstr. Madison, Wis., 124.
Murtagh, GJ. 1971. Chemical seeded preparation and the efficiency of nitrogen utilization by sod-sown oats. I. Comparison of seedbeds. Aust. J. Exp. Agric. Anim. Husb. 11:299-306.
Nuttonson, MY. 1955. Wheat-Climatic Relationships and the Use of Phenology in Ascertaining the Thermal and Photo-thermal Requirements of Wheat. Washington, D.C.: American Institute of Crop Ecology.
———. 1957. Barley-Climate Relationships and the Use of Phenology in Ascertaining the Thermal and Photo-thermal Requirements of Barley. Washington, D.C.: American Institute of Crop Ecology.
———. 1958. Rye-Climate Relationships and the Use of Phenology in Ascertaining the Thermal and Photo-thermal Requirements of Rye. Washington, D.C.: American Institute of Crop Ecology.
Pascal, JL. 1977. Mechanism of growing *Brassica* forage crops. In JFD Greenhalgh, IH McNaughton, and RF Thow (eds.), Brassica Fodder Crops. Pentlandfield, Roslin, Midlothian, Scotland: Scottish Agriculture Development Council and Scottish Plant Breeding Station, 70-75.
Patel, GA, and JF Nishimuta. 1978. Comparative nutritive value of wheat, rye and triticale forages in a beef finishing program (steers). Annu. Res. Rep. Sch. Agric. and Environ. Sci., Ala. Agric. Mech. Univ. 7:189-208.
Paulsen, GM. 1987. Wheat stand establishment. In E. Heyne (ed.), Wheat and Wheat Improvement, Am. Soc. Agron. Monogr. 13. Madison, Wis., 384-89.
Paxman, PJ, and R Hill. 1974. Goitrogenicity of kale and its relationship to thiocyanate content. J. Sci. Food Agric. 25:329-37.
Quisenberry, KS, and LP Reitz. 1967. Wheat and Wheat Improvement. Am. Soc. Agron. Monogr. 13. Madison, Wis.
Rao, SC, and TH Dao. 1987. Soil water effects on

low temperature seedling emergence of five *Brassica* cultivars. Agron. J. 79:517-19.

Rao, SC, and FP Horn. 1986. Planting season and harvest date effects on dry matter production and nutritional value of *Brassica* spp. in southern Great Plains. Agron. J. 78:327-33.

Rao, SC, SW Akers, and RM Ahring. 1987. Priming *Brassica* seed to improve emergence under different temperatures and soil moisture conditions. Crop Sci. 27:1050-53.

Sheldrick, RD, JC Fenlon, and RH Lavender. 1981. Variation in forage yield and quality of three cruciferous catch crops grown in southern England. Grass and Forage Sci. 36:179-87.

Simmons, GD. 1972. Sodseeding of small grains. In Proc. Symp. Manage. of Intensified Winter Pastures and Stocker Calves, 11-12 May, Noble Foundation, Ardmore, Okla., 12-20.

Sithamparanathan, J. 1979a. Improvement of winter forage production on high rainfall North Island hill country by altum oversowing of cereals. I. Relative performance of different winter cereal cultivars. N.Z. J. Exp. Agric. 7:281-84.

———. 1979b. Improvement of winter forage production on high rainfall North Island hill country by altum oversowing of cereals. II. Effect of oversowing date and nitrogen application. N.Z. J. Exp. Agric. 7:285-88.

Skovmand, B, PN Fox, and RL Villareal. 1984. Triticale in commercial agriculture: Progress and promise. Adv. Agron. 37:1-45.

Smith, TJ. 1979. Producing cruciferous forages—a cultural practice guide. In Proc. Symp. Sheep Harvested Feeds for the Intermt. West, Colorado State University, Fort Collins, Colo., 64-68.

Stewart, BA, DL Grunes, AC Mathers, and FP Horn. 1981. Chemical composition of winter wheat forage grown where grass tetany and bloat occur. Agron. J. 73:337-47.

Swain, FG, AM Decker, and HJ Retzer. 1975. Sodseeding annual forages into 'Midland' bermudagrass (*Cynodon dactylon* L.) pasture. I. Species evaluation. Agron. J. 57:596-98.

Swanson, AF. 1935. Pasturing Winter Wheat in Kansas. Kans. Agric. Exp. Stn. Bull. 271.

Tokumasu, S, I Kanada, and M Kato. 1985. Germination behavior of seeds as affected by different temperatures in some species of *Brassica*. J. Jap. Soc. Hort. Sci. 54:364-70.

Toosey, RD. 1972. Profitable Fodder Cropping. London: Ipswich Farming Press.

Tsunoda, S, H Hinata, and C Gomez-Campo. 1980. Brassica Crops and Wild Allies. Tokyo: Japan Scientific Societies Press.

US Department of Agriculture. 1991. Agricultural Statistics. Washington, D.C.: US Gov. Print. Off.

Vartha, EW. 1971. Overdrilled winter-growing annual grass to supplement lucerne. In Proc. N.Z. Grassl. Assoc. 33:115-23.

Varughese, G, T Baker, and E Sarri. 1987. Triticale. Mexico City, Mexico: CIMMYT.

Westerman, RL, 1981. Factors affecting soil acidity. Solutions 25:64-81.

Westover, HL, HA Schoth, and AT Semple. 1933. Growing Root Crops for Livestock. USDA Farmers Bull. 1699.

Wikse, S, and N Gates. 1987. Preventive Health Management of Livestock That Graze Turnips. Wash. State Univ. Coop. Ext. Serv. Bull. EB 1453.

Wilkinson, SR, and JA Stuedemann. 1979. Tetany hazard of grass as affected by fertilization with nitrogen, potassium and poultry litter and methods of grass tetany prevention. In VV Redding and DL Grunes (eds.), Grass Tetany, ASA Spec. Publ. 35. Madison, Wis.: American Society of Agronomy, Crop Science Society of America, and Soil Science Society of America, 93-121.

Wilson, RE, EH Jensen, and GCJ Fernandez. 1992. Seed germination response for eleven forage cultivars of *Brassica* to temperature. Agron. J. 84:200-202.

Wright, MJ, and KL Davison. 1964. Nitrate accumulation in crops and nitrate poisoning in animals. Adv. Agron. 16:197-217.

37 Summer Annual Grasses

HENRY A. FRIBOURG
University of Tennessee

EFFICIENT summer livestock production is limited by the lack of adequate amounts of high-quality pasture. During the hot, often dry summer months, most perennial cool-season forages become semidormant, and most perennial warm-season forages do not produce high-quality forage. Summer forage annuals include members of several gramineous genera such as *Sorghum* and *Pennisetum*. These sorghums and millets are valuable in the development of year-round forage systems, particularly where quality is important, as with lactating or rapidly growing animals. The more productive cultivars are characterized by rapid growth of nutritious forage in late spring and summer and are used for pasture, green chop, silage, and hay.

DISTRIBUTION AND ADAPTATION

Sorghums used for forage in the US include grain and forage sorghums and sudangrass (formerly *Sorghum sudanense* [Piper] Stapf), all now classified as *S. bicolor* (L.) Moench (Harlan and de Wet 1972). Sorghums and pearlmillet, *Pennisetum americanum* (L.) Leeke (Burton 1982), can endure considerable moisture stress. They can be grown where annual precipitation is as low as 400-650 mm, but they grow better with more moisture or, especially in arid and semiarid regions, when irrigated. They can resume vegetative growth after a dormant period induced by drought.

The most favorable temperatures for growth are 25°-30°C (Squire et al. 1984), with a minimum of about 15°C. Pearlmillet is more sensitive than sorghums to lower temperatures; early plantings of sorghums exposed during germination or the seedling stage to early morning temperatures of 5°-10°C may survive when pearlmillet seedlings are killed. Sorghums and pearlmillet are rarely grown above 2000-2700 m in the southwest US, and this elevation limit decreases farther north. In the East they are grown from Florida to 42°N. Their range extends from southern Texas to Minnesota and North Dakota in the central grassland regions. Foxtail or Italian millet, *Setaria italica* (L.) Beauv., and other millets sometimes replace them in more northern latitudes.

Census data indicate about 600,000 ha of sorghums are grown for forage in the US, and another 300,000 ha for silage. In the US and elsewhere, vegetative regrowth of grain sorghum after grain harvest can provide good pasture. Pearlmillet is grown for forage on about half of the 500,000 ha planted in millets. These data do not reflect small fields often not listed in the census that play an important role in supplying high-quality pasture or silage for good animal rations during summer.

Forage sorghums are characterized by abundant juice in the culms, which are 1.5-3.0 m or more tall. The importance of forage sorghums rapidly increased in the 1960s with the advent of higher-yielding hybrids. Like-

HENRY A. FRIBOURG is Professor of Plant and Soil Science, University of Tennessee. He holds the MS degree from Cornell University and the PhD from Iowa State University. His special interests are the ecology of forage plants and he interrelationships among forage crops, their environment, grazing animals, and management factors.

wise, improved pearlmillets led to an increase in their use. Although pearlmillet is as widely adapted as sorghum, it has been grown primarily in the Southeast because of apparent greater tolerance to pathogens and high humidity. Most other millets are either not increasing or are decreasing in use, although they remain important in areas with short growing seasons.

Some sorghums (Escalada and Plucknett 1975) and pearlmillets are short-day plants. However, many appear to be insensitive to length of darkness. Summer annual grasses are best suited to soils with moderate to high available water supply where erosion is not a problem. They compete with row crops for allocation on highly productive soils; under such conditions their potentially high yields lower the unit cost and may justify their use. Moderately well-drained and imperfectly drained soils are well suited to summer annual grasses when excessive surface water is removed. Planting may be delayed in spring by wet conditions. Green chop or grazing may be difficult or undesirable when soil is wet. Better-distributed forage production can be obtained by soil selection and staggered plantings. Small amounts of alkali in the soil reduce performance considerably, though tolerance to salinity is moderate (4-8 dSm^{-1}) (Bernstein 1964) and may exist in some genotypes for greater concentrations (Bottacin et al. 1983). Moderately acid soil conditions down to pH 5.5 do not appreciably affect production.

Although seeding can be as early as March in southern Texas, date of planting is determined by the intended use, such as early summer or late summer grazing, silage, or compatibility in a double-cropping sequence (Doyle et al. 1986; Robinson 1991). Seeding before soil temperature at 10 cm reaches 12°-13°C can be hazardous, but later or multiple plantings often are made to equalize forage production throughout the season. In climates approaching subtropical, late summer-early autumn plantings also may be made.

PLANT DESCRIPTIONS

Sorghums. Sorghum is a coarse, erect grass with considerable variability in growth characteristics. Height at maturity can range from 0.45 to over 5 m. Some cultivars tiller early, while others do not tiller until physiological maturity is reached or until meristematic tissue is removed or damaged. More tillers, larger leaf areas, and greater yields are obtained under 14-h days and warm temperatures than under cooler temperatures and 10-hr days. Culm thickness varies greatly, ranging from 5 to over 30 mm. Culms are solid, though spaces may occur in the pith. They may be comparatively dry at maturity or have sweet or insipid juice. Root primordia occur at culm nodes, and prop roots may grow from these. There is a bud at each node from which a tiller may grow. Early tillers originate from basal or epigeal nodes adjacent to elongated internodes; later tillers originate from nodes adjacent to elongated internodes and at successively longer time intervals as the season progresses. Leaf blades commonly are similar in shape to those of corn (*Zea mays* L.) but are shorter and may be wider. The blades are glabrous and waxy. The sheaths encircle the culm and have overlapping margins.

The inflorescence of sorghum is a panicle that ranges from compact in most grain sorghums to open in sudangrass and many forage sorghums. Some cultivars have seed that is completely covered by glumes not removed in threshing; most have seed that threshes free. Panicles usually are erect but sometimes are recurved. A sorghum panicle may contain as many as 6000 fertile spikelets. Sorghum is generally self-pollinated, but considerable cross-fertilization can occur, depending on environmental conditions. Seed can vary in size: grain sorghums may have 25,000-60,000 seeds/kg, but forage sorghums may have 120,000-150,000 seeds/kg. Seed may be white, yellow, red, or brown, and some of the darker-colored seed may contain considerable amounts of tannin.

Millets. Pearlmillet (Rachie and Majmudar 1980) is an erect annual grass that may grow 2-5 m tall. Leaf blades are long and pointed, with finely serrated margins, and stems are pithy. Leaf initiation is constant, and successive primordia are increasingly larger. Tillering occurs freely from axillary meristems borne at each node. In some cultivars tillering may occur from nodes 20 cm or more above ground level. The inflorescence is a thick, cylindrical spike 20-50 cm long and 2-4 cm in diameter. The caryopsis threshes free at maturity (Powers et al. 1980). Because the stigmas appear before the anthers are protruded, pearlmillet is largely cross-pollinated. The spike flowers from the tip downward over several days. Pearlmillet seed usually is yellowish and much larger than the hulled seed of other millets. Pearlmillet will average 200,000 seeds/kg, and foxtail millet, 500,000 seeds/kg.

Browntop millet, *Panicum ramosum* L., is a

rapidly growing annual, 0.6-1.2 m tall, with a yellow to brown panicle 5-15 cm long. It shatters seed in sufficient amounts to reseed itself. Proso millet, *P. miliaceum* L., has been used at high elevations and in more northern locations (Hinze et al. 1978). Foxtail millet is an annual with slender erect and leafy culms which grow from 0.3 to 1.75 m. Its inflorescence is a dense, cylindrical, bristly panicle 5-30 cm long. Japanese millet, *Echinochloa crusgalli* var. *frumentacea* (Link) W. F. Wight, is coarser and grows more rapidly under cool conditions than does foxtail millet (Muldoon 1983).

IMPORTANCE AND USE

Some sorghum cultivars may produce more dry matter per hectare than does corn; however, these types usually are low in grain content. Cattle, sheep, and horses can utilize sorghum forage for maintenance, growth, and fattening. In the US, beef cattle consume more sorghum grain, and possibly more forage, than do dairy cattle.

The feeding of sorghum or pearlmillet silage or green chop to lactating cows is a common practice in dairy intensive enterprises; grazing of these crops by growing beef cattle is done on a more limited, but increasing, scale. Feedlot rations containing 50%-65% sorghum forage have been used successfully with steers, and lambs have been fattened on rations of sorghum forage and grain supplemented with protein. Sorghum rations must be supplemented with protein, calcium (Ca), and other minerals. If sorghum grain is fed, it should be cracked, rolled, steam flaked, or reconstituted for best animal digestibility. In Australia, New Zealand, and Africa, supplementation of pastures with sodium (Na) and sulfur (S) has been beneficial (Wheeler et al. 1980).

Later-maturing sorghum cultivars, which yield more than earlier-maturing types, may not reach maturity before a killing frost. However, frosted sorghums and corn can be successfully ensiled with minimal adverse effects on their chemical composition by adding water as needed for a 60%-70% moisture silage.

Sorghum silage has largely replaced corn silage in the High Plains region of the southwest US because of high yields. In areas of higher rainfall or irrigation, corn silage is considered superior to sorghum silage because of its higher energy content. Sorghum hybrid cultivars that produce large quantities of both grain and stover have produced acceptable silage, but the stage of maturity at harvest greatly influences the composition of silage. In general, the best animal performance is obtained from silages cut after physiological maturity but before senescence is too marked. Grain content of silage is very important, and grain should constitute at least 25% of the material harvested for silage. Highest-quality silage is obtained when crop nutrients, as affected by soil fertility, are kept in balance (Fribourg et al. 1976). Silage made from bird-resistant sorghum cultivars generally is inferior in quality to that made with non-bird-resistant cultivars. Forage sorghum yield variations caused by differences in growing conditions consist mostly of differences in culm production (Creel and Fribourg 1981).

Forage sorghums are sometimes cut for hay, but curing may be difficult in humid climates. During a growing season two to five harvests can be attempted, each one having a potential yield of 2 mt/ha or more. Plants are difficult to cure because of their thick culms; even sudangrass, with its thinner culms, may be difficult to cure because of the large mass of material to be dried. Sunshine and use of forage crushers, crimpers, and tedders help to reduce drying time. Bleaching and discoloration are common side effects. Pearlmillet is seldom made into hay, although sometimes other millets are grown in a mixture with soybeans (*Glycine max* [L.] Merr.) for this purpose.

Pasturing is the cheapest method of harvesting forage, although efficient utilization of fast-growing summer annuals demands considerable attention to many details of plant and animal management. Since grazing animals selectively consume certain plant portions, they may produce more milk or meat than when fed the whole plant. However, grazing leads to waste by trampling or fouling by excreta. Plant management may be easier under a green chop system, when feed needs are better matched to livestock requirements. Both systems are used with forage sorghums and pearlmillet with equally desirable results. Summer annuals are generally more suitable economically for lactating dairy cows, for which a continuous supply of high-quality forage is essential, than for beef steers.

CULTIVARS

Since the 1950s plant breeders have created many sorghum hybrids that are well adapted to specific environments and uses. The term *sorghum-sudangrass hybrids* has been used to describe these, with considerable confusion resulting, because the phenotypic expression of genetic characteristics within the genus is a continuum from one extreme to the other (Harlan and de Wet 1972). Generally the male parent used is a fertile sudangrass. The male-sterile female parent can be another sudangrass, a sorghum with either sweet or insipid juice, or a grain sorghum. Crosses involving three or more parents also can be made. Most hybrids have been developed by private industry. Some seed companies have active breeding programs from which have issued unique gene recombinations and cultivars excellent in many characteristics. Frequently blends of tall, medium, and/or short types are sold. Blended entries differing in stature do not yield more than do pure tall types (Gorz and Haskins 1982). More recently, some companies have produced cultivars in which sudangrass parentage predominates; these cultivars usually tend to be leafier and have lower prussic acid (HCN) potential than earlier cultivars. In subtropical Queensland, Australia, 'Silk' sorghum (*S. halepense* × *S. roxburghii* × *S. arundinacea*) is well adapted for pasture and may last three to four seasons (Clewett and Young 1988). There are fewer pearlmillet cultivars than sorghum cultivars, partly because hybrids may have to be produced by intermating unrelated inbreds, which are still few but are increasing in number (Burton 1981; Hanna et al. 1988).

In addition to larger yields, future breeding goals for summer annuals will be less HCN potential in sorghums (Wheeler et al. 1980; Haskins et al. 1987); less lignin and fiber concentrations; increased palatability, sugar concentration, and cold and drought tolerance; greater leaf:stem ratios; greater resistance to leaf diseases and insect pests; faster regrowth after harvest; and increased intake and digestibility by ruminants. Since the number of available cultivars is large, use of the latest cultivar performance reports or descriptions by the USDA and state universities is advised.

CULTURE AND MANAGEMENT

Establishment. Adequate germination of sorghums and pearlmillet takes place when soil temperature is 20°-30°C; much poorer germination occurs at lower or higher soil temperatures. Poor stands usually result from very early or very late plantings. Although some late plantings are justified when staggered production is desired, they will result in low yields because growth will be hampered by cool fall temperatures, summer droughts, and long nights. Inadequate surface soil moisture also makes late planting riskier. A well-prepared, firm, moist seedbed is best, although acceptable stands may be established with stubble-planting machinery in no-till systems and for reclamation (Mislevy and Blue 1981). Seeding should be at depths of 1.5-5.0 cm, depending on soil moisture and texture. Compaction of the seedbed is desirable after seeding, particularly if the soil is dry.

Seeding Rates and Spacing. Summer annual grasses can be drilled, broadcast, or sown in rows, in level seedbeds or after listing, in double rows on beds, in rows too narrow to cultivate, or in no-till systems. The narrower the spacing, the heavier the seeding rate needed. Smaller yields result occasionally from seeding rates of less than 10-20 kg/ha. At low seeding rates (Caravetta et al. 1990) culm diameter may be increased 20%-35%, thus increasing hay-drying time. Greater seeding rates, up to 50 kg/ha, produce larger first-harvest yields, result in thinner culms, and may decrease drying time. Normal seeding rates are 20-30 kg/ha. Seeding rates should be heavy for large-seeded cultivars and decreased for smaller-seeded plants such as sudangrass or pearlmillet.

Row spacing has relatively little effect on total forage production, but grazing animals can damage drilled or broadcast seedings. Seeding rate and row spacing therefore will depend upon the intended use. Although broadcast or drilled plantings make more efficient use of solar energy and produce more in the first growth than do wide row stands, total seasonal production is usually the same unless the number of harvests is small. If the number of seeds/m^2 is kept constant, increasing the width of rows may decrease production. As the seeding rate increases, intra-row competition reduces the percentage of established plants. Since tillering is inversely related to plant spacing, new tillers in the regrowth compensate for differences in initial plant spacing. Within a row spacing, plant size is inversely related to plant number. Cell wall constituents and digestibility thus can decrease with increasing plant density, although increases in population density result in larger total digestible dry matter production. Orientation of wide rows may have some effect on yield because of insolation effects: plants in north-south rows may yield 10%

more than plants in east-west rows when harvested frequently at an immature stage of growth.

Growth Habits. Species and cultivars within species are widely different in growth habits and performance. Some hybrids produce more digestible dry matter than others when harvested at a 40 cm height; others yield more when harvested at the dough state than at other stages of growth, even though the percent digestibility within each stage is about the same for several hybrids. Plant types differ among and within hybrids, ranging from those with long, thin stalks to others with short, thick stalks, and from very compact heads to open panicles. Just as wide a range probably exists in characteristics of internal morphology, physiology, or chemical components. A gradual increase in digestible dry matter production has been observed in pearlmillet when the frequency of harvest was increased from 6 wk to 4-, 3-, and 2-wk intervals. Dry matter concentration was increased in some forage cultivars from the first to the fifth harvest.

Mechanical Damage. The wheel pressure of implements on stubble at time of cutting can seriously affect regrowth ability and decrease productivity of summer annual grasses. Node tissue may be damaged and bud development may be prevented. Passage of a tractor wheel over the row at each of five cuttings decreased dry matter yield 20% on a silt loam soil (Fribourg et al. 1975). When the tractor wheel was followed by the wheel of a loaded wagon, an additional 10% loss was sustained. These treatments increased bulk density about 20% in the top 4 cm of soil, thereby affecting soil aeration and water infiltration. The harvest method contemplated therefore should influence the planting pattern selected. Plants should not be green-chopped when the soil is very wet; however, the need for dairy cattle feed sometimes may override this consideration.

Harvesting Management. A thin-culmed sorghum or pearlmillet recovers more rapidly after cutting and can tolerate closer grazing or cutting than one with thicker culms (Fig. 37.1). Stubble height and plant height or stage of growth when harvested influence not only total yields and digestible dry matter yields but also leafiness. 'Gahi-1' pearlmillet repeatedly cut from 50-cm growth to a 25-cm stubble yielded 14 mt/ha containing 90% leaf (Fribourg 1985). Regrowth of summer annual grasses depends not only on the amount of photosynthetic area left on the stubble and the amount of water in the plant but also on the presence and regrowth of terminal, axillary, and basal meristems. As the height of defoliation is raised, regrowth from terminal meristems increases. Basal and axillary tillering increases as the point of defoliation is lowered. These relationships affect not only total production but also its seasonal distribution, since higher stubbles shift production to later periods in the summer. Low stubbles are more detrimental to some cultivars than to others; pearlmillet is more sensitive to low stubble height than are some sorghum cultivars. Since regrowth must come from terminal or axillary meristems, higher stubbles generally lead to more vigorous and leafier regrowth. Stubbles lower than 10-15 cm result in reduced regrowth or even plant death. Relationships between stubble carbohydrate content and yield are not clear. Greatest forage dry matter yields from pearlmillet and sorghum are obtained as plants approach maturity or attain heights of 80-120 cm. However, mature plants of such heights are not suitable for grazing (Fig. 37.2). Most uniform grazing and least waste are achieved at the 50-70 cm height. Best regrowth is obtained if grazing is suspended when 10-15 cm of growth and some succulent plant parts with buds are left. Rotational stocking or strip-grazing provide optimum management where two or more areas are grazed successively in a controlled manner. To avoid prussic acid (HCN) potential toxicity, young growth should not be grazed, particularly after frost or drought (see Chap. 9, Vol. 2). Stockpiled material may be of value under specific circumstances if the growth is not so short and young as to be high in HCN potential.

Fertilization. The effects of fertilizer nitrogen (N) on summer annual grasses have been studied extensively. Summer annuals will grow on low-fertility and/or moderately acid soils, but they grow best at greater soil fertility and at pH 6-7 levels. Sorghum hybrids are more sensitive to low pH (Ahlrichs et al. 1991) and to low soil phosphorus (P) and potassium (K) availability than is pearlmillet. Generally, the summer annual grasses are fertilized with 30-60 kg/ha of P and K each, even though yield responses to such applications often have not been demonstrated on soils

Fig. 37.1. Sorghum cultivars differ in response to management. Both cultivars were planted the same day and, when 50 cm tall, were cut back to 15 cm. Note the poor regrowth and high culm mortality of the thick-culmed cultivar (*left*) as compared with the numerous tillers and good regrowth of the thin-culmed cultivar (*right*).

with medium or greater amounts of available P and exchangeable K. Use soil tests to pinpoint fertility problems and ensure fertilizer efficiency.

High yields of forage may remove as much as 30-50 kg/ha of P and 150-200 kg/ha of K. When available K is abundant, luxury consumption may lead to large concentrations of K in forage dry matter. Pearlmillet grown with high Ca and K fertilizer levels and fed to lactating cows is associated with severe milkfat depression. These effects appear related to organic acid constituents of the forage, which are themselves influenced by fertilization and are genetically controlled (Keisling et al. 1990).

Positive linear responses of summer annual grasses to N fertilization of up to 200 kg/ha have been reported (Scheffer et al. 1985). At the larger rates, split applications of N are essential for uniform growth and balanced nutrition of plants. In warm, humid climates, 15-20 mt/ha of forage dry matter have resulted from applications of 400 kg/ha of N or more. If relatively large N rates are used, especially when growth is suppressed by moisture stress or other causes, concentrations of nitrate (NO_3^-) ions, HCN, and alkaloids may occur in plants at levels that can be toxic to some grazing animals (Fales and Wilkinson 1984; Krejsa et al. 1987).

Digestibility and Animal Intake. Generally, summer annual grasses are moderately to highly digestible and are readily consumed in the vegetative stage of growth by livestock. Good quality forage has a large leaf:stem ratio, a large concentration of protein and digestible nutrients, and a small concentration of fiber and lignin. Cutting and grazing man-

Fig. 37.2. Properly managed forage sorghum will provide quality forage during the summer months.

agement, nature of growth, moisture supply, night length, and sometimes temperature are critical factors in the production of summer annual forages. Positive correlations between leafiness and dry matter digestibility have been established for pearlmillet and sorghum (Fribourg et al. 1976). Leafiness is related more to stage of growth and height than to time elapsed since a previous harvest.

Many factors influence intake and digestibility: the physiologic state and genotype of both consuming animal and consumed plant parts, soil fertility, moisture availability, previous crop management, and plant stage of growth. Plant parts are selected differentially by animals even when these other factors are held constant, and selection is affected by grazing pressure and forage availability. High lignin concentration has been related to advancing maturity and decreased animal intake and digestibility. The brown midrib-inherited character has been associated in both sorghums and pearlmillet with smaller neutral detergent fiber, acid detergent fiber and hemicellulose concentrations, and greater apparent digestibility than in similar forage genotypes without the brown midrib (Akin et al. 1991; Cherney et al. 1990; Fritz et al. 1988). Young leaf blades may approach 75% dry matter digestibility, while older leaves may be only 50%-60% digestible and culms may be as low as 30%-50% digestible. Well-managed summer annual grass pastures result in animal performance similar to that from alfalfa. However, consumption by grazing animals of sorghum varies among cultivars and is related to chemical constituents and their fluctuation with time, environment, stage of growth, and the presence of bloom on leaves. The highest levels of protein and digestibility result from the most frequently harvested forage.

Green Chop and Silage. The harvesting of summer annual grasses for green chop or silage is well suited to mechanization because of the bulk and mass involved. Equipment has a low labor requirement. Because of the low labor requirements with green chop equipment, one person can cut, haul, store, and dispense the feed from large areas. Green chop may require daily cutting, but it allows for maximum management control and efficient use of resources. It also limits feed selection by animals.

Sorghum for silage should be harvested when seed is in the milk-to-dough stage but before leaf blades reach senescence. Earlier or later harvest will reduce harvested energy. When grain sorghum for silage is harvested in the milk, dough, and hard seed stages, dry matter intake increases with maturity, but no differences in milk production are noted. Milk production is greater with grain sorghums than with grainless types. Although harvesting in the hard seed stage or later stages may lead to some loss of grain passing through the digestive tract, the appearance of some grain in the feces appears to bear little relation to animal performance. The highest-quality silage is obtained from sorghum cultivars having 25%-30% or more dry matter yield as grain. Although height differences of mature forage cause few quality changes in cultivars, differences in silage fermentation and preservation have been shown among genotypes. Sorghum silage is slightly less palatable and digestible than corn silage. Silage, if low in grain, may require additives to prevent undesirable fermentation. Interplanting corn and sorghum in an attempt to hedge against unfavorable environmental conditions has not been advantageous in the southeast US.

SEED PRODUCTION

For summer annual grasses grown for seed, the crop is generally planted in rows about 1 m apart and at low seeding rates (3-6 kg/ha). Most seed production is done in the southern High Plains and the southwest US, usually under irrigation. The field should be isolated from others of the same genus to minimize the risk of unwanted hybridization. Although many sorghums are self-pollinated, they will cross freely with other types grown adjacent to them. Seed set on male-sterile heads varies from near 0% to almost 100%, depending on pollen supply and weather. Johnsongrass (*S. halepense* [L.] Pers.) is particularly troublesome in sorghum seed production; it will cross freely with grain or forage sorghums. Crop improvement associations require 200-400 m or more isolation.

Production of the male-sterile parents and of hybrid seed is done by interplanting fertile and male-sterile types in isolated crossing blocks. The maintainer line is used for pollen production. When grown in an isolated and rogued block, the seed produced is either selfed or sibbed. The male-sterile line is wind pollinated by the maintainer line, and the seed produces only male-sterile plants in the next generation. The male parent is produced by selfing or sibbing in another isolated block.

In hybrid seed production the male-sterile line is interplanted with the male restorer parent line. Seed harvested from the male-sterile plants is sold as hybrid seed. A production field usually has a 3:1 ratio of male-sterile rows to pollinator rows.

The ease and cost of producing seed of a line, cultivar, or species determine the availability and price of seed. Some crosses are difficult because the male-sterile line and the pollinator line do not mature at the same time. Staggered or multiple plantings can be used to ensure pollen availability when stigmas of male-sterile plants are receptive. A tall male-sterile parent also materially increases seed production costs. Seed production may range from 0.2 to 6.0 mt/ha. Seed shattering, limited moisture, diseases, and/or insects may cause low yields.

DISEASES AND PESTS

Unique pest management is required for each agroecosystem. Summer annual grasses grown for pasture or green chop may escape pests that will damage the same crop grown for silage or seed. Forage sorghum can be attacked by insects that feed on many warm-season crops, such as the fall armyworm (*Spodoptera frugiperda* J. E. Smith). In addition, forage sorghum is sometimes attacked by the greenbug (*Schizaphis graminum* Rondani); by the chinch bug (*Blissus leucopterus* Say), which also attacks pearlmillet (Merkle et al. 1983), especially in dry years on cultivars that have no genetic resistance; and by the sorghum webworm (*Nola sorghiella* Riley). The sorghum midge (*Contarinia sorghicola* Coquillett) can destroy a seed crop. In Asia and Africa, the sorghum stem borer (*Chilo partellus* Swinhoe) can cause much damage. Many other pests affect summer annual grasses. When resistant cultivars are not available or cultural control is not effective, the assistance of state experiment station or extension service personnel should be solicited.

Many diseases of millets and sorghums have been described (Nyvall 1989). Few cultivars with resistance to specific diseases have been developed (Monson et al. 1986). Where resistant cultivars are not available, some control may be obtained with seed treatment, rotation with other crops, or control of alternate hosts. Another way to circumvent the problem is to plant the crop during seasons when environmental conditions favor the crop more than the pathogen. In major disease outbreaks, consultation with local specialists is advised.

QUESTIONS

1. Contrast the ways in which summer annual grasses fill special forage needs in your area and in other parts of the country.
2. What are the most salient characteristics of sorghum and pearlmillet that may lead to their use for green chop? For grazing? For silage? For hay?
3. Discuss the factors that should be considered in making management decisions involving the use

of summer annual grasses for pasture or green chop.
4. Which cultivars of corn, sorghums, and pearlmillet would you plant in your locality for forage? Discuss the advantages and disadvantages of each available cultivar.
5. In what ways can sorghums, pearlmillet, and other forages fit into different farm enterprises and management strategies in your region?
6. How do climatic and edaphic factors interact with the genetic makeup and the morphology of summer annual grasses? Discuss how these interactions influence the decisions that must be made by the farm manager for best animal performance or production.
7. What important fertilizer and other soil management practices should be considered in planning for growing summer annual grasses for forage? Give specific examples applicable to the most prevalent soils in your locality.
8. Identify the most important diseases and pests of summer annual grasses in your area, and describe some of the more effective and ecologically sound means of control.

REFERENCES

Ahlrichs, JL, RR Duncan, G Ejeta, PR hill, VC Baligar, RJ Wright, and WW hanna. 1991. Pearlmillet and sorghum intolerance to aluminum in acid soil. In RJ Wright et al. (ed), Plant-Soil Interactions at Low pH, Kluwer Acad. Pub., Netherlands, 947-51.

Akin, DE, LL Rigsby, WW Hanna, and RN Gates. 1991. Structure and digestibility of tissues in normal and brown midrib pearl millet (*Pennisetum glaucum*). J. Sci. Food Agric. 56:523-38.

Bernstein, L. 1964. Salt Tolerance of Plants. USDA Agric. Info. Bull. 283.

Bottacin, A, M Saccomani, P Spettoli, and G Cacco. 1983. NaCl-induced modifications of nitrogen absorption and assimilation in salt tolerant and salt resistant millet ecotypes. In MR Saric and BC Longhman (eds.), Genetic Aspects of Plant Nutrition, Dev. Plant and Soil Sci. 8. The Hague: Martinus Nijhoff/W. Junk, 203-7.

Burton, GW. 1981. Registration of pearl millet inbreds Tift 23DBE, Tift 23DAE and Tift 0756. Crop Sci. 21:804.

———. 1982. Improving the heterotic capability of pearlmillet lines. Crop Sci. 22:655-57.

Caravetta, GJ, JH Cherney, and KD Johnson. 1990. Within-row spacing influences on diverse sorghum genotypes: II. Dry matter yield and forage quality. Agron. J. 82:210-15.

Cherney, DJR, JA Patterson, and KD Johnson. 1990. Digestibility and feeding value of pearl millet as influenced by the brown-midrib, low-lignin trait. J. Anim. Sci. 68:4345-51.

Clewett, JF, and PD Young. 1988. Silk sorghum. Queensl. Agric. J. 114:122-24.

Creel, RJ, and HA Fribourg. 1981. Interactions between forage sorghum cultivars and defoliation managements. Agron. J. 73:463-69.

Doyle, KM, M Collins, and S Kaplan. 1986. Yield and quality of annual crops seeded following pea harvest. Can. J. Plant Sci. 66:87-94.

Escalada, RG, and DG Plucknett. 1975. Rattoon cropping of sorghum: II. Effect of daylength and temperatures on tillering and plant development. Agron. J. 67:479-84.

Fales, SL, and RE Wilkinson. 1984. Mefluidide-induced nitrate accumulation in pearl millet forage. Agron. J. 76:857-60.

Fribourg, HA. 1985. Summer annual grasses. In ME Heath, RF Barnes, and DS Metcalfe (eds.), Forages: The Science of Grassland Agriculture, 4th ed. Ames: Iowa State Univ. Press, 278-286.

Fribourg, HA, JR Overton, and JA Mullins. 1975. Wheel traffic on regrowth and production of summer annual grasses. Agron. J. 67:423-26.

Fribourg, HA, WE Bryan, GM Lessman, and DM Manning. 1976. Nutrient uptake by corn and grain sorghum silage as affected by soil type, planting date and moisture regime. Agron. J. 68:260-63.

Fritz, JO, KJ Moore, and EH Jaster. 1988. In situ digestion kinetics and ruminal turnover rates of normal and brown midrib mutant sorghum × sudangrass hays fed to non-lactating Holstein cows. J. Dairy Sci. 71:3345-51.

Gorz, HJ, and FA Haskins. 1982. Performance of blends of short, medium, and tall sorghum for forage. Crop Sci. 22:223-26.

Hanna, WW, HD Wells, GW Burton, GM Hill, and WG Monson. 1988. Registration of 'Tifleaf 2' pearl millet. Crop Sci. 28:1023.

Harlan, JR, and JMJ de Wet. 1972. A simplified classification of cultivated sorghum. Crop Sci. 12:172-76.

Haskins, FA, HJ Gorz, and BE Johnson. 1987. Seasonal variation in leaf hydrocyanic acid potential of low- and high-dhurrin sorghums. Crop Sci. 5:903-6.

Hinze, G, HO Mann, EJ Langin, and A Fisher. 1978. Registration of Cope proso millet cultivars. Crop Sci. 18:1093.

Keisling, TC, W Hanna, and ME Walker. 1990. Genetic variation for Mg tissue concentration in pearl millet lines grown under Mg stress conditions. J. Plant Nutr. 13:1371-79.

Krejsa, BB, FM Rouquette, Jr., EC Holt, BJ Camp, and LR Nelson. 1987. Alkaloid and nitrate concentrations in pearl millet as influenced by drought stress and fertilization with nitrogen and sulfur. Agron. J. 79:266-70.

Merkle, OG, KJ Starks, and AJ Casady. 1983. Registration of pearl millet germplasm lines with chinch bug resistance. Crop Sci. 23:601.

Mislevy, P, and WG Blue. 1981. Reclamation of quartz sand-tailings from phosphate mining. III. Summer annual grasses. J. Environ. Qual. 10:457-60.

Monson, WG, WW Hanna, and TP Gaines. 1986. Effects of rust on yield and quality of pearl millet forage. Crop Sci. 26:637-39.

Muldoon, DK. 1983. Growth and development of two *Echinochloa* millet species in a warm temperate climate. In JA Smith and VW Hays (eds.), Proc. 14th Int. Grassl. Congr., Lexington, Ky. Boulder, Colo.: Westview, 244-47.

Nyvall, RF. 1989. Field Crop Diseases Handbook. 2d ed. New York: Van Nostrand Reinhold.

Powers, D, ET Kanemasu, P Singh, and G Kreitner.

1980. Floral development of pearl millet. Field Crops Res. 3:245-65.

Rachie, KO, and JV Majmudar. 1980. Pearl Millet. University Park: Penn State Univ. Press.

Robinson, DL. 1991. Yield, forage quality and nitrogen recovery rates of double-cropped millet and rye grass. Commun. Soil Sci. Plant Anal. 22:713-27.

Scheffer, SM, JC de Saibro, and J Riboldi. 1985. Efeito de nitrogenio, metodos de semadura e regimes de corte no rendimento e qualidade de forragem e da semente de milheto. Pesqui Agropecu Bras 20:309-17.

Squire, GR, B Marshall, AC Terry, and JL Monteith. 1984. Response to temperature in a stand of pearl millet. VI. Light interception and dry matter production. J. Exp. Bot. 35:599-610.

Wheeler, JL, DA Hedges, KA Archer, and BA Hamilton. 1980. Effect of nitrogen, sulphur and phosphorus fertilizer on the production, mineral content and cyanide potential of forage sorghum. Aust. J. Exp. Agric. Anim. Husb. 20:330-38.

APPENDIX

Common and Botanical Names of Grasses, Legumes, and Other Plants

THE ACCOMPANYING LIST gives common and scientific names of many grasses, legumes, and several other plants discussed in this book. Many forage crop plants are known by different common names in different sections of the US and the world (Chapter 2). Common names are listed alphabetically, e.g., alfalfa, followed by names used less frequently. The initial letter of the first part of the scientific name, the genus, is always capitalized (*Medicago*); the second part, the species epithet, is written entirely in lowercase (*sativa*). The genus corresponds roughly to a last name and the species ephithet to a first name, as *Zea mays* would to Brown, John. The scientific name is followed by the abbreviation of the name of the person who first named the species (Chapter 2). Thus, *Zea mays* L. indicates this species was named by Linnaeus. However, *Cynodon dactylon* (L.) Pers. was first described as *Panicum dactylon* by Linnaeus but later was transferred to another genus, *Cynodon*, by Persoon.

The following abbreviations and symbols are used: syn. = synonym, in which case there are two accepted scientific names; X before the genus name indicates a hybrid between two genera; x before the species ephithet or subspecies name refers to a hybrid between two species or two subspecies.

Scientific names continue to change as new technologies for classification are developed. For example, the taxonomy and nomenclature of the perennial grasses of the tribe Triticeae were revised by Barkworth and Dewey (1985). The Crop Science Society of America, the Weed Science Society of America, and the USDA Germplasm Resources Information Network (GRIN) database regularly update listings.

COMMON NAMES	SCIENTIFIC NAMES
aeschynomene (*see* jointvetch)	*Aeschynomene americana* L.
alemangrass	*Echinochloa polystachya* (H.B.K.) Hitchc.
alfalfa	*Medicago* L.
alfalfa	*M. sativa* L.
variegated	*M. sativa* nothosubsubsp. *varia* (Martyn) Arcang (syn. *M. media* Pers.)
yellow	*M. sativa* subsp. *falcata* (L.) Arcang. (syn. *M. falcata* L.)
alkaligrass, nuttall	*Puccinellia airoides* (Nutt.) Wats. & Coult.
alyceclover.	*Alysicarpus vaginalis* (L.) DC.
American jointvetch	*Aeschynomene americana* L.
angletongrass	*Dichanthium aristatum* (Poir.) C.E. Hubb.
antelopegrass	*Echinochloa pyramidalis* (Lam.) Hitchc. & Chase
Arizona cottontop	*Digitaria californica* (Benth.) Henr.
axillaris (perennial horse gram)	*Macrotyloma axillare* (E. Mey.) Verde.
bagpod	*Glottidium vesicarium* (Jacq.) Harper
bahiagrass	*Paspalum notatum* Flugge
balsamroot, arrowleaf	*Balsamorhiza sagittata* (Pursh) Nutt.

Common name	Scientific name
bamboo	*Bambusa bambos* Druce
barley	*Hordeum* L. or *Critesion* Rafin.
bulbous	*H. bulbosum* L.
common	*H. vulgare* L.
foxtail	*C. jubatum* (L.) Nevski
meadow	*C. brachyantherum* (Nevski) Barkw. & D.R. Dewey
beachgrass	*Ammophila* Host
American	*A. breviligulata* Fern.
European	*A. arenaria* (L.) Link
bean	*Phaseolus* L. or *Vigna* Savi
adsuki	*V. angularis* (Willd.) Ohwi & H. Ohashi
mat (moth)	*V. aconitifolia* (Jacq.) Marechal
mung	*V. radiata* (L.) Wilcz.
rice	*V. umbellata* (Thunb.) Ohwi & H. Ohashi
tepary	*P. acutifolius* var. *latifolius* Freem.
Texas	*P. acutifolius* A. Gray
beardgrass	*Bothriochloa*
silver (silver bluestem)	*B. saccharoides* (Sw.) Rydb.
yellow (Turkestan bluestem)	*B. ischaemum* (L.) Keng.
beggarweed (tickclover)	*Desmodium* Desv.
beggarweed, Florida	*D. tortuosum* (Sw.) DC.
creeping (kaimi clover)	*D. incanum* DC.
bentgrass	*Agrostis* L.
cloud	*A. nebulosa* Boiss. & Reut.
colonial	*A. tenuis* Sibth.
creeping	*A. stolonifera* L. var. *palustris* (Huds.) Farw. (syn. *A. palustris* Huds.)
velvet	*A. canina* L.
winter	*A. hiemalis* (Walt.) B.S.P.
bermudagrass	*Cynodon dactylon* (L.) Pers.
big bluejoint (big bluestem)	*Andropogon gerardii* Vitman
birdwoodgrass	*Cenchrus setigerus* Vahl
blackberry, Allegheny	*Rubus allegheniensis* Porter
bladygrass (cogongrass)	*Imperata cylindrica* (L.) Raeusch.
blowoutgrass	*Redfieldia flexuosa* (Thurb.) Vasey
bluebonnet (Texas lupine)	*Lupinus subcarnosus* Hook.
bluegrass	*Poa* L.
annual	*P. annua* L.
big	*P. ampla* Merr.
bulbous	*P. bulbosa* L.
Canada	*P. compressa* L.
Canby	*P. canbyi* (Scribn.) Piper
inland	*P. interior* Rydb.
Kentucky	*P. pratensis* L.
mutton	*P. fendleriana* (Steud.) Vasey
Nevada	*P. nevadensis* Vasey ex Scribn.
rough	*P. trivialis* L.
Sandberg	*P. secunda* J.S. Presl
Texas	*P. arachnifera* Torr.
upland	*P. glaucantha* Gaudin
winter (bulbous)	*P. bulbosa* L.
bluejoint (bluejoint reedgrass)	*Calamagrostis canadensis* (Michx.) P. Beauv.
bluestem	*Andropogon* L., *Bothriochloa* O. Kuntze, or *Schizachyrium* Nees
big (turkey foot)	*A. gerardii* Vitman
broomsedge	*A. virginicus* L.
caucasian	*B. caucasia* (Trin.) C.E. Hubb.
cane	*B. barbinodis* (Lag.) Herter

Common name	Botanical name
creeping	*S. stoloniferum* Nash
little	*S. scoparium* (Michx.) Nash
sand	*A. gerardii* var. *paucipilus* (Nash) Fern. (syn. *A. hallii* Hack)
silver	*B. saccharoides* (Sw.) Rydb.
Turkestan	*B. ischaemum* (L.) Keng.
yellow	*B. ischaemum* (L.) Keng.
bluetop (bluejoint reedgrass)	*Calamagrostis canadensis* (Michx.) Beauv.
bristlegrass	*Setaria* Beauv.
plains	*S. leucopila* (Scribn. & Merv.) K. Schumi (syn. *S. macrostachya* H.B.K.)
broadbean	*Vicia faba* L.
bromegrass	*Bromus* L.
Bieberstein	*B. biebersteinii* Roem. & Schultes
California	*B. carinatus* Hook. & Arn.
cheatgrass	*B. tectorum* L.
chess	*B. secalinus* L.
downy (cheatgrass)	*B. tectorum* L.
field	*B. arvensis* L.
fringed	*B. ciliatus* L.
Japanese (chess)	*B. japonicus* Thunb. ex Murr.
meadow	*B. riparius* Rehm. (formerly *B. erectus* Huds.)
mountain	*B. marginatus* Nees ex Steud.
nodding	*B. anomalus* Rupr.
rescuegrass (*see* rescuegrass)	
pumpelly	*B. pumpellianus* Scribn.
red	*B. rubens* L.
ripgutgrass	*B. rigidus* Roth
smooth	*B. inermis* Leyss.
soft chess	*B. mollis* L.
broomcorn	*Sorghum bicolor* (L.) Moench
broomsedge	*Agropogon virginicus* L.
broomweed, common	*Gutierrezia dracunculoides* (DC.) Blake
brownseed paspalum	*Paspalum plicatulum* Michx.
brunswickgrass	*Paspalum nicorae* Parodi
buckbrush	*Symphoricarpos orbiculatus* Moench
buffalo clover (alyceclover)	*Alysicarpus vaginalis* (L.) DC.
buffalograss	*Buchloe dactyloides* (Nutt.) Engelm.
buffelgrass	*Cenchrus ciliaris* L.
bullnettle	*Cnidoscows texanus* (Muella-Arg.) Small
burclover	*Medicago* L.
California	*M. polymorpha* L. (syn. *M. hispida* Gaertn.)
little	*M. minima* (L.) Bartal.
spotted	*M. arabica* (L.) Huds.
burroweed	*Haplopappus tenuisectus* (Greene) Blake ex Benson
butterfly pea	*Clitoria ternatea* L.
buttonclover	*Medicago orbicularis* (L.) Bartal.
cabbage	*Brassica oleracea* L. var. *capitata* L.
Caleypea (roughpea)	*Lathyrus hirsutus* L.
calopa	*Calopogonium mucunoides* Desv.
canarygrass	*Phalaris* L.
annual	*P. canariensis* L.
bulbous	*P. aquatica* L.
littleseed	*P. minor* Retz.
reed	*P. arundinacea* L.
caribgrass	*Eriochloa polystachya* Kunth
carpetgrass	*Axonopus* Beauv.
common	*A. affinis* Chase

tropical	*A. compressus* (Sw.) Beauv.
carpon desmodium (*see* desmodium)	
catclaw acacia	*Acacia greggii* A. Gray
cedar, eastern red	*Juniperus virginiana* L.
centipedegrass	*Eremochloa ophiuroides* (Munro) Hack.
centro	*Centrosema pubescens* Benth.
centurion	*Centrosema pascuorum* Mart. ex Benth.
cheat, downy (cheatgrass)	*Bromus tectorum* L.
chess	*Bromus secalinus* L.
hairy	*B. commutatus* Schrad.
Japanese	*B. japonicus* Thunb.
ripgutgrass	*B. rigidus* Roth.
soft	*B. mollis* L.
chickpea (garbanzo)	*Cicer arietinum* L.
chickweed, common	*Stellaria media* (L.) Vill.
Chinese cabbage	*Brassica pekinensis* L.
chloris	*Chloris* Sw.
weeping	*C. distichophylla* Lag.
cicer milkvetch	*Astragalus cicer* L.
clover	*Trifolium* L.
alsike	*T. hybridum* L.
arrowleaf	*T. vesiculosum* Savi
ball	*T. nigrescens* Viv.
berseem (Egyptian)	*T. alexandrinum* L.
bigflower (Mikes)	*T. michelianum* Savi
buffalo	*T. reflexum* L.
carolina	*T. carolinianum* Michx.
cluster	*T. glomeratum* L.
crimson	*T. incarnatum* L.
cup	*T. cyathiferum* Lindl.
Egyptian (berseem)	*T. alexandrinum* L.
hollyleaf	*T. gymnocarpon* Nutt.
hop	*T. agrarium* L.
Hungarian	*T. pannonicum* Jacq.
knotted	*T. striatum* L.
kura	*T. ambiguum* Bieb.
ladino	*T. repens* L.
lappa	*T. lappaceum* L.
large hop	*T. campestre* Schreb.
longstalk	*T. longipes* Nutt. ex Torr. & A. Gray
maiden	*T. microcephalum* Pursh.
Parry	*T. parryi* A. Gray
Persian	*T. resupinatum* L.
pinhole	*T. bifidum* A. Gray
pin-point	*T. gracilentum* Torr. & A. Gray
rabbitfoot	*T. arvense* L.
red	*T. pratense* L.
rose	*T. hirtum* All.
seaside	*T. wormskioldii* Lehm.
small hop	*T. dubium* Sibth.
Spanish	*T. purshianus* (Benth.) Clem. & Clem.
strawberry	*T. fragiferum* L.
striate	*T. striatum* L.
squarehead	*T. microdon* Hook & Arnott
subterranean	*T. subterraneum* L.
tomcal	*T. microdon* Hook & Arnott
tree	*T. cilolatum* Benth.
whiproot	*T. dasyphyllum* Torr. & A. Gray

white (white dutch, common) .... *T. repens* L.
whitetip .... *T. variegatum* Nutt. ex Torr. & A. Gray
zigzag .... *T. medium* L.
cocksfoot (orchardgrass) .... *Dactylis glomerata* L.
coffee .... *Coffee arabica* L.
cogongrass (bladygrass)............. *Imperata cylindrica* (L.) Raeusch.
columbusgrass .... *Sorghum* x *almum* Parodi.
cordgrass .... *Spartina* Schreb.
California .... *S. foliosa* Trin.
prairie .... *S. pectinata* Link
spike.......................... *S. foliosa* Trin.
corn (maize)...................... *Zea mays* L.
cotton............................ *Gossypium* spp.
cottontop .... *Digitaria californica* (Benth.) Henr.
couchgrass (quackgrass) .... *Elytrigia repens* (L.) Nevski
cowpea .... *Vigna unguiculata* (L.) Walp.
crabgrass .... *Digitaria* L.
hairy .... *D. sanguinalis* (L.) Scop.
southern .... *D. ciliaris* (Retz.) Koel
creeping vigna.................... *Vigna parkeri*
creosotebush .... *Larrea tridentata* (Sesse & Mocino ex DC.) Coville
crotalaria .... *Crotalaria* L.
lance .... *C. lanceolata* E. Mey.
shak .... *C. incana* L.
showy .... *C. spectabilis* Roth
slenderleaf .... *C. brevidens* var. *intermedia* (Kotschy) Polhill
smooth .... *C. pallida* Aiton
sunn hemp .... *C. juncea* L.
crownvetch .... *Coronilla varia* L.
curly mesquite .... *Hilaria belangeri* (Steud.) Nash
dalea .... *Dalea alopecuroides* Willd.
dallisgrass .... *Paspalum dilatatum* Poir.
darnel .... *Lolium temulentum* L.
Persian .... *L. persicum* Boiss. & Hohen. ex Boiss.
deertongue .... *Panicum clandestinum* L.
deervetch, big .... *Lotus crassifolius* (Benth.) Greene
desmanthus........................ *Desmanthus virgatus* (L.) Willd.
desmodium .... *Desmodium* Desv.
carpon .... *D. heterocarpon* (L.) DC.
greenleaf .... *D. intortum* (Mill) Urb.
hetero .... *D. heterophyllum* (Willd.) DC.
silverleaf .... *D. uncinatum* (Jacq.) DC.
digitgrass (fingergrass) .... *Digiteria eriantha* Steud.
dock, curly........................ *Rumex crispus* L.
dodder .... *Cuscuta* spp.
dogfennel......................... *Eupatonium capillifolium* (Lam.) Small
dogtail, crested .... *Cynosurus cristatus* L.
dropseed .... *Sporobolus* R. Br. or *Blepharoneuron* Nash
blue; pineywoods .... *S. junceus* (Michx.) Kunth
giant.......................... *S. giganteus* Nash
meadow .... *S. asper* var. *hookeri* (Trin.) Vasey
mesa.......................... *S. flexuosus* (Thurb.) Rydb.
pine .... *B. tricholepis* (Torr.) Nash
pineywoods; blue .... *S. junceus* (Michx.) Kunth
prairie .... *S. heterolepis* (A. Gray) A. Gray
sand .... *S. cryptandrus* (Torr.) A. Gray
spike.......................... *S. contractus* tall Hitchc.
tall .... *S. asper* (Michx.) Kunth

elephantgrass (napiergrass) . . . . *Pennisetum purpureum* Schumach.
falcon-pea (flatpod peavine). . . . . *Lathyrus cicer* L.
fenugreek . . . . *Trigonella foenum-graecum* L.
fescue. . . . . *Festuca* L. or *Vulpia* C. Gmelin
  Arizona . . . . *F. arizonica* Vasey
  Chewings . . . . *F. rubra* subsp. *commutata* Gaud.
  greenleaf . . . . *F. viridula* Vasey
  hard . . . . *F. ovina* var. *duriuscula* (L.) Koch
  Idaho . . . . *F. idahoensis* Elmer
  meadow . . . . *F. pratensis* Huds.
  rattail . . . . *V. myuros* (L.) C. Gmelin
  red . . . . *F. rubra* L.
  tall . . . . *F. arundinacea* Schreb.
  sheep . . . . *F. ovina* L.
  sixweeks. . . . . *V. octoflora* (Walt.) Rydb.
fir. . . . . *Pseudotsuga* spp.
flaccidgrass . . . . *Pennisetum flaccidum* Griseb.
flatpea (Wagner pea) . . . . *Lathyrus sylvestris* L.
foxtail . . . . *Alopecurus* L. or *Setaria* Beauv.
  creeping . . . . *A. arundinaceus* Poir.
  giant . . . . *S. faberi* Herrm.
  meadow . . . . *A. pratensis* L.
galletagrass . . . . *Hilaria jamesii* (Torr.) Benth.
  big . . . . *H. rigida* (Turb.) Benth.
gamagrass, eastern . . . . *Tripsacum dactyloides* (L.) L.
gambagrass . . . . *Andropogon gayanus* Kunth
garbanzo . . . . *Cicer arietinum* L.
gardener's garters . . . . *Phalaris arundinacea* var. *picta* (L.) Asch. & Graebn.
gliricida . . . . *Gliricidia sepium* (Jacq.) Kunth ex Wap.
grama . . . . *Bouteloua* Lag.
  black . . . . *B. eriopoda* (Torr.) Torr.
  blue . . . . *B. gracilis* (H.B.K.) Lag. ex Steud.
  hairy. . . . . *B. hirsuta* Lag.
  Rothrock. . . . . *B. rothrockii* Vasey
  sideoats . . . . *B. curtipendula* (Michx.) Torr.
  slender . . . . *B. filiformis* (Fourn.) Griffiths
grape . . . . *Vitis* spp.
grasses . . . . Gramineae or Poaceae
grasspea . . . . *Lathyrus sativus* L.
graythorn. . . . . *Condalia lycioides* (A. Gray) Weberb.
greasewood . . . . *Sarcobatus vermiculatus* (Hook.) Torr.
green sprangletop . . . . *Leptochloa dubia* (H.B.K.) Nees
guar . . . . *Cyamopsis tetragonoloba* (L.) Taub.
guava . . . . *Psidium guajava* L.
guineagrass . . . . *Panicum maximum* Jacq.
hairgrass . . . . *Deschampsia* Beauv.
  Bering . . . . *D. beringensis* Hult.
  tufted . . . . *D. caespitosa* (L.) Beauv.
halogeton. . . . . *Halogeton glomeratus* (Bieb.) C.A. Mey.
hardinggrass . . . . *Phalaris stenoptera* Hack.
herdgrass (timothy). . . . . *Phleum pratense* L.
hilograss . . . . *Paspalum conjugatum* Bergius
hindigrass (shedagrass,
  Kleberg bluestem) . . . . *Dichanthium annulatum* (Forssk.) Stapf
hop. . . . . *Humulus lupulus* L.
hopsage . . . . *Grayia* spp.
horsebean . . . . *Vicia faba* L.
horse gram, perennial (axillaris). . . . . *Macrotyloma axillare* (E. Mey.) Verdc.

horsenettle. . . . . *Solanum carolinense* L.
hyacinth bean . . . . *Lablab purpureus* (L.) Sweet
hymenachne . . . . *Hymenachne amplexicaulis* (Rudge) Nees
indiangrass . . . . *Sorghastrum nutans* (L.) Nash
indigo . . . . *Indigofera* L.
  hairy . . . . *I. hirsuta* L.
ipil-ipil (leucaena) . . . . *Leucaena leucocephala* (Lamb.) de Wit.
ironweed, western . . . . *Vernonia baldwinii* Torr.
Japanese millet . . . . *Echinochloa frumentacea* Link
jaragua grass . . . . *Hyparrhenia rufa* (Nees) Stapf
jimsonweed . . . . *Datura stramonium* L.
johnsongrass . . . . *Sorghum halepense* (L.) Pers.
jointvetch . . . . *Aeschynomene* L.
  American . . . . *A. americana* L.
  Australian . . . . *A. falcata* (Poir.) DC.
junegrass (prairie junegrass) . . . . *Koeleria macrantha* (Ledeb.) Schultes (formerly called *K. cristata)*
kaimi clover (creeping beggarweed) . . . . *Desmodium incanum* DC.
kale . . . . *Brassica oleracea* L.
kidneyvetch . . . . *Anthyllis vulneraria* L.
kikuyugrass . . . . *Pennisetum clandestinum* Hochst. ex Chiov.
Kleberg bluestem. . . . . *Dichanthium annulatum* (Forssk.) Stapf
kleingrass . . . . *Panicum coloratum* L.
knotgrass . . . . *Paspalum distichum* L.
koa . . . . *Acacia koa* A. Gray
koa haole (leucaena) . . . . *Leucaena leucocephala* (Lam.) de Wit
kohlrabi . . . . *Brassica oleracea* var. *gongyloides* L.
koleagrass . . . . *Phalaris tuberosa* var. *hirtiglumis* Batt. & Trab.
koroniviagrass (creeping signalgrass) . . . . *Brachiaria humidicola* (Rendle) Schweick.
kudzu. . . . . *Pueraria lobata* (Willd.) Ohwi
  tropical (puero). . . . . *P. phaseoloides* (Roxb.) Benth.
lablab bean . . . . *Lablab purpureus* (L.) Sweet
lambsquarter, common . . . . *Chenopodium album* L.
lawngrass, Japanese . . . . *Zoysia japonica* Steud.
legumes . . . . Leguminosae or Fabaceae
lentil, common . . . . *Lens culinaris* Medikus (syn. *L. esculenta* Moench)
lespedeza . . . . *Kummerowia* or *Lespedeza*
  annual; korean. . . . . *K. stipulacea* (Maxim.) Makino
  annual; striate (common). . . . . *K. striata* (Thunb.) Schindler
  sericea . . . . *L. cuneata* (Dum.-Cours.) G. Don
leucaena . . . . *Leucaena leucocephala* (Lam.) de Wit
limpograss . . . . *Hemarthria altissima* (Poir.) Stapf & C.E. Hubb.
llanos macro . . . . *Macroptilium longepedunculatum* (Benth.) Urb.
lotononis . . . . *Lotononis bainesii* Baker
lovegrass . . . . *Eragrostis* Wolf
  Boer . . . . *E. curvula* (Schrad.) Nees var. *conferta* Nees (formerly *E. chloromelas* Steud.)
  Lehmann . . . . *E. lehmanniana* Nees
  sand . . . . *E. trichodes* (Nutt.) Wood
  weeping . . . . *E. curvula* (Schrad.) Nees var. *curvula* Nees
  Wilman. . . . . *E. superba* Peyr.
lucerne (alfalfa) . . . . *Medicago sativa* L.
lupine . . . . *Lupinus* L.
  blue . . . . *L. angustifolius* L.
  Nootka . . . . *L. nootkatensis* Donn ex Sims
  sundial . . . . *L. perennis* L.

Texas (bluebonnet) .... *L. subcarnosus* Hook.
white .... *L. albus* L.
yellow .... *L. luteus* L.
maize; corn .... *Zea mays* L.
makarikarigrass .... *Panicum coloratum* L.
manilagrass .... *Zoysia matrella* (L.) Merr.
mannagrass .... *Glyceria* R. Br.
marshelder .... *Iva xanthifolia* Nutt.
marvelgrass .... *Dichanthium annulatum* (Forssk.) Stapf
mascarenegrass................... *Zoysia tenuifolia* Willd. ex Thiele
maygrass......................... *Phalaris caroliniana* Walt.
meadow grass, salt (Nuttall
alkaligrass) .... *Puccinellia nuttalliana* Hitchc.
medic............................ *Medicago* L.
black .... *M. lupulina* L.
snail .... *M. scutellata* (L.) Mill.
mesquite .... *Prosopis juliflora* (Sw.) DC.
algaroba......................... *P. pallida* (Willd.) Kunth
honey .... *P. glandulosa* Torr. var. *glandulosa*
velvet .... *P. velutina* Woot.
milkvetch .... *Astragalus* L. or *Oxytropis* DC.
cicer .... *A. cicer* L.
ruby .... *O. riparia* Litv.
sicklepod (sickle) .... *A. falcatus* Lam.
millet
broom or broomcorn............... *Panicum miliaceum* L.
browntop .... *Panicum ramosum* L.
(formerly *Brachiaria ramosa* [L.] Stapf)
foxtail (Italian)................ *Setaria italica* (L.) Beauv.
Japanese .... *Echinochloa crusgalli* var. *frumentacea* (Link)
W. F. Wight (formerly *Echinochloa frumentacea*
Link)
pearlmillet...................... *Pennisetum americanum* (L.) Leeke
(formerly *Pennisetum glaucum* [L.] R.Br.)
proso............................ *Panicum miliaceum* L.
proso (hog) .... *Panicum miliaceum* L.
mitchellgrass .... *Astrebla pectinata* (Lindl.) F. Muell.
molassesgrass .... *Melinis minutiflora* Beauv.
Mormon tea....................... *Ephedra* spp.
muhly .... *Muhlenbergia* Schreb.
bush .... *M. porteri* Scribn.
mountain .... *M. montana* (Nutt.) Hitch.
sandhill .... *M. pungens* Thurb.
spike............................ *M. wrightii* Vasey
Nadi bluegrass .... *Dichanthium cariscosum* (L.) A. Camus
napiergrass (elephantgrass) ..... *Pennisetum purpureum* Schumach.
natalgrass .... *Rhynchelytrum repens* (Willd.) C. E. Hubb.
needle-and-thread .... *Stipa comata* Trin. & Rupr.
needlegrass .... *Stipa* L.
California .... *S. californica* Merr. & Davy
green .... *S. viridula* Trin.
needlegrass (needle-and-thread) .... *S. comata* Trin. & Rupr.
Texas wintergrass .... *S. leucotricha* Trin. & Rupr.
nimblewill .... *Muhlenbergia schreberi* J.F. Gmel.
nutsedge, yellow .... *Cyperus esculentus* L.
oatgrass......................... *Arrhenatherum* Beauv. or *Danthonia* Lam. & DC.
bulbous .... *A. elatius* var. *bulbosum* (Willd.) Spenner
California .... *D. californica* Boland.

tall . . . . *A. elatius* (L.) J. S. & C. Presl
oat . . . . *Avena* L.
animated . . . . *A. sterilis* L.
cultivated . . . . *A. sativa* L. (syn. *A. bysantina* K. Koch)
slender . . . . *A. barbata* Pott ex Link
wild . . . . *A. fatua* L.
Ohia (Ohia lehua) . . . . *Metrosideros collina* A. Gray
Old World bluestems . . . . *Bothriochloa* spp.
orchardgrass (cocksfoot) . . . . *Dactylis glomerata* L.
Oriental pennisetum . . . . *Pennisetum orientale* L. Rich
palisadegrass . . . . *Brachiaria brizantha* (A. Rich.) Stapf
pampasgrass . . . . *Cortaderia selloana* (Schultes) Asch. & Graebn.
pangolagrass . . . . *Digitaria eriantha* Steud.
(formerly *D. decumbens* Stent)
panic, blue . . . . *Panicum antidotale* Retz.
panicgrass . . . . *Panicum* L.
blue . . . . *P. antidotale* Retz.
green . . . . *P. maximum* Jacq. var. *trichoglume*
paragrass . . . . *Brachiaria mutica* (Forssk.) Stapf
partridge pea . . . . *Chamaecrista fasciculata* (Michx.) E. Greene
paspalum . . . . *Paspalum* L.
brownseed . . . . *P. plicatulum* Michx.
field . . . . *P. laeve* Michx.
knotgrass . . . . *P. distichum* L.
ribbed . . . . *P. malacophyllum* Trin.
sand . . . . *P. stramineum* Nash
pea . . . . *Lathyrus* L. or *Pisum* L.
beach . . . . *L. maritimus* (L.) Bigel.
Caley . . . . *L. hirsutus* L.
field . . . . *P. sativum* L. subsp. *sativum* var. *arvense* (L.) Poir.
rough . . . . *L. hirsutus* L.
singletary . . . . *L. hirsutus* L.
Tangier . . . . *L. tingitanus* L.
peanut . . . . *Arachis hypogaea* L.
perennial . . . . *A. glabrata* Benth.
pinto . . . . *A. pintoi* Krap. & Greg., nom. nud.
rhizoma . . . . *A. glabrata* Benth.
pearlmillet . . . . *Pennisetum americanum* (L.) Leeke
(formerly *P. glaucum* [L.] R. Br.)
peavine . . . . *Lathyrus* L.
flatpod . . . . *L. cicer* L.
Tangier (Tangier pea) . . . . *L. tingitanus* L.
persimmon . . . . *Diospyros virginiana* L.
phasey bean . . . . *Macroptilium lathyroides* (L.) Urb.
pigeonpea . . . . *Cajanus cajan* (L.) Millsp.
pigweed, redroot . . . . *Amaranthus retroflexus* L.
pinegrass . . . . *Calamagrostis rubescens* Buckl.
pinto peanut . . . . *Arachis pintoi* Krap. & Greg., nom. nud.
polargrass . . . . *Arctagrostis arundinacea* (Trin.) Beal
porcupinegrass . . . . *Stipa spartea* Trin.
potato . . . . *Solanum tuberosum* L.
povertygrass . . . . *Danthonia spicata* (L.) Beauv.
pumpkin . . . . *Curcurbita* species
purpletop . . . . *Tridens flavus* (L.) Hitchc.
quackgrass (couchgrass) . . . . *Elytrigia repens* (L.) Nevski
(formerly *Agropyron repens* [L.] Beauv.)
quakinggrass . . . . *Briza* L.
big . . . . *B. maxima* L.

little	*B. minor* L.
ragweed	*Ambrosia* L.
common	*A. artemisiifolia* L.
lanceleaf	*A. bidentata* Michx.
western	*A. psilostachya* DC.
rape (winter rape)	*Brassica napus* L.
rattailgrass (smutgrass)	*Sporobolus africanus* (Poir.) Robyns & Tourn.
rattle-box	*Crotalaria* spp.
redtop	*Agrostis gigantea* Roth (syn. *A. alba* L.)
reed, common	*Phragmites australis* (Cav.) Steudel
reed canarygrass	*Phalaris arundinacea* L.
reedgrass	*Calamagrostis* Adans.
bluejoint	*C. canadensis* (Michx.) Beauv.
rescuegrass	*Bromus unioloides* (Willd.) H.B.K. (syn. *B. catharticus* Vahl. and *B. willdenowii* Kunth)
rhizoma peanut	*Arachis glabrata* Benth.
rhodesgrass	*Chloris gayana* Kunth
ribbongrass (gardener's garters)	*Phalaris arundinacea* var. *picta* (L.) Asch. & Graebn.
rice	*Oryza sativa* L.
ricegrass	*Oryzopsis* Michx. or x *Stiporyzopsis* B. L. Johnson & Rogler
bloomer	*O. bloomeri* (Bolander) Ricker
indian	*O. hymenoides* (Roem. & Schult.) Ricker
Mandan	*S. caduca* (Beal) B.L. Johnson & Rogler
ripgutgrass	*Bromus rigidus* Roth
rivergrass	*Scolochloa festucacea* (Willd.) Link
ronphagrass	*Phalaris aquatica* L. *(P. tuberosii* L.) x *P. arundinacea* L.
rose, multiflora	*Rosa multiflora* Thunb. ex Murr.
roughpea (caleypea; singletarypea)	*Lathyrus hirsutus* L.
roundleaf cassia	*Chamaecrista rotundifolia* (Pers.) Greene
rushes	*Juncus* spp.; *Eleocharis* spp.
rushgrass, longleaf (tall dropseed)	*Sporobolus asper* (Michx.) Kunth
ruzigrass	*Brachiaria ruziziensis* Germ. & Evrard
rye	*Secale cereale* L.
ryegrass	*Lolium* L.
annual (Italian)	*L. multiflorum* Lam.
dalmatian	*L. subulatum* Vis.
flax	*L. remotum* Schrank
Italian (annual)	*L. multiflorum* Lam.
perennial	*L. perenne* L.
Wimmera	*L. rigidum* Gaud.
sacaton	*Sporobolus* Munro ex Scribn.
alkali	*S. airoides* (Torr.) Torr.
big	*S. wrightii* Munro ex. Scribn.
sagebrush	*Artemisia* L.
big	*A. tridentata* Nutt.
black	*A. nova* A. Nels.
sand	*A. filifolia* Torr.
threetip	*A. tripartita* Rydb.
sainfoin	*Onobrychis* Mill.
common	*O. viciifolia* Scop.
Russian	*O. transcaucasica* Grossh.
Siberian	*O. arenaria* (Kit.) DC.
St. Augustine grass	*Stenotaphrum secundatum* (Walt.) Kuntze
saltbush	*Atriplex* spp.

saltgrass, inland .... *Distichlis spicata* subsp. *stricta* (Torr.) Thorne
sandreed .... *Calamovilfa* Hack.
  big .... *C. gigantea* (Nutt.) Scribn. & Merr.
  prairie .... *C. longifolia* (Hook.) Scribn.
scrobicgrass .... *Paspalum scrobiculatum* L.
sea-oats .... *Uniola paniculata* L.
sedge, threadleaf.................. *Carex filifolia* Nutt.
senna, sickle .... *Cassia tora* L.
serradella .... *Ornithopus sativus* Brot.
sesbania........................... *Sesbania sesban* (L.) Merr.
  hemp sesbania .... *S. exaltata* (Raf.) Rybd. ex A.W. Hill
setaria (golden timothy) .... *Setaria sphacelata* (Schum.) Stapf & C. E. Hubb. ex M.B. Moss
shadscale (shadscale saltbush) .... *Atriplex confertifolia* (Torr. & Frem.) S. Wats.
shedagrass (hindigrass, Kleberg bluestem) .... *Dichanthium annulatum* (Forssk.) Stapf
signalgrass .... *Brachiaria decumbens* Stapf
  creeping (koroniviagrass) .... *B. humidicola* (Rendle) Schweick.
siratro .... *Macroptilium atropurpureum* (Mocino & Sesse ex DC.) Urb.
sleepygrass .... *Stipa robusta* (Vasey) Scribn.
sloughgrass, American .... *Beckmannia syzigachne* (Steud.) Fernald
smilograss .... *Oryzopsis miliacea* (L.) Asch. & Schweinf.
sneezeweed, bitter................. *Helenium omarum* (Raf.) H. Rock
sorghum .... *Sorghum bicolor* (L.) Moench
sourclover .... *Melilotus indica* (L.) All.
soybean .... *Glycine max* (L.) Merr.
soybean, perennial (glycine) .... *Neonotonia wightii* (Wight & Arn.) Lackey
Spanish clover .... *Lotus purshianus* (Benth.) Clem. & Clem.
speargrass .... *Heteropogon contortus* (L.) Beauv. ex Roem. & Schult.
sprangletop, green .... *Leptochloa dubia* (Kunth) Nees
spruce .... *Picea* spp.
squirreltailgrass .... *Elymus elymoides* (Rafin.) Swezey (formerly *Sitanion hystrix* [Nutt.] J.G. Smith)
staghorn fern...................... *Nephrolepsis exaltata* Schott
stargrass .... *Cynodon nlemfuensis* Vanderyst
  giant............................ *C. aethiopicus* Clayton & Harlan
  Naivasha .... *C. plectostachyus* (K. Schum.) Pilger
stipa (needlegrass) .... *Stipa* L.
stylo .... *Stylosanthes guianensis* (Aubl.) Sw.
  Caribbean stylo .... *S. hamata* (L.) Taub.
  shrubby stylo .... *S. scabra* (Vog.)
  townsville stylo .... *S. humilis* Kunth
sudangrass .... *Sorghum bicolor* (L.) Moench (formerly *S. sudanense* [Piper] Stapf) (syn. *Sorghum* x *drummondii* [Steudel] Millsp. & Chase)
sugarbeet.......................... *Beta vulgaris* L.
sugarcane .... *Saccharum officinarum* L.
sulla (sulla sweetvetch)............ *Hedysarum coronarium* L.
sumac .... *Rhus* spp.
sunolgrass .... *Phalaris coerulescens* Desf.
sweetclover .... *Melilotus* Mill.
  annual yellow.................... *M. indica* (L.) All.
  Banat .... *M. dentata* (Waldst. & Kit.) Pers.
  Daghestan .... *M. suaveolens* Ledeb.
  Israel (hubam) .... *M. alba* var. *annua* Coe
  white .... *M. alba* Medik.

yellow . . . . *M. officinalis* Lam.
switchgrass . . . . *Panicum virgatum* L.
tanglehead. . . . . *Heterogon contortus* (L.) Beauv. ex Roem. & Schult.
tansymustard . . . . *Descurainia pinnata* (Walt.) Britt.
teosinte . . . . *Zea mays* subsp. *mexicana* (Schrad.) Iltis
thistle . . . . *Cirsium* L.
Canada. . . . . *C. arvense* (L.) Scop.
musk . . . . *C. nutans* L.
three-awn . . . . *Aristida* L.
pineland. . . . . *A. stricta* Michx.
purple. . . . . *A. purpurea* Nutt.
red . . . . *A. longiseta* Steud.
tickclover . . . . *Desmodium* Desv.
tall . . . . *D. tortuosum* (Sw.) DC.
ticklegrass (winter bentgrass). . . . . *Agrostis hiemalis* (Walt.) B.S.P.
timothy . . . . *Phleum pratense* L.
alpine . . . . *P. alpinum* L.
turf . . . . *P. bertolonii* DC.
tobosagrass . . . . *Hilaria mutica* (Buckl.) Benth.
tomato . . . . *Lycopersicon esculentum* L.
torpedograss . . . . *Panicum repens* L.
trefoil. . . . . *Lotus* L.
big . . . . *L. uliginosus* Schkuhr. (syn. *L. pedunculatus* Cav.)
birdsfoot . . . . *L. corniculatus* L.
narrowleaf birdsfoot . . . . *L. tenuis* Waldst. & Kit. ex Willd.
trisetum, yellow . . . . *Trisetum flavescens* (L.) Beauv.
triticale . . . . X *Triticosecale* Wittmack
tunisgrass . . . . *Sorghum arundinaceum* (Desv.) Stapf (formerly *S. virgatum* [Hack.] Stapf)
turkeyfoot (big bluestem) . . . . *Andropogon gerardii* Vitman
turnip . . . . *Brassica rapa* L.
uniola, broadleaf . . . . *Chasmanthium latifolium* (Michx.) Yates
urochloagrass . . . . *Urochloa mosambicensis* (Hack.) Dandy
vaseygrass . . . . *Paspalum urvillei* Steud.
veldtgrass . . . . *Ehrharta calycina* Smith
velvetbean . . . . *Mucuna* Adans.
Florida . . . . *M. pruriens* (L.) DC. var. *utilis* (Wight) (formerly *Stizolobium decringianum* Bart.)
velvetgrass. . . . . *Holcus* L.
common . . . . *H. lanatus* L.
German . . . . *H. mollis* L.
velvetleaf. . . . . *Abutilon theophrasti* Medicus
vernalgrass, sweet. . . . . *Anthoxanthum odoratum* L.
vetch . . . . *Vicia* L.
bard . . . . *V. monantha* Retz.
bigflower . . . . *V. grandiflora* Scop.
bird. . . . . *V. cracca* L.
bitter . . . . *V. ervilia* (L.) Willd.
common . . . . *V. sativa* L.
cordateleaf, common . . . . *V. sativa* subsp. *cordata* (Wulfen ex Hoppe) Asch. & Graebn.
grandiflora (bigflower) . . . . *V. grandiflora* Scop.
hairy. . . . . *V. villosa* Roth
Hungarian . . . . *V. pannonica* Crantz
narrowleaf . . . . *V. angustifolia* L.
purple. . . . . *V. benghalensis* L.
showy . . . . *V. grandiflora* Scop.

single-flowered . . . . *V. articulata* Hornem.
winter. . . . . *V. villosa* Roth
vine mesquitegrass . . . . *Panicum obtusum* H.B.K.
wheat. . . . . *Triticum aestivum* L. emend. Thell.
durum . . . . *T. turgidum* L.
einkorn. . . . . *T. monococcum* L.
emmer . . . . *T. turgidum* L.
goatgrass . . . . *T. tauschii* (Coss.) Schmal. (syn. *Aegilops tauschii* Coss.)
wheatgrass . . . . *Agropyron* Gaertn., *Elymus* L., *Elytrigia* Desv., *Pseudoroegneria* (Nevski) A. Love, *Pascopyrum,* or *Thinopyrum* A. Love
arctic . . . . *Elymus macrourus* (Turcz.) Tzvelev
bearded . . . . *Elymus trachycaulus* subsp. *subsecundus* (Link) Gould
bluebunch. . . . . *Pseudoroegneria spicata* (Pursh) A. Love (formerly *Elytrigia spicata* [Pursh] D.R Dewey)
fairway crested. . . . . *Agropy ron cristatum* (L.) Gaertn.
intermediate . . . . *Thinopyrum intermedium* (Host) Barkworth & D. R. Dewey (formerly *Elytrigia intermedia* [Host] Nevski)
pubescent. . . . . *Thinopyrum intermedium* subsp. *barbulatum* (Schur) Barkworth & D.R. Dewey (formerly *Elytrigia trichophora* [Link] Nevski)
Siberian crested. . . . . *Agropyron fragile* (Roth) Candargy
slender . . . . *Elymus trachycaulus* (Link) Gould ex Shinners subsp. *trachycaulus*
standard crested . . . . *Agropyron desertorum* (Fisch. ex Link) Schultes
tall . . . . *Thinopyrum ponticum* (Podp.) Barkworth & D.R. Dewey (formerly *Elytrigia elongata* [Host] Nevski)
thickspike . . . . *Elymus lanceolatus* (Scribn. & Smith) Gould (formerly *Elytrigia dasystachya* [Hook.] A. & D. Love)
western . . . . *Pascopyrum smithii* (Rydb.) A. Love (formerly *Elymus smithii* [Rydb.] Gould)
wildbean, trailing . . . . *Strophostyles helvola* (L.) Ell.
wildrice, cultivated . . . . *Zizania palustris* L.
wildrye . . . . *Elymus* L., *Leymus* Hochst., or *Psathyrostachys* Nevski
Altai . . . . *L. angustus* (Trin.) Pilger
beardless . . . . *L. triticoides* (Buckl.) Pilger
blue . . . . *E. glaucus* Buckl.
Canada. . . . . *E. canadensis* L.
dune . . . . *L. mollis* (Trin.) Pilger subsp. *mollis*
giant . . . . *L. condensatus* (Presl) A. Love
Great Basin . . . . *L. cinereus* (Scribn. & Merr.) A. Love
Russian . . . . *P. juncea* (Fisch.) Nevski
Siberian . . . . *E. sibiricus* L.
willow . . . . *Salix* spp.
winterfat . . . . *Ceratoides lanata* (Pursh) J. Howell
wiregrass (three-awn) . . . . *Aristida* spp.
woollyfinger. . . . . *Digitaria* spp.
Yorkshire fog . . . . *Holcus lanatus* L.
zornia. . . . . *Zornia latifolia*
zoysia. . . . . *Zoysia japonica* Steud.

REFERENCES

Barkworth, ME, and DR Dewey. 1985. Genomically based genera of the perennial Triticeae of North America: Identification and membership. Amer. J. Bot. 72:767-76.

Barnes, RF, and JB Beard. 1992. Glossary of Crop Science Terms. Madison, Wis: Crop Science Society of America.

Bayer, AG, Agrichemicals Division. 1986. Important Crops of the World and Their Weeds. Leverkusen, Germany: Bayer AG.

Cronquist, A. 1981. An Intergrated System of Classification of Flowering Plants. New York: Columbia Univ. Press.

Encke, F, G Buchheim, and S Seybold. 1984. Zander-Handworterbuch der Pflanzennamen. 13th ed. Stuttgart, Germany: Eugen Ulmer.

Hitchcock, AS (rev. by A Chase). 1951. Manual of the Grasses of the United States. USDA Misc. Publ. 200. Reprinted by Dover Press.

LH Bailey Hortorium Staff. 1976. Hortus Third. New York: Macmillan.

Patterson, DT (chairman). 1984. Composite list of weeds. Weed Sci. 32, Suppl. 2:1-137.

Terrell, EE, SR Hill, JH Wiersema, and WE Rice. 1986. A Checklist of Names of 3000 Vascular Plants of Economic Importance. USDA Agric. Handb., rev.

Voss, EG, et al. (eds.). 1983. International Code of Botanical Nomenclature. Regnum Vegetabile 111. Utrecht, Netherlands: Bohn, Scheltema, and Holkema.

Weed Science Society of America. 1989. Composite List of Weeds. Champaign, Ill.: Weed Science Society of America.

GLOSSARY

aberrant Different from the normal type of species, genus, or higher group in one or more characters, but not readily assignable to another group.
abiotic Nonliving components of the environment, such as water, solar radiation, oxygen, organic compounds, and soil nutrients.
abomasum The fourth compartment of the ruminant stomach, comprising the true stomach, in which occur digestive processes similar to those found in the nonruminant stomach.
acceptability Readiness with which animals select and ingest a forage; sometimes used interchangeably to mean either palatability or voluntary intake.
acid detergent fiber (ADF) Insoluble residue following extraction of herbage with acid detergent (van Soest); cell wall constituents minus hemicellulose.
acid detergent fiber digestibility The digestibility of acid detergent fiber (ADF) of a forage, as determined by the percentage decrease in ADF measured before and after in vitro or in vivo digestion.
acid detergent lignin (ADL) Lignin in the residue determined following extraction with acid detergent.
acid pepsin Used in second stage of in vitro forage digestion, 2 g of 0.1 g kg^{-1} pepsin in 1 L of 0.1 M HCl.
ad libitum feeding Daily feed offerings that allow free-will consumption, generally fed to have a daily excess of 15% of feed remaining.
aerobic Pertaining to life or processes occurring in free oxygen or in oxygen concentrations normal in air (21% O_2). *See also* **anaerobic.**
aflatoxin $C_{17}H_{10}O_6$ A polynuclear substance derived from molds; a known carcinogen. It is produced by a fungus occurring on peanuts, corn, and other plants, especially seeds.
aftermath Residue and/or regrowth of plants (forage) used for grazing after harvesting of a crop.
agroforestry Land use system in which woody perennials are grown for wood production in association with agricultural crops, with or without animal production, or other commercial enterprises.
agro-silvo-pastoral Land use system in which woody perennials are grown with agricultural crops, forage crops, and livestock production.
agrostology Study of grasses; their classification, management, and utilization.
air-dry weight The weight of a substance after it has been allowed to dry to equilibrium in the ambient atmosphere.
alkaloid One of a class of basic organic compounds with nitrogen in their structure; a secondary product of plant metabolism. An example is perloline, produced by tall fescue.
allele Any of a group of possible mutational forms of a gene.
allelopathy The positive or negative influence of one living plant upon another due to secretion of chemical substances. *See* **autotoxicity.**
alternate stocking The repeated grazing and resting of forage by using two paddocks in succession.
ambient temperature Air temperature at a given time; not radiant temperature.
amino acid Organic acid containing one or more amino groups (NH_2) and at least one carboxyl group (COOH). Some amino acids such as cystine and methionine contain sulfur. Many amino acids linked together in a definite pattern form a molecule of protein.
amylopectin The form of starch in which branching occurs through alpha 1-6 linkages from an amylose backbone. Amylopectins are more easily dissolved and digested than amylose. The contents of amylose and amylopectins in seed grains are genetically controlled.
amylose The form of starch in which linkages are exclusively alpha 1-4. While representative formulas usually are presented in a linear fashion, the molecules actually have a spiral form.
anaerobic Living in the absence of free oxygen; the opposite of aerobic.
animal day One day's tenure upon pasture by one animal. Not synonymous with animal unit day.

animal days per hectare Unit to express total tenure of animals upon a unit of pasture. *Usage:* Typically expressed in terms of a longer time period: e.g., animal d ha^{-1} yr^{-1}.

animal month One month's tenure upon pasture by one animal. *Usage:* Not synonymous with animal unit month.

animal performance Production per animal (weight change or animal products) per unit of time.

animal unit One mature non-lactating cow weighing 500 kg and fed at maintenance level, or the equivalent, expressed as $(weight)^{0.75}$, in other kinds or classes of animals.

animal unit day The amount of dry forage consumed by one animal unit per 24-h period. *Usage:* The term may be extrapolated to other time periods, such as a week, month, or year (e.g., animal unit month).

animal unit month (AUM) The amount of feed or forage required by an animal unit for 1 mo; tenure of one animal unit for a period of 1 mo. Not synonymous with animal month.

anoxia Oxygen deficiency.

anthesis Stage in floral development when pollen is shed.

antiquality constituents Constituents that have negative effects on forage intake or that produce negative responses in animals consuming the produce containing the constituent.

apical dominance Inhibiting effect of a terminal bud upon the development of lateral buds.

apomixis Formation of viable embryos without actual union of male and female gametes, as in Kentucky bluegrass.

apparent dry matter digestibility. *See* **digestibility, apparent.**

ash The residue remaining after complete burning of combustible matter; consists mainly of minerals in oxidized form.

atomic absorption spectroscopy Observation by means of an optical device (spectroscope) of the wavelength and intensity of electromagnetic radiation (light) absorbed by various materials. Particular elements absorb well-defined wavelengths on an atomic level. The wavelengths absorbed are in the visible and infrared regions. Theoretical interpretation of the resulting spectra leads to knowledge of atomic and molecular structure.

autotoxicity A specific type of allelopathy where the presence of adult plants interferes with the germination and development of its own seedlings.

available forage That portion of the forage, expressed as weight of forage per unit land area, that is accessible for consumption by a specified kind, class, sex, size, age, and physiological status of grazing animal.

axillary bud Meristematic apex located in the junction of the leaf and stem; gives rise to tillers in grasses and to branches and flowers in dicots.

backgrounding Intensive management of young cattle, postweaning, using forages to facilitate maximum performance before animals are moved to feedlot for finishing using more concentrates.

bioassay The use of living organisms to quantitatively estimate the amount of biologically active substances present in a sample.

biomass The weight of living organisms (plants and animals) in an ecosystem at a given point in time, expressed either as fresh or dry weight.

bloat Excessive accumulation of gases in the rumen of animals because loss through the esophagus is impaired, causing distension of the rumen.

bloom, early Initial flowering (anthesis) in the uppermost portion of the inflorescence.

bloom, full Essentially all florets in the inflorescence in anthesis.

blooming Refers to anthesis in the grass family, or to the period during which florets are open and anthers are extended.

body weight, empty Conceptually, weight of an animal when the alimentary tract is empty; equal to live weight minus gut contents. (Calculation often used to determine gain in short time period where differences in weight of gut contents may exist.)

body weight, shrunk Body weight after a period of fast (no feed and/or water, usually overnight or for 24 h) to reduce variation in gut-fill contribution to body weight.

bolus The mass of food prepared by the mouth of ruminants for swallowing. The bolus retains some structure and is regurgitated for mastication.

bomb calorimetry Process whereby a substance is completely oxidized in 25 to 30 atmospheres of oxygen to determine gross energy (GE) content based on heat released.

boot stage Growth stage when a grass inflorescence is enclosed by the sheath of the uppermost leaf.

bract A modified or reduced leaf subtending a flower or inflorescence.

bran Pericarp of cereal grain; coproduct from converting endosperm to flour.

brown midrib In maize (br) and sorghum (bmr), a single recessive gene character resulting in the dark brown coloration of the back side of the leaf midrib and under the leaf sheaths; associated with reduced lignin content of the plant.

browning Refers to the reaction between reducing sugars and free amino groups in proteins to form a complex that undergoes a series of reactions to produce brown polymers. Higher temperatures and basic pH favor the reaction. Process renders product less digestible.

browning reaction *See* **browning.**

browse (1) *n.* Leaf and twig growth of shrubs, woody vines, trees, cacti, and other nonherbaceous vegetation available for animal consumption. (2) *v.* To browse; the consumption of browse in situ by animals.

bunch-type growth habit Plant development, especially grasses, where the new tillers emerge virtually along the stem while remaining enclosed in the sheath; tillering at or near the soil surface without production of rhizomes or stolons.

bundle sheath A sheath of one or more layers of parenchymatous or of sclerenchymatous cells surrounding a vascular bundle in the leaf.

by-product Something produced in addition to the principal product.

C_3-plant A plant employing ribulose bisphosphate carboxylase as the primary CO_2-capturing enzyme, with the first product being a 3-carbon acid, also displays photorespiration.

C_4-plant A plant employing phosphoenolpyruvate carboxylase as the primary CO_2-capturing enzyme, with the first product being a 4-carbon acid, does not display photorespiration.

calorie (gram calorie) The amount of heat required to raise the temperature of 1g of water one degree Celsius. One kilocalorie (kcal) = 1,000 calories; one megacalorie (Mcal) = 1,000,000 calories.

cannula A tubular device inserted into a body cavity, duct, or vessel (e.g., esophagus or rumen); mainly used to divert digesta or to allow sampling of digesta of animals.

canopy The aerial portion of plants in their natural growth position; usually expressed as percent of ground so occupied or as leaf area index.

carbohydrate Compound of carbon, hydrogen, and oxygen in the ratio of CH_2O, as in sugar, starch, and cellulose.

carbohydrates, nonstructural Soluble carbohydrates found in the cell contents, as contrasted with structural carbohydrates in the cell walls. Assumed to be available to support life processes.

carbohydrates, structural Carbohydrates found in the cell walls (e.g., hemicellulose, cellulose); assumed to not be available to support life processes.

carrying capacity The maximum stocking rate, i.e., animals ha^{-1}, that will achieve a target level of animal performance, in a specified grazing method, that can be applied over a defined period of time without deterioration of the ecosystem. *Usage:* Carrying capacity is not static from season to season or from year to year and may be defined over fractional parts of years. *Average carrying capacity* refers to a long-term carrying capacity averaged over years; *annual carrying capacity* refers to a specific year. *Synonym:* grazing capacity.

caryopsis Small, one-seeded, dry fruit with a thin pericarp surrounding and adhering to the seed; the seed (grain) or fruit of grasses.

cecum Intestinal pouch located at the junction of large and small intestines of nonruminants. Functions somewhat similar to a rumen. Usually it is much larger in the herbivorous horse than in the nonherbivorous monogastrics.

cell wall constituents Compounds that make up or constitute the cell wall, including cellulose, hemicellulose, lignin, and minerals (ash).

cell wall content The proportion of plant material made up of cell walls as opposed to cell contents, usually determined by solubility differential.

cellulase Enzymes that digest cellulose to hexose units.

cellulose A carbohydrate formed from glucose that is linked by beta 1, 4 bonds, a major constituent of plant cell walls. A colorless solid; insoluble in water.

chasmogamy Opening of a mature flower in the normal way to ensure pollination and fertilization, either self- or cross-pollinated.

chemostatic A theory for regulation of feed intake based on blood levels of components that signal the hypothalamus gland.

chilling injury Temporary reduction in photosynthesis and plant growth of sensitive plants following exposure to temperature just above freezing.

chlorosis Yellowing or blanching of leaves

and other parts of chlorophyll-bearing plants; usually caused by a mineral deficiency.

chromic oxide A completely indigestible chemical (Cr_2O_3) used as an indicator to estimate forage intake.

cleistogamy The condition of having flowers, often small and inconspicuous, which are self-pollinated before the flower opens, or the flower may never open.

climate A characteristic condition or pattern of the various elements of weather for a given geographic area or region of the earth.

clone Progeny produced asexually from a single original individual by vegetative propagation, usually by cuttings or natural propagation of axillary buds, bulbs, tubers, or rhizomes.

coated seed Seed for planting purposes to which a substantial amount of foreign material is used to cover each seed to make the seed uniform in size and shape and free-flowing, or to serve as a carrier of fertilizer, pesticides, nitrogen-fixing microorganisms, coloring, or other additives. *Synonym:* pelleted seed.

companion crop A crop such as a small grain that is sown with another crop, especially one that will emerge and develop slowly, such as a forage crop. Preferred to the term *nurse crop.*

concentrate All feed, low in fiber and high in total digestible nutrients, that supplies primary nutrients (protein, carbohydrate, and fat); for example, grains, cottonseed meal, wheat bran.

continuous grazing *Usage:* Not a recommended term, because animals do not graze continuously. *See* **continuous stocking.**

continuous stocking A method of grazing livestock on a given unit of land where animals have unrestricted and uninterrupted access throughout the time period when grazing is allowed. *Usage:* Specify the length of the grazing period. *See also* **rotational stocking; set stocking.**

controlled grazing *Usage:* Not an acceptable term; sometimes erroneously used for increased grazing management.

cool-season grass Grass species that grow best during cool, moist periods of the year; they commonly have temperature optimums of 15 to 25°C (59 to 77°F.)

coproduct *See* **by-product**.

cored hay samples Samples taken by forcing a hollow cylinder into a bale or stack of hay, the core of which is 3-5 cm in diameter, and which represent the total mass.

coumarin A white, crystalline compound with a vanilla-like odor that gives sweetclover its characteristic odor. An antiquality component of sweetclover.

coumestrol Estrogenic factor occurring naturally in forage crops, especially in ladino clover, strawberry clover, and alfalfa.

creep grazing The practice of allowing juvenile animals to graze areas that their dams cannot access at the same time.

crimped Rolled with corrugated rollers, especially fresh forage, to break stems and facilitate drying.

cropland Land devoted to the production of cultivated crops, including forage crops.

cropland pasture Cropland on which grazing occurs but is generally of limited duration.

crop residue Portion of plants remaining after seed harvest; said mainly of grain crops such as corn stover or of small grain straw and stubble.

crude fiber Coarse, fibrous portions of plants, such as cellulose, that are partially digestible and relatively low in nutritional value. In chemical analysis, it is the residue obtained after boiling plant material with dilute acid and then with dilute alkali. Term is being replaced with more specific NDF (neutral detergent fiber) and ADF (acid detergent fiber).

cubing Process of forming hay into high-density cubes to facilitate transportation, storage, and feeding.

cultivar (1) A variety, strain, or race that has originated and persisted under cultivation or was specifically developed for the purpose of cultivation. (2) For cultivated plants, the equivalent of botanical variety, in accordance with the International Code of Nomenclature of Cultivated Plants—1980.

cuticle A waxy layer secreted by epidermal cells on the outer surface on plants.

cutin A waxy, somewhat waterproof material that provides an outer covering on plants.

cyanogenesis The release of hydrocyanic acid (HCN) in the process of chemical change. *Cyanogenetic* is the adjective form.

days per hectare, animal Total tenure of animals on pasture, expressed as animal days per hectare, usually per unit of time (month, year, etc.).

deferment The postponement or purposeful delay of grazing to achieve a specific management objective. A strategy aimed at providing a discontinuance of livestock grazing on an area for an adequate period of time to provide for plant reproduction, establishment of new plants, or restoration of plant vigor; a return to environmental conditions appropriate for grazing; or the accumulation of forage for later use.

deferred grazing The deferment or purposeful delay of grazing an area within a nonsystematic rotation with other land units.

defoliation (1) Application of a chemical or cultural practice to make leaves fall from a plant prematurely. (2) Removal of the leaves (tops) from a plant by cutting or grazing.

density (1) The number of individuals per unit area. (2) The relative closeness of individuals to one another.

desert Land on which the vegetation is absent or sparse, is usually shrubby, and is characterized by an arid, hot to cool climate.

desiccant A drying agent.

determinate Growth habit characterized by termination of vegetative growth by differentiation of the shoot apex into an inflorescence. In a broader sense, vegetative growth also can be terminated or arrested by competitive sink demand (e.g., fruit loading) or other physiological stresses.

diet The feed regularly offered to or consumed by an animal.

digestibility, apparent Digestibility determined by animal feeding trials, calculated as feed consumption minus excretion (feces), expressed as percent; does not account for endogenous excretions in the feces.

digestibility, true Actual digestibility or availability of feed, forage, or nutrient as represented by the balance between intake and fecal loss of the same ingested material with endogenous excretions in feces accounted for; in vitro digestibility without adjustment to in vivo base.

digestible dry matter (DDM) Dry weight of feed consumed minus dry weight of feces, expressed as percentage of feed dry matter consumed.

digestible energy (DE) Feed-intake gross energy minus fecal energy, expressed as calories per unit feed dry matter consumed.

digestible energy intake Feed consumption expressed as units of digestible energy.

digestible nutrients Portion of nutrients consumed that is digested and taken into the animal body. This may be either apparent or true digestibility; generally applied to energy and protein.

digestible protein Feed protein minus feces protein (N × 6.25), expressed as a percentage of amount in feed.

digestion The conversion of complex, generally insoluble foods to simple substances which are soluble in water.

dough stage Seed development stage at which endosperm development is pliable, like dough (e.g., soft, medium, hard); usually used when 50% of seeds on an inflorescence are in this stage of development.

dry matter (DM) The substance in a plant remaining after oven drying to constant weight at a temperature slightly above the boiling point of water.

dry matter accumulation Total (aboveground) plant development up to some point in time, including both new and old growth.

dry matter disappearance (DMD) (1) Grazing: forage present at the beginning of a grazing period plus growth during the period, minus forage present at the end of the period. (2) Digestibility: loss in dry weight of forage exposed to in vitro digestion.

dry matter intake, daily Amount of dry matter ingested by an animal on a daily basis.

ecological system Energy-driven complex of one or more organisms and their environment. *Synonym:* ecosystem.

ecology The study of communities of living things and the relationships between organisms and their environment.

ecosystem A living community and all the factors in its nonliving environment. *See* **ecological system.**

ecotype A variety or strain within a given species that maintains its distinct identity by adaptation to a specific environment.

endophyte An organism that lives at least part of its life cycle within a host plant, as a parasite or symbiont.

ensilage Silage.

enzyme A protein with an active site where one or more substrates can be changed to one or more products; generally a reaction that would not occur if the substrates were merely mixed.

epinasty Increased growth on upper surface of a plant organ or part (especially the leaf) causing it to bend downward.

ether extract Fats, waxes, oils, and similar plant components which are extracted with warm ether in chemical analysis.

exotic plant An introduced plant that is not fully naturalized or acclimated.

extensive grazing management Grazing management that utilizes relatively large land areas per animal and a relatively low level of labor, resources, or capital.

facilitated recurrent selection A type of recurrent selection in which genetic male sterility is maintained in the population to maintain heterozygosity and genetic diversity and to permit the recombination and shifting of gene frequencies.

fecal index Indirect method of estimating digestibility of dry matter by determining concentration of an indigestible indicator in feces.

feeding value Characteristics that make feed valuable to animals as a source of nutrients; the combination of chemical, biochemical, physical, and organoleptic characteristics of forage that determine its potentials to produce animal meat, milk, wool, or work. Considered by some as synonymous with nutritive value. *See* **forage quality.**

fermentation Anaerobic chemical transformation induced by activity of enzyme systems of microorganisms such as yeast that produce carbon dioxide and alcohol from sugar.

fescue foot Syndrome characterized by red and swollen skin at junction of the hoof in cattle grazing tall fescue, along with loss of appetite and emaciation. Sloughing off of hoofs, tail tips, and ear tips may occur in advanced stages.

fescue toxicity (1) Syndrome caused by ingestion of endophyte-infected tall fescue plants, eliciting a toxic response in the animal. (2) The state of being toxic to the animal.

fescue toxicosis The collective animal syndromes associated with exposure to tall fescue. These include fescue foot, fat necrosis, what is called summer syndrome, and other related disorders.

fiber A unit of matter characterized by a length at least 100 times its diameter or width. In forages it generally means cell walls, especially those low in digestibility.

fibrous Being finely lined in appearance or composed of fibers.

first-last grazing A method of utilizing two or more groups of animals, usually with different nutritional requirements, to graze sequentially on the same land area. First grazers can be more selective and consume a higher-quality diet. *Synonyms:* leader-follower; preference-follower; top-and-bottom grazing. *See also* **forward creep.**

fistula A surgical opening, duct, or passage from a cavity or hollow organ of the body.

flag leaf The uppermost leaf on a fruiting (fertile) grass culm; the leaf immediately below the inflorescence or seed head.

flash grazing *See* **mob grazing.**

flowering stage The physiological stage of a grass plant in which anthesis (blooming) occurs, or in which flowers are visible in nongrass plants.

fluorescence The emission of electromagnetic radiation from a body, resulting from the absorption of incident radiation and persisting only as long as the stimulating radiation is continued.

flushing Improving the nutrition of female breeding animals, such as by providing high-quality forage or energy concentrates prior to and at the beginning of the breeding season as a means of stimulating ovulation.

fodder Coarse grasses such as corn and sorghum harvested with the seed and leaves green or live, cured and fed in their entirety as forage.

foliage The green or live leaves of growing plants; plant leaves collectively. Often used in reference to aboveground development of forage plants.

forage (1) *n.* Edible parts of plants, other than separated grain, that can provide feed for grazing animals or that can be harvested for feeding, including browse, herbage, and mast. *Usage:* Generally, the term refers to more-digestible material (e.g., what is called pasturage, hay, silage, dehy, and green chop) in contrast to less-digestible plant material, known as roughage. (2) *v.* To search for or to consume forage (of animals); *compare* **browse; graze.**

forage accumulation The increase in forage mass per unit area over a specified period of time.

forage allowance The relationship between the weight of forage dry matter per unit area and the number of animal units or forage intake units at any one point in time; a forage-to-animal relationship. The inverse of grazing pressure.

forage crop A crop of cultivated plants or plant parts, other than separated grain, produced to be grazed or harvested for use as feed for animals.

forage intake unit An animal with a rate of forage consumption equivalent to 8 kg dry matter per day.

forage mass The total dry weight of forage per unit area of land, usually above ground level and at a defined reference level.

forage nutritive value *See* **nutritive value.**

forage quality Characteristics that make forage valuable to animals as a source of nutrients; the combination of chemical and biocharacteristics of forage that determines its potential to produce meat, milk, wool, or work. Considered by some as synonymous with feeding value and nutritive value.

forb Any herbaceous broadleaf plant that is not a grass and is not grasslike.

forest grazing The combined use of forestland or woodland for both wood production and animal production by grazing of the coexisting indigenous forage or vegetation that is managed like indigenous forage.

forestland Land on which the vegetation is dominated by forest or, if trees are lacking, the land bears evidence of former forest and has not been converted to other vegetation.

forward creep A method of creep grazing in which dams and offspring rotate through a series of paddocks with offspring as first grazers and dams as last grazers. Specific form of first-last grazing.

fractionation To separate into components, as by distillation, crystallization, or physical separation.

frontal grazing A grazing method by which forage within a land area is allocated by means of moving periodically a sliding fence such that livestock can advance to gain access to new increments of ungrazed forage.

fructan A storage carbohydrate polymer of fructose that includes one glucose molecule. Synthesized from sucrose, the polymer is water soluble and stored in vacuoles.

gene The physical and functional unit of heredity; a segment of chromosome, plasmid, or DNA molecule that encodes a functional protein or RNA molecule.

genome The hereditary material of a cell comprising an entire chromosomal set found in each nucleus or organelle of a given species.

genotype Genetic makeup of an individual or group.

genus A taxonomic category that designates a closely related and definable group of plants, including one or more species. The name of the genus becomes the first word of the binomial employed in literature.

germ Biology: a small organic structure or cell from which a new organism may develop. Seed: refers to the embryo.

germplasm The living substance of the cell nucleus that determines the hereditary properties of organisms and that transmits these properties to the next generation.

gluten A mixture of plant proteins occurring in cereal grains, chiefly corn and wheat; substance in wheat flour that gives cohesiveness to dough.

grain, percent Threshed grain weight × 100 (threshed grain weight + stalk weight).

grain grade Market standard established to describe the amount of contamination, grain damage, immaturity, test weight, and marketable traits.

grain maturity Storage after which no further dry matter is accumulated in the grain.

grain-to-stalk ratio Threshed grain weight/stalk weight.

grass Member of the plant family Poaceae.

grasslike Vegetation that is similar to grass in appearance and is usually a member of the plant family Cyperaceae (sedges) or Juncaceae (rushes).

grassland Any plant community in which grasses and/or legumes compose the dominant vegetation. Land on which the vegetation is dominated by grasses.

grass tetany (hypomagnesemia) Condition of cattle and sheep marked by tetanic staggers, convulsions, coma, and frequently death; characterized by a low level of blood magnesium.

grazable forestland Forestland that produces, at least periodically, sufficient understory vegetation that can be grazed. Forage is indigenous or, if introduced, it is managed as though it were indigenous. *Synonym:* grazable woodland, woodland range, and forest range.

graze *v.* To graze. The consumption of forage in situ by animals; *compare* **browse, forage.** *Usage*: Use with the animal as the subject, not the object; that is, in technical terminology, cattle graze, but people do not graze cattle.

grazer Animal that grazes in situ grass as herbage. Animals on experimental pastures which may or may not remain on specified pasture treatments for the entire grazing period or season, but which are of a kind or physiological condition not necessarily represented on all pasture treatments for the entire grazing period or season.

grazing, high-intensity/low-frequency A grazing system in which the forage on individual pastures is removed by grazing in a relatively short period (high-intensity) and the pasture is not grazed again for a relatively long period (low-frequency).

grazing capacity *Synonym:* carrying capacity.

grazing cycle The time elapsed between the beginning of one grazing period and the beginning of the next grazing period in the same paddock where the forage is regularly grazed and rested.

grazing event The length of time that an animal grazes without stopping.

grazing land Any vegetated land that is grazed or that has the potential to be grazed by animals.

grazing land management The manipulation of the soil-plant-animal complex of the grazing land in pursuit of a desired result.

grazing management The manipulation of animal grazing in pursuit of a defined objective.

grazing management unit The grazing land area used to support a group of grazing animals for a grazing season. It may be a single area or it may have a number of subdivisions.

grazing method A defined procedure or technique of grazing management designed to achieve a specific objective(s).

grazing period The length of time that livestock or wildlife have access to in situ forage in a specific land area.

grazing pressure The relationship between the number of animal units or forage intake units and the weight of forage dry matter per unit area at a given time; an animal-to-forage relationship. The inverse of forage allowance.

grazing season The time period during which grazing can normally be practiced each year or portion of each year.

grazing system A defined, integrated combination of animal, plant, soil, and other environmental components and the grazing method(s) by which the system is managed to achieve specific results or goals.

groat The caryopsis (kernel) of oats after the husk (lemma and palea) has been removed.

gross energy (GE) The amount of heat that is released when a substance is completely oxidized in a bomb calorimeter containing 25 to 30 atmospheres of oxygen.

haylage Product resulting from ensiling forage with about 45% moisture in the absence of oxygen.

head components Components of the inflorescence of grains and grass crops; generally, grain versus vegetative structures but may include individual vegetative structures such as rachis, peduncle, and pedicel.

heading The stage of development of a grass plant between initial emergence of the inflorescence from the boot and the time the inflorescence is fully exerted.

hemicellulose Polysaccharides that are associated with cellulose and lignin in the cell walls of green plants. It differs from cellulose in that it is soluble in alkali and, with acid hydrolysis, gives rise to uronic acid, xylose, galactose, and other carbohydrates, as well as glucose.

herbaceous Nonwoody vegetation.

herbage The biomass of herbaceous plants, other than separated grain, generally above ground but including edible roots and tubers.

high-quality protein A protein containing the appropriate proportions of amino acids for a particular dietary usage.

histochemistry The chemistry of cells and tissues.

hybrid First-generation progeny resulting from the controlled cross-fertilization between individuals that differ in one or more genes.

in situ In the natural or original position.

in vitro In glass, outside the living body.

in vitro digestible dry matter (IVDDM) *See* **in vitro dry matter (digestibility) disappearance (IVDMD).** The procedures and the numerical values are the same for IVDDM and IVDMD.

in vitro dry matter (digestibility) disappearance (IVDMD) A gravimetric measurement of the amount of dry matter lost upon filtration following the incubation of forage in test tubes with rumen microflora, usually expressed as a percentage: weight dry matter sample—weight residue/weight dry matter sample.

in vivo In a living organism, such as in the animal or in the plant.

in vivo nylon bag technique System of determining dry matter disappearance of forage contained in a fine-mesh nylon bag, after placing the bag in the rumen of a fistulated animal for a specified period of time, usually 48 h.

indigenous Originating or produced naturally in a particular land or region or environment; native.

inoculant A seed or soil additive, especially for legume seed, composed of specific nitrogen-fixing bacteria that facilitate dinitrogen fixation in the subsequent crop.

intensive grazing management Grazing management that attempts to increase production or utilization per unit area or production per animal through a relative increase in stocking rates, forage utilization, labor, resources, or capital.

intermittent grazing A method that imposes grazing for indefinite periods at irregular intervals.

introduced species A species not part of the original fauna or flora of the area in question, that is, brought by human activity from another geographical region; exotic.

kernel A mature ovule of a grass plant which has the ovary wall fused to it. Same as caryopsis.

ketosis A pathological accumulation of ketone bodies in an organism.

legume Members of the plant family Fabaceae.

lesion A wound or injury; a circumscribed pathological alteration of tissue.

ley A biennial or perennial hay or pasture portion of a rotation, including cultivated crops.

lignify *v.* To make woody. The thickening, hardening, and strengthening of plant cells by the deposition of lignin on and in the walls of plant cells.

lignin An organic chemical of very low digestibility which strengthens and hardens the walls of cells, especially wood cells.

lipid Organic compounds which contain long-chain aliphatic hydrocarbons and their derivatives, such as fatty acids, alcohols, amines, amino alcohols, and aldehydes; includes waxes, fats, and derived compounds.

lodging, root Stalk fall at ground level without stalk breakage, because of weak root system, root damage, or soil condition.

lodging, stalk Stalk breakage above the ground level.

maillard browning reaction *See* **browning.**

marshland Flat, wet, treeless land usually covered by water and dominated by marsh grasses, indigenous rushes, sedges, or other grasslike plants.

mast Fruits and seeds of shrubs, woody vines, trees, cacti, and other non-herbaceous vegetation available for animal consumption.

mastication Initial chewing prior to swallowing; in ruminants, chewing the cud after regurgitation of a bolus.

meadow Area covered with grasses and/or legumes, often native to the area, grown primarily for hay but with secondary grazing potential.

mesophyll The leaf cells which contain chloroplasts and are located between the upper and lower epidermis.

metabolic body weight ($W^{0.75}$) Basal metabolic rate (energy expenditure per unit body weight per unit time; i.e., kcal heat/weight/day) varies as a function of a fractional power of body weight, usually determined to be body weight raised to the 0.75 power. Loss of protein from body also varies by a similar fractional power of body weight. Thus, the metabolic mass to body weight is presumed to be related by the same exponential power of body weight.

metabolizable energy (ME) Digestible energy (DE) less the energy lost as methane from the rumen and energy lost in urine by ruminant animals.

microclimate The local, rather uniform climate of a specific place or habitat, compared with the climate of the entire area of which it is a part.

middlings A coproduct of flour milling that contains varying proportions of endosperm, bran, and germ.

milk stage In grain (seed), the stage of development following pollination in which the endosperm appears as a whitish liquid that is somewhat like milk.

mixed grazing Grazing by two or more species of grazing animals (e.g., sheep and cattle) on the same land unit, not necessarily at the same time but within the same grazing season.

mob grazing In the management of a grazing unit, grazing by a relatively large number of animals at a high stocking density for a short time period.

moisture equilibrium The condition reached by a sample when it no longer takes up moisture from, or gives up moisture to, the surrounding atmosphere.

mycotoxin A toxin or toxic substance produced by a fungus.

native pasture Native vegetation (predominantly herbaceous) used for grazing in untilled areas. The term *tame* or *introduced* is used instead of *native* for pastures that include mainly nonnative species.

native species A species indigenous to an area, not introduced from another environment or area.

naturalized pasture Plants introduced from other countries which have become established in and more or less adapted to a given region by long continued growth there. The name is appropriate for pastures made up of plants such as white clover, bluegrass, and bermudagrass.

near infrared reflectance spectroscopy (NIRS) A method of forage quality analysis based on spectrophotometry at wavelengths in the near infrared region.

net energy (NE) Metabolizable energy minus the energy lost in the heat increment.
neutral detergent fiber (NDF) That portion of a forage that is insoluble in neutral detergent; synonymous with cell wall constituents.
neutral detergent fiber digestibility The digestibility of neutral detergent fiber determined as the difference in NDF in a forage before and after in vivo or in vitro digestion.
nitrate Any compound containing the NO_3^- radical.
nitrate poisoning *See* **nitrate toxicity**.
nitrate toxicity Conditions in animals resulting from ingestion of feed high in nitrate. The toxicity actually results from nitrite, which results when nitrate is reduced to nitrite in the rumen. Nitrite affects the O_2 carrying capacity of the blood.
nitrite A compound containing the radical NO_2^-; can be organic or inorganic.
nitrogen-free extract (NFE) The highly digestible portion of a plant, consisting mostly of carbohydrates, that remains after the protein, ash, crude fiber, ether extracts, and moisture content have been determined.
NMR oil content The amount of oil in a whole seed sample measured by use of the nuclear magnetic resonance spectroscopy technique.
nonenzymatic browning *See* **browning**.
nonnutritive fiber That portion of fiber in a feed which is not digestible and hence is of no nutritive value.
nonselective grazing Utilization of forage by grazing animals so that all forage species and/or all plants within a species are grazed.
nuclear magnetic resonance (NMR) A type of radio frequency or microwave spectroscopy, based on the magnetic field generated by the spinning of the electrically charged nucleus of certain atoms. This nuclear magnetic field is caused to interact with a very large magnetic field of the instrument magnet. The electron distribution around each atom can distinguish between the three kinds of hydrogen atoms in ethyl alcohol. The direct relation of NMR spectral bank height to the concentration of chemical compound allows NMR to be used for quantitative analysis and the following of rates of reaction and equilibrium of chemicals in solution.
nutrient, animal Food constituent or group of food constituents of the same general chemical composition required for support of animal life.
nutritive value Relative capacity of a given feed to furnish nutrition for animals; may be prefixed by low, high, moderate, etc.
nutritive value index (NVI) Daily digestible amount of forage per unit of metabolic body size relative to a standard forage.
omasum The third chamber of the ruminant stomach, where the contents are mixed to a more or less homogeneous state.
opaque-2 An endosperm mutant of maize associated with suppressed prolamine production in the endosperm, resulting in increased lysine content of the protein fraction.
orts Rejected feedstuffs left under conditions of ad libitum stall feeding.
overgrazing The grazing of a number of animals on a given area that, if continued to the end of the planned grazing period, will result in less than satisfactory animal performance and/or less than satisfactory pasture forage production.
overstocking The placing of a number of animals on a given area that will result in overuse if continued to the end of the planned grazing period. Not to be confused with overgrazing, as an area may be overstocked for a short period, but the animals are removed before the area is overused. However, continued overstocking will lead to overgrazing.
paddock A grazing area that is a subdivision of a grazing management unit and is enclosed and separated from other areas by a fence or barrier.
palatability Preference based on plant characteristics eliciting a choice between two or more forages or parts of the same forage, conditioned by the animal and environmental factors that stimulate a selective intake response.
parenchyma A tissue of higher plants consisting of living cells with thin walls that are agents of photosynthesis and storage; cells generally very high in digestibility.
pasturage Vegetation on which animals graze, including grasses or grasslike plants, legumes, forbs, and shrubs. Not a recommended term.
pasture A type of grazing management unit enclosed and separated from other areas by fencing or other barriers and devoted to the production of forage for harvest primarily by grazing.
pasture, carrying capacity *See* **carrying capacity**.
pasture, permanent Composed of perennial

or self-seeding annual plants kept indefinitely for the purpose of grazing.

pasture, rotation A fenced pasture area used for a few seasons and then plowed for other crops.

pasture, supplemental A crop used to provide grazing for supplemental use, usually during periods of low production of permanent or rotational pastures.

pasture, tame Grazing lands planted with introduced or domesticated forage species which may receive periodic cultural treatments such as renovation, fertilization, and weed control.

pasture, temporary A field of crop or forage plants grazed for only a short period, usually not more than one crop season.

pastureland Land devoted to the production of indigenous or introduced forage for harvest primarily by grazing. Pastureland generally must be managed to arrest successional processes.

perennial A plant or group of plants that persists for several years, usually with new growth from a perennating part.

pericarp The ripened and variously modified walls of a plant ovary, especially those contributing the outer layer in a cereal caryopsis.

perloline A plant alkaloid found to interfere with cellulose digestion by rumen microoranisms; commonly associated with tall fescue.

phloem A conducting tissue present in vascular plants, chiefly concerned with the transport of sugars and other organic food materials in the plant. When fully developed, the phloem consists of sieve tubes and parenchyma, generally with companion cells.

photoperiod Period of a plant's daily exposure to light.

photorespiration Respiratory activity due to O_2 reaction instead of CO_2 in the photosynthetic pathway that takes place in cool-season plants during the light period; no useful form of energy is derived.

photosensitization A noncontagious disease resulting from the abnormal reaction of light-colored skin to sunlight after a photodynamic agent has been absorbed through the animal's system. Grazing certain kinds of vegetation or ingesting certain molds under specific conditions causes photosensitization.

photosynthesis Process by which carbohydrates are produced from CO_2 and water, chloroplasts or chlorophyll-bearing cell granules, and the energy of sunlight.

pith A usually continuous central strand of spongy tissue in the stems of most vascular plants which probably functions chiefly in storage.

prairie Nearly level or rolling grassland that was originally treeless; usually characterized by fertile soil.

prebloom The plant stage or period immediately preceding blooming.

preservative An additive used to protect against decay, discoloration, or spoilage.

protected variety (cultivar) A plant variety (cultivar) that is released and granted a certificate of plant variety protection under the legal statutes of the US or some other country. The owner of a protected variety has the right during the term of the protection to exclude others from selling the variety, offering it for sale, reproducing it, importing it, exporting it, or using it in producing a hybrid or different variety.

protein, crude An estimate of protein content based on a determination of total nitrogen (N) content multiplied by 6.25 because proteins average about 16% N.

protein fraction Refers to solubility of protein as originated by Osborne, who designed a sequential extraction scheme based on water, dilute salt, 70% alcohol, and dilute acid or alkali.

protein quality Refers to the balance of essential amino acids in the protein, as well as the biological availability of the protein. In general, many cereals are low in lysine relative to animal needs and thus have low quality.

proximate analysis Analytical system for feedstuffs that includes the determination of ash, crude fiber, crude protein, ether extract, moisture (dry matter), and nitrogen-free extract.

prussic acid A poison produced as a glucoside by several plant species, especially sorghums. Also called hydrocyanic acid.

pubescence A general term for hairs or trichomes.

pubescent Covered with fine, soft, short hairs, or trichomes.

pulses The edible seeds of various leguminous crops (such as peas, beans, lentils); collectively, legume plants that produce edible seeds.

put-and-take animal *See* **grazer.**

put-and-take stocking The use of variable animal numbers during a grazing period or grazing season, with a periodic adjustment in animal numbers in an attempt to maintain desired sward management criteria, i.e., a desired quantity of forage, degree of defoliation, or grazing pressure.

range Land supporting indigenous vegetation that is grazed or that has the potential to be grazed and is managed as a natural ecosystem. Includes grazable forestland and rangeland.

rangeland Land on which the indigenous vegetation (climax or natural potential) is predominantly grasses, grasslike plants, forbs, or shrubs suitable for grazing or browsing use and is managed as a natural ecosystem. If plants are introduced, they are managed as indigenous species. Rangelands include natural grasslands, savannas, shrublands, most deserts, tundra, alpine communities, coastal marshlands, and wetland meadows.

range management The science of maintaining maximum forage production, generally with natural vegetation, without jeopardy to other resources or uses of the land.

ration The total amount of feed (diet) allotted to one animal for a 24-h period.

ration grazing Confining animals to an area of grazing land to provide the daily allowance of forage per animal.

reducing sugars Sugars that have the ability to donate electrons to copper cations to produce copper metal. Some common reducing sugars are glucose, fructose, and maltose, as opposed to nonreducing sugars such as sucrose, raffinose, melibiose, and stachyose.

residue biomass The biomass that remains following removal or utilization of part of the biomass by grazing, harvesting, burning, etc.

resins Sticky to brittle plant products from essential oils that sometimes posses marked odors; more common with woody vegetation than with herbaceous vegetation. Used in medicines, varnishes, etc.

resistance (1) The ability of a plant or crop to grow and produce even though heavily inoculated or actually infected or infested with a pest. (2) The ability of a plant to survive a period of stress such as drought, cold, or heat.

respiration The process by which tissues and organisms exchange gases with their environment; generally associated with oxidation of sugars to release energy for the plant to grow and reproduce.

rest To leave an area of grazing land ungrazed or unharvested for a specific time, such as a year, a growing season, or a specified period required within a particular management practice.

rest period The length of time that a specific land area is allowed to rest.

reticulum The second stomach in ruminants.

rind The epidermis and sclerenchyma tissue on the outer surface of stems of corn, sorghum, and other grass plants.

ripe Fully grown and developed; mature.

rotational deferred grazing A grazing management system which uses a systematic rotation of deferment among land areas within a grazing management unit.

rotational stocking A grazing method that utilizes recurring periods of grazing and rest among two or more paddocks in a grazing management unit throughout the period when grazing is allowed.

roughage Animal feeds that are relatively high in crude fiber and low in total digestible nutrients and protein.

rumen First compartment of the stomach of a ruminant or cud-chewing animal.

ruminant A suborder of mammals having a complex multichambered stomach; uses forages primarily as feedstuffs.

rumination Regurgitation and remastication of food in preparation for true digestion in ruminants.

saponin Any of various plant glucosides that form soapy colloidal solutions when mixed and agitated with water.

savanna Grassland with scattered trees or shrubs; often a transitional type between true grassland and forestland and accompanied by a climate with alternating wet and dry seasons.

scarification Process of scratching or abrading of the seed coat of seed of certain species to allow uptake of water and gases as an aid to seed germination.

sclerenchyma Strengthening tissue made up of cells with heavy lignified cell walls; supports and protects the softer tissue of the plant.

seasonal grazing Grazing restricted to one or more specific seasons of the year.

sedge A plant of the family Cyperaceae. Sedges generally have a 3-sided stem.

seed *n.* Ripened (mature) ovule consisting of an embryo, a seedcoat, and a supply of food that, in some species, is stored in the endosperm. *v.* To sow, as to broadcast or drill small-seeded grasses and legumes or other crops.

seed size A measure, usually expressed as weight per unit number of seed.

senescence The gradual degradation of a plant or plant organ, usually due to old age, accompanied or preceded by a loss of protein and some minerals.

sequence grazing The grazing of two or more land units in succession that differ in forage species composition. Sequence grazing takes advantage of differences among forage species and species combinations grown in separate areas for management purposes to extend grazing seasons or enhance forage quality.

set stocking The practice of allowing a fixed number of animals on a fixed area of land during the time when grazing is allowed.

shrub A perennial woody plant smaller than a tree and having several stems arising at a point near the ground.

shrubland Land on which the vegetation is dominated by shrubs.

silage Forage preserved in a succulent condition by organic acids produced by partial anaerobic fermentation of sugars in the forage.

silage, additive Material added to forage at the time of ensiling to enhance favorable fermentation processes.

silage preservative *See* **silage, additive**.

silvo-pastoral Preferred term is *forest grazing*.

sod-seeding Mechanically placing seed, usually legumes or small grains, directly into a grass sod.

species A taxonomic category ranking immediately below a genus and including closely related, morphologically similar individuals which actually or potentially interbreed. The name of the species becomes the second word of the binomial employed in literature.

sprigging Vegetative propagation by planting stolons or rhizomes (sprigs) in furrows or holes in the soil.

stalk diameter The diameter of a stalk, usually at a designated node or internode.

stalk girdle Ring around the stem or stalk made by an insect removing material; may result in lodging of the stalk. Girdle may be either internal or external.

stalk tunnels Longitudinal tunnels in plant stalks produced by insects. Examples of stalk-tunneling insects are the sugarcane borer and lesser corn stalk borer.

starch fractions Refers to amylose and amylopectin. In most cereal starches, there is usually 20% to 30% amylose and 70% to 80% amylopectin in the starch. However, waxy cereals contain 100% amylopectin (branch-chain starch), whereas high-amylose corn cultivars have been developed that contain 80% amylose (linear-chain starch) and only 20% amylopectin in the starch.

starch granules The fundamental unit in which starch is deposited in the storage tissue of many higher plants. Granules are insoluble in cold water and have a characteristic size and shape depending on the plant species which produced them.

steppe Vast semi-arid grass-covered plain, usually lightly wooded; semi-arid grassland characterized by short grasses occurring in scattered bunches with other herbaceous vegetation and occasional woody species.

stocker Young cattle, post weaning, generally being grown on forage diets to increase size before going to concentrate feed in feedlots.

stocking density The relationship between number of animals and area of land at any instant of time, or grazing management unit utilized over a specified time period.

stocking density index The reciprocal of the fraction: land available to the animals at any one time/land available to the animals for the entire grazable period.

stocking plan The number and kind of livestock assigned to one or more given management areas or units for a special period.

stocking pressure *See* **grazing pressure**.

stocking rate The relationship between the number of animals and the grazing management unit utilized over a specified time period.

stockpiling forage To allow forage to accumulate for grazing at a later period. Forage is often stockpiled for autumn and winter grazing after or during dormancy or semidormancy, but stockpiling may occur at any time during the year as a part of a management plan. Stockpiling can be described in terms of deferment and forage accumulation.

stover The matured, cured stalks of such crops as corn or sorghum from which the grain has been removed. A type of roughage.

stress A stimulus or succession of stimuli of such magnitude as to tend to reduce the growth rate or survival of an organism.

strip grazing Confining animals to an area of grazing land to be grazed in a relatively short period of time, where the paddock size is varied to allow access to a specific land area.

structural carbohydrates *See* **carbohy-**

drates, structural.

stubble The basal portion of the stems of herbaceous plants left standing after harvest.

substrate (1) A substance that is acted upon in a chemical reaction. (2) A culture medium.

supplement Nutritional additive (salt, protein, phosphorus, etc.) intended to improve nutrition balance and remedy deficiencies of the diet.

supplemental feeding Supplying concentrates or harvested feed to correct deficiencies of the pasture diet. Often erroneously used to mean emergency feeding.

surfactants Compounds that are active at the interface between nonpolar (oil) and polar (water) molecules (i.e., soaps, detergents).

sustained yield The continuation of desired forage or animal production.

sward A population of herbaceous plants characterized by a relatively short habit of growth and relatively continuous ground cover, including both above- and belowground parts.

tannin Broad class of soluble polyphenols that occur naturally in many forage plants. They have a common property of condensing with protein to form a leatherlike substance that is insoluble and of impaired digestibility.

tedding A mechanical fluffing of a cut forage in the field to aid in drying.

terminal Of or relating to an end or extremity; growing at the end of a branch or stem. Sometimes called the growing point, but the preferred term is *shoot apex.*

tester animals Animals of like kind and similar physiological condition used in grazing experiments to measure animal performance or pasture quality; usually assigned to a treatment for the duration of the grazing season, versus "grazer" animals, which may be assigned temporarily.

total digestible nutrients (TDN) Sum total of the digestibility of the organic components of plant material and/or seed; for example, crude protein + NFE + crude fiber + fat.

total nonstructural carbohydrates (TNC) *See* **carbohydrates, nonstructural.**

toxicity Injury, impairment, or death resulting from poison or toxin.

toxin A substance that is a specific product of the metabolic activities of a living organism and is usually very unstable, notably toxic when introduced into the tissues, and typically capable of inducing antibody formation.

toxoid Toxin that has been treated to be rendered nontoxic but which will still induce the formation of antibodies.

trichome A filamentous outgrowth; an epidermal hair structure on a plant.

tropical Related to or having characteristics of the tropics.

tundra Land areas in arctic and alpine regions devoid of large trees, varying from bare ground to various types of vegetation consisting of grasses, sedges, forbs, dwarf shrubs and trees, mosses, and lichens.

undergrazing Utilizing pasture forage with grazing animals at a rate less than that required for optimum animal performance and/or forage production.

ungrazed (1) The status of grazing land that is not grazed by animals. (2) The status of plants or plant parts that are not grazed by animals.

variable stocking The practice of allowing a variable number of animals on a fixed area of land during the time when grazing is allowed.

varietal protection Legal statutes that give to the developers or owners of a variety (cultivar) the exclusive right to control the seed production and marketing of that cultivar. *See also* **protected variety.**

variety *See* **cultivars.**

vascular bundle An elongated strand containing xylem and phloem, the conducting tissues of plants that transport food and water.

vascular tissue Conducting tissue with vessels or ducts.

vegetation Plant mass or plant parts in general.

vegetative Non-reproductive plant parts, (leaf and stem) in contrast to reproductive plant parts (flower and seed) in developmental stages of plant growth. The nonreproductive stage in plant development.

vegetative cover A soil cover of plants, irrespective of species.

vegetative propagation *See* **vegetative reproduction.**

vegetative reproduction (1) In seed plants, reproduction by means other than seeds. (2) In lower forms, reproduction by vegetative spores, fragmentation, or division of the plant body.

vegetative state Stage prior to the appearance of fruiting structures.

veld Grasslands of eastern and southern

Africa that are usually level and mixed with trees and shrubs; or grasslands similar to the African veld.

vigor Indicative of active growth; relative absence of disease or other stresses.

vitality The capacity to live and develop; power of enduring or continuing.

voluntary intake Ad libitum (free will) intake achieved when an animal is offered an excess of a single feed or forage.

weather The state of the atmosphere, mainly with respect to its effect upon life and human activities. As distinguished from climate, weather consists of the short-term (minutes to months) variations of the atmosphere.

weathering The loss of quality in a crop due to the effects of weather on the product or process.

xylem The portion of the conducting tissue which is specialized for the conduction of water and minerals.

REFERENCES

Barnes, RF, and JB Beard. 1992. Glossary of Crop Science Terms. Madison, Wis.: Crop Science Society of America.

Heath, ME, RF Barnes, and DS Metcalfe. (eds.) 1985. Forages: The Science of Grassland Agriculture. 4th ed. Ames, Iowa: Iowa State Univ. Press, 619-30.

The Forage and Grazing Terminology Committee. 1992. Terminology for grazing lands and grazing animals. J. Prod. Agric. 5:191-201.

Parker, SP. 1984. McGraw-Hill Dictionary of Scientific and Technical Terms. New York: McGraw-Hill Book Co.

Vallentine, JF. 1990. Grazing Management. San Diego, Calif.: Academic Press, 456-73.

INDEX

The index lists common names of grasses, legumes, and other plants. The corresponding botanical names are listed in the appendix.